HAAg
hepatitis A antigen
HAV
hepatitis A virus
HBcAg
hepatitis B core antigen
HBeAg
hepatitis B e antigen
HBsAg
hepatitis B surface antigen
HBV
hepatitis B virus
HCT
hematocrit
HDL
high-density lipids
HHNK
hyperosmolar hyperglycemic nonketosis
HIV
human immunodeficiency virus
IDDM
insulin dependent diabetes mellitus
IF
interstitial fluid
IgE
immunoglobulin E
IgG
immunoglobulin G
ITP
idiopathic thrombocytopenic purpura
LATS
long-acting thyroid stimulator
LDH
lactate acid dehydrogenase
LDL
low-density lipoproteins
LH
luteinizing hormone
LH-RH
luteinizing hormone releasing hormone
MBD
minimal brain dysfunction
MI
myocardial infarction
MRI
magnetic resonance imaging
NIDDM
noninsulin dependent diabetes mellitus
NMRI
nuclear magnetic resonance imaging
NSAID
nonsteroidal anti-inflammatory drug
OGTT
oral glucose tolerance test
OP
oncotic or osmotic pressure
PA
pulmonary artery

PCOD
polycystic ovary disease
PEEP
positive end-expiratory pressure
PMN
polymorphonuclear neutrophil
PSGN
poststreptococcal glomerulonephritis
PTE
pulmonary thromboembolism
PTH
parathormone
RBC
red blood cell
RIND
reversible ischemic neurologic deficit
RPGN
rapidly progressive glomerulonephritis
RTA
renal tubular acidosis
SA node
sinoatrial node
SAH
subarachnoid hemorrhage
SBE
subacute bacterial endocarditis
SGOT
serum glutamic-oxaloacetic transaminase
SGPT
serum glutamic-pyruvic transaminase
SIADH
syndrome of inappropriate ADH secretion
SLE
systemic lupus erythematosus
SRS-A
slow reacting substance of anaphylaxis
TIA
transient ischemic attack
t-PA
tissue-type plasminogen activator
TRH
thyrotropin releasing hormone
TSH
thyroid-stimulating hormone
TSI
thyroid-stimulating immunoglobulin
UTI
urinary tract infection
UTO
urinary tract obstruction
VLDL
very low-density lipoproteins
vWD
von Willebrand's disease
WBC
white blood cell

PATHOPHYSIOLOGY

WITH PRACTICAL APPLICATIONS

PATHOPHYSIOLOGY

WITH PRACTICAL APPLICATIONS

Phyllis L. Chowdry

MESA STATE COLLEGE

Wm. C. Brown Publishers

Dubuque, Iowa•Melbourne, Australia•Oxford, England

Book Team

Editor *Colin H. Wheatley*
Developmental Editor *Jane DeShaw*
Production Editor *Michelle M. Campbell*
Publishing Services Coordinator *Barbara J. Hodgson*
Photo Editor *Shirley M. Lanners*
Permissions Editor *Karen L. Storlie*
Visuals/Design Developmental Consultant *Donna Slade*

WCB
Wm. C. Brown Publishers
A Division of Wm. C. Brown Communications, Inc.

Vice President and General Manager *Beverly Kolz*
National Sales Manager *Vincent R. Di Blasi*
Assistant Vice President, Editor-in-Chief *Edward G. Jaffe*
Director of Marketing *John C. Calhoun*
Marketing Manager *Christopher T. Johnson*
Advertising Manager *Amy Schmitz*
Director of Production *Colleen A. Yonda*
Manager of Visuals and Design *Faye M. Schilling*
Publishing Services Manager *Karen J. Slaght*
Permissions/Records Manager *Connie Allendorf*

WCB
Wm. C. Brown Communications, Inc.

Chairman Emeritus *Wm. C. Brown*
Chairman and Chief Executive Officer *Mark C. Falb*
President and Chief Operating Officer *G. Franklin Lewis*
Corporate Vice President, President of WCB Manufacturing *Roger Meyer*

Cover design and interior illustrations by Diphrent Strokes, Inc.

Interior design by Diphrent Strokes, Inc. (Sheree Goodman).

Cover illustration by Diphrent Strokes, Inc. (Felipe Passalacqua).

The credits section for this book begins on page 514 and is considered an extension of the copyright page.

Library of Congress Catalog Card Number: 91-78216

ISBN 0-697-08538-4

Printed in the United States of America by Wm. C. Brown Communications, Inc., 2460 Kerper Boulevard, Dubuque, IA 52001

10 9 8 7 6 5 4 3 2

Brief Contents

Expanded Contents

Preface

This textbook is intended for use in a one semester pathophysiology course that serves sophomore and junior level students with paramedical interests. The first unit deals with fluid and electrolyte balance, and the remaining eight units are concerned with disease processes involving the major systems of the body. Each unit includes a brief review of anatomy and physiology that precedes a discussion of the pathophysiology of a body system.

ORGANIZATION

The overall organization of this text is based on the premise that some degree of fluid and/or electrolyte imbalance is an accompanying feature of most disease processes and that evaluation of laboratory test results is an essential aspect of understanding disease processes. Fluid/electrolyte balance is a unifying theme and examples of electrolyte and fluid imbalances are included throughout the text. This theme is introduced in the first unit in which the topics of transmembrane movements, fluid compartments, electrolyte imbalance, and acid-base imbalance are discussed.

The last chapter of each unit is a Practical Applications chapter in which there are case studies and evaluations of related laboratory test results. In many of these cases, electrolyte and acid-base imbalances are a part of the picture. The purpose of these chapters is to integrate unit concepts by using examples, with the added perspective of laboratory analysis. This, in some cases, leads to a description of aspects of a disease process that have not been discussed in preceding chapters. These chapters expand the concepts regarding the pathophysiology of disorders and, consequently, are more than simply review chapters.

Unit I deals with the mechanisms whereby fluid imbalance occurs and with derangements of sodium, potassium, and calcium balance. Osteoporosis is included in the discussion of calcium homeostasis. Basic concepts related to acidosis/alkalosis are discussed including homeostatic mechanisms and arterial blood gases. The end of the unit chapter describes sources of bicarbonate and hydrogen ions in the blood, focuses on the effects of fluid/electrolyte loss by way of vomiting or diarrhea, and includes brief case studies with laboratory test results that show the consequent fluid/electrolyte imbalance.

Units II and III deal with the respiratory and the genitourinary systems and are a natural extension of unit I. Chapters 6 and 7 correlate respiratory anatomy with clinical significance, discuss pleural effusion based on principles outlined in unit I, and include disease processes involving obstruction to airflow. The remainder of unit II deals with conditions in which there is diminished lung capacity, as well as various other pulmonary disorders. Chapter 9 gives examples of pulmonary diseases and expands the topic of acid-base balance. Unit II includes the topic of fluid balance in terms of pleural effusion and illustrates the concept of respiratory control of acid-base balance.

Unit III describes disorders of the male and female reproductive systems and disorders of the excretory system. There is an overview of normal nephron function including renal control of acid-base balance, a discussion of laboratory tests indicative of renal function, the mechanisms of action of various diuretics that affect electrolyte and acid-base balance, and disorders involving the entire urinary tract. The remainder of unit III deals with glomerular injury, primary and secondary renal diseases, and renal failure in which dialysis is a principal means of management. Chapter 16 integrates information from preceding chapters with the use of case studies. There are examples of hepatorenal syndrome, reduced renal perfusion, nephrosclerosis, glomerulonephritis, diabetic nephropathy, and renal transplant rejection. These examples are discussed in terms of mechanisms that lead to abnormal alkaline phosphatase levels, acidosis, electrolyte imbalance, abnormal anion gap, anemia, and elevations of the blood urea nitrogen and creatinine levels.

Units IV and V focus on the cardiovascular system. Unit IV deals with blood disorders, and the discussion includes erythropoiesis, oxygen delivery to tissues, and disorders involving inadequate oxygenation and abnormal red blood cell/hemoglobin production. Reasons for variations in numbers of white blood cells precede an emphasis on malignant proliferation of white blood cells. A description of the processes of coagulation and fibrinolysis leads to the topic of bleeding disorders, with emphasis on disseminated intravascular coagulopathy. Chapter 20 deals with blood disorders and the evaluation of laboratory test results, including tests involving red blood cells and white blood cells. The coagulation tests include platelet count, prothrombin time, activated partial thromboplastin time, thrombin time, and fibrin split products. Chapter 20 includes cases of acute myelogenous leukemia and leukemia and prostate cancer associated with disseminated intravascular coagulopathy. The laboratory test results associated with these cases are blood cell tests, uric acid levels, potassium and calcium levels, coagulation tests, and enzyme levels.

Unit V deals with blood pressure disorders, vascular disorders, disorders of the heart including congestive heart failure, and congenital heart defects. Chapter 25 discusses congestive heart failure in terms of mechanisms for fluid and osmolal homeostasis and failure of homeostasis in relation to a specific example of congestive heart failure. The mechanisms leading to edema, pleural effusion, and electrolyte imbalance are discussed in relation to this case. A case of myocardial infarction is evaluated in terms of blood lipid levels and cardiac enzymes. The pathogenesis and manifestations of septic shock, with an example, is the final topic, which includes acid-base imbalance.

Unit VI deals with the immune system and includes specific immune responses and the role of lymphocytes, hypersensitivity reactions, organ transplant rejection, AIDS, and inherited immunodeficiency syndromes. Autoimmune disorders include rheumatoid arthritis, polyarteritis nodosa, lupus erythematosus, scleroderma, and Sjögren's syndrome. Chapter 30 deals with immunologic tests and evaluates cases of autoimmune hemolytic anemia, AIDS, and lupus erythematosus.

Unit VII deals with endocrine disorders that include the endocrine pancreas. Chapter 33 focuses on diabetes mellitus and hyperosmolar, hyperglycemic non-ketosis, with emphasis on laboratory findings and the pathophysiology of electrolyte and acid-base imbalance.

Unit VIII deals with disorders of the gastrointestinal tract, liver, and gallbladder. The effects of AIDS on the esophagus, intestinal tract, and liver are included. Chapter 37 describes liver function tests and discusses laboratory test results correlated with disorders involving both the liver and the pancreas. Electrolyte and acid-base imbalance are a part of the picture.

Unit IX is concerned with disorders of the nervous system. Topics include the flow of cerebrospinal fluid, fluid accumulation in the cerebral hemispheres, and alterations in cerebral blood flow that lead to stroke. Cerebral disorders are discussed including epilepsy and intracranial tumors followed by examples of brain injury. Degenerative disorders are discussed and include Alzheimer's disease, Creutzfeldt-Jakob disease, and AIDS dementia. There are selected examples of sensory and motor disorders, i.e., head and face pain and a sensory disorder of the inner ear, motor neuron abnormalities, degenerative muscle changes, defective neuromuscular transmission, motor pathway abnormalities, and demyelination. Chapter 41 focuses on laboratory data and diagnostic tests related to cases of seizure, transient ischemic attack, and stroke. Blood gas values and fluid and electrolyte disturbances are included in the discussion.

Finally, there are two appendices with Appendix A, answers to chapter review questions, and Appendix B, a list of normal laboratory values.

In summary, the scope of this text is fluid/electrolyte balance and pathologic processes associated with the major systems of the body. Fluid and electrolyte (including acid-base) balance is a unifying theme and end of the unit chapters are Practical Applications chapters. These chapters include case studies with laboratory findings. A discussion of AIDS appears in chapters 29, 30, 34–36, and 39 involving systems markedly affected by the HIV infection.

PEDAGOGICAL AIDS

An introduction to each unit provides both an overview of the organization of the chapters and the general concepts included in each chapter. Chapter outlines indicate major topics with all subheadings, which clearly show the flow of ideas and aid in finding a particular topic.

Each chapter includes phonetic pronunciations, and key words, when first introduced and defined, are in bold type. Definitions of words are clarified further by footnotes that indicate derivation. Boxed comments are included throughout the text, which in some cases, are for special emphasis and in others are added as a note of interest. A chapter summary provides a review of the main ideas in each chapter.

Review questions at the end of each chapter are intended for self-study practice, with answers in the appendix. There is also a list of selected readings at the end of each chapter for further information related to selected topics within the unit. There is a comprehensive index with extensive cross referencing.

The case studies included in each chapter at the end of the unit should be used for practice with interpretation of laboratory findings. Reference to fluid/electrolyte imbalance associated with various disease processes throughout the text is intended as a pedagogical aid as well. Fluid and electrolytes seem to involve elusive concepts, and repetition in various situations should provide a better grasp of the operative mechanisms when an imbalance occurs.

SUPPLEMENTARY MATERIALS

An instructor's manual is available for use with this text. This includes a test item file with an average of 30–50 questions for each chapter.

WCB TestPak is a computerized testing service offered free upon request to adopters of this textbook. It provides a call-in/mail-in test preparation service. A complete test item file is also available on computer diskette for use with IBM compatible, Apple IIe or IIc, or Macintosh computers.

ACKNOWLEDGMENTS

I am indebted to Dr. David Smith for his help in obtaining photographs of X-rays and to Dr. M. G. Klein, who expended considerable effort to provide additional photographs for illustrations. I owe a word of thanks to Dr. Mary Turley and to Susan Dickson for their efforts on my behalf. Many students have shared insights and experiences that have contributed to this effort and Shirley Roe has been a special source of encouragement and help. Janet Coleman has spent many hours as a research assistant and has prepared X-ray photographs. Reviewers for this writing project have been key players in determining the final form of the manuscript.

Lori Kashuba, RN
University of Alberta Hospitals School of Nursing

Kay Doyle, Ph.D.
University of Massachusetts Lowell

Patricia M. O'Mahoney-Damon, Ph.D.
University of Southern Maine

Marlene Reimer, RN, MN, CNN(C)
The University of Calgary

Louis Giacinti
Milwaukee Area Technical College

James F. Berry
Elmhurst College

Dr. Katharina Burns, M.D., Ph.D., Dip H.S.A.
Burns Healthcare and Wellness Consulting Inc.

Stephen R. Overmann
Southeast Missouri State University

Louis D. Trombetta
St. John's University

Dr. Ralph W. Stevens III
Old Dominion University

My department chairman, Gary McCallister, has encouraged and supported my efforts in every way. No good thing would have come from this effort without the guidance and help of the editors and production staff at Wm. C. Brown Publishers. Finally, a special word of thanks to Sally Reeves, who typed and corrected my manuscript with perpetual patience and good cheer.

PATHOPHYSIOLOGY

WITH PRACTICAL APPLICATIONS

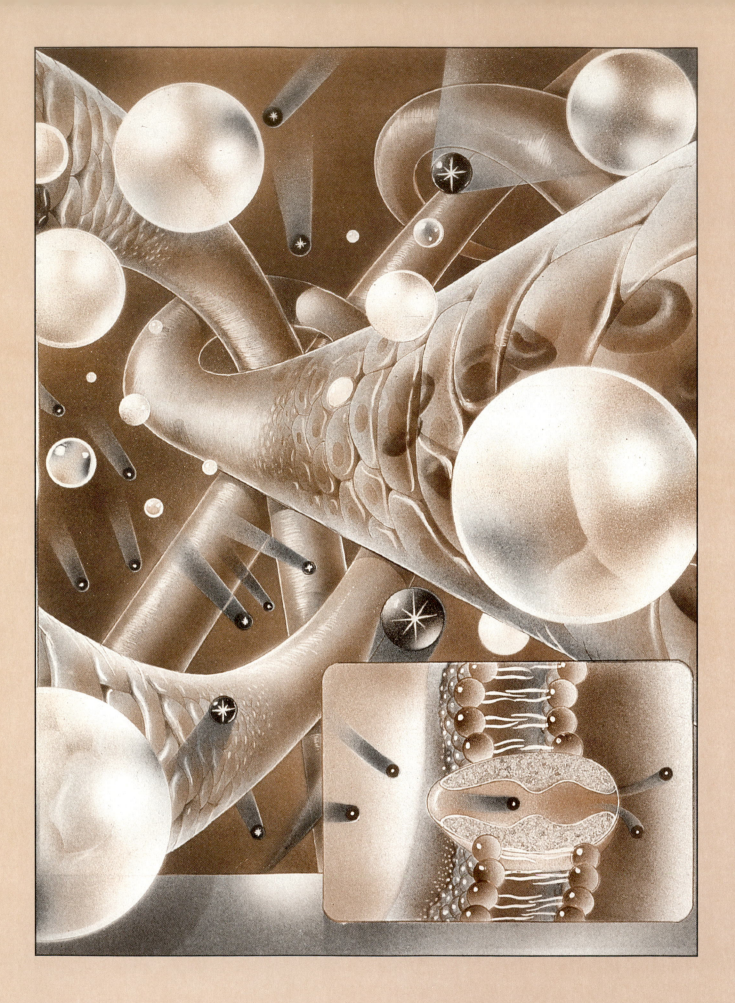

Unit I

FLUID AND ELECTROLYTES

The premise for the content and organization of unit I is that some degree of fluid and/or electrolyte imbalance is an accompanying feature of most disease processes. Unit I is the basis for including references to electrolyte derangements, and to a lesser extent fluid imbalance, throughout the text. Chapter 1 deals with the mechanisms by which transmembrane movements occur and sets the stage for subsequent discussions of exchanges between cells and extracellular compartments.

The normal distribution of body water and the mechanisms for maintaining that distribution are discussed in the first part of chapter 2. This is followed by a description of fluid imbalances, both excess and deficit. The main purpose of chapter 2 is to establish the importance of fluid balance and the mechanisms whereby imbalances occur.

The emphasis of chapter 3 is imbalance involving sodium, potassium, and calcium ions. The purpose of this chapter is to establish some principles involving the physiological role of these ions and to provide the basis for incorporating electrolyte imbalance into the remainder of the text. There is also introductory reference to acidosis and alkalosis which leads into the following chapter on acid-base balance.

Chapter 4 deals with basic concepts related to acidosis/alkalosis including definitions, causes, homeostatic mechanisms, and arterial blood gases. The principles in this chapter are expanded and illustrated in chapter 5 and throughout the text.

The purpose of chapter 5 is to provide an overview of fluid, electrolyte, and acid-base balance. The physiological consequences of vomiting and diarrhea are discussed as a means of integrating the topics of preceding chapters. Specific examples, accompanied by laboratory data, are included to illustrate the principles discussed.

Chapter 1

Introduction to Pathophysiology

The study of physiology involves a consideration of how the body maintains itself by establishing balance or **homeostasis** (ho″me-o-sta′sis), whereas pathophysiology focuses on loss of homeostasis as the result of disease. The purpose of this unit is to introduce the subject of pathophysiology with a discussion of a type of imbalance that affects all body systems and frequently accompanies disease processes. Imbalances of total body water volume, fluid shifts, and changes in the chemistry of these fluids are discussed as an introduction to subsequent units dealing with disease processes associated with specific systems of the body.

Fluid and electrolyte imbalance develops as the result of exchanges across cell membranes in response to other physiological derangements. This chapter deals with mechanisms by which movements across cell membranes occur. The chapters which follow are concerned with water balance and the chemistry of body fluids.

CELL MEMBRANE STRUCTURE

The cell membrane encloses cell contents and provides selective control over exchanges between the cell and its environment. The chemical constituents of the cell membrane include varying proportions of lipids and proteins with lesser amounts of carbohydrate.

LIPIDS

Cholesterol is an important lipid constituent in many animal cell membranes (figure 1.1). Phospholipids make up a major class of cell membrane lipids. The basic structure of a phospholipid is glycerol bonded to two fatty acid chains and to a phosphate containing group (figure 1.2). A phospholipid molecule has a hydrophilic (water loving)

homeostasis: Gk. *homoios*, unchanging; Gk. *stasis*, standing

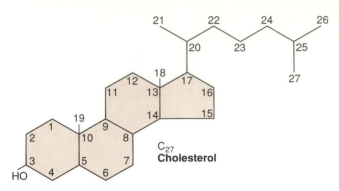

Figure 1.1 Structure of cholesterol. The numbers indicate carbon atoms.

portion that attracts water and a hydrophobic region that repels water. The ionized, electrically charged phosphate group is the hydrophilic region and is referred to as the polar head of the molecule (figure 1.3). Water molecules are polar. They have a positive and negative region; thus they are attracted to the polar head of a phospholipid molecule (figure 1.4). The hydrophobic region is the nonpolar tail consisting of fatty acid chains.

Phospholipids surrounded by an aqueous environment become oriented so there is no contact between water molecules and the nonpolar tails. A common arrangement is the lipid bilayer in which the polar heads are oriented toward the interior and exterior surfaces with the nonpolar tails sandwiched in the middle (figure 1.5).

PROTEINS

Membrane-associated proteins may be either **extrinsic** (i.e., on the surface), or they may be **intrinsic** and an integral part of the membrane. The intrinsic proteins penetrate the membrane from either surface or span the

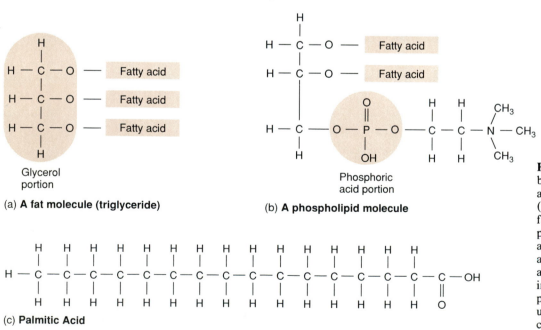

(a) **A fat molecule (triglyceride)**

(b) **A phospholipid molecule**

(c) **Palmitic Acid**

Figure 1.2 (*a*) Glycerol bonded to three fatty acids is a fat or lipid molecule. (*b*) Glycerol bonded to two fatty acid chains and to a phosphate containing group is a phospholipid. (*c*) Palmitic acid is an example of a fatty acid. Fatty acids that are incorporated into phospholipid molecules are usually 16 or 18 carbon chains.

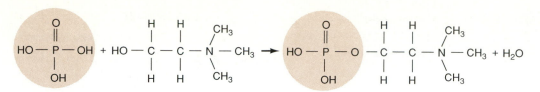

(a) **Phosphoric acid + Choline → Phosphatidyl choline**

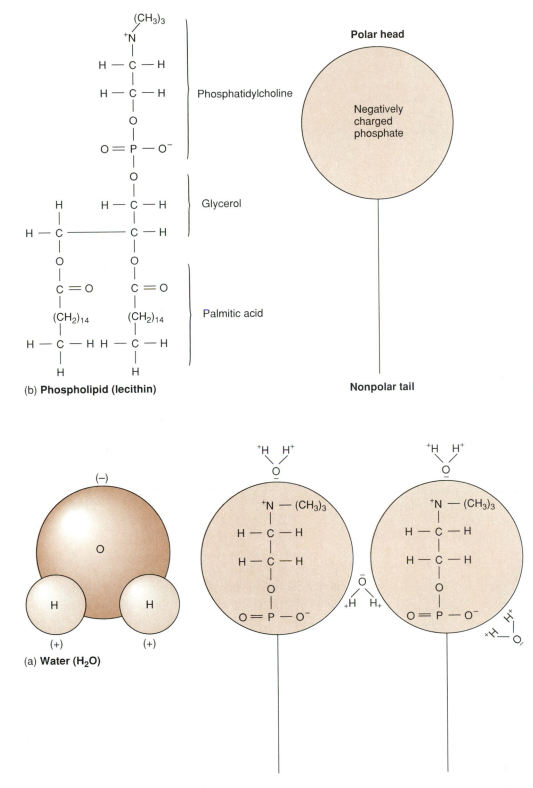

Phosphatidylcholine

Glycerol

Palmitic acid

(b) **Phospholipid (lecithin)**

Polar head

Negatively
charged
phosphate

Nonpolar tail

Figure 1.3 (*a*) Choline is an
alcohol that reacts with
phosphoric acid. (*b*) The
polar head of a phospholipid
molecule consists of a
negatively charged phosphate
group bonded to an alcohol.
The alcohol may also have
electrical charge. The head is
hydrophilic and the nonpolar
tail is hydrophobic.

(a) **Water (H₂O)**

(b)

Figure 1.4 (*a*) A model of a
water molecule showing the
charged areas and thus its
polar characteristics.
(*b*) Water molecules are
attracted to the polar regions
of phospholipid molecules.

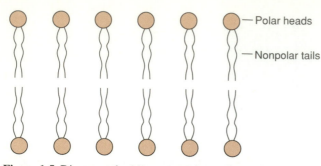

Figure 1.5 Diagram of a bilayer membrane formed by phospholipid molecules.

thickness of the membrane one or more times. Some membrane proteins are bound on the external surface to carbohydrate side chains and are called glycoproteins. Membrane proteins carry out functions, such as transport in which protein molecules behave as pumps, gates, or channels. Proteins are receptors for hormones and neurotransmitters and serve an enzymatic function.

Fluid Mosaic Membrane Model

The fluid mosaic model of membrane structure based on experimental observations is shown in figure 1.6. There is a lipid bilayer with the hydrophilic phospholipid heads forming the inner and outer membrane surface. The middle region made up of phospholipid tails is in a disordered fluid state that allows lateral diffusion of lipids and some proteins. Both cholesterol and the fatty acid chains of phospholipids play a role in maintaining fluidity. Proteins are interspersed throughout the lipid bilayer in a mosaic pattern. Carbohydrates are attached to lipids or proteins on the external surface of the membrane.

Transmembrane Movements

The proteins and the lipid bilayer of a cell membrane act as a barrier to transmembrane movements and control the rate of exchange between the cell and its environment. Some molecules cross a cell membrane by diffusion, while others are carried across.

Simple Diffusion

Diffusion is a process whereby molecules move from an area of high concentration to an area of low concentration. The process is due to random molecular collisions and the fact that a greater number of collisions occur in a high concentration region. The result is a tendency to disperse, and the overall effect is to equalize concentrations of both areas. Molecules tend to diffuse regardless of whether the matter is in a solid, liquid, or gaseous state (figure 1.7).

Diffusion occurs across a cell membrane, although the rate of diffusion is affected by such factors as lipid solubility, electrical charge, and molecular size. The following are useful generalizations:

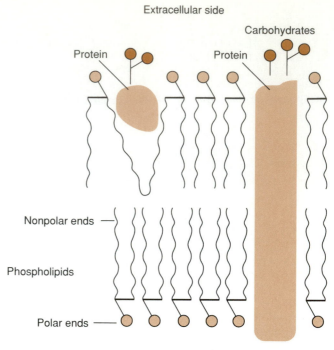

Figure 1.6 Fluid mosaic model of membrane structure. Hydrophilic phospholipid heads line the interior and exterior membrane surfaces. The middle region is made up of phospholipid tails. Proteins are interspersed throughout the bilayer, and carbohydrates are attached to the external surface.

1. The rate of diffusion through a membrane markedly decreases with increasing molecular size.
2. Lipid soluble substances dissolve in the nonpolar interior of a membrane and cross by diffusion; thus, membranes are more permeable to lipid soluble substances, and the hydrophobic interior of a membrane is a barrier to water soluble molecules.
3. Ions are relatively insoluble in lipids, and membranes are relatively impermeable to both ions and polar molecules.

A cell membrane is freely permeable to water and this is a notable exception to the generalizations listed above. Water molecules diffuse across a membrane 10^9 times faster than sodium and potassium ions do. There is evidence that membrane protein molecules may be constructed so they form a hydrophilic channel along the central axis to provide quick passage for water. Proteins may also form passages that allow selected ions to cross the membrane (figure 1.8).

Facilitated Diffusion

Evidence indicates that proteins embedded in the cell membrane function as carriers to aid certain molecules in crossing the membrane. The direction of movement is the same as in unassisted diffusion, that is, from an area of

(a) Two pieces of metal clamped together

Over a period of time, there is some diffusion of atoms

(b) Instant tea added to water

Tea dissolves, and with time, becomes evenly dispersed

(c)

Figure 1.7 Diffusion occurs in (*a*) solids, (*b*) liquids, and (*c*) gases.

high concentration to that of a lower concentration. The rate of transfer is faster than it would be in simple diffusion. A model of how facilitated diffusion, also called **carrier-mediated diffusion,** may occur is shown in figure 1.9. Intracellular transfer of certain water soluble molecules, such as sugars and amino acids, is facilitated by carrier-mediated diffusion.

ACTIVE TRANSPORT

Simple and facilitated diffusion involve the movements of ions or molecules down a concentration gradient, that is, from high to low concentrations with a tendency to establish equal concentrations on both sides of a membrane. The concentration of chemical substances in the cytoplasm of cells in the body is, however, different from that of extracellular fluids. A concentration gradient is established by active transport by using energy to pump a substance across a membrane to maintain a high concentration on one side of that membrane (figure 1.10).

Sodium-Potassium Pump

As will be discussed further in chapter 3, animal cells have a high potassium (K^+) and low sodium (Na^+) content relative to the surrounding extracellular fluids. The system that maintains this concentration gradient is called the

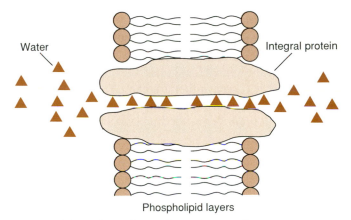

Water

Integral protein

Phospholipid layers

Figure 1.8 Membrane integral proteins may form a hydrophilic channel for quick passage of water molecules.

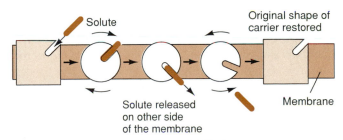

Solute

Original shape of carrier restored

Solute released on other side of the membrane

Membrane

Figure 1.9 A diagram of facilitated diffusion.

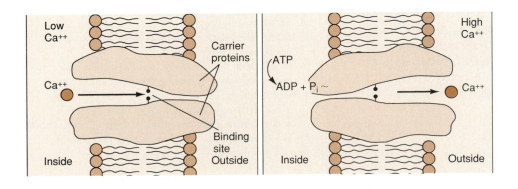

Low Ca++

Carrier proteins

ATP

ADP + P$_i$ ~

High Ca++

Ca++

Ca++

Binding site

Inside

Outside

Inside

Outside

Figure 1.10 A diagram of active transport showing the involvement of integral proteins and ATP as a source of energy. The direction of movement is from an area of low to an area of high concentration.

sodium-potassium (Na^+/K^+) pump because energy is utilized and because the transport of the two ions is linked.

The essential features of this active transport system appear to include the following. The carrier molecule is an integral membrane protein and is an enzyme. The enzyme is called ATPase, and it causes the breakdown of adenosine triphosphate (ATP) to adenosine diphosphate (ADP) plus inorganic phosphate with the release of energy. Sodium and potassium ions bind to specific receptor sites on the carrier protein, which leads first to phosphorylation and then to dephosphorylation of the carrier. This, in turn, brings about a conformational change in the carrier molecule resulting in the extrusion of three sodium ions in exchange for inward transport of two potassium ions. Active transport makes it possible to maintain a concentration gradient for sodium ions external to the cell and an internal potassium ion concentration gradient.

Cotransport

There are transport processes that involve the binding of an ion plus another molecule to a membrane protein; thus, the term **cotransport** is used. The ion flows down a concentration gradient and the second molecule is carried by this flow in the same (sometimes the opposite) direction and against a concentration gradient. Figure 1.11 shows a model for the cotransport of sodium and glucose in which the flow of sodium drives the transport of glucose.

SUMMARY

• • •

The mechanisms by which ions or molecules may cross a membrane include (1) simple diffusion, (2) passage through channels, (3) facilitated diffusion, and (4) active transport systems. The explanation of these mechanisms is based on a model of the cell membrane with a lipid bilayer penetrated by cholesterol and protein molecules. Diffusion is a passive process by which substances move from an area of high concentration to an area of low concentration. Particles may dissolve in the nonpolar middle region of the membrane or may pass through protein channels in the process of membrane crossing. In general, lipid soluble substances diffuse across a membrane most efficiently. Water is a notable exception because it rapidly diffuses across a membrane. Facilitated diffusion involves attachment to carrier protein molecules, which assist in the high to low concentration movement. Active transport requires energy to bring about transmembrane movements against a concentration gradient, i.e., from low to high concentrations. Two examples of active transport are the cotransport of sodium and glucose and the Na^+/K^+ pump.

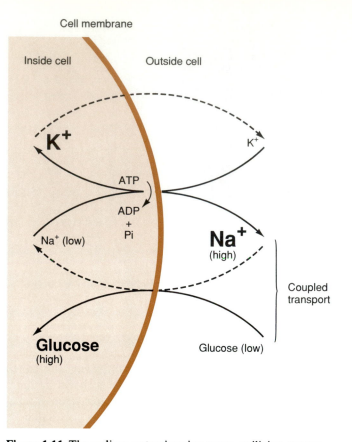

Figure 1.11 The sodium-potassium ion pump, utilizing energy from the breakdown of ATP, maintains a high intracellular concentration of potassium (K^+) and a high extracellular concentration of sodium (Na^+). The passive intracellular diffusion of Na^+ (dotted lines) provides the energy for glucose movements into the cell (against a concentration gradient). This is an example of cotransport.

REVIEW QUESTIONS

• • •

1. Is the surface of a cell membrane, which is made up of the heads of phospholipid molecules, polar or nonpolar?
2. Is the surface of a membrane hydrophilic or hydrophobic?
3. Is the middle region of the bilipid layer of the cell membrane hydrophilic or hydrophobic?
4. Are lipid soluble or water soluble substances more likely to dissolve in the middle region of the bilipid layer?
5. To which category(ies) of substances is a cell membrane relatively impermeable: polar molecules, nonpolar molecules, ions?
6. In which of the following processes are membrane proteins involved: crossing through channels, facilitated diffusion, cotransport, Na^+/K^+ pump?
7. What drives the transport of glucose against a concentration gradient in the process of cotransport?
8. What accounts for the remarkable permeability of cell membranes to water?

9. What is the basic reason for the movements of sodium and potassium as controlled by the Na^+/K^+ pump?
10. What is the source of energy for the Na^+/K^+ pump?

SELECTED READING

• • •

Harvey, B. J. et al. 1988. Intracellular pH controls cell membrane Na^+ and K^+ conductances and transport in frog skin epithelium. *Journal of General Physiology* 92:767–91.

Pool, R. 1989. New microscope images ions ins and outs (news). *Science* 243(4891):609.

Reuter, H. 1987. Modulation of ion channels by phosphorylation and second messengers. *News in Physiological Sciences* 2:168–71.

Simons, K. et al. 1988. Lipid sorting in epithelial cells. *Biochemistry* 27(7):6197–6202.

Testa, I. et al. 1988. Abnormal membrane cation transport in pregnancy-induced hypertension. *Scandinavian Journal of Clinical Laboratory Investigation* 48(1):7–13.

Yeagle, P. L. et al. 1988. Effects of cholesterol on (Na^+, K^+)—ATPase ATP hydrolyzing activity in bovine kidney. *Biochemistry* 27(17):6449–52.

Chapter 2

Fluid Balance

The focus of this chapter is the distribution of body fluid, whereas discussion of electrolytes as important chemical constituents of body fluids is reserved for the chapter that follows. The first part of this chapter describes fluid compartments and factors that control exchanges between compartments. The last section deals with fluid imbalance, specifically water excess and dehydration.

FLUID COMPARTMENTS

The concept of fluid compartments separated by a semipermeable membrane is used to describe the distribution of water in the body. Fluid within and outside of cells, i.e., intracellular and extracellular fluid, represents two major compartments. Extracellular fluid is either plasma in blood vessels (figure 2.1) or a watery medium called **interstitial** (in″ter-stish′al) fluid, which surrounds cells of the body. Figure 2.2 shows the two major compartments which may be visualized as three separate spaces: (1) **intravascular** within blood vessels, (2) interstitial, and (3) intracellular. The membrane that separates blood plasma and interstitial fluid is the capillary wall made up of a single layer of cells (figure 2.3). The partition between the cytoplasm of cells and interstitial fluid is the cell membrane. Water moves freely across the membranes of these compartments (figure 2.4). Fluids in other localized parts of the body, such as cerebrospinal and intraocular fluid, have unique functions and for now will not be included in this discussion.

intravascular: L. *intra,* within; L. *vasculum,* small vessel
interstitial: L. *interstitium,* space between

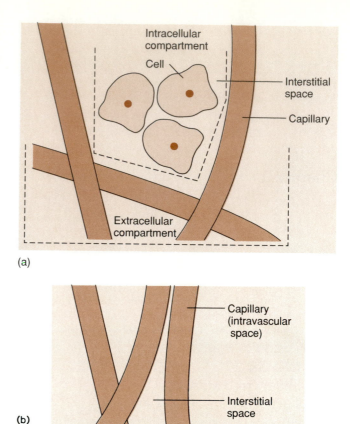

(a)

(b)

Figure 2.2 (*a*) Intracellular and extracellular compartments are the major fluid compartments of the body. (*b*) The extracellular compartment is made up of the intravascular space within blood vessels and interstitial space around blood vessels and cells.

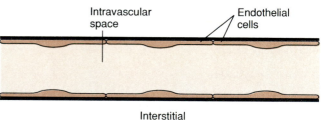

Figure 2.3 The capillary wall is made up of a single layer of cells. This constitutes a semipermeable membrane that separates the intravascular and interstitial spaces.

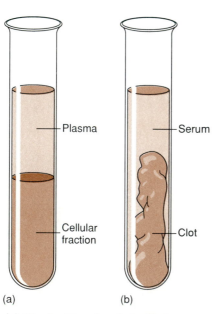

(a)   (b)

Figure 2.1 (*a*) Blood with a chemical added to prevent clotting. The tube has been centrifuged to separate the cellular and the liquid fraction. (*b*) Blood that has been allowed to clot.

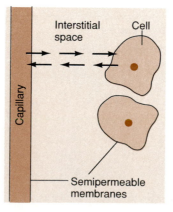

Figure 2.4 Fluid compartments separated by a semipermeable membrane. The intravascular and interstitial spaces are separated by capillary walls. The interstitial and intracellular spaces are separated by cell membranes. Fluid and electrolyte shifts occur between these compartments.

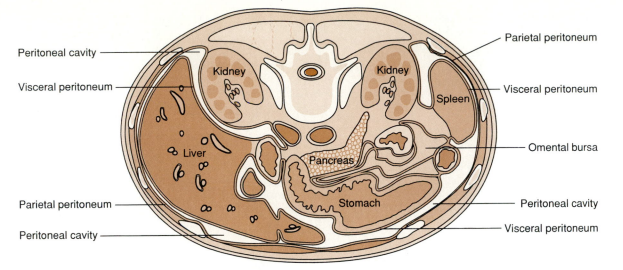

Peritoneal cavity

Visceral peritoneum

Kidney

Kidney

Parietal peritoneum

Visceral peritoneum

Spleen

Liver

Omental bursa

Pancreas

Parietal peritoneum

Stomach

Peritoneal cavity

Peritoneal cavity

Visceral peritoneum

Figure 2.5 A transverse section through the abdominal cavity showing the peritoneal cavity. There is a membrane called the parietal peritoneum that lines the body wall. The membrane covering the surfaces of organs or viscera is called the visceral peritoneum. The peritoneal cavity is an example of a third-space compartment when fluid accumulates in the space.

Fluid shifts may occur as the result of disease or injury. When there is an accumulation of fluid in a tissue or in a body cavity, it is called a **third-space compartment.** Third-space compartments play a major physiological role in certain disease processes. For example, liver disease may lead to significant accumulations of fluid in the peritoneal cavity (figure 2.5), and since the fluid remains trapped in the space, this essentially represents a fluid loss for the body (chapter 36).

CAPILLARY-INTERSTITIAL FLUID DYNAMICS

There is a relatively free exchange of substances between blood plasma in capillaries and fluid in the interstitial spaces. The evidence of this is that, with the exception of protein, constituents on both sides of the capillary membrane are about equal (table 2.1). The factors that control these exchanges are discussed in the following paragraphs.

DIFFUSION

The capillary wall is made up of a single layer of endothelial cells with small openings or pores between the cells. Water, small molecules, and ions diffuse in both directions through this barrier, although the small pore size limits the movements of such large protein molecules as albumin. Diffusion involves movements of substances from an area of high to low concentrations and, if separated by a membrane, tends to establish equal concentrations on both sides of the membrane (chapter 1). Figure 2.6 illustrates diffusion of salt and water across a semipermeable

TABLE 2.1 Electrolyte composition of plasma and interstitial fluid

	Plasma (meq/liter)	Interstitial Fluid (meq/liter)
Na^+	140.0	145.5
K^+	4.5	4.8
Ca^{++}	5.0	2.8
Mg^{++}	1.5	1.0
Cl^-	104.0	116.6
HCO_3^-	24.0	27.4
Protein	15.0	8.0

From H. J. Carroll and M. S. Oh, Water, Electrolyte, and Acid-Base Metabolism. *Copyright © 1989 J. B. Lippincott Company, Philadelphia, PA. Reprinted by permission.*

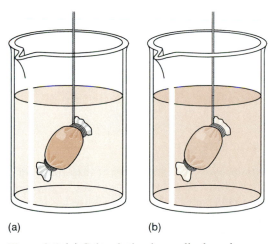

(a)   (b)

Figure 2.6 (*a*) Salt solution in a cellophane bag suspended in a beaker of water. (*b*) Diffusion of salt and water across the semipermeable membrane results in equal concentrations on both sides of the membrane.

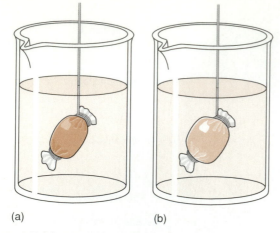

Figure 2.7 (*a*) A protein solution in a cellophane bag suspended in a beaker of water. (*b*) Protein molecules are too large to cross the membrane. Water diffuses into the bag.

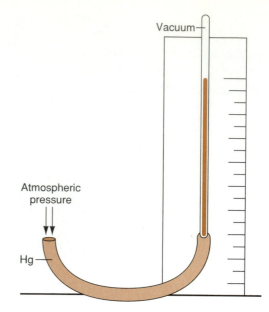

Figure 2.8 A column of mercury in a tube supported by atmospheric pressure. The pressure is usually expressed in millimeters (the height of the column).

membrane. Water alone diffuses when the membrane is impermeable to the solute (figure 2.7). The basic concepts shown in these figures are essential to understanding both fluid and electrolyte balance in the body. The overall tendency of intravascular/interstitial and interstitial/intracellular diffusion is to establish an equilibrium with equal concentrations on both sides of the semipermeable membrane. Concentration differences develop as the result of various factors that will be discussed in this and the following chapters.

FILTRATION

Filtration is the net flow of water due to the overall effect of pressures on both sides of a membrane. Fluid is filtered out of the arterial ends of capillaries in response to changes in hydrostatic and oncotic pressures. **Hydrostatic pressure** (HP) is fluid pressure, either blood pressure in the capillaries or pressure exerted by interstitial fluid outside the capillaries. **Oncotic pressure** is a drawing force resulting from the presence of protein in solution in plasma or interstitial fluid. The unit of measurement for pressure is millimeters (mm) of mercury (Hg), and it is used to identify the pressure that would support a column of mercury, with the height of that column expressed in millimeters (figure 2.8).

HYDROSTATIC PRESSURE

Blood pressure or hydrostatic pressure at the arterial end of a capillary is about 30 mm Hg and at the venous end is about 10 mm Hg (figure 2.9). This pressure forces fluid out of the capillary. Indirect measurements show that the hydrostatic pressure in interstitial spaces is less than atmospheric pressure, on the order of −6 mm Hg. These measurements imply that this is a pulling force or a kind of suction; thus, interstitial hydrostatic pressure draws fluid out of capillaries as well (figure 2.10).

hydrostatic: Gk. *hydor*, water; Gk. *statos*, standing

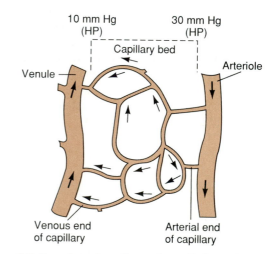

Figure 2.9 Branches of small arteries, called arterioles, form microscopic capillary beds. Blood from the capillaries drains into small veins or venules. The blood hydrostatic pressure (HP) at the arterial end of the capillary is about 30 mm Hg. At the venous end the hydrostatic pressure is about 10 mm Hg.

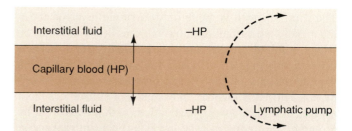

Figure 2.10 Hydrostatic pressure of blood tends to cause fluid to leak out of capillaries. The HP of interstitial fluid is negative or is like a suction, and it also draws fluid out of capillaries. The lymphatic pump removes excess fluid as it accumulates.

16

The removal of interstitial fluid by the lymphatic system contributes to the negative pressure (pull) of interstitial hydrostatic pressure. There is a network of microscopic lymphatic capillaries in the interstitial spaces. Interstitial fluid with its dissolved substances, including protein, diffuses into this special capillary system. The fluid carried by lymphatic vessels is called **lymph.** Lymphatic capillaries coalesce to form larger vessels and ultimately drain into two major vessels: the **thoracic duct** and the **right lymphatic duct.** The thoracic duct empties into the left subclavian vein, while the right lymphatic duct returns lymph to the general circulation by way of the right subclavian vein (figure 2.11). This constitutes the lymphatic circulation whereby fluid that leaks out of capillaries is returned to blood vessels. The process is described as the **lymphatic pump.**

OSMOTIC-ONCOTIC PRESSURE

Osmotic pressure is an important force in controlling water movements. It may be defined as pressure that develops when there is net movement of water across a membrane into a second compartment. It is a water drawing force expressed in mm Hg and is directly proportional to solute concentration. It is useful to think in terms of potential pressure. For example, a thistle tube filled with a protein solution has a specific osmotic pressure potential that depends on the amount of protein dissolved in the water. If, however, a membrane is placed over the open end of that tube and it is inverted in a beaker of pure water, a real pressure develops in the tube as water moves into the tube to increase the total volume (figure 2.12).

The osmotic pressure of plasma or of interstitial fluid depends on the concentration of such substances as urea, glucose, amino acids, electrolytes, and proteins in solution. Since the composition of plasma and interstitial fluid is essentially the same except for protein, the term **oncotic pressure** is used to identify pressure or potential pressure, which is the result of protein concentration differences. The oncotic pressure of plasma at the arterial end of a capillary is about 28 mm Hg. The oncotic pressure of interstitial fluid is about 5 mm Hg. Some limited leakage of protein into interstitial spaces accounts for the oncotic pressure of interstitial fluid. The first value represents a tendency to draw water into the capillary, and the second value is a force to draw water out of the capillary (figure 2.13).

CAPILLARY-INTERSTITIAL FLUID EXCHANGES

Along the length of a capillary there are opposing inward and outward forces. At the arterial end, the capillary hydrostatic pressure (30 mm Hg), the negative interstitial fluid pressure (−6 mm Hg), and interstitial fluid oncotic pressure (5 mm Hg) are forces that cause fluid to move out of the capillary. The force that favors inward movement is plasma oncotic pressure (28 mm Hg). The sum of these two tendencies, 41 mm Hg outward and 28 mm Hg

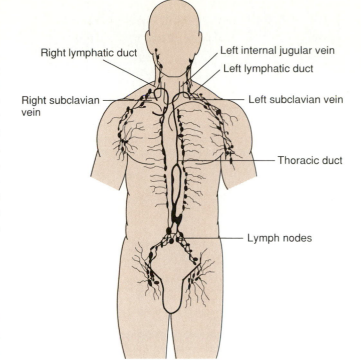

Figure 2.11 Interstitial fluid diffuses into special lymphatic capillaries that drain into larger lymph vessels. A major vessel, the thoracic duct, drains into the left subclavian vein and the right lymphatic duct empties into the right subclavian vein.

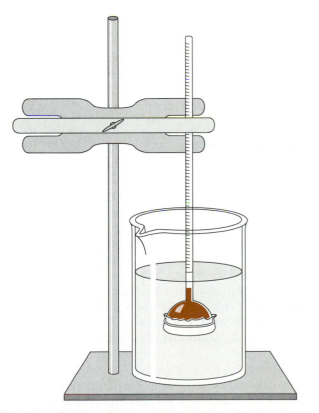

Figure 2.12 Thistle tube inverted in a beaker of water. The tube contains dissolved protein, and with time, water moves into the tube and the water level rises.

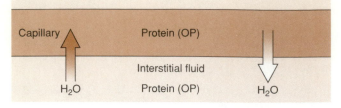

Figure 2.13 Opposing oncotic pressure (OP) inside and outside of a capillary. There is a higher protein concentration in plasma as compared to interstitial fluid; thus, it has a greater oncotic pressure.

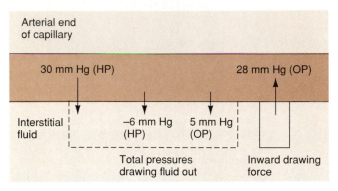

Figure 2.14 Opposing forces that influence fluid shifts across the capillary wall. Hydrostatic pressures (a total of 36 mm Hg) and interstitial oncotic pressure (5 mm Hg) favor outflow. Oncotic pressure (28 mm Hg) inside the capillary is an inward drawing force. The net effect is outflow of fluid (41 mm out minus 28 mm in) with a net pressure of 13 mm.

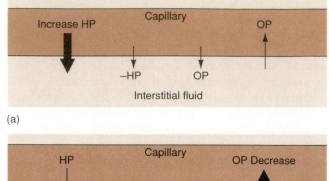

(a)

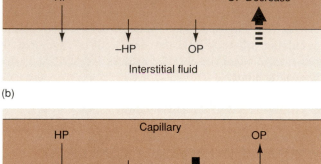

(b)

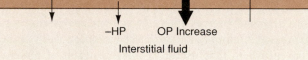

(c)

Figure 2.15 Increased capillary fluid loss resulting in edema may occur under conditions in which there is (*a*) increased capillary hydrostatic pressure, (*b*) decreased oncotic plasma pressure (loss of protein), and (*c*) increased oncotic pressure in interstitial fluid.

inward, indicates that there is a net outward filtration pressure of 13 mm Hg with capillary fluid loss (figure 2.14). There is evidence that pressures are such that some fluid is filtered out of the capillary along its entire length.

ABNORMAL CAPILLARY DYNAMICS

The following observations will be useful in interpreting underlying causes of disturbances in water balance. There is some loss of protein and fluid from capillaries with provision for the return of both to the general circulation by way of lymphatic vessels. You have seen that water movements across the capillary membrane are the result of opposing forces. What happens when there are changes in those forces? In theory, one would predict that there would be increased capillary fluid loss when there is (1) increased capillary hydrostatic pressure, (2) decreased plasma oncotic pressure, and (3) increased interstitial fluid oncotic pressure (figure 2.15).

Edema

An accumulation of fluid in interstitial or tissue spaces is identified as edema, and is apparent because of swelling or puffiness. This may occur throughout the body or may

be localized in a specific region. If interstitial fluid increase is extreme, there will be pitting edema in which touching the skin leaves a depressed imprint of the finger.

It is not always possible to identify what causes disturbances of water balance with certainty, but the following examples illustrate governing principles (table 2.2). Malnutrition leads to a decreased synthesis of protein by the liver, marked lowering of plasma albumin (protein), which probably accounts for edema. The logical steps are (1) decreased plasma protein, (2) decreased plasma oncotic pressure, (3) decreased capillary water drawing force, and (4) accumulation of fluid outside capillaries (figure 2.16). Various liver diseases that result in decreased protein synthesis lead to edema.

Capillary wall damage and increased capillary permeability to protein occurs in response to exposure to certain chemicals, venoms, bacterial toxins, or to the process of inflammation. With protein loss there is predictably a decrease in capillary oncotic pressure and subsequent edema.

Excess fluid and protein in the interstitial spaces is collected by lymphatic vessels and is returned to the blood. Any obstruction to this flow of lymph, possibly due to a tumor or surgical removal of lymph nodes, can lead to edema.

TABLE 2.2 Possible causes of increased capillary hydrostatic pressure or decreased plasma oncotic pressure

Disorder	Mechanism of Edema Formation
Starvation; severe protein deficiency; skin of legs and arms show marked pitting edema	Causes are not completely clear. Frequently, there is decreased oncotic pressure
Glomerulonephritis (an inflammatory kidney disease)	Edema at the onset of the disease is caused by fluid retention; thus, there is increased capillary hydrostatic pressure
Nephrotic syndrome (refers to a clinical picture of urinary protein loss and decreased plasma protein)	There is generalized edema as the result of lowered plasma oncotic pressure
Congestive heart failure (the heart fails as a pump)	There is increased pulmonary venous and capillary pressure because of increased blood volume (increased hydrostatic pressure)
Administration of large amounts of blood or salt solutions leads to pulmonary edema (especially in individuals with heart or kidney disease)	Pulmonary edema is the result of increased blood volume; thus, increased hydrostatic pressure

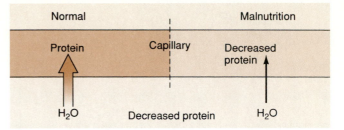

Figure 2.16 In malnutrition, there is decreased synthesis of protein by the liver, resulting in a lowered protein level in plasma. A decrease in oncotic pressure causes a diminished tendency to draw water into capillaries. There are probably other factors that contribute, but the net result is edema.

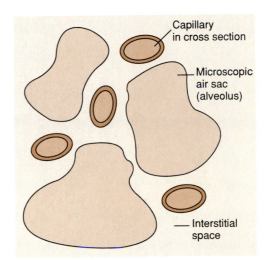

Figure 2.17 A diagram representing a cross section of lung tissue. There is a network of air sacs with capillaries in close proximity. Interstitial space is limited.

Pulmonary Edema

Pulmonary edema is a special kind of edema in which fluid accumulates in interstitial spaces and in the microscopic air sacs or **alveoli** (al-ve′ ō lī′) of the lungs (figure 2.17). The problem deserves particular attention because the process is life threatening and because the lungs in some ways are uniquely susceptible to edema. Pulmonary capillaries are more permeable to protein as compared to capillaries elsewhere in the body, and this implies a tendency toward increased oncotic pressure, or water drawing force, in the interstitium. Lung tissue is largely made up of a network of air sacs and interstitial space is limited. If excessive amounts of fluid leak into the interstitium, the alveoli will fill with fluid and interfere with gas exchange across the alveolar walls.

Two factors protect the lungs from fluid accumulations: (1) a high rate of lymph flow away from the lungs and (2) pulmonary capillary pressure that is lower than systemic capillary pressure. The principles of water balance in the lungs, however, are the same as in peripheral tissues. Any condition that causes increased capillary hydrostatic pressure, decreased capillary oncotic pressure, or increased capillary permeability may lead to pulmonary edema.

In heart failure when the left ventricle fails as a pump, pulmonary circulation is slowed, there is a buildup of blood in the pulmonary vessels, and blood pressure increases (figure 2.18). Pulmonary edema is characteristic of various forms of heart failure (figure 2.19).

Noxious gases, inflammation (such as occurs in pneumonia), or respiratory burns may cause increased permeability of pulmonary capillaries with loss of fluid and protein. The causes of pulmonary edema in such cases may be a combination of factors: excess fluid and protein loss, lowered plasma oncotic pressure, and increased oncotic pressure in interstitial spaces.

pulmonary: L. *pulmo,* lung
alveolus, pl. alveoli: L. *alveolus,* small cavity

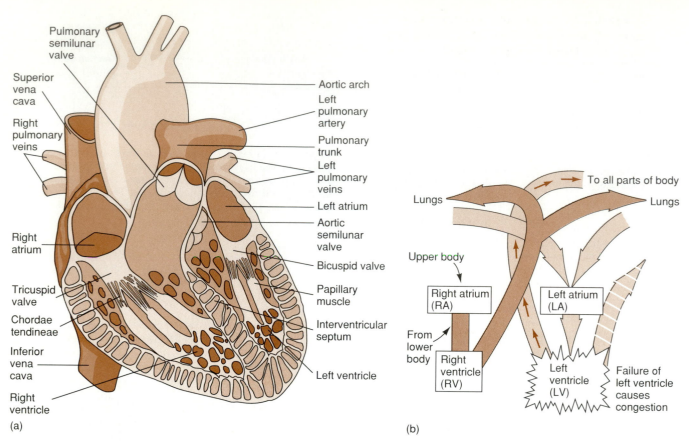

(a)

(b)

Figure 2.18 (*a*) Blood is returned to the right atrium by way of the superior and inferior vena cavae. Blood flows into the right ventricle and is then pumped to the lungs through the pulmonary artery and its branches. Blood is returned to the left atrium from the lungs by the pulmonary veins. The left ventricle receives blood from the left atrium. The left ventricle

pumps blood to all parts of the body through the aorta and its branches. (*b*) Diagrammatic scheme of blood flow. If the left ventricle fails as a pump, there is slowed blood flow from the left atrium and from the lungs. The result is increased volume and increased blood pressure in pulmonary capillaries. This leads to pulmonary edema.

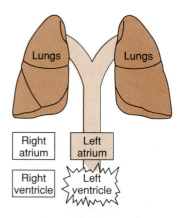

Figure 2.19 Diagram showing areas of blood congestion (shaded) when the left ventricle fails as a pump.

INTERSTITIAL FLUID-CELLULAR DYNAMICS

In addition to fluid exchange between the intravascular and interstitial compartments, there is rapid water movement between cells and the interstitium as well. Electrolyte composition is a major factor in this exchange with osmotic pressure as the drawing force (table 2.3).

Fluid imbalance may occur and result in either cellular dehydration or cellular hydration. The following is an example of cell water loss. Glucose does not cross a cell membrane freely by simple diffusion and (as in diabetes mellitus) may reach high concentrations in extracellular fluids. The concentration gradient outside the cells draws water out of cells by creating an osmotic pull (figure 2.20).

Cellular hydration can occur as well. Sodium ions predominate in extracellular fluid (plasma and interstitial fluid) and, for that reason, play a major role in determining osmotic pressure. If there is sodium loss for any reason, the result is decreased solute concentration, decreased osmotic pressure, and the flow of water into cells

TABLE 2.3 A comparison of electrolyte
concentrations in interstitial fluid
and intracellular fluids

	Interstitial Fluid (meq/liter)	Muscle Cells (meq/liter)
Na+*	145.5	12
K+***	4.8	150
Ca++	2.8	0.0000001
Mg++	1.0	7
Cl−*	116.6	3
HCO3−	27.4	10
PO4−3**	2.3	116
Protein	8.0	40

From H. J. Carroll and M. S. Oh, Water, Electrolyte, and Acid-Base
Metabolism. *Copyright © 1989 J. B. Lippincott Company,
Philadelphia, PA. Reprinted by permission.*
*The principle electrolytes in interstitial fluid
**The main intracellular ions

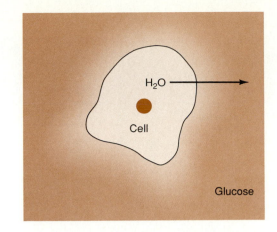

Figure 2.20 An increase in glucose concentration in
extracellular fluid draws water out of cells. Cellular
dehydration is the result.

(figure 2.21). If low sodium levels (perhaps caused by diuretics, vomiting, or low sodium intake) are combined with intake of excess water, the problem is exaggerated.

EFFECTS OF IMBALANCE ON COMPARTMENTS

Thus far there has been discussion of filtration and diffusion resulting in fluid shifts into the intravascular, intracellular, and interstitial spaces. From a practical point of view, which area will actually be most affected by fluid losses or gains? There are several factors to be considered. How quickly does the change occur? Is the fluid pure water, or does it have dissolved substances in it? Is the solute concentration equal to, less than, or greater than that of plasma? Are all three areas equally susceptible to loss or gain?

The intravascular compartment is likely to be affected by volume changes first. This is reasonable when you consider the following. Fluid taken by mouth is absorbed into the bloodstream; fluids are frequently given to a patient by intravenous infusion; the kidney directly filters blood; and injury and disease often cause blood loss. In contrast to this, the interstitial spaces are less available and represent a reservoir for fluid exchange, while the cells are last to be affected by imbalances.

Involvement of interstitial space depends on how quickly blood volume changes occur. If the change is abrupt, then the immediate predominant effect will be on the intravascular compartment. Consider the case of an individual who, as the result of a neurotic disorder, drinks large amounts of water in a short period of time, perhaps a liter in an hour. Contrast this with an individual who drinks the same amount over a 24-hour period. In the first

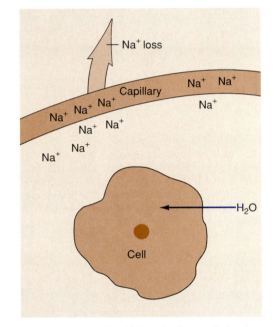

Figure 2.21 Sodium ion (Na+) loss leads to cellular hydration.

instance, there will be immediate intravascular expansion and, with time, distribution into interstitial space. In the second instance, both intravascular and interstitial volume will expand together unless increased urinary output compensates for volume overload (figure 2.22). If instead of fluid excess there is loss, as in the case of hemorrhage, and this loss occurs within minutes, then only the intravascular compartment is immediately affected. Both compartments share the loss equally if it occurs over a period of hours. Significant shifts of fluid between the intravascular and interstitial areas require several hours.

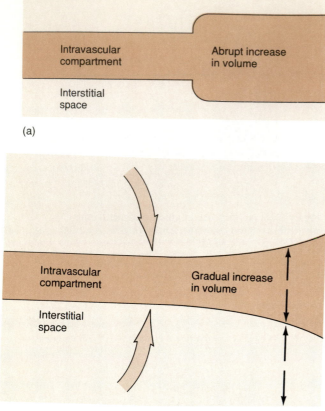

(a)

(b)

Figure 2.22 (*a*) When there is an increase in volume over a short period of time, there is immediate intravascular expansion with distribution to the interstitial spaces over a period of hours. (*b*) A gradual increase in intravascular volume leads to expansion of both intravascular and interstitial compartments.

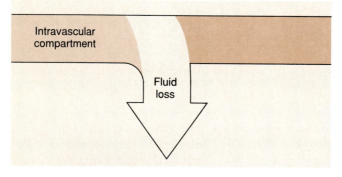

Figure 2.23 A loss of hypotonic fluid has a concentrating effect on extracellular fluid.

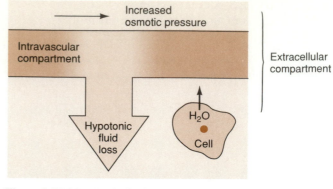

Figure 2.24 Hypotonic fluid loss from the extracellular compartment causes increased osmotic pressure due to the concentrating effect. Increased osmotic pressure draws water out of cells.

TABLE 2.4 Summary of how the intravascular, intracellular, and interstitial compartments are affected by fluid imbalance

Imbalance	Compartment
Order of involvement (fluid loss or fluid excess)	a. Intravascular b. Interstitial c. Intracellular
Imbalance occurs suddenly	Intravascular compartment affected first
Imbalance develops slowly	Extracellular compartments affected equally
Fluid excess is pure water	Water diffuses freely into all three compartments
Loss of hypotonic fluid	Osmotic pressure becomes higher in the extracellular compartment; water drawn out of cells

The effect of fluid loss or excess depends on solute concentration as well as on rate of change. The loss of **hypotonic fluid** (which has a solute concentration less than that of plasma) has a concentrating effect on extracellular fluid, resulting in an increase of osmotic pressure (figure 2.23). Osmotic pressure depends on the number of particles in solution, and this kind of fluid loss draws water out of cells in response to the increased extracellular osmotic pressure (figure 2.24). Cells are unaffected if there is no change in extracellular solute concentration.

Cellular involvement in fluid excess depends on the solute concentration of the fluid as compared to plasma. Table 2.4 summarizes the general principles governing compartmental involvement in fluid imbalance.

hypotonic: Gk. *hypo*, less than; Gk. *tonos*, stretching

FLUID BALANCE DISORDERS

Normally the total volume of body water is maintained within relatively narrow limits, and in adults, that total volume is approximately 60% of body weight. Table 2.5 shows that intracellular fluid is approximately 67% of total body water, plasma volume is about 8%, and interstitial fluid constitutes about 25% of the total volume.

TABLE 2.5 Normal water distribution and volume in an individual weighing 70 kg

	Volume (liters)	Percent of Lean Body Mass	Percent of Total Body Water
Total body water	42	60	
Intracellular fluid	28	40	67
Extracellular fluid	14	20	33
Plasma	3.5	5	8.3
Interstitial fluid	10.5	15	25

From A. I. Arieff and R. A. DeFronzo, Fluid, Electrolyte, and Acid-Base Disorders. *Copyright © Churchill Livingstone, New York, 1985. Reprinted by permission.*

THIRST CENTER

Overall water balance is maintained by mechanisms that control thirst (ultimately determining ingestion) and renal control of urine volume. Neurons that stimulate thirst are located in the hypothalamus, a region of the brain making up the floor and a part of the walls of the third ventricle (figure 2.25). The thirst center is limited to neurons in a small area of the hypothalamus, lateral to the region where the optic nerves cross. Mechanisms that determine thirst or the desire for water are complex and are not completely understood.

A primary stimulus for thirst is cellular dehydration, which may be the result of inadequate intake of water or the result of an increase in extracellular solute concentration. Thirst is also caused by a loss of extracellular fluid with the decreased volume stimulating specialized neurons called stretch receptors. These receptors, which respond to changes in volume, are located in the left atrium

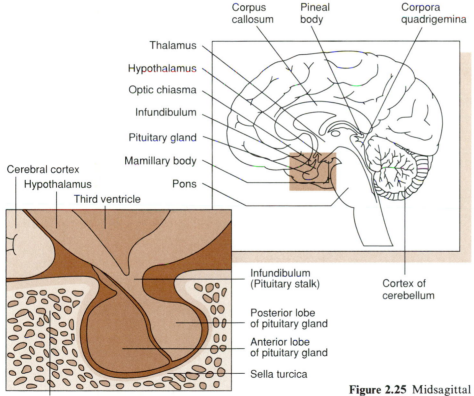

Figure 2.25 Midsagittal section through the brain showing the region of the hypothalamus.

of the heart, in the aortic arch, at the bifurcation of the carotid arteries, and in the pulmonary veins (figure 2.26). There are factors other than cellular dehydration and decreased fluid volume that influence fluid intake. **Prostaglandins,** a family of fatty acids that affect a wide variety of physiological processes, probably have a role in regulating thirst. Prostaglandins are 20-carbon fatty acids that contain a 5-carbon ring (figure 2.27). Another possible factor in thirst is increased plasma levels of the hormone **angiotensin II.** This hormone has been implicated in causing thirst and plays a role in reabsorption of water by the kidney as well.

RENAL CONTROL

Kidney function is a major homeostatic mechanism for maintaining water balance, and one aspect of that function involves hormonal control. This topic is discussed further in chapter 11.

• • •

The term osmolality is a combined form of the words osmotic and molal. Osmotic pressure is determined by solute concentration, and a molal solution is a gram molecular weight of a substance dissolved in a liter of water; hence, the term osmolality refers to solute concentration.

• • •

WATER EXCESS

Fluid excess initially affects the extracellular compartment and is the result of inadequate renal output as compared to intake. If the retained water is either pure or hypotonic as compared to plasma, the effect is to dilute the extracellular fluid, which causes a decrease in osmolality. Increased volume, depending on the magnitude, may lead to pulmonary and/or cerebral edema and generalized edema. When interstitial fluid is dilute as compared to the intracellular salt concentration, water is drawn into the cells by an osmotic pull. As a result, there is increased cellular volume resulting in possible changes in cellular function (figure 2.28). Evidence of those changes may be particularly apparent in the central nervous system.

If the total body water is increased by intravenous infusion of an isotonic solution (solute concentration the same as plasma), the result will be increased extracellular volume with little or no effect on cellular volume (figure 2.29). For now we will designate the solute as being a salt that will diffuse across a membrane. In the case of an isotonic solution, both water and solute will move across the capillary wall to establish equal concentrations in plasma and in interstitial fluid. The net result is no change in solute concentration and no effect on cells.

bifurcation: L. *bis,* twice; L. *furca,* two pronged fork
isotonic: Gk. *iso,* same; Gk. *tonos,* stretching

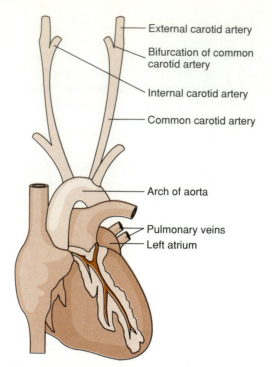

Figure 2.26 Specialized neurons called stretch receptors are located in the left atrium of the heart, the aortic arch, at the bifurcation of the carotid arteries, and in the pulmonary veins. Stimulation of these receptors by a loss of extracellular volume is a factor in the sensation of thirst.

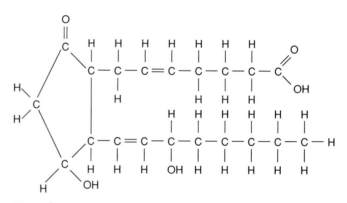

Figure 2.27 Prostaglandin E₂ (PGE₂) is a 20-carbon fatty acid that contains a 5-carbon ring.

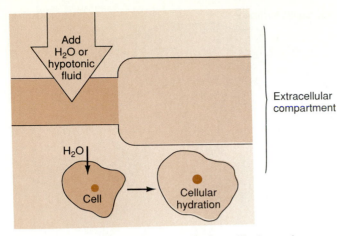

Figure 2.28 Hypotonic fluid excess leads to dilution and expansion of the extracellular compartment. When interstitial fluid has a lesser solute concentration as compared to intracellular fluid, water is osmotically drawn into cells. The result is cellular hydration.

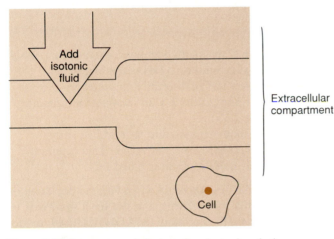

Figure 2.29 Intravenous infusion of an isotonic solution causes expansion of the extracellular compartment with little effect on cellular volume.

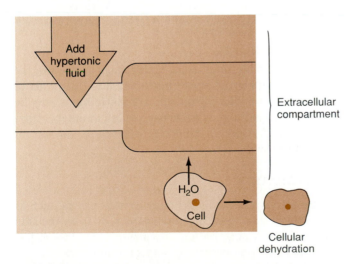

Figure 2.30 Intravenous infusion of a hypertonic solution causes expansion of the extracellular compartment and causes an increase in solute concentration. The resulting increase in osmotic pressure draws water out of cells and leads to cellular dehydration.

In the case of infusion of a hypertonic solution, diffusion of both solute and water into the interstitial fluid occurs, and an equal concentration on both sides of capillary walls is established. That concentration, however, is higher than normal and creates an osmotic drawing force that pulls water out of the cells (figure 2.30).

The results of overhydration or water intoxication may be summarized as follows:

1. If the fluid is hypotonic, there will be an increase in extracellular volume, a dilutional effect, and subsequent cellular hydration.
2. If the fluid retained is isotonic to plasma, there will be expansion of the extracellular compartment and no effect on cells.
3. If the excess fluid is hypertonic, there will be increased extracellular volume and cellular dehydration.

The following discussion and examples are limited to pure water or hypotonic fluid excess.

Signs and Symptoms

The clinical manifestations of pure water overhydration are associated with lowered sodium ion levels and depend on the magnitude and rate of change. If the fluid excess develops over a period of time and the sodium ion level is about 130 meq/liter, the symptoms are likely to include **anorexia** (an″o-rek′se-ah) or loss of appetite, exhaustion, muscle cramps, and **dyspnea** (disp′ne̅-ah) upon exertion. If the sodium level is lower than 130 meq/liter and down to about 120 meq/liter, there may be nausea, vomiting, and abdominal cramps. When the sodium level is 115 meq/liter or below, such neurological effects as lethargy, confusion, delirium, and convulsions are apparent. Nausea, vomiting, seizures, violent behavior, and coma are the consequences of a sudden lowering of serum sodium ion level to the range of about 125 meq/liter. When death occurs, it is usually attributable to cerebral edema.

In addition to the preceding observations, there are some nonspecific signals that point to overhydration, and these symptoms include marked salivation and **lacrimation,** watery diarrhea, neck vein distention, and a history of weight gain.

hypertonic: Gk. *hyper*, greater than; Gk. *tonos*, stretching
anorexia: Gk. *anorexia*, absence of appetite
dyspnea: Gk. *dys*, hard; Gk. *pnein*, to breathe
lacrimation: L. *lacrimare*, to shed tears

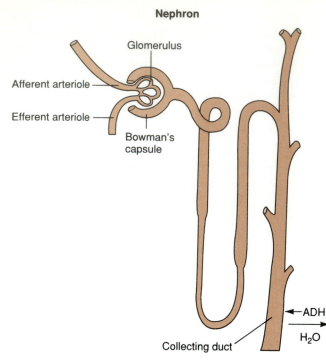

Nephron

Glomerulus

Afferent arteriole

Efferent arteriole

Bowman's capsule

← ADH

H₂O →

Collecting duct

Figure 2.31 Urine is formed in microscopic nephrons in the kidney. Antidiuretic hormone (ADH) increases the permeability of collecting ducts to water. The overall effect is to promote water retention and to decrease the volume of urine.

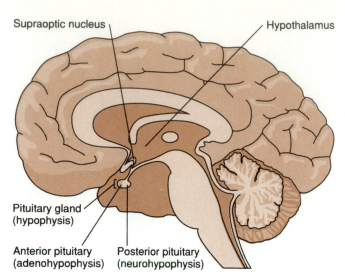

Supraoptic nucleus

Hypothalamus

Pituitary gland (hypophysis)

Anterior pituitary (adenohypophysis)

Posterior pituitary (neurohypophysis)

Figure 2.32 Sagittal section of the brain showing the location of neurons that make up the supraoptic nuclei of the hypothalamus. These neurons synthesize ADH which is stored in the posterior lobe of the pituitary gland.

Excessive Intake of Water

Patients with histories of psychiatric disorders sometimes show a pattern of compulsive water drinking called **psychogenic polydipsia** (pol''ē-dip'sē-ah) or **primary polydipsia**. An individual suffering from psychogenic polydipsia may ingest 10–15 liters of water per day, and acute water intoxication is a probable outcome. A second example in which acute water intoxication has been described is child abuse in which forced water ingestion is used as a form of punishment.

Inappropriate ADH Secretion

Antidiuretic hormone (ADH), also called **vasopressin** or **arginine vasopressin (AVP)**, promotes water retention by increasing water permeability of the collecting ducts in the kidney (figure 2.31). In addition to this, it is a powerful vasoconstrictor causing constriction of arterioles in particular (chapter 31). The hormone is produced by the neurons of the supraoptic nuclei of the hypothalamus and is stored in axon end-plates that terminate in the posterior pituitary gland (figure 2.32).

The following is an overview of events involving ADH:

1. Increased body fluid osmolality triggers ADH release. ADH is carried by blood to the kidney where it promotes water reabsorption. Finally, retained water dilutes body fluids to correct the original imbalance.

polydipsia: Gk. *polys*, many; Gk. *dipsa*, thirst

2. On the other hand, decreased extracellular fluid osmolality inhibits ADH secretion, which results in less reabsorption of water by the kidney and ultimately an increase in osmolality (figure 2.33).
3. Low blood pressure due to fluid loss may stimulate the release of ADH. This occurs if there is a 7% or greater blood loss.

At times ADH is secreted when volume depletion or increased osmolality is absent, and the term **syndrome of inappropriate ADH secretion** (SIADH) is used to describe the phenomenon. Inappropriate ADH secretion has been observed in such conditions as head trauma, bronchogenic carcinoma, encephalitis, pneumonia, and diseases of the central nervous system. Various types of tumors secrete substances with ADH-like activity as well.

Principles of Restoring Balance

If there are no life-threatening neurological symptoms, the mode of treatment of water intoxication is restriction of fluids. If neurological symptoms are alarming, it may be necessary to judiciously administer small amounts of hypertonic (5%) sodium chloride to increase the serum sodium ion level.

DEHYDRATION

Water loss may occur by the more obvious routes, such as the urinary tract, gastrointestinal tract, or from the skin in the form of perspiration. **Insensible water loss** is less obvious. This loss occurs by way of water vapor in the breath and by way of evaporation of water that diffuses through the skin. Table 2.6 summarizes water turnover in the average adult in a day. The effects of a fluid deficit depend on volume and rate of loss and on the amount of

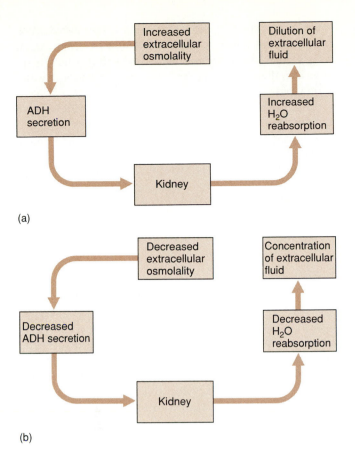

(a)

(b)

Figure 2.33 (a) An increase in extracellular osmolality stimulates ADH secretion. ADH promotes water retention by the kidney. The overall effect is to dilute extracellular fluid with restoration of normal osmolality. (b) Dilution of extracellular fluid leads to a decrease in osmolality. This leads to inhibition of ADH secretion. There is a decrease in water reabsorption from the collecting ducts of nephrons in the kidney. The result is increased urine volume, which has a concentrating effect on extracellular fluid.

TABLE 2.6 Water gain and loss
in an average adult per day

Source of Water	Volume in ml	Loss	Volume in ml
Drink	1,500	Urine	1,500
Food	800	Skin	600
Metabolism	294	Lungs	400
		Feces	100

From E. Goldberger: A Primer of Water, Electrolyte and Acid-Base Syndromes, 7th edition. Philadelphia: Lea & Febiger, 1986, p. 42. Reprinted by permission.

electrolytes lost with the water. Isotonic or hypertonic losses will be discussed in chapter 5. Simple dehydration or hypotonic loss will be considered in the following paragraphs.

The effects of hypotonic fluid loss are predictable on the basis of the principles described in the first part of this chapter. Water loss that exceeds solute loss has a concentrating effect, causing increased osmolality and **hypernatremia** (increased sodium ion concentration). When this

occurs, water is osmotically drawn out of cells into the extracellular compartment. The initial effect is hypernatremia with subsequent cellular dehydration. Volume depletion stimulates the release of the two hormones ADH and aldosterone. ADH mediates renal water retention, and aldosterone favors sodium ion and water retention (chapter 11). To summarize, the immediate effect of water loss is hypernatremia and the physiological responses to correct the imbalance are (1) movement of cellular water to the extracellular compartment and (2) hormonal responses that increase extracellular volume.

Signs and Symptoms
A number of observations may be made in cases of dehydration, and these observations depend on the severity of the deficit as well as the rate at which the loss has occurred. There may be weight loss, dry mucous membranes, increased body temperature, sunken eyeballs, general weakness, and if it is an acute episode, hypotension and **tachycardia** (tak″e-kar′dē-ah). Central nervous system effects, due to fluid loss from brain cells, are **somnolence,** confusion, and coma. In infants there may be depression of the fontanelle and decreased **skin turgor** or resiliency. Skin turgor is not likely to be decreased in young adults because of natural elasticity of the skin. In the elderly, skin turgor is difficult to assess because of loss of natural elasticity.

Sweat
Excessive sweating causes dehydration, and because sweat is normally a hypotonic fluid, hypernatremia may occur. The sodium ion concentration of sweat varies depending on the rate of secretion. When the rate of secretion is low, the concentration of sodium ions is low because the ions are reabsorbed before the secretion reaches the surface. Sodium ion concentration of sweat increases with increased rate of sweating. Increases in any insensible fluid losses from the skin or respiratory tract cause dehydration and hypernatremia as well.

Osmotic Diuresis
The term osmotic **diuresis** refers to a situation in which a solute, cleared from the blood by the kidney and appearing in the glomerular filtrate, is not reabsorbed but remains in the glomerular filtrate to be eliminated by the kidney. This high solute concentration creates an osmotic pull that draws water in that direction. Figure 2.31 shows the functional unit of the kidney, the nephron, with a mass of capillaries (a glomerulus) encased by the upper nephron (Bowman's capsule). Plasma filters from the glomerulus into Bowman's capsule and down through the tubules to

hypernatremia: Gk. *hyper,* greater than; L. *natrium,* sodium
tachycardia: Gk. *tachys,* swift; Gk. *kardia,* heart
somnolence: L. *somnulentia,* sleepiness
diuresis: Gk. *diourein,* to pass urine

be ultimately collected as urine. This fluid is called glomerular filtrate and its composition is changed throughout the length of the tubule as substances are either reabsorbed back into the blood or added from the blood to the filtrate. If there is a high concentration of a solute in the filtrate that is (1) not reabsorbed into the blood from the tubule or (2) the concentration is so high that it overwhelms the reabsorptive capacity of the tubule, the presence of this solute in the filtrate osmotically draws water into the tubule, causing increased urine flow with the potential for a serious water loss.

An example of this situation is uncontrolled **diabetes mellitus** in which large amounts of glucose appear in the glomerular filtrate (chapter 32). Glucose is normally reabsorbed from the kidney tubule, but there is a limit to that reabsorptive capacity. The presence of this solute in the glomerular filtrate osmotically draws water into the tubules and causes increased urine flow with the potential of causing excessive water loss. A second example of osmotic diuresis is when **mannitol,** a polysaccharide, is used as a diuretic in cases of cerebral edema. It is a nonreabsorbable solute in the kidney tubules, and its presence creates an osmotic drawing force that promotes water loss. Osmotic diuresis will be discussed further in unit III.

Diabetes Insipidus

Diabetes insipidus may be described as a deficiency of ADH characterized by excretion of excessive amounts of dilute urine and by excessive thirst (chapter 31). Since the normal function of ADH is to favor water reabsorption from kidney tubules into the blood, a deficiency of this hormone implies that water is not reabsorbed but excreted. Diabetes insipidus is the opposite extreme of inappropriate secretion of ADH (discussed earlier), which is characterized by an excess of ADH and by overhydration.

Principles of Restoring Balance

The objective in treating fluid loss is to increase blood volume by replacing fluid. The specific circumstance dictates the route and rate of administration and the type of fluid. There are fluids available that have varied electrolyte composition, and the choices range from sodium chloride and other salt solutions to whole blood.

WATER BALANCE IN SPECIAL CASES

Both the very old and the very young are especially vulnerable to fluid imbalance. With increasing age comes a decreased renal capacity to save water and to put out a concentrated urine. This becomes a significant fact when fluid intake is limited, especially if there is increased insensible water loss, as would be the case with fever. On the other side, the elderly are especially susceptible to water intoxication as well. In some cases, this is the result of increased ADH secretion associated with the stress of surgery or with such illnesses as pneumonia, meningitis, or subdural hematoma (chapter 38). Another general

cause of susceptibility to water intoxication is reduction in renal blood flow due to heart failure, liver disease, or drug induced hypotension. When renal blood flow is diminished, there is a decrease in volume of urine, and water is retained.

Infants belong in a special category as well because an infant has a greater surface area compared to its weight. This means there will be increased insensible fluid losses. Add to this the fact that this little individual has less renal concentrating ability than an adult, and this describes vulnerability to dehydration.

SUMMARY

• • •

Key terms from this chapter are summarized in table 2.7. There are two major fluid compartments in the body, intracellular and extracellular, with intravascular and interstitial areas constituting the extracellular compartment. Water movements into the intravascular, interstitial, and intracellular areas are controlled by a summation of forces (hydrostatic and oncotic pressures) on both sides of the separating membrane. Hydrostatic pressure is the pressure of the volume of fluid, and oncotic pressure is a water drawing force caused by protein dissolved in fluid. Edema,

TABLE 2.7 Definitions of key terms from this chapter

Osmotic pressure	A water drawing force expressed in mm of Hg. The magnitude of the force is determined by the number of particles in solution
Oncotic pressure	A water drawing force expressed in mm of Hg. The magnitude of the force is determined by the concentration of protein in solution
Hydrostatic pressure	The result of the weight of a particular volume of fluid. The force is expressed in mm of Hg
Isotonic	A solution that has the same concentration as a second solution
Hypotonic	A solution of a lesser concentration than a second solution
Hypertonic	A solution of a higher concentration than a second solution
Osmolality	Solute concentration. Hyperosmolality means an increase in solute concentration; hence, an increased osmotic pressure. Hypoosmolality means decreased solute concentration and decreased osmotic pressure
Osmotic diuresis	Increased excretion of fluid due to increased solute concentration in the glomerular filtrate

an accumulation of fluid in interstitial spaces, occurs with an intravascular increase of hydrostatic pressure, a decrease of oncotic pressure, or as the result of increased interstitial oncotic pressure. Intravenous infusion of hypotonic fluid causes extracellular volume expansion and cellular dehydration. Increased osmolality of extracellular fluid causes cellular dehydration. Total body water is influenced by the hormones aldosterone and ADH, both of which favor water retention by the kidney.

Thirst, primarily stimulated by cellular dehydration, and urinary output control total body water. Retention of excessive amounts of water or hypotonic fluid causes an increased extracellular volume and cellular hydration; isotonic fluid causes increased extracellular volume with little or no effect on cells; hypertonic fluid increases extracellular volume and causes cellular dehydration. The manifestations of water intoxication depend on magnitude and rate of change and are associated with lower plasma sodium ion levels. The symptoms range from anorexia and behavioral changes to death. Some causes of water intoxication are psychogenic polydipsia and inappropriate ADH secretion. If the imbalance is not life-threatening, the mode of treatment is water restriction.

The pathophysiology of dehydration depends on the amount of electrolytes lost with the fluid, and only hypotonic loss or simple dehydration is included in this chapter. Hypotonic fluid loss causes increased osmolality of extracellular fluid and hypernatremia, with cellular dehydration as a subsequent event. The evidence of dehydration depends on extent and rate of loss and includes weight loss, dry mucous membranes, hypotension, tachycardia, confusion, and coma. Some causes of dehydration are excessive sweating, osmotic diuresis, and diabetes insipidus. Treatment for fluid loss is restoration of fluid with various salt or dextrose solutions or blood. Infants and the elderly are especially vulnerable to fluid imbalance.

REVIEW QUESTIONS

• • •

1. What are four types of extracellular fluid?
2. What is the semipermeable membrane that separates the intravascular compartment from interstitial spaces?
3. What is the semipermeable membrane that separates interstitial fluid from the intracellular compartment?
4. What is the quantitatively important cation in plasma?
5. What is the reason for a drawing force in oncotic pressure?
6. Oncotic pressure inside capillaries favors water movement (into or out of) capillaries.
7. There is a net (loss or gain) of water from the arterial end of capillaries.
8. What is a mechanism for the return of excess interstitial fluid to the general circulation?

9. Given the following:

    Plasma oncotic pressure = 26 mm Hg; plasma hydrostatic pressure = 24 mm Hg; interstitial fluid hydrostatic pressure = −4.3 mm Hg; interstitial fluid oncotic pressure = 4 mm Hg.

    In which direction is there net movement of water, and what is the magnitude of the force involved?
10. If there is a decreased concentration of sodium in extracellular fluid, there will be a net movement of water (into, out of) cells.
11. List two factors that make lungs especially susceptible to edema.
12. In general, a/an (increased, decreased) interstitial fluid oncotic pressure favors net water movement out of capillaries.
13. What is the evidence of a 30–50% increase of interstitial fluid?
14. If there is sodium loss for any reason, you would expect cellular (hydration, dehydration).
15. Inappropriate secretion of ADH causes (water retention, water loss).
16. A response to water excess is (increased, decreased) ADH secretion.
17. Why may blood volume remain unchanged in dehydration?
18. The term hypotonic fluid loss refers to fluid which has (equal, greater, lesser) solute concentration than plasma.
19. What is the process called when the kidney must eliminate a large solute load?
20. Give two reasons that the elderly are at risk for dehydration.

SELECTED READING

• • •

Ballermann, B. J., and B. M. Brenner. 1986. Role of atrial peptides in body fluid homeostasis. *Circulation Research* 58(5):619–30.

Baylis, P. H. et al. 1986. Development of a cytochemical assay for plasma vasopressin: Application to studies on water loading normal man. *Clinical Endocrinology* 24:383–93.

Erkert, J. D. 1988. Dehydration in the elderly. *Journal of American Academy of Physician Assistants* 1(4):261–67.

Guyton, A. C. 1963. A concept of negative interstitial pressure based on pressures in implanted perforated capsules. *Circulation Research* 12:399–413.

Jeffries, A. L., G. Coates, and H. O'Brodovich. 1984. Pulmonary epithelial permeability in hyaline-membrane disease. *New England Journal of Medicine* 311:1075–80.

Kimura, T. et al. 1986. Effects of acute water load, hypertonic saline infusion, and furosemide administration on atrial natriuretic peptide and vasopressin release in humans. *Journal of Clinical Endocrinology and Metabolism* 63(5):1003–10.

Nielson, O. M., and H. C. Emgell. 1986. The importance of plasma colloid osmotic pressure for interstitial fluid volume and fluid balance after elective abdominal vascular surgery. *Annals of Surgery* 203(1):25–29.

Oian, P. et al. 1986. Transcapillary fluid balance in pre-eclampsia. *British Journal of Obstetrics and Gynecology* 93(3):235–39.

Raison, J. et al. 1986. Extracellular and interstitial fluid volume in obesity with and without associated systemic hypertension. *American Journal of Cardiology* 57(4):223–26.

Raymond, K. H., and M. D. Lifschits. 1986. Effect of prostaglandins on renal salt and water excretion. *American Journal of Medicine* 80 (suppl 1A):22–29.

Skorecki, K. L., and B. M. Brenner. 1981. Body fluid homeostasis in man: A contemporary overview. *American Journal of Medicine* 70(1):77–88.

Stromberg, D. D., and C. A. Wiederkielm. 1976. Interstitial fluid oncotic pressure in rabbit subcutaneous tissue. *American Journal of Physiology* 231(3):888–91.

Chapter 3

Electrolyte Balance

Sodium ions are the quantitatively important cations in extracellular fluid, are accompanied by anions (mainly chloride (Cl⁻) and bicarbonate (HCO₃⁻) ions), and are considered an indicator of total solute concentration of plasma osmolality. Consequently, sodium ions are osmotically important in determining water movements, and a discussion of one must include the other. The first topic of this chapter is hypernatremia, and this represents a continuation of the last part of chapter 2 in which dehydration was discussed.

HYPERNATREMIA

The normal range for blood levels of sodium is approximately 137–143 meq/liter. Hypernatremia refers to an elevated serum sodium ion level, in the range of 145–150 meq/liter or greater. The effects of increased levels of sodium ions are the results of diffusion and osmosis as described in chapter 2. It is useful to keep three things in mind:

1. Sodium ions do not cross cell membranes as efficiently and quickly as water does.
2. Cells pump sodium ions to the outside by means of active transport.
3. Increases in extracellular sodium ion levels do not change intracellular sodium ion concentration.

Hypernatremia causes two things to occur. Water is osmotically drawn out of cells, resulting in some degree of cellular dehydration, and there is an increase in extracellular fluid volume (figure 3.1).

The central nervous system deserves special attention in terms of a shift of cellular water due to hypernatremia. A unique anatomical feature of capillaries in the central nervous system is that there are tight junctions between the endothelial cells that make up the capillary walls. These junctions restrict diffusion from capillaries to the interstitium of the brain. Consequently, when there is an increased level of sodium ions in the blood, there is not an accompanying increase of sodium ions in brain interstitial fluid. As the result of an osmotic gradient, water shifts from the interstitium and cells of the brain and enters the capillaries (figure 3.2). The brain tends to shrink, and the capillaries dilate and tend to rupture. The result is cerebral hemorrhage, blood clots or **thromboses,** and neurological dysfunctions.

• • •

It has long been recognized that there is either no diffusion or slow diffusion of substances from the blood into the brain and cerebrospinal fluid. This phenomenon is called the blood-brain barrier, and the tight junctions between brain capillary endothelial cells appear to be a factor in this phenomenon.

• • •

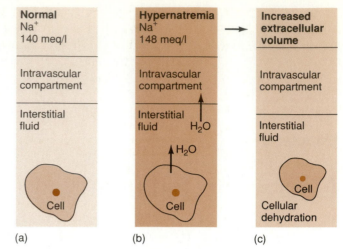

Figure 3.1 (*a*) Normal sodium ion concentration in extracellular fluid. (*b*) Increases in extracellular sodium ion levels do not change intracellular sodium ion concentrations. Consequently, water is osmotically drawn out of cells. (*c*) Hypernatremia causes some degree of cellular dehydration with dilution of the extracellular compartment.

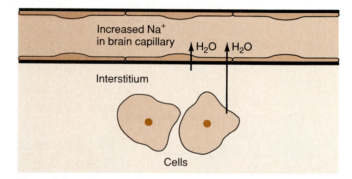

Figure 3.2 An increase in sodium ion concentration in the blood of brain capillaries draws water out of brain cells and the surrounding interstitium.

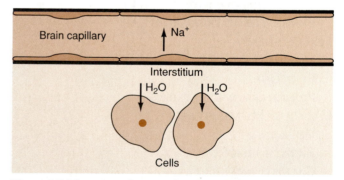

Figure 3.3 There is an increase in solute concentration inside of brain cells in response to hypernatremia. This occurs within about a day and is a protective mechanism against dehydration of brain cells.

A mechanism that protects the brain from shrinkage has been identified, and since this is a significant factor when treatment is considered, it deserves special mention. In some unknown way, the **intracellular osmolality** of brain cells increases within about a day in response to extracellular hyperosmolality. Unidentified substances called **idiogenic osmoles** accumulate inside brain cells. These substances may be potassium or magnesium ions released from binding sites or taurine and other amino acids released from the breakdown of protein. Idiogenic osmoles create an osmotic force that draws water back into the brain and protects cells from dehydration (figure 3.3).

MANIFESTATIONS

The manifestations of hypernatremia may be predicted on the basis of the foregoing discussion. The symptoms are primarily neurological and depend on the extent of the sodium ion elevation and the rate at which it occurs, with increased rate paralleling severity. Varying degrees of central nervous system disturbance may be observed including hyperirritability, muscle twitchings and weakness, lethargy, coma, and convulsions (perhaps because of cerebral hemorrhages). A strong sense of thirst, polyuria, and a low grade fever may be present. Dehydration is almost always an accompanying feature of hypernatremia, but unless fluid loss is profound, the effects on circulation are not marked.

ETIOLOGY

Hypernatremia is basically caused by two things: water loss or sodium ion overload. The cause is almost always water deficit due to loss or inadequate intake (table 3.1).

Absence of Thirst or Access to Water

Damage or destruction of neurons of the thirst center in the hypothalamus perhaps incurred by cerebral tumor or meningitis, results in impaired sensation of thirst and leads to inadequate fluid intake. Infants or comatose individuals who do not have access to water suffer inadequate hydration, and the problem is compounded under conditions of increased insensible water loss, i.e., fever, hyperventilation, or in the case of infants, large surface area compared to weight.

Water Loss

Diabetes insipidus (see chapters 2 and 31), caused by inadequate ADH or renal insensitivity to ADH, results in large urinary fluid losses and subsequent hypernatremia. Increased fluid loss also occurs as the result of osmotic diuresis, a situation in which a high solute load is delivered to the kidney for elimination. The solute in the glomerular filtrate creates an osmotic gradient that draws

idiogenic: Gk. *idio*, peculiar; Gk. *genic*, producing
osmoles: Gk. *osmo*, pushing
polyuria: Gk. *poly*, many; Gk. *ouron*, urine
hyperkalemia: L. *kalium*, potassium

TABLE 3.1 Causes of hypernatremia

Cause	Comments
Coma	Inadequate fluid intake
Essential hypernatremia	A disorder in which thirst is impaired; cause is unclear
Fever	Increased insensible fluid loss
Hot environment, especially with strenuous exercise	Sweat, hypotonic fluid loss
Vomiting	Often a hypotonic fluid loss
Diarrhea	Often a hypotonic fluid loss
Pituitary diabetes insipidus	Deficiency of ADH; excessive urinary loss
Nephrogenic diabetes insipidus	Renal tubules insensitive to ADH; excessive urinary loss
Uncontrolled diabetes mellitus	Glucose in glomerular filtrate; osmotic diuresis
Mannitol used as a diuretic	Mannitol in glomerular filtrate; osmotic diuresis
Large amounts of protein and amino acids given by nasogastric tube	Urea is a product of protein metabolism; urea causes osmotic diuresis
Excessive intravenous infusion of hypertonic sodium salt solutions	Administration of excessive sodium ions

water into the kidney tubules; thus, urine volume increases (chapter 2). In uncontrolled diabetes mellitus, glucose appears in the urine in such high amounts that it is not possible for the kidney tubules to reabsorb it into the blood. The presence of this solute increases the osmotic pressure inside the kidney tubules, draws water into the tubules resulting in increased urine volume, and causes fluid loss as the final outcome. A third possible cause of water loss is the case of a patient who requires high protein feedings provided by a stomach tube. Protein is metabolized by the liver with urea produced as a waste product. Urea is then eliminated by the kidney. High levels of urea in the glomerular filtrate create an osmotic gradient just as glucose does, and diuresis or increased urinary output is the result. Excess fluid loss leads to hypernatremia.

Sodium Excess

Less frequently, hypernatremia may be caused by retention or intake of excess sodium. Intravenous infusion of hypertonic sodium ion solutions is a case in point. The hormone aldosterone promotes sodium and water retention by the kidney, and when there are high levels of aldosterone, a mild hypernatremia may ensue. Aldosteronism is discussed further in chapter 32.

PRINCIPLES OF TREATMENT

Fluid deficit is usually the underlying cause of hypernatremia; thus, the objective of therapy is to restore normal hydration to decrease sodium ion concentration. There is a particular point of concern with rehydration and that is the matter of cerebral adaptation to hyperosmolality described earlier. In response to hypernatremia, idiogenic osmoles accumulate inside brain cells within 24 hours. This establishes an osmotic gradient and draws water into the brain to protect against water loss. If this adaptation has occurred and treatment involves a rapid infusion of dextrose, for example, there is danger of cerebral edema with fluid being drawn into brain tissue. Consequently, rapid infusions can cause cerebral edema and, subsequently, convulsions.

• • •

The brain adapts to increased blood levels of Na^+ by accumulating osmotically active substances inside brain cells. These substances minimize water movement out of the brain. Skeletal muscle cells do not respond to increased blood osmolality in this way, and the reasons for this lack of response are unclear.

• • •

In general, treatment is best handled by giving a slow infusion of glucose solution for the purpose of diluting high plasma sodium ion concentration. Ideally, the goal is to avoid overloading with fluid and to remove excess sodium. Diuretics may be used to induce sodium and water diuresis, but if kidney function is not normal, peritoneal dialysis may be required (chapter 11).

HYPONATREMIA

Hyponatremia is defined as a serum sodium ion level that is lower than normal. Since this ion is an indicator of osmolality, hyponatremia implies an increased ratio of water to sodium in extracellular fluid, which means extracellular fluid is more dilute than intracellular fluid. The result is a shift of water into cells until equilibrium is reestablished. The resulting cellular hydration is particularly significant in brain cells. It was noted earlier that the brain adapts to hypernatremia with the accumulation of idiogenic osmoles to reverse water movement back into the cells. In hyponatremia the opposite occurs, the brain cells lose osmoles, creating a higher extracellular solute concentration. The effect of this phenomenon is to protect against cerebral edema by drawing water out of brain tissue.

A general physiological response to decreased osmolality is suppression of thirst and suppression of ADH secretion, both of which favor a correction of the problem by decreasing water ingestion and increasing urinary output.

MANIFESTATIONS

The symptoms of hyponatremia are primarily neurological, probably because of the net flux of water into the brain. The symptoms are more alarming, even with a moderate decrease in sodium ion level, if the onset is sudden. The following summarizes some frequently observed symptoms of hyponatremia. When the serum sodium ion level is about 125 meq/liter or less, the patient may experience nausea, vomiting, and loss of taste. Weakness and muscle cramps occur with a lower sodium ion level. Seizures and coma occur when serum sodium ion concentration is 110 meq/liter or less.

• • •

Although the terms serum and plasma are sometimes used interchangeably, there is a difference. Blood collected in a tube and allowed to stand will form a clot that separates from a watery fraction called serum. If blood is collected in a tube to which heparin or other anticoagulant substance has been added, the unclotted blood separates into a cellular fraction and a liquid part called plasma.

• • •

ETIOLOGY

Hyponatremia is produced by loss of sodium ions or water excess. The water excess may be due to ingestion or renal retention.

Dilutional Effect
An indirect cause of hyponatremia is a dilutional effect when there is isotonic fluid loss. Loss of fluid with a solute concentration comparable to plasma does not cause lower sodium ion levels (chapter 5), but volume depletion stimulates thirst and leads to increased water ingestion. Isotonic fluid loss can cause hyponatremia, not because of sodium ion loss, but because of increased intake of water.

There are other reasons for the occurrence of water excess. Antidiuretic hormone enhances renal water retention, and in a number of situations, this hormone may be secreted in the absence of normal stimuli, i.e., in the absence of increased osmolality or volume depletion (chapter 2). Ingested water is retained in the presence of ADH, which may cause a dilutional hyponatremia. Psychogenic polydipsia or compulsive water drinking has the same effect (chapter 2). In both acute and chronic renal failure, the failing kidney is unable to excrete water, and water retention leads to hyponatremia.

Hyponatremia and Potassium Ion Loss
The relationship between potassium ion loss and hyponatremia deserves a special note. Potassium ions are the predominant intracellular cations, and when they are lost from the body, replacement occurs by the diffusion of intracellular potassium into extracellular fluid. **Electrical**

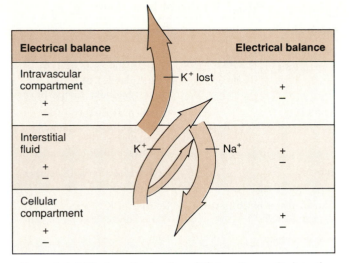

Figure 3.4 Diffusion of potassium (K) ions from cells into the extracellular compartment is a compensatory response to extracellular potassium ion loss. Intracellular electrical balance is maintained by an accompanying diffusion of sodium ions into cells. Hyponatremia is the result because sodium ion levels are measured in blood, not in the cellular compartment.

TABLE 3.2 Causes of hyponatremia

Cause	Comments
Psychogenic polydipsia	Excessive ingestion of water
Syndrome of inappropriate secretion of ADH	ADH causes renal water retention
Addison's disease	Aldosterone deficiency; several other factors
K+ losses from extracellular fluid	K+ move out of cells to replace losses; Na+ move into cells to maintain electrical neutrality

balance is maintained by the diffusion of sodium ions into the cells in exchange for potassium ions (figure 3.4); hence, hyponatremia may ensue. Total body sodium ion content would be unchanged, but extracellular fluid level is decreased, and this is hyponatremia.

Diuretic therapy is a common cause of hyponatremia and, although there are different types of diuretics and the underlying mechanisms of diuresis vary, loss of sodium and potassium often occurs in addition to fluid loss (chapter 11). A shift of sodium ions into cells in response to extracellular and cellular loss of potassium ions, as well as direct renal sodium ion loss, contribute to hyponatremia (table 3.2).

PRINCIPLES OF TREATMENT

The treatment of choice for hyponatremia depends on the severity of symptoms, whether the condition is acute or chronic, and on the underlying cause. If the condition is chronic, treating the underlying disorder is a priority. If the condition is acute, the goal is to correct cellular overhydration and increase serum sodium levels. This may be accomplished with the administration of hypertonic sodium chloride solution alone or in combination with a diuretic. If the patient is asymptomatic, water restriction may be adequate.

HYPERKALEMIA

Normal serum potassium ion levels are in the range of 3–5 meq/liter, as compared to an average of 142 meq/liter for sodium ions. Intracellular levels of potassium are much higher, about 140–150 meq/liter, with some variation depending on cell type. This high intracellular level is maintained by active transport. The amount of total body potassium ion content is maintained by a balance between ingestion and renal elimination, and cellular-extracellular exchange of this ion. The latter is an important aspect of potassium ion homeostasis because excess potassium in extracellular fluids quickly enters cells until the kidney, over a period of hours, eliminates the excess from the body. Potassium ions also shift from cells to extracellular fluid as replacement when there is loss of these ions from the body.

Hyperkalemia refers to an elevated serum potassium ion level. Since it is the serum concentration that is measured and reported by clinical laboratories, it is important to recognize that this measurement may not reflect the intracellular content of potassium.

One aspect of the consequences of hyperkalemia involves acidosis, a condition in which there is an increase of hydrogen ions (H+) in body fluids. Chapter 4 deals with acidosis/alkalosis in more detail, but it is appropriate to point out here that when there are changes in either potassium ions or hydrogen ion levels, the one causes a compartmental shift of the other. When hyperkalemia develops, potassium ions move into the cells in exchange for hydrogen ions, which then diffuse into extracellular fluid (to maintain electrical neutrality). As a result, the physiological response to hyperkalemia causes acidosis (figure 3.5). The reverse occurs as well. When there is an increase of extracellular hydrogen ions (acidosis), the body is protected from deleterious effects as the ions diffuse into cells where they are effectively tied up by protein molecules. This causes a shift of potassium ions out of the cells (figure 3.6). A summary of these ideas is that hyperkalemia causes acidosis and acidosis causes hyperkalemia.

MANIFESTATIONS

Muscle contraction is affected by disturbances of potassium ion concentration. Hyperkalemia blocks the transmission of nerve impulses along muscle fibers and can cause muscle weakness and paralysis, commonly involving muscles of the extremities and the trunk. Hyperkalemia

hyperkalemia: L. *kalium*, potassium

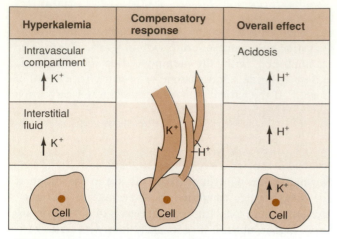

Hyperkalemia	Compensatory response	Overall effect
Intravascular compartment ↑K⁺		Acidosis ↑H⁺
Interstitial fluid ↑K⁺	K⁺ —H⁺	↑H⁺
Cell	Cell	↑K⁺ Cell

Figure 3.5 Diffusion of potassium ions into cells in exchange for the efflux of hydrogen ions into the extracellular compartment is a compensatory response to hyperkalemia. The net effect is acidosis, which is an increased hydrogen ion level in extracellular fluid.

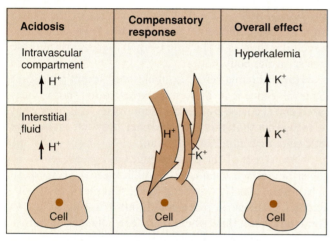

Acidosis	Compensatory response	Overall effect
Intravascular compartment ↑H⁺		Hyperkalemia ↑K⁺
Interstitial fluid ↑H⁺	H⁺ —K⁺	↑K⁺
Cell	Cell	Cell

Figure 3.6 Acidosis may lead to hyperkalemia. Intracellular hydrogen ion diffusion is accompanied by extracellular potassium ion diffusion.

causes arrhythmias and heart conduction disturbances, even cardiac arrest, although cardiac effects may be absent if the onset of potassium ion increase is slow. The patient may also experience a decreased vibratory sense, diminished sense of position, and diminished cutaneous sensory perception.

ETIOLOGY

Serum potassium ion levels are determined by ingestion, urinary excretion, and cellular-extracellular shifts. The basis for hyperkalemia is increased input, impaired excretion, or impaired uptake by cells (table 3.3).

TABLE 3.3 Causes of hyperkalemia

Cause	Comments
Hyperkalemic periodic paralysis	Inherited disorder in which there are sudden shifts of cellular K⁺ to extracellular compartment
Acidosis	Compensatory shift of H⁺ into cells in exchange for movement of K⁺ to extracellular compartment
Burns	Cell destruction with release of K⁺
Transfusion of blood that has been stored	Release of K⁺ from hemolyzed red blood cells
Spironolactone	Diuretic that is an aldosterone antagonist; interferes with reabsorption of Na⁺ and secretion of K⁺
Too rapid intravenous infusion of KCl	Special risk of hyperkalemia if there is impaired renal secretion of K⁺
Use of K⁺ containing salt substitutes	Excessive ingestion
Potassium salts of antibiotics	Additional source of K⁺
Acute oliguric renal failure	Impaired secretion of K⁺

Increased Input

There are a number of possible sources for increased input of potassium ions including intravenous potassium chloride (KCl) infusion, use of potassium containing salt substitutes, and blood transfusions in which some hemolysis of red blood cells occurs with release of potassium ions. Damaged and dying cells release potassium ions so that situations in which tissue damage has occurred, burns, crushing injuries, or ischemia (is-kē'mē-ah), lead to hyperkalemia. Intravascular hemolysis resulting from increased fragility of red blood cells causes elevated serum levels of potassium as well.

• • •

Exercise is accompanied by an elevation of plasma potassium levels, with the elevation proportional to the amount of exertion involved. It is believed that the hyperkalemia is caused by the release of potassium from muscle cells. The potassium levels quickly decrease when exercise stops. The mechanisms for the sudden rise and fall of potassium levels are not clear.

• • •

ischemia: Gk. *ischein*, to check; Gk. *haima*, blood

Cellular-Extracellular Shifts

There is a predisposition toward hyperkalemia in the type of diabetes mellitus in which there is insulin deficiency (chapter 31). The basis for this observation is that cellular uptake of potassium ions, which provides immediate protection from potassium overload, is enhanced by such hormones as insulin, aldosterone, and epinephrine. Insulin deficiency represents decreased protection if the body is challenged by an excess of potassium ions.

Hyperkalemia occurs in the case of low levels or absence of aldosterone. This hormone is secreted by the adrenal cortex and enhances renal excretion of potassium in exchange for the reabsorption of sodium ions. It promotes cellular uptake of potassium to protect against increased extracellular levels of potassium ions. In the absence of aldosterone, there is loss of sodium ions in the urine, renal retention of potassium, and ensuing hyperkalemia. These and other hormones will be discussed further in unit VII.

There is an inherited disorder called **hyperkalemic periodic paralysis** in which the serum potassium level rises periodically. It is assumed that the cause of this periodic increase is a shift of potassium from muscle to blood in response to ingestion of potassium, exercise, and perhaps other stimuli. The reason for the shift is not clear and the attacks, characterized by muscle weakness, may be short or long and vary in the degree of severity. There is a comparable condition which causes **periodic hypokalemia.**

As noted earlier, hyperkalemia is a compensatory response to acidosis with a reciprocal movement of potassium ions out of cells as hydrogen ions move into an intracellular position.

Renal Insufficiency

The hormone aldosterone has a primary role in promoting the conservation of sodium ions and secretion of potassium ions by the nephrons of the kidney. In Addison's disease in which there has been destruction or atrophy of the adrenal cortex, there is aldosterone deficiency, so the kidney is unable to secrete potassium at a normal rate. Hyperkalemia is the result. A second example of renal insufficiency is the **oliguric** (ōl''ĭ-gu'rik) phase of renal failure. The kidney loses the ability to secrete potassium ions and hyperkalemia is one of a multitude of problems that develop.

Spironolactone (spī-rō''nō-lak'tōn) is a diuretic that is antagonistic to the effects of aldosterone. Since aldosterone normally favors sodium ion retention and potassium ion secretion by the kidney, this diuretic causes some rise in serum potassium level by interfering with potassium secretion. The increase may not be significant, but the kidney's ability to respond to a potassium ion load is impaired. A patient taking this diuretic is at risk for hyperkalemia if potassium is administered.

oliguric: Gk. *oligo*, small; Gk. *ouron*, urine

PRINCIPLES OF TREATMENT

Correcting the underlying cause of the disturbance is an aspect of treatment, while the specific procedures are dictated by the severity of the symptoms. In principle, there are three approaches to the problem of acute hyperkalemia: (1) counteracting the effects of potassium ions at the level of the cell membrane, (2) promotion of potassium ion movements into cells, and (3) removal of potassium ions from the body.

Infusion of either calcium gluconate or sodium chloride solutions almost immediately counteract the effects of potassium ions on the heart and is effective for 1–2 hours. Sodium bicarbonate ($NaHCO_3$) also reverses hyperkalemic effects on the heart, is effective within minutes, and, if acidosis is a factor in the hyperkalemia, $NaHCO_3$ raises the pH of body fluids. Insulin given with glucose is effective in lowering potassium ion level in about 30 minutes, and the duration of action is up to 6 hours. The basis for using the combination of these two substances is that insulin promotes the shift of potassium ions into cells and glucose prevents insulin-induced hypoglycemia.

A **cation-exchange resin** (Kayexalate) removes potassium ions from the body by exchanging potassium for sodium. It may be administered orally or rectally and is effective for up to 6 hours. Peritoneal dialysis or hemodialysis (chapter 11) effectively clears the blood of high levels of potassium as well.

• • •

Kayexalate is sodium polystyrene sulfonate, a cation-exchange resin that exchanges sodium for potassium. When it is given by mouth, a laxative is given at the same time to avoid fecal impaction. Oral administration may lead to nausea and vomiting. The resin may be instilled as an enema or may be introduced into the rectum in a dialysis bag for easy recovery. The exchange time is about 30–60 minutes and the procedure may be repeated every 6 hours.

• • •

Long-term management of hyperkalemia includes treatment of disorders contributing to the problem and dietary restriction of potassium.

HYPOKALEMIA

Hypokalemia is defined as a serum potassium ion level that is below normal or less than 3 meq/liter. Serum concentrations of potassium will decrease if there is an intracellular flux of these ions and will also be low if potassium ions are lost from the gastrointestinal or urinary tract. The distinction between redistribution and loss of total body stores of potassium should be made.

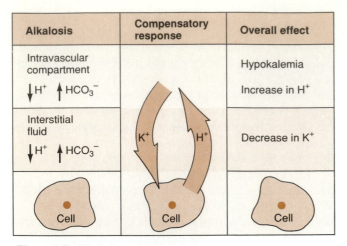

Alkalosis	Compensatory response	Overall effect
Intravascular compartment $\downarrow H^+$ $\uparrow HCO_3^-$		Hypokalemia Increase in H^+
Interstitial fluid $\downarrow H^+$ $\uparrow HCO_3^-$	K^+ H^+	Decrease in K^+
Cell	Cell	Cell

Figure 3.7 Alkalosis results in a compensatory response whereby potassium ions diffuse into cells in exchange for hydrogen ions, which diffuse into the extracellular compartment. The net effect is to increase the level of hydrogen ions extracellularly and to cause a decrease in the number of potassium ions in that compartment. The net effect is hypokalemia.

The relationship between hyperkalemia and acidosis was described earlier. It is also true that alkalosis causes and is caused by hypokalemia. Alkalosis is defined as a decrease of hydrogen ions or an increase of bicarbonate in extracellular fluids (chapter 4) and is more or less the opposite of acidosis. Alkalosis elicits a compensatory response causing hydrogen ions to shift from cells to extracellular fluid to correct the acid-base imbalance. The hydrogen ions are exchanged for potassium, which moves into cells; thus, serum concentration of potassium is decreased and alkalosis causes hypokalemia (figure 3.7). Conversely, when potassium ions are lost from the cellular and extracellular compartments, sodium and hydrogen ions enter body cells in a ratio of 2:1 as replacement. This loss of hydrogen ions from extracellular fluid causes alkalosis.

Hypokalemia affects kidney function as well (chapter 11). In general, sodium ions are reabsorbed into the blood when potassium ions are secreted into the urine by kidney tubules. If adequate numbers of potassium ions are not available for this exchange, hydrogen ions are secreted instead. It is useful to think of this as a reciprocal relationship with sodium ions being exchanged for potassium/ hydrogen ions (figure 3.8). Hypokalemia promotes renal loss of hydrogen ions and thus alkalosis. If hypokalemia is prolonged, it impairs renal urine concentrating ability and **polyuria** ensues.

MANIFESTATIONS

Hypokalemia reduces the responsiveness of muscles to nerve impulse stimulation and result in muscle weakness and paralysis. The lower extremities are usually affected first with progressive involvement of the trunk, upper extremities, and respiratory muscles. Smooth muscle is affected as well as skeletal muscle, and the result of this is

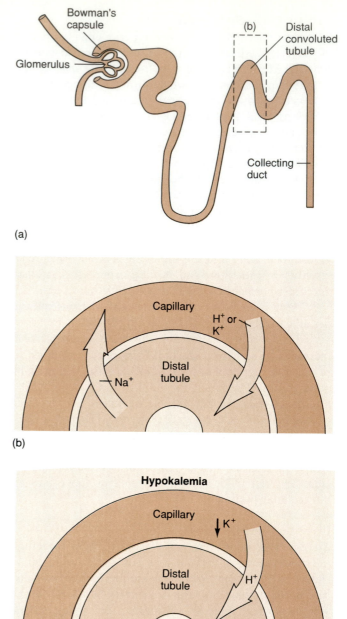

(a)

(b)

Hypokalemia

(c)

Figure 3.8 (*a*) Microscopic nephron in the kidney where urine is formed. (*b*) Enlarged view of distal convoluted tubule and an associated capillary. Potassium and hydrogen ions are secreted in the distal tubule in exchange for sodium ions. (*c*) In hypokalemia, the kidney selectively secretes hydrogen ions in preference to potassium ions. This loss of hydrogen ions may lead to alkalosis.

anorexia, nausea, vomiting, constipation, and urinary retention. Cardiac arrhythmias are also characteristic.

ETIOLOGY

There are various causes of hypokalemia (table 3.4). A mysterious disorder called hyperkalemic periodic paralysis was described briefly earlier in this chapter. There is a similar inherited condition called **hypokalemic periodic**

TABLE 3.4 Causes of hypokalemia

Cause	Comments
Aldosterone excess	Favors renal Na$^+$ reabsorption and K$^+$ excretion
Diarrhea	Diarrheal fluid contains high amounts of K$^+$
Diuretics	In general causes K$^+$ loss
Distal renal tubular acidosis	Kidney tubule defect in which K$^+$ are secreted, and H$^+$ are retained by the body
Hypokalemic periodic paralysis	Cause unknown; periodic influx of K$^+$ into cells
Bartter's syndrome	Syndrome in which aldosterone is sometimes elevated; probably a renal tubular defect so that K$^+$ are lost

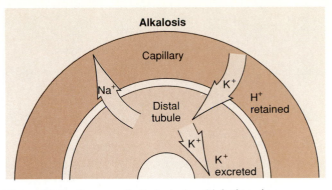

Figure 3.9 In the case of alkalosis, in which there is a decrease in hydrogen ions in extracellular fluid, the kidney retains hydrogen ions to correct the alkalosis and eliminates potassium ions. This may lead to hypokalemia.

paralysis that is characterized by movement of potassium ions into muscle cells causing hypokalemia. Precipitating factors include a high carbohydrate diet, high sodium intake, and rest after exercise. The result is attacks of weakness and paralysis. Mechanisms and underlying causes are not known.

A second example of hypokalemia resulting from intracellular potassium ion movements is the injection of insulin, which promotes such a shift. A third example is the physiological response to alkalosis. To correct the problem of alkalosis, there are actually two compensatory responses, both of which contribute to hypokalemia. (1) There is a reciprocal exchange of K$^+$ and H$^+$ in cells with potassium ions moving into cells and hydrogen ions moving out. (2) There is renal retention of hydrogen ions, which corrects the problem of alkalosis, but the hydrogen ions are exchanged for potassium ions eliminated from the body (figure 3.9).

Gastrointestinal losses by way of vomiting and diarrhea are possible causes of hypokalemia, and this topic is discussed further in chapter 5. In addition, diuretics in general augment urinary potassium ion excretion and may cause hypokalemia (see chapter 11).

PRINCIPLES OF TREATMENT

Hypokalemia is treated by replacement of potassium either by diet, oral potassium salt supplements, or intravenous administration of a potassium salt solution. If excessive renal loss is a part of the picture, the diuretic spironolactone is an antagonist of aldosterone and inhibits renal potassium secretion.

CALCIUM HOMEOSTASIS

Calcium (Ca^{++}) plays an important role in many physiological processes including muscle contraction, nerve impulse transmission, hormone secretion, and blood clotting. The range of normal values for serum calcium (9–10.5 mg/dl) has narrow limits, and a number of factors contribute to maintaining that level precisely. The overall picture is one of balance between input and output, but there are various factors involved.

• • •

The normal range for serum calcium may be expressed as follows: 4.5–5.3 meq/liter or 9–10.5 mg% or 9–10.5 mg/dl. An equivalent is the mass of an element chemically equivalent to 1 gram of hydrogen. A milliequivalent is 0.001 of an equivalent. Milligrams percent refers to the number of mg per 100 ml of serum, and a deciliter is equal to 100 ml.

• • •

VITAMIN D

Vitamin D is involved in maintaining serum calcium levels. The source of vitamin D is either dietary or is synthesized by the body. Cholesterol, as a precursor, is chemically converted in the skin by exposure to sunlight into a product which is converted to an active form of vitamin D by the liver and kidneys (figure 3.10). Vitamin D enhances serum calcium ion levels by (1) directly promoting bone resorption with the release of calcium salts, (2) potentiating the effects of parathormone (PTH) on bone resorption, (3) increasing absorption of calcium ions from the intestine, and (4) reabsorption by the kidney tubules (figure 3.11).

Vitamin D promotes bone build-up as well, both directly and by increasing extracellular levels of calcium ions.

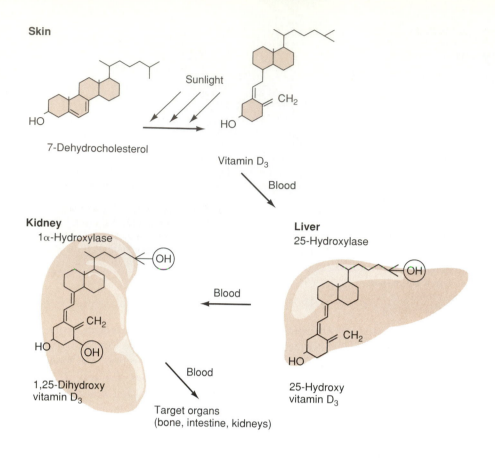

Skin

7-Dehydrocholesterol

Sunlight

HO

CH_2

Vitamin D_3

Blood

Kidney
1α-Hydroxylase

OH

Blood

CH_2

HO

OH

1,25-Dihydroxy
vitamin D_3

Blood

Target organs
(bone, intestine, kidneys)

Liver
25-Hydroxylase

OH

CH_2

HO

25-Hydroxy
vitamin D_3

Figure 3.10 The active form of vitamin D, 1,25-dihydroxy vitamin D_3, is produced by the kidneys. Vitamin D_3 is formed in the skin from the precursor, 7-dehydrocholesterol. This in turn is converted to the inactive 25-hydroxy vitamin D_3 in the liver. The final step in the pathway occurs in the kidneys with the production of the active form of vitamin D_3.

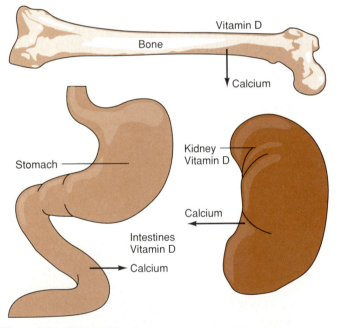

Vitamin D

Bone

Calcium

Stomach

Kidney
Vitamin D

Calcium

Intestines
Vitamin D

Calcium

Figure 3.11 Vitamin D is important in maintaining serum calcium levels. When calcium intake is inadequate, vitamin D promotes resorption of bone with release of calcium salts. It also promotes absorption of calcium from the intestines into blood, and the reabsorption of calcium into the blood from the kidneys.

PARATHORMONE

Parathormone is secreted into the bloodstream by the parathyroid glands and is an essential part of the calcium ion homeostatic mechanism (chapter 31). Parathormone increases calcium ion absorption from the intestine by enhancing the synthesis of the active form of vitamin D, favors reabsorption of calcium and excretion of phosphate (PO_4^{-3}) by kidney tubules, and enhances bone resorption with the release of calcium salts. Parathormone secretion is stimulated by decreased serum levels of calcium and is inhibited by increased serum levels of calcium.

CALCITONIN

Calcitonin or thyrocalcitonin is a hormone secreted by the thyroid gland. It decreases serum calcium levels by interfering with bone resorption and favoring bone uptake of calcium, and by promoting excretion of calcium by the kidney. The effects of calcitonin are weak when compared to PTH, and either a deficiency or excess of this hormone may not cause significant changes in serum calcium levels.

The calcium in bone is in the form of calcium phosphate salts, and about 0.5% of this calcium is easily released into extracellular fluid. There are some ex-

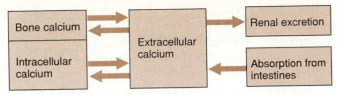

Figure 3.12 There are exchangeable stores of calcium in bone and in cells, and this total pool is in equilibrium with calcium in extracellular fluid. Serum calcium levels are also maintained by a balance between renal excretion and intestinal absorption.

changeable stores of calcium in cells as well and this total pool is in dynamic equilibrium with extracellular calcium. Serum calcium level is maintained ultimately by a balance between absorption from the intestines and renal excretion (figure 3.12). Exchangeable stores of calcium act as a buffer against disequilibrium.

FORMS OF SERUM CALCIUM

There are three forms of serum calcium: (1) ionized form (Ca^{++}), about 45% of the total, (2) protein (mainly albumin) bound 40%, (3) calcium combined with substances other than protein, about 15%. The significance of these distinctions is that only the ionized form is physiologically active, and the protein-bound form is a reservoir for calcium. There is a dynamic equilibrium between the two forms, i.e., ionized calcium and protein-bound calcium. If there is a loss of calcium ions, dissociation of the protein-calcium complex replaces the deficit (figure 3.13). It is total calcium, not the ionized form that is routinely measured (normal range, 9–10.5 mg/dl), and these values require some interpretation if there is an increase/decrease of serum albumin. Consider the following observations:

1. An increase in serum protein results in an increase in total serum calcium, but the ionized form of calcium (the physiologically active form) may be unchanged.
2. There is a decrease in total serum calcium when serum protein is lowered, but the ionized form may remain unchanged.
3. A combination of normal protein concentration and low total serum calcium level indicates a low level of ionized calcium and a physiological hypocalcemia.

HYPERCALCEMIA

A serum calcium level of 10.5 mg/dl or above constitutes hypercalcemia, and there are serious physiological consequences. In general, high levels of calcium interfere with nerve impulse transmission and muscle contraction, may cause kidney stones and attendant kidney damage, and at high levels (17 mg/dl), may precipitate out in any body tissue.

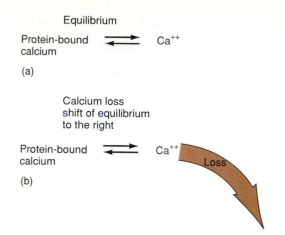

Figure 3.13 (*a*) There is an equilibrium between protein-bound calcium and ionized calcium. (*b*) In the case of a decrease of ionized calcium, there is a shift of the equilibrium to the right, so dissociation of protein and calcium releases ionized calcium.

MANIFESTATIONS

Neurological symptoms of hypercalcemia include lethargy, depression, insomnia, decreased memory, and headache. Delirium and coma may follow high levels of calcium. There are some muscle-related symptoms, such as muscle weakness, muscle aches and pain, nausea and vomiting, abdominal pain, constipation, and cardiac arrhythmias. The patient may experience itching, and there may be evidence of renal tubular disturbances evidenced by polyuria and even oliguria in later stages. A characteristic finding is **band keratopathy,** a condition in which gray or white crescent-shaped opacities develop in the cornea of the eye, usually in the lateral or medial margins. There may also be clear, glasslike deposits in the conjunctivae near the **palpebral** fissure.

ETIOLOGY

Overactive parathyroid glands result in increased secretion of PTH. Parathormone increases serum calcium levels because of its effects on absorption from the intestine, bone resorption, and reabsorption of calcium by kidney tubules. Thyroid hormone increases bone resorption so that in hyperthyroidism, serum calcium levels are increased as well (chapter 31).

Large doses of vitamin D cause hypercalcemia because of increased intestinal absorption of calcium and because of increased bone resorption. When a patient is confined to bed for a period of weeks, bone resorption occurs at a greater rate than bone formation. The consequence of an imbalance between the two processes is a weakening of bones and increased serum calcium levels.

Hypercalcemia occurs in many types of malignancies and is not limited to those cases where there has been **metastasis** (mĕ-tas′tah-sis), or spread of the malignancy, to

palpebra: L. *palpebra,* eyelid
metastasis: Gk. *metastasis,* removal

TABLE 3.5 Causes of hypercalcemia

Cause	Comments
Vitamin D excess	Promotes calcium absorption from intestine, reabsorption from kidney
Hyperparathyroidism	Excessive production of PTH
Hyperthyroidism	Causes mild hypercalcemia; causes bone resorption
Immobilization	Patients who are confined to bed; bone resorption
Thiazide diuretics	Cause small increases in serum calcium in normal individuals; increased renal reabsorption
Malignancies: bronchogenic carcinoma carcinoma of kidney carcinoma of breast multiple myeloma	Probably secretion of PTH-like substances by malignant tissue or cause uncertain
Excessive and prolonged ingestion of milk and antacids	Alkalosis causes hypercalcemia

TABLE 3.6 Physiological functions of calcium

1. Regulation of membrane permeability
2. Control of excitability of nerves and muscles (decreased Ca^{++} results in increased excitability)
3. Essential for normal skeletal muscle contractions
4. Essential for normal contractility and rhythmicity of heart
5. Required for blood coagulation
6. Essential for normal enzyme activity

bone. In some instances, malignant tumors secrete hormones that cause bone resorption, and there may be various other contributing factors.

Milk-alkali syndrome refers to hypercalcemia resulting from excessive and prolonged ingestion of milk and alkaline antacids for the treatment of peptic ulcer. Sodium bicarbonate is an example of an absorbable alkali, and at one time, calcium carbonate was used as an antacid also. Metabolic alkalosis occurs because of increased levels of plasma bicarbonate. The alkalosis promotes hypercalcemia in two ways: (1) alkalosis causes increased kidney reabsorption of calcium and (2) decreases the capacity of bone to take up additional calcium. Milk, as a source of calcium, is not essential for the development of the syndrome, but if large quantities are ingested, it contributes to hypercalcemia.

Table 3.5 summarizes some causes of hypercalcemia.

PRINCIPLES OF TREATMENT

The approach to the treatment of hypercalcemia depends on the severity of the symptoms and the serum calcium levels, whether mild (less than 12 mg/dl), moderate (12–15 mg/dl), or severe (greater than 15 mg/dl). Treating underlying disease and correcting fluid and electrolyte deficits are aspects of the overall regimen.

Intravenous or oral administration of phosphate decreases plasma calcium levels, perhaps in part by interfering with bone resorption. Calcitonin reduces the activity of bone destroying cells (osteoclasts) and reduces bone resorption or breakdown. A cytotoxic antibiotic, mithramycin, is effective in reducing serum calcium levels, although the mechanism by which this occurs is not clear.

Glucocorticoids (chapter 32) inhibit intestinal absorption of calcium and, in malignant conditions, directly inhibit bone resorption.

Loop diuretics, such as furosemide or ethacrynic acid, promote excretion of calcium by the kidney tubules. These diuretics are useful in treating hypercalcemia provided that fluid and other electrolyte losses are replaced. The specific action of different categories of diuretics is discussed in chapter 11. Infusion of saline (NaCl) or sodium sulfate (Na_2SO_4) increases urinary calcium excretion.

Substances that form complexes with calcium immediately remove calcium. **Ethylenediaminetetracetate (EDTA)** is an example of such a substance used in extreme cases of hypercalcemia.

HYPOCALCEMIA

Hypocalcemia is defined as a low serum calcium level less than 9 mg/dl. Calcium is essential in all tissues and has various functions (table 3.6). The physiological response to low serum calcium levels is increased secretion of PTH by the parathyroid glands. Parathormone increases serum calcium by favoring bone resorption, intestinal absorption of calcium, and renal reabsorption.

MANIFESTATIONS

The earliest and most obvious symptoms of hypocalcemia are neuromuscular, and these symptoms include numbness and tingling (especially of fingers and around the lips), muscle spasms, and **tetany.** Tetany refers to a series of muscle contractions so close together that the contractions fuse and are indistinguishable. The calcium level at which tetany occurs, usually less than 7 mg/dl, varies due to the many factors that influence neural excitability. **Latent (la'tent) tetany,** tetany which is not clinically obvious, may be demonstrated in two ways. Tapping the facial nerve just anterior to the earlobe elicits a unilateral contraction of facial muscles, a response called **Chvostek's (vos'teks) sign. Trousseau's (troo-soz')** sign may be observed as muscle spasms in the arm and contractions of the fingers after a blood pressure cuff has been put on the arm and

tetany: Gk. *tetanos,* spasm
latent: L. *latens,* concealed

Figure 3.14 Trousseau's sign in hypocalcemia. Inflating a blood pressure cuff above a patient's systolic pressure elicits a carpal spasm, contraction of thumb and fingers.

pumped up above normal systolic blood pressure (figure 3.14).

Hypocalcemia impairs cardiac contractility and, if the low calcium level develops quickly, may cause enlargement of the heart and heart failure. Chronic hypocalcemia may cause depression, emotional lability, or severe psychoses. Loss of hair, scaly skin, and thickened nails are the results of hypocalcemia, and occasionally diarrhea occurs. Hypoparathyroidism leading to hypocalcemia may cause calcium deposits to form in the lens of the eye and result in cataracts.

ETIOLOGY

The role of vitamin D in calcium homeostasis was discussed earlier in this chapter. Since vitamin D promotes absorption of calcium from the intestines and is essential for the resorptive action of PTH on bone, inadequate amounts of this vitamin lead to hypocalcemia. Nutritional deficiency of vitamin D is a possible cause, but there are other causes, such as the following: (1) impaired intestinal absorption, for example, as the result of surgical removal of a part of the stomach or subtotal gastrectomy, (2) liver or kidney dysfunction resulting in interference with the formation of the active metabolite of vitamin D, (3) intestinal or bone unresponsiveness to the action of vitamin D, or (4) inadequate exposure to sunlight so that there is a reduction in the formation of the active vitamin D metabolite in the skin.

Parathormone is essential in maintaining normal blood levels of calcium by means of bone resorption, renal reabsorption of calcium, and intestinal absorption of calcium.

gastrectomy: Gk. *gaster,* belly; Gk. *ektome,* excision
lability: L. *labilis,* unstable

TABLE 3.7 Causes of hypocalcemia

Cause	Comments
Hypoparathyroidism	May be caused by surgical procedures involving the neck; PTH deficiency
DiGeorge's syndrome	Congenital absence of parathyroid glands
Multiple endocrine deficiency-autoimmune-candidiasis syndrome	Inherited condition in which hypoparathyroidism is present
Vitamin D deficiency	Decreased absorption of calcium from intestines

Hypocalcemia is the result of the loss of the parathyroid glands or loss of function of those glands, or unresponsiveness of target tissues (kidney tubules, bone, gastrointestinal tract) to the influence of PTH.

Hypocalcemia occurs in conjunction with a number of disease processes. Most patients with pancreatitis have some degree of hypocalcemia, but the relationship between decreased calcium levels and the disease process is unclear (chapter 36). It has been observed in pancreatitis that calcium deposits form in soft tissues and that there is a defect in PTH secretion, both of which may contribute to hypocalcemia.

Hypocalcemia occurs in renal failure, and in this condition, an important contributing factor in the decreased blood calcium levels is a reduction in the formation of an active vitamin D metabolite. It is likely that altered bone response to PTH is a factor as well. Table 3.7 is a list of causes of hypocalcemia.

PRINCIPLES OF TREATMENT

The specific steps taken to bring blood levels of calcium back to a normal range depend on onset, severity, and contributing factors. Calcium salts may be administered intravenously or given orally, and vitamin D may be given as well.

• • •

Acute hypocalcemia is treated by intravenous infusion of calcium gluconate or calcium chloride solutions. Calcium chloride infusions are associated with an increased incidence of thrombophlebitis. If there is leakage of the solution during infusion, soft tissue necrosis occurs. Consequently, calcium gluconate is preferred. Calcium must not be added to bicarbonate-containing solutions, because calcium salts precipitate in the presence of bicarbonate.

• • •

Osteoporosis

Osteoporosis is a condition in which there is reduced bone mass due to demineralization associated with enlarged spaces in the bone. The weakened skeletal structure results in fractures following minimal trauma. The incidence of osteoporosis is higher in whites as compared to blacks, and the disease is a major health problem in Western countries.

Osteoporosis involves two phases of bone loss affecting two types of bone. Cortical or compact bone forms the outer portion of bone. Spongy bone, also called trabecular bone, forms the interior meshwork. There is a slow, prolonged phase of bone loss that occurs in both males and females. Both sexes begin to experience slow cortical bone loss at about age 40, and the loss increases in females at menopause. In addition to this, women experience accelerated cortical bone loss immediately after menopause, with a slow decline of the process over a period of 8 to 10 years. A transient accelerated phase occurs in men with androgen deficiency. Over a lifetime, women lose approximately 35% of their cortical bone and 50% of the trabecular bone, whereas men lose about two-thirds of these amounts.

Factors in Bone Homeostasis

A summary of factors that contribute to a balance between bone formation and bone resorption is listed in table 3.8. Homeostasis of bone involves a dynamic process known in the adult as remodeling. Remodeling consists of a balance between the activity of bone destroying and bone forming cells. Osteoclasts destroy both trabecular and cortical bone, and osteoblasts fill in the cavity created by osteoclasts. In the slow phase of bone loss associated with the aging process, there is impaired bone formation due to a decline in the function of osteoblasts, although osteoclasts create resorption cavities of normal depth. Osteoclasts create deeper resorption cavities in accelerated postmenopausal bone loss, while osteoblastic function is normal.

Pathogenesis

There are various sources of evidence regarding the pathogenesis of osteoporosis (table 3.9). In most cases, no other disease causing bone loss is apparent. The condition is either postmenopausal or is associated with the aging process and is called senile osteoporosis. The postmenopausal type of loss is found in females between the ages 51–65 and is characterized by fractures of bones, such as the ver-

osteoporosis: Gk. *osteon*, bone; Gk. *poros*, passage
osteoclast: Gk. *osteon*, bone; Gk. *klan*, to break
osteoblast: Gk. *blastos*, germ (associated with bone production)

TABLE 3.8 A summary of factors that affect homeostasis of skeletal structure

Factor	Effect	Action
Parathormone (PTH)	Maintains serum calcium level	Stimulates bone resorption; increases renal reabsorption of Ca; decreases renal reabsorption of phosphate (limits deposition of Ca and phosphate in bone)
Calcitonin	Reduces serum calcium level	Inhibits bone resorption
Vitamin D	Maintains serum calcium level	Increases intestinal absorption of Ca and P
Sex steroids	Required for skeletal growth	Stimulate protein synthesis
Thyroid hormone	Required for normal bone growth	Stimulates protein synthesis and growth in general
Growth hormone	Required for bone growth	Promotes cellular uptake of amino acids and growth in all body organs
Glucocorticoids	Increased bone resorption	Inhibit bone cell formation
Genetic factors	Determine bone mass	Control cell activities

tebrae and the distal forearm. A probable sequence of events is accelerated bone resorption, decreased secretion of parathyroid hormone, impaired production of the active form of vitamin D, and decreased intestinal calcium absorption. The fact that all women do not develop osteoporosis indicates there are factors other than a decrease in estrogen that cause bone loss.

Senile osteoporosis occurs in both males and females past the age of 70 in a female to male ratio of 2–3:1. Fractures of hip and vertebrae are common in these individuals. Two probable factors in senile osteoporosis are increased parathyroid hormone and decreased osteoblast function.

Manifestations

The first evidence of osteoporosis is often a fracture, and in the case of vertebral compression, may occur when the individual coughs or bends. Each compression fracture causes a loss of height, and the patient may develop kyphosis (ki-fo′sis), a curvature of the spine. Osteoporosis

TABLE 3.9 Observations that relate to the pathogenesis of osteoporosis

Observation	Probable Implication
Female runners with amenorrhea have a decrease in vertebral bone density	Accelerated bone resorption is associated with decreased sex hormone levels
There is increased secretion of PTH with aging	Leads to increased bone loss
Intestinal absorption of Ca decreases after age 70	Contributes to decreased bone formation
Hyperthyroidism is associated with bone loss	Excessive levels of thyroid hormone stimulate bone resorption
Glucocorticoids can inhibit bone cell proliferation, and cause increased bone resorption	Cause osteoporosis
Calcium plus estrogen in perimenopausal women prevents accelerated bone loss	The combination is more effective than calcium alone
Prolonged immobility causes osteoporosis	Leads to increased bone loss
Alcohol abuse is associated with osteoporosis	Toxic effects on osteoblasts
Obesity appears to protect against osteoporosis	Higher estrogen levels and increased skeletal loading
More rapid trabecular bone loss occurs after surgical removal of ovaries as compared to menopause	Abrupt decline of estrogen
Greater number of osteoclasts present soon after menopause	Cause accelerated phase of bone loss
Estrogen receptors have been found in osteoblasts	Reason for association between low estrogen levels and osteoporosis
Adrenal androgens decline with aging	Role in osteoporosis is uncertain

can affect the entire skeleton, and there may be fractures involving various bones, with a high incidence of hip fracture.

PRINCIPLES OF TREATMENT

Prevention is the best approach to treatment of this disease and includes the following measures:

1. The calcium content of the diet should be 800 mg/day for adults and 1500 mg/day for children.
2. Physical activity is recommended.
3. Cigarette smoking and heavy alcohol consumption should be avoided.
4. The daily calcium intake of postmenopausal women should be 1,000–1,500 mg/day.

5. Evidence indicates that the most effective measure to prevent bone loss is estrogen/progesterone replacement therapy for postmenopausal women (chapter 10).

Treatment for osteoporosis includes calcium, estrogen, and calcitonin. All three substances promote the maintenance of bone mass or contribute to a decrease in the rate of bone loss. A total of 1,000–1,500 mg of calcium taken in three to four doses throughout the day is recommended. The inclusion of vitamin D in the regimen improves intestinal absorption of calcium. A contraindication to calcium therapy is possible formation of calcium containing kidney stones.

SUMMARY

• • •

Hypernatremia is an elevated serum sodium ion level that causes cellular dehydration and increased extracellular fluid volume. It is usually caused by a water deficit and, less often, by sodium overload. The symptoms are primarily neurological and range from lethargy to coma. Brain cells accumulate idiogenic osmoles to protect against central nervous system dehydration. Hyponatremia refers to a low serum sodium ion level. This results in cellular hydration and primarily neurological symptoms. It is caused by sodium ion loss or water excess. Brain cells lose idiogenic osmoles as a defense against overhydration.

Hyperkalemia is an elevated level of potassium in the blood caused by (1) increased input of potassium, (2) impaired excretion of potassium, or (3) impaired cellular uptake of potassium. Hyperkalemia causes acidosis and blocks nerve impulse transmission. Hypokalemia is a low serum potassium ion level and may be caused by (1) intracellular flux of potassium, (2) gastrointestinal loss, or (3) urinary loss. Hypokalemia causes alkalosis and decreased muscle responsiveness.

Blood calcium levels are maintained by a balance between dietary intake and excretion, moderated by the influence of hormones and vitamin D. The ionized form of calcium is physiologically active. Bone represents a reservoir of exchangeable stores of calcium. Hypercalcemia is a high blood level of calcium. The possible causes include (1) increased secretion of PTH, (2) excess vitamin D, (3) immobilization, and (4) malignancy. High levels of calcium interfere with nerve impulse transmission and may cause kidney stones. There are neurological and muscle related symptoms including cardiac arrhythmias. There are various causes of hypocalcemia including vitamin D or PTH deficiency. The predominant effect of hypocalcemia is to increase the excitability of nerve fibers and cause abnormal muscle contractions.

Osteoporosis is a condition in which there is reduced bone mass due to demineralization associated with enlarged spaces in the bone. Both males and females begin to experience slow cortical bone loss at about age 40, and the loss increases in females at menopause. Osteoporosis may be categorized as postmenopausal or senile. Factors that contribute to bone loss are decreased levels of estrogen and various factors associated with the aging process that lead to increased bone resorption or decreased formation.

REVIEW QUESTIONS

• • •

1. Hypernatremia causes cellular (hydration, dehydration).
2. In the case of hypernatremia, why isn't there an increased sodium level in interstitial spaces of the brain?
3. What is a compensatory response of brain cells to hypernatremia?
4. List two neurological symptoms resulting from hypernatremia.
5. Indicate the reasons for caution in treating hypernatremia.
6. A physiological response to hyponatremia is water movement (into or out of cells).
7. Isotonic fluid loss does not directly cause hyponatremia. Indirectly, however, there is what effect?
8. List two specific causes of hyponatremia.
9. Hyperkalemia means that there is an increased amount of intracellular potassium. (true, false)
10. How does hyperkalemia cause acidosis?
11. List two predominant symptoms of hyperkalemia.
12. A diuretic, spironolactone, causes hyperkalemia if the patient's body is challenged by a potassium overload. Why?
13. What are three substances that reverse hyperkalemic effects on the heart?
14. Hypokalemia promotes renal loss of hydrogen. Why?
15. Vitamin D enhances serum calcium levels in what three ways?
16. Parathormone (increases, decreases) serum calcium levels.
17. List three forms of calcium in the body. Which of these forms are physiologically active?
18. In general, hypercalcemia affects what two physiological responses?
19. List three neurological symptoms of hypercalcemia.
20. List three muscle related symptoms of hypercalcemia.
21. List four possible causes of hypercalcemia.
22. What two general effects are caused by hypocalcemia?
23. List three neuromuscular manifestations of hypocalcemia.
24. What are the effects of hypocalcemia on the heart?

SELECTED READING

• • •

Arieff, A. I. 1986. Hyponatremia, convulsions, respiratory arrest, and permanent brain damage after elective surgery in healthy women. *New England Journal of Medicine* 314:1529–34.

Balogh, D. et al. 1986. Sodium balance and osmolarity in burn patients. *Intensive Care Medicine* 12:100–103.

Buckalow, Jr., V. M. 1986. Hyponatremia: Pathogenesis and management. *Hospital Practice* 21(11A):49–58.

Drinkwater, B. L. et al. 1984. Bone mineral content of amenorrheic and eumenorrheic athletes. *New England Journal of Medicine* 311(5):277–80.

Hammond, D. et al. 1986. Hypodipsic hypernatremia with normal osmoregulation of vasopressin. *New England Journal of Medicine* 315(7):433–36.

Joyce, S. M., and R. Potter. 1986. Beer potomania: An unusual cause of symptomatic hyponatremia. *Annals of Emergency Medicine* 15(6):745–47.

Laragh, J. H. 1985. Atrial natriuretic hormone, the renin-aldosterone axis, and blood pressure-electrolyte homeostasis. *New England Journal of Medicine* 313(21):1330–39.

Mundy, G. R. et al. 1984. The hypercalcemia of cancer: Clinical implications and pathogenic mechanisms. *New England Journal of Medicine* 310(26):1718–35.

Sawyer, W. T. et al. 1985. Hypokalemia, hyperglycemia, and acidosis after intentional theophylline overdose. *American Journal of Emergency Medicine* 3:408–11.

Sterns, R. H., J. E. Riggs, and S. S. Schochet. 1986. Osmotic demyelination syndrome following correction of hyponatremia. *New England Journal of Medicine* 314:1535–42.

Tieder, M. et al. 1985. Hereditary hypophosphatemic rickets with hypercalciuria. *New England Journal of Medicine* 312(10):611–16.

Chapter 4

Acid-Base Balance

The topic of acid-base balance is mainly concerned with two ions, hydrogen (H^+) and bicarbonate (HCO_3^-). These are included in a separate chapter because they represent a special case of electrolyte balance. Derangements of hydrogen and bicarbonate ion concentrations in body fluids are common in disease processes. The hydrogen ion has special physiological significance because the concentration of this ion must be maintained within narrow limits in order to be compatible with life.

ACIDS

An acid may be defined as a proton (H^+) donor with acidic properties associated with those protons. Carbonic acid (H_2CO_3), phosphoric acid (H_3PO_4), pyruvic acid, and lactic acid are examples of physiologically important acids. These acids are dissolved in body fluids, and in solution, some of the molecules dissociate or break apart to release hydrogen ions (figure 4.1).

• • •

The element hydrogen is electrically neutral and is composed of a proton (positively charged) and an electron (negatively charged). A hydrogen ion has lost its electron and is composed only of a proton.

• • •

WEAK ACIDS

Different acids contribute varying numbers of hydrogen ions, depending on their natural tendency to dissociate in solution. Those that have a strong tendency to dissociate

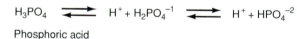

$$H_2CO_3 \rightleftharpoons H^+ + HCO_3^{-1}$$

Carbonic acid

$$H_3PO_4 \rightleftharpoons H^+ + H_2PO_4^{-1} \rightleftharpoons H^+ + HPO_4^{-2}$$

Phosphoric acid

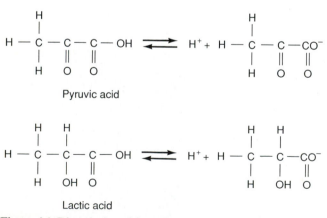

Pyruvic acid

Lactic acid

Figure 4.1 Dissociation of four physiologically important acids.

TABLE 4.1 Derivation of the formula for the ionization constant (K_a) of an acid*

1. $H_2CO_3 \rightleftharpoons H^+ + HCO_3^-$

2. In an equilibrium, the rate of the forward reaction R_f = the rate of the reverse reaction R_r.

3. In an equilibrium, the rate of each reaction is proportional to the product of the reactants.

4. Therefore, $R_f = k[H_2CO_3]$ and $R_r = k'[H^+][HCO_3^-]$

5. If $R_f = R_r$, then $k[H_2CO_3] = k'[H^+][HCO_3^-]$

6. Then, rearranging the equation,

$$\frac{k}{k'} = \frac{[H^+][HCO_3^-]}{[H_2CO_3]}$$

7. Finally, k and k' are both constants so that a new constant K_a may be substituted.

$$K_a = \frac{[H^+][HCO_3^-]}{[H_2CO_3]}$$

Formula for the ionization constant of an acid

*This is based on the law of mass action which states that, in an equilibrium, the rate of each reaction is proportional to the product of the concentration of the reactants (step 4). The symbols k and k' represent proportionality constants, and the brackets indicate moles/liter.

are strong acids, whereas weak acids dissociate to a limited extent.

PHYSIOLOGICALLY SIGNIFICANT EQUILIBRIA

An equation, such as $H_2CO_3 \rightleftharpoons H^+ + HCO_3^-$, represents an equilibrium between two reactions: (1) the breakdown of carbonic acid on the left and (2) the formation of carbonic acid by the combination of H^+ and HCO_3^- on the right. An equilibrium is a balanced and a dynamic process in which the rate of the forward reaction is equal to the rate of the reverse reaction. The concentrations, however, on either side of the arrows are not necessarily equal. The relationship between the reactant(s) and the products is expressed by the following formula:

$$K_a = \frac{[H^+][HCO_3^-]}{H_2CO_3}$$

The derivation for this formula, which is the ionization constant of an acid, is shown in table 4.1. This formula indicates that there is a constant numerical value (K_a) that is equal to the quotient of the (1) product of the concentration of hydrogen and bicarbonate ions and (2) the reactant, carbonic acid. Table 4.2 lists some observations about the significance of the ionization constant equation. Essentially, if there is an increase in concentration on the left side of the equilibrium, the result is an increased rate of the forward reaction. This is called a **shift to the right**. The reverse reaction rate increases if there is an increase in concentration of products on the right side of the equilibrium; thus, there is a **shift to the left**.

TABLE 4.2 Some observations about the ionization constant equation of an acid

$$K_a = \frac{[H^+][HCO_3^-]}{[H_2CO_3]}$$

1. This equation shows the relationships between reactant and products of the following equilibrium (ionization of H_2CO_3):

$$H_2CO_3 \rightleftharpoons H^+ + HCO_3^-$$

2. The numerical value of K_a remains the same for a particular acid at a constant temperature.

3. If the value of the numerator decreases (removal of H^+ and HCO_3^-), then dissociation of more H_2CO_3 molecules occurs to reestablish the equilibrium. This maintains relative numbers in the numerator as compared to the denominator in the ionization constant equation. It is a shift to the right in the equilibrium.

4. If H_2CO_3 is removed from the equilibrium, hydrogen ions combine with HCO_3^- to form H_2CO_3 to reestablish equilibrium (shift to left).

The following equilibria are physiologically important.

$$H_2O + CO_2 \rightleftharpoons H_2CO_3 \rightleftharpoons H^+ + HCO_3^-$$

Carbon dioxide is a metabolic waste product that diffuses from cells into blood and dissolves in plasma. A chemical reaction between carbon dioxide and water takes place, and about 70% of the carbon dioxide is converted to bicarbonate according to the reaction previously shown. About 7% of the carbon dioxide remains in solution in plasma and is measured in terms of **partial pressure** (pCO_2). About 23% of the carbon dioxide combines with hemoglobin.

The equilibria already discussed are useful as points of reference to visualize the following. When carbon dioxide diffuses into the air sacs of the lungs and is eliminated by expiration, several things occur (figure 4.2). The equilibria shift to the left, that is, H_2CO_3 breaks down to form more CO_2 and H_2O, and more hydrogen ions combine with bicarbonate to form H_2CO_3. The new equilibria have the same relative number, although fewer, molecules to compensate for the loss of carbon dioxide. The loss of carbon dioxide causes a decrease in the number of hydrogen ions, and an increase in breathing rate would exaggerate this effect. In contrast to this, a diminished breathing rate causes an increase in carbon dioxide in the blood, which results in a shift to the right of the equilibria. More carbonic acid forms, which dissociates to hydrogen ions and bicarbonate. There is an increase of all molecules, although relative numbers are the same. The result is an increase in hydrogen ions in plasma.

These ideas may be summarized by saying that (1) hyperventilation, or increased breathing rate, increases the elimination of carbon dioxide, decreases hydrogen ion concentration, and increases pH of the blood

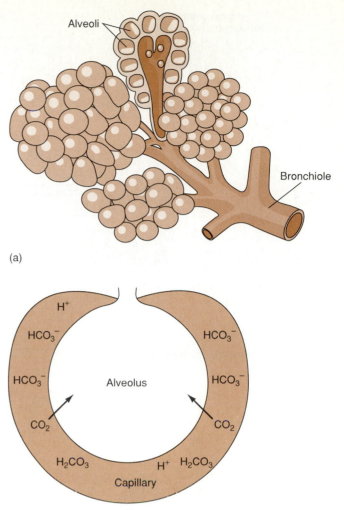

(a)

(b)

Figure 4.2 (*a*) Air passages end in microscopic sacs called alveoli. (*b*) Each alveolus is surrounded by a network of capillaries. Carbon dioxide diffuses freely from blood into the alveoli.

(figure 4.3); and (2) hypoventilation, or decreased breathing rate, causes carbon dioxide to accumulate in the blood, increases hydrogen ion concentration, and decreases pH (figure 4.4). The effect of increasing carbon dioxide is also brought about if there is interference with diffusion of carbon dioxide into the air sacs because of a disease or injury.

HYDROGEN ION CONCENTRATION

The pH scale (figure 4.5) is a convenient means of expressing hydrogen ion concentration in water solutions and is based on the fact that water ionizes to a limited extent ($H_2O \rightleftharpoons H^+ + OH^-$) to form 10^{-7} mol/liter of hydrogen ions and 10^{-7} mol/liter of hydroxide ions (OH^-). The hydrogen ion is an acid, and the hydroxide ion is a base. Since

hypoventilation: Gk. *hypo,* under; L. *ventus,* wind

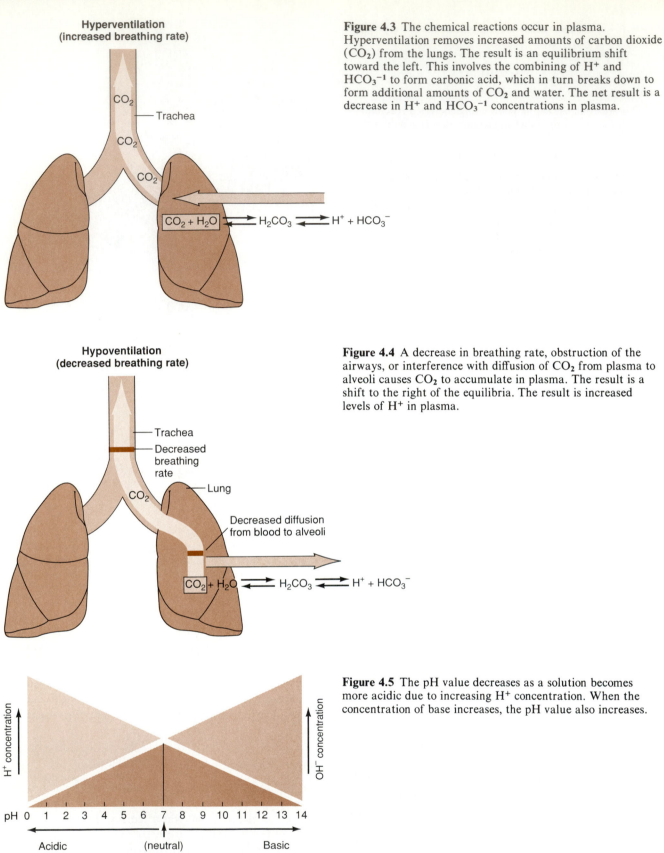

**Hyperventilation
(increased breathing rate)**

CO₂

Trachea

CO₂

CO₂

$$CO_2 + H_2O \rightleftharpoons H_2CO_3 \rightleftharpoons H^+ + HCO_3^-$$

Figure 4.3 The chemical reactions occur in plasma. Hyperventilation removes increased amounts of carbon dioxide (CO_2) from the lungs. The result is an equilibrium shift toward the left. This involves the combining of H^+ and HCO_3^{-1} to form carbonic acid, which in turn breaks down to form additional amounts of CO_2 and water. The net result is a decrease in H^+ and HCO_3^{-1} concentrations in plasma.

**Hypoventilation
(decreased breathing rate)**

Trachea

Decreased breathing rate

Lung

CO₂

Decreased diffusion from blood to alveoli

$$CO_2 + H_2O \rightleftharpoons H_2CO_3 \rightleftharpoons H^+ + HCO_3^-$$

Figure 4.4 A decrease in breathing rate, obstruction of the airways, or interference with diffusion of CO_2 from plasma to alveoli causes CO_2 to accumulate in plasma. The result is a shift to the right of the equilibria. The result is increased levels of H^+ in plasma.

H⁺ concentration

OH⁻ concentration

pH 0 1 2 3 4 5 6 7 8 9 10 11 12 13 14

Acidic   (neutral)   Basic

Figure 4.5 The pH value decreases as a solution becomes more acidic due to increasing H^+ concentration. When the concentration of base increases, the pH value also increases.

TABLE 4.3 The pH scale*

	H⁺ Concentration (molar)	pH	OH⁻ Concentration (molar)
	1.0	0	10^{-14}
	0.1	1	10^{-13}
	0.01	2	10^{-12}
Acids	0.001	3	10^{-11}
	0.0001	4	10^{-10}
	10^{-5}	5	10^{-9}
	10^{-6}	6	10^{-8}
Neutral	10^{-7}	7	10^{-7}
	10^{-8}	8	10^{-6}
	10^{-9}	9	10^{-5}
	10^{-10}	10	0.0001
Bases	10^{-11}	11	0.001
	10^{-12}	12	0.01
	10^{-13}	13	0.1
	10^{-14}	14	1.0

From Stuart Ira Fox, Human Physiology, 3d ed. Copyright © 1990 Wm. C. Brown Communications, Inc., Dubuque, Iowa. All Rights Reserved. Reprinted by permission.
The product of the H⁺ and OH⁻ concentration expressed in moles/liter is equal to 10^{-14}. An increasing H⁺ concentration is associated with a decrease on the pH scale.

these ions are present in equal amounts, pure water is neutral. The pH value is defined as the negative logarithm of hydrogen ion concentration expressed in moles/liter ($pH = -\log[H^+]$), or it is the exponent of 10 expressed as a positive number (table 4.3). In neutral water the pH is 7, and less than 7 on the pH scale represents an increase in hydrogen ion concentration. The normal pH range for blood is 7.35–7.45 and the extreme pH range that is compatible with life is 6.7–7.9.

BASES

An acid is a proton donor, and a base is a proton acceptor. Figure 4.6 shows an example in which H_2CO_3 acts as an acid and HCO_3^- acts as a base. In the second example, dihydrogen phosphate ($H_2PO_4^{-1}$) acts as an acid and, in the reverse reaction, biphosphate (HPO_4^{-2}) is a base.

ACIDOSIS/ALKALOSIS

Acidosis and alkalosis refer to a physiological condition in which there is a relative increase in hydrogen ions, in the case of acidosis, and a relative increase in bicarbonate in the case of alkalosis. The normal ratio of HCO_3^- to H_2CO_3, which is a source of hydrogen ions in the body, is 20:1. Although HCO_3^- is not the only base (proton acceptor) in extracellular fluids, it is the quantitatively important one (figure 4.7). Deviations from this ratio are used to identify acid-base imbalance. Acidosis is a decrease in

$$CO_2 + H_2O \rightleftharpoons H_2CO_3 \rightleftharpoons H^+ + \boxed{HCO_3^-}$$

When this equilibrium shifts to the left, HCO_3^- acts as a base by combining with the H⁺ to form H_2CO_3.

$$H_2PO_4^{-1} \rightleftharpoons H^+ + \boxed{HPO_4^{-2}}$$

The biphosphate ion acts as a base when it combines with H⁺ to form $H_2PO_4^{-1}$.

Figure 4.6 The bicarbonate and biphosphate ions are examples of physiologically important bases.

Normal base/acid ratio

$$\frac{Base}{Acid} \quad \frac{20}{1}$$

Acidosis

$$\frac{Base}{Acid \uparrow} \quad or \quad \frac{Base \downarrow}{Acid} = Ratio < 20{:}1$$

Alkalosis

$$\frac{Base \uparrow}{Acid} \quad or \quad \frac{Base}{Acid \downarrow} = Ratio > 20{:}1$$

Figure 4.7 The normal pH is associated with a base to acid ratio of 20:1. Acidosis occurs when that ratio decreases due either to an increase in acid or a decrease in base. Alkalosis is the result of increased base or decreased acid that leads to an increase in the base/acid ratio.

the normal 20:1 base to acid ratio and can occur when there is either an increase in the number of hydrogen ions or a decrease in the number of bicarbonate ions. An increase of base to acid ratio is alkalosis, and there are two ways this may occur. There may be an increase in the amount of bicarbonate or a decrease in the number of hydrogen ions. Acidosis is actually too much acid or too little base, and alkalosis is base excess or acid deficit.

SOURCES OF HYDROGEN IONS

Intracellular metabolism of glucose leads to the formation of intermediate products with the release of energy and the production of carbon dioxide and water as waste products (figure 4.8). Carbon dioxide diffuses into blood where the reaction $CO_2 + H_2O \rightleftharpoons H_2CO_3 \rightleftharpoons H^+ + HCO_3^-$ occurs in red blood cells. Carbon dioxide is responsible for the formation of both hydrogen ions and bicarbonate, and as has already been noted, bicarbonate is one of the forms of carbon dioxide in plasma. Food and medication are sources of small amounts of hydrogen ion. In addition, there are intermediate products of metabolism, such as lactic acid, pyruvic acid, and acetoacetic acid, that are proton donors.

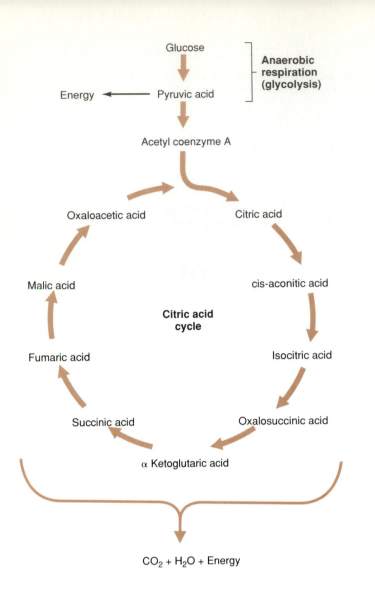

Glucose

Energy ← Pyruvic acid

Anaerobic respiration (glycolysis)

Acetyl coenzyme A

Oxaloacetic acid

Citric acid

Malic acid

cis-aconitic acid

Citric acid cycle

Fumaric acid

Isocitric acid

Succinic acid

Oxalosuccinic acid

α Ketoglutaric acid

Aerobic respiration

$CO_2 + H_2O$ + Energy

Figure 4.8 The breakdown of glucose by way of anaerobic and aerobic metabolism leads to the production of the waste products, CO_2 and H_2O.

• • •

Carbonic acid is called a volatile acid because it breaks down to carbon dioxide, which in turn can be eliminated by respiration. Nonvolatile acids are metabolic acids that are not eliminated in this way.

• • •

SOURCES OF BICARBONATE IONS

The reaction between carbon dioxide and water has been mentioned several times. This reaction occurs as carbon dioxide diffuses into red blood cells, and it is mediated by the enzyme **carbonic anhydrase.** Most of the hydrogen ions combine with hemoglobin, which acts as a buffer, and bicarbonate ions diffuse into plasma in exchange for chloride ions (figure 4.9). Although hydrogen ions are free to diffuse out of the cell, most of the diffusion involves bicarbonate. The direction of bicarbonate diffusion reverses in pulmonary capillaries. There is an influx into red blood cells as carbon dioxide diffuses into alveoli. Under conditions of lowered pCO_2, the predominant reaction within the red blood cells is the formation of carbonic acid with dissociation to carbon dioxide.

• • •

The hydration of carbon dioxide also occurs in plasma, although the reaction rate is slow due to the absence of carbonic anhydrase in plasma. Electrical neutrality is maintained in red blood cells by a shift of chloride ions (Cl^-) into the cell as HCO_3^- diffuses out. The process is called the chloride shift.

• • •

FLUID AND ELECTROLYTES

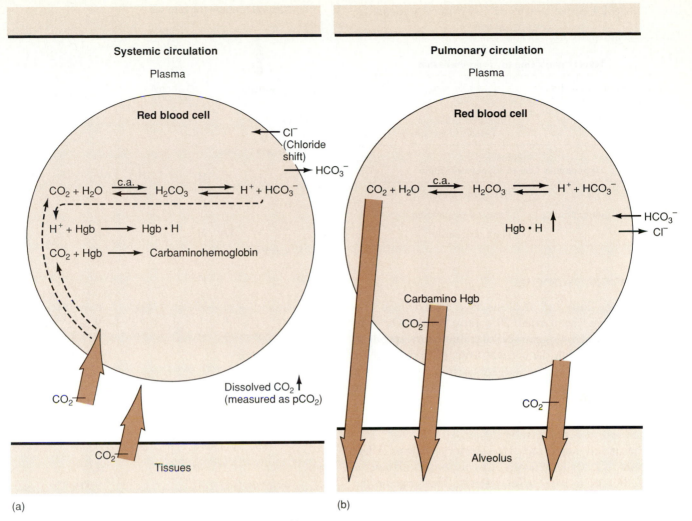

Systemic circulation

Plasma

Red blood cell

Cl⁻ (Chloride shift)

$$CO_2 + H_2O \xrightleftharpoons[]{c.a.} H_2CO_3 \rightleftharpoons H^+ + HCO_3^- \longrightarrow HCO_3^-$$

$H^+ + Hgb \longrightarrow Hgb \cdot H$

$CO_2 + Hgb \longrightarrow$ Carbaminohemoglobin

Dissolved CO_2↑ (measured as pCO_2)

CO_2

CO_2

Tissues

(a)

Pulmonary circulation

Plasma

Red blood cell

$$CO_2 + H_2O \xrightleftharpoons[]{c.a.} H_2CO_3 \rightleftharpoons H^+ + HCO_3^-$$

HCO_3^-

$Hgb \cdot H$ ↑

Cl⁻

Carbamino Hgb

CO_2

CO_2

Alveolus

(b)

Figure 4.9 (*a*) Carbon dioxide diffuses into plasma and into red blood cells. Within red blood cells, the hydration of carbon dioxide is catalyzed by the enzyme carbonic anhydrase. Bicarbonate thus formed diffuses into plasma. (*b*) Bicarbonate diffuses back into red blood cells in pulmonary capillaries and reacts with hydrogen ions to form carbonic acid. The acid breaks down to carbon dioxide and water.

Another source of HCO_3^- is the following process in which **parietal** (pah-ri′ě-tal) cells of the gastric mucosa secrete hydrogen ions into the lumen of the stomach, while bicarbonate ions diffuse into the blood to maintain electrical neutrality. The original source of those ions is the hydration of carbon dioxide. The direction of ion movement is reversed in pancreatic epithelial cells. Hydrogen ions are secreted into the blood, and bicarbonate ions diffuse into pancreatic juice. If these two processes are balanced, there is no net change in the amount of bicarbonate in blood (figure 4.10). Loss of gastric or pancreatic juice can change that balance (chapter 5).

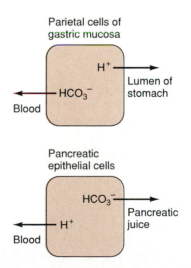

Parietal cells of gastric mucosa

H^+

Lumen of stomach

HCO_3^-

Blood

Pancreatic epithelial cells

HCO_3^-

Pancreatic juice

H^+

Blood

Figure 4.10 Cells of the gastric mucosa secrete hydrogen ions into the lumen of the stomach in exchange for the diffusion of bicarbonate ions into blood. The direction of the diffusion of these ions is reversed in pancreatic epithelial cells.

TABLE 4.4 Some causes of respiratory acidosis and alkalosis

Acidosis Hypercapnia Due to Hypoventilation	
Depression of respiratory center	Collapse of lung
Obstruction of airway	Impaired diffusion of CO_2 (blood to air sacs)

Alkalosis Hypocapnia Due to Hyperventilation	
Anxiety	Salicylate poisoning
Respiratory center lesions	Assisted respiration
Fever	High altitude

RESPIRATORY ACIDOSIS

The terms acidosis and alkalosis have been defined as a deviation from the normal 20:1 ratio for HCO_3^- and H_2CO_3. A pH of 7.4 of extracellular fluids corresponds to a 20:1 ratio in which the concentration of bicarbonate is 24 meq/liter and H_2CO_3 is 1.2 meq/liter. Acidosis is a decrease in pH, which means there is a relative increase of hydrogen ions. This may be caused by an increase in hydrogen ions (H_2CO_3) or a decrease in bicarbonate, both of which lead to a decrease in the ratio.

The underlying cause of respiratory acidosis is **hypercapnia** (hī''per-kap′nē-ah), i.e., an accumulation of carbon dioxide in extracellular fluids. This may be the result of obstruction of air passages, decreased respiration, or decreased gas exchange between pulmonary capillaries and air sacs of lungs; thus, respiratory acidosis is caused by hypercapnia due to hypoventilation (table 4.4).

RESPIRATORY ALKALOSIS

Respiratory alkalosis is a condition in which the normal 20:1 ratio of HCO_3^- and H_2CO_3 is increased, and the pH is above 7.45. The cause is hyperventilation, which leads to eliminating excessive amounts of carbon dioxide and a decrease in hydrogen ions (table 4.4). There is a shift to the left of these equilibria, $CO_2 + H_2O \rightleftharpoons H_2CO_3 \rightleftharpoons H^+ + HCO_3^-$.

METABOLIC ACIDOSIS

Metabolic acidosis occurs when there is a decrease in the normal 20:1 ratio of HCO_3^- and H_2CO_3 with a subsequent decrease in pH. Any acid-base imbalance not attributable to carbon dioxide is classified as metabolic, even though such an imbalance may not be the direct result of metabolism.

Unregulated diabetes mellitus, in which there is improper utilization of glucose, causes what is called **ketoacidosis.** The body metabolizes fat rather than carbohydrate, and this leads to excessive production of three

hypercapnia: Gk. *hyper,* above; Gk. *kapnos,* smoke

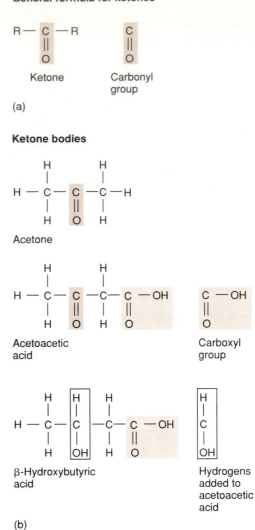

General formula for ketones

(a)

Ketone bodies

Acetone

Acetoacetic acid

Carboxyl group

β-Hydroxybutyric acid

Hydrogens added to acetoacetic acid

(b)

Figure 4.11 (*a*) Ketones have a carbonyl group. (*b*) Acetone, acetoacetic acid, and β-hydroxybutyric acid are ketone bodies produced in ketoacidosis. A carboxyl group is characteristic of organic acids. β-hydroxybutyric acid is produced by the reduction (addition of hydrogen) to acetoacetic acid.

ketones: **acetone, acetoacetic acid,** and **β-hydroxybutyric acid** (figure 4.11). The last two are physiologically strong acids and contribute excessive numbers of hydrogen ions to body fluids (figure 4.12). Table 4.5 lists some possible causes of metabolic acidosis, which will be discussed in later chapters.

METABOLIC ALKALOSIS

Metabolic alkalosis is defined as a condition in which there is an elevation of pH due to an increased HCO_3^-/H_2CO_3 ratio. This may be caused by an increase of bicarbonate or a decrease in hydrogen ions, but the imbalance is not due to carbon dioxide. As in the case of metabolic acidosis, metabolic alkalosis is an increase in pH that has a nonrespiratory origin. Either vomiting or diarrhea can cause metabolic alkalosis if there is excessive loss of hydrogen ions as compared to bicarbonate (chapter 5). Table 4.5 lists some causes of metabolic alkalosis.

FLUID AND ELECTROLYTES

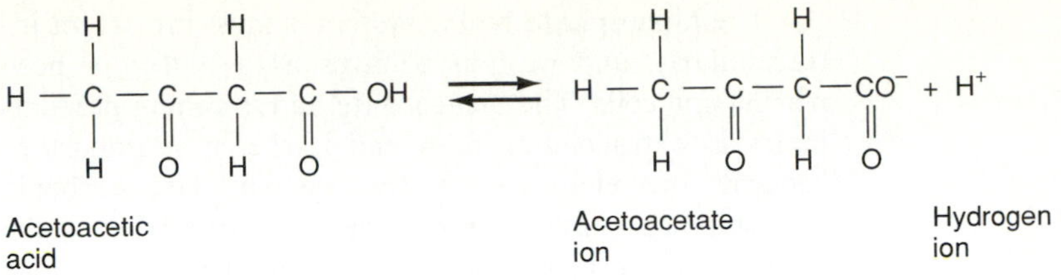

Figure 4.12 Organic acids ionize to release hydrogen ions that cause a decrease in the pH of body fluids.

TABLE 4.5 Some causes of metabolic acidosis and alkalosis

Acidosis	
Excessive H^+ or Decreased HCO_3^-	
Diabetes mellitus	Metabolic production of acids
Chronic renal failure	Retention of H^+
Strenuous exercise	Anaerobic conversion of pyruvic to lactic acid
Vomiting	Loss of excessive HCO_3^-
Diarrhea	Loss of excessive HCO_3^-
Alkalosis	
Decreased H^+ or Increased HCO_3^-	
Excessive use of antacids	$NaHCO_3$
Vomiting	Loss of excessive H^+
Diarrhea	Loss of excessive H^+

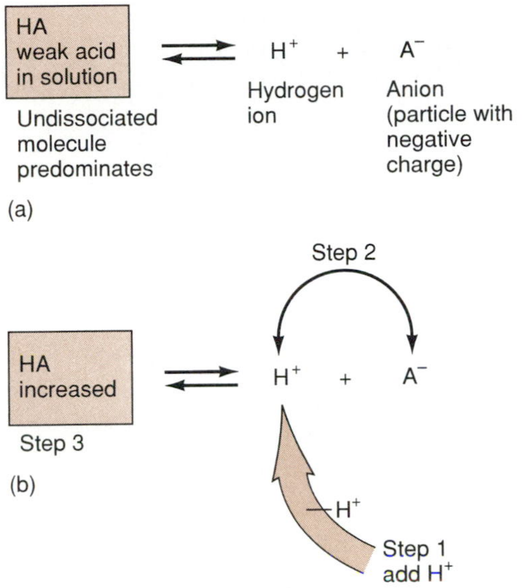

Figure 4.13 (*a*) A weak acid in solution dissociates to a limited extent to release hydrogen ions and anions. (*b*) When acid (H^+) is added to the solution (step 1), hydrogen ions combine with anions to form additional molecules of undissociated HA.

Responses to Acidosis/Alkalosis

There are several homeostatic mechanisms that protect the body against life-threatening changes in hydrogen ion concentration. Those responses include (1) buffering systems in body fluids, (2) respiratory response, (3) renal control, and (4) intracellular shifts of ions. Buffering systems provide an immediate response to fluctuations in pH, and the predominant buffer in extracellular fluid is the bicarbonate system.

Buffers

A **buffer** is a combination of chemicals in solution that resists any significant change in pH. A buffer system is made up of either a weak acid and a salt of that acid, or a weak base and a salt of that base. The key to the functioning of a buffer system is that a weak acid dissociates to a limited extent, so the equilibrium is shifted to the left with undissociated molecules predominating (figure 4.13). Hydrogen ions added to this equilibrium combine with the anions to form additional weak acid, the equilibrium remains far to the left, and hydrogen ions are effectively tied up. A salt of that acid ionizes completely in solution and contributes anions to react with excess hydrogen ions that may be added to the solution (figure 4.14). Three physiologically important buffer systems are shown in figure 4.15.

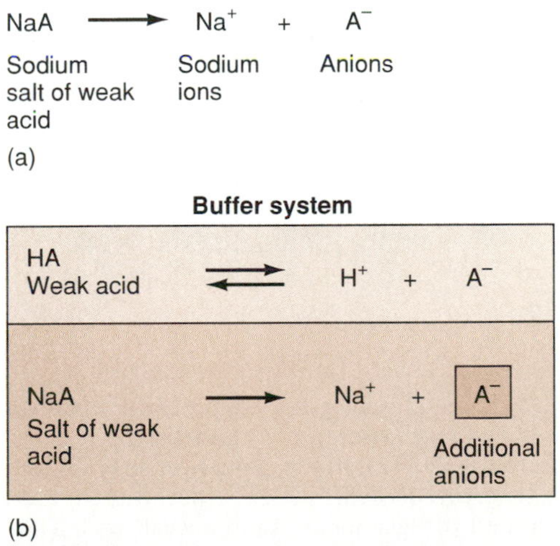

Figure 4.14 (*a*) Salts ionize completely in solution. (*b*) A weak acid and a salt of that acid constitute a buffer system. The weak acid dissociates to a limited extent and the salt ionizes completely. The anions contributed by the salt are available to react with hydrogen ions that may be added to the solution. When acid is added to the solution, the weak acid equilibrium shifts to the left.

Bicarbonate buffer system

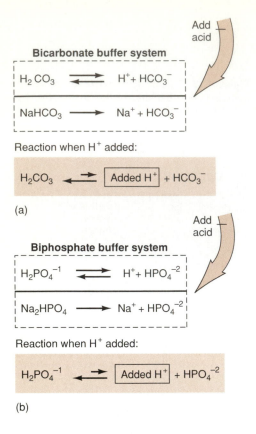

Bicarbonate buffer system

$$H_2CO_3 \rightleftharpoons H^+ + HCO_3^-$$

$$NaHCO_3 \longrightarrow Na^+ + HCO_3^-$$

Reaction when H^+ added:

$$H_2CO_3 \longleftarrow \boxed{Added\ H^+} + HCO_3^-$$

(a)

Biphosphate buffer system

$$H_2PO_4^{-1} \rightleftharpoons H^+ + HPO_4^{-2}$$

$$Na_2HPO_4 \longrightarrow Na^+ + HPO_4^{-2}$$

Reaction when H^+ added:

$$H_2PO_4^{-1} \longleftarrow \boxed{Added\ H^+} + HPO_4^{-2}$$

(b)

Hemoglobin and other proteins as buffers

(Protein) R — C — OH $\rightleftharpoons$ H^+ + R + CO^- (with ||O below)

(Protein) R — C — O — Na $\longrightarrow$ Na^+ + R – CO^- (with ||O below)

Reaction when H^+ added:

R — C — OH $\longleftarrow$ $\boxed{Added\ H^+}$ + R — CO^- (with ||O below)

(c)

Figure 4.15 Three physiologically important buffer systems. There is an excess of base, i.e., (*a*) bicarbonate, (*b*) biphosphate, and (*c*) protein anions, in body fluids. As there is an increase in hydrogen ion concentration, these bases react with the hydrogen ions to form a weak acid. This reaction removes free H^+ because the molecular form of a weak acid predominates.

The biphosphate buffer system is most important intracellularly, and protein behaves as a buffer in both plasma and cells. The bicarbonate buffer system predominates in extracellular fluid and is the most important because the elements of the system, i.e., carbonic acid and bicarbonate, can be independently varied by the kidney and by respiration. Hyperventilation eliminates additional carbon dioxide and pulls the equilibria, $CO_2 + H_2O \rightleftharpoons H_2CO_3 \rightleftharpoons H^+ + HCO_3^-$, to the left. Hypoventilation increases carbon dioxide and pushes the equilibria to the right. The kidney determines the excretion or retention of hydrogen and bicarbonate ions (figure 4.16).

RESPIRATORY RESPONSE

Clusters of neurons in the medulla oblongata and pons constitute the respiratory center. Dorsally located medullary neurons stimulate inspiration and determine breathing rhythm. Ventrolaterally located neurons in the medulla make up both an inspiratory and expiratory group, which are particularly important when respiration increases. The **pneumotaxic** (nū''mō-tak'sik) region in the dorsal area of the pons is composed of neurons that send impulses to the inspiratory area, and the primary effect is to limit inspiration (figure 4.17).

Chemoreceptor

A **chemosensitive** area in the medulla responds to increases in carbon dioxide and hydrogen ions by stimulating the respiratory center. The stimulating effect is diminished within 1–2 minutes, and the effect is weak if increased carbon dioxide levels persist for several days. In addition, there are **chemoreceptors,** neurons stimulated by changes in the partial pressure of oxygen or carbon dioxide (pO_2, pCO_2), or pH. These receptors are called the **carotid** and **aortic bodies.** There are two carotid bodies located laterally at the bifurcations of the common carotid arteries, and the aortic bodies are found in the arch of the aorta (figure 4.18). Afferent fibers from these neurons carry impulses to the dorsal respiratory area with increased signals in response to low oxygen levels or increased carbon dioxide/hydrogen ion levels. Normally, it is increases of carbon dioxide or hydrogen ions that stimulate the chemosensitive region of the medulla to signal for increased respiration. The overall effect is hyperventilation as a compensatory response to increased carbon dioxide/hydrogen ions and hypoventilation in response to decreased carbon dioxide/hydrogen ions.

ELECTROLYTE SHIFTS

In general acidosis, either respiratory or metabolic, may be associated with hyperkalemia that occurs when hydrogen ions diffuse into cells. The hydrogen ions are buffered by protein and potassium ions diffuse out of cells.

apneustic: Gk. *a*, not; G. *pneusis*, breathing
afferent: L. *afferre*, to carry to

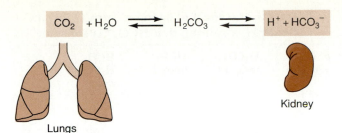

$$CO_2 + H_2O \rightleftharpoons H_2CO_3 \rightleftharpoons H^+ + HCO_3^-$$

Lungs

Kidney

Figure 4.16 Hyperventilation removes hydrogen ions by pulling the equilibria to the left. Conversely, hypoventilation results in a shift to the right with an increase in hydrogen ion concentration. The kidney exerts control over acid-base balance by either eliminating or retaining hydrogen and bicarbonate ions.

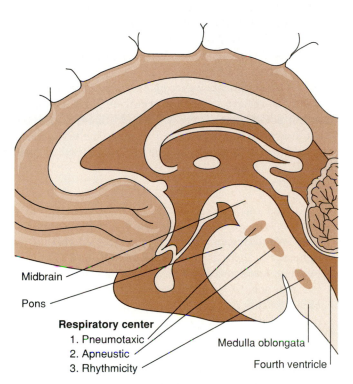

Midbrain

Pons

Respiratory center
1. Pneumotaxic
2. Apneustic
3. Rhythmicity

Medulla oblongata

Fourth ventricle

Figure 4.17 Sagittal view of the brain. Clusters of neurons in the pons and medulla oblongata constitute the respiratory center. The apneustic center promotes inspiration, the pneumotaxic center inhibits inspiration, and the rhythmicity center controls automatic breathing.

Alkalosis is associated with hypokalemia with a reciprocal exchange of K^+ and H^+ in the opposite direction (figure 4.19). The contributing factors are complex and not completely defined, and the kidney plays a role in determining serum potassium ion levels as well.

RENAL RESPONSE

The kidney compensates for acid-base imbalance within 24 hours and is responsible for long-term control. Unit III deals with the details of kidney function, but in general, the kidney retains bicarbonate ions and eliminates hydrogen ions in response to acidosis. The response is reversed in alkalosis.

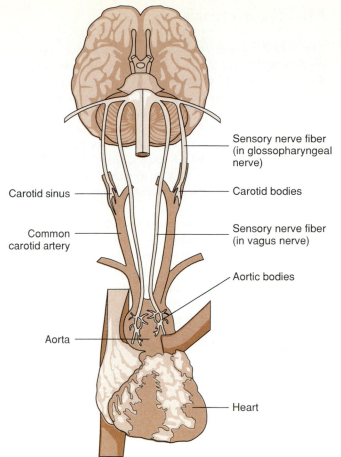

Carotid sinus

Common carotid artery

Aorta

Sensory nerve fiber (in glossopharyngeal nerve)

Carotid bodies

Sensory nerve fiber (in vagus nerve)

Aortic bodies

Heart

Figure 4.18 Chemoreceptors in the medulla and in the carotid and aortic bodies are stimulated by changes in pCO_2, pH, or pO_2. The normal stimuli for increased respiration is increased levels of CO_2 or H^+. Marked low levels of O_2 may also stimulate respiratory rate.

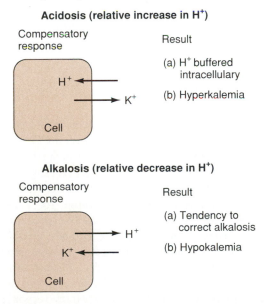

Acidosis (relative increase in H^+)

Compensatory response

Result

H^+

K^+

Cell

(a) H^+ buffered intracellulary

(b) Hyperkalemia

Alkalosis (relative decrease in H^+)

Compensatory response

Result

H^+

K^+

Cell

(a) Tendency to correct alkalosis

(b) Hypokalemia

Figure 4.19 There are intracellular/extracellular shifts that are compensatory responses to acid-base imbalance. In acidosis, hydrogen ions diffuse into cells in exchange for extracellular diffusion of potassium ions. The processes are reversed in alkalosis; thus, acidosis may cause hyperkalemia and alkalosis may lead to hypokalemia.

Manifestations

The manifestations of acid-base imbalance vary according to the rate of onset. In the case of respiratory acidosis, depressed respiration may be obvious. There may be symptoms of both respiratory acidosis and lack of oxygen (hypoxia) because the two conditions occur together. In acute respiratory acidosis, breathlessness, stupor, or coma may be observed. Carbon dioxide causes dilation of cerebral and other blood vessels, and dilation of superficial vessels of the face may be apparent. Confusion, loss of memory, and somnolence occur in chronic respiratory acidosis.

The manifestations of acute respiratory alkalosis are numbness and **paresthesias** (par″es-the′ze-ahs), particularly of the lips and extremities. The term paresthesia refers to tingling, crawling, or burning sensations of the skin. The patient may feel light-headed, confused, or even lose consciousness. In the chronic condition, the symptoms are those of the underlying disease process.

The manifestations of metabolic acidosis include the following:

1. Obvious hyperventilation
2. Decreased cardiac contractility
3. Peripheral vasodilation
4. **Anorexia** (an″o-rek′se-ah)
5. Nausea
6. Vomiting
7. Stupor
8. Coma

In acute metabolic acidosis, there may be intense, deep sighing respiration called **Kussmaul respiration.** No specific symptoms are associated with metabolic alkalosis. In severe cases, there may be apathy, confusion, or stupor and possibly increased neuromuscular excitability.

Evaluation

Arterial blood gases are measured to evaluate acid-base status, and the laboratory values include pCO_2, HCO_3^-, and pH. The pCO_2 is (1) a measure of the partial pressure of CO_2 dissolved in plasma, (2) the normal range of values is 35–45 mm Hg, and (3) is changed by hyperventilation and hypoventilation. The normal range for HCO_3^- is 24–33 meq/liter, and since renal control predominates, deviations from normal indicate metabolic acidosis and alkalosis. The pH indicates hydrogen ion concentration. The normal range for blood is 7.35–7.45, and this value represents the overall acid-base balance reflecting, not only primary imbalance, but also compensatory responses.

paresthesia: Gk. *para,* beyond; Gk. *aisthesis,* perception
anorexia: Gk. *a,* not; Gk. *orexis,* appetite

Table 4.6 The effect of changes in pCO_2 on the blood pH, assuming that HCO_3^- is constant

pCO_2 (mm Hg)	H_2CO_3 (meq/L)	HCO_3^- (meq/L)	$HCO_3^-/$ H_2CO_3 ratio	pH
20	0.6	24	40:1	7.70
30	0.9	24	26.7:1	7.53
40	1.2	24	20:1	7.40
50	1.5	24	16:1	7.30
60	1.8	24	13.3:1	7.22

From Stuart Ira Fox, Human Physiology, *3d ed. Copyright © 1990 Wm. C. Brown Communications, Inc., Dubuque, Iowa. All Rights Reserved. Reprinted by permission.*

Base excess is an additional laboratory test that provides information about metabolic acid-base imbalance (table 4.6). This test is a measure of base or acid added to the blood due to metabolic disturbance. Normal values for base excess are between +2.5 and −2.5 meq/liter. Positive values indicate either an excess of base or deficit of acid. The reverse is true for negative values, i.e., base deficit or acid excess. There will be additional discussion about interpretation of laboratory values, including electrolytes and arterial blood gases, in the last chapter of each unit.

• • •

Bicarbonate levels generally reflect metabolic acid-base balance because the kidney is the predominant influence on these levels. Carbon dioxide, and so respiration, has some effect on bicarbonate by way of the reaction $CO_2 + H_2O \rightleftharpoons H_2CO_3 \rightleftharpoons H^+ + HCO_3^-$. This reaction, however, is of primary importance in terms of hydrogen ion concentration. A shift to the right of these equilibria increases H^+ and HCO_3^- in a ratio of 1:1. Increased hydrogen ions have, however, a more significant effect on the normal base-acid ratio of 20:1 due to the relative numbers of ions involved.

• • •

Principles of Treatment

Treatment of acid-base imbalance involves correction of the underlying cause, with other measures determined by the severity of the condition as well as whether the disorder is acute or chronic. If gastrointestinal losses have caused acidosis, the administration of intravenous sodium bicarbonate is appropriate. In respiratory acidosis, assisted ventilation may be required. Dilute hydrochloric acid or an acidifying salt, such as arginine hydrochloride, corrects severe metabolic alkalosis. Rebreathing air from a paper bag alleviates respiratory alkalosis caused by anxiety hyperventilation.

SUMMARY

• • •

The equation for an ionization constant (K_a) of a weak acid shows that in equilibrium ($H_2CO_3 \rightleftharpoons H^+ + HCO_3^-$) the relative number of molecules on either side of the arrows remains the same. Carbon dioxide dissolves in plasma and the following reaction occurs, $CO_2 + H_2O \rightleftharpoons H_2CO_3 \rightleftharpoons H^+ + HCO_3^-$. When carbon dioxide diffuses into the air sacs of the lungs, the equilibria shift to the left. If carbon dioxide accumulates in the plasma, the equilibria shift to the right; hence, hyperventilation decreases hydrogen ion concentration as bicarbonate combines with hydrogen ions. Hypoventilation increases hydrogen ion concentration as carbonic acid ionizes to release hydrogen and bicarbonate ions. Carbonic acid in the equilibria is a proton donor and behaves as an acid. Bicarbonate is a base because it is a proton acceptor.

Acidosis/alkalosis are defined in terms of a normal (20:1) base to acid ratio. Acidosis occurs when there is (1) an increase of hydrogen ions or a decrease of bicarbonate (decrease of the 20:1 ratio), and (2) there is a lowered pH. Alkalosis is a relative increase in bicarbonate (increase of the 20:1 ratio) and an elevation of pH. Respiratory acidosis is caused by hypercapnia due to hypoventilation, whereas respiratory alkalosis is caused by hypocapnia resulting from hyperventilation. Metabolic acid-base imbalances are the result of nonrespiratory mechanisms.

Homeostatic mechanisms that regulate hydrogen ion concentration are buffer systems, respiratory response, renal response, and intracellular exchange of K^+ and H^+. The treatment of acid-base imbalance involves correction of the underlying cause and may include administration of intravenous sodium bicarbonate and assisted ventilation in the case of acidosis. Metabolic alkalosis may require acidifying salts, or respiratory alkalosis may be treated by rebreathing air from a paper bag.

REVIEW QUESTIONS

• • •

1. What are two physiologically important acids?
2. Which would have a larger (K_a) ionization constant, a weak or strong acid? Explain.
3. Does a decrease in pH represent an increase or decrease in hydrogen ion concentration?
4. What does hypoventilation mean?
5. What happens to these equilibria during hyperventilation?

$$CO_2 + H_2O \rightleftharpoons H_2CO_3 \rightleftharpoons H^+ + HCO_3^-$$

6. Identify the acid and the base in this equilibrium.

$$H_2PO_4^{-1} \rightleftharpoons H^+ + HPO_4^{-2}$$

7. An increase in the HCO_3^-/H_2CO_3 ratio is (acidosis, alkalosis).

8. What are two possible sources of hydrogen in the body?
9. Where in the body does carbon dioxide react with water to form carbonic acid?
10. What are two sources of bicarbonate in the body?
11. What causes respiratory acidosis?
12. What causes hypocapnia?
13. What are three possible causes of hypoventilation?
14. Distinguish between respiratory and metabolic acidosis.
15. What instantaneous homeostatic mechanism operates to protect against fluctuation in pH?
16. What is the most effective long-term protection against changes in pH?
17. In the bicarbonate buffer system what ion effectively ties up excess hydrogen?
18. What are important intracellular buffers?
19. What is the respiratory response to alkalosis?
20. What is the significance of a laboratory report showing increased bicarbonate?

SELECTED READING

• • •

Chazan, J. A. et al. 1989. Acid-base abnormalities in cardiopulmonary arrest: Varying patterns in different locations in the hospital (letter). *New England Journal of Medicine* 320(9):597–98.

Giammarco, R. A. 1987. The athlete, cocaine and lactic acidosis: A hypothesis. *American Journal of Medical Science* 294:412.

Grossman, L. et al. 1983. Lactic acidosis in a patient with acute leukemia. *Clinical and Investigative Medicine* 6:85.

Kilderberg, P. 1983. Acid-base status of biological fluids: Amount of acid, kind of acid, anion-cation difference, and buffer value. *Scandinavian Journal of Clinical Laboratory Investigation* 43:103–9.

Ledley, F. D. et al. 1984. Benign methylmalonic aciduria. *New England Journal of Medicine* 311:1015–18.

McCartney, N., G. J. Heigenhauser, and N. L. Jones. 1983. Effects of pH on maximal power output and fatigue during short-term dynamic exercise. *Journal of Applied Physiology: Respiratory, Environmental and Exercise Physiology* 55:225–29.

Morris, L. R., M. B. Murphy, and A. Kitabchi. 1986. Bicarbonate therapy in severe diabetic ketoacidosis. *Annals of Internal Medicine* 105(6):836–40.

Oh, M. S. et al. 1988. Electrolyte case vignette: A case of unusual organic acidosis. *American Journal of Kidney Diseases* 11:80.

Relman, A. S. 1986. Blood gases: arterial or venous? *New England Journal of Medicine* 315:188.

Chapter 5

Practical Applications: Loss of Fluid and Electrolytes

The focus of each chapter so far has been homeostasis of water and specific ions. The purpose of this chapter is to integrate these topics and to illustrate the physiological consequences of fluid and electrolyte loss. The discussion that follows deals with two types of fluid loss that lead to derangement of electrolytes and to acid-base imbalance.

VOMITING

The consequences of vomiting large quantities of gastric juice are the result of both fluid loss as well as electrolyte loss. Gastric juice contains amounts of hydrochloric acid as well as sodium and potassium ions. A loss of this fluid leads to alkalosis, volume depletion, and loss of total body sodium and potassium ion stores (table 5.1).

METABOLIC ALKALOSIS

Figure 5.1 shows the microscopic appearance of gastric mucosa, and figure 5.2 identifies the cells that make up the gastric mucosa. Hydrogen ions appear in gastric juice as the result of a reaction in the parietal cells of the stomach involving carbon dioxide and water, resulting in the formation of hydrogen ions in the parietal cells (chapter 4).

$$CO_2 + H_2O \rightleftharpoons H_2CO_3 \rightleftharpoons H^+ + HCO_3^-$$

Hydrogen ions pass into the lumen of the stomach by means of active transport, and bicarbonate diffuses into interstitial fluid and then into blood. Chloride ion secretion is coupled with hydrogen ion secretion, and water diffuses into the stomach as well (figure 5.3). In vomiting, the net result is that bicarbonate is added to the blood as hydrogen ions are lost from the body.

• • •

When a patient is given acetazolamide (Diamox), a diuretic that acts as a carbonic anhydrase inhibitor, the secretion of gastric HCl is repressed. The reason this occurs is that carbonic anhydrase catalyzes the reaction between carbon dioxide and water.

• • •

TABLE 5.1 Electrolyte composition of gastric contents expressed in meq/liter*

	Na⁺	K⁺	Cl⁻	HCO₃⁻
Average	60	9	84	0
Range	30–90	4.3–12	52–124	0

From M. Maxwell and C. Kleeman, Clinical Disorders of Fluid and Electrolyte Metabolism. Copyright © 1972 McGraw-Hill, Inc., New York, NY. Reprinted by permission of McGraw-Hill, Inc.
There is variation in these values.

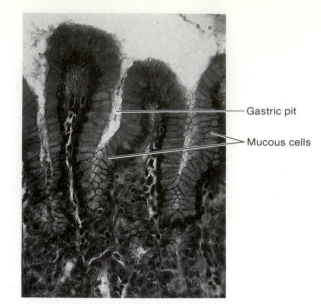

Figure 5.1 Microscopic view of gastric mucosa. © Edwin Reschke

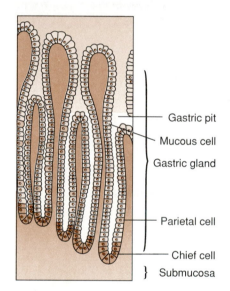

Figure 5.2 Mucous cells of the gastric mucosa produce mucus, chief cells produce digestive enzymes, and parietal cells produce hydrochloric acid.

Under normal conditions, hydrogen ions in gastric juice are carried into the duodenum with **chyme** (kīm) or partially digested food material. The ions are neutralized by the secretion of pancreatic cells (figure 5.4), that is, high in bicarbonate (figure 5.5). As in gastric parietal cells, a chemical reaction occurs in pancreatic epithelial cells, so carbon dioxide and water form hydrogen ions and bicarbonate. There is a difference, however, because pancreatic cells secrete hydrogen ions into venous blood, and bicarbonate appears in pancreatic juice emptied into the duodenum (figure 5.6). As a result, there is a balance between bicarbonate and hydrogen ions, the former being

chyme: Gk. *chymos*, juice

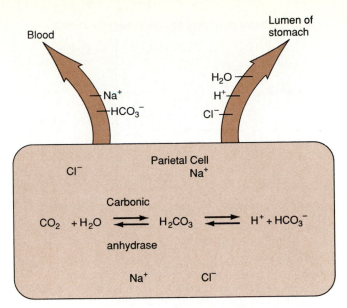

Figure 5.3 A reaction between carbon dioxide and water in parietal cells is catalyzed by the enzyme carbonic anhydrase. Hydrogen and chloride ions are secreted by parietal cells into the lumen of the stomach. At the same time, sodium and bicarbonate ions diffuse into blood.

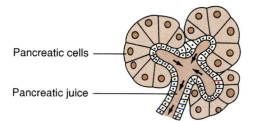

Figure 5.4 Cells of the pancreas secrete pancreatic juice, which is high in bicarbonate.

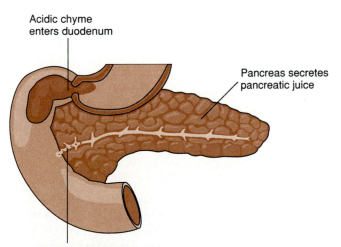

Figure 5.5 The pancreas releases alkaline pancreatic juice into the duodenum where it neutralizes acidic chyme from the stomach.

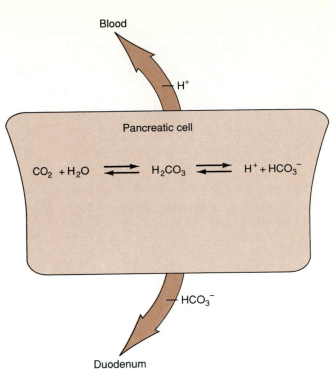

Figure 5.6 Pancreatic cells secrete hydrogen ions into the blood and bicarbonate ions into pancreatic juice. Pancreatic juice enters the duodenum.

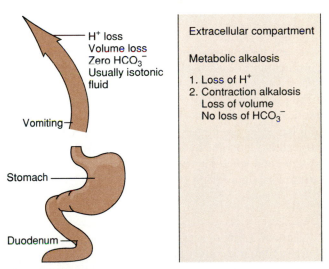

Figure 5.7 Prolonged vomiting is associated with metabolic alkalosis due to loss of hydrogen ions. Loss of volume with no accompanying bicarbonate ions has a concentrating effect on extracellular fluid and is the cause of contraction alkalosis. Both factors contribute to metabolic alkalosis.

added to the blood from gastric cells and the latter being added to the blood from pancreatic cells. Loss of gastric contents upsets the balance between these two ions, so hydrogen ions are lost as bicarbonate is added to the blood. Alkalosis is the outcome (figure 5.7).

TABLE 5.2 A comparison of electrolyte composition of gastric juice and serum*

	Na⁺	K⁺	Cl⁻	HCO₃⁻
Serum	142	4.5	102	26
Gastric juice	60	9	84	0

From M. Maxwell and C. Kleeman, Clinical Disorders of Fluid and Electrolyte Metabolism. Copyright © 1972 McGraw-Hill, Inc., New York, NY. Reprinted by permission of McGraw-Hill, Inc.
Average values expressed in meq/liter.

There is a second factor in the development of metabolic alkalosis. Loss of vomitus from the body has the net effect of decreasing the extracellular volume without changing the amount of bicarbonate (since there is no bicarbonate in gastric juice). This has a concentrating effect resulting in **contraction alkalosis,** a term used because the extracellular volume contracts around a fixed amount of bicarbonate. The increased concentration of bicarbonate may not be marked, but loss of volume by vomiting is a contributing factor to alkalosis.

• • •

Contraction alkalosis usually does not increase serum HCO₃⁻ more than 5 meq/liter.

• • •

HYPOVOLEMIA

There are three important factors related to fluid loss by vomiting. (1) The extracellular compartment is the source of fluid lost by vomiting; (2) the loss may be substantial in the case of persistent vomiting; and (3) the fluid, for practical purposes, is isotonic compared to serum. Potassium ion levels are higher in gastric juice as compared to extracellular fluid, the hydrogen ion concentration is highly variable, and when highly acidic, gastric juice has fewer sodium ions than serum (table 5.2).

• • •

A loss of 3 liters of gastric juice per day is considered to be moderately severe.

• • •

Volume depletion indirectly stimulates the adrenal cortex to release the hormone aldosterone into the blood (chapter 32), which acts on the kidney tubules to cause sodium ion and water retention by the body. A second hormone, ADH, is secreted by the posterior lobe of the pituitary and favors reabsorption of water by collecting ducts of the kidney. Volume loss stimulates thirst as well (chapter 2). All three mechanisms are compensatory responses to restore volume and lead to less than ideal consequences.

hypovolemia: Gk. *hypo,* under; L. *volumen,* volume

Effects of volume depletion due to vomiting

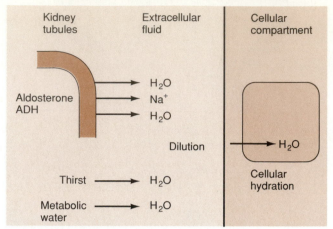

Figure 5.8 Volume depletion may lead to dilution of extracellular fluid and ultimately to cellular hydration. The extracellular compartment may remain fluid deficient. The hormones aldosterone and ADH, thirst, and the accumulation of metabolic water are factors that contribute to the dilution.

The patient's kidney retains sodium ions and water. Salt-free metabolic water accumulates, and if the individual is able to drink water, the result is a dilution of extracellular fluid. A sufficient volume of water may diffuse into cells, leaving those cells swollen, while the extracellular compartment remains fluid deficient (figure 5.8).

POTASSIUM IONS

Low or normal serum potassium levels may be observed in cases of vomiting. Two contributing factors are (1) direct loss in the vomitus, resulting in decreased total body potassium, and (2) reduced dietary intake. A major factor is loss of potassium ions by way of the kidney in response to elevated aldosterone levels (chapters 3 and 11). The influence of aldosterone on sodium ion retention favors the reciprocal loss of potassium and hydrogen in the urine. Volume depletion then stimulates the release of aldosterone from the adrenal cortex, which causes the kidney to restore volume by returning sodium ions and water to the extracellular compartment. This is accompanied by the elimination of potassium and hydrogen ions in the urine; thus, the loss of hydrogen ions in the urine contributes to the problem of alkalosis.

• • •

A predictable compensatory response to alkalosis is the excretion of bicarbonate in the urine. In view of this response, it is remarkable to see an example in which metabolic alkalosis is associated with output of urinary hydrogen ions.

• • •

TABLE 5.3 Factors that determine serum potassium levels in the case of vomiting

Mechanism	Comment
Direct loss in vomitus	Gastric juice is essentially isotonic to serum. This loss affects total body contents
Aldosterone secretion is a compensatory response to volume depletion	Favors Na^+ and water retention; reciprocal loss of K^+ and H^+
ADH secretion is a compensatory response to volume depletion	Favors H_2O retention; dilutional effect on ions
Volume loss stimulates thirst center	Water ingestion dilutes extracellular ions
Hypokalemia causes H^+ to move into cells as K^+ enter extracellular fluid	Response to hypokalemia contributes to alkalosis
Response to alkalosis is for H^+ to move out as K^+ move into cells	Alkalosis contributes to hypokalemia

Effects of vomiting on potassium levels

1. Direct loss of potassium (K^+)

2. Hormonal influences

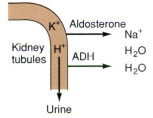

3. Opposing cellular effects

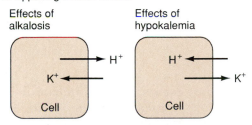

Figure 5.9 Serum potassium ion levels are usually low or normal in the case of persistent vomiting. Contributing factors are (*1*) direct loss of K^+ in the vomitus, (*2*) hormonal influence that promotes urinary loss of K^+, and (*3*) cellular ion exchanges. At the cellular level, alkalosis promotes intracellular flux of K^+, and hypokalemia favors extracellular diffusion. In addition to these factors, water ingestion retention has a dilutional effect.

In addition to the events already described, serum potassium ion levels are affected by two opposing tendencies for cellular ion exchange:

1. Hydrogen ions move out of cells in response to alkalosis and are exchanged for potassium ions, which move to an intracellular position (chapter 3).
2. In response to hypokalemia, potassium ions diffuse out of cells. Hydrogen ions, in a reciprocal shift, move into cells.

The factors that determine serum potassium ion levels in vomiting are summarized in table 5.3. Potassium ions are lost from the body directly. Hormonal influences cause dilution of extracellular ions and cause renal excretion of potassium ions. Finally, there are opposing tendencies for cellular K^+/H^+ exchange (figure 5.9). On the basis of this information, it is not surprising that laboratory reports of serum potassium ion levels in case of persistent vomiting vary, showing either low or normal values.

SODIUM IONS

As in the case of potassium ions, the effect of vomiting on extracellular sodium ion levels is variable. ADH and water ingestion cause a dilutional effect, and aldosterone favors sodium ion and water retention. As a result, normal or low levels of sodium ions may be observed.

CHLORIDE IONS

Gastric fluid contains substantial amounts of chloride ions, and consequently, total body loss of chloride ions is a predictable outcome of vomiting. Decreased chloride ion levels are significant because low levels of this ion influence exchange of ions in the kidney.

The following is an overview of some interrelated processes that occur in the kidney (chapter 11). When sodium ions are reabsorbed in the kidney, either they are accompanied by anions or there is a reciprocal exchange with K^+ and H^+ in order to maintain electrical balance. Normally, all of the chloride present in the glomerular filtrate will be reabsorbed into the blood with sodium ions. If there are any remaining sodium ions to be reclaimed by the body, they will be exchanged for K^+ and H^+. When chloride is depleted, sodium ion reabsorption requires an increase in hydrogen ion secretion, which enhances bicarbonate reabsorption. Figure 5.10 summarizes these events and shows that (1) low blood levels of chloride ion requires that bicarbonate accompany the reabsorption of sodium, and (2) this leads to increased hydrogen ion secretion. The condition of metabolic alkalosis associated with low chloride levels is called **hypochloremic alkalosis.** As shown in figure 5.10, the body's response to hypochloremic alkalosis is to reabsorb bicarbonate, which maintains the alkalotic condition.

Relationship between chloride (Cl⁻) and reabsorption/ elimination of other ions in kidney tubules

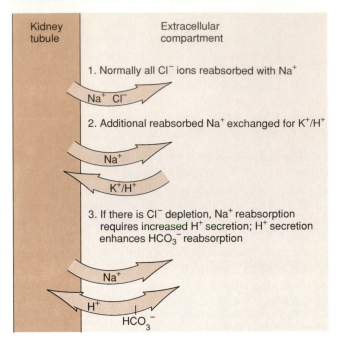

Figure 5.10 Chloride ions (Cl⁻) are normally reabsorbed with sodium ions. In the absence of Cl⁻, Na⁺ are reabsorbed in exchange for K⁺/H⁺. In the case of Cl⁻ depletion, there is increased H⁺ secretion accompanying Na⁺ reabsorption. Increased H⁺ secretion enhances HCO₃⁻ reabsorption. The net result is that chloride depletion causes alkalosis, a condition called hypochloremic alkalosis.

• • •

Molal solution = 1 mole of solute dissolved in 1 liter of water

Molar solution = 1 mole of solute diluted to a total volume of 1 liter of water

Osmolality refers to the osmotic pressure exerted by a molal concentration of particles. There is a relationship between the terms milliosmole (mOsm) and milliequivalent. Milliequivalent (meq) denotes 0.001 of the atomic weight, expressed in grams, divided by the valence of the ion. A milliosmole identifies the osmotic pressure exerted by 0.001 of the gram atomic weight of the ion. If the ion is monovalent, then 1 meq exerts 1 mOsm of pressure.

• • •

SERUM OSMOLALITY

It is useful to evaluate serum osmolality in cases of fluid and/or electrolyte loss because this parameter indicates total solute concentration of serum. One meq/liter of a monovalent ion exerts 1 mOsm of pressure. If the serum sodium value is 142 meq/liter then, knowing that the

TABLE 5.4 Consequences of prolonged vomiting

Imbalance	Comment
Metabolic alkalosis	Loss of H⁺ with the addition of HCO₃⁻ to the blood; contraction alkalosis due to decreased extracellular volume with no change in HCO₃⁻
Hypovolemia, essentially isotonic fluid loss; causes dehydration	Stimulates thirst, and ADH and aldosterone secretion
Low or normal serum K⁺	Total body loss of K⁺, reduced dietary intake; aldosterone promotes urinary K⁺ loss; alkalosis promotes K⁺ cellular uptake
Serum Na⁺ levels are variable	Aldosterone favors Na⁺ and water retention; ADH favors water retention; dilutional effect
Chloride ions lost; hypochloremia	Hypochloremia indirectly causes renal secretion of H⁺ and retention of HCO₃⁻
Serum osmolality decreases	Water retention causes decrease

sodium ion is accompanied by an anion, the osmolality can be estimated by multiplying that value by two. Laboratory reports may show both sodium ion concentration and serum osmolality because some other serum constituents, such as glucose, may increase osmolality.

In the case of vomiting, the fluid loss is isotonic. Changes in osmolality are not due to direct loss of sodium ions, but to water ingestion and fluid retention by the kidney. The compensatory responses to vomiting decrease serum osmolality.

URINARY OUTPUT

The following are expected findings in terms of urinary excretion during persistent vomiting. Volume depletion stimulates the secretion of ADH and aldosterone, which favor sodium ion and water retention by the kidney. The net effect is (1) decreased urinary output, (2) low sodium ion excretion, (3) increased potassium and hydrogen ion elimination, and (4) a urine low in chloride ions.

OVERVIEW

Table 5.4 summarizes the derangements caused by vomiting. The effects of prolonged vomiting can be summarized by identifying two stages in a series of events. (1) There is an early stage in which fluid loss causes dehydration, and serum electrolytes are normal because the fluid loss is isotonic (figure 5.11). (2) There is a second

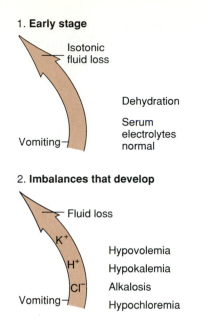

1. Early stage

Isotonic
fluid loss

Dehydration

Serum
electrolytes
normal

Vomiting

2. Imbalances that develop

Fluid loss

K$^+$

Hypovolemia

H$^+$

Hypokalemia

Cl$^-$

Alkalosis

Vomiting

Hypochloremia

Figure 5.11 In the early stages of vomiting there may be dehydration with normal electrolyte values. In prolonged vomiting, K$^+$, H$^+$, and Cl$^-$ losses occur.

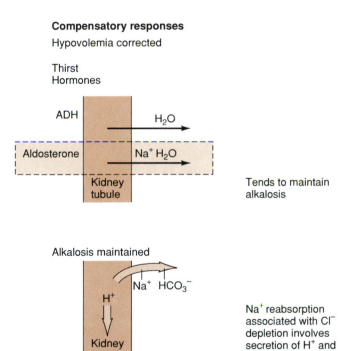

Compensatory responses

Hypovolemia corrected

Thirst
Hormones

ADH

H$_2$O

Aldosterone

Na$^+$ H$_2$O

Kidney
tubule

Tends to maintain
alkalosis

Alkalosis maintained

Na$^+$ HCO$_3^-$

H$^+$

Kidney
tubule

Na$^+$ reabsorption
associated with Cl$^-$
depletion involves
secretion of H$^+$ and
retention of HCO$_3^-$

Figure 5.12 Hypovolemia and alkalosis are two primary imbalances that develop in association with prolonged vomiting. Compensatory responses correct volume depletion and may maintain alkalosis.

stage in which compensatory responses cause fluid retention, decreased serum osmolality, and variable serum electrolyte levels. When the body must respond to both alkalosis and fluid loss, there is a selective response to correct fluid loss at the expense of sustaining alkalosis (figure 5.12). Adequate nutrition and removal of wastes is a cellular life and death matter that makes maintenance of ex-

Problem	Treatment
Hypovolemia	Fluid replaces volume
Total body loss of Na$^+$	Na$^+$ replacement
Hypochloremia	Cl$^-$ replacement
Renal elimination of K$^+$	K$^+$ replacement
Alkalosis and renal elimination of H$^+$ retention of HCO$_3^-$	Cl$^-$ replacement means that there is Cl$^-$ available to accompany renal reabsorption of Na$^+$; when Cl$^-$ is reabsorbed, HCO$_3^-$ is eliminated and H$^+$ is retained

tracellular volume of primary importance. The response to conserve sodium ions and water, which is promoted by aldosterone, is a factor in correcting the volume deficit. Sodium retention associated with chloride depletion, however, tends to favor alkalosis (as shown in figure 5.12).

PRINCIPLES OF TREATMENT

The basis for correcting fluid and electrolyte imbalance is recognition of the underlying cause. The basic problem in persistent vomiting is fluid loss and metabolic alkalosis. The goal in treatment is to provide water and to intervene so the kidney can excrete bicarbonate. Sodium ions play a major role. Based on maintaining electrical balance, there are three ways this ion is conserved by the kidney: (1) sodium ions exchanged for K$^+$ and H$^+$, (2) sodium ions accompanied by bicarbonate, and (3) sodium ions accompanied by chloride ions. In addition, hydrogen ion secretion is associated with bicarbonate retention.

If hypochloremia occurs and adequate amounts of chloride ions become available, three things happen:

1. Sodium reabsorption is balanced by the accompanying chloride ion
2. Bicarbonate becomes available for elimination
3. Hydrogen ions are retained rather than secreted (table 5.5).

Renal potassium ion depletion needs to be corrected; thus, infusion of a fluid with potassium and chloride ions with a lesser amount of sodium is appropriate treatment (figure 5.13).

• • •

If potassium ions are not replaced in a patient with hypochloremic metabolic alkalosis, correction of alkalosis may be delayed. The reason is that potassium ion depletion favors hydrogen ion secretion in exchange for sodium ions.

• • •

Practical Applications: Loss of Fluid and Electrolytes

Treatment of persistent vomiting

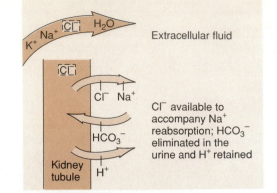

Figure 5.13 Treatment of persistent vomiting involves infusion of a fluid that contains potassium and chloride ions and a lesser amount of sodium. This corrects the volume deficit and potassium ion loss. Chloride ions are important because they are needed to accompany reabsorbed sodium ions. Bicarbonate can then be eliminated in the urine in exchange for hydrogen ions.

On a practical level, the effects of gastric fluid loss can be unpredictable because there are wide variations in electrolyte composition of gastric contents. One may or may not see the alkalosis described. Consider the following examples:

> Patient 5.1 was on gastric suction for 3 days with about 2,000 cc removed in an 8-hour period. Her laboratory test results were as follows (normal values of the reporting lab are in parentheses): Arterial blood pH 7.48 (7.35–7.45), pCO_2 58 mm (35–45), HCO_3^- 44 meq/liter (22–30), base excess +l6 meq/liter (+2.5 to −2.5)
> The afternoon of the same day:
> Arterial blood pH 7.52, HCO_3^- 50 meq/liter, base excess +21, Na^+ 148 meq/liter (137–47), K^+ 3.1 meq/liter (3.4–5.0), Cl^- 85 meq/liter (100–10).

The pH is high and bicarbonate and base excess elevated, indicating metabolic alkalosis. The second set of results show further increases in those values. Sodium ions are above normal and a low chloride value reflects loss in gastric contents. The pCO_2 is high, indicating respiratory compensation for alkalosis. This patient showed signs and symptoms of alkalosis including disorientation, loss of voice, and tingling sensations in the hands.

> Patient 5.2 was a 58-year-old female admitted to the hospital for nausea and vomiting. She had a history of gallstones and peptic ulcer. Her laboratory test values were as follows:
> Arterial blood pH 7.45, (7.37–7.45), pCO_2 55.8 mm (35–45), base excess +2.6 (+2.5 to −2.5), Cl^- 92 meq/liter (100–10), Na^+ 3.5 meq/liter (137–47), K^+ 3.5 meq/liter (3.4–5.0).

There is a slight elevation of base excess, an indication of alkalosis, and the high pCO_2 may reflect some respiratory compensation. It is more likely that respiratory movements were limited by pain.

Patient 5.3 was a 38-year-old female admitted to the hospital with a diagnosis of gastric outlet obstruction and was experiencing nausea and vomiting. She had a history of vagotomy, antrectomy, and gastroduodenostomy (see chapter 34). Her laboratory tests showed the following results:
Arterial blood pH 7.34, pCO_2 50.1 mm, base excess −2.4, Na^+ 129 meq/liter, K^+ 2.7 meq/liter, Cl^- 95 meq/liter.

The pCO_2 is high, and a likely contributing factor is pain with respiratory movements. The electrolyte values are low, although the chloride level depression is minimal.

DIARRHEAL FLUID LOSS

The example of intestinal fluid loss will reinforce and clarify the concepts about the multiple mechanisms by which the body responds to the challenge of fluid and electrolyte depletion. The fluid that passes through the intestinal tract undergoes progressive changes as the result of reabsorption of both water and electrolytes. Normally, the epithelial cells of the intestinal tract secrete bicarbonate into the lumen in exchange for hydrogen ions which are added to the blood.

The fluid which enters the colon is high in bicarbonate and that level is further enhanced by bicarbonate secretion into the lumen by epithelial cells of the colon. Acids formed by bacterial fermentation neutralize much of that bicarbonate, so that ultimately, in the normal stool, there is a decrease in the measurable anions, bicarbonate and chloride, but the total cation concentration is about equal to that found in the small intestine. Potassium ions represent an exception because the level in the stool is higher. Table 5.6 compares electrolyte levels in serum and intestinal contents.

The mechanism by which watery fecal material is generated is passive diffusion of water to an area of higher solute concentration (the colon). An increased solute concentration may have several underlying causes. Possible contributors to an increased osmotic concentration of solute are (1) substances that stimulate electrolyte secretion, such as bacterial enterotoxins, (2) ingested substances that are not efficiently absorbed and appear in the colon as in the case of infant lactose intolerance, (3) inflammation of the bowel, causing electrolyte secretion, and (4) inherent malabsorption in the small intestine.

The effect of diarrhea on fluid and electrolyte balance, including hydrogen ions, depends on the cause and severity of the problem and how prolonged it is. In general, sodium ion concentration increases and potassium ion concentration decreases as there is an increase in stool volume. A probable explanation for this observation is that active transport couples potassium ion secretion with sodium ion absorption. A decrease in one event inhibits the other. A loss of about 3 liters of fluid per day is considered moderately severe diarrhea; and with this magnitude of loss, total fecal solute concentration is similar to plasma, although there is a difference in individual constituents. This represents an isotonic loss. In severe diarrhea, loss of sodium ions and chloride ions greatly exceeds loss of potassium.

VOLUME AND ELECTROLYTE LOSS

The consequences of vomiting and diarrhea are not exactly the same, but the principles that govern their overall effects are the same. A moderately severe diarrheal fluid loss is an isotonic loss from which one would expect no effect on serum osmolality or no direct effect on serum sodium ion levels. Reduced circulatory volume causes ADH and aldosterone secretion, which promote both sodium ion and water retention by the kidney. If fluid losses are replaced primarily by water, retention of this water leads to dilutional hyponatremia and decreased serum osmolality.

As stool volume increases, fecal potassium ion levels decrease. With moderately severe losses, potassium ion concentration is similar to plasma levels. If diarrhea is prolonged, a total body potassium deficit develops due to intracellular loss. This situation is exacerbated by the stimulation of renal potassium ion excretion by aldosterone.

ACID-BASE IMBALANCE

In general, bicarbonate losses parallel increased diarrheal fluid losses. If the problem is secretory diarrhea, where there is stimulation of active secretion, the intestine or colon increases secretion of bicarbonate and chloride ions. Generally, there is a greater bicarbonate loss compared to chloride ions, and the result is hyperchloremic metabolic acidosis. A mixed acid-base disorder may occur if both upper and lower portions of the gastrointestinal tract are irritable. In vomiting, hydrogen ions are lost and bicarbonate is added to the blood, whereas bicarbonate is lost as the result of diarrhea. The degree of both problems determines whether acidosis or alkalosis prevails.

The metabolic consequences of diarrhea depend on the severity of the problem. Clinically significant depletion of body potassium only occurs in severe and chronic conditions, and acidosis is usually not marked. Consider the following examples:

Patient 5.4 was a 32-year-old male admitted to the hospital because he had diarrhea for several days. Laboratory tests showed the following values:
Arterial blood pH 7.34 (7.35–7.45), Na^+ 130 meq/liter (137–47), K^+ 3.5 meq/liter (3.4–5.0).

In this case, the sodium ion level is low, potassium ion level is normal, and pH is close to normal range.

Patient 5.5 was a 22-year-old female admitted to the hospital after vomiting for 3 days and having diarrhea for 2 days. Her laboratory tests upon admission showed the following:
HCO_3^- 16 meq/liter (22–30), Na^+ 140 meq/liter (137–47), K^+ 3.3 meq/liter (3.4–5.0), base excess −8.7 meq/liter (+2.5 to −2.5).

Notice acidosis as indicated by the bicarbonate and base excess values.

SUMMARY

* * *

Fluid and electrolyte loss must be considered in terms of accompanying compensatory responses. The following are key points in defining the problem in vomiting: (1) it is an isotonic loss which leads to volume depletion; (2) there is a high concentration of hydrogen and chloride ions in the vomitus; (3) hydrogen ions are added to stomach, bicarbonate ions diffuse into the blood; and (4) hydrogen ions are added to blood as bicarbonate enters the intestine.

Volume depletion causes contraction alkalosis, elicits hormonal responses, which lead to renal retention of sodium and water. Sodium ion retention, in turn, promotes renal K^+ and H^+ excretion. Direct loss of hydrogen ions in the vomitus causes alkalosis, which promotes diffusion of potassium ions into the cells.

Key events in the kidney are (1) sodium ions may be accompanied by anions, either Cl^- or HCO_3^-; (2) sodium ions are exchanged for K^+ and H^+; (3) when chloride ion levels are low, sodium is accompanied by bicarbonate; (4) the mode of formation dictates that secretion of hydrogen ions accompanies bicarbonate reabsorption.

Blood levels of sodium ions are not directly affected by isotonic gastric loss but are influenced by renal retention of water. Hypokalemia is the result of renal loss and diffusion of potassium into cells in response to alkalosis. Loss of hydrogen ions directly causes alkalosis, sustained by sodium and bicarbonate ion retention by the kidney. Bicarbonate retention is promoted by low chloride ion levels. The response of the body is to replace volume and sustain alkalosis.

The problem in diarrheal fluid loss is defined as an isotonic loss that leads to (1) volume depletion with bicarbonate and chloride loss, (2) acidosis, (3) renal retention of sodium and water, and (4) a possible total body deficit for potassium ions.

REVIEW QUESTIONS

• • •

1. Metabolic alkalosis is a consequence of prolonged vomiting. What are two contributing factors?
2. The serum K^+ level may be low or normal when there is prolonged vomiting. What are two factors involved?
3. Prolonged vomiting promotes (normal, high, low) serum Cl^- levels.
4. What is the effect of vomiting on serum osmolality? Why?
5. Vomiting is a/an (isotonic, hypertonic, hypotonic) fluid loss.
6. What are two compensatory responses to hypovolemia?
7. What ions may accompany or be exchanged for sodium as this ion is conserved by the kidney?
8. What are four main imbalances that develop as the result of vomiting?
9. What is a likely effect on acid-base balance if there is both vomiting and diarrhea?
10. What is the mechanism of developing hyperchloremic metabolic acidosis in association with diarrhea?

SELECTED READING

• • •

Arrambide, K. A. et al. 1989. Loss of absorptive capacity for sodium chloride as a cause of diarrhea following partial ileal and right colon resection. *Digestive Diseases Science* 34(2):193–201.

Mitchell, J. E. et al. 1983. Electrolyte and other physiological abnormalities in patients with bulimia. *Psychological Medicine* 13:273–78.

Oster, J. R. 1987. The binge-purge syndrome: A common albeit unappreciated cause of acid-base and fluid-electrolyte disturbances. *Southern Medical Journal* 80:58–67.

Scully, R. E., ed. 1985. Case records of Massachusett's General Hospital (laxative abuse). *New England Journal of Medicine* 313(21):1341–46.

Wang, F. et al. 1986. The acidosis of cholera. *New England Journal of Medicine* 315(25):1591–96.

Unit II

Respiratory System

Chapter 6 includes a review of respiratory system anatomy correlated with the clinical significance of certain anatomical features. These correlations include mucus secretion and insensible water loss, smoking and inhibition of the mucociliary escalator, the absence of cartilage in bronchioles and obstruction to airways, mast cells and allergic reactions, inflammation and destruction of alveolar walls, and inhibition of the phagocytic activity of macrophages. The lungs and the pleura are discussed in terms of loss of compliance and pleural effusions.

Chapter 7 includes a brief review of factors in air movements and emphasizes those diseases in which there is obstruction to airflow. The description of the disease processes provides a foundation for the discussion of arterial blood gases in chapter 9.

Chapter 8 includes a discussion of conditions in which there is diminished lung capacity and various other pulmonary disorders including bronchogenic carcinoma. This chapter both describes additional pulmonary diseases and provides a basis for a discussion of blood gases in chapter 9.

Chapter 9 includes examples of selected respiratory disorders to give an overview of disease processes discussed in the unit. This chapter expands the topic of acid-base balance (chapter 4) and introduces hypoxemia, which is discussed further in unit IV.

Chapter 6

Respiratory System: Structure Related to Disease Processes

This chapter introduces disease processes in the respiratory system by describing the major anatomical features, with emphasis on aspects significant under abnormal conditions.

AIRWAYS

The term **respiration** refers to those processes by which cells are supplied with oxygen as the waste product carbon dioxide is removed. These processes include moving air into and out of the lungs (**ventilation** or breathing), gas exchange between air sacs in the lungs and blood or **external respiration,** and the exchange of oxygen and carbon dioxide at the cellular level called **internal respiration** (figure 6.1). The major organs of respiration include the air passageways: nose, pharynx, larynx, trachea, bronchi (brong'kī), and bronchioles (figure 6.2). The external nose has two openings, the nostrils or **external nares** (na'rēz), that lead posteriorly to **internal nares,** which open into the upper part of the **pharynx** (far'inks) or throat. The pharynx also communicates with the passages from the middle ear called the **Eustachian (u-sta'ke-an) tubes** and with the oral cavity. In addition to this, the pharynx opens into both the **esophagus** and into the **larynx.**

The upper surface of the larynx or voice box is covered by a flap of leaf-shaped cartilage called the **epiglottis** which is attached anteriorly to cartilage and extends to the base of the tongue. In the process of swallowing, the epiglottis prevents food or foreign objects from entering the air passages.

naris, pl. es: L. *naris,* nostril
epiglottis: Gk. *epi,* on or upon; Gk. *glottis,* mouth of windpipe

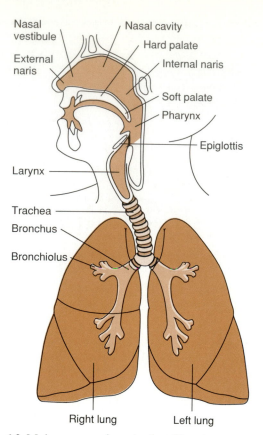

Figure 6.2 Major organs of respiration. The air passageways of the respiratory system include the nose, pharynx, larynx, trachea, bronchi, and bronchioles.

• • •

The trachea is about 2.5 cm in diameter and about 11 cm long from the larynx to the level of the sixth thoracic vertebra.

• • •

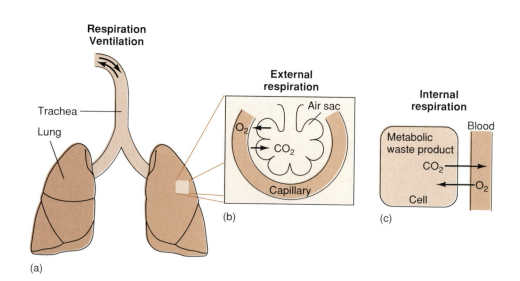

Figure 6.1 Respiration is the (*a*) act of breathing, (*b*) exchange of CO_2 and O_2 between blood and air sacs, and (*c*) the exchange of CO_2 and O_2 between cells and blood. Respiration removes the waste product CO_2 and supplies O_2 to cells.

The entire system of air passages, the trachea, bronchi, and smaller branches, is referred to as the bronchial tree because of its extensive branching (figure 6.3). The bronchi send branches called **secondary bronchi** to each of the five lobes of the lung where there is further repeated branching into **tertiary bronchi.** The following identifies the smaller airways formed from tertiary bronchi. Listed according to decreasing size, these airways include bronchioles, **terminal bronchioles, respiratory bronchioles, alveolar ducts, alveolar sacs,** and finally thin-walled air sacs, the **alveoli.** The distinction between bronchi and bronchioles is not only decreasing size, but also the disappearance of supporting cartilage plates in the bronchioles.

• • •

The right primary bronchus is wider and shorter than the left and forms a less acute angle. It is more likely that a foreign object would lodge in the right rather than in the left bronchus.

• • •

Each lobe of the lung is made up of small compartments called **lobules** separated by connective tissue and serviced by a terminal bronchiole. The smaller division of terminal bronchioles are called respiratory bronchioles, and these are distinguished by their microscopic size and also by the fact that there are air sacs or alveoli that open into the lumen. The respiratory bronchioles form smaller passages called alveolar ducts, each of which is surrounded by an alveolar sac (figure 6.4b). Alveoli are individual air sacs that communicate by way of small

bronchus, pl. bronchi: Gk. *bronchos,* windpipe
alveolus, pl. alveoli: L. *alveus,* tube or cavity

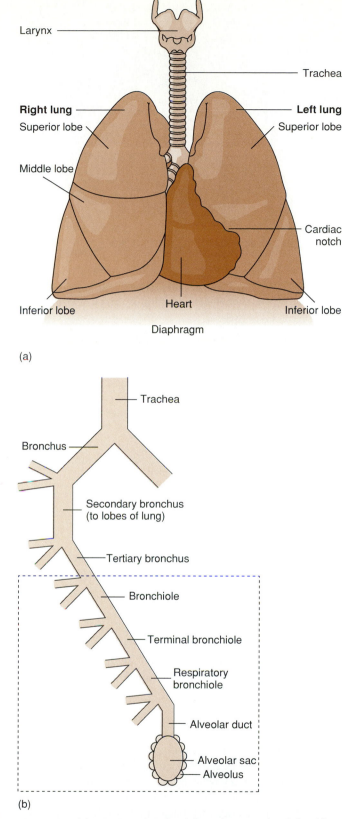

(a)

(b)

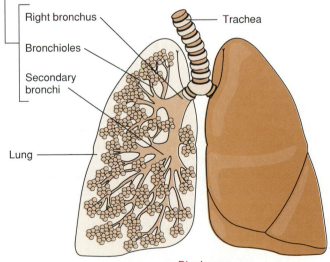

Figure 6.3 The bronchial tree includes the trachea, bronchi, and bronchioles.

Figure 6.4 (*a*) The lung has five lobes, three on the right side and two on the left. (*b*) The bronchi divide into branches called secondary bronchi that go to the five lobes of the lung. Repeated branching forms subdivisional of decreasing size that ultimately end in an alveolar duct and an alveolar sac surrounded by microscopic, thin-walled alveoli.

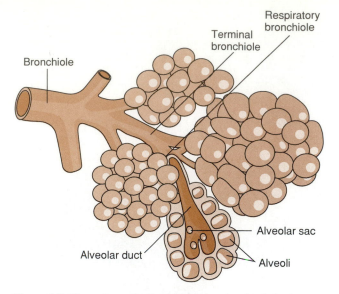

Figure 6.5 The acinus, the basic functional unit of the lung, includes a terminal bronchiole, respiratory bronchiole, ducts, and alveoli.

TABLE 6.1 Definitions of the smaller airways

Term	Definition
Terminal bronchiole	Small air passage that supplies a lobule of the lung
Respiratory bronchiole	Microscopic size; associated with alveoli that open directly into lumen
Alveolar duct	Surrounded by an alveolar sac
Alveolar sac	Cluster of air sacs with a common opening
Alveolus	Single air sac; alveoli communicate by way of pores of Köhn

openings called pores of Köhn. The basic functional unit of the lung is the **acinus** (as' in us) and includes a terminal bronchiole, respiratory bronchiole, ducts, and air sacs (figure 6.5). Table 6.1 lists terms and definitions of the air passages.

AIRWAY SUPPORT

Cartilage provides support and a degree of rigidity to the air passageways. The anterior and lateral walls of the trachea and the primary bronchi are supported by a series of incomplete rings of cartilages with the circle completed posteriorly by bundles of muscle fibers (figure 6.6).

acinus, pl. acini; L. *acinus,* grapestone

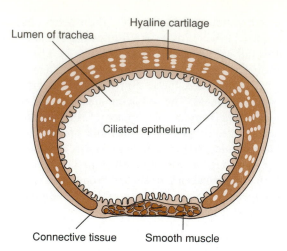

Figure 6.6 The anterior and lateral walls of the trachea and primary bronchi are supported by incomplete rings of cartilage. Bundles of muscle fibers complete the circle posteriorly.

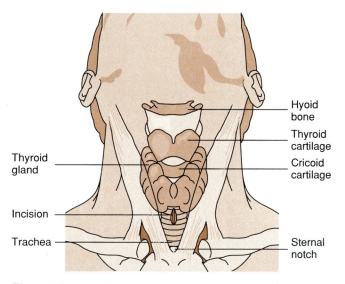

Figure 6.7 A small incision into the trachea, a tracheotomy, may be necessary to open a blocked airway in an emergency situation. The thyroid and cricoid cartilages may be felt or palpated when the patient's neck is extended.

The cartilaginous support of the trachea is shown in figure 6.7. An incision into the trachea may be required if the air passage is blocked.

Cartilage rings completely encircle the primary bronchi superior to the point where these branches enter the lung, and at this point there is a layer of muscle internal to the cartilage. The smaller pulmonary branches of the bronchi are encircled by smooth muscle and are supported by small plates of cartilage that are absent in the bronchioles.

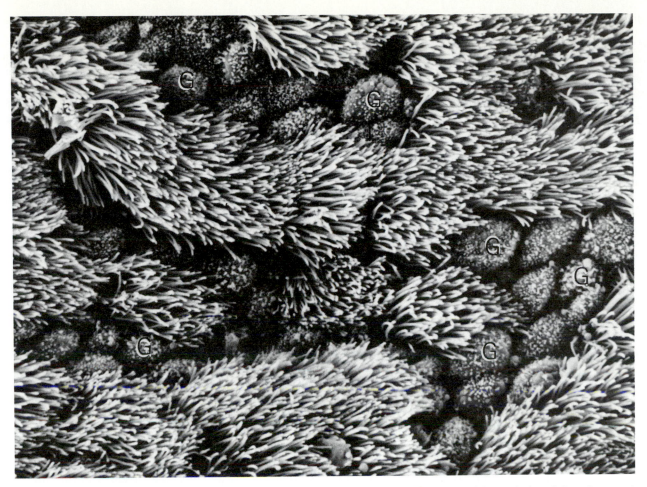

Figure 6.8 Scanning electron micrograph of the surface of the lumen of a bronchus. The nonciliated cells are goblet cells (G). Their surface is characterized by small blunt microvilli that give a stippled appearance to the cell. The cilia of the many ciliated cells occupy the remainder of the micrograph. *From M. H. Ross, E. J. Reith, and L. J. Romrell:* Histology: A Text and Atlas. *1989 Williams & Wilkins, Baltimore. Reproduced with permission.*

• • •

In an extreme emergency in which there is suffocation, a small incision may be made in the trachea and a tube inserted to establish airflow. The term tracheotomy refers to an incision in the trachea, and tracheostomy means an insertion of a tube. Since there is a risk of bleeding and nerve damage associated with a tracheotomy, this is done only if it is impossible to introduce a tube into the trachea by way of nose or throat.

• • •

AIRWAY WALLS

The entire passageway is lined with a mucous membrane that has a surface covering of mucus. Mucous membranes are made up of a surface layer of epithelial cells interspersed with mucus secreting **goblet cells.** The epithelial lining of the trachea down to the respiratory bronchioles

tracheotomy: Gk. *trachys*, rough; Gk. *tomos*, cutting
tracheostomy: Gk. *stoma*, mouth

is made up of ciliated columnar cells. The cilia are cylindrical projections that move rhythmically to propel mucus and trapped foreign particles toward the pharynx (figure 6.8). This mechanism is called the **mucociliary escalator** (mu″ko-sil′e-er″e).

• • •

Kartagener's syndrome is a genetic disorder in which recurrent respiratory infections are due to immotile cilia. The syndrome is further characterized by permanently dilatated bronchi, heart transposed to the right side, and sinusitis.

• • •

The structure of the alveolar wall has special significance because it is a barrier that both oxygen and carbon dioxide must cross. The wall is made up of two types of closely spaced epithelial cells, a basement membrane, and a narrow connective tissue space. Of the two types of cells, **Type I** is flattened and for the most part constitutes the

dilatate: L. *dilatare*, to enlarge

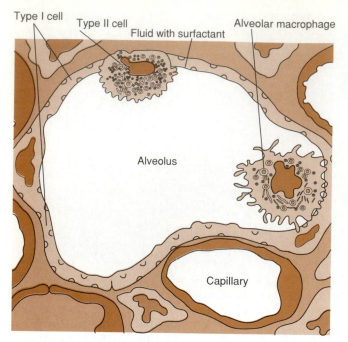

Figure 6.9 The alveolar wall is made up of, for the most part, epithelial cells called Type I cells. Type II cells secrete surfactant, a phospholipid, that is important in preventing the collapse of alveoli.

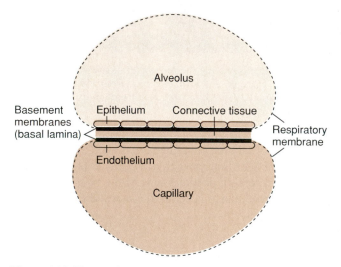

Figure 6.10 The respiratory membrane (alveolar/capillary walls) is made up of the epithelium, basement membranes, and connective tissue of the wall of the alveolus and capillary basement membrane and endothelium. Basement membrane is extracellular material to which cells are attached.

lining that allows gas exchange between capillary blood and air sacs. The thicker **Type II cells** are interspersed among the other cells and secrete a phospholipid called **surfactant** that decreases surface tension and prevents collapse of the alveoli. There are also special types of cells present, such as **macrophages** (dust cells), which ingest foreign particles and **mast cells** that contain histamine and other substances involved in allergic reactions (figure 6.9).

The term **respiratory membrane** refers to the walls that separate alveoli from capillary blood. Alveolar/capillary walls constitute a 0.4 micron thick barrier (figure 6.10). The total surface area for diffusion of gases between capillaries and alveoli is about 70 square meters.

PROTECTIVE MECHANISMS

Table 6.2 summarizes protective mechanisms in air passages. A factor that prevents collapse of air sacs is significant. The alveoli have a tendency to collapse, both because of elasticity of the walls and also because of an inner lining of a thin film of fluid that creates surface tension. Surface tension is the result of intermolecular attractions that tend to draw alveolar walls together. Type II cells secrete a phospholipid mixture of substances called surfactant that decreases surface tension and minimizes the possibility of alveolar collapse.

Lymphatic vessels are scattered throughout lung tissue and are important in removing accumulations of fluid to protect against pulmonary edema. Rhythmic breathing movements and possibly contraction of smooth muscle in lymph vessel walls promote the flow of fluid through the pulmonary lymph system away from the lungs.

micron: 0.001 mm

TABLE 6.2 A summary of some protective mechanisms in the air passageways

Cause	Effect
Blood supply and mucus covering	Warms and humidifies air as it enters airways
Mucus	Provides moist lining and traps foreign particles
Cilia	Sweep mucus and trapped particles toward pharynx
Macrophages	Phagocytosis—engulf and destroy foreign particles
Type II cells	Produce surfactant, which decreases surface tension and prevents collapse of alveoli
Lymphatic vessels	Prevent pulmonary edema by removing excess fluid

RESPIRATORY SYSTEM

The purpose of this first section has been to review the structure of the airways as a basis for understanding disease processes that will be considered in the next several chapters. The following is a synopsis of some observations, listed in table 6.3, which have clinical significance.

The air passages are covered with a blanket of mucus, secreted both by goblet cells and mucous glands. Goblet cell secretion is stimulated by irritants in the airways, and the mucous glands are stimulated by nerve impulses and possibly by direct irritation. Water evaporates from the surface of this mucus blanket and saturates the incoming air. Vapor pressure of mucus water increases when an individual has a fever, and there is greater water loss. Water lost in this way constitutes insensible water loss (chapter 2) and may be significant in case of fever and increased breathing rate.

Mucus secretion is increased in response to irritation or infection. In the case of chronic irritation, there is an increase in the number of mucus secreting goblet cells. If the excess is large, mucus may plug the bronchi and cause increased susceptibility to infection.

The mucociliary escalator, which is important in removing foreign particles, may be rendered ineffective under certain conditions. Smoking, for example, inhibits ciliary action as well as causes **hypertrophy** or an increase in size of mucous glands. In such chronic inflammatory conditions as bronchitis, the cilia may be decreased in number and this, combined with excessive secretion of mucus, contributes to ineffective ciliary action. Certain viruses cause destruction of cilia as well. Excessive drying of the airways interferes with the action of cilia.

It has been noted that the number of cartilage plates decreases in the small bronchi and that cartilage is absent in bronchioles. In such diseases as asthma, contractions of the smooth muscle, which encircles these airways, decrease the lumen size and cause obstructed airflow. This type of smooth muscle contraction is called **bronchospasm.**

Destruction of both capillaries and alveolar walls occurs in certain diseases, such as emphysema. One of the consequences of these degenerative processes is a decrease in capillary/alveolar surface area for the exchange of carbon dioxide and oxygen. Another aspect of the problem is that the flow of inspired air normally stops at the alveolar ducts with subsequent diffusion into the alveoli. If the alveoli are enlarged because of alveolar wall destruction, then diffusion into air sacs from alveolar ducts requires more time. This decreases the efficiency of air/blood gas exchange.

Macrophages are found within alveoli and occasionally in alveolar walls. These cells defend the body against bacteria as well as inhaled dust particles. Both **hypoxia,**

hypertrophy: Gk. *hyper*, more than; Gk. *trophe*, nutrition

TABLE 6.3 Clinically significant observations

Observations	Significance
Mucus secretion	Humidifies air; insensible water loss
Increased mucus secretions	Caused by chronic irritation; plugs bronchi
Inflammation destroys cilia	Ineffective mucociliary escalator
No cartilage in bronchioles	Bronchospasms obstruct airflow
Destruction of capillaries and alveoli	Decreased surface area for gas exchange
Alveolar macrophages	Cigarette smoke and hypoxia decrease bactericidal effects of macrophages
Macrophage release of the enzyme elastase	Destruction of alveolar wall
Mast cells in connective tissue of airways	In allergic reactions, mast cells release histamine and slow reacting substance
Type II alveolar cells produce surfactant	Surfactant decreases surface tension

an inadequate oxygen supply, and cigarette smoke decrease the bactericidal effects of macrophages. Asbestos damages macrophages as well.

Inflammation of the lung leads to destruction of alveolar walls. The mechanism involved is the release of the enzyme **elastase** by macrophages, which enzymatically destroys the elastic fibers of the alveolar wall.

Mast cells are found in connective tissue underneath the smooth muscle of the airways and respond in allergic reactions by releasing chemicals that affect the airways. Two of these chemicals, **histamine** and **slow reacting substance,** cause contraction of bronchial smooth muscle.

Surfactant decreases surface tension and helps prevent collapse of the air sacs. A deficiency of surfactant in the lungs of premature infants is called **hyaline membrane disease** or **respiratory distress syndrome** and leads to inadequate inflation of the air sacs.

LUNGS

The lungs consist of three lobes on the right side and two on the left side. The right lung is larger than the left lung (figure 6.4). The apex of the lung on each side is just above the level of the clavicle with the inferior surface bounded by the diaphragm. There is a concavity on the left side for the heart called the **cardiac notch.** The **mediastinum** (mē''de-as-tī'num) consists of tissues between the lungs including the heart, esophagus, thymus, blood vessels, and lymphatic vessels. Primarily the middle right lobe and both

mediastinum: L. *mediastinus*, in the middle

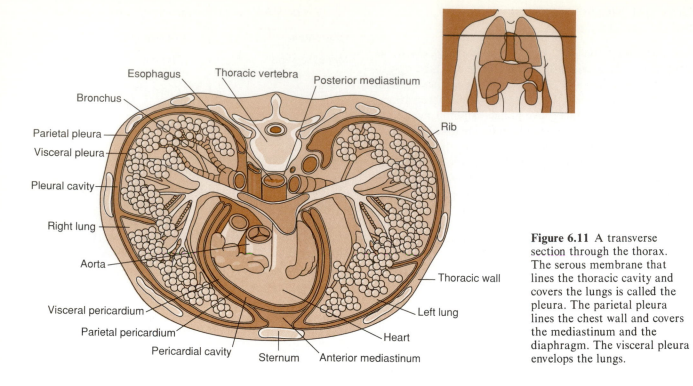

Esophagus
Thoracic vertebra
Posterior mediastinum
Bronchus
Parietal pleura
Visceral pleura
Pleural cavity
Right lung
Aorta
Visceral pericardium
Parietal pericardium
Pericardial cavity
Sternum
Anterior mediastinum
Heart
Left lung
Thoracic wall
Rib

Figure 6.11 A transverse section through the thorax. The serous membrane that lines the thoracic cavity and covers the lungs is called the pleura. The parietal pleura lines the chest wall and covers the mediastinum and the diaphragm. The visceral pleura envelops the lungs.

upper lobes are visible anteriorly, while the two lower lobes are almost completely in a posterior position. The hilus (hī′lus) is the area where there is entrance and exit of pulmonary vessels, lymphatics, bronchi, and nerves.

BRONCHOPULMONARY SEGMENTS

Bronchopulmonary segments are lung compartments separated by connective tissue compartments. There are 10 segments on the right and 9 on the left. Lobules are smaller divisions of the lungs separated by connective tissue, and each one is supplied by a branch of a terminal bronchiole. Ultimately, the basic functional unit of the lungs is the acinus, consisting of a microscopic respiratory bronchiole, alveolar duct, alveolar sac, and alveoli (figure 6.5).

LUNG COMPLIANCE

The term **compliance** refers to the elastic properties or the stretchability of a tissue. The most important components of the lung responsible for elasticity are **elastin** and **collagen;** both proteins are scattered throughout lung tissue. The geometric arrangement of these fibers combined with the surface tension of the alveolar walls are mainly responsible for the lung's capacity for stretching and recoil. Muscles, tendons, and connective tissue give the chest wall elasticity, although the chest wall is less compliant than the lungs. A low compliance means the lungs and chest wall do not expand easily and more effort is required to breathe.

PLEURA

The airways of the respiratory system are lined by a mucous membrane described earlier in this chapter. A mucous membrane is one that lines a cavity that opens to the exterior of the body and secretes mucus to maintain a moist surface. A serous membrane lines a cavity that does not open to the exterior. The **serous membrane** that lines the thoracic cavity and covers the lungs is called the **pleura.** The pleura is made up of a single layer of squamous epithelial cells and a thin layer of connective tissue with blood vessels and lymph vessels. The pleura forms a double layer, and the part that envelops the lungs is called the **visceral pleura.** The part that lines the chest wall and covers the mediastinum and the diaphragm is called the **parietal pleura** (pah-rī′ĕ-tal) (figure 6.11). The visceral pleura is continuous with the parietal pleura at the hilus with a double fold of pleura called the **pulmonary ligament** extending from that point almost down to the diaphragm. The innervation and blood supply to the two membranes are different. The visceral pleura is mainly supplied by branches of the bronchial artery, a part of the pulmonary circulation, and is not supplied by sensory nerves. The parietal pleura, in contrast, receives blood by way of branches of the **intercostal** and **mammary arteries** from the systemic circulation and is endowed with sensory nerves. There is a potential space between the two membranous layers, but because of the surface tension of a thin film of 10–30 ml of fluid, the membranes are normally held tightly together.

pleura: Gk. *pleura,* a side
visceral: L. *viscera,* inner parts of body
parietal: L. *parietalis,* a wall

CLINICAL SIGNIFICANCE RELATED TO LUNGS AND PLEURA

Some observations related to the lungs and pleura that have clinical significance are summarized in table 6.4.

The lungs have the property of compliance or elasticity because of elastin and collagen throughout lung tissue. A decrease in lung compliance increases the work of breathing and may be caused by any disease process that causes edema, an obstruction of bronchioles, or **fibrosis.** Fibrosis is a process by which fibrous connective tissue replaces normal tissue. The spread of cancer throughout lung tissue, for example, causes fibrous tissue **hyperplasia,** which is excessive tissue formation, and it also causes disruption of lymph vessels. The result is a decrease in lung compliance due to both increase in tissue and to obstruction of interstitial fluid drainage.

• • •

A condition called diffuse-interstitial pulmonary fibrosis causes diminished lung compliance. This disorder may be the result of various disease processes, such as pneumonia or reactions to drugs. In some cases, it is idiopathic (cause unknown). Typically, there is widespread distribution of interstitial fibrosis, and ultimately, the lungs become shrunken and stiff.

• • •

Irritation or inflammation of the pleura results in pain that originates only in the parietal pleura because the visceral pleura is not supplied by sensory nerves. The pain is usually sharp, limited to one side, and made worse by breathing or coughing. This pain may be referred to the abdomen, shoulder, or neck because the innervation is supplied by branches of the intercostal nerves, spinal nerves, and the phrenic nerve.

The potential intrapleural space may be the site of accumulation of excessive amounts of fluid under abnormal conditions. The contributing factors are the same as those influencing vascular/interstitial fluid exchanges (chapter 2). Normally, the 10–35 ml of pleural fluid is in a constant state of flux, with a turnover rate of about 35–75 percent/hour. Fluid is filtered out of the capillaries of the parietal pleura and is reabsorbed by the pulmonary capillaries of the visceral pleura as well as by lymphatic vessels. Hydrostatic pressure is higher in parietal pleural capillaries compared to those of the visceral pleura, and this favors outflow (figure 6.12).

The term **pleural effusion** is used to describe excessive accumulations of fluid in the pleural space, and it may occur because of changes in hydrostatic or oncotic pressure or because of changes in permeability due to inflammation or other damage.

TABLE 6.4 Clinical significance of some observations regarding the lungs and pleura

Observation	Significance
Bronchopulmonary segments	Can surgically remove a segment
Parietal pleura supplied by sensory nerves	Inflammation results in pain that originates in parietal pleura
Parietal capillaries of systemic origin; visceral pleural capillaries of pulmonary origin	Fluid diffuses out of parietal capillaries and is reabsorbed by visceral pleural capillaries
Lymphatic vessels in region of visceral pleura	Remove protein and some fluid
Elasticity (compliance) conferred to lungs by elastic fibers and to chest wall by muscles, tendons, connective tissue	Edema, fibrosis causes a decrease of lung compliance, resulting in increased difficulty in breathing

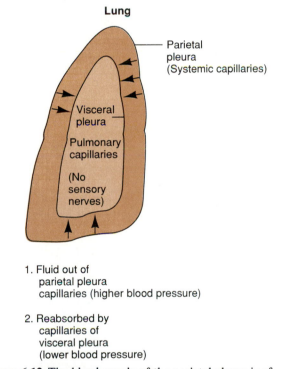

1. Fluid out of parietal pleura capillaries (higher blood pressure)

2. Reabsorbed by capillaries of visceral pleura (lower blood pressure)

Figure 6.12 The blood supply of the parietal pleura is of systemic origin with a higher blood pressure as compared to vessels that supply the visceral pleura. The blood supply of the visceral pleura is a part of the pulmonary circulation.

TABLE 6.5 Possible causes of pleural effusion

Increased Hydrostatic Pressure

Disease Process	Mechanism
Left ventricular heart failure	Increased hydrostatic pressure in visceral pleural vessels due to decreased cardiac output
Right ventricular heart failure	Increased parietal pleural capillary pressure
Pulmonary venous thrombosis	Increased pressure in pulmonary capillaries due to obstruction

Changes in Oncotic Pressure

Nephrotic syndrome; loss of protein by the kidney	Decreased plasma oncotic pressure
Cirrhosis; decreased protein synthesis by the liver	Decreased protein in plasma
Pleural tumor; (may be result of exposure to asbestos)	Block lymphatic removal of protein; increased pleural oncotic pressure

Increased Permeability

Bacterial pneumonia; malignancies	Inflammation increases permeability of capillaries

Pleural effusion

Capillary of pleura

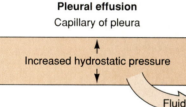

(a)

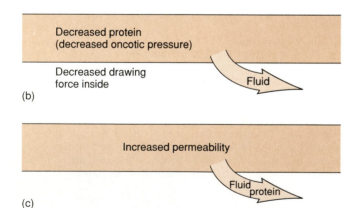

(c)

Figure 6.13 Pleural effusion occurs under circumstances of increased hydrostatic pressure, decreased oncotic pressure, or increased permeability of capillary walls.

• • •

The process of fluid accumulation is called **transudation** if hydrostatic or oncotic pressure is the driving force. The process is called **exudation** if change in permeability of the membrane is responsible. Typically, the resulting fluid, a **transudate,** contains less protein than an **exudate.**

• • •

Table 6.5 lists some possible causes of pleural effusion and shows that the underlying mechanisms involve imbalance of hydrostatic or oncotic pressures, either in capillaries or the pleural space. The examples include (1) increased capillary hydrostatic pressure, favoring outflow of fluid, (2) decreased plasma oncotic pressure, causing a diminished drawing force into capillaries, (3) increased capillary wall permeability, resulting in loss of fluid and protein into the pleural space, and (4) impaired removal of protein by lymph vessels, leading to increased oncotic pressure in pleural fluid (figure 6.13).

• • •

The term **pleurisy** or **pleuritis** means an inflammation of the pleura. This is accompanied by fever, pain, breathing difficulties, and often, pleural effusion. Trauma to the chest resulting in bruising of the parietal pleura can cause pleurisy, which is also caused by a variety of lung diseases.

• • •

SUMMARY

• • •

The major divisions of the airways, listed in the order of decreasing size, are the (1) trachea, (2) bronchi, (3) bronchioles, (4) alveolar ducts, and (5) alveoli. The passages down to the bronchioles are supported either by rings or plates of cartilage, and the inner lining is a ciliated mucous membrane. Capillaries surround the alveoli and the alveolar/capillary wall constitutes the respiratory membrane where gas exchange occurs. Protective mechanisms include the mucociliary escalator; macrophages, which are phagocytic; and surfactant, which decreases surface tension of the alveoli. In disease processes, bronchospasms obstruct airflow in bronchioles because these airways are supported only by smooth muscle. Gas exchange is limited when degenerative changes cause destruction of alveolar walls and capillaries. Hypoxia and cigarette smoke interfere with the phagocytic activity of

transudation: L. *trans,* through; L. *sudare,* to sweat
exudation: L. *ex,* out; L. *sudare,* to sweat

macrophages. Inflammation stimulates the release of elastase by macrophages, causing the enzymatic breakdown of elastic fibers in the alveolar walls.

There is normally a small amount of fluid in the potential space between the visceral and parietal pleura. The formation and reabsorption of the pleural fluid is a dynamic process governed by the forces controlling vascular/interstitial fluid. Pleural effusion is the accumulation of fluid in the intrapleural space and occurs when there is (1) increased plasma hydrostatic pressure, (2) decreased plasma oncotic pressure, (3) increased capillary wall permeability, or (4) increased pleural fluid oncotic pressure.

The lungs as well as the chest wall have elastic properties which are responsible for the characteristic of compliance. Any disease process that causes a decrease in compliance increases the work of breathing.

REVIEW QUESTIONS

• • •

1. What are two distinguishing features of bronchi and bronchioles?
2. What lung compartments do terminal bronchioles service?
3. What is the difference between terminal bronchioles and respiratory bronchioles?
4. Describe an alveolar duct.
5. What is the definition of acinus?
6. What is the source of mucus in the airways?
7. Describe the respiratory membrane.
8. What is the source and function of surfactant?
9. What is the function of macrophages?
10. What is the significance of the presence of mast cells in the lungs?
11. What is the significance of the fact that there are more lymphatic vessels in the lungs than in any other organ?
12. List three conditions in which the mucociliary escalator is rendered ineffective.
13. What airways would bronchospasms obstruct?
14. What is the significance of enlarged alveoli in emphysema?
15. What is the source and significance of the enzyme elastase in the lung?
16. What components of the lung are responsible for elasticity?
17. What is the significance of low compliance of the lungs and chest wall?
18. What is the difference in blood supply to the parietal and visceral pleura?
19. List three general conditions that would cause a decrease in compliance.
20. What are three factors that favor the movement of fluid into the intrapleural space?

SELECTED READING

• • •

Baer, R. E. et al. 1987. The increased expiratory muscle use in upright dogs: Role of cardiovascular receptors. *Respiratory Physiology* 70:359–68.

Bloch, H. 1987. Phenomena of respiration: Historical overview to the twentieth century. *Heart Lung* 16:419–23.

Jobe, A., and M. Ikegami. 1987. Surfactant for the treatment of respiratory distress syndrome. *American Review of Respiratory Disease* 136(5):1256–75.

Lee, R. M. et al. 1987. Assessment of postmortem respiratory ciliary motility and ultrastructure. *American Review of Respiratory Disease* 136:445–47.

Marchal, F. et al. 1987. Measurement of ventilatory system compliance in infants and young children. *Respiratory Physiology* 68:311–18.

Redline, S. et al. 1987. Influence of upper airway sensory receptors or respiratory muscle activation in humans. *Journal of Applied Physiology* 63:368.

Uddman, R. et al. 1987. Neuropeptides in the airways: A review. *American Review of Respiratory Disease* 136:S3–8.

Chapter 7

Obstructive Pulmonary Disorders

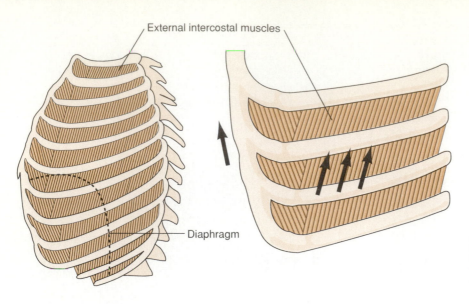

External intercostal muscles

Diaphragm

Figure 7.1 Contraction of external intercostal muscles and contraction of the diaphragm increase the size of the thoracic cavity. This leads to inspiration.

This chapter focuses on those alterations in structure and function of the respiratory system that lead to some degree of obstruction with an increased resistance to airflow. The obstructed airways may be large or small; thus, the trachea and larger bronchi may be involved, or the obstruction may be limited to the smaller bronchioles and alveoli. **Restrictive** pulmonary diseases, discussed in chapter 8, are characterized by inadequate expansion of the lungs with a decrease in total lung capacity. The distinction between obstructive and restrictive pulmonary disorders is useful for purposes of discussion, although one problem does not exclude the other, and there may be both airflow resistance and diminished lung capacity.

Chronic obstructive pulmonary disease (COPD), the most common form of pulmonary disease in humans, is the main topic of this chapter. The following discussion deals first with normal airflow and then with diseases in which there is increased resistance to airflow, i.e., chronic bronchitis, emphysema, and asthma.

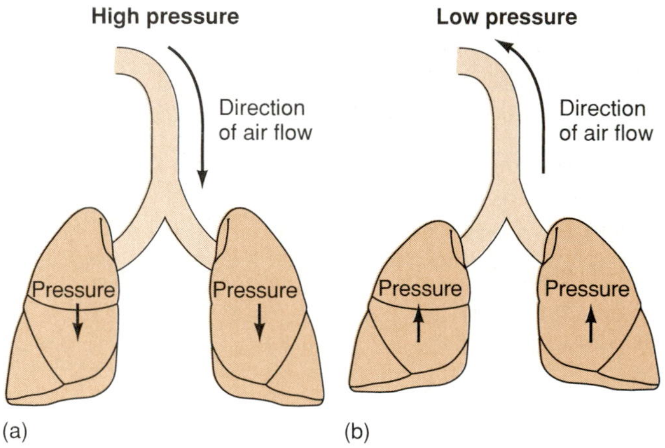

RESPIRATORY AIRFLOW

The inflow and outflow of air between the atmosphere and the lungs is called **ventilation** and is the result of pressure changes created by chest movements leading to variations in the size of the thoracic cavity. The diaphragm is a dome-shaped sheet of muscle separating the thoracic and abdominal cavities. In quiet inspiration, the predominant event is downward contraction of the diaphragm along with contraction of external intercostal muscles (figure 7.1) to raise the rib cage. The result is an increase in the size of the thoracic cavity. In quiet expiration, muscles relax and thoracic structures return to original position. In forced expiration, there is both contraction of internal intercostal muscles to depress the rib cage and abdominal muscle contraction.

intercostal: L. *inter*, between; L. *costa*, rib

High pressure　　　　**Low pressure**

Direction of air flow　　Direction of air flow

Pressure　Pressure　　Pressure　Pressure

(a)　　　　　　　　(b)

Figure 7.2 Air flows from a direction of high pressure to low pressure.

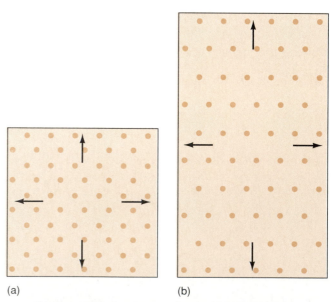

(a)　　　　　　　　(b)

Equal number of gaseous molecules in both containers

Figure 7.3 Air pressure, due to molecules randomly bouncing against the walls of a container, is less in a larger container.

Figure 7.4 (*a*) Intrapulmonic pressure is equal to atmospheric pressure prior to inspiration. (*b*) Intrapulmonic pressure decreases as the diaphragm contracts and the chest wall expands. Air flows into the lungs, from high to lower pressure. (*c*) Intrapulmonic pressure increases with the relaxation and recoil of the diaphragm and muscles of the chest wall. Air flows out, from an area of high to an area of lower pressure.

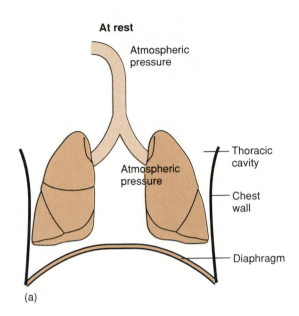

At rest

Atmospheric pressure

Atmospheric pressure

Thoracic cavity

Chest wall

Diaphragm

(a)

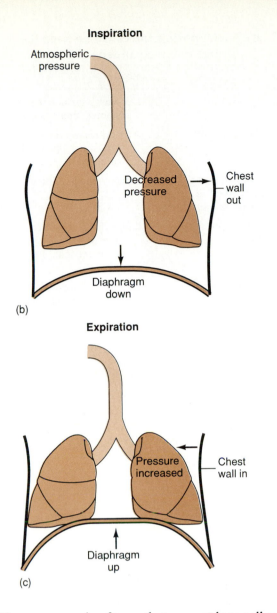

Inspiration

Atmospheric pressure

Decreased pressure

Chest wall out

Diaphragm down

(b)

Expiration

Pressure increased

Chest wall in

Diaphragm up

(c)

Air moves in the direction of low pressure and the purpose of respiratory movements is to establish a pressure gradient (figure 7.2). An increase in the size of the thoracic cavity causes decreased intrapulmonary or intraalveolar pressure as compared to atmospheric pressure. There is a pressure/volume relationship in accordance with **Boyle's law,** which states that with a fixed mass of gas, the volume and pressure are inversely proportional at a constant temperature (figure 7.3). At rest, intrapulmonic and atmospheric pressure are equal. Inspiration occurs as a lower intrapulmonic pressure is established by increased thoracic size, and expiration is the result of increased intrapulmonic pressure (figure 7.4).

• • •

The pleural space is a closed system with no opening to the outside. When an infant is born, the lungs fill the thoracic cavity without stretching. As the child grows, the rib cage grows faster than the lungs. The increased space causes the pressure in the intrapleural space to drop below the pressure in the lungs, so the lungs stretch and expand.

• • •

There are opposing forces that prevent lung collapse, and these forces are summarized in table 7.1. There are elastic fibers throughout lung tissue. The natural recoil tendency of these fibers combined with the surface tension of the thin film of fluid of the alveolar walls would cause lung collapse, if unopposed. The opposing forces are surfactant (which decreases surface tension), outward recoil tendency of chest wall, air in the airways, and the subatmospheric pressure in the pleural space (figure 7.5). Intrapleural pressure is normally less than intrapulmonary pressure (figure 7.6). Preceding inspiration, intrapleural pressure is about 756 mm Hg compared to intrapulmonary pressure of 760. The intrapleural pressure is sometimes indicated as −4 mm Hg to show that it is 4 mm below atmospheric pressure.

TABLE 7.1 Opposing forces which keep the lungs in an expanded state

Force	Effect
Surfactant	Decreases surface tension; opposes lung collapse
Elasticity of chest wall	At rest, elastic recoil of chest is outward; opposes lung collapse
Air in airways; at rest the pressure is equal to atmospheric pressure	Opposes lung collapse
Subatmospheric pressure in pleural space	Favors lung expansion because pressure in alveoli is greater than pleural pressure
Elastic fibers in lung tissue	Recoil, if unopposed, would cause lung collapse

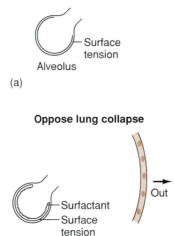

Lung collapse

Elastic fibers

Surface tension

Alveolus

(a)

Oppose lung collapse

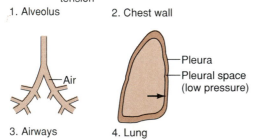

Surfactant
Surface tension

1. Alveolus

Out

2. Chest wall

Air

3. Airways

Pleura
Pleural space (low pressure)

4. Lung

(b)

Figure 7.5 (a) Factors that favor a tendency for lung collapse are the elasticity of lung tissue and the surface tension within alveoli. (b) Opposing forces are (1) surfactant that decreases surface tension, (2) the outward recoil tendency of the chest wall, (3) air in airways, and (4) pressure in the pleural space that is lower than intrapulmonary pressure.

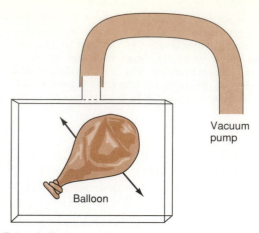

Vacuum pump

Balloon

Figure 7.6 A balloon expands as air is removed from the closed container. In the same way, the lungs remain expanded when they are surrounded by a lower pressure.

• • •

Introduction of air into the pleural space with subsequent collapse of the lung is called a pneumothorax. The cause may be injury due to accident or surgery or may be due to lung disease.

• • •

The walls of airways present resistance to the flow of air and respiratory muscle contractions help overcome that resistance. The pulmonary diseases discussed in the following section are characterized by obstruction due to thickened muscle, muscle contractions, and mucus secretions. The narrowing of the airways increases resistance to airflow and increases the work of breathing.

CHRONIC BRONCHITIS

Bronchitis is an inflammation of the airways in which there is excessive production of mucus accompanied by a persistent cough. Inflammation, regardless of cause, involves redness, swelling, and an accumulation of white blood cells. The term chronic implies that the condition has continued over a period of time. For purposes of comparison, chronic bronchitis is sometimes defined precisely as a cough with mucus production occurring on most days for at least 3 consecutive months for more than 2 years.

ETIOLOGY

Cigarette smoking, atmospheric pollution, and infection are three factors that contribute to chronic bronchitis, although exact causes are uncertain. It is not believed that infection is an initiator of the disease process but that it is a contributor because (1) inflammation increases susceptibility to infections, and (2) infection prolongs and ex-

pneumothorax: Gk. *pneuma*, air; Gk. *thorax*, thorax

TABLE 7.2 Structural changes resulting in obstruction of airflow in chronic bronchitis

Change	Effect
Increased size of mucous glands, mainly in larger bronchi	Increased mucus production
Goblet cells often increased in number (mucous glands absent in bronchioles)	Bronchioles blocked by mucus
In some cases, hyperplasia of bronchial smooth muscle (increase in number of muscle fibers)	Narrowed lumen
Inflammation and scarring	Narrowed lumen
Squamous epithelium may replace columnar cells if mucous membrane becomes ulcerated and then heals (squamous cell metaplasia)	Possibility of cancerous transformation

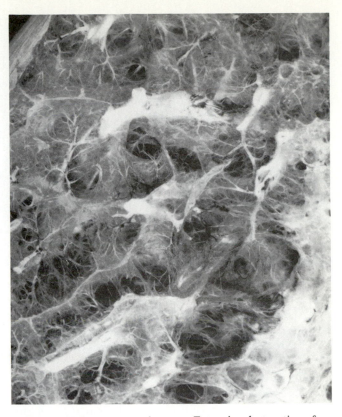

Figure 7.7 Pulmonary emphysema. Extensive destruction of pulmonary architecture leaves only fibrotic strands and blood vessels traversing the distended airspaces.
From R. S. Cottran, V. Kumar, and S. L. Robbins: Robbins Pathologic Basis of Disease. 1984 W. B. Saunders, Philadelphia. Reproduced with permission.

aggerates the inflammatory process. Other factors, such as an inherited low resistance to infection or respiratory illnesses in infancy, may predispose an individual to this condition.

PATHOLOGY

Table 7.2 lists some characteristic structural changes observed in this disease process. Excessive mucus production by both goblet cells and mucous glands, thickening and scarring of airway walls, and contraction of smooth muscle all narrow the air passageways and interfere with free flow of air. Evidence indicates that tobacco smoke and other irritants are responsible for these changes and that they are reversible in the early stages.

MANIFESTATIONS

The symptoms that characterize chronic bronchitis are (1) a productive cough, (2) **dyspnea** or breathlessness, and (3) wheezing, either persistent or during acute attacks. Although the airway obstruction is reversible in early stages, in the later stages it is irreversible. The final outcome is total disability, and ultimately, there is respiratory failure (chapter 9).

MANAGEMENT

If the patient is a smoker, an obvious and important first step in the management of chronic bronchitis is to eliminate that habit. Even in later stages, the cough will improve when the individual stops smoking, even though pulmonary function may not improve. Clearing excess mucus and improving airflow may be accomplished by use of (1) expectorants, (2) bronchodilators, such as aminophylline (ah-me″no-fil′in), or (3) **postural drainage** in

which the patient lies with head lower than the feet. Antibiotics are used to treat infections, and sometimes long-term antibiotic treatment is used to decrease the number of acute episodes of severe symptoms. Dyspnea may require hospitalization and oxygen therapy.

• • •

Bronchiectasis (brong″ke-ek′tah-sis) is a permanent dilatation or enlargement of the bronchi and bronchioles. Obstruction or repeated infection are the most common factors contributing to the condition. Chronic bronchitis may be complicated by bronchiectasis. Manifestations include a persistent cough, production of large amounts of fetid sputum, and, in severe cases, dyspnea.

• • •

EMPHYSEMA

Evidence of emphysema is sometimes seen at autopsy in cases of chronic bronchitis, indicating that there is a close relationship between these two conditions. Emphysema is defined in anatomical terms as a condition in which there is enlargement of air spaces accompanied by destruction of alveolar walls (figure 7.7).

dyspnea: Gk. *dys*, hard; Gk. *pnein*, to breathe

bronchiectasis: Gk. *bronchia*, bronchial tubes; Gk. *ektasis*, extension

Obstructive Pulmonary Disorders

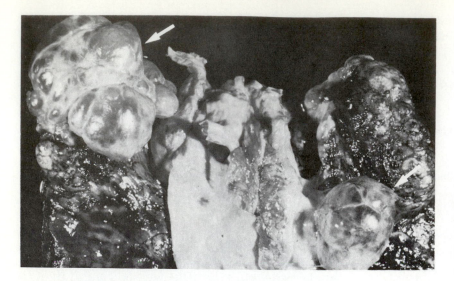

Figure 7.8 Bullous emphysema with large subpleural bullae (arrows). *From R. S. Cottran, V. Kumar, and S. L. Robbins:* Robbins Pathologic Basis of Disease. *1989 W. B. Saunders, Philadelphia. Reproduced with permission.*

CLASSIFICATION

Emphysema may be classified in several different ways, one of which is the designation of four types depending on the involvement of the lobules. A lobule is surrounded by connective tissue and is composed of terminal bronchioles and acini. Each acinus consists of respiratory bronchioles, alveolar ducts, and alveoli (chapter 6). Four types of emphysema are (1) **centriacinar** (sen-tri-uh-sin'ar) in which the respiratory bronchioles are mainly involved (2) **panacinar** in which both acini that open directly into respiratory bronchioles and distal acini are enlarged, (3) **paraseptal** in which there is destruction of predominantly distal acini adjacent to the pleura and next to lobular septa, and (4) no particular pattern of involvement called **irregular.**

PATHOLOGY

Chronic bronchitis frequently precedes emphysema, and they coexist. There are cases, however, in which either condition may exist alone. Cigarette smoking is the most important constant in emphysema, and it is likely that chronic bronchitis and emphysema frequently occur together because of this common factor, sometimes combined with atmospheric pollution. The sequence of events leading to alveolar wall destruction is uncertain. The inflammation and obstruction of the small airways that occur in chronic bronchitis may indirectly damage alveoli. It has been theorized that air trapped in alveoli by obstructed airways exerts pressure on capillaries, cuts off blood supply, and causes injury (figure 7.8).

centriacinar: L. *centrum*, center
panacinar: Gk. *pan*, all
paraseptal: Gk. *para*, alongside
septum, pl. a: L. *septum*, partition
bulla, pl. ae: L. *bulla*, bubble
proteolytic: Gk. *protos* (proteins), first; Gk. *lysis*, loosening

• • •

Distended spaces more than 1 cm in diameter may develop in any form of emphysema. These are called bullae (bul'lē) and are the result of progressive destruction of connective tissue walls that separate the lobules of the lung. Bullae are usually subpleural and rupture may lead to pneumothorax.

• • •

Chronic bronchitis does not always precede emphysema, and the enzyme elastase may initiate the destruction of alveolar walls. In pulmonary inflammations, macrophages and leukocytes release the enzyme elastase into the lung. This enzyme destroys elastic fibers (the protein elastin) found in alveolar walls, surrounding bronchioles, and pulmonary capillaries. There is normally a glycoprotein called **alpha-1-antitrypsin** (α-1-AT) synthesized by the liver and carried in the blood. Alpha-1-antitrypsin inactivates several protein-hydrolyzing enzymes including trypsin and elastase. Some individuals have an inherited deficiency of α-1-AT and have diminished protection against the proteolytic activity of elastase. Panacinar type emphysema is associated with α-1-AT deficiency. In view of the fact that macrophages and white blood cells release elastase, it is significant that smokers have an increased number of both of these cells. Table 7.3 summarizes the role of elastase and α-1-AT in maintaining the integrity of the alveolar walls.

MANIFESTATIONS

Manifestations of emphysema appear when the function of one-third or more of the lung tissue is affected. It is likely that dyspnea will be the first symptom of emphysema, and it may or may not be accompanied by coughing or wheezing. The patient may sit forward in a hunched position to aid in squeezing air out of the lungs with each expiration. There may be an increased anterior-posterior chest dimension (**barrel chest**), and weight loss is common.

TABLE 7.3 The role of elastase and α-1-AT in maintaining the integrity of the alveolar walls

Cause	Response	Effect
Pulmonary inflammation	Macrophages and white blood cells release elastase in the lungs	Elastase destroys elastic fibers of alveolar walls
Smoking	Increased number of macrophages and leukocytes	Increase in elastase
α-1-AT synthesized in liver; carried in blood		α-1-AT inactivates elastase; protects against proteolytic effect of elastase
Inherited α-1-AT deficiency		Increased susceptibility to alveolar wall destruction

TABLE 7.4 Distinguishing features of extrinsic and intrinsic types of asthma

Extrinsic	Intrinsic
Allergens trigger asthmatic attacks	No demonstrable allergen
Frequently childhood onset	Usually begins during adult years
Asthma tends to be paroxysmal	Asthma tends to be chronic and continuous
Family history of other allergies	Less family history of other allergies
Genetic factors, genes may predispose	Genetic factors, different genes may predispose

TABLE 7.5 Some possible allergens that cause asthmatic attacks

Food	Environmental	Occupational (handling)
Chocolate	Pollen	Grain
Eggs	Dust	Flour
Milk	Animal dander	Coffee
Wheat	House dust mite	
	Spores	

MANAGEMENT

The overall approach to the treatment of emphysema is first, as in chronic bronchitis, the avoidance of irritating factors, such as smoking and environmental pollution. A second concern is the prevention and treatment of infection. Measures are taken to control cough and bronchial secretions and to maintain adequate oxygenation.

ASTHMA

The term **asthma** is used for a condition in which there is obstruction to airflow due to widespread narrowing of the airways. The obstruction is not due to cardiovascular disease, and there are periodic fluctuations in severity. The characteristic feature of asthma is its reversibility.

ETIOLOGY

Evidence indicates that **bronchospasms** or contraction of bronchial smooth muscle is the important factor in severe episodes of asthma. The precise mechanisms are not clear. In some cases, an asthmatic attack may be triggered by an allergic reaction to certain stimuli called **allergens.** Asthma is divided into two main categories: (1) an **extrinsic type** in which there is a response to an allergenic stimulus and (2) an **intrinsic type** in which there is no demonstrable allergenic stimulus. Table 7.4 lists some distinguishing features of both intrinsic and extrinsic types of asthma. The list shows that extrinsic asthma typically (1) appears at a younger age, (2) tends to be paroxysmal (par″ok-siz′mal), (3) is associated with a family history

paroxysm: Gk. *paroxysmos*, sudden recurrence

of other allergies, and (4) there may be specific genes which predispose to the condition. Table 7.5 lists some possible allergens.

Another difference between the two types of asthma involves **immunoglobulin E** (IgE) blood level, which is significant in terms of allergic reactions. IgE is an **antibody,** a protein produced in response to and capable of combining with a foreign protein or antigen. The antigen is called an allergen when an allergic reaction ensues. Allergen triggers the following series of events. First the allergen stimulates lymphocytes and plasma cells to produce IgE, which attaches to mast cells and basophils. Lymphocytes and basophils are both types of white blood cells, and plasma cells develop from lymphocytes (chapter 18). After a second exposure to an allergen, **mast cells** release chemicals called **primary mediators.** These include **histamine** that causes bronchoconstriction, increased capillary permeability, and increased nasal and bronchial secretions. Chemotactic (ke″mo-tak′tik) factors for leukocytes, specifically eosinophils and neutrophils, are released as well.

A second series of events involves mast cells and leukocytes that release chemicals called **secondary mediators.** Secondary mediators include a group of substances which were first called **slow reacting substance of anaphylaxis** (SRS-A) and are made up of derivatives of arachidonic acid called leukotrienes (lu′kōtri′ēnes). These

chemotactic: Gk. *chemeia*, chemistry; Gk. *taxis*, arrangement

TABLE 7.6 The role of primary and secondary mediators in symptoms of extrinsic asthma

Primary Mediators		
Stimulus	Response	Effect
Allergen	Triggers production of IgE	IgE attaches to mast cells and basophils
Second exposure to allergen	Mast cells release primary mediators	Bronchoconstriction, increased capillary permeability
Secondary Mediators		
Allergen (slower response)	Mast cells release secondary mediators	Leukotrienes, bronchoconstriction; prostaglandins, vasodilation; platelet activator, aggregation of platelets
Overall Effect		
Bronchoconstriction and continued inflammatory response		

TABLE 7.7 A summary of morphological findings in status asthmaticus

Morphology	Significance
Thickening of basement membrane of alveolar wall	Increases thickness of respiratory membrane
Hypertrophy of mucous glands	Increased mucus production
Increase in number of goblet cells; decrease of epithelial cells	Increased mucus; less effective mucociliary escalator
Hyperplasia of smooth muscle of airways	Contributes to increased resistance to airflow
Infiltration of medium and small bronchi by eosinophils	Release enzymes to inactivate histamine and leukotrienes

are released more slowly than histamine but are more powerful bronchoconstrictors. Additional secondary mediators include **prostaglandins** that are potent vasodilators and **platelet activators** that induce aggregation of platelets.

The overall effect of primary and secondary mediators is to cause bronchoconstriction and increased capillary permeability, followed by a period of continued reactivity. Increased capillary permeability and the presence of blood cells, including platelets, neutrophils, and eosinophils, cause further inflammatory responses. Table 7.6 summarizes the role of primary and secondary mediators.

Factors other than allergens stimulate asthmatic attacks in susceptible individuals, i.e., those individuals that have **hyperreactive** bronchi. Exercise, especially in cold air, triggers bronchoconstriction, sometimes even in **latent asthmatics** who have a family history but lack symptoms of asthma. Respiratory infections in asthmatic children cause attacks. Emotional episodes can precipitate asthmatic attacks, although in most individuals psychological factors are probably not primary. Elevations of IgE are less common in intrinsic asthma, but it does not exclude the possibility of an allergic component in the disease process.

PATHOLOGY

Morphological changes which have been identified microscopically in **status asthmaticus** are summarized in table 7.7. Status asthmaticus is a severe form of asthmatic episode that may last for days or weeks and may lead to death. The structural changes are typical regardless of whether the asthma is an intrinsic or extrinsic type. A thickening of basement membrane increases the dimensions of the capillary/alveolar wall (respiratory membrane) where diffusion of gases occurs. The alveolar wall is made of epithelial cells with underlying basement membrane and connective tissue (chapter 6), so a thick basement membrane interferes with diffusion of gases. Impaired diffusion may only be significant during exercise if the morphological changes are not marked. Excessive mucus production is the result of enlarged mucous glands and an increase in the number of goblet cells. Mucus plugs and hyperplasia of airway smooth muscle cause resistance to airflow. Eosinophils are attracted by chemicals released by mast cells and seem to inhibit allergic reactions.

The factors that contribute to airflow obstruction characteristic of asthma are bronchoconstriction, edema of the mucous membranes, and viscous mucus plugs. The central airways are narrowed by hyperplasia of the smooth muscle, while the distal airways are distended by trapped air (figure 7.9).

MANIFESTATIONS

A typical asthmatic attack begins suddenly with breathlessness characterized by difficulty in forcing air out of the lungs. Usually, there is wheezing with expiratory effort, although in very severe attacks there may be no audible wheeze. The attack may last several hours and is followed by prolonged coughing with mucus production. In severe cases, there may be inspiratory as well as expiratory difficulty, and there may be jugular vein distention during expiration due to increased intrathoracic pressure.

• • •

Forced expiration causes high intrapleural pressure, which pushes against the outside walls of the airways. The result is dynamic compression in which the walls of airways collapse or are squeezed together. This occurs more readily if the airway walls are diseased.

• • •

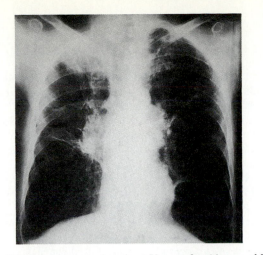

Figure 7.9 Posterior-anterior chest X-ray of a 66-year-old male. This individual has chronic obstructive pulmonary disease (COPD) as demonstrated by hyperexpanded lungs and white patchy areas in both upper lung fields indicating pneumonia.

MANAGEMENT

Control of environmental and other factors is an aspect of long-term management of asthma. Acute asthmatic episodes are treated with bronchodilators, such as isoproterenol, to relieve bronchoconstriction. **Corticosteroids** inhibit inflammatory responses and are used if symptoms are not controlled by bronchodilators or if the disease process becomes life-threatening. **Disodium cromoglycate (cromolyn),** which inhibits the release of histamine and SRS-A in the lung, may be effective as a prophylactic measure.

COMPLICATIONS OF COPD

A common complication of COPD is **cor pulmonale** (pulmo-nal′ē) or right-sided heart failure. Airway obstruction causes inadequate oxygenation or **hypoxemia** (hi″pok-se′me-ah) as well as respiratory acidosis. Both of these factors cause constriction of the pulmonary arteries. The constriction decreases the blood vessel lumen size and increases resistance to blood flow. Consequently, the right ventricle must work harder to pump blood to the lungs, and this ultimately causes heart failure (chapter 23).

• • •

The term "blue bloater" is sometimes used to describe a patient who suffers from severe bronchitis, often associated with emphysema. This individual typically has right-sided heart failure, edema, and severe hypoxemia. Oxygen deficiency causes cyanosis, a bluish tinge to the skin and mucous membranes. Severe emphysema is characterized by weight loss and dyspnea upon exertion. Severe hypoxemia occurs only late in the course of the disease. These individuals are called "pink puffers."

• • •

SUMMARY

• • •

Chronic obstructive pulmonary disease is the most common pulmonary disease in humans and includes chronic bronchitis, emphysema, and asthma. Chronic bronchitis is a prolonged inflammation of the airways, which causes excessive mucus production and a persistent cough. Cigarette smoking, air pollution, and infection are three major etiological factors. Characteristically, there is hypertrophy of mucous glands, an increased number of goblet cells, and hyperplasia of smooth muscle.

Emphysema is defined as a condition in which there is destruction of alveolar walls and is classified on the basis of the nature of involvement of the lobules of the lung. Chronic bronchitis may lead to emphysema. Macrophages release elastase, an enzyme that destroys fibers in the lung, but α-1-AT deactivates this enzyme to provide protection against its proteolytic action. Any circumstances that lead to increased elastase or decreased α-1-AT make the individual susceptible to emphysema.

Asthma is a condition in which there is reversible narrowing of the airways with periodic increases in severity. Bronchospasms are an important factor in airflow obstruction. The extrinsic type of asthma is triggered by allergens with the typical onset during childhood with an increased level of IgE. IgE combines with mast cells, which release primary mediators and secondary mediators. The overall effect is bronchoconstriction, vasodilation, and a prolonged inflammatory response. In the intrinsic type of asthma, the onset is during adulthood. There is no elevation of IgE levels, and factors other than allergens trigger an asthmatic response. Such nonspecific factors as exercise, cold air, and emotional upset may cause an asthmatic attack. The onset of symptoms in asthma is sudden with difficulty in expiration, a productive cough, and wheezing. Morphologically, the basement membrane in alveolar walls is thickened, and there is hypertrophy of mucous glands, increased numbers of goblet cells, hyperplasia of smooth muscle, and infiltration of walls of small bronchi by eosinophils. Obstruction to airflow in asthma is caused by bronchoconstriction, edema, hyperplasia of smooth muscle, and mucus.

The most common complication of COPD is cor pulmonale or right-sided heart failure caused by pulmonary hypertension due to hypoxemia and acidosis.

cor pulmonale: L. *cor,* heart; L. *pulmonis,* lung
hypoxemia: Gk. *hypo,* under; oxygen; Gk. *haima,* blood

1. Identify the disorder described by each of the following definitions.
   a. COPD in which there is enlargement of air space by destruction of alveolar walls
   b. COPD in which there is reversible and paroxysmal narrowing of the airways
   c. COPD characterized by inflammation of the airways accompanied by cough and excessive mucus production
2. What are two common causes in chronic bronchitis?
3. List three structural changes that occur in chronic bronchitis.
4. List three factors that contribute to obstruction of airflow in chronic bronchitis.
5. What is the source(s) of elastase?
6. What effect does elastase have in the lungs?
7. What is the source of α-1-AT?
8. What is one effect of α-1-AT in the body?
9. How is smoking involved in upsetting the balance between elastase and α-1-AT?
10. What is the first symptom of emphysema?
11. What is the most important factor in airflow obstruction in asthma?
12. Identify each of the following with either intrinsic or extrinsic asthma.
    a. Adult onset
    b. Family history of allergies
    c. No demonstrable allergenic stimulus
    d. IgE level increased
13. What cells do IgE molecules attach to?
14. What cells release primary and secondary mediators?
15. Give two examples of primary mediators and indicate their effects.
16. What is the significance of increased capillary permeability, vasodilation, and the presence of blood cells in response to primary and secondary mediators?
17. What are the symptoms of asthma?
18. List four morphological changes that occur in asthma.
19. What is the cause of airflow obstruction in asthma?
20. What causes pulmonary hypertension in COPD that leads to cor pulmonale?

Berk, J. L. 1987. Cold-induced bronchoconstriction: Role of cutaneous reflexes vs. direct airway effects. *Journal of Applied Physiology* 63(2):659–64.

Burr, M. L. 1987. Why is chest disease so common in South Wales? Smoking, social class, and lung function: A survey of elderly men in two areas. *Journal of Epidemiology and Community Health* 91(2):140–44.

Ingram, R. H., Jr. 1987. Site and mechanism of obstruction and hyperresponsiveness in asthma. *American Review of Respiratory Disease* 136(4 Pt 2):562–64.

Littenberg, B., and E. H. Gluck. 1986. A controlled trial of methylprednisolone in the emergency treatment of acute asthma. *New England Journal of Medicine* 314:150–52.

Murciano, D., M. Aulier, Y. Lecocquic, and Pene Fariente. 1984. Effects of theophylline on diaphragmatic strength and fatigue in patients with chronic obstructive pulmonary disease. *New England Journal of Medicine* 311:349–53.

Pingleton, S. K. 1990. Adult asthma. *Physician Assistant* 14:23–36.

Shekleton, M. E. 1987. Coping with chronic respiratory difficulty. *Nursing Clinical of North America* 22:569–81.

Weinberg, H. 1988. Long-term management of asthma. *Physician Assistant* 12:30–42.

Chapter 8

Restrictive and Other Pulmonary Disorders

The first section of the following discussion focuses on several pathological conditions characterized by diminished lung capacity, classified as restrictive pulmonary disorders. The separation of pulmonary disorders on the basis of obstructive or restrictive is imprecise but is convenient for discussion purposes. The last part of this chapter is concerned with various other pulmonary diseases including bronchogenic carcinoma.

RESTRICTIVE DISEASES

In general, any condition resulting in a loss of expandability of the lungs and a decreased compliance leads to a diminished lung capacity. The following paragraphs deal with three examples of restrictive pulmonary disorders.

HYALINE MEMBRANE DISEASE

Hyaline membrane disease is also called **respiratory distress syndrome** in the newborn. It is characterized by life-threatening respiratory distress that mainly affects premature infants (figure 8.1). The basic defect is a deficiency of surfactant, which causes inadequate inflation of alveoli. It is believed that hypoxia either before or after birth is an underlying cause and that maternal diabetes or Caesarean section predisposes infants to this condition through prematurity or fetal hypoxia. A **hyaline membrane** composed of cellular debris and protein fibers forms a lining within the alveoli, and this creates a barrier for alveolar/capillary gas exchange. The gross appearance of the lungs is solid and liverlike with diminished compliance. Hypoxia, caused by inadequate inflation and formation of a hyaline membrane, leads to constriction of pulmonary arterioles and increased pulmonary artery pressure. The infant typically has breathing difficulties accompanied by cyanosis and tachycardia (tak″-e-kar′ de-ah).

There is evidence that a corticosteroid given to the mother when premature delivery is anticipated, promotes maturation of the fetal lung and helps to prevent hyaline membrane disease. Treatment of hyaline membrane disease involves the judicious use of oxygen, correction of low blood pH with intravenous bicarbonate, and possibly the use of artificial respiration. There are similarities between hyaline membrane disease and adult respiratory distress syndrome.

ADULT RESPIRATORY DISTRESS SYNDROME

Respiratory failure is defined as the inability of the lungs to provide adequate amounts of oxygen to the body (chapter 9). This occurs in an acute form of **adult respiratory distress syndrome (ARDS)**. Identifying features of ARDS are (1) a preceding acute illness or trauma, (2) a latent period of 24–72 hours, (3) dyspnea, (4) increased

tachycardia: Gk. *tachys*, swift; Gk. *kardia*, heart

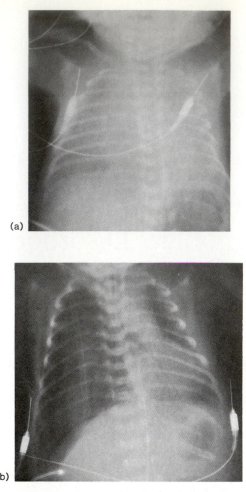

(a)

(b)

Figure 8.1 (*a*) Chest X-ray of 1-hour-old baby with hyaline membrane disease. The whitish area indicates inadequate aeration. (*b*) The same baby, 1-day-old and on a ventilator. Ventilation has caused a pneumothorax. The white, more or less vertical lines are the margins of the ribs. On the left, the pneumothorax is demonstrated by the dark space on the outer edge of the patient's lung.

breathing rate or **tachypnea** (tak″ip-ne′ah), (5) severe hypoxemia, and (6) chest radiographs showing bilateral **infiltrates** or fluid accumulation. Table 8.1 lists some illnesses or injuries that lead to ARDS. The term "**shock lung**" is sometimes used for ARDS when it is caused by nonthoracic trauma.

Pathogenesis

The pathological processes in ARDS include an increased permeability of alveolar/capillary walls and a decrease in the synthesis of surfactant. The mechanisms responsible for the changes are uncertain. Pulmonary edema is the result of increased permeability as fluid leaks into the alveoli and interstitium. Surfactant is an essential factor in preventing alveolar collapse (chapter 6) and **atelectasis**

tachypnea: Gk. *tachys*, swift; Gk. *pnoie*, breathing
atelectasis: Gk. *ateles*, imperfect; Gk. *ektasis*, extension

TABLE 8.1 Some injuries or illnesses that may lead to ARDS	
Pneumonia	Septicemia
Aspiration of gastric juice	Drug overdose
Hypotensive shock	High concentrations of oxygen
Cerebral trauma	Massive blood transfusion
Acute pancreatitis	

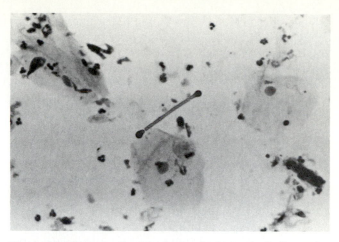

Figure 8.2 Microscopic appearance of asbestos body (center) in sputum. Exposure to asbestos may lead to pulmonary fibrosis, called asbestosis, or it may lead to lung cancer years after first exposure.
Courtesy of Dr. M. G. Klein.

(at′e-lek′tah-sis) or collapse occurs due to decreased surfactant. **Interstitial fibrosis** is an increase in fibrous connective tissue, and when this occurs, it makes the lungs stiff and noncompliant. A hyaline membrane composed of fibrinogen and cellular debris forms as an inner lining of some of the alveoli. These factors all contribute to hypoxemia: edema, atelectasis, hyaline membrane formation, and decreased lung compliance.

Treatment

The principles of management of ARDS involve the treatment of the underlying cause with correction of hypoxemia. Corticosteroids help prevent the development of the syndrome but are not useful after the syndrome has occurred. The mortality rate for ARDS is 50% or more.

SARCOIDOSIS

Sarcoidosis (sar′′koi-dō′sis) may affect several systems of the body. The lung is almost always involved, and most patients have respiratory symptoms. The etiology of the disease is unknown and it is characterized by lesions which are **noncaseating** (ka′′se-a′ting) **epithelioid granulomas.**

• • •

The term caseating refers to necrosis of tissue, which becomes cheeselike. Epithelioid means that the tissue resembles epithelial tissue, and granuloma refers to a growth in which there are red granules formed by new capillaries. The lesion in sarcoidosis is described as noncaseating because in some ways it resembles the caseated lesion in tuberculosis.

• • •

In subacute form, the symptoms of the disease appear abruptly and include such general symptoms as fever, malaise, anorexia and in many instances, dyspnea and cough. The symptoms of the chronic form are respiratory complaints that develop over a period of months, with other symptoms depending on the organs that are involved. Most affected individuals have an abnormal chest X-ray with enlarged lymph nodes in the region of the hilus and trachea.

The lesions in sarcoidosis sometimes stimulate the production of fibrous connective tissue in the interstitium

of the lungs, which interferes with normal function. The development of **pulmonary interstitial fibrosis** occurs in a number of other conditions as well, i.e., rheumatoid arthritis, chronic pulmonary edema, irradiation pneumonitis, and pneumoconiosis (nu′′mō-kō′′nē-ō′sis) (figure 8.2). Respiratory function is characterized by a decreased pCO_2 and normal pO_2 at rest, and hypoxemia with exercise.

• • •

Pneumoconiosis is a condition in which pulmonary dust deposits have stimulated fibrosis. Silica, asbestos, and coal dust are examples of such particles.

• • •

Prognosis for sarcoidosis is good and may clear spontaneously in about half the cases. Corticosteroids are used to treat the chronic form of the disease.

OTHER PULMONARY DISORDERS

The following section deals with some disease processes that primarily affect the lung and represent frequently occurring health problems.

PULMONARY THROMBOEMBOLISM

The term **thrombus** means blood clot, and the term **embolism** refers to an occlusion of a blood vessel. A pulmonary thromboembolism is a blood clot that occludes a blood vessel of the lungs. The size of both clot and blood vessel varies from microscopic to relatively large. The clot usually originates from the deep veins of the leg and then travels to the pulmonary artery and other parts of the pulmonary tree. There are three factors in the formation of a blood clot: (1) blood flow stasis, (2) an increased tendency for blood to coagulate, and (3) damage to a blood vessel wall. Risk of blood clot formation is increased by immobility, obesity, increasing age, contraceptive pills, and pregnancy. Pulmonary thromboembolism is a common cause of death after surgery and after childbirth. Extravascular sources of emboli include air emboli and fat emboli subsequent to long bone fracture. Both of these emboli may pass through pulmonary capillaries and be carried to the brain.

Blood clots are lysed by the **fibrinolytic system** of the body (chapter 19), and the event of pulmonary thromboemboli formation may be asymptomatic. The opposite extreme is sudden death due to occlusion of the pulmonary artery. The severity of the symptoms depends on the magnitude of occlusion. Typically, there is a sudden onset of breathlessness accompanied by tachycardia and hypotension. **Pulmonary infarction** may occur. This is necrosis of lung tissue due to inadequate blood supply and is usually accompanied by pleuritic pain and pleural effusion.

The effect of an embolus is that the area beyond the occlusion is not perfused. It receives no blood, and gas exchange in that area does not occur. After several hours there is loss of surfactant, and there is subsequent alveolar collapse. Surfactant is important in maintaining normal permeability, and a regional pulmonary edema occurs in response to loss of surfactant. Arterial hypoxemia develops, and if there is hyperventilation, pCO_2 is lowered as well.

Treatment involves administration of oxygen and anticoagulant therapy. In a case where massive thromboembolism has occurred, external cardiac massage and administration of oxygen is required.

Pneumonia

The term **pneumonia** refers to an inflammation of the lung in which the patient typically has an increased temperature, a rapid pulse, rapid breathing, pleuritic pain, and appears flushed. **Pneumonitis** is a synonym, although the use of this term is sometimes limited to an inflammation of a segment of the lung. A variety of agents, such as bacteria, viruses, and fungi may cause pneumonia. *Pneumocystis carinii* is a protozoan that causes pneumonia, especially in the immunosuppressed patient (chapter 29).

Inflammatory Response

The inflammatory response in the lungs involves (1) mast cell release of such vasoactive substances as histamine and bradykinin, (2) vasodilation with increased blood flow, and (3) disruption of tight junctions between Type I alveolar epithelial cells. The infective agent can pass from alveolus to alveolus by way of pores of Köhn and, ultimately, cause pleuritis. These events are followed by exudation of fluid

necrosis: Gk. *nekrosis*, killing
infarction: L. *infarcire*, to stuff

into alveoli along with red blood cells, fibrin, and neutrophils. Neutrophils phagocytize bacteria and disintegrate. As the disease process is resolved, the alveolar exudate is enzymatically digested and coughed up or is ingested by macrophages. The consequence of fluid filled alveoli is interference with capillary/alveolar gas exchange leading to hypoxia. Hyperventilation forestalls an increase in pCO_2.

Factors

Several factors contribute to the development of pneumonia postoperatively. An inactive cough mechanism due to pain or sedation allows bacteria to be protected by mucus and to multiply. Respiratory defenses may be impaired by anesthetics as well. In general, anything that suppresses either respiratory defenses or drainage of the lung predisposes to pneumonia. Severe illness, increased age, or an ineffective immune system puts an individual at risk. Aspiration of gastric juice or vomitus occurs in some situations, particularly in cases where the individual is under sedation, after anesthesia, and especially with infants and the elderly. Aspiration of strongly acidic gastric juice damages the airway epithelium and may be followed by infection. Pulmonary edema, atelectasis, and hemorrhage are the results, and the condition is called **aspiration pneumonia.**

Cystic Fibrosis

Cystic fibrosis is a condition with prominent respiratory symptoms. It is inherited as an autosomal recessive disorder, rare in Asians and Africans, more common in Caucasians. The basic defect is unknown, but the disease process involves the production of viscid exocrine gland secretions. Typically, there is pancreatic insufficiency, inflammation, and fibrosis of the hepatobiliary (hep″ah-to-bil′e-ār″e) system. Intestinal obstruction occurs, and there is increased sodium, potassium, and chloride ions in sweat. Bronchial mucous glands hypertrophy, secretions are thick, and there is inflammation of bronchioles. The changes are complicated by repeated infections, and there is airway obstruction with hypoxia.

Treatment includes improving pulmonary drainage, antibiotic therapy, and the use of bronchodilators. The average survival in cystic fibrosis is about 20 years.

Neoplasms of Lung

Lung cancer is a major cause of death throughout the world. In the United States, it is the leading cause of cancer death in males 35 years and older and the second leading cause of cancer death in females in the age group of 35–75 years. There is a peak incidence of lung cancer in both sexes in the age category of 55–65 years.

viscid: L. *viscidus*, sticky
hepatobiliary: Gk. *hepar*, liver; L. *bilis*, bile
neoplasm: Gk. *neos*, new; Gk. *plasma*, anything formed

Malignant vs. Benign

The term cancer refers to growths or **malignant** tumors as opposed to **benign** (be-nīn') tumors, and there are several distinguishing features for both (table 8.2). A tumor or mass of tissue of either type is the result of uncontrolled cell division. The study of tumors is called **oncology** (ong-kol'ō-jē). Growth rates vary, but a malignant tumor often grows so rapidly that central parts of the tumor become necrotic because of inadequate blood supply. These growths may produce tissue destroying enzymes and tend to penetrate adjacent tissues. Not only is invasiveness characteristic, but also **metastasis,** which means that the growth becomes established at sites far from the original lesion. Malignant tumors are almost never enclosed by a fibrous capsule.

The degree of cellular **differentiation** or specialized function is yet another difference. The cells constituting benign tumors are well differentiated. They resemble normal cells, while in cancerous growths there are various degrees of differentiation. In some types of cancer, the cells are immature, vary in size and shape, and many cells are atypical. In other cases, the cells are well differentiated and they may secrete products normally produced by the tissue in which the neoplasm resides.

Benign tumors characteristically grow slowly, are noninvasive, and do not metastasize.

BRONCHOGENIC CARCINOMA

Cancers are grouped in three major categories: (1) **carcinomas,** originating in epithelial tissue; (2) **sarcomas,** originating in undifferentiated connective tissue called **mesenchyme;** and (3) **leukemias** and **lymphomas,** originating from white blood cells and cells of lymphoid tissues, respectively.

Cell Types

The type of neoplasm called bronchogenic carcinoma represents more than 90% of the tumors that develop in the lower respiratory tract. Bronchogenic carcinoma includes four major cell types: (1) squamous cell or epidermoid carcinoma, (2) small cell (oat cell) carcinoma, (3) adenocarcinoma, and (4) large cell carcinoma.

Squamous cell carcinoma most commonly develops in major bronchi and obstructs the lumen of the passage. The neoplasm tends to grow through the bronchial wall and invade both blood and lymph vessels. Metastasis to other parts of the body is common.

Adenocarcinoma is made up of cuboidal or columnar cells that form glandular structures (figure 8.3). The tumor most commonly arises in the peripheral parts of the lung and may be asymptomatic until the tumor becomes large.

The tumor in small cell carcinoma is made of small malignant cells with scanty cytoplasm. The neoplasm usually develops in the wall of a bronchus and spreads to

benign: L. *benignus,* kind
oncology: Gk. *onkos,* bulk; Gk. *logos,* description
metastasis: Gk. *metastasis,* removal

TABLE 8.2 Some common characteristics of malignant and benign tumors

Malignant	Benign
Uncontrolled growth	Uncontrolled growth
Rapid growth rate	Slow growth rate
Outgrows blood supply, center becomes necrotic; develops blood supply	Does not become necrotic or develop blood supply
Invasive, produce tissue destroying enzymes	Noninvasive
Metastasis; cells disorganized, do not adhere to each other	No metastasis; cells adhere to each other
Almost never encapsulated	Nearly all enclosed by a fibrous capsule
Range from undifferentiated to differentiated	Well differentiated cells

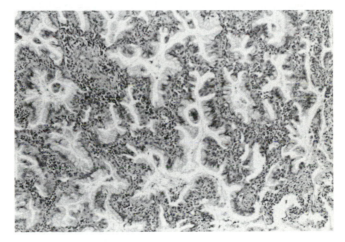

Figure 8.3 Adenocarcinoma of the lung. *Courtesy of Dr. M. G. Klein.*

lymph nodes, blood vessels, and distant organs. Metastasis has usually occurred at the time of diagnosis. Large cell carcinoma is made of undifferentiated cells that have abundant cytoplasm and lack the characteristics of squamous cell carcinoma or adenocarcinoma. This type of tumor grows rapidly and metastasizes widely.

Etiology

Cigarette smoking is the leading cause of lung cancer. Evidence indicates that exposure to various carcinogens (kar-sin'ō-jens) increases the risk of lung cancer. Exposure to radiation due to radon and exposure to asbestos are two notable examples. There is some evidence suggesting that family members of lung cancer patients are at increased risk of getting the disease, possibly due to inherited factors.

carcinogen: Gk. *karkinoma,* cancer; Gk. *genes,* become

Restrictive and Other Pulmonary Disorders

Manifestations

Respiratory symptoms of bronchogenic carcinoma include difficult breathing or dyspnea, which may be the result of obstruction, paralysis of the diaphragm, pulmonary collapse, or pleural effusion (chapter 6). Harsh sounds called **stridor** may occur during expiration due to obstruction of the airways. There may be a productive cough and **hemoptysis** (he-mop′ti-sis). Hemoptysis refers to blood tinged sputum that indicates destruction of blood vessels. There may be chest pain as the result of invasion of pleura, the chest wall, or mediastinum. Pneumonia is a possible complication and is caused by obstruction and inefficient drainage. General symptoms, such as anorexia and weight loss, are associated with lung cancer. There may be symptoms associated with metastasis to distant sites, such as lymph nodes, the brain, bone, and liver. Endocrine disturbances may occur due to abnormal production of hormones, such as adrenocorticotrophic hormone, parathormone, and antidiuretic hormone (chapters 31 and 32). Proliferation of fibrous connective tissue called **finger clubbing** causes a thickening of the fingers and is common in lung cancer.

Diagnosis

Bronchogenic carcinoma is sometimes first detected because of a shadow on a chest X-ray (figure 8.4). In combination with history and physical examination the chest X-ray is a major tool in diagnosis. Cytological examination of sputum with a special staining technique reveals abnormal cells and aids in diagnosis. Direct examination of the airways is made possible by the use of a flexible tube with light-transmitting fibers that provides a magnified image, a procedure called **bronchoscopy** (brong-kos′ko-pē′).

Staging

The American Joint Committee on Cancer has developed a system for identifying the anatomic involvement in bronchogenic carcinoma. This system, widely used in the United States, uses the letters T, N, and M to stand for primary tumor, lymph nodes, and metastases. Numbers are added to the letters to show increasing involvement. A primary tumor diagnosed by the presence of malignant cells in bronchopulmonary secretions, but which can not be demonstrated by an X-ray or bronchoscopy, is designated **TX.** The symbol **TD** means that there is no evidence of primary tumor. Carcinoma **in situ** (in si′tu), in its original location, is indicated by **TIS.** The symbols **T1–T4** are used for (1) a noninvasive tumor of 3 cm or less, (2) tumor greater than 3 cm or any size that invades the visceral pleura or is associated with atelectasis or obstructive pneumonitis, (3) tumor of any size that invades the chest wall or diaphragm with no involvement of heart, trachea,

hemoptysis: Gk. *haima,* blood; Gk. *ptysis,* spit
in situ: L. *in,* in; L. *situs,* position

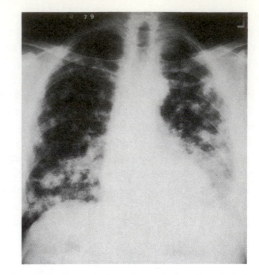

Figure 8.4 Posterior-anterior chest X-ray of an elderly male with a history of cancer of the colon that has metastasized to the lung. Diffuse white, patchy areas are the evidence of that metastasis.

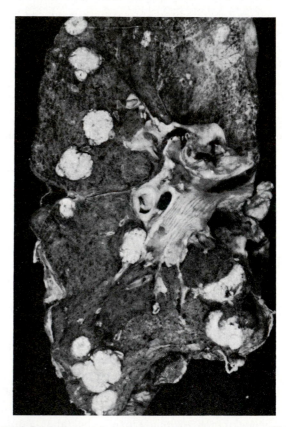

Figure 8.5 Metastases in lung from a breast carcinoma. This pattern of metastatic growth shows multiple discrete nodules scattered throughout all lobes. The lesions tend to occur in the periphery of the lung parenchyma.
From R. S. Cottran, V. Kumar, and S. L. Robbins: Robbins Pathologic Basis of Disease. *1989 W. B. Saunders, Philadelphia. Reproduced with permission.*

TABLE 8.3 Stages of lung cancer according to the American Joint Committee on Cancer

Stage I	Tumor with no lymph node involvement or metastasis; atelectasis or obstructive pneumonitis does not affect entire lung
Stage II	Tumor with limited lymph node involvement and no metastasis
Stage III$_a$	Tumor that invades the chest wall or the diaphragm; spread to lymph nodes
Stage III$_b$	Mediastinal, hilar, or supraclavicular lymph node involvement; any size tumor
Stage IV	Any tumor with varying lymph node involvement; metastasis

or esophagus, and (4) tumor of any size that invades the mediastinum, the heart and associated vessels, trachea, esophagus, or vertebral bodies; or is associated with pleural effusion.

Lymph node involvement is indicated by **NO,** the absence of metastasis to regional lymph nodes, and by **N1** through N3 to show increasing invasion of lymph nodes. The symbol **MO** is used for no metastasis and **M1** for distant metastasis (figure 8.5). Table 8.3 lists the stages of lung cancer according to this system.

Principles of Treatment

With the exception of small or oat cell type bronchogenic carcinoma, surgery is the most effective treatment. Radiation and multiple drug chemotherapy are used as well. The prognosis for patients with lung cancer is poor. The average 5-year survival rate is less than 10%.

SUMMARY

• • •

Examples of restrictive pulmonary disease in which lung capacity is diminished include hyaline membrane disease, adult respiratory distress syndrome, and sarcoidosis. Hyaline membrane disease occurs in premature infants, and the primary problem is insufficient surfactant. Acute illness or trauma precedes ARDS, and this condition is characterized by (1) decreased surfactant, (2) pulmonary edema, (3) an alveolar hyaline membrane, and (4) interstitial fibrosis. Sarcoidosis is associated with noncaseating epithelioid granulomas involving the lungs and other systems in the body. Respiratory symptoms, such as dyspnea and cough, are common.

Other pulmonary problems include thromboemboli, pneumonia, and cystic fibrosis. Cystic fibrosis causes viscid exocrine gland secretions including bronchial secretions. Airway obstruction and hypoxemia are the result.

Bronchogenic carcinoma is classified according to the predominant cell type of the tumor, i.e., squamous cell, small cell, adenocarcinoma, and large cell carcinoma.

Squamous cell carcinoma usually develops in major bronchi, whereas adenocarcinoma, composed of cuboidal or columnar cells, arises in the peripheral parts of the lung. The tumor in both small and large cell carcinoma is composed of cells that lack the characteristics of either squamous cell or adenocarcinoma. Widespread metastasis is characteristic of both malignant cell types. Cigarette smoking is the leading cause of lung cancer. Respiratory symptoms are common and may be accompanied by endocrine disturbances and evidence of metastasis to other parts of the body. Diagnosis is made by chest X-ray, cytological examination of sputum, and bronchoscopy. The TNM system of staging lung cancer describes the primary tumor, the lymph node involvement, and metastasis. Lung cancer is treated surgically, and by both radiation and chemotherapy. The prognosis is poor.

REVIEW QUESTIONS

• • •

1. What is the basic defect in hyaline membrane disease?
2. What is a hyaline membrane, and where is it formed in hyaline membrane disease?
3. List four factors that cause hypoxia in hyaline membrane disease.
4. List six identifying features of ARDS.
5. What is responsible for decreased pulmonary compliance in ARDS?
6. Describe the lesions in sarcoidosis.
7. List three contributing factors in blood clot formation.
8. What is pulmonary infarction?
9. List three effects that an embolus has on the area beyond the point of occlusion.
10. Why does the inflammatory response promote pulmonary edema?
11. List two factors that contribute to the development of postoperative pneumonia.
12. List two major features of cystic fibrosis.
13. List five differences between malignant and benign growth.
14. List four major cell types that make up almost all of the bronchogenic carcinomas.
15. List four possible symptoms of bronchogenic carcinoma.
16. What are three approaches in diagnosing bronchogenic carcinoma?
17. What cell types are characteristic in adenocarcinoma?
18. What are four major categories of cancer?
19. What does T1, ND, and MD mean?
20. Describe stage IV of bronchogenic carcinoma.

Selected Reading

• • •

D'Costa, M. et al. 1987. Lecithin/sphingomyelin ratios in tracheal aspirates from newborn infants. *Pediatric Research* 22(2):154–57.

Durant, J. R. 1987. Immunotherapy of cancer: The end of the beginning? *New England Journal of Medicine* 316: 939–40.

Fleischer, R. 1987. A decade of cancer prevention and detection. *Physician Assistant* 11:15–33.

Fowler, A. A. et al. 1987. The adult respiratory distress syndrome. Cell populations and soluble mediators in the air spaces of patients at high risk. *American Review of Respiratory Disease* 136(5):1225–31.

Holt, J. A. et al. 1990. Automated rapid assessment of surfactant and fetal lung maturity. *Laboratory Medicine* 21:359–66.

Hull, R. D. et al. 1986. Continuous intravenous heparin compared with intermittent subcutaneous heparin in the initial treatment of proximal-vein thrombosis. *New England Journal of Medicine* 315:1109–14.

LeGrys, V. A. 1990. Trends and methodology in sweat testing for cystic fibrosis. *Laboratory Medicine* 21:155–58.

Perry, M. C. et al. 1987. Chemotherapy with or without radiation therapy in limited small-cell carcinoma of the lung. *New England Journal of Medicine* 316:912–18.

Sefrin, P. 1987. Catecholamines in the serum of multiple trauma patients-mediators of ARDS? *Progress in Clinical and Biological Research* 236A:477–86.

Simberhoff, M. S. et al. 1986. Efficacy of pneumococcal vaccine in high risk patients: Results of a Veteran's Administration cooperative study. *New England Journal of Medicine* 315:1318.

Singer, F. R., and J. S. Adams. 1986. Abnormal calcium homeostasis in sarcoidosis. *New England Journal of Medicine* 315:755–56.

Chapter 9

Practical Applications: Arterial Blood Gases

The preceding chapters in this unit describe normal anatomy and function of the respiratory system, with emphasis on abnormal function. The purpose of this chapter is to discuss further the diffusion of gases between alveoli and pulmonary capillaries and to consider the effects of pulmonary disorders on blood levels of oxygen and carbon dioxide. Practical applications involve examples of disease processes discussed in relation to blood gases and other laboratory data.

This chapter deals with the process of gas exchange, the consequences of variations in both air flow and flow of blood in different regions of the lungs, and with abnormalities in the diffusion of gases.

GAS EXCHANGE

Oxygen is carried in the blood in two forms: (1) combined with the hemoglobin of red blood cells and (2) dissolved in the plasma of blood. Both carbon dioxide and oxygen dissolved in plasma are measured in terms of partial pressure expressed in mm of mercury, and the gases in alveoli are measured in the same way. The partial pressure of oxygen (pO_2) in pulmonary capillary blood is about 40 mm Hg, and pCO_2 is 46 mm Hg; while in alveolar air the pO_2 equals 100 mm Hg, and pCO_2 is 40 mm Hg. A pressure gradient is the driving force for diffusion; so oxygen diffuses into blood, and there is a net movement of carbon dioxide into alveolar air, both moving from areas of high to low pressure (figure 9.1). At the end of the capillary, the gases have equilibrated, so the pO_2 in blood and alveolar air are the same, and the pCO_2 on both sides of the respiratory membrane are equal.

The uptake of alveolar oxygen by capillary blood is determined by (1) a pO_2 pressure difference between plasma and alveolar air and (2) the combining of hemoglobin with oxygen, which removes oxygen from plasma. This second step is necessary for diffusion to continue (figure 9.2). Each hemoglobin molecule can bind up to four molecules of oxygen to form oxyhemoglobin. The alveolar pO_2 is usually not high enough to cause all of the hemoglobin molecules to be completely saturated with oxygen, so arterial hemoglobin is normally about 97% saturated.

A unique feature of hemoglobin-oxygen binding is that it is determined by pO_2, even though the relationship is not linear (chapter 17). As shown in figure 9.3, the percent saturation of hemoglobin remains relatively constant over a rather wide range of pO_2 values. With a pO_2 of 60 mm Hg, there is approximately 90% saturation of hemoglobin, as compared to a pO_2 of 100 mm Hg and 97% hemoglobin saturation. A normal range for arterial pO_2 values is 75–100 mm Hg, depending on age and assuming that the individual is breathing room air.

Changes in levels of carbon dioxide in both blood and alveolar air affect diffusion of carbon dioxide from blood

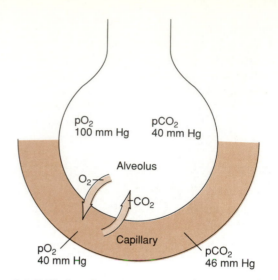

Figure 9.1 Diffusion of gases across a respiratory membrane involves movements down a pressure gradient, from high to low pressure.

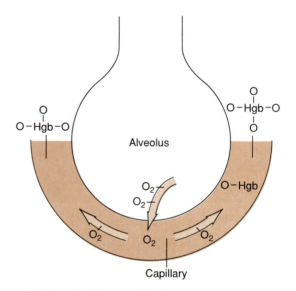

Figure 9.2 The chemical combination of oxygen and hemoglobin promotes continued diffusion of oxygen into capillaries.

into the air sacs. Hyperventilation increases the movement of air through the air passages and removes alveolar carbon dioxide, resulting in diffusion of this gas from blood into the air sacs. Hypoventilation causes an accumulation of carbon dioxide in alveolar air with subsequent inhibition of blood to alveolar diffusion.

RESPIRATORY FAILURE

The term respiratory failure means that pulmonary gas exchange is inadequate to meet the body's need for oxygen and the need to eliminate carbon dioxide. Blood gas values used to define respiratory failure are pO_2 less than 60 mm

Hg and pCO$_2$ greater than 50 mm Hg. These values are obtained while the individual is at rest and breathing room air. Respiratory failure occurs first upon physical exertion, progresses to failure at rest, and may be either acute or chronic. Hypoxemia affects the metabolism of all cells and, if severe, can lead to irreversible damage within minutes. Chronic, moderately low levels of pO$_2$ lead to pulmonary arteriolar vasoconstriction, increased pulmonary vascular resistance, and ultimately, right ventricular failure (cor pulmonale discussed in chapter 7). Hypercapnia causes respiratory acidosis with direct effects on the central nervous system (chapter 4).

VENTILATION/PERFUSION RATIO

The adequate exchange of gases depends on both ventilation and perfusion. Ventilation refers to flow of air, and perfusion refers to blood flow. Normally, the total flow of blood to the lungs is about 5 liters/min, and the volume of inspired air is about 5 liters/min. The overall ventilation/perfusion ratio ($\dot{V}a/\dot{Q}c$) is approximately 1. The symbol $\dot{V}a$ means alveolar gas flow, and the symbol $\dot{Q}c$ means capillary blood flow. At any moment those volumes may not be equal, although an equal balance between ventilation and perfusion is the ideal.

There are regional differences in ventilation and perfusion in the normal lung, with the upper areas being less well perfused as compared to the lower areas. Normally, these variations are not significant in terms of overall gas exchange.

VENTILATION/PERFUSION IMBALANCE

Mismatching of ventilation and perfusion leads to abnormal blood gas values. One important observation related to the ventilation/perfusion imbalance is that an increase in the $\dot{V}a/\dot{Q}c$ ratio in one region of the lung may not compensate for a decrease in this ratio in another area. Compensation is more effective for carbon dioxide as compared to oxygen. This means that hyperventilation involving nearly normal areas of the lung increases elimination of carbon dioxide, but increased alveolar pO$_2$ does not appreciably affect oxygen content of blood. The oxyhemoglobin dissociation curve illustrates that relatively large changes in pO$_2$ have small effects on the uptake of oxygen by hemoglobin. Figure 9.3 shows that in the upper, relatively flat portion of the curve, changes in pO$_2$ between 60–100 mm Hg are related to oxyhemoglobin saturation of 90% or more.

The following discussion is related to the effects of ventilation/perfusion imbalance on arterial blood gases.

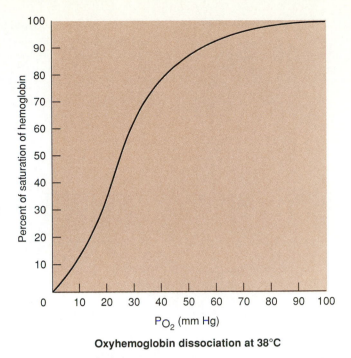

Figure 9.3 The oxyhemoglobin dissociation curve showing the percent of oxyhemoglobin saturation and the blood oxygen content, indicated for different values of pO$_2$.

DECREASED VENTILATION

A decrease in the $\dot{V}a/\dot{Q}c$ ratio occurs when there is hypoventilation, i.e., an inadequate flow of air through the air passages. The result is a lowered pO$_2$, a decrease oxyhemoglobin saturation, and an increase in pCO$_2$ (figure 9.4).

Decreased ventilation occurs in a number of situations as shown in table 9.1. Neuromuscular diseases that cause respiratory muscle weakness are associated with hypoventilation. Any condition or substance that depresses the respiratory center, such as narcotics, anesthesia, stroke, and meningitis, also lead to hypoventilation. Other possible causes include rigid thorax, severe pulmonary disease, and atelectasis (at″e-lek′tah-sis). Atelectasis refers either to collapse of the lungs or inadequate expansion of the lungs at birth. Decreased surfactant (chapter 8) or pneumothorax, air in the pleural cavity as the result of injury, or rupture of a lung abscess cause pulmonary collapse. Hypoventilation may occur during sleep and involve a condition called the sleep apnea syndrome.

Sleep Apnea Syndrome
The sleep apnea syndrome is associated with many disorders, including COPD, and involves recurrent episodes of hypoxemia and hypercapnia during sleep. The term apnea (ap-ne′ah) means to stop breathing, and sleep apnea syndrome is a group of disorders in which there are repeated periods of apnea and/or hypoventilation. There

apnea: Gk. *a*, not; Gk. *pnein*, breathe

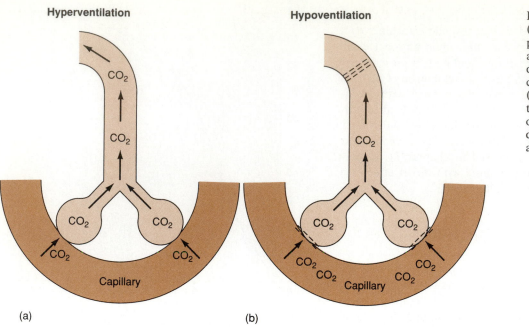

Hyperventilation

Hypoventilation

CO_2

CO_2

CO_2

CO_2

CO_2

CO_2

CO_2

CO_2

CO_2

CO_2

Capillary

CO_2

CO_2

Capillary

(a)

(b)

Figure 9.4
(*a*) Hyperventilation promotes removal of alveolar CO_2 and the diffusion of CO_2 from capillaries into air sacs. (*b*) Hypoventilation slows the movement of CO_2 out of alveoli and inhibits CO_2 diffusion from blood to alveoli.

TABLE 9.1 Causes of hypoventilation and the associated blood gas pattern

Causes of Hypoventilation

Respiratory muscle weakness

Depression of respiratory center

Rigid thorax (kyphoscoliosis)

Severe obstructive/restrictive pulmonary disease

Atelectasis

Sleep apnea syndrome

Blood Gases

Arterial pO_2 decreased

Arterial pCO_2 increased

TABLE 9.2 Blood gas values associated with uneven pulmonary capillary blood flow

Causes of Uneven Blood Flow

Thrombosis (blood clot)

Embolism (foreign particle in blood, such as blood clot, air, clumps of bacteria)

Blood Gases

Decreased oxyhemoglobin saturation

Decreased pCO_2 due to increased breathing rate

may be 200 events per night during sleep with a duration of at least 10 seconds. There are two types of apnea: (1) central apnea in which there is no airflow and no respiratory effort, evidently due to decreased respiratory drive and (2) obstructive apnea, more common in adults, in which airflow is absent in spite of respiratory effort. There is obstruction to the airways which may be caused by the tongue, obesity, narrowing of the airways, or enlarged tonsils. Individuals with obstructive apnea may have little evidence of airway obstruction during the day. There is cyclical hypoxemia and hypercapnia, depending on the degree of hypoventilation through the night. The periodic hypoxemia may lead to cardiac arrhythmias and pulmonary hypertension.

There are a number of clinical symptoms associated with the syndrome including snoring, morning headaches, sleep walking, daytime sleepiness, personality changes, and insomnia. About half the affected individuals are obese and about 10% have cardiopulmonary abnormalities. Modalities of treatment include surgical correction of narrowing of the upper airway, a permanent tracheostomy, and mechanical ventilation during sleep.

DECREASED PERFUSION

A pattern of uniform alveolar ventilation associated with variations in pulmonary capillary blood flow occurs in situations in which there is occlusion of a pulmonary vessel (table 9.2). This occurs, for example, in the case of pulmonary thromboembolism in which a blood clot interferes with blood flow. The result is an area in the lung with decreased blood flow and a high $\dot{V}a/\dot{Q}c$ ratio. Thrombi create an alveolar dead space, so called because the alveoli are ventilated but receive no blood supply. The result is constriction of airways in the affected area. A second consequence of diminished blood flow is, over a period of hours, loss of alveolar surfactant that leads to atelectasis

thromboembolism: Gk. *thrombos*, clot; Gk. *embole*, to throw in

RESPIRATORY SYSTEM

within 24–48 hours. Normally, there is constriction of blood vessels in the area where collapse of air sacs has occurred. Blood flow to this area may, however, be maintained because the embolic occlusion that blocks the flow of blood causes increased pressure that overcomes the vasoconstriction; thus, there may be continued perfusion to collapsed air sacs. The overall picture is one of imbalance of the $\dot{V}a/\dot{Q}c$ ratio, and this is a major factor in the development of hypoxemia. Although a normal pO_2 value may be associated with pulmonary embolism, massive embolism is usually characterized by hypoxemia, hypocapnia due to an increased breathing rate, and respiratory alkalosis.

Most pulmonary thrombi originate in the deep veins of the legs, which are then carried to the pulmonary circulation. There is increased risk of thrombi after surgery or with prolonged confinement to bed.

> Patient 9.1 was a female with a diagnosis of pulmonary embolism and infarction that occurred 3 days after surgery. Laboratory test results are as follows (normal values in parentheses).
> Arterial blood pH 7.45 (7.35–7.45), pCO_2 31 mm Hg (35–45), pO_2 68 mm (75–80).

The next patient below experienced dyspnea and hypoxemia 4 days after hip replacement surgery. She received 2 units of packed red blood cells and 2 units of blood postoperatively.

> Patient 9.2 was a 60-year-old female whose pulmonary arteriogram indicated she had a pulmonary embolism. Laboratory findings:
> Arterial blood pH 7.51 (7.35–7.45), pCO_2 34.4 mm Hg (35–45), HCO_3 27.4 meq/liter (22–30), pO_2 41 mm Hg (75–80), oxygen Hgb saturation 81.9% (94–95).

Most patients with pulmonary embolism have, when breathing room air, a pO_2 of less than 80 mm Hg. Frequently, the patient is hyperventilating, which lowers the pCO_2 and results in respiratory alkalosis. Both of these patients with pulmonary embolism have low values for pO_2 as well as pCO_2, and the pH value for the second patient indicates alkalosis.

IMPAIRED DIFFUSION

There may be impaired alveolar/capillary diffusion of gases in cases in which there is a thickening of the respiratory membrane, fluid accumulations, or the formation of fibrous connective tissue. Impaired diffusion also occurs when the disease process causes a decrease in surface area for gas exchange, i.e., destruction of pulmonary capillaries or alveolar walls. Individuals who experience impaired diffusion may hyperventilate; thus, blood levels of carbon dioxide may or may not be low, depending on breathing rate. The pCO_2 varies because carbon dioxide diffuses from the blood into the alveoli, even in cases of severe alveolar/capillary block. Table 9.3 lists the effects of impaired diffusion on blood gases.

TABLE 9.3 Blood gases associated with impaired alveolar/capillary diffusion

Causes

Thickened respiratory membrane: sarcoidosis, scleroderma of lungs

Decreased surface area for gas exchange: emphysema, pulmonary vascular disease

Blood Gases	Interpretation
Low pO_2	Impaired gas exchange
pO_2 and Hgb saturation normal (resting condition)	Patient tends to hyperventilate
pCO_2 normal	Diffusion of CO_2 into alveoli occurs even in cases of severe block
pCO_2 decreased	Patient hyperventilating

Adult respiratory distress syndrome (ARDS) is a disease process in which impaired diffusion of gases is the result of several factors. The initial event in ARDS is injury to the alveolar walls and capillary endothelium, resulting in increased capillary permeability and alveolar edema. Although not every case involves the following series of events, possible complications include (1) deposition of fibrin, (2) formation of a hyaline membrane lining the alveolar walls, (3) loss of surfactant, (4) stiff noncompliant lung, and (5) thickening of alveolar walls caused by fibrosis. Consider the following cases.

> Patient 9.3 was an adult male, an alcoholic, involved in a car accident. He sustained multiple injuries with blood loss. He developed ARDS in the intensive care unit, and blood tests showed the following results.
> Arterial blood pH 7.35 (7.35–7.45), pCO_2 27 mm Hg (35–45), pO_2 47 mm Hg (75–100), base excess −8.9 meq/liter (−2.5 to +2.5).
> The patient was put on positive end-expiratory pressure (PEEP). Two hours later:
> Arterial blood pH 7.3, pCO_2 38.2 mm, pO_2 185 mm. One hour later: Arterial blood pH 7.35, pCO_2 41.3 mm, pO_2 126 mm, base excess −2 meq/liter.

The base excess indicates metabolic acidosis due to lactic acid. Normally, a small amount of pyruvic acid undergoes (1) reduction to produce lactic acid, (2) lactic acid is then reoxidized to pyruvic acid by the liver, and (3) is metabolized to CO_2 and water. Shock or severe hypoxia causes an increased production of lactic acid and, in the case of ethanol intoxication, there is decreased metabolism of this acid. In patient 9.3, an accumulation of lactic acid is predictable.

There is hypoxemia and hypocapnia with a pH at the lower limit of normal. The pH represents the combined effects of respiratory alkalosis and metabolic acidosis. The hypoxemia is the result of pulmonary edema, atelectasis, and the formation of intraalveolar hyaline membrane. These factors all contribute to decreased diffusion and

$\dot{V}a/\dot{Q}c$ imbalance. Hyperventilation is the reason for the lowered pCO_2. In a later stage of ARDS, hypercapnia is predictable when hyperventilation no longer eliminates accumulating carbon dioxide. Hypoxemia caused by ARDS responds to **positive end-expiratory pressure** (PEEP) mechanical ventilation. This type of ventilation maintains pressure at the end of exhalation and minimizes both airway closure and atelectasis. Patient 9.4 was treated by a PEEP ventilator, and there was an improvement in pCO_2, pO_2, and base excess.

> Patient 9.4 was an elderly male who had undergone a gastrectomy (chapter 34). He developed staphylococcal pneumonia postoperatively and subsequently developed ARDS. Laboratory findings:
>
> pO_2 41 mm Hg (75–100), pCO_2 38 mm Hg (35–45), arterial blood pH 7.41 (7.35–7.45), base excess 0.5 meq/liter (+2.5 to −2.5) HCO_3^- 24.3 meq/liter (22–30), oxygen Hgb saturation 76.6% (90–100).
>
> The patient was given oxygen with a face mask. Two hours later:
>
> pO_2 67 mm Hg (75–100), pCO_2 35.9 mm Hg (35–45), arterial blood pH 7.42 (7.35–7.45), base excess 0.2 meq/liter (+2.5 to −2.5), HCO_3^- 23.6 meq/liter (22–30), oxyhemoglobin saturation 93.4% (90–100).

The characteristic gas exchange abnormality in ARDS is a low pO_2, related to the absence of oxygenation of blood in the areas where alveoli are filled with fluid. In other less affected regions of the lung, ventilation/perfusion mismatching occurs and contributes to hypoxemia as well. Laboratory data indicate the absence of metabolic acidosis and a normal pCO_2 in patient 9.4. Early in the course of the disease process, the critical value is pO_2. The patient is usually hyperventilating at this stage and the pCO_2 may be in the normal range, as seen in the case of patient 9.4.

> Twenty-four hours later: Patient on PEEP ventilation, pO_2 92 mm Hg, pCO_2 51.8 mm Hg, arterial blood pH 7.32.

In ARDS, after 24–48 hours, hyaline membranes form on the inner lining of the air sacs. The membranes consist of edema fluid with high protein content, fibrin, and remnants of necrotic epithelial cells. The consequent loss of surfactant leads to atelectasis, and the result is a noncompliant lung. An increase in pCO_2 indicates alveolar hypoventilation.

OBSTRUCTIVE/RESTRICTIVE PATTERNS

Increased resistance to airflow occurs in asthma and emphysema (figure 9.5) and, there is uneven distribution of air because the severity of the narrowing varies among the many thousands of air passages. In the case of asthma, obstruction is due to bronchospasms, viscous mucus secretions, and edema and inflammation of the airway mucosa. There are regions of the lung with blood flow to underventilated areas, i.e., a decreased $\dot{V}a/\dot{Q}c$ ratio, re-

Figure 9.5 Typical posture of a patient with chronic obstructive pulmonary disease.

sulting in hypoxemia and little change in pCO_2. The pCO_2 decreases if the patient is hyperventilating. Increased breathing rate may not compensate during a prolonged attack and there may be an increase in the blood content of carbon dioxide (table 9.4).

Destruction of alveolar walls and their associated capillaries occurs in emphysema. The result is both impaired diffusion and abnormal $\dot{V}a/\dot{Q}c$ ratios. There may also be collapse of weakened airways during expiration, which leads to resistance to airflow. Overall, in the late stages of COPD, there is inadequate alveolar ventilation with both hypoxemia and hypercapnia.

Restrictive pulmonary disorders involve limited expansion of the lungs (table 9.5), and the effects on arterial blood gases are determined by complications that may include impaired diffusion, obstruction, or hypoventilation.

The following are cases of COPD that show abnormalities of arterial blood gases.

> Patient 9.5 was a middle-aged male admitted to the hospital with a diagnosis of pneumonia and with a 12-year history of COPD. He had smoked for 35 years and was currently smoking one pack a day. He complained of chest pain and cough for 6 days. The following laboratory values were obtained at admission.
>
> Arterial blood pH 7.4 (7.35–7.45), pCO_2 32 mm Hg (35–45), pO_2 59 mm Hg (75–80).

The pH is within normal range, the pCO_2 is somewhat decreased, and the pO_2 is low. In COPD, there is obstruction to airflow caused by narrowing of airways, decreased elasticity, and collapse of airways during forced expiration. In emphysema, alveoli are enlarged, which decreases surface area for diffusion. Fluid in the interstitium

TABLE 9.4 Examples of an obstructive pattern for abnormal blood gases

Emphysema*

Blood Gases	Interpretation
Increased pO_2, decreased pCO_2	Large part of lung not well ventilated, but there is good blood flow
Normal pO_2, pCO_2	Blood flow to poorly ventilated alveoli reduced in proportion to air flow
Decreased pO_2, normal pCO_2	Uneven ventilation/blood flow ratios

Asthma

Blood Gases	Interpretation
Decreased pO_2	May be caused by hypoventilation, uneven ventilation, or impaired diffusion
Normal pCO_2	Regions with increased ventilation/perfusion ratios have compensated for regions with decreased ratios

May show a pattern for both hypoventilation and obstruction

TABLE 9.5 Restrictive pulmonary diseases and abnormal blood gas values

Causes

Limited expansion of the lungs:

Scleroderma	Pleural effusion
Interstitial fibrosis	Pneumothorax
Abdominal mass with elevation of diaphragm	

Blood Gases

Increased pCO_2 only if there is severe restriction of thoracic movements resulting in hypoventilation

Abnormalities are due to complications of impaired diffusion, airway obstruction, or hypoventilation

and alveoli due to pneumonia interferes with gas exchange. This leads to an overall reduction of oxygen uptake and carbon dioxide elimination due to a decrease in the ventilation/perfusion ratio. A compensatory response to increasing pCO_2 is hyperventilation, and this causes normal areas of the lungs to eliminate carbon dioxide. Increased ventilation in these areas does little to improve oxygenation because there is not a linear relationship between increased ventilation and hemoglobin oxygen saturation. The disease processes limit increased ventilation in affected areas. The kidney is involved in maintaining a normal pH by controlling bicarbonate retention or loss.

The following is a similar case in which the pCO_2 is within normal range, and the pO_2 is low. The pH is somewhat elevated, and this may represent overcompensation by the kidney.

Patient 9.6 was an 80-year-old female admitted to the hospital with diagnosis of COPD. She was not in pain but her breathing was labored. Laboratory test results upon admission are as follows.

Arterial blood pH 7.48 (7.35–7.45), pO_2 53 mm Hg (75–80), pCO_2 37.2 mm Hg (35–45).

Patient 9.7 was a 77-year-old male with a diagnosis of COPD. He was wheezing and having difficulty breathing when admitted to the hospital. He was lethargic and increased anterior-posterior chest diameter was noted. Laboratory test results follow.

Arterial blood pH 7.17, pCO_2 59 mm, pO_2 60 mm, and 81.9% oxygen saturation.

Aminophylline was administered intravenously, and oxygen was administered.

Arterial blood pH changed to 7.13, pCO_2 80 mm, pO_2 163 mm, and 98.2% oxygen saturation.

After sodium bicarbonate was given intravenously:

Arterial blood pH 7.24, pCO_2 78 mm, pO_2 150 mm, 98.4% oxygen saturation.

Oxygen must be administered judiciously to individuals who have chronically elevated pCO_2 levels. The respiratory center becomes insensitive to increased carbon dioxide as a stimulus for respiration, and at that point, the respiratory drive depends on low pO_2. The desensitization of the respiratory center to carbon dioxide is called **carbon dioxide narcosis,** and if the respiratory center responds only to low oxygen levels, the administration of oxygen can remove any respiratory stimulus. After being given oxygen, patient 9.7 showed an increased pCO_2 due to increased respiration. There was respiratory acidosis and sodium bicarbonate was administered in an effort to return the pH to a normal range. Aminophylline was given because it is a bronchodilator. In evaluating a chronic condition, the laboratory values should be considered in terms of change from what is normal for that individual. The pH may be normal in COPD because of compensatory renal retention of bicarbonate.

Patient 9.8 was a 65-year-old female with hoarseness, weight loss, and severe dyspnea. Cytological examination of sputum revealed malignant cells. The diagnosis was adenosquamous carcinoma, stage III (T1, MX, N1) with bilateral involvement. Laboratory findings:

WBC 12,500/mm³ (4,800–10,800), Hct 46% (37–47), pH 7.37 (7.35–7.45), pCO_2 46 mm Hg (35–45), pO_2 72 mm Hg (75–100) Na+ 122 meq/liter (137–47), Cl− 80 meq/liter (100–10).

A small percentage of cases of bronchogenic carcinoma are identified as adenosquamous carcinoma. This term is used because both cell types are present in great numbers (chapter 8). A chest X-ray of this patient showed a right pleural effusion and no obvious pulmonary mass. In most cases in which pleural effusion is associated with lung cancer, the underlying cause is directly attributable to the tumor. Sixteen hundred milliliters of fluid were withdrawn from the right pleural cavity of this patient by way of a chest tube, and her breathing became less labored. A bronchodilator, aminophylline, was administered. The arterial blood gases, obtained upon hospital admission, reflect inadequate ventilation, most likely due to restricted respiratory movements. The hyponatremia may be dilutional, the result of the syndrome of inappropriate antidiuretic hormone (chapter 3).

Two days later: Arterial blood pH 7.01, pCO_2 140 mm Hg, HCO_3^- 25 meq/liter (22–30).

Patient 9.8 went into acute respiratory failure. The increase in pCO_2 is associated with a normal bicarbonate level, and consequently, there was a fall of pH. When respiratory failure develops slowly, there is time for renal compensatory changes, resulting in an increase in bicarbonate and a normal or near normal pH.

SUMMARY

• • •

A pressure gradient is the driving force for diffusion. Oxygen moves into blood and carbon dioxide crosses the respiratory membrane into alveolar air. Respiratory failure refers to inadequacy of pulmonary gas exchange leading to hypoxemia and hypercapnia. Equally matched ventilation and perfusion is the ideal for efficient alveolar/capillary gas exchange, and mismatching results in abnormal blood gas values. In the case of $\dot{V}a/\dot{Q}c$ imbalance, hyperventilation involving normal areas of the lung increases the elimination of carbon dioxide, but it does not appreciably affect oxygen content of blood.

Hypoventilation, inadequate airflow through the airways, leads to a lowered pO_2, a decrease in oxyhemoglobin saturation, and an increase in pCO_2. The underlying cause of hypoventilation is inadequate respiratory movement. The sleep apnea syndrome, involving repeated periods of apnea and/or hypoventilation, may be included in this category.

Pulmonary thromboembolism interferes with blood flow and leads to an increased $\dot{V}a/\dot{Q}c$ ratio. Massive embolism causes hypoxemia, hypocapnia due to an increased breathing rate, and respiratory alkalosis. Impaired alveolar/capillary diffusion is the result of thickening of the respiratory membrane, fluid accumulations, and decreased surface area for gas exchange. In such cases, there may be hypoxemia and normal pCO_2, although increased breathing rate causes variations in blood gas values.

Obstructive/restrictive patterns of blood gases depend on the severity of the disease process. In the late stages of COPD, there is inadequate alveolar ventilation with hypoxemia associated with hypercapnia. Arterial blood gases in restrictive pulmonary disorders are determined by complications that may include impaired diffusion, obstruction, or hypoventilation.

REVIEW QUESTIONS

• • •

1. Give an example of a pressure gradient involving the diffusion of oxygen.
2. What is a limiting factor for the diffusion of oxygen into blood?
3. When there are some unventilated respiratory units, why does increased ventilation in other regions of the lung not prevent hypoxemia?
4. Does hyperventilation decrease blood pCO_2 when there are areas of the lung unventilated? Explain.
5. What is the basic cause of a decrease in the ventilation/perfusion ratio?
6. In general, what effect does a decrease in the ventilation/perfusion ratio have on arterial blood gases?
7. Identify three causes of hypoventilation.
8. What are two possible complications of periodic hypoxemia associated with the sleep apnea syndrome?
9. List two respiratory consequences of pulmonary thromboembolism.
10. What is a characteristic blood gas pattern associated with massive embolism?
11. What acid-base imbalance typically accompanies a pulmonary embolism, and what is the cause?
12. What are two general causes of impaired alveolar/capillary diffusion?
13. Why, in the case of impaired diffusion, may the pCO_2 be either normal or decreased?
14. List three factors in obstruction associated with asthma.
15. What are two possible causes of a low pO_2 in asthma?
16. What is a typical blood gas pattern in late stages of COPD?
17. What is the mechanism in restrictive pulmonary disease that may cause hypoventilation?
18. Why is metabolic acidosis associated with hypoxia?
19. What is the renal response to an elevation of pCO_2?
20. What two mechanisms cause the sleep apnea syndrome?

SELECTED READING

• • •

Bach, J. R. et al. 1987. Intermittent positive pressure ventilation via nasal access in the management of respiratory insufficiency. *Chest* 92:168–70.

Fiaccadori, E. et al. 1987. Skeletal muscle energetics, acid-base equilibrium and lactate metabolism in patients with severe hypercapnia and hypoxemia. *Chest* 92:883–87.

Hoflin, F., and A. Thomi. 1985. Pulmonary embolism caused by high ski boots. *New England Journal of Medicine* 312:1645.

Kupfer, Y. et al. 1987. Disuse atrophy in a ventilated patient with status asthmaticus receiving neuromuscular blockade. *Critical Care Medicine* 15:795–96.

Lane, R. et al. 1987. Arterial oxygen saturation and breathlessness in patients with chronic obstructive airways disease. *Clinical Science* 72:693–98.

Pepe, P. E. et al. 1984. Early application of positive end-expiratory pressure in patients at risk for the adult respiratory-distress syndrome. *New England Journal of Medicine* 311:281–86.

Rose, C. E. et al. 1983. Synergistic effects of acute hypoxemia and hypercapnic acidosis in conscious dogs. *Circulatory Research* 53:202.

Udwadia, Z. F. et al. 1987. Radiation necrosis causing failure of automatic ventilation during sleep with central sleep apnea. *Chest* 92:567–69.

Warner, G. et al. 1987. Effect of hypoxia-inducted periodic breathing on upper airway obstruction during sleep. *Journal of Applied Physiology* 62:2201–11.

GENITOURINARY SYSTEM

Unit III combines the reproductive and the excretory systems. The first part of chapter 10 includes a brief review of both the anatomy of the female reproductive system and the menstrual cycle. This is followed by a discussion of nonmalignant gynecologic disorders, malignant gynecologic disorders, and anatomy and disorders of the breast. Both the nonmalignant and the malignant gynecologic disorders include dysfunctions of the uterus, fallopian tubes, and ovaries. A discussion of two benign disorders of the female breast, galactorrhea and fibrocystic breast disease, is followed by a consideration of the etiology, diagnosis, manifestations, prognosis, and treatment of breast cancer. The second part of chapter 10 reviews male reproductive anatomy and nonmalignant and malignant diseases of the male reproductive system. These diseases include disorders of the penis, the testes, and the prostate gland. The emphasis is on testicular tumors and prostatic adenocarcinoma.

Chapter 11 introduces excretory function with a review of anatomy and normal function of the excretory system. The topics related to normal function include (1) urine formation, (2) the countercurrent mechanism, (3) hormonal influences, and (4) control of acid-base balance. The remainder of the chapter deals with evaluation and control of renal function. The topics of evaluation and control include symptoms, diagnostic tests, imaging techniques, and diuretic control.

Chapter 12 deals with injury of the urinary tract by infection, obstruction, and nephrotoxins. The first section describes urethritis, cystitis, and pyelonephritis as consequences of infection. The second part of the chapter discusses obstruction to urinary flow as caused by nephrolithiasis, prostatic hypertrophy, and malignancies. The last part of the chapter deals with chemical agents that damage the kidney.

Chapter 13 focuses on glomerular damage due to immune responses, which is a continuation of the discussion of urinary tract infections in the preceding chapter. The first part of the chapter deals with the nature of immunological injury, and the remainder of the chapter deals with acute glomerulonephritis, rapidly progressive glomerulonephritis, and Goodpasture's syndrome. The last topic of the chapter is the nephrotic syndrome, characterized by increased glomerular permeability to proteins.

Chapter 14 deals with diseases of the kidney, both primary and those secondary to other systemic disease processes. The primary renal diseases are polycystic kidney disease, renal tubular acidosis, nephrogenic diabetes insipidus, and renal cell carcinoma. The secondary renal diseases discussed in the last part of this chapter include the following: nephrogenic diabetes insipidus, hemolytic-uremic syndrome, diabetic nephropathy, malignant hypertension, and scleroderma.

Chapter 15 describes decline in renal function, a consequence of excretory dysfunction and renal disease. The first part of the chapter deals with acute and chronic renal failure. The last part of the chapter discusses hemodialysis and peritoneal dialysis as a means of treating renal failure.

Chapter 16 illustrates previously discussed principles and introduces new information related to the significance of laboratory findings. This chapter integrates information from preceding chapters by using case histories. The chapter gives examples of acute and chronic renal failure, diabetic nephropathy, and renal transplant rejection, each with laboratory data. The laboratory test results are evaluated in terms of contributing factors. The discussion includes metabolic acidosis, BUN and creatinine levels, anion gap, electrolyte balance, alkaline phosphatase level, and red blood cell count. The discussion is based on concepts in unit I involving electrolyte and acid-base balance. The discussion is also related to chapter 11, the topics of renal control of acid-base balance and laboratory evaluation of renal function. There is a summary of reasons for laboratory test results associated with renal failure.

Chapter 10

Disorders of the Reproductive Systems

The term genitourinary (jen″ĭ-to-u′rĭ-nar-e) refers to the relationship between the reproductive and excretory systems, both of which are discussed in this unit. This introductory chapter focuses on disorders of the male and female reproductive systems. The remainder of the chapters in the unit deal with various aspects of the process of excretion. The following section includes a brief review of gynecologic anatomy and physiology and a discussion of both malignant and nonmalignant disorders of the female reproductive system.

ANATOMY OF THE FEMALE REPRODUCTIVE SYSTEM

External female genital organs are known collectively as the **vulva** or **pudendum.** The vulva includes the structures shown in figure 10.1. The **mons pubis** is a prominence over the symphysis pubis that is covered with hair after puberty. There are two folds, the **labia majora** and the **labia minora,** that extend back from the mons. The external portion of the labia majora is covered with hair, while the inner surfaces are smooth and have sebaceous glands. The labia minora are smaller, inner folds that lack hair. The labia minora surrounds the **clitoris,** a small erectile structure that is homologous to the male penis. Posterior to the clitoris, the female urethra opens to the exterior. The area between the labia minora is called the **vestibule.** The opening or orifice of the vagina is bordered by the hymen and occupies most of the vestibule. The **vestibular bulb** refers to elongated masses of erectile tissue located on either side of the vaginal orifice. Mucus secreting glands called **paraurethral** (pair″ah-u″re′thrahl) **glands** are located on either side of the urethral orifice. **Greater vestibular glands** or **Bartholin's glands** on either side of the vaginal orifice secrete a lubricating mucus into the space between the hymen and labia minora.

• • •

Bacterial infection may lead to either a Bartholin's gland cyst or abscess. An abscess, which becomes swollen and painful, is the result of dilation of the glandular ducts due to obstruction. A chronic bacterial infection leads to a fluid-filled cyst. The most common cause of either condition is gonorrhea, although other types of bacterial infections also occur.

• • •

genito: L. *genitalis,* belonging to birth
gynecologic: Gk. *gynaikos,* woman
pudendum: L. *pudere,* to be ashamed
mons: L. *montanus,* mountain
labia: L. *labium,* lip
vestibule: L. *vestibulum,* porch

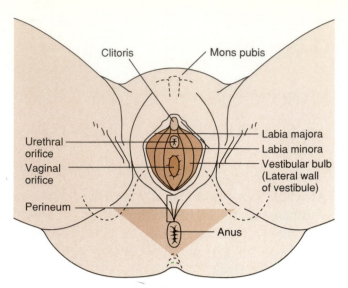

Figure 10.1 External female genitalia.

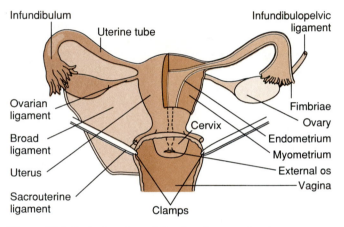

Figure 10.2 Posterior view of the female reproductive organs.

Internal female organs of reproduction are shown in figures 10.2 and 10.3. Paired **ovaries** are the female gonads that produce secondary oocytes, which upon maturation, become ova or egg cells. **Uterine tubes,** also called **fallopian tubes** or **oviducts,** transport ova from the ovaries to the uterus. The distal end of each tube is funnel-shaped and surrounded by fingerlike projections called **fimbriae** (fim′bre-ee). The **uterus** is between the urinary bladder and the rectum and is the site of implantation of a fertilized egg. The lower cylindrical portion of the uterus is the neck or the **cervix.** The vagina leads from the uterus to the vestibule of the vulva. The vaginal orifice is surrounded by a mucous membrane called the hymen.

MENSTRUAL CYCLE

The menstrual cycle involves monthly changes in the **endometrium** of a nonpregnant uterus. Table 10.1 summarizes the events of the cycle.

endometrium: Gk. *endon,* within; Gk. *metra,* womb

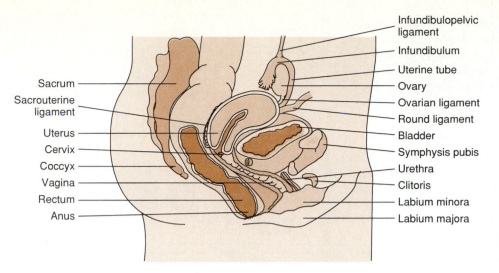

Figure 10.3 Midsagittal section of female reproductive organs.

TABLE 10.1 A summary of hormonal control of the menstrual cycle

Hormone	Source	Effect
Gonadotropin releasing hormone (GnRH)	Hypothalamus	Stimulates release of FSH and LH
Follicle-stimulating hormone (FSH)	Anterior pituitary	Stimulates initial development of ovarian follicles
Luteinizing hormone (LH)	Anterior pituitary	Stimulates further follicle development and ovulation; stimulates production of estrogens and progesterone
Estrogens	Ovaries	Maintain endometrium; inhibit GnRH
Progesterone	Ovaries	With estrogens, prepares endometrium for implantation and mammary gland for lactation

MENSTRUAL PHASE

A menstrual cycle requires an average of 28 days. Menstruation usually occurs during the first 5 days of the cycle. The major events of the menstrual phase are as follows. Sudden reduction of estrogens and progesterone cause a discharge of blood due to degeneration of the endometrial lining. Primary follicles begin to develop and start to produce low levels of estrogens. During the last 2 days of menstruation, the primary follicles develop into secondary

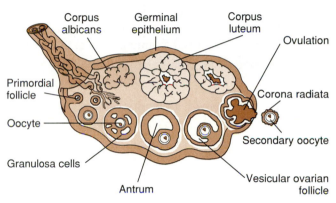

Figure 10.4 A schematic diagram of an ovary showing the various stages of ovum and follicle development.

follicles in which there is a secondary oocyte and a stratified layer of epithelial cells called **granular cells** or **granulosa cells.** The secondary follicles also produce estrogens. Cells outside the follicles condense into a layer around the follicle called the theca. Granulosa-theca ovarian tumors are discussed later in this chapter. Figure 10.4 shows various stages of the development of the ovum and the follicle.

PREOVULATORY PHASE

The period between menstruation and ovulation is called the preovulatory phase, which involves approximately 6–13 days. During this time, one secondary follicle matures to form a **Graafian follicle.** This structure is relatively large, contains an immature ovum, and secretes increasing amounts of estrogens. There is a high level of follicle-stimulating hormone during this phase. Secretion of luteinizing hormone increases immediately preceding ovulation.

A high level of estrogens during the preovulatory phase promotes repair of the endometrial lining and inhibits the hypothalamic secretion of gonadotropin releasing hormone (GnRH). Inhibition of GnRH leads to diminished secretion of FSH. High levels of estrogens also stimulate the anterior pituitary to secrete increased amounts of luteinizing hormone, and this triggers ovulation. Following the release of the ovum from the Graafian follicle, the follicle collapses and develops into the **corpus luteum.**

POSTOVULATORY PHASE

The postovulatory phase completes the menstrual cycle and involves approximately day 15 to day 28 of the cycle. After ovulation, luteinizing hormone stimulates the development of the corpus luteum, and this body secretes increasing amounts of progesterone and estrogens, which leads to inhibition of GnRH and luteinizing hormone secretion. During this phase, there is a gradual increase in the level of FSH and a gradual decrease in the level of LH. The corpus luteum degenerates to form the **corpus albicans.** Finally, the decreased production of progesterone and estrogens results in the initiation of another menstrual cycle. Figure 10.5 summarizes the events of the menstrual cycle in relation to the ovarian cycle.

NONMALIGNANT GYNECOLOGIC DISORDERS

Numerous infectious agents invade the female reproductive tract including the herpes simplex virus, *Neisseria gonorrhea,* the fungus *Candida albicans,* the protozoan *Trichomonas vaginalis,* and *Chlamydia trachomatis.* The vulva, which is the portal of entry for such agents, is a common site of lesions. The herpes simplex virus is the most common pathogen. Vaginal inflammation is often caused by *Candida albicans* and *Trichomonas vaginalis,* with frequent involvement of the vulva and cervix as well. Postpartum bacterial sepsis leads to an acute inflammation of the uterine lining called acute **endometritis,** with a streptococcus or an anaerobic *Bacteroides* species as likely causes. Chronic endometritis may be the result of recent abortion or delivery or the presence of an intrauterine contraceptive device. The fallopian tubes may be infected from the lower genital tract with a resulting inflammatory condition called **salpingitis.**

The term **pelvic inflammatory disease** refers to infection of the uterus, fallopian tubes, and peritoneum that leads to inflammation of both these and nearby structures. The source of the infection may be the lower genital tract or, less commonly, such conditions as appendicitis or inflammatory bowel disease.

In addition to infectious diseases, there are various other nonmalignant disorders of the female reproductive tract that are discussed in the following paragraphs.

corpus luteum: L., yellow body
corpus albicans: L., white body

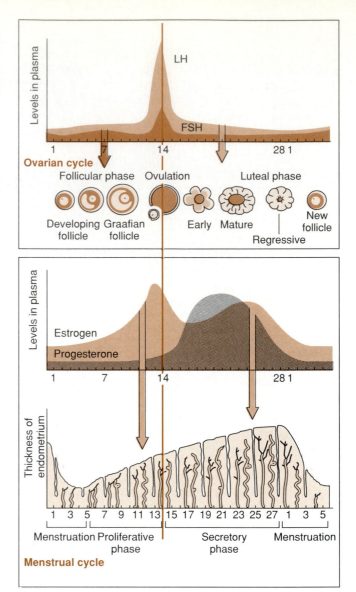

Figure 10.5 Phases of the menstrual cycle in relation to ovarian changes and hormone secretion, and the relationship between changes in the ovaries and the endometrium of the uterus during different phases of the menstrual cycle.

DYSMENORRHEA

Dysmenorrhea, which refers to painful menstrual cramps, affects about 50% of young females and may be accompanied by nausea, vomiting, and diarrhea. About 75% of the cases are the result of excessive prostaglandin production or absorption. Endometrial tissue is one source of prostaglandins, and it has been shown that two types of endometrial prostaglandins are increased in women who suffer from dysmenorrhea. Mechanisms that cause pain include the sensitization of pain fibers by prostaglandins and the effects of **cyclic endoperoxides,** intermediate products in prostaglandin synthesis. The endoperoxides cause pain directly and also increase myometrial contractility, which leads to ischemia and pain.

dysmenorrhea: Gk. *dys,* bad or painful

Prostaglandin inhibitors treat dysmenorrhea successfully in 80% of the cases. Nonsteroidal anti-inflammatory drugs (NSAIDs) inhibit the production of endoperoxides and prostaglandins. Examples of drugs that have the property of inhibiting prostaglandin synthesis include aspirin, indomethacin, tolmetin, ibuprofen, and naproxen.

ENDOMETRIOSIS

The term endometriosis refers to the implantation of normal endometrial tissue outside of the uterus. The sites of implantation include the pelvic peritoneum and the pelvic organs as well as other sites, such as the lungs and abdominal wall. How this ectopic implantation occurs is unclear. One theory suggests that endometrial tissue, during menstruation, is regurgitated through the fallopian tubes onto the ovaries and other pelvic organs. There may be spread by way of lymphatic or blood vessels. A second theory proposes **in situ** formation of the tissue from undifferentiated peritoneal cells. Chronic irritation, either infection or menstrual blood, is a possible stimulus. It is likely that hereditary tendencies and defects in cellular immunity are factors as well, and the term for such an event is **coelomic metaplasia.**

Endometriosis affects women during their reproductive years, with an incidence of 4–17% in the general population. There is a much higher incidence among infertile women. The lesions have a central area that is red, purple, or brown, depending on the amounts and age of the blood. Fibrosis occurs and leads to dense adhesions around the fallopian tubes, the intestinal tract, and elsewhere.

The symptoms of endometriosis depend on the location and extent of the lesions. If the ectopic endometrial tissue responds to the cyclic ovarian hormonal changes, there may be dysmenorrhea, infertility, and premenstrual staining. Cyclic swelling and bleeding within the implants and excessive prostaglandin production are reasons for secondary dysmenorrhea. Adhesions and changes in tubal motility due to prostaglandins may lead to infertility. Adhesions may cause pelvic pain, pain during defecation, dysuria, and dyspareunia (dis''pah-roo'ne-ah) or painful intercourse.

Endometriosis is often treated with oral contraceptive pills, which suppress endometrial tissues. Obstructing lesions may be excised surgically. In extreme cases, an oophorectomy or removal of ovaries may be required.

LEIOMYOMA

Leiomyomas (li''o-mi-o'mahs), also called **fibroids,** are benign tumors of smooth muscle. This is a common uterine tumor that occurs in 20–30% of women past the age of 30 and occurs more commonly in blacks than in whites. The tumors are sensitive to the stimulation of estrogens and enlarge during pregnancy and shrink after menopause. They may be replaced by fibrous tissue and calcification

dyspareunia: Gk. *dyspareunos,* badly mated

may occur. Each tumor may be a few millimeters or may weigh many kilograms. In most cases, the tumors are asymptomatic. They may contribute to infertility and abortion and cause bleeding. In some cases, there may be pressure on the bladder causing urinary frequency or pain. A hysterectomy may be required to treat the condition.

ENDOMETRIAL HYPERPLASIA

Endometrial hyperplasia is a condition in which the endometrium is thickened due to an increase in the number of normal cells of the uterine lining. It usually affects women during menopause. Mild hyperplasia is characterized by hyperplasia of the endometrial glands and stroma. In moderate hyperplasia, the endometrial glands are of irregular size. Atypical **hyperplasia** is characterized by abnormal cells. Moderate and atypical hyperplasia are believed to be premalignant conditions in which the transition to a malignancy may require a period of years.

TUBAL ECTOPIC PREGNANCY

A tubal ectopic pregnancy is the implantation of a fertilized ovum in one of the fallopian tubes rather than in the uterus. This abnormality is more likely to occur if the fallopian tubes have been affected by inflammation. Although many patients do not present with typical symptoms, the implantation may lead to amenorrhea, pain, and vaginal bleeding. Excision of the implanted tissue may be successful, although there is a high rate of repeat ectopic pregnancy that accompanies the procedure. If the tube ruptures and there is massive bleeding, a **salpingectomy** is required.

POLYCYSTIC OVARY DISEASE

The clinical features of polycystic ovary disease (PCOD) include enlarged ovaries, increased cellularity or hyperplasia of ovarian stroma, and a layer of small follicles beneath a thickened capsule. The normal ovary at the time of puberty shows a lesser degree of enlargement and has many large follicles scattered throughout the tissue. PCOD is associated with multiple follicular cysts lined with granulosa and theca cell layers. Androgen levels are elevated, and there is a high ratio of luteinizing to follicle-stimulating hormone. The normal function of follicle-stimulating hormone (FSH) is to promote the development of mature follicles. The function of luteinizing hormone (LH) is to stimulate the rupture of the follicle to release an oocyte; thus, the hormone imbalance associated with PCOD interferes with ovulation.

ectopic: Gk. *ektopos,* displaced
salpingo: Gk. *salpinx,* tube; Gk. *ektome,* excision

Polycystic ovary syndrome, as first described by Stein and Leventhal in 1935, involved infertility, secondary amenorrhea, hirsutism, and obesity. These manifestations were observed in a group of young women who had pale, cystic ovaries.

· · ·

The manifestations of polycystic ovary disease may include infertility, amenorrhea, obesity, and **hirsutism** (her′soot-izm) or abnormal hairiness. PCOD affects young women and, in some cases, may be hereditary.

Hyperthecosis is a condition that may be associated with PCOD and is characterized by an increase in the number of theca cells. There is a marked decrease in the number of follicles and ovaries are normal in size. The ovaries may be solid or partly cystic and abnormal stromal cells produce androgens that lead to masculinization, including clitoral hypertrophy, frontal balding, and a deepening of the voice. Except for the process of masculinization associated with hyperthecosis, the condition is similar to PCOD.

Etiology
The cause of PCOD may be a primary or a secondary hypothalamic defect in the secretion of releasing factors that control pituitary LH and FSH. Excess androgen production by the adrenal glands is a possible cause. Obesity may be a factor in PCOD due to the fact that adipose tissue converts androgens to estrogens and the estrogens, in turn, stimulate the inappropriate release of pituitary LH and inhibit the secretion of FSH. Increased levels of LH cause hyperplasia of both ovarian stroma and theca cells, and these cells produce androgens, thus perpetuating the cycle. The maturation failure of follicles is the result of decreased FSH levels.

Treatment
Treatment of PCOD involves measures to correct abnormal hormonal responses. (1) Estrogen-progestogen oral contraceptives suppress ovarian androgen secretion. Weight loss contributes to a decrease in peripheral estrogen production. (2) Clomiphene citrate, which is an antiestrogenic drug, may be administered to induce ovulation and restore normal menstrual periods. (3) Ovarian wedge resection, a procedure whereby a triangular wedge of tissue is removed, reduces hormonal secretions and leads to more normal function of the pituitary and hypothalamus. (4) Surgical removal of the ovaries may be required for treatment of hyperthecosis.

hirsutism: L. *hirsutus*, shaggy

MALIGNANT GYNECOLOGIC DISORDERS

Cancer of the female reproductive tract is a significant cause of mortality. A cancerous condition is named for the type of cell from which the malignancy originates. For example, a **sarcoma** originates from mesenchymal cells, a **carcinoma** originates from epithelial cells, and an **adenocarcinoma** is a type of carcinoma that shows evidence of glandular formation. The degree of cellular differentiation, i.e., characteristics of specialized function, is significant. Undifferentiated malignant cells, in general, are associated with a less optimistic prognosis. The term **neoplasm** refers to any new, abnormal growth and this uncontrolled growth may be either benign or malignant. **Hyperplasia** (hi″per-pla′ze-ah) describes an abnormal increase in the number of cells in a tissue, **dysplasia** (dis-pla′se-ah) is an alteration of size and shape of adult cells, and **metaplasia** (met″ah-pla′ze-ah) refers to abnormal changes in mature cells. **In situ** identifies a growth that is confined to the site of origin in contrast to the process of **metastasis** in which the growth has invaded other tissues. The following paragraphs deal with malignancies that affect the uterus, the fallopian tubes, and the ovaries.

CARCINOMA OF THE CERVIX

The incidence of carcinoma of the cervix in the United States has been declining as the result of screening and successful early treatment. In contrast, the incidence in developing countries remains high. Squamous cell carcinoma makes up about 90% of all cervical neoplasms, and adenocarcinomas and mixed tumors constitute most of the remainder of cervical malignancies.

Dysplasia of the cervix is a condition in which the majority of epithelial surface cells resemble cancer cells. These abnormal cells vary in size and shape and have a relative abundance of cytoplasm. The sequence of maturation from the basal layer to the surface is disorderly. Dysplasia is graded as slight, moderate, severe, or **carcinoma in situ.** Carcinoma in situ indicates that all cells from the basement membrane to the surface have the appearance of cancer cells. The abnormal cells are more uniform in size and shape than in dysplasia, there is scant cytoplasm, and a maturation sequence is not apparent.

Although the etiology is unknown, studies show that there is a link between squamous cell carcinoma of the cervix and (1) multiple sexual partners at an early age, (2) low socioeconomic conditions, (3) smoking, and (4) venereal infections. Viral studies show a link between cervical neoplasia, papilloma viruses, and preexistent infections.

hyperplasia: Gk. *hyper*, above; Gk. *plassein*, to form
dysplasia: Gk. *dys*, bad
metaplasia: Gk. *meta*, after
in situ: L., in the normal place or confined to site of origin

Staging

Physical examination and X-ray studies are the basis for staging of cervical carcinoma. Stage O is a preinvasive stage, and stage Ia is a microinvasive stage associated with a small probability of lymph node involvement. Stage Ib is the gross malignant involvement of the cervix involving lesions larger than 7 mm. In the more advanced clinical stages, there is spread to other structures. In stage II, for example, the vagina is involved; in stage III, there is spread to the lower third of the vagina and to the pelvic wall. Stage IV represents the invasion of the bladder or rectum or spread to distant organs.

Symptoms

Cytologic examination of a cervical smear, the **Papanicolaou** (pap″ah-nik″la′o͞oz) smear, aids in the identification of the disease in early, preinvasive stages. This is an important screening test because cervical carcinoma may be asymptomatic. In most cases of cervical carcinoma, abnormal findings are limited to the pelvis. The common symptoms are vaginal bleeding, perceived as abnormal menstrual bleeding or bleeding following intercourse, and possibly a vaginal discharge. The cervix may appear to be normal or may be hard, or there may be evidence of an invasive lesion in the form of ulcer or a tumor.

Metastasis of squamous cell cervical carcinoma most commonly involves direct infiltration of the vaginal mucosa and then spread to adjacent ligaments. This leads to ureteral obstruction. When the condition is untreated, bilateral ureteral obstruction occurs and is followed by renal failure. Spread by way of blood and the lymphatic system may occur as well. In advanced stages of the disease, it is common for metastasis to occur and to involve pelvic organs, lungs, and bone, particularly the vertebrae. Liver, brain, and skin are less commonly affected. In general, adenocarcinomas of the cervix metastasize more aggressively than squamous cell carcinomas. The evidence of metastatic disease includes lymphadenopathy, swollen legs, ascites (chapter 36), hepatomegaly, pleural effusion, or rarely, skin metastases.

Treatment

The treatment for adenocarcinoma and squamous cell carcinoma of the cervix is usually similar, and the primary treatment is either surgery or radiation therapy. Early stages of the disease can be treated, with equal success, by surgery or radiation therapy. Surgery may be preferred in young patients because ovarian function is affected by radiation therapy.

The extent of invasion by the tumor determines whether the tumor is excised or a hysterectomy is performed and whether or not pelvic lymph nodes are removed. A **radical hysterectomy** involves the removal of the uterus, adjacent ligaments, and the upper one-third to one-half of the vagina. Postoperative radiation therapy is indicated if the patient is at high risk for recurrence or if pelvic lymph nodes are involved.

Chemotherapy may be administered under several circumstances:

1. When there is a recurrence of the tumor after radiation therapy
2. May be combined with radiation therapy when there is a high probability of failure with radiation alone
3. May be administered prior to radiation therapy in the treatment of stage IIb or stages more advanced than stage IIb

ENDOMETRIAL CARCINOMA

Endometrial carcinoma, involving the inner lining of the uterus, has become an increasingly common malignancy, with an incidence in the United States that is now higher than cervical carcinoma. The disease has a peak incidence among women who are 50–70 years of age and is more likely to affect females who have no children. The frequency of this type of cancer increases with age in association with such risk factors as hypertension, diabetes mellitus, and obesity. Stimulation of the endometrium by either exogenous or endogenous estrogens unopposed by progesterone is believed to be a cause. It appears that estrone is the significant estrogenic hormone that promotes endometrial cancer. Estrone is produced by the conversion of adrenal or ovarian androgens by adipose tissue. This may explain the link between obesity and endometrial carcinoma. Other possible sources of estrogens include an estrogen-secreting ovarian tumor and exogenous estrogen prescribed for postmenopausal symptoms. In certain cases, there is increased susceptibility to endometrial cancer in families. This is sometimes associated with susceptibility to breast cancer.

Most endometrial tumors are adenocarcinomas, graded 1 to 3 according to the degree of differentiation. These tumors may occasionally be squamous cell carcinomas or may be of an adenosquamous type, as well as other rare cytologic patterns.

Staging

The degree of spread of the cancer determines the staging of endometrial carcinoma. Stage I represents confinement of the disease process to the body of the uterus, spread to the cervix is stage II. In stage III, there has been metastasis to the pelvic region; and in stage IV, metastasis has progressed to involve the rectum and areas outside the pelvis.

Symptoms

Abnormal bleeding is a symptom of endometrial carcinoma, and the cancer should be suspected in younger women with recurrent heavy menses as well as in older

endometrium: Gk. *endon,* within; Gk. *metra,* uterus

women with postmenopausal bleeding. The cancer frequently appears as a diffuse thickening of the endometrium that spreads over the entire surface and then ulcerates. The uterine muscle, the myometrium, may be invaded. The pattern of spread involves lymphatic vessels first. Later, there is dissemination by way of blood vessels to the lungs, liver, and bones and other parts of the body.

Treatment

The basic treatment for all stages of endometrial carcinoma is total abdominal hysterectomy and surgical removal of fallopian tubes and ovaries, a **salpingo-oophorectomy** (o″of-o-rek′to-me). All stages except well-differentiated stage I are treated postoperatively by radiation therapy.

Prognosis for endometrial carcinoma is influenced by various factors. The prognosis for a well differentiated tumor is favorable, compared to a poor prognosis for an undifferentiated tumor. If the cancer is confined to the endometrium, nearly 90% of patients survive 5 years. If the inner half of the myometrium is involved, survival is about 70% at 5 years and is less than 15% if metastasis has occurred outside the uterus.

FALLOPIAN TUBE ADENOCARCINOMA

Adenocarcinoma of the fallopian tubes is the most aggressive and least common neoplasm of the female genital tract. The cause is usually metastasis from another site, frequently the genital tract. The etiology of primary lesions is unknown, although there is evidence to suggest a relationship between chronic inflammation and malignancy.

Symptoms include pain and vaginal discharge, both of which are characteristic of tubal inflammation as well as of adenocarcinoma. An uncommon symptom is sudden copious watery discharge accompanied by relief of pain. Only about one patient in five lives for 5 years after the diagnosis of fallopian tube carcinoma.

OVARIAN NEOPLASMS

The ovary is a common site for both benign and malignant tumors. The majority of ovarian malignancies are primary, although about 6% are the result of metastasis from sites, such as the colon, stomach, or breast. Women with no children or with breast cancer are at increased risk for ovarian cancer. The early symptoms of the disease include abdominal discomfort, loss of appetite, vaginal discharge, and abnormal uterine bleeding. Modalities of treatment depend upon the specific type of tumor involved.

The following are types of ovarian cells that may give rise to neoplasms:

1. Mesothelium covers the surface of the ovary, as well as other organs in the abdominal cavity.
2. Oocytes are germ cells that have the potential for reproducing tissues of all germ layers.
3. Ovarian stromal cells adjacent to oocytes specialize to form the granulosa and theca cells that produce female sex hormones. Unspecialized ovarian stromal cells produce collagen and respond to hormonal stimuli to produce steroid hormones. The ovarian stroma also contains smooth muscle fibers.
4. Large cells with abundant cytoplasm called hilum cells are associated with nonmyelinated nerve fibers in the hilum of the ovary.

Major categories of ovarian neoplasms include tumors of the surface epithelium, germ cell tumors, stromal tumors, and lipid cell tumors (including hilum cell tumors).

Surface Cell Epithelial Tumors

Surface epithelial tumors are the most common group of ovarian tumors and cause more deaths than any of the other malignancies of the female genital tract. The ovarian mesothelial covering is considered the tissue of origin.

There are three main groups of epithelial tumors categorized on the basis of the appearance of the neoplastic cells. Those tumors in which the cells resemble cells of the fallopian tubes are called **serous tumors.** The cells of **mucinous tumors** resemble those of the endocervix. **Endometrioid tumors** have a cellular resemblance to the endometrium. In all three categories the tumors may be benign **cystadenomas** that have low malignant potential or may be **adenocarcinomas,** also called **cystadenocarcinomas** (sis-tad″e-no-kar″si-no′mahs).

Papillary cystadenocarcinoma is the most common primary cancer of the ovary, making up about 40% of ovarian malignancies. Cysts are common, and there is usually abundant papillary ingrowths, both within the cysts and projecting from the surface. The most malignant tumors are either solid or have many solid areas. Metastasis to lymph nodes and to distant sites occurs.

The majority of endometrioid tumors are malignant, and they make up about 20% of all cancers of the ovary. Microscopically, they resemble adenocarcinoma of the endometrium. This type of carcinoma is often a combination of cysts filled with brown or bloody fluid and solid areas. The cysts are lined with papillae that give a velvety appearance. In about one-third of the cases, there is an association between the tumor and endometriosis.

Serous cystadenoma of the ovary is the most common type of benign tumor of the ovary. This neoplasm usually occurs during the reproductive years, and in one out of five cases, involves both ovaries. Serous tumors form fluid-filled cysts with smooth outer surfaces and papillary ingrowths within the cyst.

Mucinous cystadenoma is found in an older age group and is less common as compared to serous tumors. The tumor usually involves only one ovary, the cysts are filled with a sticky mucin, and sparse papillary ingrowths are present. A possible complication of this condition is the escape of mucin into the peritoneal cavity resulting in multiple tumor implants on the peritoneal surface. The final outcome is the binding together of abdominal viscera. The occurrence of borderline and malignant mucinous tumors is less common than the serous types.

Treatment for epithelial ovarian cancer in an early stage is surgery, i.e., salpingo-oophorectomy and hysterectomy. The 5-year survival rate is 90%. In 60% of the cases, however, metastasis has occurred. When the pelvis is involved, surgery is followed by radiation therapy and/or chemotherapy. Spread of the malignancy to the abdominal cavity or to distant sites is treated by reduction of the tumor mass followed by chemotherapy. There is a 20–30% 5-year survival rate in the case of distant metastasis.

Tumor Markers

Research has been directed toward the identification of any substance, or tumor marker, that may appear in the serum in direct association with epithelial ovarian cancer. Two-thirds of the patients with this disease are in an advanced stage at the time of diagnosis, and a reliable tumor marker would be useful for early detection. At the present time, there is no ideal marker for the disease.

Elevations of **carcinoembryonic antigen** (CEA) have been shown in 20–70% of patients with epithelial ovarian cancer. CEA levels, however, are dependent on the tumor cell type and the degree of differentiation. Patients with mucinous tumors or undifferentiated tumors are more likely to have positive values. CEA monitoring may be most helpful in patients with mucinous tumors.

A protein called CA-125 has been studied as a possible tumor marker. This protein is present in coelomic (se-lom′ik) epithelial tissue during embryonic development and is found in certain adult tissues, although it is not found in the ovary. It is possible to monitor CA-125 in serum and body fluids by immunoassay using a monoclonal antibody (chapter 28), OC-125, that binds to the CA-125 protein. Studies have shown elevated CA-125 levels in up to 80% of patients with ovarian cancer. The studies also have shown a good correlation between CA-125 levels and either regression or progression of the disease. CA-125 is, however, not specific for epithelial ovarian cancer and may be elevated in various diseases including liver disease, pancreatitis, renal failure, and gynecologic cancers other than ovarian cancers.

Germ Cell Tumors

Germ cell tumors originate from cells that produce the oocytes, and the tumor cells retain the capacity to produce various types of tissues in tumor development. Most tumors are **benign cystic teratomas** that make up 20% of all ovarian tumors in adults and half of all ovarian tumors in children. Most patients are between the ages of 20 and 40, although this type of tumor occurs at all ages.

The cysts that develop are filled with hair and cellular secretions, particularly sebaceous gland secretions. Various types of tissue protrude from the inner surface of the cysts, i.e., salivary gland, fat, smooth muscle, cartilage, bone, neural tissue, and others. **Solid teratomas** are nearly always found in children and young women. They are malignant, and usually involve one ovary.

Dysgerminomas are germ cell tumors that are large encapsulated masses of soft, gray-white tissue that are often hemorrhagic and necrotic. This type of tumor usually affects young women and, in about 10% of cases, involves both ovaries. **Embryonal carcinoma** is an uncommon germ cell tumor that affects young patients. The tumor causes precocious puberty in prepubertal girls. Most of these tumors are unilateral.

The treatment for germ cell tumors depends on the specific type of tumor. Dysgerminomas are successfully treated with oophorectomy and radiation therapy, even in the case of metastasis. There is a 5-year survival rate of 70–90% for these patients. Germ cell tumors in an early stage, other than dysgerminomas, are treated by removal of the affected ovary. In the case of metastasis, chemotherapy is used for treatment.

Sex Cord Stromal Tumors

Sex cord stromal tumors arise from specialized stromal cells that produce ovarian steroid hormones. A specific type of sex cord stromal tumor is a **granulosa-theca** cell tumor. This type of tumor may be composed of granulosa or theca cells, or a combination of the two. Granulosa cell tumors are usually a combination of cysts with multiple hemorrhagic areas and solid, yellow-brown areas. Theca cell tumors, or **thecomas,** are solid and yellow or yellow-white and benign. Typically, granulosa-theca cell tumors produce estrogenic hormones and, less commonly, may produce androgens. There is often endometrial hyperplasia with the estrogen-secreting tumors. Uterine bleeding is the most common symptom.

About 5% of ovarian tumors are large fibrous tumors called **fibromas.** These tumors are sex cord stromal tumors that do not produce hormones. **Sertoli-Leydig cell tumors** arise from the same female sex cord stromal cells as granulosa-theca cell tumors. This type of tumor often produces androgens that have masculinizing effects, although many of the tumors either produce estrogens or are hormonally inert. Nearly all Sertoli-Leydig cell tumors are benign.

Stromal tumors in general have a better prognosis than some of the other types of tumors because the hormone-secreting cells are relatively well differentiated. Treatment for this type of ovarian cancer is surgery.

Gonadoblastoma

Gonadoblastoma is a rare type of ovarian tumor that contains both immature germ cells and sex cord-stromal cells. The patients are usually phenotypic females that have a Y chromosome. Most gonadoblastomas are benign.

coelomic: Gk. *koiloma*, cavity (abdominal)

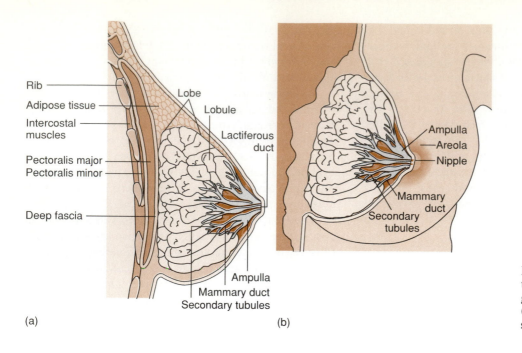

Labels on figure (a): Rib, Adipose tissue, Intercostal muscles, Pectoralis major, Pectoralis minor, Deep fascia, Lobe, Lobule, Lactiferous duct, Ampulla, Mammary duct, Secondary tubules

Labels on figure (b): Ampulla, Areola, Nipple, Mammary duct, Secondary tubules

(a)   (b)

Figure 10.6 The structure of the breast and mammary glands. (*a*) A sagittal section. (*b*) An anterior view partially sectioned.

Lipid Cell Tumors

A hilum cell tumor is an example of lipid cell tumors. These occur in the form of soft yellow or brownish nodules that are usually benign.

MAMMARY GLAND ANATOMY

Mammary glands are located within the breast and are modified **sudoriferous** or sweat glands (figure 10.6). The glands are composed of 15–20 lobes or compartments separated by adipose tissue. Each lobe is composed of several smaller compartments called **lobules.** The lobules contain milk secreting cells called **alveoli.** Milk is carried by a series of secondary tubules to **lactiferous ducts** that terminate at the nipple. The lactiferous ducts are enlarged near the nipple to form milk storage areas called **ampulla.**

DISORDERS OF THE BREAST

The following discussion includes both benign and malignant disorders of the female breast.

GALACTORRHEA

Galactorrhea is a condition in which milklike material with fat globules is discharged from the nipples. The cause is unknown in most cases although galactorrhea may occur at the cessation of oral contraceptive medication or may represent a persistent discharge after pregnancy. A blue or brown nipple discharge, usually involving one breast, is

mammary: Gk. *mamma*, breast
sudoriferous: L. *sudor*, sweat; L. *facere*, to make
alveoli: L. *alveolus*, small cavity

likely to be a sign of benign changes in the breast. A bloody discharge is most likely evidence of a benign intraductal papilloma, which is a benign tumor. The pituitary hormone prolactin is required for normal lactation, and a prolactin-secreting pituitary tumor is a possible cause of galactorrhea.

FIBROCYSTIC BREAST DISEASE

The term fibrocystic breast disease refers to local or diffuse lumps in the breast as the result of (1) fibrosis, (2) cyst formation, (3) an increase in the number of epithelial cells called epithelial hyperplasia, or (4) adenosis in which there is an increase in the number of glandular units in the breast. Breast biopsy may be required to confirm the absence of cancer and, of such biopsies, about 30% represent an increased risk for carcinoma.

BREAST CANCER

Breast cancer affects one out of nine women in the United States and is the most common cause of death in females between the ages of 35 and 50.

Most of the breast carcinomas originate from epithelial cells of the ducts, and the tumors are called **intraductal tumors** if they are confined to the duct lumen with no invasion of stroma or are called **infiltrating duct carcinomas** if stromal tissue has been invaded. Early metastasis and a less favorable prognosis is associated with the infiltrating type tumor. The majority of breast cancers are infiltrating adenocarcinomas that form glandular structures from ductal or lobular epithelial cells.

Etiology

The etiology of breast cancer is unknown, although various risk factors have been identified:

1. Early menarche or late menopause, both of which involve prolonged exposure to estrogen
2. First pregnancy after the age of 30
3. Obesity or tall stature, perhaps due to dietary or hormonal factors
4. Western culture, possibly related to high dietary intake of fat and dairy products
5. Family history of breast cancer (Risk is highest if a close relative has bilateral breast cancer during premenopausal years.)
6. Exposure to ionizing radiation
7. Prolonged postmenopausal estrogens
8. Benign breast disease with atypical epithelial hyperplasia

Diagnosis

Self-examination is an important aspect of early detection of breast cancer, and the appearance of a palpable breast mass requires further evaluation by a physician. A mass that is not malignant may vary with the menstrual cycle and is likely to be most obvious in the week preceding menses. If the mass remains unchanged through a menstrual period, cytological evaluation may be performed by means of needle aspiration or surgical biopsy of the tissue. Periodic mammograms may be recommended, depending on the risk factors involved. Although false negative readings may occur, mammography makes it possible to detect deep, small malignancies within breast tissue.

Breast tumor tissue may be analyzed for cytoplasmic **estrogen receptors** and **progesterone receptors.** These receptors are proteins that bind to and transfer the hormones to the nuclei of cells. The presence or absence of the receptors is significant in terms of prognosis and the treatment of the disease.

Manifestations

A painless mass in the breast is the most common early indication of breast cancer. Most carcinomas are hard and have an irregular border. Possible manifestations of the disease include breast pain, skin dimpling, and nipple discharge. Skin edema, ulceration, and enlarged lymph nodes are manifestations of more advanced breast cancer.

Prognosis

The course of the disease in patients with breast cancer varies with the individual. The disease may be slowly progressive or may lead to death within a few months.

The stage of the disease at the time of diagnosis is an important factor in the prognosis. The clinical stages of breast cancer follow:

1. *Stage I* Tumor is less than 2 cm, and there is no evidence of lymph node involvement or of distant metastasis.
2. *Stage II* Tumor is less than 5 cm; any palpable lymph nodes are not fixed; there is no distant metastasis.
3. *Stage III* Tumor is more than 5 cm, or the tumor has invaded the skin or is attached to the chest wall; supraclavicular lymph nodes are involved; no distant metastasis.
4. *Stage IV* Distant metastasis.

The 5-year survival rate for the disease in stage I is in the range of 82–94%, whereas stage IV is associated with a 5-year survival rate of 10%.

The number of affected axillary lymph nodes is a factor in the prognosis. Increased numbers of positive lymph nodes represent increased risk of recurrence of the disease. A high degree of tumor differentiation is associated with a lesser degree of risk of recurrence. One indication of differentiation is the presence of hormone receptors, and this is indicative of a more hopeful outlook. When receptor positive tumors metastasize, bone and soft tissue is likely to be affected, whereas receptor negative tumors tend to metastasize to viscera.

Treatment

Surgery is the initial treatment for breast cancer and, in the case of stage I or stage II disease, usually involves a modified radical mastectomy with preservation of chest muscles and removal of axillary lymph nodes, or there may be excision of the tumor followed by radiation of the breast. More advanced tumors are treated by surgery and radiation therapy for local control.

It has been recognized in recent years that there is micrometastasis during early stages, i.e., spread of the cancer on a microscopic level. Based on the assumption that there is early micrometastasis, chemotherapy or endocrine manipulation are postsurgical modalities of treatment for those patients at high risk for early recurrence. The overall survival rate of premenopausal women is significantly improved by chemotherapy involving three to five drugs, whereas evidence indicates that postmenopausal women receive the most benefit from treatment with tamoxifen, a drug that blocks the effects of estrogen.

The approach to treatment of metastatic breast cancer is determined by the presence or absence of estrogen receptors and the status of the patient with regard to menopause. The treatment may involve chemotherapy or endocrine manipulation. The latter includes oophorectomy, the antiestrogenic drug tamoxifen, androgens, and aminoglutethimide (inhibits steroid synthesis) plus hydrocortisone.

A bilateral oophorectomy may be the initial treatment for a patient who (1) has metastatic disease, (2) has a tumor that is either estrogen receptor positive or whose receptor status is unknown, and (3) is premenopausal. The elimination of ovarian estrogen and estrogen precursors is thought to remove a source of stimulation of breast cancer growth. There is a positive response to this treatment in about 50% of those who have estrogen receptors and in less than 10% of those who are estrogen receptor negative. In some cases, the antiestrogenic drug tamoxifen is preferred to oophorectomy. Antiestrogenic drugs block the

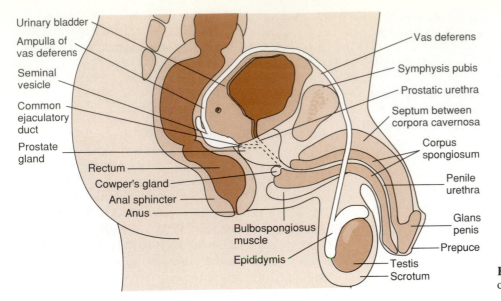

Figure 10.7 Midsagittal view of male reproductive organs.

Labels in figure: Urinary bladder; Ampulla of vas deferens; Seminal vesicle; Common ejaculatory duct; Prostate gland; Rectum; Cowper's gland; Anal sphincter; Anus; Bulbospongiosus muscle; Epididymis; Vas deferens; Symphysis pubis; Prostatic urethra; Septum between corpora cavernosa; Corpus spongiosum; Penile urethra; Glans penis; Prepuce; Testis; Scrotum

effects of nonovarian estrogens by inactivating estrogen receptors. These drugs may be administered to those patients who suffer a relapse following a positive response to oophorectomy. If there is an initial response to oophorectomy followed by a relapse, the drug aminoglutethimide plus hydrocortisone is an additional option. Aminoglutethimide blocks the production of both adrenal androgens and the enzymatic conversion of androgens to estrogens in peripheral tissues. The patients who do not respond to hormonal manipulations are treated with chemotherapy.

Tamoxifen is used as the initial treatment for postmenopausal women with metastatic breast cancer. If the tumor is estrogen receptor positive, the response rate is approximately 50%. If, however, the tumor is estrogen receptor negative, the response rate is less than 10%. In those cases in which there is an initial response followed by a relapse, the other approaches to endocrine manipulation may be used.

Chemotherapy is the initial treatment for patients with estrogen receptor negative tumors or for any patients that have a life-threatening condition.

ANATOMY OF THE MALE REPRODUCTIVE SYSTEM

External male genitalia include a scrotal sac, which encloses a pair of testes (tes'tēz), and the penis (figure 10.7). The expanded tip is called the **glans penis** which is covered by a circular fold of skin called the **prepuce** (pre'pūs). The penis contains three cylindrical bodies, each held together by a sheath of connective tissue. The two dorsal cylindrical bodies are called the **corpora cavernosa.** The single ventral body is called the **corpus spongiosum** through which the urethra passes. The distal end of the corpus spongiosum forms the glans of the penis. The three cylindrical bodies contain spongelike spaces that fill with blood and cause an erection of the penis during sexual arousal.

The **testes,** also called **testicles** (figure 10.8), are divided into compartments or lobules, each of which contains coiled **seminiferous tubules** where sperm production or **spermatogenesis** occurs (figure 10.9). Seminiferous tubules from all compartments form a network of tubules toward the posterior part of the testis called the **rete testis.** These tubules empty into **vasa efferentia** that leave the testis and enter the **epididymis.** The epididymis, located in the scrotum, is an elongated structure that fits tightly against the posterior surface of the testis. The epididymis consists of coiled tubules where sperm maturation occurs. In the lower region, tubules called **vas deferens** transport sperm away from the epididymis. The vas deferens pass through the body wall into the pelvic cavity. The vas deferens, between the epididymis and the body wall, is next to blood and lymph vessels and nerves. These vessels and nerves are enclosed in a sheath of connective tissue and are called the **spermatic cord.** Inside the cavity, the vas deferens crosses over the top of the ureter and down the posterior surface of the urinary bladder. The terminus of the vas deferens is enlarged to form an **ampulla.** Both vas deferens are joined by the duct of a seminal vesicle to form a short **ejaculatory duct,** which passes through the prostate gland and empties into the prostatic urethra. This is continuous with the penile urethra that leads to the outside of the body.

Accessory glands, i.e., seminal vesicles, the prostate gland, and bulbourethral glands (Cowper's glands), contribute alkaline secretions to seminal fluid. There are two **seminal vesicles** lateral to the vas deferens and located on the posterior surface of the urinary bladder. The duct of each seminal vesicle joins a ductus deferens to form an ejaculatory duct that, in turn, enters the urethra below the point of exit from the urinary bladder. The **prostate gland** is the largest of the three glands, is located in front

vas deferens, pl. vasa deferentia

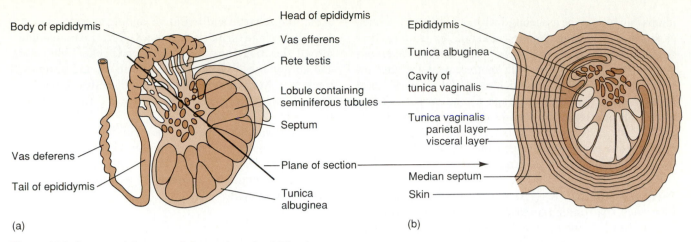

Figure 10.8 Structural features of the testis and epididymis. (*a*) A longitudinal view. (*b*) A transverse view.

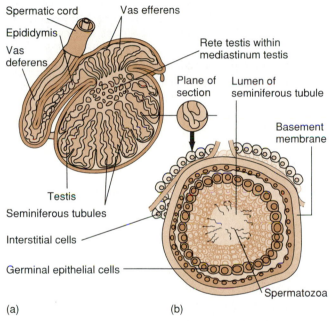

Figure 10.9 A diagrammatic representation of seminiferous tubules. (*a*) A sagittal section of a testis. (*b*) A transverse section of a seminiferous tubule.

of the rectum, and surrounds the urethra just below the bladder. The paired **bulbourethral glands** are below the prostate on either side of the urethra.

NONMALIGNANT MALE REPRODUCTIVE DISORDERS

The remainder of the chapter deals with various disorders of the male reproductive system. The following paragraphs focus first on nonmalignant disorders.

DISORDERS OF THE PENIS

This section deals with disorders of the penis involving infection, edema, and prolonged erection.

Balanitis and Posthitis

The terms balanitis (bal″ah-ni′tis) and posthitis (pos-thi′tis) refer to inflammation of the glans penis and the foreskin respectively. They usually occur together and are the result of poor hygiene. The foreskin becomes edematous and erythematous, and local ulcers may occur. In mild cases, retraction of the foreskin, cleansing, and application of an antibiotic ointment may be sufficient. Severe inflammation may require circumcision after the swelling diminishes.

Phimosis

Phimosis (fi-mo′sis) is a condition in which the prepuce is tight and cannot be retracted over the glans penis. The condition may be congenital or acquired. Urinary retention may occur if the prepuce is adherent. Immediate relief may be obtained by dilatation with a hemostat or a dorsal slit procedure may be required, i.e., an incision in the dorsum of the prepuce. Circumcision will provide permanent relief.

Paraphimosis

The term paraphimosis (par″ah-fi-mo′sis) is used to describe a situation in which the prepuce cannot be pulled back to cover the glans penis. Permanent retraction of the prepuce may occur after masturbation, intercourse, or may follow catheter insertion. The tissue becomes edematous and tender, and infarction is a possibility. It may be necessary to inject a local anesthetic at the base of the penis and squeeze out the edema fluid in order to pull the prepuce forward. If this procedure is unsuccessful, an emergency dorsal slit procedure may be required.

Priapism

Priapism (pri′ah-pizm) refers to prolonged erection due to failure of drainage of venous blood. The condition may be

balanitis: Gk. *balanos*, an acorn
posthitis: Gk. *posthe*, foreskin
paraphimosis: Gk. *para*, beside; Gk. *phimosis*, closure
priapism: L. *priapismus*, abnormal erection

idiopathic or may be associated with other disorders, such as sickle cell anemia, tumor, leukemia, or trauma. Persistent priapism leads to fibrosis of the corpora cavernosa and impotence. The approach to treatment includes instillation of heparin, fibrinolytic therapy (chapter 19), local application of ice, and drainage of the corpora. There should be an attempt to irrigate the corpora free of clots with saline. If these efforts are unsuccessful, a surgical procedure is carried out to provide internal drainage of the obstructed corpora cavernosa.

Peyronie's Disease

Peyronie's (pa-ron-ez') disease involves fibrosis of the cavernous sheath. This causes a thickening that usually begins in the septum between the two corporal bodies, involving the dorsal side of the penis. The thickened areas or plaques vary in size. The erectile tissue is usually not involved. The manifestations of this disorder include a mass in the penis, curvature on erection, pain due to erection, and difficulty in achieving penetration during intercourse. The condition is treated if it interferes with intercourse. Vitamin E or para-aminobenzoate, oral agents, have been used for treatment. If the problem persists, it is possible to surgically macerate the plaque and insert a prosthesis.

DISORDERS OF TESTES

The disorders of the testes included in this section are (1) failure of testes to descend into the scrotum, (2) inflammation of epididymis, (3) inflammation of the testes, (4) twisted spermatic cord, (5) fluid-filled mass in testes, (6) dilatation of veins of testes, and (7) fluid-filled cyst of epididymis.

Cryptorchidism

Cryptorchidism (krip-tor'ki-dizm) is a condition in which one or both testes fail to descend into the scrotum during the course of fetal development. A majority of those that are undescended at birth descend during the first year following birth. An undescended testis does not mature after puberty. There is an increased incidence of malignancy associated with cryptorchidism, and this risk remains, even after surgical correction.

Epididymitis

The most common cause of scrotal swelling is epididymitis, an inflammation of the epididymis that may be either acute or chronic. The cause in most cases is reflux of urine into the epididymis, although the lymphatic system may also be the origin of the infection. The inflammation frequently occurs after strenuous physical activity, urethral instrumentation, or prostatectomy.

The symptoms of the acute condition include sudden onset of pain in the scrotum, extreme sensitivity of the epididymis, and swelling of the testis due to congestion. The patient may or may not be febrile. There may also be

cryptorchidism: Gk. *kryptos,* hidden; Gk. *orchis,* testis

urethral discharge and irritative voiding symptoms, such as urgency and urinary frequency. Treatment includes scrotal elevation and support to provide relief, analgesics and antipyretics to control pain and fever, and antibiotic treatment. Most cases of acute epididymitis are resolved within 3 weeks, either spontaneously or after antibiotic treatment.

In chronic epididymitis, the epididymis is enlarged, but tenderness is not always present. Antibiotics may be less effective in the chronic than in the acute condition, and anti-inflammatory agents, such as ibuprofen, may be used for treatment.

Orchitis

Orchitis is a rare condition in which the testis becomes inflamed. This inflammation may follow epididymitis or a viral infection. The most common cause is mumps, in which case it is likely to be unilateral. There is sudden onset of pain, the testis becomes enlarged and tender, and the scrotum may become erythematous and edematous. Urinary tract symptoms are absent. If the inflammation is associated with mumps, there may be a few red blood cells and some protein in the urine. Infiltration of the spermatic cord with a local anesthetic provides pain relief. Modalities of treatment are bed rest, ice packs, and analgesics. Of patients with mumps orchitis, 25–35% experience loss of spermatogenesis.

Testicular Torsion

Testicular torsion involves a twisted spermatic cord, a condition that is common in young men and in children. A congenital anatomic abnormality that allows the testis to rotate around the spermatic cord may be the cause. The testis can rotate during violent exercise or following trauma as well. The result is twisting of both the spermatic cord and its blood vessels, thus causing veins and arteries to become obstructed. The testis then becomes congested and the interrupted blood supply, i.e., infarction, may lead to irreversible damage to the testis.

Typically, there is onset of acute testicular pain during physical activity or at rest, the testis swells, and the urinalysis is clear. Epididymitis is associated with testicular torsion in postpubertal males. Elevation of the testis usually increases the pain of testicular torsion, whereas elevation relieves pain in the case of epididymitis. Ultrasound is useful in distinguishing between epididymitis and testicular torsion.

Treatment involves immediate surgery to confirm the nature of the problem and to untwist the testis and to secure the testis to the scrotal wall. A high incidence of testicular infarction is associated with failure to treat the condition within 4–6 hours.

Hydrocele

A hydrocele (hi''dro-sēl) is a fluid-filled mass that surrounds the testis. It does not result in tenderness and is

hydrocele: Gk. *hydor,* water; Gk. *kele,* tumor

the result of an accumulation of fluid between the two layers of the tunica vaginalis. This condition develops most commonly in older men and usually is asymptomatic. A hydrocele may be the result of a congenital defect or may be a complication of other conditions, such as orchitis, epididymitis, tumor, or trauma. The occurrence of a hydrocele in a younger man suggests underlying disease. Treatment is not required if the mass is small and is causing no problems. Surgical excision of the hydrocele sac relieves discomfort associated with a large mass.

Varicocele

A varicocele (var'ĭ-ko-sēl'') is a dilatation of the veins leading away from the testes. Possible causes include thrombosis or compression of the vena cava. On the left side, the spermatic vein empties into the left renal vein, and a thrombus may cause an obstruction. A varicocele may be asymptomatic or may cause discomfort or a sense of heaviness. A varicocele is more common on the left than the right side, and about 15% of males have some degree of a left varicocele.

Support of the scrotum usually provides relief of symptoms, and treatment is aimed at correcting the cause.

Spermatocele

The term spermatocele (sper'mah-to-sēl) refers to a dilatation of the epididymis that causes a mass that is usually at the head of the epididymis. This dilatation is a cyst filled with a clear or milky fluid containing spermatozoa. Typically, the mass formed is small and painless and may be a sequel to acute epididymitis. Surgical excision is required if the cyst becomes large or painful.

DISORDERS OF THE PROSTATE

Enlargement of the prostate gland associated with obstruction to outflow of urine is a common occurrence, particularly in older males. The gland encases the portion of the urethra that lies between the neck of the bladder and the urogenital diaphragm; hence, enlargement leads to stricture of prostatic urethra. The following paragraphs deal with common causes of prostatic enlargement.

Benign Prostatic Hypertrophy

Benign prostatic hypertrophy (BPH) involves a process of overgrowth that leads to prostatic enlargement. The gland may be palpated through the rectal wall, which allows evaluation in terms of size, consistency, and symmetry. The normal prostate is the size of a golf ball and has a rubbery consistency. Characteristically, the enlargement due to BPH is up to two to three times that of normal and has a firm uniform consistency.

The prostate gland is made up of an inner layer of urethral and submucosal glands in proximity to the urethra, and the remaining tissue is composed of prostatic exocrine glands. The glands surrounding the urethra are subject to the formation of nodules composed of collagen, smooth muscle, and hyperplastic glands. The enlarged glands displace and compress the more peripheral true prostatic glands.

The etiology of BPH is not clear, although abnormal testosterone metabolism is a possible cause. The incidence of BPH increases after age 40. It is estimated that 90% of males past the age of 80 are affected. The majority of cases, however, do not require surgery.

BPH may lead to symptoms of obstruction to the flow of urine or to symptoms of irritation. Obstruction causes voiding problems that include (1) hesitancy to initiate voiding, (2) straining to void, (3) diminished force of urinary stream, (4) dribbling after micturition, and (5) urinary retention. Irritative symptoms may be caused by incomplete emptying of the bladder or by urinary tract infection. These symptoms include nocturia, dysuria, hematuria, urgency, and urinary frequency.

The treatment for BPH is surgical removal of all hyperplastic tissue, a partial or complete prostatectomy. Surgery is required in cases of intolerable symptoms or evidence of injury to the urinary tract. Indications for surgery include urinary retention and dilation of the ureters, secondary renal failure, and recurrent gross hematuria.

Prostatitis

Prostatitis is an inflammation of the prostate gland that may occur in either an acute or a chronic form. Both forms may be caused by bacterial infection, although the etiology is unknown in many cases of chronic prostatitis. The inflammation causes extreme tenderness and enlargement of the gland.

Acute prostatitis is characterized by fever, extreme prostatic tenderness, and enlargement of the gland. It is accompanied by a urinary tract infection and bacteremia. The diagnosis is based on an elevated white blood cell count with a shift to the left (see chapter 18) with a urinalysis showing many white blood cells and bacteria. Blood and urine cultures may be taken to identify the bacteria. The bladder is drained if there is a bladder obstruction. The patient with acute prostatitis is usually hospitalized and treated with parenteral antibiotics, frequently broad spectrum antibiotics.

Chronic prostatitis is a common condition that may either be of bacterial or nonbacterial origin. There is always an accompanying urinary tract infection in the case of bacterial origin. The majority of cases involve gram negative organisms. Prostatitis in the young patient is often preceded by a venereal infection of the urethra and a resulting urethral stricture. Nonbacterial prostatitis may be diagnosed on the basis of a negative urine culture and the demonstration of increased numbers of white blood cells in the prostatic fraction of voided urine. The urine sample is obtained by collecting the first 10 ml of urine following prostatic massage.

varicocele: L. *varicosus*, dilated veins; Gk. *kele*, tumor
micturition: L. *mic*, to urinate

bacteremia: Gk. *bakterion*, little rod; Gk. *haima*, blood

A patient with chronic prostatitis may be asymptomatic or, more typically, have irritative voiding symptoms, or may even have some of the obstructive symptoms previously discussed. There may be gross hematuria and septic complications, i.e., bloodstream infection, including cystitis, pyelonephritis, and epididymitis.

Antibiotic therapy is used to treat irritative symptoms and to eradicate the infecting organism in the prostate. There is a high rate of relapse, usually within a year, if the organism is not eradicated. In the case of nonbacterial prostatitis, a trial therapy with tetracycline or a related drug may be effective against *Chlamydia trachomatis, Neisseria gonorrhoeae,* and T-strain *Mycoplasma.* The patient's partner should be treated as well.

General symptomatic measures may be effective if antibiotic treatment fails to provide relief. Sitz baths provide relief for those patients suffering from perineal discomfort or low back pain. Increased sexual activity provides increased prostatic drainage. Oral antispasmodic agents, such as imipramine or phenoxbenzamine, relieve urinary frequency and urgency.

Surgery may be necessary when symptoms are extreme or when infection does not respond to antibiotic treatment. Surgery is required when there is a significant obstruction to urine flow.

MALIGNANT MALE REPRODUCTIVE DISORDERS

The following paragraphs discuss carcinoma of the penis, various tumors of the testes, and carcinoma of the prostate gland.

SQUAMOUS CELL CARCINOMA OF THE PENIS

Squamous cell carcinoma may appear first as an eroded ulcer or as a raised erythematous plaque and frequently originates on the inner surface of the prepuce or glans of an uncircumcised patient. The lesion is usually painless although secondary infections may cause discomfort. Phimosis may occur, and there may be a bloody discharge from the prepuce. Biopsy and histologic examination of the tissue establishes the diagnosis. The malignancy invades the connective tissue and spreads to the lymphatic system. The penis may be partially amputated, and if the neoplasm has invaded the corpora and lymph nodes, lymph nodes may be dissected as well. Chemotherapy is advisable if metastasis has occurred.

TESTICULAR TUMORS

Most masses within the scrotum are benign and are the result of inflammation, trauma, or congenital defect. Testicular cancer also presents as a scrotal mass and is a life-threatening condition. This type of cancer is the number one cause of malignancy in males between the ages of 15

TABLE 10.2 Staging of testicular tumors

Stage IA	Disease confined to testes
Stage IB	Microscopic lymph node involvement
Stage II	Metastasis to lymphatics below the diaphragm
Stage III	Metastasis to lymphatics above and below the diaphragm and to viscera

and 34. Although the prognosis has improved with current approaches to treatment, it was once the fifth leading cause of death within this young age group. The etiology is unknown. Blacks are rarely affected, and it is more likely to occur in males with cryptorchidism than in those with normally descended testes.

A testicular tumor seldom causes pain and typically is first noticed as a hard nodule, often discovered by the patient. The tumor may secrete hormones that cause gynecomastia. The first symptoms in up to 20% of the cases are associated with metastasis to lymph nodes or the lungs.

Diagnosis and Classification

The diagnosis is confirmed by histologic examination of the tissue. The clinical staging of testicular tumors is indicated in table 10.2. Most testicular tumors are of germ cell origin. Nongerm cell tumors are rare and usually are benign. The germ cell tumors are categorized as (1) seminomas (2) embryonal cell carcinomas (3) teratomas, and (4) choriocarcinomas.

Seminoma

Seminomas represent 30–40% of the cases of testicular tumors. These germinal cell tumors tend to occur in an older age group than other testicular tumors. There is diffuse enlargement of the testes with involvement of the right testis slightly more frequent than the left. The tumor replaces the entire testis and metastasizes to lymph nodes around the aorta. Seminoma responds to radiation therapy and chemotherapy and, with treatment, there is a 5-year survival rate of over 90%.

Embryonal Cell Carcinoma

Embryonal cell carcinomas are less differentiated and more aggressive than other germinal cell testicular tumors. This type of tumor represents about 20% of the testicular tumors. It is generally smaller than seminomas and can replace part or all of the testis. It is common for hemorrhage and necrosis to occur. There is an adult and an infantile type of embryonal cell carcinoma. The adult type has a 5-year survival rate of about 30%, whereas the infantile type has a 75% 5-year survival rate. The infantile type is the most common testicular tumor in infancy and childhood.

Teratoma

Teratomas arise from primitive cells that have the potential for developing into any of the embryonic germ layers, i.e., ectoderm, mesoderm, or endoderm. This type of tumor constitutes 4–9% of testicular tumors. There is a mature solid type teratoma that consists of islands of various mature, benign types of tissues. These islands are interspersed by malignant areas. Immature solid teratomas contain primitive neuroectoderm and primitive malignant cells. Cystic teratomas are uncommon and benign. There is about a 70% 5-year survival rate with teratomas.

Choriocarcinoma

The least common and the most malignant type of testicular tumor is choriocarcinoma. Less than 1% of testicular tumors are of this type, and virtually all patients die within 5 years. Widespread metastasis occurs, and there is poor response to treatment.

Mixed Cell Type Tumors

About 40% of testicular tumors show a combination of growth patterns. About 25% of testicular tumors, for example, are a combination of teratoma and embryonal carcinoma.

Tumor Markers

The assay of two hormones, **alpha-fetoprotein** and **human chorionic gonadotropin,** is useful in evaluating a patient with a testicular tumor. The blood level of alpha-fetoprotein is elevated in embryonal carcinoma and, less frequently, in other testicular tumors. Human chorionic gonadotropin is typically elevated in choriocarcinoma. Elevations of these hormones after surgical removal of the tumor indicate a recurrence or metastasis. Normal blood levels of these hormones occur in about 25% of the cases in which the malignancy has persisted.

In addition to these two hormones, **alkaline phosphatase** is a tumor marker and is elevated in 50% of cases of seminoma. A marker detected on red blood cells, B5, is raised in most cases of testicular tumor and seems to be elevated in all cases of seminoma.

Treatment

Surgical removal of the testes, orchiectomy, is the first step in treatment of testicular tumor. In addition to orchiectomy, the treatment for stage I nonseminomas includes surgical removal of retroperitoneal (re″tro-per″i-to-ne′al) lymph nodes, i.e., lymph nodes located behind the parietal peritoneum. In the case of microscopic involvement of lymph nodes, combination chemotherapy is indicated. A greater than 90% 5-year survival rate is expected for stage I and stage II disease. Treatment for seminomas is orchiectomy followed by radiation therapy. Radiation therapy is less effective with bulky stage II or stage III disease, and combination chemotherapy may be instituted.

PROSTATIC ADENOCARCINOMA

Adenocarcinoma is a glandular epithelial type malignancy. Adenocarcinoma of the prostate rarely occurs in men in their forties, but the incidence increases with age. Fifty percent of males past the age of 80 are affected, even though most of these cases are not clinically apparent. The disease is the second leading cause of cancer deaths in American men and accounts for 20,000 deaths annually. The etiology is unknown, although the fact that mortality from prostate cancer is low in Japan but increases among Japanese living in the United States suggests that there may be environmental/dietary factors. The incidence of the disease in families suggests that there may be genetic factors as well.

Symptoms/Diagnosis

In the majority of patients, the cancer is not detected until it is locally advanced or has spread to other parts of the body. The tumor usually arises from the posterior aspect of the gland and may be palpated as a hard nodule during rectal examination. There are no laboratory methods for detecting small tumors of the prostate; thus, rectal examination is the best currently available means of detection of early prostate cancer. Symptoms do not occur until the disease is in an advanced stage.

Elevated serum levels of the **acid phosphatase** enzyme is useful in identifying prostatic cancer in an advanced state or in a metastatic stage. Acid phosphatase is produced by various tissues including the prostate, bone marrow, red blood cells, liver, and kidney. The prostate gland produces a unique acid phosphatase, which can be identified and which is persistently elevated in the serum in advanced stages of prostatic cancer. A diagnosis of prostatic cancer is made by microscopic examination of prostate tissue obtained by needle biopsy, either by way of the perineum or the rectum. A bone scan, involving the intravenous administration of a radioisotope and X-ray evidence of increased uptake of that isotope by the bone, is a means of diagnosing malignancy of the bone. A bone scan is performed to diagnose the spread of prostatic cancer to the bone.

Staging and Treatment

Stage A prostatic carcinoma involves either a small, localized differentiated tumor or a diffuse and/or undifferentiated tumor. If there is a negative bone scan, the tumor is classified as A_1, and the patient is monitored by yearly rectal examinations. Stage A_2 identifies more extensive involvement of prostatic tissue or a more poorly differentiated cancer. The prognosis for this category is less promising, and there already may be spread to lymph nodes; thus, it is aggressively treated with radiation therapy.

Prostatic cancer is classified as stage B if the tumor is confined to the prostate gland and is identified by rectal examination. These patients are asymptomatic, have a negative bone scan, and have no elevation of serum acid phosphatase. If the tumor is limited to one lobe of the gland, it is considered to be stage B_1, and this represents minimal likelihood of accompanying lymph node involvement. When there is involvement of more than one lobe of the prostate, it is considered to be stage B_2, and there is a greater chance that there has been spread to lymph nodes. Although the rate at which the stage B disease progresses varies greatly, it metastasizes and leads to death. Aggressive treatments are appropriate, except in the case of advanced age or a concurrent serious illness. Modalities of treatment include (1) radical prostatectomy, (2) external irradiation, or (3) interstitial irradiation. The latter procedure involves a lower abdominal incision exposing the prostate and the placement of radioactive seeds through hollow needles into the gland. Surgical removal of the prostate leads to impotence in 90% of the cases. There is stricture of the urethra in close to 10% of the cases. There is a mortality rate of about 1% of the cases. Radiation causes impotence in about 50% of the patients and results in permanent bladder or rectal inflammation.

Prostate cancer in stage C involves a tumor that extends outside the prostate and can be palpated rectally, but clinical evidence of metastases is absent. The symptoms may be typical of benign prostatic hypertrophy or of prostatitis. There is a negative bone scan and normal serum enzyme levels. About 50% of the cases have lymph node involvement in stage C. There is evidence that radical prostatectomy is followed by a recurrence of the disease in most cases. Aggressive radiation therapy may provide a cure and, in 90% of the cases, controls the tumor. In an older patient who has a tumor larger than 8 cm in diameter, treatment is delayed until symptoms appear and is then aimed at symptomatic relief. Most prostatic cancers require male sex hormones and androgens for growth. One approach to management of the disease is to decrease these hormonal levels. Such measures result in a temporary effect and do not cure the disease. Surgical removal of testicles, i.e., **orchiectomy** (or″ke-ek′tô-me), is one method of achieving androgen deprivation, but this procedure has the disadvantage of causing permanent impotence. Estrogen therapy suppresses the pituitary release of luteinizing hormone that normally stimulates release of testosterone. This method of testosterone suppression may be accompanied by unpleasant side effects, such as fluid retention and breast development, called **gynecomastia** (jin″ĕ-ko-mas′te-yah). A third approach to androgen suppression involves the use of luteinizing hormone releasing hormone (LH-RH) agonists. Agonists mimic the effects of LH-RH that is produced by the hypothalamus (chapter 31). Luteinizing hormone releasing hormone normally stimulates the secretion of LH by the anterior pituitary, which stimulates testosterone production by the testes. It is paradoxical that these agonists depress testosterone levels although for the first week cause an increase in testosterone. The mechanism of action is not entirely clear, but with prolonged administration, the pituitary becomes less responsive to LH-RH, and this leads to a decrease in testosterone levels. The agonists do not lead to gynecomastia, and the effects are potentially reversible.

Prostatic cancer that has metastasized, either to pelvic lymph nodes or to distant sites, is considered to be stage D. There may be indications of a pelvic mass; or there may be metastatic symptoms including arthritic pain, anemia, clotting abnormalities, hematuria, and symptoms of acute spinal cord compression. In about 80% of the patients, the serum acid phosphatase level is elevated. A bone scan shows increased uptake of radioactive isotope in areas of pain. Involvement of the spinal cord may lead to extremity weakness, paresthesias or tingling sensations, and permanent lower body paralysis or paraplegia. A few of these patients who have limited lymph node involvement may be treated with surgery or surgery combined with radiation therapy. In cases that there has been metastasis to distant sites, the treatment is palliative, i.e., measures are taken to alleviate symptoms. These include orchiectomy or hormonal therapy as discussed in the preceding paragraph. Although survival is not changed by hormonal therapy, there is amelioration of symptoms, such as obstruction to urine flow, bleeding, and bone pain in about 80% of the patients. Radiation therapy relieves ureteral obstruction and chronic hematuria and plays an important role in the treatment of localized bone pain.

SUMMARY

• • •

Nonmalignant gynecologic disorders include infections caused by *Neisseria gonorrhoeae,* the herpes simplex virus, the fungus *Candida albicans,* the protozoan *Trichomonas vaginalis,* and *Chlamydia trachomatis.* Dysmenorrhea or painful menstrual cramps affects about half of young females, and in most cases, excessive prostaglandin production or absorption is the cause. Endometriosis is a condition in which there is implantation of normal endometrial tissue outside of the uterus. It is possible that endometrial tissue is regurgitated through the fallopian tubes onto the ovaries and other pelvic organs. Fibrosis occurs and dense adhesions develop around the fallopian tubes, the intestinal tract, and elsewhere. Leiomyomas, also called fibroids, are benign tumors of smooth muscle. These are common uterine tumors that vary considerably in size. Bleeding may occur or the condition may be asymptomatic. Endometrial hyperplasia refers to a thickening of the uterine lining. Moderate and atypical hyperplasia are believed to be premalignant conditions. A tubal ectopic pregnancy is the implantation of a fertilized ovum in one of the fallopian tubes. The result may be pain and bleeding. Polycystic ovary disease is characterized by enlarged ovaries, hyperplasia of ovarian stroma, and a layer of small follicles beneath a thickened capsule. The condition is associated with multiple follicular cysts, elevated androgen levels, a high LH to FSH ratio, and the absence of ovulation. Hyperthecosis may be associated with PCOD and is characterized by increased numbers of theca cells, decreased numbers of follicles, and high levels of androgens that lead to masculinization.

Malignant gynecologic disorders are significant contributors to mortality. Carcinoma of the cervix, in the majority of cases, is squamous cell carcinoma. Risk factors for squamous cell carcinoma of the cervix include (1) multiple sexual partners at an early age, (2) low socioeconomic conditions, (3) smoking, and (4) venereal infections. The Papanicolaou smear is important as a screening test for cervical carcinoma because the condition may be asymptomatic. Endometrial carcinoma occurs more commonly among women over the age of 50 and appears to be associated with prolonged exposure to estrogens. Most endometrial tumors are adenocarcinomas. A symptom of endometrial carcinoma is abnormal bleeding. Fallopian tube adenocarcinoma is the least common neoplasm of the female genital tract and is the most aggressive. It is usually caused by metastasis from another site. About one patient out of five survives 5 years after diagnosis. The ovary is a common site for both benign and malignant tumors. Most of the ovarian malignancies are primary. Types of ovarian cells that may give rise to neoplasms include (1) mesothelium of the ovary, (2) oocytes, (3) ovarian stromal cells that specialize to form granulosa and theca cells, and (4) hilum cells. Surface cell epithelial tumors are the most common group of ovarian tumors. Three main groups of epithelial tumors are (1) serous tumors, (2) mucinous tumors, and (3) endometrioid tumors. In all three categories, the tumors may be benign cystadenomas, that have low malignant potential; or they may be adenocarcinomas, also called cystadenocarcinomas. Germ cell tumors originate from cells that produce the oocytes, and the tumors are characterized by the production of various types of tissues. Most of these tumors are benign cystic teratomas. Dysgerminomas are germ cell tumors that are often hemorrhagic and necrotic. Embryonal carcinoma is an uncommon germ cell tumor that affects young patients and causes precocious puberty. Granulosa-theca cell tumors produce estrogenic hormones or, in some cases, produce androgens. Fibromas are sex cord stromal tumors that do not produce hormones. Sertoli-Leydig cell tumors arise from sex cord stromal cells and often produce androgens that have masculinizing effects. These tumors are nearly always benign. Gonadoblastoma is a rare type of ovarian tumor that contains both immature germ cells and sex cord-stromal cells. The patients are usually phenotypic females that have a Y chromosome. Hilum cell tumors are usually benign.

Disorders of the breast include galactorrhea, a condition in which milklike material with fat globules is discharged from the nipples. The etiology is unknown in most cases, although galactorrhea may occur at the cessation of oral contraceptive medication or may represent a persistent discharge after pregnancy. A prolactin-secreting pituitary tumor is a possible cause of galactorrhea. Fibrocystic breast disease involves local or diffuse lumps in the breast due to fibrosis, cyst formation, epithelial hyperplasia, or adenosis. Breast cancer is the most common cause of death in females between the ages of 35–50. The majority of breast cancers are infiltrating adenocarcinomas that form glandular structures from ductal or lobular epithelial cells. Risk factors for breast cancer include early menarche or late menopause, late first time pregnancy, Western culture, family history, radiation exposure, and benign breast disease with atypical epithelial hyperplasia. The presence or absence of estrogen or progesterone receptors is significant in terms of prognosis and treatment of the disease. Based on the assumption that there is early micrometastasis, chemotherapy or endocrine manipulation are postsurgical modalities of treatment for patients at high risk for recurrence.

Nonmalignant disorders of the penis include balanitis and posthitis, inflammation of the glans penis and the foreskin respectively. Phimosis is a condition in which the prepuce cannot be retracted over the glans penis. Paraphimosis describes a situation in which the prepuce cannot be pulled back to cover the glans penis. Priapism is prolonged erection due to failure of drainage of venous blood. Peyronie's disease involves fibrosis of the cavernous sheath that causes a thickening on the dorsal side of the penis. The condition is characterized by a mass in the penis, curvature on erection, and pain due to erection.

Disorders of the testes include cryptorchidism, involving failure of one or both testes to descend into the scrotum during fetal development. Epididymitis is the most common cause of scrotal swelling. A common cause is believed to be reflux of the urine into the epididymis. Testicular torsion, a twisted spermatic cord, is common in young men and in children. A fluid-filled mass surrounding the testis is called a hydrocele. This may be the result of a congenital defect, or a complication of inflammation of the genitalia or tumor, or trauma. Dilatation of the veins leading away from the testes is called a varicocele. The cause may be thrombosis or compression of the vena cava. The term spermatocele refers to a fluid-filled cyst at the head of the epididymis that may be a sequel to acute epididymitis.

Disorders of the prostate gland include benign prostatic hypertrophy that may lead to obstruction of urine flow. Inflammation associated with tenderness and enlargement of the prostate gland is called prostatitis.

Malignant male reproductive disorders include carcinoma of the penis, tumors of the testes, and carcinoma of the prostate gland. Squamous cell carcinoma of the penis may appear first as an eroded ulcer or as a raised erythematous plaque and frequently originates on the inner surface of the prepuce or glans of an uncircumcised patient. The malignancy invades the connective tissue and spreads to the lymphatic system. Testicular tumors are usually benign and are the result of inflammation, trauma, or congenital defects. Testicular cancer presents as a scrotal mass and is the number one cause of malignancy in males between the ages of 15–34. Seminoma is a testicular tumor of germinal cell origin tending to occur in an older age group. Embryonal cell carcinoma is an aggressive type of germinal cell testicular tumor. Teratoma is a type of testicular tumor in which the cells have the potential for developing into any of the embryonic germ layers. Choriocarcinoma is the least common and the most malignant type of testicular tumor. About 40% of testicular tumors show a combination of growth patterns and are classified as mixed cell tumors. Prostatic adenocarcinoma is a glandular epithelial type malignancy. This is the second leading cause of cancer deaths in American men.

21. Define the term priapism.
22. What is the name of a disorder of the penis in which fibrosis causes thickened areas or plaques?
23. What is the term for failure of testes to descend into the scrotum?
24. What is the most common cause of scrotal swelling?
25. What is inflammation of the testis called?
26. What is the term for a twisted spermatic cord?
27. List three symptoms of benign prostatic hypertrophy.
28. Identify a germ cell tumor of the testes.
29. What is the term for a tumor that arises from primitive cells that have the potential for developing into any of the embryonic germ layers?
30. What serum enzyme is likely to be elevated in advanced prostatic cancer?

REVIEW QUESTIONS

• • •

1. What chemical(s) produced by the body cause(s) dysmenorrhea in most cases?
2. Define endometriosis.
3. What term describes the implantation of a fertilized egg in a fallopian tube?
4. What is the term for a condition in which the endometrium becomes thickened due to an increase in the number of endometrial cells?
5. List four clinical features of polycystic ovary disease.
6. What are two symptoms of polycystic ovary disease?
7. Define the term hyperplasia.
8. Define the term dysplasia of the cervix.
9. Define carcinoma in situ.
10. Identify three risk factors for cancer of the cervix.
11. What is a likely cause of endometrial cancer?
12. Identify three types of ovarian cells that may be involved in a malignancy of the ovary.
13. What is the term for the discharge of milklike material from the nipples?
14. Identify three causes of lumps in the breast.
15. What is the term for a breast carcinoma that originates from epithelial cells of the ducts and invades stromal tissue?
16. List five risk factors for breast cancer.
17. A high degree of tumor differentiation as indicated by the presence of estrogen receptors indicates a (more hopeful or less hopeful) prognosis.
18. What is the principle of surgical removal of ovaries as a treatment of breast cancer?
19. What are the terms for inflammation of the glans penis and the foreskin?
20. Define the terms phimosis and paraphimosis.

SELECTED READING

• • •

Bosl, G. J., et al. 1991. Serum tumor markers and patient allocation to good-risk and poor-risk clinical trials in patients with germ cell tumors. *Cancer* 67:1299–304.

Cramer, B. M., et al. 1991. MR imaging in the differential diagnosis of scrotal and testicular disease. *Radiographics* 1:9–21.

Dodin, S. et al. 1991. Bone mass in endometriosis patients treated with GnRH agonist implant or danazol. *Obstetrics and Gynecology* 77:410–15.

Estreich, S. et al. 1990. Sexually transmitted diseases in rape victims. *Genitourinary Medicine* 66:433–38.

LaVecchia, C. et al. 1991. Genital and urinary tract diseases and bladder cancer. *Cancer Research* 51:629–31.

Lawhead, R. A. 1990. Vulvar self-examination: What your patients should know. *Physician Assistant* 14:55–62.

Olt, G. J., et al. 1990. Gynecologic tumor markers. *Seminars Surgical Oncology* 6:305–13.

Remennick, L. I. 1990. Induced abortion as cancer risk factor: A review of epidemiological evidence. *Journal of Epidemiology and Community Health* 44:259–64.

Shafik, A. 1990. Inguinal pelviscopy: A new approach for examining pelvic organs. *Gynecologic and Obstetric Investigation* 30:159–61.

Chapter 11

Excretory Function: Evaluation and Control

The major functions of the kidney are to eliminate wastes and excess water from the body and maintain a proper balance of electrolytes in body fluids. This chapter presents an overview of both the anatomy of the excretory system and the process of urine formation, as a basis for the topics related to dysfunctions discussed in the chapters that follow. The emphasis in this chapter is on the relationship between the kidney, its blood supply, and the factors that determine urine composition. The last part of the chapter deals with evaluation of renal function and the effects of diuretics.

STRUCTURE AND NORMAL FUNCTION

The topics discussed in the following paragraphs include (1) structure of the kidney and urine flow, (2) urine formation, (3) hormonal influences on urine formation, and (4) control of acid-base balance.

KIDNEYS

The kidneys are about 10–12 cm long and are located retroperitoneally (re″tro-per″i-to-ne′al-le), that is, posterior to the peritoneal cavity (figures 11.1 and 11.2). They are on either side of the vertebral column, extend down to the third lumbar vertebra, and are partially protected by ribs 11 and 12. The **costal margin,** which is the medial edge of ribs 7–10, is a useful point of reference. The kidneys are located below the costal margin in line with the middle of the clavicle.

• • •

The term palpable comes from the Latin word *palpare,* which means to touch softly. Although normal kidneys are usually not palpable, abnormalities may be apparent when an examiner applies pressure with one hand on the back beneath the costal margin, with the other hand placed anteriorly and below the costal margin.

• • •

Internal Structure

The general outline of the kidney is bean-shaped. The **renal hilus** in the medial area is where the ureter and renal vein exit and where the renal artery enters. Figure 11.3 is a coronal section of kidney that shows three main areas: the outer **cortex,** the **medulla** in the middle region, and the medially located **pelvis.** The cortex is composed of approximately one million microscopic units called **nephrons** along with their tubules. The medulla is the region of triangular-shaped **pyramids** formed by tubules and blood vessels separated by **renal columns.** Blood vessels pass through the renal columns. **Renal papillae** (papil′laē) at

costal: L. *costa,* rib

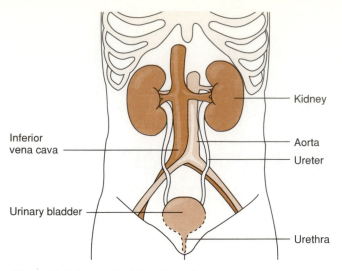

Figure 11.1 Anatomical locations of excretory system structures. Ureters carry urine, formed in the kidney, to the bladder. The urethra provides passage to the outside.

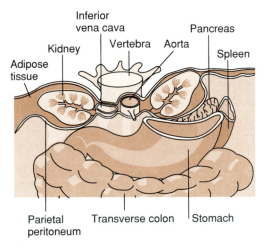

Figure 11.2 Retroperitoneal location of kidneys. The parietal peritoneum is a membrane that lines the interior of the abdominal cavity.

the tips of the pyramids open into compartments called **minor calyces** (cal′yces), and several of these compartments combine to form a **major calyx.** Finally, the renal pelvis is formed by the major calyces.

• • •

The term hilus comes from the Latin word *hilum,* which means a little thing. It refers to a depression that provides passage for vessels or ducts. The pulmonary hilus is the area in the lungs where there is passage of blood vessels, lymphatics, bronchi, and nerves. The renal hilus is where blood vessels and the ureter either enter or exit the kidney.

• • •

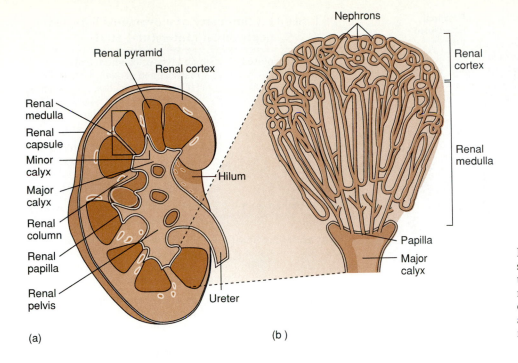

Renal pyramid
Renal cortex
Renal medulla
Renal capsule
Minor calyx
Major calyx
Renal column
Renal papilla
Renal pelvis

Nephrons
Renal cortex
Renal medulla
Hilum
Papilla
Major calyx
Ureter

(a)

(b)

Figure 11.3 (*a*) A coronal section of the kidney showing the three main regions: cortex, medulla, and pelvis. (*b*) An enlarged diagram of the cortex and a pyramid to show nephrons.

Urine Flow

Urine is formed by nephrons that empty into collecting ducts in the medulla. Urine drops out of the renal pyramids of the medulla by way of openings in the papillae into minor calyces. These, in turn, open into the major calyces and into the renal pelvis.

A ureter carries urine away from each kidney and leads to the posterior lateral surface of the **bladder.** The bladder is a muscular organ located just posterior to the pubic symphysis that stores up to about 800 ml of urine.

Backflow of urine during **micturition** (mik'tu-rish'un) is prevented by contraction of muscles of the bladder, which compresses the ureters. Backflow is also prevented by folds of mucous membrane that cover the ureteral openings. The **urethra** exits the inferior surface of the bladder and provides passage for the urine to the outside. There is an **internal urethral sphincter** at the point of urethral exit from the bladder that opens as the bladder contracts. There is an **external urethral sphincter** as the urethra passes through the pelvic floor, and this sphincter is under voluntary control. Table 11.1 defines a list of terms related to the kidney.

Nephron

The basic functional unit of the kidney is a microscopic structure called the nephron. It consists of a cuplike capsule called a Bowman's capsule continuous with a long tubule. The configuration of the tubule and the names of the segments are indicated in figure 11.4. The distal part of the tubule is a collecting duct that carries urine to the pyramids.

The function of nephrons is to filter the blood. Each nephron has both a rich supply of blood and is in close contact with blood vessels. The renal artery, a branch of the abdominal aorta, enters the hilus of the kidney and

TABLE 11.1 Terminology related to the kidney

Calyx	Gk. *kalyx,* cup; a compartment in the renal pelvis
Papillae	L., nipple; small projection like a nipple; open at tips of pyramids into calyces
Renal	L. *renalis,* of the kidneys
Micturition	L. *mictus,* making water; the act of urinating
Afferent arterioles	L. *ad,* to; L. *ferre,* to bear; carry blood to the glomerulus
Efferent arterioles	L. *effere,* to carry away; carry blood away from glomerulus
Glomerulus	L. *glomus,* ball; mass of capillaries surrounded by Bowman's capsule

subdivides to ultimately form afferent arterioles. Each afferent arteriole carries blood to a mass of capillaries called a **glomerulus** encased by a Bowman's capsule. Blood is carried away from the glomerulus by an efferent arteriole, and this forms a second capillary system called the **peritubular capillaries.** These capillaries are wrapped around the tubules of the nephron as shown in figure 11.4.

URINE FORMATION

The events leading to the formation of urine involve filtration, reabsorption, and secretion. Both passive diffusion and active transport are involved. The overall effect is the filtering of blood from glomeruli into Bowman's capsule. There are subsequent exchanges of constituents between the filtrate and interstitial fluid along the length of the tubule until the final product, urine, appears in the collecting duct.

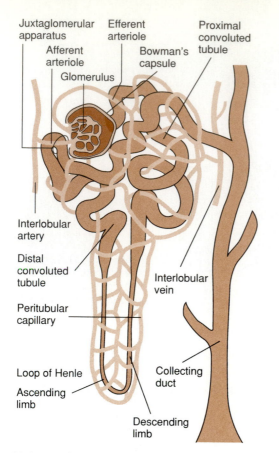

Figure 11.4 A nephron, made up of a Bowman's capsule, a tubule, and an associated collecting duct. An afferent arteriole leads into a glomerulus with an efferent arteriole carrying blood away. Peritubular capillaries surround the nephron tubules.

• • •

About 20% of the cardiac output passes through the kidney. The rate of blood flow per unit mass of tissue exceeds that of other organs in the body. About 180 liters of fluid per day are filtered into Bowman's capsules, and since only approximately 1.5 liters of urine per day is produced, most of this filtered volume is reabsorbed.

• • •

Filtration/Diffusion

The principles involved in the glomerular-to-capsule movements are the same as capillary/interstitial fluid exchanges (chapter 2). Filtration refers to the net movement of fluid under pressure, and the main driving force is capillary blood pressure or hydrostatic pressure. The opposing forces are plasma oncotic pressure due to plasma proteins and hydrostatic pressure within Bowman's capsules. The capillary/capsular wall restricts the passage of plasma proteins, so the filtrate in Bowman's capsules is relatively protein free. The net effect is the filtration of plasma into Bowman's capsules with the composition of the filtrate similar to that of interstitial fluid.

TABLE 11.2 Summary of movements between nephron and interstitial fluid

Proximal convoluted tubule	Single layer of cuboidal cells; reabsorption of Na^+, Cl^-, H_2O, HCO_3^-, K^+, glucose, amino acids, urea, HPO_4^{-2}, and SO_4^{-2}; secretion of H^+, foreign substances
Descending limb	Squamous cells; permeable to H_2O, relatively impermeable to solutes; H_2O drawn out of tubule osmotically
Thin ascending limb	Squamous cells; permeable to solutes, impermeable to H_2O
Thick ascending limb	Cuboidal cells; low permeability to H_2O; Na^+ and Cl^- reabsorbed; creates high osmolality of interstitial fluid
Distal convoluted tubule	Cuboidal cells influenced by aldosterone, ADH; fluid hypotonic; secretion of K^+, H^+, and NH_3; reabsorption of Na^+, Cl^-, HCO_3^-, K^+, and H_2O
Collecting duct	Influenced by aldosterone, ADH; reabsorption of urea, H_2O, Na^+, and Cl^-; K^+/H^+ reabsorbed, or secreted

Modification of Filtrate

Changes in the composition of the glomerular filtrate begin in the proximal convoluted tubule. The term **reabsorption** refers to movement out of the tubule into interstitial fluid and into peritubular capillaries. The term **secretion** is used to identify movement in the opposite direction, that is, from interstitial fluid into tubular cells and, ultimately, into tubular fluid. The processes of reabsorption and secretion in the tubules may be summarized as follows. The overall effect in the proximal convoluted tubule is reabsorption of 80% of the electrolytes, almost all of the glucose and amino acids, and reabsorption of water. The filtrate remains isotonic as compared to plasma.

Table 11.2 summarizes reabsorption and secretion. The following are some observations about these events. Sodium ions are carried out of the proximal convoluted tubule by active transport, and Cl^- and HCO_3^- follow to maintain electrical neutrality. The efflux of sodium ions creates an osmotic gradient that draws water out of the tubule. Hydrogen ions and foreign substances, such as penicillin or organic acids, are secreted into the proximal convoluted tubule.

The descending limb is composed of squamous cells and is thin-walled, and more permeable to water than to solutes. Water is drawn out of the tubule because of the high osmolality of the medullary interstitial fluid. The first segment beyond the loop of Henle is the thin ascending limb, which becomes more thick walled with a lining of

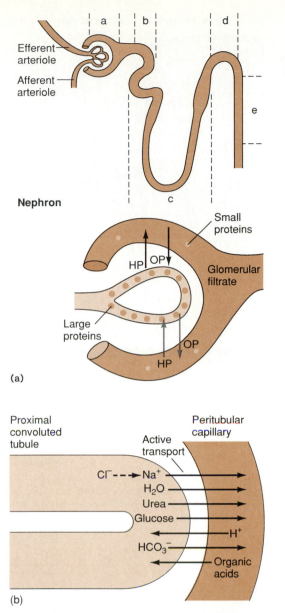

(a)

(b)

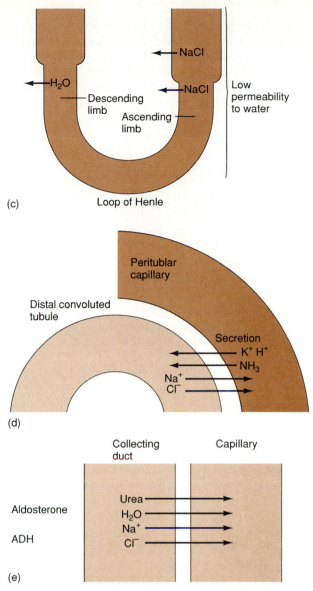

(c)

(d)

(e)

Figure 11.5 Nephron. (*a*) Enlarged view of glomerulus and Bowman's capsule. Plasma is filtered into Bowman's capsule to form a glomerular filtrate. The forces favoring the filtration are capillary hydrostatic pressure (HP) and capsular oncotic pressure (OP). Opposing forces are capillary oncotic pressure and capsular hydrostatic pressure. (*b*) Enlarged view of a portion of the proximal convoluted tubule in close proximity to a peritubular capillary. Sodium is actively transported out of the tubule and the chloride ions passively follow. Water is drawn osmotically out of the tubule. Hydrogen ions are secreted into the tubule. (*c*) There is osmosis of water out of the descending limb of the loop of Henle and both Na^+ and Cl^- diffuse out of the ascending limb. (*d*) There is secretion of ammonia and potassium and/or hydrogen in the distal convoluted tubule. Sodium and chloride are reabsorbed from the tubule as well. (*e*) The hormones ADH and aldosterone influence the reabsorption of water and sodium from collecting ducts.

cuboidal cells. Both the thin and thick segments of the ascending limb have low water permeability. Substantial amounts of sodium and chloride ions are reabsorbed in this region, and this creates high osmolality in the medullary interstitial fluid.

The distal convoluted tubule is sensitive to both aldosterone and ADH, and these hormones control sodium ion and water reabsorption. For the most part, chloride ions are the anions that accompany sodium ions because bicarbonate reabsorption occurs mainly in the proximal convoluted tubule. There is secretion of potassium ions, hydrogen ions, and ammonia (NH_3) in this segment of the tubule with reciprocal K^+/H^+ exchange. Water and sodium ion reabsorption is also controlled by ADH and aldosterone in the collecting duct. This segment is permeable to urea, and reabsorption of urea contributes to high osmolality in the interstitial fluid. Figure 11.5 shows the overall picture of events in each segment of the nephron.

COUNTERCURRENT MECHANISM

The mechanism by which the kidney is able to conserve water and eliminate urine with a high solute concentration is called the **countercurrent mechanism.** There are three main elements in this mechanism: (1) nephrons with loops of Henle located deep within the medulla, (2) reabsorption of substances that creates high osmolality in the medullary interstitial fluid, and (3) a unique peritubular capillary system, which maintains that high osmolality.

Types of Nephrons

Nephrons are classified on the basis of the location of their glomeruli (figure 11.6). About 80% of the nephrons originate in the outer zone of the cortex with the loop of Henle extending a short distance into the medulla. These are called **cortical nephrons.** The remainder of the nephrons arise from glomeruli located deep within the cortex close to the medulla. Consequently, they are called **juxtamedullary nephrons.** The loop of Henle of each of these nephrons penetrates deep into the medulla. It is the juxtamedullary nephrons that are important in the countercurrent mechanism.

Peritubular Capillaries

There is a difference between peritubular capillaries associated with the cortical nephrons as compared to those associated with juxtamedullary nephrons. In both types of nephrons, the capillaries are wrapped around the convoluted tubules; and in the case of the cortical nephrons, are wrapped around the loop of Henle as well. Each juxtamedullary loop of Henle is supplied by a capillary that makes a hairpin loop parallel to the loop of Henle (figure 11.7). This type of capillary is called the vasa recta (va'sa rek'ta) and is located deep within the renal medulla. The functional significance of this arrangement is discussed in the following paragraphs.

Medullary Osmolality

The juxtamedullary loops of Henle, the associated vasa recta, and collecting ducts in the renal medulla make it possible for the kidney to conserve water and produce a concentrated urine by creating high osmolality in the medullary interstitial fluid and maintaining that osmolality.

Table 11.3 lists key events that lead to a high solute concentration in the medullary interstitial fluid. High osmolality is established by efflux of sodium and chloride ions out of the ascending limb of the loop of Henle and the diffusion of urea from the collecting duct into interstitial fluid; thus, sodium, chloride, and urea create a concentration gradient in the medullary interstitial fluid.

juxtamedullary: L. *juxta,* near; *medius,* middle

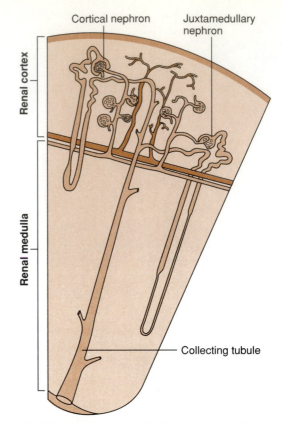

Figure 11.6 Nephrons are classified on the basis of the location of their glomeruli.

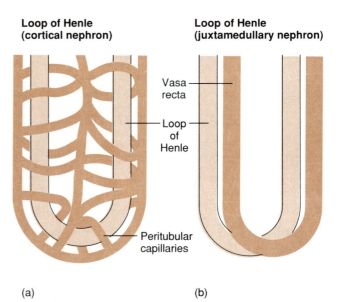

Figure 11.7 (*a*) In cortical nephrons, the peritubular capillaries are wrapped around the loop of Henle. (*b*) The capillary associated with a juxtamedullary nephron makes a hairpin loop parallel to the loop of Henle.

TABLE 11.3 Events that lead to high osmolality of interstitial fluid in the renal medulla

1. Active transport of Cl⁻ out of ascending limb into medullary interstitial fluid
2. Diffusion of Na⁺ out of ascending limb
3. Ascending limb is impermeable to H_2O; H_2O does not diffuse out
4. Descending limb is permeable to H_2O; H_2O diffuses out and concentrates tubular fluid
5. New fluid flows through loop of Henle; process repeated
6. Collecting duct is highly permeable to urea; urea diffuses into medullary interstitial fluid

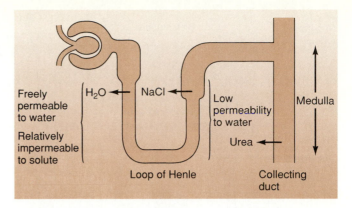

Figure 11.8 Mechanisms by which an increased solute concentration is established in medullary interstitial fluid. Water diffuses out of the descending tubule and is removed by peritubular capillaries. The tubular fluid thus becomes more concentrated as it moves down to the loop of Henle. Medullary interstitial fluid has a high solute concentration due to the presence of Na⁺ and Cl⁻ and urea.

There are several contributing factors in the events that lead to a high solute concentration in the medullary interstitial fluid. There is variation in permeability to water along the length of the loop of Henle. The descending limb is freely permeable to water, while the ascending limb is relatively impermeable. There is active transport of chloride ions out of the ascending lumen with subsequent efflux of sodium ions to maintain electrical neutrality. Urea, which is an end product of protein metabolism, contributes to the solute concentration of medullary interstitial fluid as well. Figure 11.8 provides an overview of the movements of substances that occur in the medulla.

Vasa Recta

The mechanism for maintaining a high solute concentration in medullary interstitial fluid involves the vasa recta. The key element is the hairpin configuration of the vasa recta in a region of high solute concentration. The following are factors that contribute to the removal of water from the interstitium of the medulla:

1. The descending and ascending segments of the vasa recta are close to each other and are parallel. Blood flows in opposite directions in these two segments; thus, the term countercurrent is used.
2. In the descending vasa recta, there is a net efflux of water out of the vessel and an influx of urea, sodium, and chloride ions due to interstitial/capillary concentration gradients. The overall effect is to create a high solute concentration within the vasa recta loop.
3. Osmolality is high in the ascending segment. The ascending and descending limbs are in proximity, and there is a high solute concentration in the ascending limb compared to both the interstitium and the descending limb.
4. Water diffuses into the ascending limb, and solute diffuses out in response to the concentration gradient.
5. Water is carried away from the medulla, and solute is left in the interstitial fluid (figure 11.9).

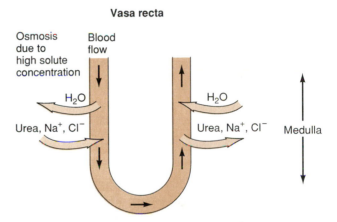

Figure 11.9 The countercurrent exchange mechanism is important in maintaining a high solute concentration in medullary interstitial fluid. Blood flows in opposite directions and in close proximity. As blood flows down, water diffuses out and solutes diffuse into the capillary. The process is reversed as the blood flows upward. The net result is that water is carried away and interstitial solute concentration is maintained.

Overview

In summary, high osmolality in the medulla is established by the efflux of sodium and chloride ions out of the ascending limb of the loop of Henle and by the diffusion of urea from the collecting ducts into interstitial fluid. High osmolality in the medulla is maintained by vasa recta. These capillaries remove water that is reabsorbed from collecting ducts and thus prevent dilution of interstitial fluid. This function is made possible by the hairpin configuration of the capillaries. Finally, water is reabsorbed under hormonal influence by osmosis (figure 11.10).

vasa recta: L. *vasa*, vessel; L. *rectus*, straight

143

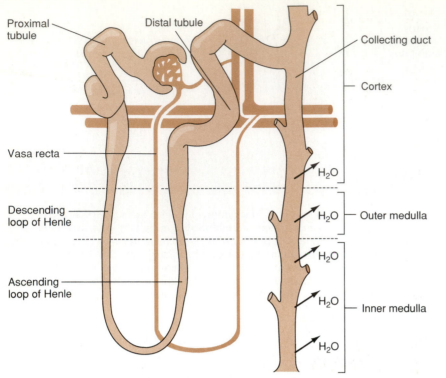

Figure 11.10 Osmolality is high in the medullary interstitial fluid. The collecting ducts are impermeable to salt, but under hormonal influence, are permeable to water. Water is drawn out of collecting ducts by osmosis and carried away by capillaries; thus, a concentrated urine is formed.

HORMONAL INFLUENCES

Hormones influence the composition and volume of the urine produced by nephrons. A discussion of four of these hormones follows.

Antidiuretic Hormone

Antidiuretic hormone (ADH) increases permeability to water in the distal convoluted tubules and collecting ducts. Under the influence of this hormone, water is reabsorbed from these tubules, and in its absence, there is an increased volume of urine.

There are specialized neurons that have cell bodies in the hypothalamus that synthesize ADH and control its secretion. ADH is synthesized within the neurons and is packaged in granules that migrate by way of axons to be stored in the posterior lobe of the pituitary, also called the **neurohypophysis** (nu″ro-hi-pof′ĭ-sis) (chapter 31).

Stimuli for ADH Secretion

There are neurons in the hypothalamus that are specialized as **osmoreceptors.** These neurons are affected by changes in osmolality of the extracellular fluid in the brain (figure 11.11). Increased osmolality, mainly caused by increased sodium ion concentration, draws water out of the osmoreceptors and causes subsequent shrinkage. This stimulates nerve impulse transmission to the neurohypophysis, which causes ADH to be secreted into the blood.

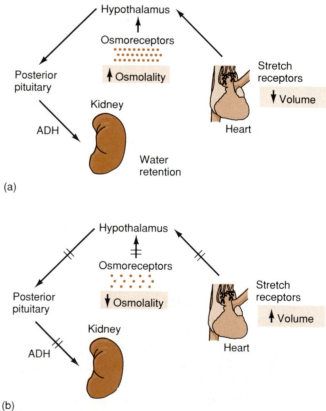

Figure 11.11 (a) ADH secretion is stimulated by increased osmolality of extracellular fluid in the brain or by a decrease in blood volume in the left atrium of the heart. (b) Decreased osmolality or stimulation of stretch receptors in the left atrium inhibit the secretion of ADH.

Decreased extracellular osmolality causes influx of water into the osmoreceptors that inhibits ADH secretion.

There are nonosmotic factors that influence ADH secretion as well. Specialized neurons called **stretch receptors** are located within the left atrium of the heart, and these receptors respond to variations in blood volume in that chamber. Increased volume causes the atrial walls to stretch, which stimulates the stretch receptors to transmit impulses to the hypothalamus. ADH secretion is inhibited. A decrease in volume causes a decrease in nerve impulse transmission, and ADH secretion increases.

Various other factors, such as pain, stress, and fever, stimulate ADH secretion, while ethanol is a common inhibitor (chapter 2).

Effects of ADH

Cells that compose the walls of the distal convoluted tubules and collecting ducts are responsive to ADH. The hormone combines with cell **membrane receptor sites,** initiating a series of chemical events that lead to an increase in water permeability. Under the influence of ADH, water reabsorption is augmented, increasing extracellular volume and producing a concentrated urine. An increased volume of dilute urine is produced in the absence of ADH.

ADH is also called vasopressin because it is a powerful vasoconstrictor. It is secreted in response to hemorrhage and contributes to maintaining blood pressure by way of vasoconstriction.

• • •

The **half-life** of a substance is the time required for half of the plasma concentration to be removed by metabolism or elimination from the body. ADH is metabolized by the kidney and liver and has a half-life of about 10 minutes.

• • •

Aldosterone

There is a significant structural feature of the nephron that relates to hormone production. The distal convoluted tubule is close to both the glomerulus and proximal convoluted tubule of the same nephron. There are specialized smooth muscle cells of the afferent and efferent arterioles called **juxtaglomerular cells** that store the enzyme renin (ren'in). The segment of the distal tubule in contact with the **juxtaglomerular cells** is made of specialized cells called the **macula densa.** The combination of the two is called the **juxtaglomerular apparatus** (figure 11.12). The juxtaglomerular cells release renin into the bloodstream in response to various stimuli including a decrease in plasma sodium ion concentration, a decrease in chloride ions in the tubular fluid at the macula densa, and a decrease in blood pressure in the afferent arterioles. Renin acts on a plasma protein called **angiotensinogen** (an"je-o-ten'sin-o-jen) to form **angiotensin I** (figure 11.13). In the presence

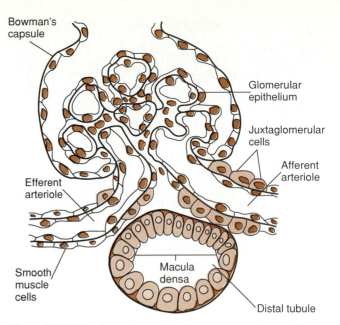

Figure 11.12 The juxtaglomerular apparatus consists of the macula densa and the juxtaglomerular cells. Juxtaglomerular cells store the enzyme renin.

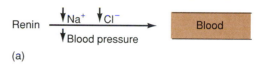

Juxtaglomerular cells

(a)

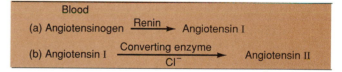

(b)

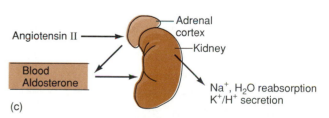

(c)

Figure 11.13 (*a*) The release of renin leads to (*b*) the production of angiotensin II, which in turn (*c*) stimulates the release of aldosterone. Aldosterone promotes sodium ion and water retention and secretion of K⁺/H⁺.

of converting enzyme and chloride ions, angiotensin I is changed to angiotensin II. The effects of angiotensin II are shown in table 11.4. Angiotensin II stimulates the release of **aldosterone** from the adrenal cortex. This hormone is carried by the bloodstream and acts on the distal convoluted tubule to promote reabsorption of sodium ions. Water is osmotically drawn out of the tubule by the efflux

TABLE 11.4 Physiological effects of angiotensin II

Target Tissue	Effect
Adrenal cortex	Stimulates release of aldosterone
Arterioles	Vasoconstriction
Posterior pituitary	Release of ADH
Hypothalamus	Stimulates thirst

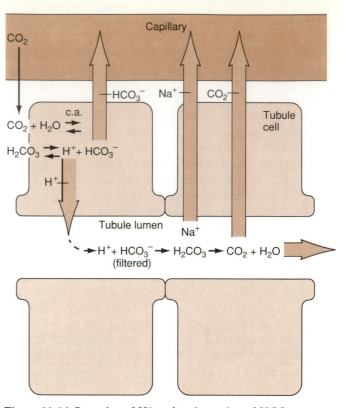

Figure 11.14 Secretion of H^+ and reabsorption of HCO_3^-. Carbon dioxide diffuses into tubular cells and reacts with water, catalyzed by the enzyme carbonic anhydrase, to form H^+ and HCO_3. Bicarbonate diffuses into extracellular fluid and hydrogen ions are secreted into tubular fluid. Bicarbonate, filtered from glomeruli, reacts with H^+ to form CO_2 and water. Carbon dioxide freely diffuses into extracellular fluid.

of sodium. The net effect is to increase extracellular fluid volume. The combined effects of angiotensin II and aldosterone increase blood pressure. There is a reciprocal exchange of potassium and hydrogen ions for sodium ions, so aldosterone enhances urinary excretion of both potassium and hydrogen.

• • •

Excessive amounts of aldosterone lead to a mild metabolic alkalosis due to loss of H^+. This is further aggravated by K^+ losses.

• • •

Other Hormonal Effects

Atrial natriuretic factor (ANF) is a hormone synthesized in and secreted by the atria of the heart. The stimuli for secretion of this hormone is increased plasma sodium ion concentration or increased extracellular volume. ANF promotes sodium ion excretion with diuresis (chapter 21).

Parathormone (PTH) is produced by the parathyroid glands and influences the renal handling of biphosphate and calcium ions. PTH increases biphosphate excretion and promotes reabsorption of calcium (chapter 31).

CONTROL OF ACID-BASE BALANCE

The kidney responds to acid-base imbalance within 24 hours and is responsible for long-term control. An overview of the homeostatic response is that hydrogen ions are eliminated and bicarbonate is reabsorbed in response to acidosis. The process is reversed in alkalosis.

Secretion of Hydrogen Ions

An important renal mechanism for correcting acid-base imbalance involves the secretion of hydrogen ions. The steps in this process are shown in figure 11.14. Carbon dioxide diffuses into the tubule cells and is enzymatically hydrated to form carbonic acid. Hydrogen ions formed by the ionization of carbonic acid are actively transported into the tubular lumen as sodium ions diffuse into the renal cell. The bicarbonate, also formed from the ionization of carbonic acid, diffuses into interstitial fluid and is accompanied by sodium ions. The secretion of one hydrogen ion leads to the reabsorption of one bicarbonate ion.

• • •

Bicarbonate is lost from the body in the process of buffering acids. It is lost in the form of exhaled CO_2 as the result of the following reaction:

$$H^+ + HCO_3^- \rightleftharpoons H_2CO_3 \rightleftharpoons CO_2 + H_2O$$

The kidney replenishes the body's store of HCO_3^- by the hydration of CO_2 in tubular cells ($CO_2 + H_2O \rightleftharpoons H_2CO_3 \rightleftharpoons H^+ + HCO_3^-$). This bicarbonate is newly formed and is not the same bicarbonate that was filtered into the Bowman's capsule.

• • •

Urinary Excretion of Hydrogen Ions

Hydrogen ions are eliminated in the urine in the following ways (figure 11.15). Hydrogen ions, secreted into tubular fluid, combine with bicarbonate to form carbonic acid. Catalyzed by the enzyme carbonic anhydrase, carbonic acid breaks down to carbon dioxide and water. Carbon dioxide diffuses into extracellular fluid. The water produced in the tubular fluid represents one form in which hydrogen ions are eliminated.

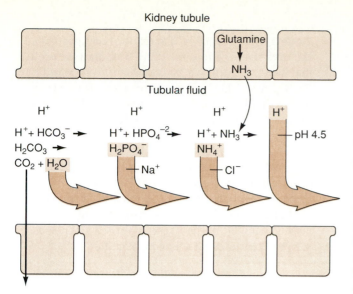

Kidney tubule

Tubular fluid

H^+ H^+ H^+ H^+

$H^+ + HCO_3^-$ → $H^+ + HPO_4^{-2}$ → $H^+ + NH_3$ →   pH 4.5

H_2CO_3 → $H_2PO_4^-$ NH_4^+

$CO_2 + H_2O$

Na^+ Cl^-

Figure 11.15 Mechanisms for eliminating H^+ in the urine, i.e., water, dihydrogen phosphate, ammonium, and free hydrogen ions. Hydrogen ions are buffered to prevent the formation of highly acidic urine.

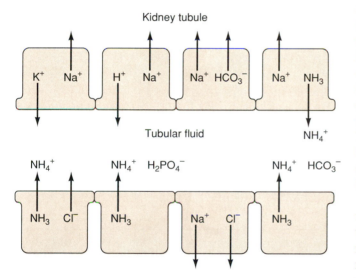

Kidney tubule

K^+ Na^+ H^+ Na^+ Na^+ HCO_3^- Na^+ NH_3

Tubular fluid NH_4^+

NH_4^+ NH_4^+ $H_2PO_4^-$ NH_4^+ HCO_3^-

NH_3 Cl^- NH_3 Na^+ Cl^- NH_3

Figure 11.16 Interrelationships of ions as they are secreted or reabsorbed in the kidney tubules. The movements are such that electrical neutrality is maintained.

A second type of reaction is the buffering of hydrogen ions with biphosphate in the tubular fluid. The product of this reaction is dihydrogen phosphate, a weak acid that dissociates to a limited extent. Dihydrogen phosphate is eliminated in the urine accompanied by sodium.

A third mechanism for the excretion of hydrogen ions involves ammonia. Ammonia is formed in renal tubular cells from amino acids, notably glutamine. Glutamine is enzymatically converted, first into glutamic acid, and then into alpha ketoglutaric acid. Two molecules of ammonia are products of these reactions. The ammonia diffuses into tubular fluid and combines with hydrogen ions to form

ammonium. The ammonium ion does not diffuse out of the tubule and is eliminated in urine, accompanied by the chloride anion.

A fourth mechanism for the elimination of hydrogen ions is secretion into tubular fluid and urinary excretion in the form of free ions. Hydrogen ions decrease the pH of urine. The ability of the nephron to secrete free hydrogen ions is limited, so the minimum pH of urine is 4.5. This limitation protects the renal tubules from highly acidic conditions.

Hydrogen ions (1) are secreted into the renal tubule and are limited to a concentration represented by pH 4.5, (2) are buffered by bicarbonate in the tubular fluid with the formation of water and carbon dioxide, (3) are combined with biphosphate in the tubular fluid to form a weak acid, dihydrogen phosphate, and (4) are tied up in the tubular fluid by the conversion of ammonia (NH_3) to ammonium (NH_4^+).

Interrelationships of Ions

There is a principle of maintaining electrical neutrality as ions are reabsorbed or secreted in the renal tubules (figure 11.16). The following generalizations are useful in understanding kidney function. The reabsorption of sodium ions is coupled with (1) reabsorption of chloride ion, (2) reabsorption of bicarbonate, (3) secretion of potassium ions, or (4) secretion of hydrogen ions. Any of these four events maintain electrical balance. The distal convoluted tubule is less permeable to water and ions, so the reabsorption of sodium in this segment of the tubule creates an electrical potential that favors K^+ and H^+ secretion. There is a reciprocal relationship between potassium and hydrogen ions and, although the exchange is not necessarily one to one, when one is secreted the other tends to be retained.

Increased acidity of the tubular fluid favors the formation of ammonium. The formation of this cation in the tubular fluid allows either sodium or potassium to be reabsorbed, accompanied by either bicarbonate or chloride ions. The ammonium ion is accompanied by an anion, chloride, dihydrogen phosphate, or bicarbonate.

RENAL PATHOLOGY

The major focus of this unit is the effects of disease processes involving the excretory system, especially the kidney. This chapter provides a perspective in terms of relating locus of kidney damage and consequences of that injury. In general the renal effects of disease or toxins are initially selective, i.e., effects on (1) renal blood flow, (2) glomerular filtration, (3) cortical convoluted tubules, (4) medullary tubules, or (5) the renal pelvis. Progressive effects lead ultimately to kidney failure in which all aspects of function are affected.

Evaluation and Control of Renal Function

The discussion that follows involves evidence of renal disorders, i.e., symptoms and diagnostic tests, as well as diuretic control of urinary output.

Symptoms

There are a number of symptoms of renal dysfunction. The following is a summary of some symptoms that may be observed. In subsequent chapters, there is further discussion of symptoms caused by specific disease processes.

Volume and Appearance of Urine

Blood flow to the kidneys is about 1200 ml/min with glomerular filtrate forming at a rate of 125 ml/min. The total filtrate per day is about 180 liters, with all being reabsorbed except about 1 liter. A healthy adult, assuming average fluid intake, produces between 0.5–2.5 liters of urine per day. Changes in total output and frequency of urination are significant observations. **Nocturia** (nok-tū′re-ah), or urination at night, indicates a diminished renal concentrating ability.

The appearance of urine provides some information about function. Concentrated urine is darker, blood in the urine may be visible, and a frothy urine indicates excessive protein or the presence of bile salts.

Edema and Pain

Edema or water retention is a sign of certain renal diseases and is evidenced by puffiness around the eyes, in the entire face, or in the lower parts of the body. Renal pain occurs at the costovertebral (kos″to-ver′tĕ-bral) angle, which is the angle formed by the twelfth rib and the spinal column. The pain may radiate anteriorly and down the abdominal wall. Pain originating in the bladder is usually in the suprapubic region. Obstruction of the ureters causes severe pain called **renal colic,** which often radiates into the lower abdomen, testes, or labia.

Diagnostic Tests

There are diagnostic tests that show both structural abnormalities and specific functional deficiencies. The following are tests that indicate renal functional capacity.

nocturia: L. *nox*, night; Gk. *ouron*, urine
suprapubic: L. *supra*, above; L. *pubes*, private parts

Urea is synthesized by a series of reactions called the urea cycle. The urea cycle was the first cyclic metabolic pathway to be discovered and was first identified in 1932 by Hans Krebs and Kurt Henseleit. A breakdown product of amino acids, the ammonium ion (NH_4^+) is converted to urea in the liver. Liver disease leads to toxic levels of NH_4^+ in the blood. There are inherited disorders in which there is a partial block of urea cycle reactions. The outcome is death immediately after birth in cases where there is a complete absence of urea cycle enzymes. Partial enzyme deficiencies lead to mental retardation. Mild forms of these inherited deficiencies may be controlled by a low protein diet.

• • •

Blood Urea Nitrogen

Urea is a waste product of protein metabolism. It is a nitrogen containing substance (figure 11.17) produced in the liver and is carried by the blood to the kidney. Urea is removed from the blood by glomerular filtration. It diffuses freely across cell membranes, and consequently, some reabsorption occurs in the nephron tubules by passive diffusion. Blood levels of urea are expressed as urea nitrogen, also called blood urea nitrogen (BUN). The concentration of urea in the blood is determined by (1) protein ingestion, (2) protein metabolism, and (3) excretion in the urine. Hormonal influence is a factor because certain hormones promote either protein **anabolism** (synthesis) or **catabolism** (breakdown). Androgens and growth hormone have an anabolic effect, which may lower blood levels of urea. Corticosteroids and thyroxine both promote protein catabolism, which tends to increase the BUN.

• • •

The concentration of urea is the same in all body fluids including saliva, plasma, perspiration, and cerebrospinal fluid.

• • •

The normal range of values for BUN is 6–22 mg/dl. Pathological processes that cause deviations from normal fall into three categories:

1. Any process resulting in fluid imbalance, either dehydration or edema
2. Conditions in which there is excessive protein catabolism
3. Decreased renal function.

The first two may be associated with renal disease.

anabolism: Gk. *anabole*, a rising up
catabolism: Gk. *katabole*, a throwing down

Urea cycle

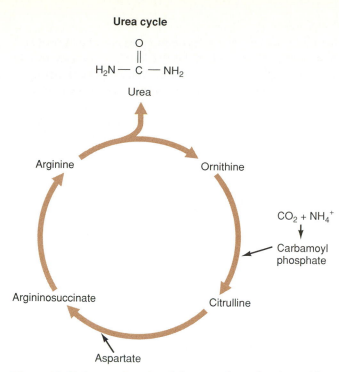

Figure 11.17 Ammonia, a breakdown product of amino acids, is used in the synthesis of urea.

Nephron

Water deficit

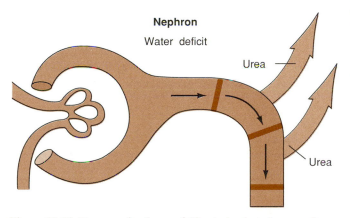

Figure 11.18 Decreased volume of filtrate leads to increased solute concentration, which favors reabsorption of urea. Decreased volume of filtrate also results in a decrease in the rate of flow, which in turn promotes reabsorption of urea.

Water retention or deprivation influences blood levels of urea by way of affecting urine flow. Low urine volume increases solute concentration and creates an increased tubular/interstitial concentration gradient (figure 11.18). This favors urea reabsorption. A low urine volume also means decreased flow, which results in additional time for reabsorption to occur. Urinary excretion of urea is most efficient with increased urine flow.

A number of conditions, including diabetes mellitus, hyperthyroidism, excessive glucocorticoid production, and neoplasms, lead to increased protein breakdown. When such a condition is accompanied by renal dysfunction, the result is chronic elevated levels of blood urea.

Creatine

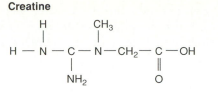

Creatine phosphate (phosphocreatine)

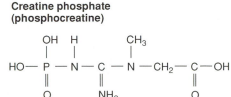

Creatinine

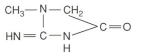

Figure 11.19 Structures of three nitrogen-containing compounds: creatine, creatine phosphate, and creatinine.

Various pathological processes in renal disease cause an elevated BUN. These processes may involve blood flow to the kidney, glomerular/tubular injury, or obstruction leading to increased hydrostatic pressure.

• • •

Creatine is synthesized from three amino acids: glycine, arginine, and methionine. Creatine is converted to phosphocreatine, which temporarily stores high-energy phosphate groups. A reaction that keeps ATP at high levels in muscle involves the transfer of phosphate from phosphocreatine to ADP to form ATP. The reaction is catalyzed by the enzyme creatine kinase. Creatine kinase leaks into blood when injury to heart muscle occurs due to myocardial infarction; thus, an elevated blood level of creatine kinase has diagnostic significance in suspected myocardial infarction.

• • •

Creatinine

Creatine and **creatinine** are both nitrogen-containing compounds (figure 11.19); the latter is eliminated in the urine. Creatine is synthesized in the liver from amino acids and is carried by the blood to the muscles and the brain. Small amounts appear in the urine and are reabsorbed. Creatinine is a product of chemical reactions in muscle involving creatine. Creatinine appears in the urine by filtration and secretion. Determination of blood creatinine levels is useful in evaluating kidney function because those levels are unaffected by dietary protein or urine volume. Since creatinine is affected by muscle mass, the normal values are somewhat different for men and women, i.e., 0.6–1.5 mg/dl for men and 0.1 mg/dl lower for women. Creatinine levels are not increased in the early stages of renal failure.

BUN/Creatinine Ratio

The preceding discussion outlined the fact that blood levels of urea may be elevated by reason of (1) prerenal conditions involving water imbalance, (2) renal dysfunction, or (3) a postrenal disorder, such as ureteral obstruction (figure 11.20). Both creatinine and urea appear in the urine mainly by glomerular filtration, but urea is partially reabsorbed as well (figure 11.21). The BUN/creatinine ratio is useful in evaluating the cause of an elevated BUN. In general, a high ratio is the result of decreased urine flow, which does not affect creatinine, but allows increased urea reabsorption. A ratio greater than 20:1 is characteristic of dehydration, ureteral obstruction, and decreased renal blood flow.

Elevated BUN

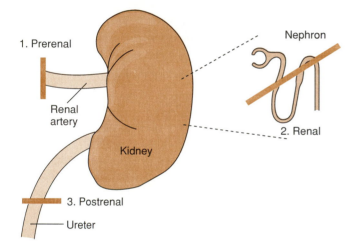

Figure 11.20 There are three general reasons for an elevated BUN: (*1*) prerenal in which there is decreased blood flow to the kidney, (*2*) renal dysfunction, or (*3*) postrenal in which there is obstruction to outflow of urine.

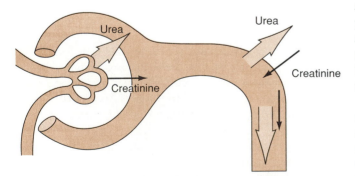

Figure 11.21 Creatinine appears in the urine by filtration and secretion. Urea is both filtered and reabsorbed.

Renal Clearance Tests

Several tests have been devised to measure the removal of a substance from plasma by the kidney. The rate of removal is expressed as volume of cleared plasma per unit of time, usually per minute. Solute removal is accomplished by glomerular filtration and/or secretion by cells of the convoluted tubules and collecting ducts. A clearance test is an indicator of **glomerular filtration rate** if the solute, the removal of which is being measured, is solely or predominantly filtered into Bowman's capsule and is not also secreted.

There are two categories of clearance tests. One involves the intravenous administration of a substance known to be filtered by the kidney, correlated with assay of that substance as it appears in the urine. The other type of test involves measurement of blood and urine concentrations of an **endogenous substance,** that is substance normally produced by the body.

Inulin and **para-aminohippuric acid** are two exogenous substances, i.e., not normally produced by the body, which may be used for clearance tests. The procedure includes four main steps:

1. A blood and urine sample are collected.
2. An exogenous substance is administered intravenously.
3. Urine flow is promoted by oral or intravenous fluids.
4. Urine (sometimes collected by catheterization) and blood samples are taken, and the substance is measured in all samples.

• • •

Inulin is a starchlike polymer of fructose that is freely filtered from the glomerulus but is neither reabsorbed nor secreted. Para-aminohippuric acid is filtered freely and is secreted by nephron tubules as well.

• • •

The appearance of either creatinine or urea in the urine, both endogenous substances, is an indicator of renal clearance and glomerular filtration rate. The rate of creatinine production is constant, so the blood level of creatinine is also constant. Creatinine is primarily eliminated by glomerular filtration with limited tubular secretion. Urea is both filtered and reabsorbed, so urinary concentration of urea is not solely the result of glomerular filtration.

The creatinine clearance test requires the determination of creatinine concentration in one blood sample and creatinine in a 24-hour urine collection. The steps in evaluating creatinine clearance are shown in table 11.5, and the basic premise is that the rate of formation is equal to the rate of excretion. Range of normal values for creatinine clearance in men is 90–139 ml/min and in women

endogenous: Gk. *endon*, within

TABLE 11.5 Steps in evaluating creatinine clearance using normal values

1. Urinary creatinine (U cr) = 120 mg/ml
2. Flow rate of urine ($\dot{V}$) = 2 ml/min
3. (U cr)($\dot{V}$) = (120 mg/ml)(2 ml/min) = 240 mg/min
4. Creatinine excreted = amount cleared from blood = 240 mg/min
5. Plasma creatinine (P cr) = 2 mg/ml
6. Creatinine clearance is the volume required per minute to supply creatinine at the rate excreted in urine; glomerular filtration rate (GFR)
7. Therefore, (GFR)(P cr) = (U cr)($\dot{V}$)
   (GFR)(2 mg/ml) = (120 mg/ml) (2 ml/min)
8. Then, GFR = 120 ml/min

80–125 ml/min. The value for the creatinine clearance test is the plasma volume per minute required to supply creatinine at the rate excreted. It is equivalent to the glomerular filtration rate, assuming that creatinine is only filtered by the nephron.

The significance of creatinine clearance should be considered in terms of several facts. Plasma creatinine concentration is not directly proportional to early kidney damage, because there may be compensatory increases in secretion. Moderate impairment of filtration is indicated by creatinine clearance values of 28–42 ml/min and less than 28 ml/min indicates marked impairment. Age is a factor because creatinine clearance decreases with age, starting at age 20, and elderly patients with decreased muscle mass may have lower values.

Concentration Test

A decrease in the kidney's ability to produce urine with a higher solute concentration as compared to plasma indicates failing renal function. The urine concentration test measures this ability. It involves restriction of water ingestion for 14–16 hours and collection of urine at 1, 2, and 4 hour intervals. Specific gravity of the urine samples is determined (figure 11.22). The specific gravity of plasma is 1.010. A urine sample with a specific gravity of less than 1.020 in this test indicates decreased renal function. The specific gravity approaches 1.010 late in the course of renal disease. Normal specific gravity values are between 1.016 and 1.022 with normal fluid intake.

IMAGING TECHNIQUES

There are a number of techniques that make it possible to visualize the excretory system, i.e., size and location of kidneys, location of obstructions, and lesions. Some of these techniques are summarized in table 11.6. The intravenous

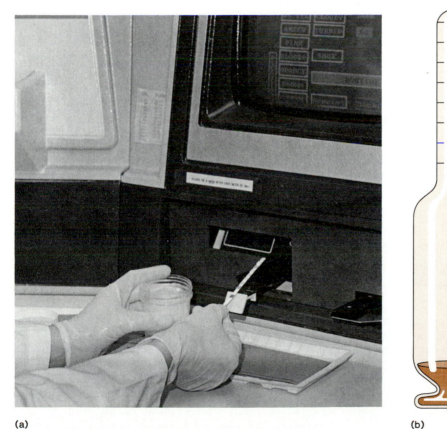

(a)

(b)

Figure 11.22 (*a*) A reagent test strip is dipped into a urine sample. The instrument scans the strip and provides a reading of total dissolved solids (specific gravity) and an analysis of the chemical composition of the urine. (*b*) A urinometer provides for a direct reading of specific gravity of a urine sample determined by the level at which it floats in the urine. A reading is taken at the bottom of the meniscus. A urinometer is calibrated to read 1.000 in pure water at 25°C.

Excretory Function: Evaluation and Control

TABLE 11.6 Imaging techniques to show abnormalities of the excretory system

X-ray of Abdomen
Often shows position, shape, and size of kidneys; areas of calcification

Excretory Urogram (Intravenous Pyelogram)
Intravenous injection of iodinated contrast material; X-rays after injection

Retrograde Pyelogram
Contrast material introduced by means of a catheter; X-rays show renal pelvis, ureters, and bladder

Computerized Axial Tomography (CAT scan)
X-ray source rotates around area to be examined; computer draws a picture of the area

Nuclear Magnetic Resonance Scan
Signals are sent out by atoms in the body when exposed to a strong magnetic field; signals are converted to images

TABLE 11.7 Generic and brand names of different categories of diuretics

Category	Generic Name	Brand Name
Osmotic diuretic	Mannitol	Osmitrol
Carbonic anhydrase inhibitor	Acetazolamide	Diamox
Loop diuretics	Ethacrynic acid	Edecrin
	Furosemide	Lasix
	Bumetanide	Bumex
Potassium sparing	Triamterene	Dyrenium
	Amiloride	Midamor
	Spironolactone	Aldactone
Benzothiadiazides	Chlorothiazide	Diuril

pyelography tests require the injection of one of several iodine-containing substances which is quickly cleared from the blood by glomerular filtration and tubular secretion. The iodine is relatively opaque to X-rays and appears as a shadow on X-ray film. If the image is not clear, it indicates a decrease in renal function. In the case of a retrograde pyelogram, the iodinated contrast material is introduced directly by catheter rather than intravenously. **Computer axial tomography scans (CAT scans)** provide computer generated pictures from a series of X-rays, and the nuclear magnetic resonance procedure does not require X-rays.

• • •

The prefix pyelo refers to the pelvis of the kidney. Roentgenogram (rent-gen′ō-gram″) is another term for X-ray film.

• • •

DIURETIC CONTROL

The term diuresis means an increase in urine flow. **Water diuresis** is caused by ingestion of a large volume of fluid. **Osmotic diuresis** is caused by a high concentration of nonreabsorbable solute in renal tubules.

• • •

Organomercurial diuretics were developed as the result of a search for drug therapy for syphilis in the 1920s. They were the first known diuretics to act on the loop of Henle. This category of diuretics has fallen out of favor due to poor absorption from the gastrointestinal tract, slow onset of action, and potential for toxic effects on the kidney.

• • •

TABLE 11.8 Classification, examples, and major actions of diuretics

Osmotic Diuretics

Mannitol; increased loss of Na^+, K^+, Cl^-, HCO_3^-, Ca^{+2}, and Mg^{+2}

Carbonic Anhydrase Inhibitors

Acetazolamide; inhibits secretion of H^+; increased loss of Na^+, K^+, and HCO_3^-

Thiazide Diuretics

Chlorothiazide; inhibits reabsorption of Na^+ and Cl^-; increased loss of Na^+, Cl^-, K^+, and HCO_3^-

Loop Diuretics

Ethacrynic acid and furosemide; inhibit reabsorption in ascending limb; increased excretion of Na^+, Cl^-, K^+, and Ca^{+2}; increased loss of HCO_3^- with furosemide

Potassium Sparing Diuretics

Spironolactone; competitive antagonist of aldosterone; K^+ and H^+ excretion decreased; mild Na^+ and Cl^- loss; triamterene, amiloride; increased excretion of Na^+ and Cl^-

Mannitol

Diuretics are agents that increase urine volume and are useful therapeutically in circumstances in which urine flow is inadequate or in the treatment of edema. Tables 11.7, 11.8, and 11.9 list several categories of diuretics, summarize the major action, and list therapeutic use of each. **Mannitol,** a low molecular weight sugar-alcohol, is an example of an osmotic diuretic. It is characterized by being freely filtered at the glomerulus with negligible reabsorption. The net result is increased osmolality of the filtrate, decreased water reabsorption, and increased urine volume.

retrograde: L. *retrograde,* to go backward

TABLE 11.9 Therapeutic uses of various diuretics

	Therapeutic Use	Comments
Mannitol	Intravenous use; acute renal failure; cerebral edema; decreases CSF pressure; decreases intraocular pressure	Contraindicated if anuria develops due to severe renal disease, marked pulmonary congestion, or intracranial hemorrhage
Acetazolamide	Oral, parenteral administration; decreases intraocular pressure; acute mountain sickness	Limited usefulness
Chlorothiazide	Oral, intravenous; hypertension; edema due to heart, liver, renal disease	Antidiuretic effect in diabetes insipidus
Furosemide	Oral, parenteral; edema due to heart, liver, or renal disease	Peak diuresis greater than other diuretics
Spironolactone	Oral use; hypertension; use with other diuretics to minimize K$^+$ loss	Relatively weak diuretic

Mannitol is used to reestablish urine flow in acute renal failure and is useful in decreasing intraocular and cerebrospinal fluid pressure. The presence of this solute in extracellular fluid draws water out of intraocular and cerebrospinal fluid.

Carbonic Anhydrase Inhibitors

The carbonic anhydrase inhibitors have only limited usefulness. Carbonic anhydrase has been identified as an enzyme that catalyzes the following reaction within renal tubule cells:

$$CO_2 + H_2O \rightleftharpoons H_2CO_3 \rightleftharpoons H^+ + HCO_3^-$$

Acetazolamide and the other carbonic anhydrase inhibitors interfere with this reaction. The effect is to inhibit reciprocal exchange of H$^+$ and Na$^+$ by the tubule. There is excretion of sodium accompanied by bicarbonate and loss of potassium. Metabolic acidosis may occur.

The distal tubule and collecting duct have a limited capacity for sodium reabsorption. It is for this reason that loop diuretics, also called high-ceiling diuretics, are potent. They act by inhibiting sodium reabsorption in the ascending limb of the loop of Henle. The reabsorptive capacity of the distal tubule is overwhelmed by the amount of sodium presented from the loop. There is low diuretic potency associated with carbonic anhydrase inhibitors because they exert their effect on the proximal tubule, and this is followed by normal reabsorption in the loop of Henle.

Other Diuretics

As table 11.8 shows, the remainder of the diuretics in the list exert their effects by promoting increased electrolyte loss, which must be accompanied by water. The loop diuretics inhibit reabsorption of sodium in the ascending limb of the loop of Henle and are potent agents that promote large fluid losses. Spironolactone is unique because it is an aldosterone antagonist. It acts by competitively inhibiting the binding of aldosterone to receptor proteins in the collecting ducts. There is consequent inhibition of sodium reabsorption and a reduction in entry of K$^+$ and H$^+$ into the ducts.

Generalizations

Some generalizations in regard to diuretic effects will be useful. Table 11.9 indicates therapeutic uses of various diuretics. The underlying mechanism for increasing urine volume is usually the inhibition of sodium and water reabsorption. In most cases, the locus of action is a segment of the tubule ahead of the point where potassium ions are secreted in the distal convoluted tubule (figure 11.23). Those diuretics that are potassium-sparing are exceptions to this. The fact that the filtrate presented to the distal tubule is loaded with sodium is significant because the secretion of potassium is coupled with reabsorption of sodium in this segment of the tubule. Sodium reabsorption creates an electrical gradient, i.e., the lumen becomes negative with respect to the interstitium. This gradient enhances the diffusion of potassium into the lumen. For this reason, hypokalemia is a common side effect of diuretic use.

Chloride is often the predominant excreted anion, and this means bicarbonate is reabsorbed. The result is hypochloremic metabolic alkalosis (chapter 4). Table 11.10 summarizes the effect of four diuretics and identifies possible acid-base imbalance, and this is further illustrated in figure 11.24.

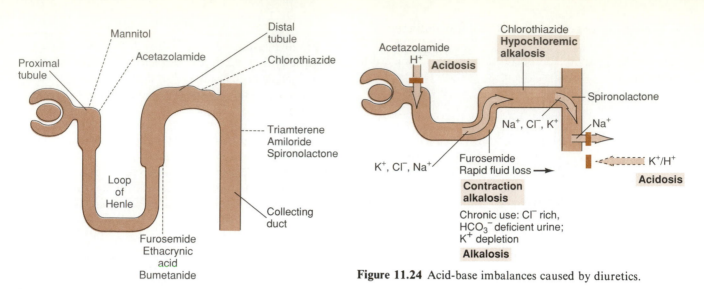

Figure 11.23 Major sites of action of various diuretics.

Figure 11.24 Acid-base imbalances caused by diuretics.

TABLE 11.10 Summary of the effects of four diuretics

1. Acetazolamide	Na^+, Cl^-, HCO_3^-, K^+, and uric acid loss; metabolic acidosis
2. Furosemide	Na^+, Cl^-, and K^+ loss; hypochloremic metabolic alkalosis
3. Chlorothiazide	Na^+, Cl^-, and K^+ loss; hypochloremic metabolic alkalosis
4. Spironolactone	Mild Na^+ and Cl^- loss; decreased excretion of K^+ and H^+; metabolic acidosis

• • •

Any combination of three mechanisms may be involved when a diuretic causes metabolic alkalosis. (1) There may be excessive secretion of hydrogen ions and consequent regeneration of bicarbonate by kidney tubules. (2) Excretion of chloride ions is accompanied by bicarbonate reabsorption. (3) Potassium depletion results in intracellular movement of hydrogen ions.

• • •

SUMMARY

• • •

The nephron is the functional unit of the kidney and consists of (1) a glomerulus, (2) Bowman's capsule, (3) proximal convoluted tubule, (4) loop of Henle, (5) distal convoluted tubule, and (6) collecting duct. Glomerular blood is filtered into Bowman's capsule, and the resulting filtrate is similar in composition to interstitial fluid. Reabsorption and secretion along the length of the tubule determines the final volume and composition of urine.

The kidney is able to produce a concentrated urine by means of the countercurrent mechanism. Juxtamedullary nephrons and a special capillary system called the vasa recta are the structural elements of the mechanism. The major events in the countercurrent mechanism are (1) reabsorption of sodium and chloride ions in the region of the ascending limb and (2) urea reabsorption from the collecting ducts. The result is high osmolality in the medullary interstitial fluid. The interstitial osmolality is maintained by the vasa recta, a capillary that extends into the medulla and has a loop configuration. In the descending vasa recta, there is efflux of water and influx of urea, sodium, and chloride ions. In the ascending segment, there is a reversal of these movements. The result is the removal of water and deposition of solute in the interstitium. The high solute concentration in the medullary interstitial fluid promotes water reabsorption from collecting ducts. The net effect is to concentrate the urine in the collecting ducts.

Hormonal effects on the kidney include the following. Antidiuretic hormone is released in response to increased osmolality of extracellular fluid and promotes water reabsorption by kidney tubules. Aldosterone release occurs as the result of a fall in blood pressure or decreased sodium ion concentration. Aldosterone promotes renal retention of both sodium and water. Atrial natriuretic factor promotes sodium ion excretion with diuresis. Parathormone favors renal biphosphate excretion and reabsorption of calcium.

The kidney provides long-term control of acid-base balance. Hydrogen ions are secreted indirectly as the result of a reaction within the tubule cells, $CO_2 + H_2O \rightleftharpoons H_2CO_3 \rightleftharpoons H^+ + HCO_3^-$. The hydrogen ions thus formed are secreted as bicarbonate is reabsorbed. Hydrogen ions are buffered in the urine by combining with ammonia, biphosphate, or bicarbonate.

In general, the reabsorption of sodium ion is coupled with (1) reabsorption of chloride, (2) reabsorption of bicarbonate, (3) secretion of potassium ions, or (4) secretion of hydrogen ions. The excretion of ammonium ions in the urine provides for reabsorption of either sodium or potassium. Ammonium ion is accompanied by an anion. There is a reciprocal relationship in the secretion of K^+ and H^+.

Some nonspecific symptoms of renal disease are changes in the volume of urinary output, nocturia, edema, and pain. Diagnostic tests include BUN, creatinine, creatinine clearance, and the urine concentration test. Blood urea nitrogen is influenced by protein ingestion and metabolism. Decreased blood flow to the kidney, glomerular/tubular injury, or obstruction lead to an elevated BUN.

Creatinine is formed from creatine and blood creatinine is affected by muscle mass, not by diet. A high BUN/creatinine ratio usually is caused by decreased urine flow that does not affect creatinine, but allows increased urea reabsorption.

Creatinine clearance is an indicator of glomerular filtration rate because this substance is filtered, not reabsorbed and is secreted only to a limited extent. Both endogenous and exogenous substances may be used for clearance tests, provided they have these characteristics. The urine concentration test shows the kidney's ability to produce urine that is more concentrated than plasma.

Imaging techniques make it possible to visualize the excretory system and include pyelograms, CAT scan, and nuclear magnetic resonance scan.

Diuretic effects are usually based on the inhibition of sodium and water reabsorption. There are varying degrees of electrolyte loss with hypokalemia as a common side effect accompanied by acid-base imbalance.

REVIEW QUESTIONS

• • •

1. Define the term renal hilus.
2. What makes up the renal medulla?
3. Define the term calyx.
4. Blood flows through two capillary systems between the renal artery and renal vein. What are these capillaries?
5. What is/are the force(s) favoring glomerular filtrate formation?
6. What forces oppose filtrate formation?
7. What process is most important in the proximal convoluted tubule?
8. Compare the descending and ascending limbs in terms of water permeability.
9. In what segment of the nephron is most of the HCO_3^- reabsorbed?
10. In what segments of the nephron are H^+ secreted?
11. Which nephrons are involved in the countercurrent mechanism?
12. Summarize the events that create high osmolality in the medullary interstitial fluid.

13. What mechanism maintains high osmolality in interstitial fluid around the collecting ducts?
14. High extracellular osmolality (stimulates/inhibits) secretion of ADH.
15. What is the source of renin?
16. What is the source of secreted H^+?
17. What two events in the kidney compensate for acidosis?
18. How are H^+ buffered in urine?
19. List four things that may occur to maintain electrical neutrality when Na^+ are reabsorbed.
20. What chemical reaction in renal tubular cells allows either Na^+ or K^+ to be reabsorbed?
21. What is the source of urea?
22. What processes determine the amount of urea excreted?
23. List a hormone that favors anabolism and one that favors catabolism.
24. Identify three general causes of an abnormal BUN.
25. Why does edema affect BUN values?
26. Ureteral obstruction would cause (elevated, lowered) BUN. Why?
27. What is the source of creatinine?
28. List three kinds of problems indicated by an increased BUN/creatinine ratio.
29. What does a creatinine clearance of 120 ml/min mean?
30. What imaging technique does not require X-rays?
31. What is the mechanism of action of osmotic diuretics?
32. What is the relationship between potassium depletion and metabolic alkalosis?
33. Carbonic anhydrase inhibitors cause metabolic (acidosis/alkalosis).
34. What is the mechanism of action of most diuretics?
35. Identify a specific aldosterone antagonist.
36. Why is hypokalemia a common adverse side effect of diuretic therapy?
37. What is the significance of Cl^- being the predominant excreted anion in diuretic therapy?
38. What is the primary site of action of spironolactone?
39. Why are the loop diuretics especially potent?
40. The thiazide diuretics cause (acidosis, alkalosis).

SELECTED READING

• • •

Brenner, B. M., K. F. Badr, and I. Ichikawa. 1980. Hormonal influences on glomerular filtration. *Mineral Electrolyte Metabolism* 4:49–56.

Chan, A. Y. et al. 1988. Functional response of healthy and diseased glomeruli to a large, protein-rich meal. *Journal of Clinical Investigation* 81:245–54.

Copley, J. B. 1991. What can be learned from urinalysis? *Physician Assistant* 15:21–26.

Garcia-Cuerpo, E. et al. 1988. Bone metaplasia in the urinary tract; a new radiological sign. *Journal of Urology* 139:104.

Glazer, G. M. 1988. MR imaging of the liver, kidneys and adrenal glands. *Radiology* 166:303–12.

Karniski, L. P., and P. S. Aronson. 1987. Formate: A critical intermediate for chloride transport in the proximal tubule. *News in Physiological Sciences* 2:160–63.

Katz, F., R. Eckert, and M. Gebott. 1972. Hypokalemia caused by surreptitious self-administration of diuretics. *Annals of Internal Medicine* 76:85–90.

Lacasse, J. et al. 1987. Cultured juxtaglomerular cells: Production and localization of renin. *Canadian Journal of Physiology and Pharmacology* 65:1409–15.

Laragh, J. H. 1985. Atrial natriuretic hormone. The renin-aldosterone axis, and blood pressure-electrolyte homeostasis. *New England Journal of Medicine* 313:1330–39.

Lifschitz, M. D., and H. H. Stein. 1983. Hormonal regulation of renal salt excretion. *Seminars in Nephrology* 3:196–202.

Medical Letter on Drugs and Therapeutics. 1981. Amiloride-a potassium-sparing diuretic. Medical Letter Inc., New Rochelle, NY. 23:109.

Medical Letter on Drugs and Therapeutics. 1983. Bumetanide (Bumex)-a new "loop" diuretic. Medical Letter Inc., New Rochelle, NY. 25:61.

Medical Letter on Drugs and Therapeutics. 1984. Indapamide (Lozal)-a new antihypertensive agent and diuretic. Medical Letter Inc., New Rochelle, NY. 26:17.

Peterson, L. N. et al. 1988. Plasma AVP and renal concentrating defect in chloride depletion metabolic alkalosis. *American Journal of Physiology* 254:F15–24.

Sands, J. M. et al. 1987. Vasopressin effects on urea and H_2O transport in inner medullary collecting duct subsegments. *American Journal of Physiology* 253:F823.

Schwartz, D., N. C. Katsube, and P. Needleman. 1986. Atriopeptins in fluid and electrolyte homeostasis. *Federation Proceedings* 45:2361–64.

Sullivan, L. P. 1975. *Physiology of the Kidney.* Lea and Febiger, Philadelphia.

Walser, M. 1986. Roles of urea production, ammonium excretion and amino acid oxidation in acid-base balance. *American Journal of Physiology* 250:F181–88.

Chapter 12

Urinary Tract: Infection, Obstruction, and Nephrotoxins

The excretory system is susceptible to injury by a variety of agents and pathological processes. There are three areas of focus in this chapter related to urinary tract injury: (1) the effects of infection, (2) obstruction, and (3) nephrotoxins.

INFECTIONS

Urinary tract infections are common and occur more frequently in females than in males. The organisms responsible for infection are usually from the patient's lower intestinal tract and are gram-negative bacilli, such as *Escherichia coli, Klebsiella, Proteus, Pseudomonas,* and others. Sometimes gram-positive cocci belonging to the genera *Staphylococcus* or *Streptococcus* cause infections. Other bacterial, as well as viral and fungal organisms, may be etiological agents under special circumstances. For example, *Chlamydia trachomatis, Neisseria gonorrhoeae,* and the herpes simplex virus are sexually transmitted organisms that may infect the urinary tract.

PREDISPOSING FACTORS

Obstruction to urine flow predisposes to urinary tract infection. Causes of obstruction include (1) kidney stones, (2) hypertrophy of the prostate gland, (3) tumors, or (4) diminished nerve impulses to the bladder. The latter is called **neurogenic bladder dysfunction** and may be the result of spinal cord injury, multiple sclerosis, or other neuropathies.

• • •

The term neuropathy refers to a disease process involving the nervous system. An example is neuropathy associated with long-standing diabetes mellitus. It is characterized by damage to Schwann cells and axons and by degeneration of myelin. The affected nerves include autonomic nerves supplying the bladder.

• • •

The probability of infection is increased by other factors as well. For example, women are at risk because the female urethra is only about 4 cm long and is close to the anus, which increases the opportunity for introduction of bacteria into the urethra. Catheterization or any procedure that involves introducing an instrument into the urinary tract carries a high risk of infection. Upper urinary tract infections occur more commonly during pregnancy.

URETHRITIS AND CYSTITIS

Invading organisms usually gain entry by way of the urethra, causing urethritis or an inflammation of the urethra. Symptoms of urethritis are frequency of urination

and **dysuria** (painful urination). Cystitis, or inflammation of the bladder, may follow with accompanying symptoms of dysuria, frequency of urination, and suprapubic pain. Urinalysis shows pyuria or pus, bacteria, and possibly red blood cells in the urine. Treatment involves appropriate antibiotic therapy, depending on the organism causing the infection.

ACUTE PYELONEPHRITIS

Pyelos refers to the pelvis of the kidney. Pyelonephritis is an inflammation of the kidney involving the renal pelvis and may include the interstitium and tubules as well (figure 12.1). The acute form of pyelonephritis usually represents an upward progression of a urinary tract infection. A predisposing factor is **vesicoureteral reflux** (ves"ĭ-ko-u-re'ter-al). This reflux is the backflow of urine from the bladder into ureters and renal pelvis. This occurs more commonly in children who either have a urinary tract infection or an anatomical abnormality of the ureter-bladder (ureterovesical) junction. The reflux of urine occurs during micturition.

Pyelonephritis causes the development of abscesses on the cortical surface of the kidney and within the kidney as well. The result is destruction of tubules.

Symptoms of acute pyelonephritis include chills and fever, nausea, vomiting, diarrhea, and tenderness with pressure at the costovertebral angle (chapter 11). The white blood cell count is usually elevated and urinalysis shows pyuria, white blood cell casts (figure 12.2), and bacteriuria. The symptoms usually disappear within a few days, after appropriate antibiotic therapy or sometimes without treatment.

CHRONIC PYELONEPHRITIS

Chronic pyelonephritis is a prolonged inflammatory condition involving the renal pelvis and associated calyces, combined with damage to tubules and interstitium. Bacterial infection plays a role in most cases, but there is evidence that other forms of kidney damage may also lead to chronic pyelonephritis.

There are few symptoms in the early stages, but these often include hypertension with intermittent bacteriuria and white blood cell casts in the urine. The progression of the disease is slow and may continue for many years. Ultimately, there is scarring of the tubules and interstitium as well as the surface of the kidney. Finally, the glomeruli are damaged, and a declining glomerular filtration rate leads to renal failure as described in chapter 15.

Treatment for chronic pyelonephritis involves correction of any underlying conditions, such as obstruction or infection and control of hypertension.

dysuria: Gk. *dys,* bad; Gk. *ouron,* urine
cystitis: Gk. *kystis,* bladder; Gk. *itis,* inflammation

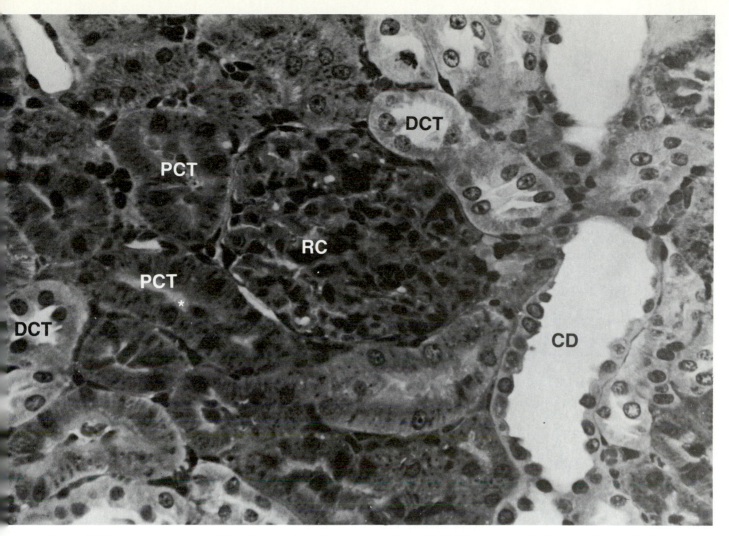

Figure 12.1 Light micrograph of the kidney cortex. CD, collecting duct; DCT, distal convoluted tubule; PCT, proximal convoluted tubule; RC, renal corpuscle (the portion of the nephron consisting of the glomerulus and Bowman's capsule). ×600.

From D. T. Moran and J. C. Rowley: Visual Histology. *© 1988 Lea & Febiger, Philadelphia. Reproduced with permission.*

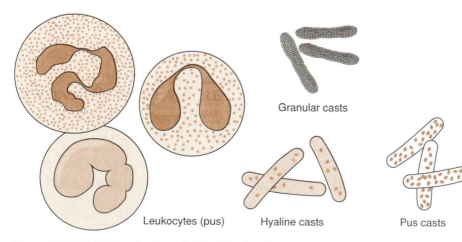

Granular casts

Leukocytes (pus) Hyaline casts Pus casts

Figure 12.2 White blood cells and white blood cell casts, seen microscopically, appear in the urine in acute pyelonephritis. Casts are made up of protein that precipitates in the nephron tubule.

Urinary Tract: Infection, Obstruction, and Nephrotoxins

OBSTRUCTION

Obstruction to the outflow of urine has serious immediate consequences and may lead to irreversible damage with either acute or chronic renal failure (chapter 15). Obstruction may occur anywhere in the urinary tract and is caused by three types of mechanisms: (1) luminal obstruction, (2) anatomical abnormalities, and (3) external compression. The following discussion deals with the effects and complications of urinary tract obstruction.

EFFECTS OF OBSTRUCTION

The location, degree, and duration of obstruction determine subsequent effects. Table 12.1 summarizes the consequences of ureteral obstruction. The immediate effect is dilatation and increased pressure in the renal pelvis and calyces, which leads to progressive renal damage. This is called **hydronephrosis** and occurs as the result of obstruction anywhere between the renal pelvis and the urethra. The pressure in the renal pelvis is transmitted to the tubules, and this inhibits glomerular filtration. Increased pressure in the pelvis also compresses medullary blood vessels resulting in **ischemia** (is-kē'me-ah) or decreased blood supply. The physiological response to the ischemia is a transient increase in renal blood flow, perhaps caused by vasodilation mediated by prostaglandins. Vasoconstriction occurs in response to prolonged obstruction, resulting in decreased renal blood flow and a fall of the glomerular filtration rate.

Tubular function (chapter 11) is affected by increased pressures, so the following occurs. Initially, a decreased flow rate for filtrate allows time for increased sodium ion and water reabsorption. Continued high pressure causes defects in tubular function, so there is decreased sodium ion reabsorption, loss of urine concentrating ability, and loss of ability to secrete K^+ and H^+. Laboratory data show hyperkalemia and hyperchloremic acidosis. Sodium ion balance depends on sodium ion and water ingestion.

Further complications of obstruction include hypertension, formation of stones, and secondary infection. Either complete or prolonged obstruction leads to oliguria or anuria and irreversible damage. The final outcome is renal failure unless there is intervention.

Severe pain is the outstanding feature of acute obstruction. If there is a gradual elevation of pressure within the bladder and renal pelvis, the individual may experience only a sense of abdominal or back discomfort. Obstruction of the upper urinary tract usually causes flank pain. If the lower urinary tract is obstructed, there may be suprapubic pain radiating to the testicles or labia.

hydronephrosis: Gk. *hydor,* water; Gk. *nephros,* kidney; Gk. *osis,* condition

TABLE 12.1 Primary and secondary effects of ureteral obstruction

Dilation of ureter and renal pelvis

Secondary increase in tubular hydrostatic pressure; inhibits glomerular filtration

Compression of medullary blood vessels

Medullary ischemia leads to transient increase in renal blood flow

Prolonged ischemia results in vasoconstriction and decreased GFR

Loss of tubular ability to conserve Na^+, concentrate urine, secrete K^+ and H^+

Oliguria or anuria if complete obstruction

Secondary infections

Stones

Hypertension

Complete or prolonged obstruction causes irreversible damage and renal failure

• • •

Nephron tubules atrophy within 4–6 weeks after complete obstruction. Correction of the obstruction before that time results in significant recovery of function.

• • •

NEPHROLITHIASIS

Nephrolithiasis refers to the formation of a kidney stone or calculus and is a common cause of obstruction (figure 12.3). The chemical composition of these stones is listed in table 12.2. Stones usually develop within the renal pelvis, on the surfaces of renal papillae, or in the bladder. If a stone is carried into the ureter, it can cause obstruction, pain, and hematuria.

Multiple factors contribute to the process of stone formation including diet, inherited tendencies, and inadequate fluid intake. In some cases, the individual has high blood levels of the constituents that make up the stone. Ultimately, for a combination of reasons, the glomerular filtrate becomes supersaturated and solute precipitates to form a crystalline deposit.

• • •

There are inhibitors of stone formation in the urine, such as inorganic pyrophosphate, glycoproteins, and citrate.

• • •

nephrolithiasis: Gk. *nephros,* kidney; Gk. *lithos,* stone
calculus: L. *calx,* stone

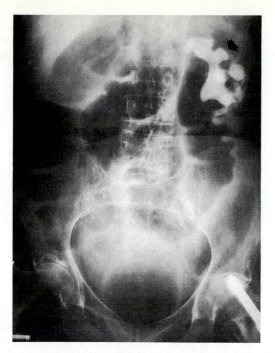

Figure 12.3 This is a film of the abdomen of a 90-year-old female. The large opaque structure in the area of the patient's left kidney (arrow) is a "staghorn" kidney stone.

TABLE 12.2 Types of kidney stones listed in decreasing order of most common occurrence

1. Calcium oxalate
2. Calcium oxalate and calcium phosphate
3. Calcium phosphate
4. Magnesium ammonium phosphate
5. Uric acid
6. Cystine

Management

The severity of the problem and the nature of contributing factors determine the specific measures for managing nephrolithiasis. Obstruction, infection, pain, and bleeding may require surgical removal of the stone called lithotomy. Lithotripsy (lith′ō-trip″sē) is an alternative to surgery. Lithotripsy means to crush a stone. There are three types of procedures: (1) **extracorporeal shock-wave lithotripsy,** (2) **percutaneous lithotripsy,** and (3) **endoscopic lithotripsy.** The first of the three is noninvasive and involves placing the patient in a tub of water. Shock-wave energy, generated by an underwater high-voltage discharge, is focused by a reflector. The shock waves are relatively harmless to soft tissue, but shatter a kidney stone. Both percutaneous and endoscopic lithotripsy involve the use of ultrasound to disintegrate a kidney stone. In percutaneous lithotripsy, a cystoscopelike instrument is introduced into the renal pelvis by way of a small incision. The endoscopic procedure involves introduction of an ultrasonic transducer by way of a cystoscope into the uri-

lithotripsy: Gk. *lithos*, stone; L. *tritus*, grind

Urinary Tract: Infection, Obstruction, and Nephrotoxins

TABLE 12.3 The effect of pH on kidney stones

Stone Composition	pH and Solubility
Calcium oxalate	Solubility unchanged within pH range of urine
Calcium phosphate	Insoluble in alkaline urine
$MgNH_4PO_4$	Insoluble in alkaline urine
Uric acid	Insoluble if pH < 5.5
Cystine	Maximum solubility in range pH 4.5–7.0

nary tract and up to the ureter. In both endoscopic and percutaneous lithotripsy, ultrasound is the force that breaks up the stone.

Measures may be taken to control stone formation. High fluid intake to promote the production of at least 2 liters of urine per day is essential. Specific measures depend on the pathogenesis of the stone formation. Solubility of stone constituents is related to urine pH as shown in table 12.3. Controlling urine pH is one aspect of preventing stone formation.

The following are possible courses of action in preventing the formation of calcium-containing stones:

1. Treat underlying conditions, such as hyperparathyroidism, that contribute to high blood calcium levels.
2. Restrict dietary calcium.
3. Treat with thiazides, which are diuretics that reduce urinary calcium by promoting tubular reabsorption.
4. In some cases, calcium oxalate crystals are deposited around a nucleus of uric acid crystals. If there is a high level of uric acid excretion, it is appropriate to limit meat, fish, and poultry in the diet. These meats are sources of purine, which is metabolized to uric acid.
5. Oxalate excess may contribute to the formation of calcium oxalate stones. Dietary restriction of oxalate containing foods and alkali therapy may be appropriate.

• • •

Some common high oxalate foods include cocoa, chocolate, black tea, spinach, rhubarb, nuts, and beets. Other foods with a lesser amount of oxalate are roasted coffee, green beans, carrots, strawberries, and sweet potatoes.

• • •

Magnesium ammonium phosphate stones (**struvite**) are formed as the result of infection, mainly caused by the *Proteus* species, which produce the urea splitting enzyme called **urease.** Urea is converted to ammonia, which forms ammonium, and this raises urine pH. Struvite stones precipitate out in alkaline conditions and may become large enough to require surgery. Antibiotic treatment of infection is an aspect of management.

Uric acid precipitates out in a strongly acidic urine, so alkali is given to increase urine pH. **Allopurinol** interferes with the production of uric acid and is useful in decreasing urinary excretion of this substance.

Cystinuria is an inherited disorder in which there is impaired reabsorption of certain amino acids. The result is excessive excretion of **lysine, arginine, ornithine,** and **cystine.** Since cystine is the least soluble of these amino acids, cystine stones tend to form. Maintaining a large urinary volume and alkalinizing the urine are two approaches to management. Administration of **penicillamine** is useful because it combines with cystine to form a product that is more soluble than cystine.

PROSTATE GLAND

Enlargement of the prostate gland (chapter 10) is a common cause of urinary obstruction in males. The gland is located inferior to the urinary bladder and encircles the superior end of the urethra (figure 12.4). It is this proximity to the urethra that makes obstruction likely. Inflammation of the prostate or prostatitis, which may be caused by infection, is one cause of enlargement. A second cause is the aging process in which most men after the age of 50 experience some increase in prostatic tissue called hyperplasia (hī″per-plā′ze-ah). A third cause of prostatic enlargement and possible urinary tract obstruction is prostatic carcinoma. This is the second most common malignancy and the third most common cause of death in males. Most of these malignancies are adenocarcinomas, a type of cancer that has a glandular growth pattern. Common symptoms of prostatic carcinoma are dysuria, difficulty in voiding, and increased urinary frequency.

BLADDER CARCINOMA

Bladder carcinoma affects three times as many males as females and occurs more commonly after age 40. There is a high incidence of this type of malignancy among cigarette smokers and workers in the dye and chemical industries. A major symptom of the disease is painless hematuria. A small number of affected individuals suffer ureteral obstruction.

OTHER CAUSES OF OBSTRUCTION

There are other causes of urinary tract obstruction in addition to renal stones, prostatic hypertrophy, and malignancies. These include neurogenic bladder dysfunction, mentioned in the first section of this chapter. Loss of neural bladder control interferes with normal urine flow and may

hyperplasia: Gk. *hyper,* beyond; Gk. *plasis,* a forming
hematuria: Gk. *haima,* blood; Gk. *ouron,* urine

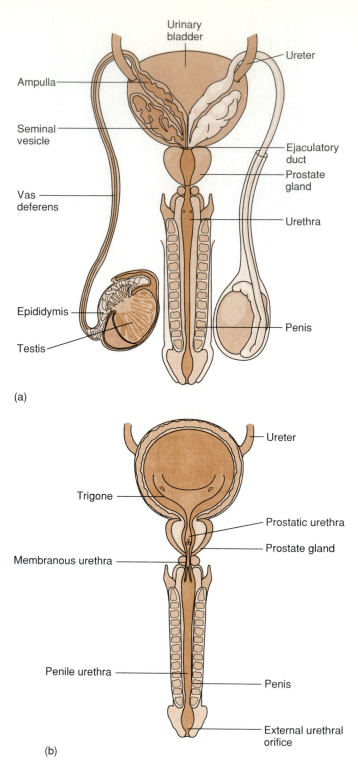

(a)

(b)

Figure 12.4 (*a*) The prostate gland in relation to other genitourinary structures. (*b*) A section of male urethra. The portion of the urethra that passes through the prostate gland is about 2.5 cm long.

TABLE 12.4 Substances in the urine due to the
Fanconi syndrome

Amino acids	Potassium
Glucose	Uric acid
Phosphate	Calcium
Bicarbonate	Protein

TABLE 12.5 A partial list of nephrotoxins

Gentamycin	Methoxyflurane (anesthetic)
Cephalosporins	X-ray contrast media
Amphotericin B	Carbon tetrachloride
Lead	Ethylene glycol (antifreeze)
Mercury	Radiation
Lithium	

be caused by injury or neuropathies. Vesicoureteral reflux, also mentioned earlier, causes a backflow of urine into ureters and renal pelvis.

NEPHROTOXINS

Chemical agents that damage the kidney are called nephrotoxins. The kidneys are particularly susceptible to injury by toxins, in part due to a high degree of exposure by way of a large blood supply. The kidneys receive 20% of the cardiac output with most of that flow going to the renal cortex. Another factor, in addition to high ratio of blood flow to small organ size, is the fact that the function of the tubules is to concentrate substances within the lumen in preparation for excretion. This concentrating function increases the exposure of renal cells to any toxic substance present.

RENAL INJURY

Toxic effects depend not only on the nature of the substance, but also on the concentration and period of renal exposure. Substances toxic to the kidney frequently cause tubular dysfunctions or, with necrosis of renal cells, either acute or chronic renal failure.

The **Fanconi syndrome** represents a dysfunction in which there is failure to reabsorb solutes from the glomerular filtrate. Proximal tubular dysfunction predominates with increased amounts of the substances appearing in the urine shown in table 12.4. A variety of conditions, such as metabolic and inflammatory diseases, can cause this syndrome. Nephrotoxins that may lead to the syndrome include heavy metals, such as lead, mercury, and cadmium, and certain antibiotics, such as **gentamycin** and the **cephalosporins.**

Injury that causes tubular unresponsiveness to ADH impairs the concentrating ability of the kidney. This condition is called **nephrogenic diabetes insipidus** and is characterized by polyuria. The toxic effects of lithium and the anesthetic, **methoxyflurane,** are two possible causes.

A third type of tubular damage that may occur is renal tubular acidosis. **Amphotericin B** causes injury to the distal tubule, resulting in **renal tubular acidosis, type I.** Hydrogen ion secretion is impaired, causing acidosis and hypokalemia. There are other types of RTA which will be discussed in chapter 14. When the concentration of the toxic substance is high, necrosis of tubular cells called **acute tubular necrosis** may occur. Mercury and carbon tetrachloride cause tubular necrosis and may cause acute renal failure (chapter 15) with sudden cessation of renal function. Damage by toxins, if severe or prolonged, may cause chronic renal failure. Table 12.5 is a partial list of nephrotoxins.

The chapters that follow deal with other aspects of renal damage due to immunological injury and renal disease.

SUMMARY

• • •

Urinary tract infections are common, especially in females, and are usually caused by intestinal tract organisms. Obstruction and catheterization are two important predisposing factors for infection. Urethritis, cystitis, and pyelonephritis may occur. Chronic pyelonephritis may progress to renal failure.

Obstruction to the outflow of urine causes hydronephrosis, which causes a decrease in the glomerular filtration rate, hypertension, loss of ability to conserve sodium, concentrate urine, and to secrete K^+ and H^+. Complete or prolonged obstruction causes irreversible damage and renal failure. Causes of obstruction include nephrolithiasis, prostatic hypertrophy, malignancies, neurogenic bladder dysfunction, and vesicoureteral reflux.

The kidneys are particularly susceptible to injury by toxins. Toxic effects include loss of reabsorptive ability by the proximal tubule, impaired concentrating ability, and loss of hydrogen ion secretory function. If there is a high concentration of the nephrotoxin, acute tubular necrosis occurs, and renal failure is a possible outcome.

Review Questions

• • •

1. What evidence of cystitis will urinalysis show?
2. Name a condition which predisposes individuals to pyelonephritis.
3. List four symptoms of pyelonephritis.
4. Define hydronephrosis and describe what causes the condition.
5. List five effects of ureteral obstruction.
6. Define lithotripsy.
7. List three general measures for the control of stone formation.
8. What are the three most common types of kidney stones?
9. Why is difficulty in voiding a symptom of prostatic carcinoma?
10. Define the Fanconi syndrome.

Selected Reading

• • •

Baggio, B. 1986. An inheritable anomaly of red-cell oxalate transport in "primary "calcium nephrolithiasis correctable with diuretics. *New England Journal of Medicine* 314:599–604.

Brunham, R. C. et al. 1984. Mucopurulent cervicitis-the ignored counterpart in women of urethritis in men. *New England Journal of Medicine* 311:1–6.

Das, G. et al. 1987. Extracorporeal shockwave lithotripsy: First 1,000 cases at London Stone Clinic. *British Medical Journal (Clinical Research)* 295:891–93.

Dick, W. H. 1987. Nephrolithiasis: Clues for the stone detective. *Indiana Medicine* 80:838–43.

Ettinger, B. et al. 1986. Randomized trial of allopurinol in the prevention of calcium oxalate calculi. *New England Journal of Medicine* 315:1386–89.

Frevele, G. 1989. Urinary tract injuries due to blunt abdominal trauma. *Physician Assistant* 13:123–36.

Gaddipati, J. et al. 1987. Prostatic and bladder cancer in the elderly. *Clinical Geriatric Medicine* 3:649–67.

Githler, A. 1989. Urinary tract infections in the elderly. *Physician Assistants in Primary and Hospital Care* (supplement) 6:22–25.

Hanno, P. M. 1989. Cystitis: A management guide to recurrent cases. *Physician Assistant* 13:25–32.

Kanel, K. T., and W. N. Kapoor. 1989. Acute pyelonephritis. *Physician Assistant* 13:34–42.

Manning, F. A., M. R. Harrison, and C. Rodeck. 1986. Catheter shunts for fetal hydronephrosis and hydrocephalus. *New England Journal of Medicine* 315:336–40.

Monsour, M. et al. 1987. Renal scarring secondary to vesicoureteric reflux. Critical assessment and new grading. *British Journal of Urology* 70:320–24.

Piper, J. et al. 1985. Heavy phenacetin use and bladder cancer in women aged 20 to 49 years. *New England Journal of Medicine* 313:292–94.

Rittenberg, M. H. et al. 1987. Ureteroscopy under local anesthesia. *Urology* 30:475–78.

Rosenfeld, J. 1987. Renal abnormalities in women admitted with pyelonephritis. *Delaware Medical Journal* 59:717–19.

Roxe, D. M. 1980. Toxic nephropathy from diagnostic and therapeutic agents. *American Journal of Medicine* 69:759.

Scriver, C. R. 1986. Cystinuria. *New England Journal of Medicine* 315:1155–56.

Scully, R. E., ed. 1985. Case records of the Massachusetts General Hospital (chronic pyelonephritis). *New England Journal of Medicine* 313:312–18.

Stark, R. P., and D. G. Maki. 1984. Bacteriuria in the catheterized patient. *New England Journal of Medicine* 311:560–64.

Chapter 13

Glomerular Diseases

This chapter deals with a group of renal diseases in which the pathological processes mainly affect the glomerulus.

GLOMERULAR INJURY

The mechanisms for glomerular injury are complex and most often are initiated by an immune response. Table 13.1 lists some definitions of terms discussed further in chapter 26 and involved in the immune response.

Antigens are substances, usually proteins, that cause production of antibodies and, in turn, combine specifically with that antigen. Antigen/antibody complexes may be deposited within glomerular capillary walls, or a circulating antibody may combine with previously deposited antigen to form a complex. Table 13.2 lists examples of antigens that may be associated with glomerular disease. As shown in table 13.3, the damage to the glomerular capillary wall may be the result of (1) deposition of an immune complex and complement, (2) cells attracted to the area, and (3) the formation of fibrin. The nature of the glomerular damage is increased permeability because of proteolytic enzymes and obstruction due to both cellular proliferation and thrombi.

The effect of platelet secretion of cationic proteins deserves a special note. The biochemical constituents of the glomerular basement membrane include **polyanionic** (pŏl″ē-ăn″ĭ-ŏn′ik) protein-saccharide molecules, specifically **proteoglycans** (prō″tē-ō-glī′kan) and **sialglycoprotein** (sī″al-glī″ko-pro-te-in). These are negatively charged molecules and create a filtration charge barrier to anions (figure 13.1). Albumin is anionic and normally is not filtered by the glomerulus. Secretion of cationic proteins by the platelets can neutralize this charge barrier. The result is the appearance of albumin in the urine.

TABLE 13.1 Definitions of terms; factors in glomerular damage

Immunoglobulin. Proteins produced by the body which are antibodies

Antibody. Protein produced in response to antigen; can combine with that antigen

Antigen. Substance, often a protein, that elicits antibody production

Complement. A group of serum proteins

Immune complex. Circulating antigen/antibody aggregate

Polyanionic binding site. Areas of negative charge along glomerular basement membrane conferred by protein/saccharide molecules

Macrophage. A large phagocytic cell

Neutrophil. Type of white blood cell

Basophil. Type of white blood cell

Platelet. Blood cell involved in clotting

Fibrin. Insoluble protein that forms long fibers

TABLE 13.2 Antigens in glomerular disease

Hepatitis B virus	Tumor antigens
Drugs	DNA
Bacteria	Thyroglobulin

TABLE 13.3 Mechanisms of glomerular injury

Deposition of immune complexes

Deposition of complement. Membrane attack complex; attracts neutrophils and macrophages

Neutrophils. Release proteolytic enzymes; membrane damage activates macrophages and platelets

Macrophages. Release proteolytic enzymes; factors that promote cellular proliferation; formation of fibrin

Platelets. Release (1) serotonin and histamine, which increase permeability, (2) promote cellular proliferation, (3) activate clotting system, and (4) secrete cationic proteins

Fibrin. Form thrombi or promotes crescent-shaped accumulation of cells within or around glomeruli

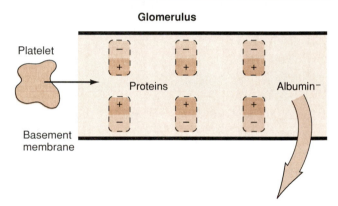

Figure 13.1 The glomerular basement membrane is negatively charged and presents a barrier to anionic albumin molecules. Cationic proteins secreted by platelets neutralize this charge resulting in filtration of albumin.

ACUTE GLOMERULONEPHRITIS

Acute glomerulonephritis is commonly caused by infection by certain strains of group A beta-hemolytic Streptococci; thus, the condition is also called **poststreptococcal glomerulonephritis (PSGN)**. Occasionally, other types of infection lead to this condition, but most often it follows a throat infection (pharyngitis) or skin infection (pyoderma) caused by group A streptococci. This usually occurs in children and less commonly in adults but only in a small percentage of those infected.

SYMPTOMS

Typically, there is an acute onset of symptoms after a latent period of about 10 days following pharyngitis and about 21 days following pyoderma. The symptoms include macroscopic hematuria causing a dark colored urine, edema, and decreased urinary output or **oliguria.**

thrombus, pl. thrombi: Gk. *thrombos*, clot
oliguria: Gk. *oligo*, small; Gk. *ouron*, urine

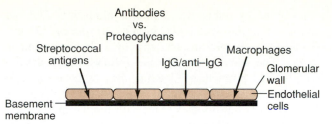
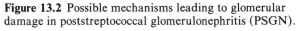

Figure 13.2 Possible mechanisms leading to glomerular damage in poststreptococcal glomerulonephritis (PSGN).

LABORATORY DATA

The evidence of some degree of impaired renal function in acute glomerulonephritis is an elevated BUN with plasma creatinine typically less than 3 mg/dl (chapter 11). The urinalysis shows protein, white blood cells, red blood cells, and **casts.** Casts are microscopic cylindrical structures formed by protein that has precipitated in the tubules (figure 12.2). Casts may have cells embedded within them. The urine usually has a high specific gravity and is concentrated.

PATHOLOGY

Kidney tissue may be obtained by biopsy, and this involves administering a local anesthetic and insertion of a needle and withdrawal of tissue. The tissue obtained in this way is observed, after appropriate staining procedures, with a light microscope, an electron microscope, or a fluorescent microscope. It is thus possible to see the histological nature of damage to glomeruli.

The fluorescent microscope makes it possible to identify immunoglobulins, complement, or other proteins deposited in the glomerulus. Antibodies that are specific for one of these proteins are labeled with a substance, such as **fluorescein,** that fluoresces when exposed to ultraviolet light. Kidney tissue is incubated with labeled antibodies, and these attach to a specific protein in the glomeruli. The formation of such an antigen/antibody complex identifies the presence of a particular protein, such as an immunoglobulin.

Microscopic findings provide evidence that immune reactions are responsible for the disease process in PSGN. This means that an antigen has stimulated antibody production and possibly the subsequent reactions listed in table 13.3. The nature of the antigen is uncertain. The following are hypotheses regarding the mechanisms leading to glomerular damage:

1. There is direct glomerular deposition of streptococcal antigens.
2. Circulating antibodies against proteoglycans in the glomerular basement membrane cause injury.
3. A streptococcal enzyme induces changes in a naturally occurring antibody function in the blood called **immunoglobulin G (IgG),** which

stimulates the production of antibodies. The resulting IgG/anti-IgG complex is deposited in the glomerulus.
4. There may be activation of a particular kind of white blood cell, a T-lymphocyte. It is theorized that activated T-lymphocytes release products that attract macrophages that damage the glomerulus (figure 13.2).

PROGNOSIS

The course of PSGN seems to be different in children and adults. Most children recover spontaneously with diuresis occurring within 1 week, and blood levels of creatinine returning to normal in 3 to 4 weeks. Proteinuria, however, may persist for a period of years. As many as 50% of adults develop both hypertension and progressive renal insufficiency that may occur many years after the initial acute episode.

TREATMENT

There is no specific treatment effective against PSGN. It is possible that antibiotics administered within 36 hours prevent glomerulonephritis. Antibiotics do not appear to prevent renal disease after that but are administered to eliminate the initial infection. Loop diuretics remove excess extracellular fluid, and dialysis may be required if diuretic therapy is ineffective.

RAPIDLY PROGRESSIVE GLOMERULONEPHRITIS

Rapidly progressive glomerulonephritis (RPGN) may develop from any form of glomerulonephritis, it may be a manifestation of systemic illness, or it may be idiopathic (cause unknown). The idiopathic form mainly affects adults in the middle age group. The disease process in RPGN typically advances within weeks or months to the end-stage of renal failure in most untreated patients.

SYMPTOMS

The onset of symptoms may be either abrupt or insidious. If the onset is abrupt, the symptoms are similar to PSGN. An insidious onset is more common, with malaise, weakness, nausea, vomiting, edema, and decreased urinary output.

MORPHOLOGIC FEATURES

The major identifying feature of RPGN is the proliferation of glomerular/capsular epithelial cells in the form of a **crescent.** The crescents fill Bowman's space and compress the glomerulus. The glomerular capillaries collapse and are bloodless, and fibrin can be identified within the capsule. There may be no antibody involvement, or there may be deposits of antigen/antibody complexes or deposits of antibodies against the glomerular basement membrane. Interstitial fibrosis, an increase in fibrous connective tissue, occurs as the disease progresses. Marked tubular atrophy occurs as well. The initiating events and the mechanisms leading to injury are uncertain, but they are probably immune in nature.

LABORATORY DATA

A patient with RPGN shows a marked decrease in the glomerular filtration rate and a plasma creatinine level usually greater than 5 mg/dl. The urinalysis indicates proteinuria, hematuria, and red blood cell casts. Anemia is common.

PROGNOSIS

The overall outlook for RPGN depends on the severity of the disease when treatment is initiated. The prognosis is poor if the patient has anuria, a plasma creatinine level greater than 6–8 mg/dl, crescents in more than 80% of the glomeruli, and shows antibodies against the glomerular basement membrane. The prognosis is relatively good, however, if the oliguric RPGN has developed from streptococcal glomerulonephritis.

TREATMENT

Glucocorticoid therapy and plasma exchange are two modalities of treatment that may lead to dramatic improvement of renal function.

• • •

Plasmapheresis (plaz″mah-fĕ-re′sis) or plasma exchange is the removal of blood from the body, the separation of cells and the liquid fraction, and reinjection of cells suspended in saline. Presumably, the procedure works by removing circulating antibodies. Removal of mediators of inflammation, such as fibrinogen and complement, may be a factor as well.

• • •

GOODPASTURE'S SYNDROME

Rapidly progressive glomerular nephritis may be accompanied by pulmonary hemorrhage associated with **anti-glomerular basement membrane antibodies (anti-GBM antibodies)** circulating in the plasma. This condition is called **Goodpasture's syndrome** or anti-GBM antibody disease. It occurs most commonly in young males and sometimes

plasmapheresis: Gk. *plasma*; Gk. *aphairesis,* a taking away

is preceded by influenza or smoke inhalation. The syndrome is characterized by (1) glomerular disease, (2) anti-GBM antibody production, and (3) pulmonary hemorrhage. The glomerular disease follows the course of RPGN. The pulmonary bleeding may be fatal and may occur in the absence of manifestations of renal disease. The anti-GBM antibodies react with the basement membrane of lung capillaries as well as the glomeruli. The disease typically has a rapid downhill course unless treated by high dose steroids and/or plasmapheresis and, in extreme cases, renal transplant. It should be noted that renal disease and pulmonary hemorrhage may also occur in association with underlying diseases, such as systemic lupus erythematosus or vasculitis.

NEPHROTIC SYNDROME

The nephrotic syndrome is defined by a combination of findings and may be either idiopathic or the result of various disease processes. The causes of the secondary form include (1) infections, (2) malignancies, (3) medications, and (4) systemic diseases.

Table 13.4 lists characteristic findings in the nephrotic syndrome. Increased glomerular permeability to proteins is a common feature. Loss of albumin is at least in part, responsible for edema. Decreased plasma albumin causes decreased oncotic pressure, which favors net movement of fluid into interstitial spaces (see chapter 2). The vascular fluid loss stimulates increased secretion of aldosterone and ADH, which promotes sodium and water retention.

There is a clinically significant loss of nonalbumin protein as well. Certain coagulation factors are small proteins and are excreted in the urine, which leads to an increased incidence of venous thrombosis.

Loss of binding proteins for vitamin D lead to vitamin D deficiency. Loss of protein in the form of IgG antibody fraction may be the reason for an increased incidence of infection.

Modalities of treatment include (1) treatment of underlying disease, (2) a protein-rich diet, (3) anticoagulants, (4) administration of vitamin D, and (5) judicious use of diuretics if edema is severe.

TABLE 13.4 Findings in the nephrotic syndrome

Observation	Cause
Heavy proteinuria (mainly albumin)	Loss of >3.5 grams of protein/day in an adult
Hypoalbuminemia	Low plasma albumin level
Hyperlipidemia	Increased level of lipids in plasma
Lipiduria	Lipids in urine
Edema	

Edema usually does not become apparent in the nephrotic syndrome until plasma albumin falls below 2.5–3.0 g/dl. When the plasma albumin is above this level, there may be other causes of edema, such as heart failure or renal failure.

. . .

SUMMARY

. . .

Glomerular injury is often initiated by an immune mechanism. Possible mechanisms for injury include (1) deposition of immune complexes, (2) deposition of complement, (3) release of proteolytic enzymes by neutrophils, (4) release of proteolytic enzymes by macrophages, (5) release of factors by platelets, and (6) formation of fibrin.

Acute glomerulonephritis is commonly caused by infection due to certain strains of group A beta-hemolytic Streptococci. Glomerular damage results in impaired renal function from which children usually recover spontaneously. Adults tend to develop hypertension and progressive renal insufficiency. There is no specific treatment.

Rapidly progressive glomerulonephritis may develop from any form of glomerulonephritis, a systemic disease, or may be idiopathic. The major identifying histological feature is the proliferation of epithelial cells in the form of crescents. These fill Bowman's space and compress the glomerulus, causing the capillaries to collapse. The result is a decrease in GFR, proteinuria, hematuria, and red blood cell casts in the urine. If untreated, the disease progresses rapidly to end-stage renal failure. Goodpasture's syndrome may accompany RPGN. The syndrome is characterized by (1) glomerular disease, (2) anti-glomerular basement membrane antibodies, and (3) pulmonary hemorrhage. The glomerular disease follows the course of RPGN, or there may be pulmonary bleeding, either with or without manifestations of renal disease.

The nephrotic syndrome is characterized by (1) heavy proteinuria, (2) hypoalbuminemia, (3) edema, (4) hyperlipidemia, and (5) lipiduria. This syndrome may be idiopathic or caused by a variety of disease processes.

REVIEW QUESTIONS

. . .

1. What are two substances that may be deposited in glomerular capillary walls and cause damage?
2. What is the nature of glomerular damage caused by neutrophils?
3. What attracts neutrophils and macrophages to the glomeruli?
4. What is the role of macrophages in glomerular damage?
5. How are platelets involved in glomerular damage?
6. What is the role of fibrin in glomerular damage?
7. What is a filtration charge barrier to anions?
8. What is a common cause of acute glomerulonephritis?
9. What are typical symptoms of acute glomerulonephritis?
10. What would you expect the urinalysis to show in this condition?
11. What is an identifying morphologic feature of rapidly progressive glomerulonephritis?
12. What is Goodpasture's syndrome?
13. What is the significance of the presence of anti-glomerular basement membrane antibodies in this syndrome?
14. Define the nephrotic syndrome.

SELECTED READING

. . .

Abrahamson, D. R. 1987. Structure and development of the glomerular capillary wall and basement membrane. *American Journal of Physiology* 253:F783–94.

Fukatsu, A. et al. 1988. The glomerular distribution of type IV collagen and laminin in human membranous glomerulonephritis. *Human Pathology* 19:64–68.

Gregory, M. C. et al. 1988. Renal deposition of cytomegalovirus antigen in immunoglobulin-A nephropathy. *Lancet* 1:11–14.

Heller, J. et al. 1987. Hemodynamics of the recently opened glomeruli. *Renal Physiology* 10:47–53.

Kawaguchi, K. et al. 1987. Glomerular alterations associated with obstructive jaundice. *Human Pathology* 18:1149–54.

Martinez-Maldonado, M. et al. 1987. Pathogenesis of systemic hypertension and glomerular injury in the spontaneously hypertensive rat. *American Journal of Cardiology* 60:471–521.

Rifai, A. et al. 1987. IgA molecular form-pathophysiology correlates: Immune complex formation and glomerular deposition. *Advances of Experimental Medicine and Biology* 216B:1515–22.

Sado, Y. et al. 1987. Experimental autoimmune glomerulonephritis in rats by soluble isologous or homologous antigens from glomerular and tubular basement membranes. *British Journal of Experimental Pathology* 68:695–704.

Scully, R. E. 1985. Case records of the Massachusetts General Hospital (glomerulonephritis). *New England Journal of Medicine* 312:1042–52.

Scully, R. E., ed. 1987. Case records of the Massachusetts General Hospital (nephrotic syndrome). *New England Journal of Medicine* 316:860–68.

Tipping, P. G. et al. 1988. Glomerular procoagulant activity in human proliferative glomerulonephritis. *Journal of Clinical Investigation* 81:119–25.

Welch, T. R. et al. 1986. Major histocompatibility-complex extended haplotypes in membranoproliferative glomerulonephritis. *New England Journal of Medicine* 314:1476–79.

Wilson, C. B. 1987. Immune aspects of renal diseases. *Journal of American Medical Association* 258:2957–61.

Glomerular Diseases

169

Chapter 14

Primary and Secondary Renal Diseases

This chapter deals with a scope broader than immunological factors related to glomerular disease (chapter 13). It includes a discussion of kidney diseases in general, primary disorders as well as disorders secondary to other systemic disease processes. Renal failure (chapter 15) is a possible outcome for each of these disorders.

POLYCYSTIC KIDNEY DISEASE

Polycystic kidney disease is an inherited condition in which multiple fluid-filled cysts develop within the kidney, compress nephrons, and interfere with kidney function (figure 14.1). The adult type of polycystic disease is the most common type and is transmitted as an autosomal dominant characteristic. The disease is characterized by cysts, varying in size from millimeters to centimeters, that develop within nephrons throughout the kidney tissue and cause enlargement of that organ. Cysts may also form in the liver, pancreas, spleen, and other organs.

The precise nature of the disease process is unclear. It is possible that the genetic defect causes abnormal basement membranes or that there is hyperplasia of tubular cells causing obstruction with subsequent cyst formation.

Manifestations of the disease usually appear between ages 30–50 (figure 14.2) and include flank pain, hematuria, and hypertension. Walls of cerebral arteries may become weakened, resulting in cerebral hemorrhage. Diagnosis is made by ultrasonography or computer tomography scanning. Table 14.1 summarizes some facts about the disease. There is continued decline in renal function for many years, and ultimately, end-stage renal failure occurs in about 70% of the affected individuals by age 65.

Treatment involves a low protein diet and control of hypertension. Patients in the final stages of renal failure are candidates for dialysis and kidney transplant.

There is both an infantile and juvenile form of polycystic kidney disease, both of which are rare. They are inherited as autosomal recessive traits. Cysts frequently

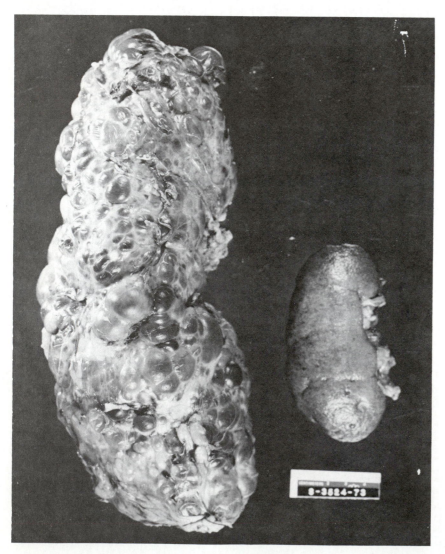

Figure 14.1 Adult polycystic disease showing a polycystic kidney (*left*) compared to a normal kidney (*right*). *A. C. Ritchie. 1990.* Boyd's Textbook of Pathology. *Lea & Febiger, Philadelphia. Reprinted by permission.*

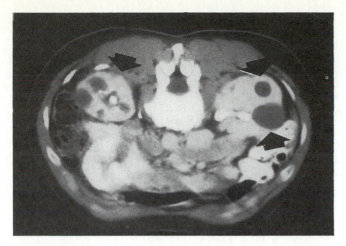

Figure 14.2 CAT scan of a 39-year-old female with polycystic kidney disease. The view is of a transverse section. The patient was given an oral contrast media to outline the intestines (white area, top part of film) and a venous iodine-based contrast media was injected to highlight the kidneys. The darker areas (arrows) are cysts.

TABLE 14.1 Manifestations, laboratory data, and possible complications of the adult type of polycystic kidney disease

Manifestations	Lab Data	Complications
Flank pain	Increased BUN	Gross hematuria
Hypertension	Increased creatinine	Infection
	Hematuria	Severe pain
	Mild proteinuria	Nephrolithiasis
	Increased red blood cells	Renal cell carcinoma
		Brain hemorrhage

develop in the liver as well as the kidney. The usual cause of death in these individuals is cirrhosis of the liver or renal failure.

RENAL TUBULAR ACIDOSIS

Renal tubular acidosis (RTA) is a collection of tubular disorders in which there is either inadequate reabsorption of bicarbonate or impaired secretion of hydrogen ions (figure 14.3). These disorders are characterized by hyperchloremic acidosis. The proximal tubule is the site of reabsorption of most of the bicarbonate and K^+ and H^+ secretion takes place mainly in the distal tubule (chapter 11); thus, there is a proximal and a distal RTA, depending on the site of the transport defect. There are subgroups of the distal defect as well.

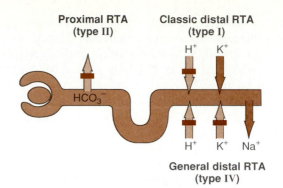

Figure 14.3 The defect type II RTA is impaired bicarbonate reabsorption. There is decreased H^+ and increased K^+ secretion in type I RTA and a decrease in both H^+ and K^+ secretion in the distal tubule in type IV.

TABLE 14.2 Defects in three types of renal tubular acidosis

Renal Tubular Acidosis

1. Classic Distal RTA (type I). Decreased H^+ secretion, increased K^+ secretion
2. Proximal RTA (type II). Impaired HCO_3^- reabsorption
3. Generalized Distal RTA (type IV). Decreased K^+ and H^+ secretion; increased Na^+ reabsorption

CLASSIC DISTAL RTA

The defects in three types of renal tubular acidosis are shown in table 14.2. The defect in classic distal RTA is decreased hydrogen ion secretion in the distal convoluted tubules. It is characterized by urine that is never below a pH of 5.5, even in metabolic acidosis. This condition occurs in all age groups and may be inherited as an autosomal dominant characteristic. It is not only genetic, but may also be caused by a variety of disease processes, as well as drugs and toxins.

Decreased hydrogen ion secretion is accompanied by increased potassium ion secretion (chapter 11); thus, symptoms of RTA may be those associated with acidosis or hypokalemia. Acidosis causes loss of calcium from bone, a release of albumin-bound calcium, and subsequent increase of calcium filtration at the glomerulus. The net effect is weakening of bones with pain, fractures, and **hypercalciuria.** Increased levels of urine calcium lead to nephrolithiasis and a condition called nephrocalcinosis in which calcium is deposited on renal papillae.

Alkali therapy prevents weakening of bones and progressive kidney damage.

GENERALIZED DISTAL RTA (TYPE IV)

The defect in type IV RTA is diminished secretion of both potassium and hydrogen ions. There may be a defect in the cells of the distal convoluted tubules, with renal disease or urinary tract obstruction as an underlying cause.

Decreased aldosterone is also a possible cause of type IV RTA. Aldosterone promotes sodium ion reabsorption, and in the case of diminished aldosterone, there is enhanced urinary sodium excretion. This in turn inhibits K^+ and H^+ secretion (chapter 3).

In contrast to type I RTA, generalized distal RTA results in acidosis combined with hyperkalemia rather than hypokalemia. In addition to this, the kidney can produce an acid urine during periods of severe acidosis. Symptoms are associated with acidosis or hyperkalemia.

Treatment involves administration of **9-α-fludrocortisone** (a synthetic mineralocorticoid) if aldosterone is deficient, or alkali may be given. Dietary potassium restriction, potassium wasting diuretics, or cation-exchange resins are measures for treating hyperkalemia (chapter 3).

• • •

Mineralocorticoids are steroid hormones produced by the adrenal cortex. Cholesterol is a precursor of this class of hormones, and aldosterone is an example.

• • •

Proximal RTA (Type II)

Proximal or **type II RTA** involves impaired bicarbonate reabsorption at the proximal convoluted tubule. Proximal RTA may be primary or may be caused by other disease processes.

The primary type of defect usually occurs in males, and there may be X-linked recessive inheritance. It is manifested in the first 18 months of life, usually with a history of vomiting and retarded growth. The defect is usually transient, and the prognosis is good. Possible causes of secondary RTA include multiple myeloma, heavy metals, carbonic anhydrase inhibitors (such as acetazolamide), and hyperthyroid states. Proximal RTA is usually associated with the Fanconi syndrome (chapter 12).

The proximal tubule is the site of reabsorption of 85–90% of bicarbonate. This reabsorption occurs in exchange for hydrogen ions, or the bicarbonate is accompanied by sodium ions. Since bicarbonate is excreted to some extent with potassium ions, hypokalemia occurs when there is increased bicarbonate excretion in proximal RTA.

Treatment includes therapy for any underlying disease, administration of bicarbonate, and potassium supplementation.

Nephrogenic Diabetes Insipidus

Antidiuretic hormone is secreted by the posterior lobe of the pituitary and causes reabsorption of water by the collecting ducts of the kidney. **Diabetes insipidus** is a condition in which there is impaired secretion of ADH or a defective renal response to ADH (chapter 2). The latter condition is called **nephrogenic diabetes insipidus,** and it causes a large volume of urine to be produced. The disorder may be inherited, but the acquired form is more common. Renal unresponsiveness to ADH may have a variety of causes including (1) renal disease, (2) obstructive uropathy, (3) chronic pyelonephritis, (4) analgesic nephropathy, (5) hypercalcemia, and (6) hypokalemia.

Treatment of nephrogenic diabetes insipidus is directed toward removing the primary cause. Dietary restriction of salt and protein reduces the solute load of the kidney and decreases urine volume.

Renal Cell Carcinoma

Renal cell carcinoma is a malignancy that originates in the proximal tubule epithelium (figure 14.4). The tumor is usually a single, unilateral growth with an average diameter of 7 cm, although there is wide variation in size. Approximately 3% of adult malignancies are of this type. Males are affected more frequently than females, with a 2:1 ratio, and most individuals are in the range of 50–70 years of age.

There is an increased incidence of renal cell carcinoma in individuals who have autosomal dominant polycystic kidney disease. Cysts develop in the kidneys of long-term dialysis patients, a condition called acquired cystic disease, and this condition also carries an increased risk of renal cell carcinoma. There are various risk factors for the malignancy including (1) cigarette smoking, (2) obesity, (3) exposure to cadmium, (4) abuse of phenacetin-containing analgesics, and (5) occupation. Leather tanners, shoe workers, and individuals exposed to asbestos are at greater risk for renal cell carcinoma than the general population.

This type of carcinoma may be asymptomatic until it reaches an advanced stage, which may be indicated by pain, hematuria, and flank mass. Metastasis occurs and, although the lung is the most common site, the spread of the malignancy may also involve soft tissue, bone, liver, skin, or the central nervous system.

Systemic symptoms of this disease include (1) anemia, (2) pyrexia, (3) cachexia (kah-kek′sē-ah), (4) fatigue, (5) weight loss, and (6) amyloidosis. The anemia may be the result of either hematuria or hemolysis. Weakness and emaciation (cachexia) may be apparent. Amyloidosis, the deposition of amyloid in tissues, is observed in a small number of cases.

Table 14.3 indicates stages and 5-year survival rates for renal cell carcinoma. Radical nephrectomy with the removal of both the kidney and adrenal gland is the standard treatment for localized renal cell carcinoma.

nephrogenic: Gk. *nephros*, kidney; Gk. *genesthai*, to be produced
cachexia: Gk. *kakos*, bad; Gk. *hexis*, state

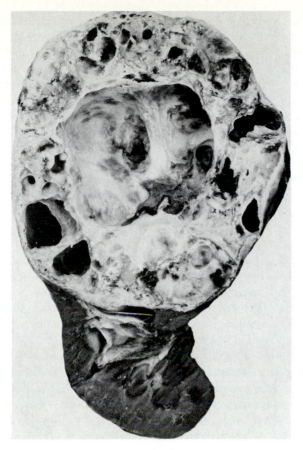

Figure 14.4 Renal cell carcinoma showing the large, well-demarcated tumor distorting the kidney.
A. C. Ritchie. 1990. Boyd's Textbook of Pathology. *Lea & Febiger, Philadelphia. Reprinted by permission.*

TABLE 14.3 Stages and 5-year survival rates for renal cell carcinoma

	Description	5-year Survival (Percent)
Stage I	Tumor within renal capsule	60–75
Stage II	Invasion through renal capsule	47–65
Stage III	Spread to lymph nodes, renal vein, or vena cava	5–15
Stage IV	Spread to distant sites	<5

• • •

Tumor-produced factors may cause systemic symptoms in renal cell carcinoma. Hypercalcemia is often associated with this malignancy and is believed to be caused by a tumor-produced factor that promotes bone resorption.

• • •

TABLE 14.4 A summary of manifestations and laboratory findings in hemolytic-uremic syndrome

Manifestations	
Pallor	Hepatosplenomegaly (enlarged spleen and liver)
Hypertension	Jaundice
Edema	Anuria (no urine produced)
Convulsions	Acute renal failure
Congestive heart failure	

Laboratory Findings	
Fragmented red blood cells	Azotemia (nitrogenous wastes in blood)
Leukocytosis (increased white blood cells)	Proteinuria
Thrombocytopenia (decreased number of platelets)	Hematuria

HEMOLYTIC-UREMIC SYNDROME

Hemolytic-uremic syndrome refers to a combination of findings in which there are both hematological and renal abnormalities. The syndrome is characterized by (1) hemolytic anemia due to fragmentation of red blood cells, (2) the formation of platelet-fibrin thrombi within small blood vessels, and (3) acute renal failure.

There is a childhood and adult form of hemolytic-uremic syndrome. In children, usually under the age of 4, the syndrome is frequently preceded by a viral infection or a bacterial gastroenteritis. In adults, the condition may be idiopathic, follow childbirth, or be associated with the use of oral contraceptives. In some cases, there may be a genetic predisposition to the syndrome.

Although the pathogenesis is not clear, there is occlusion of small blood vessels by thrombi and a thickening of blood vessel walls. This occurs in glomeruli and arterioles in the kidney as well as small blood vessels in other organs, such as liver, pancreas, and spleen. Fragmentation of red blood cells occurs as they pass through partially blocked blood vessels. Table 14.4 summarizes the usual manifestations and laboratory findings in this syndrome.

The mortality rate is about 30% and treatment is symptomatic. The steps in managing the syndrome include blood transfusion, maintaining fluid/electrolyte balance, control of blood pressure, and dialysis for anuric patients.

hemolytic: Gk. *haima*, blood; Gk. *lysis*, dissolving

Diabetic Nephropathy

Diabetes mellitus is a chronic disease of impaired glucose metabolism that causes high blood sugar levels or **hyperglycemia.** It affects the kidneys as well as other organs. General aspects of this disease are discussed in chapter 32.

Injury to the kidney associated with diabetes mellitus is called **diabetic nephropathy.** This disease process may be manifested up to 20 years after the onset of diabetes and is a common cause of renal failure. The major effect involves the glomerulus, and the result is **intercapillary glomerulosclerosis.** The term glomerulosclerosis means hardening of the glomerulus in which there is thickening of the glomerular basement membrane and an increase in intercapillary cells between the afferent and efferent arterioles. In addition to glomerulosclerosis, there is a thickening of both afferent and efferent arterioles. The inner lining or **intima** of renal arteries becomes thickened by fibrous connective tissue, and this impedes blood flow to kidney tissue. Both glomerular disease and renal vascular damage lead to interstitial nephritis in which there is edema, tubular atrophy, and interstitial fibrosis.

Proteinuria is an early manifestation of diabetic nephropathy. It progressively becomes worse and may develop into the nephrotic syndrome. Other manifestations of diabetic nephropathy include dysuria, flank pain, and hematuria due to necrosis of papillae. Difficulties in voiding may be caused by a neurogenic bladder (chapter 12) due to neuropathy. There is an increased incidence of pyelonephritis in diabetes and decreased levels of renin and aldosterone with hyperkalemia. End-stage renal disease (chapter 15) usually develops within 5 years of onset of proteinuria.

Treatment of diabetic nephropathy involves control of blood sugar and blood pressure and dietary restrictions. Dialysis and renal transplant are options as the disease progresses to end-stage renal failure.

Malignant Hypertension

A number of disease processes injure renal blood vessels and affect renal function. Prolonged hypertension not only damages peripheral blood vessels, but it also has adverse effects on renal circulation. The interrelationships between hypertension and renal function are discussed in chapter 21. Specifically, only the renal effects of malignant hypertension will be noted here.

Malignant hypertension refers to very high blood pressure (> 200/125 mm Hg) that causes variable degrees of renal function impairment. The hypertension seems to be responsible for injury and necrosis of the endothelial cells of afferent arterioles. There is subsequent

nephritis: Gk. *nephros,* kidney; Gk. *itis,* inflammation

leakage of fibrin into the arteriolar walls, and the condition is called **fibrinoid necrosis.** Rupture of arterioles or glomerular capillaries cause small hemorrhages on the cortical surface called **petechia** (pe-te′ke-ah). Thickening of the intima or inner lining of renal arteries causes a narrowing of the lumen that leads to ischemia. Renal ischemia and death of renal tissue ultimately causes a condition called nephrosclerosis. This term means hardening of the kidneys. There is a reduction in both renal blood flow and glomerular filtration rate. Manifestations of this renal damage include hematuria, proteinuria, and a sudden increase in BUN and creatinine levels. Renal failure occurs unless antihypertensive therapy is instituted.

Scleroderma

Scleroderma, also called **systemic sclerosis,** is characterized by fibrosis or replacement of normal tissue with fibrous connective tissue. The disease process affects skin as well as lungs, heart, kidney, and gastrointestinal tract. Affected individuals are usually 30–50 years of age, and the disease occurs in females twice as often as in males. The etiology and the mechanisms involved in the disease process are uncertain. There is some evidence of involvement of the immune system.

Most affected individuals experience **Raynaud's phenomenon,** which involves intermittent attacks of vasoconstriction of arteries of the fingers and toes (chapter 22). This may be precipitated by exposure to cold and lead to elevated levels of renin and decreased renal blood flow. Scleroderma also causes the skin to become smooth and drawn, as well as impaired esophageal motility, pulmonary fibrosis, cardiac disease, and malabsorption.

Vascular damage is the outstanding feature in any organ affected by scleroderma. Small renal arteries, afferent arterioles, and glomeruli are possible sites of vascular injury. Fibrinoid necrosis occurs in which there is deposition of fibrin causing a narrowing of the lumen of blood vessels and resulting in death of tissue. A narrowing of the lumen of renal arteries leads to ischemia with subsequent renin release followed by hypertension (chapters 11 and 21).

The course of the disease is variable and may progress slowly. Renal disease is common with proteinuria, azotemia, and hypertension. Renal failure is a leading cause of death in scleroderma. Treatment of the renal disease involves control of blood pressure, and dialysis may be required.

sclerosis: Gk. *sklerosis,* hardening
petechia, pl. petechiae: L. *petigo,* eruption
vascular: L. *vasculum,* small vessel

SUMMARY

• • •

Eight disease processes are discussed which may lead to renal failure. Polycystic kidney disease is an inherited condition in which multiple fluid-filled cysts develop within the kidneys and other organs. The adult form is most common, but there is an infantile and juvenile form as well. Renal tubular acidosis is a collection of tubular disorders in which there is either impaired reabsorption of bicarbonate (proximal or type II RTA) or the distal type which involves impaired hydrogen ion secretion. All types cause metabolic acidosis. Renal cell carcinoma is the most common of the primary renal neoplasms, typically affects older males, and may metastasize to any part of the body.

The hemolytic-uremic syndrome is characterized by hemolytic anemia, thrombi, and acute renal failure. In children, this syndrome usually follows a viral infection or bacterial gastroenteritis. In adults, it may be idiopathic, follow childbirth, or be associated with the use of oral contraceptives. Diabetic nephropathy is caused by diabetes mellitus and is a common cause of renal failure. Proteinuria is an early manifestation of the disease process. Malignant hypertension damages blood vessels, and ultimately causes renal failure. Scleroderma is characterized by fibrosis of the skin, lungs, heart, kidney, and gastrointestinal tract. The outstanding feature is vascular damage, and associated renal disease is common. Nephrogenic diabetes insipidus is a defective renal response to ADH and may be either inherited or acquired. It causes a large volume of urine to be formed.

REVIEW QUESTIONS

• • •

1. What is the nature of the inheritance of adult polycystic kidney disease?
2. What are three common manifestations of adult polycystic kidney disease?
3. At what age do these manifestations usually show up?
4. What is the nature of the inheritance of infantile and juvenile polycystic kidney disease?
5. What is the basic defect in classic distal (type I) RTA?
6. Symptoms of type I RTA are the result of acidosis and (hypokalemia/hyperkalemia).
7. What effect does type I RTA have on renal handling of calcium?
8. What is the defect in generalized distal RTA, type IV?
9. List three possible causes of type IV RTA.
10. What is the defect in proximal RTA, type II?
11. Define nephrogenic diabetes insipidus.
12. What is the most common primary renal neoplasm?
13. What are three characteristics of the hemolytic-uremic syndrome?
14. What is a common underlying cause of this syndrome in children?
15. What are two predisposing factors for the hemolytic-uremic syndrome in adults?
16. What is an early manifestation of diabetic nephropathy?
17. Describe intercapillary glomerulosclerosis involved in diabetic nephropathy.
18. What characteristic vascular damage occurs with malignant hypertension?
19. What organs are typically involved in scleroderma?
20. What individuals are most likely to be affected by scleroderma?

SELECTED READING

• • •

Baines, A. D. 1987. Is there a role for renal alpha 2–adrenoceptors in the pathogenesis of hypertension? *Canadian Journal of Physiology and Pharmacology* 65:1638–43.

Batlle, D. C., A. von Riotte, and W. Schlueter. 1987. Urinary sodium in the evaluation of hyperchloremic metabolic acidosis. *New England Journal of Medicine* 316:140–44.

Best, B. G. 1987. Renal carcinoma: A ten-year review 1971–1980. *British Journal of Urology* 60(20):100–102.

Cole, B. R. et al. 1987. Polycystic kidney disease in the first year of life. *Journal of Pediatrics* lll: 693–99.

D'Elia, J. A. et al. 1987. Hemolytic-uremic syndrome and acute renal failure in metastatic adenocarcinoma treated with metomycin: Case report and literature review. *Renal Failure* 10:107–13.

Feldt-Rasmussen, B. et al. 1986. Kidney function during 12 months of strict metabolic control in insulin-dependent diabetic patients with incipient nephropathy. *New England Journal of Medicine* 314:665–70.

Saifuddin, A. et al. 1987. Adult polycystic kidney disease and intracranial aneurysms. *British Medical Journal (Clinical Research)* 295:526.

Scally, J. K. 1987. Renal cell carcinoma—an unusual presentation and sequel. *Clinical Radiology* 38:653–54.

Scully, R. E., ed. 1986. Case records of the Massachusetts General Hospital (hemolytic uremic syndrome). *New England Journal of Medicine* 314:1032–40.

Shaw, S. G. et al. 1987. Atrial natriuretic peptide protects against acute ischemic renal failure in the rat. *Journal of Clinical Investigation* 80:1232–37.

Shulman, N. B. 1987. End-stage renal disease in hypertensive blacks. *Journal of Clinical Hypertension* 3(3 Suppl): 85S–88S.

Skott, D., and J. P. Briggs. 1987. Direct demonstration of macula densa-mediated renin secretion. *Science* 237:1618–20.

Taguma, Y. et al. 1985. Effect of captopril on heavy proteinuria in azotemic diabetics. *New England Journal of Medicine* 313:1617–20.

Ying, C. Y. et al. 1984. Renal revascularization in the azotemic hypertensive patient resistant to therapy. *New England Journal of Medicine* 311:1970–75.

Chapter 15

Decline in Renal Function

Renal disease and excretory dysfunction are the topics of the preceding chapters. The most serious consequence of these pathological processes is renal failure, either an acute or chronic form. The discussion that follows deals first with acute renal failure.

ACUTE RENAL FAILURE

The kidney's major function is to maintain homeostasis of the volume and constituents of body fluids. A failure of this function is a life-threatening matter. Acute failure implies rapid onset, and early evidence of that failure is a reduction of glomerular filtration rate and increasing **azotemia** (az″o-te′me-ah). This term means an accumulation of nitrogenous substances, including urea and creatinine, in the blood.

ETIOLOGY

The causes of acute renal failure fall into three main categories: **prerenal, renal,** and **postrenal** (table 15.1). The common feature of the prerenal category is decreased renal perfusion or blood supply to the kidney. Direct damage to the kidney due to ischemia, toxins, or other causes is the second category, and the postrenal type of failure is due to obstruction (figure 15.1).

MANIFESTATIONS AND TREATMENT

Manifestations and treatment of acute renal failure are summarized in table 15.2. Both sodium and water are retained in acute renal failure. Hyponatremia may result from ingestion of a greater amount of water as compared to sodium ions. Blood creatinine levels tend to rise slowly because secretion of this substance tends to increase as filtration decreases (chapter 11). Urea, which is an end product of protein metabolism, is both filtered and reabsorbed. BUN levels rise more rapidly than creatinine levels

perfusion: L. *perfusio,* to pour over

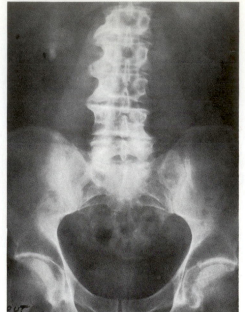

(a)

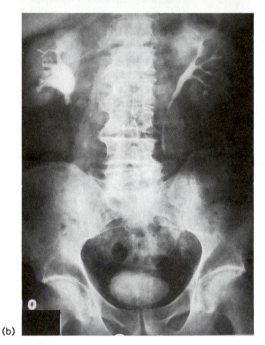

(b)

Figure 15.1 (*a*) X-ray of abdomen of an elderly male. Opaque structures on both the right and left sides (arrows at upper left corner) indicate kidney stones. (*b*) Intravenous pyelogram with the film taken 10 minutes after injection of contrast media (same patient). The contrast media is cleared by the kidneys and the white area in the patient's right kidney is evidence of obstruction due to kidney stones.

TABLE 15.1 Some causes of acute renal failure

Prerenal (Renal Perfusion)	Renal (Kidney Damage)	Postrenal (Obstruction)
Hemorrhage	Ischemia	Stones
Vomiting	Toxins	Tumors
Burns	Renal disease	Enlarged prostate
Diarrhea		
Vasodilation		
Congestive heart failure		
Pulmonary embolism		

TABLE 15.2 Manifestations and management of acute renal failure

Manifestations	Management
Hyponatremia	Salt and water restriction
Edema	Salt and water restriction
Hypertension	Salt and water restriction
Hyperkalemia	K^+ intake restricted
Hypocalcemia	May not require treatment
Hypermagnesemia	Restrict Mg^{++} containing antacids
Elevated BUN	Sufficient nutrition to prevent protein catabolism
Elevated creatinine	
Metabolic acidosis	Corrected slowly
Anemia	Transfusion if hematocrit <30%
Nausea and vomiting	May not require treatment
Susceptibility to infection	Treated promptly

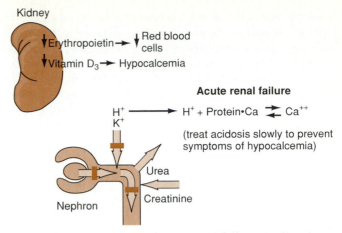

Figure 15.2 Consequences of acute renal failure. As filtration rate decreases, creatinine secretion initially increases while urea reabsorption increases. As a result, BUN levels tend to rise more rapidly.

because reabsorption of urea increases as renal blood flow decreases (figure 15.2). An impaired ability to secrete potassium causes hyperkalemia, and a failure to excrete acids results in metabolic acidosis. Hypocalcemia occurs in part because of decreased production of **1,25-dihydroxycholecalciferol** or vitamin D_3 (chapter 3) and because of skeletal resistance to the influence of PTH. Acidosis increases the level of the physiologically active form of calcium, the ionized form; thus, acidosis is treated slowly to avoid symptoms of hypocalcemia. Increased magnesium levels are usually the result of the administration of magnesium-containing antacids. Increased magnesium ion levels lead to blockage of neuromuscular transmission, which causes depression of reflexes, central nervous system depression, and may cause cardiac arrest. **Erythropoietin** (ĕ-rith″ro-poi′ē-tin) is a hormone normally produced by the kidney that stimulates red blood cell formation (chapter 17). Decreased erythropoietin combined with an increase in red blood cell fragility causes anemia.

In addition to the treatment listed in table 15.2, dialysis may be instituted. The procedures in dialysis are discussed in the latter part of this chapter. The prognosis depends on the severity and the underlying cause of renal failure.

CHRONIC RENAL FAILURE

Chronic renal failure is the result of permanent damage to nephrons resulting in (1) a decrease in the number of nephrons and/or (2) a marked decrease in the glomerular

erythropoietin: Gk. *erythros*, red; Gk. *poiesis*, production

TABLE 15.3 Fluid/electrolyte imbalances associated with a glomerular filtration rate of less than 25 ml/min

Polyuria. Loss of concentrating ability
Sodium. Increased excretion as GFR falls; hypernatremia if water restricted, hyponatremia if water excess; if low Na^+ diet, unable to reduce excretion, may cause hyponatremia
Potassium. Increased tubular secretion; increased fecal loss; rarely hyperkalemia if GFR > 5 ml/min
Calcium. Hypocalcemia in advanced stage
Magnesium, Phosphorus. Excretion of Mg, P increases with decreased GFR; Mg slightly elevated if patient taking Mg-containing antacids
Acidosis. Common at advanced stage; causes hyperkalemia (chapter 3)

filtration rate of individual nephrons. There are many mechanisms of injury including renal artery obstruction, intrinsic renal disease, and obstruction to urine flow.

MANIFESTATIONS

Symptoms are absent in most cases of early chronic renal failure. Urea and creatinine blood levels rise as there is a continuing decline in glomerular filtration rate. Since creatinine is secreted as well as filtered, blood creatinine levels rise more slowly than the BUN (chapter 11). Symptoms usually appear when the filtration rate falls to 25% of normal. In the early stages of failure, the GFR is in the range of 30–10 ml/min, and 10–5 ml/min in late stages. When renal failure is terminal, the GFR is less than 5 ml/min.

Table 15.3 summarizes the effects of a declining GFR on fluid and electrolyte balance. There is obligatory water loss as the kidney loses its concentrating capacity (chapter 11). In general, the kidney adapts to a decrease in GFR by increasing excretion of electrolytes. In the case of sodium ions, the kidney may be unable to adjust if the patient is put on a low-sodium diet. If there is continued accelerated excretion, the result is excessive loss of sodium. Potassium balance is maintained by increased secretion and by increased fecal loss. Hyperkalemia rarely occurs if the GFR is above 5 ml/min. Acidosis causes reciprocal exchange of intracellular potassium for extracellular Na^+/H^+ and leads to hyperkalemia (chapter 4).

• • •

For every decrease of 0.1 pH unit, serum K^+ will increase by about 0.6 meq/liter.

• • •

Calcium is a special case because of factors that affect metabolism. The kidney normally produces 1,25-dihydroxycholecalciferol (vitamin D_3), which promotes calcium absorption from the gastrointestinal tract. Decreased amounts of this metabolite of vitamin D results in decreased intestinal absorption of calcium and consequent fecal loss. Although parathormone (PTH) blood levels increase as GFR falls, bone and kidney tubules exhibit resistance to the influence of this hormone that normally raises blood levels of calcium. The overall effect of these factors is hypocalcemia, although tetany is rare. The ionized form of calcium in the blood is the physiologically active form, and this fraction is increased by acidosis. If acidosis is reversed quickly, it causes a sudden decrease in the level of ionized calcium and may precipitate tetany (chapter 3).

The kidney in chronic renal failure loses the ability to produce ammonia, and this contributes to acidosis (chapter 11). Manifestations of acidosis are most likely if the kidney is challenged by an excessive load of acid or by a loss of alkali.

Uremic Syndrome

An array of symptoms appear in chronic renal failure, collectively referred to as the **uremic syndrome.** These symptoms appear as the GFR falls below 20 ml/min and are more prominent as, in the final stage of failure, the GFR is less than 5 ml/min.

Manifestations of uremia are summarized in table 15.4. Anemia is common and is probably caused by decreased erythropoietin and increased red blood cell fragility. There is typically an increased bleeding tendency due to defective platelet function and capillary fragility. **Ecchymoses** (ek''ĭ-mō'sēz), or purplish discolorations of the skin, are the evidence of this and are the result of slight injuries. The skin of a uremic individual has a pallor due

uremia: Gk. *ouron,* urine; Gk. *haima,* blood
ecchymosis, pl. es: Gk. *ekchymosis,* small hemorrhagic spot

TABLE 15.4 A summary of some manifestations of uremia

Abnormal skin color	Emotional lability
Anemia	Insomnia
Bleeding tendencies	Loss of abstract thinking ability
Pruritis	Paresthesia and hypalgesia
Uremic frost	Skeletal defects
Serous membrane effusions	Diminished immune response
Anorexia, nausea, vomiting	Retention of electrolytes and water
Erosive gastritis	
Uremic colitis	Increased blood triglycerides

to anemia and a grayish-bronze color because of the kidney's failure to excrete pigments. **Pruritis** (proo-ri'tus) or itchiness is characteristic of kidney failure. An accumulation of urates in the form of a white powdery substance on the skin is described as **uremic frost.**

Serous membranes (chapter 6), pleural membranes of the lungs, and the membrane covering the heart, may become thickened and inflamed, and pleural effusion (chapter 6) and pericardial effusion may occur. Gastrointestinal manifestations of uremia include anorexia, nausea, and vomiting. **Uremic stomatitis,** or small ulcerations in the mouth, occur as well as hemorrhages and ulcerations of the stomach and colon. The two latter conditions are called erosive gastritis and uremic colitis. Neuromuscular disturbances include emotional lability, insomnia, and loss of abstract thinking ability. In the absence of intervention, coma, convulsions, and death ensue. A sensation of numbness in the lower extremities followed by **paresthesia** and **hypalgesia** are characteristic. Paresthesia refers to a tingling or burning sensation, and hypalgesia is a decreased sensitivity to pain.

The uremic syndrome includes skeletal defects. In children, growth is retarded markedly, and this is accompanied by bone deformities due to vitamin D deficiency. The common bone defects in adults are the result of increased PTH, which causes increased bone resorption. A further complication of advanced chronic renal failure is the appearance of calcium phosphate deposits in arteries, around joints, and in such organs as the lungs and the heart.

There is a diminished immune response. Increased blood triglyceride levels may be observed, and it is possible that these fatty substances contribute to atherosclerosis (chapter 22).

pruritis: L. *pururire,* to itch
stomatitis: Gk. *stoma,* mouth; Gk. *itis,* inflammation
paresthesia: Gk. *para,* beside; Gk. *aisthesis,* perception
hypalgesia: Gk. *hypo,* under; Gk. *algesis,* sense of pain

In the late stages, there is general failure of renal excretion leading to sodium and water retention accompanied by hypertension. Other substances retained include (1) potassium, (2) hydrogen ions, (3) phosphate, (4) magnesium, and (5) nitrogenous wastes.

TREATMENT OF RENAL FAILURE

There are three approaches to the treatment of renal failure: (1) early treatment of symptoms and underlying disease, (2) dialysis, and (3) renal transplant. Table 15.5 summarizes some early measures which should be taken before dialysis or renal transplant is implemented. The goal is to control symptoms and to minimize the complications of uremia. Sodium and fluid intake should be restricted to relieve hypertension. Dietary protein, a source of urea, should be restricted, and measures should be taken to correct electrolyte imbalances. Administration of calcium, vitamin D, and phosphate-binding agents maintain blood calcium levels and suppress PTH secretion.

PRINCIPLES OF DIALYSIS

Dialysis is a procedure whereby blood is exposed to **dialysate,** i.e., a fluid with solute concentration similar to plasma. The blood and dialysate are separated by a semipermeable membrane, and time is allowed for equilibration. Water and solutes diffuse across the membrane from an area of high to low concentration (chapter 1). The factors that determine the rate of exchange are (1) pore size and area of the membrane, (2) nature of solute, (3) concentration gradient, (4) hydrostatic pressure across the membrane, and (5) rate of flow of blood and dialysate.

TABLE 15.5 Early measures to control symptoms in chronic renal failure

---

Restriction of intake of Na$^+$ and fluid
Restriction of foods high in phosphate and K$^+$
Reduction of dietary protein
Correction of electrolyte imbalances
Supplements of calcium, vitamin D, and phosphate binding
    agents

HEMODIALYSIS

Hemodialysis and peritoneal dialysis are the two main types of dialysis. In hemodialysis, blood is pumped to a dialyzing apparatus or artificial kidney. There are different types of apparatus, but in each case, there is a semipermeable membrane positioned so blood flows on one side and dialysate flows on the other. Three basic designs are (1) the coil dialyzer, (2) flat plate dialyzer in which there are layers of flat sheets of membrane, and (3) the hollow fiber dialyzer in which there are thousands of fine capillaries through which blood flows, with a flow of dialysate surrounding the fibers (figure 15.3). Heparin is added to the blood as it enters the dialyzer to prevent clotting, and protamine sulfate neutralizes heparin activity as blood returns to the body's circulation. The dialysate has an electrolyte composition roughly the same as plasma so that elevated levels of electrolytes in the blood result in diffusion from the blood to dialysate. Metabolic wastes accumulate in blood and are absent in dialysate; thus, the wastes diffuse into dialysate. Since blood is pumped into the dialyzer, the hydrostatic pressure of the blood can be controlled. When the hydrostatic pressure of the blood is higher than that of the dialysate, there is net movement of water into the dialysate, a process called **ultrafiltration.** The dialysate is constantly replaced as wastes and excess water are carried away. Total dialysis time varies with individual patients, but most require several sessions with a total of 10–15 hours/week.

Hemodialysis requires vascular access (figure 15.4). Surgical creation of an **arteriovenous fistula** is frequently done to provide that access. The term fistula means a passage or canal and, in this case, involves an anastamosis or joining of an artery and a vein. The radial artery and cephalic vein in the forearm or the brachial artery and cephalic vein in the upper arm are frequently used. There are various techniques for joining the two vessels, and it may be necessary to graft bovine carotid artery or human umbilical cord artery to join the two vessels. The reason for the procedure is to enlarge the vein, so this vessel can supply adequate blood flow to the dialyzer. The vein provides venipuncture access for two needles, the venous or return needle directed away from the fistula and the arterial needle directed toward it (figure 15.5).

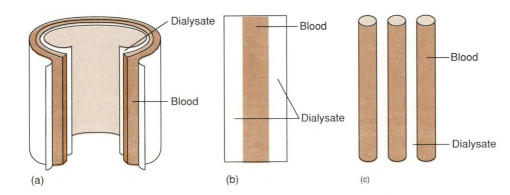

Figure 15.3 Types of dialyzers. (*a*) Coil dialyzer. (*b*) Plate dialyzer. (*c*) Hollow fiber dialyzer.

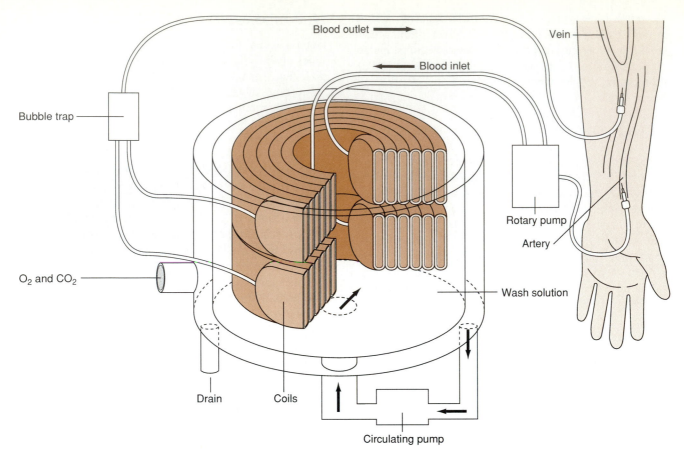

Figure 15.4 A diagram of a hemodialysis machine.

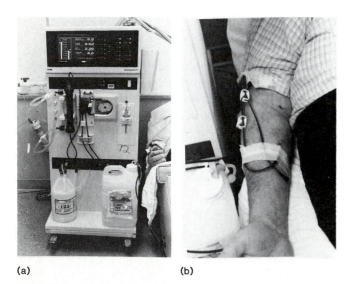

(a)   (b)

Figure 15.5 (*a*) Dialysis machine. (*b*) Patient undergoing dialysis.

PERITONEAL DIALYSIS

Peritoneal dialysis involves the introduction of dialysate into the peritoneal cavity which is lined by a membrane (figure 15.6). The peritoneal membrane covers both the visceral organs and the inner surface of membrane lining the abdominal wall. The parietal (pah-ri'ĕ-tal) membrane is continuous with the visceral membrane and forms a closed space that normally contains small amounts of fluid. Two or more liters of dialyzing fluid may be introduced into this space with little or no discomfort. A chronic indwelling catheter may be inserted into a small incision just below the umbilicus to provide instillation and drainage of dialysate.

The exchange of water and solutes primarily takes place between dialysate and peritoneal capillaries. In hemodialysis, the net flux of water into dialysate (ultrafiltration) is achieved by hydrostatic pressure. Concentrations of glucose are added to the dialysate for peritoneal dialysis. The fluid is hypertonic to plasma, and it is this osmotic gradient that is responsible for ultrafiltration in peritoneal dialysis.

There are various techniques for carrying out peritoneal dialysis procedures, and these may be either manual or automated with varying periods of time involved.

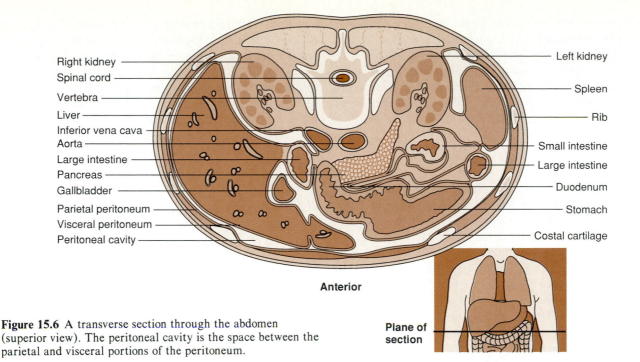

Right kidney | **Left kidney**
Spinal cord | **Spleen**
Vertebra | **Rib**
Liver | **Small intestine**
Inferior vena cava | **Large intestine**
Aorta | **Duodenum**
Large intestine | **Stomach**
Pancreas | **Costal cartilage**
Gallbladder
Parietal peritoneum
Visceral peritoneum
Peritoneal cavity

Anterior

Plane of section

Figure 15.6 A transverse section through the abdomen (superior view). The peritoneal cavity is the space between the parietal and visceral portions of the peritoneum.

Intermittent Peritoneal Dialysis

Intermittent peritoneal dialysis is usually performed by nursing staff. If done manually, this type of dialysis involves (1) the instillation of about 2 liters of dialysate over a 10 minute period, (2) a dwell time of about 30 minutes, and (3) drainage by gravity over a 20-minute period. One session may last for 36–48 hours, and in acute renal failure, one or two treatments per week may adequately remove wastes and maintain electrolyte levels. If the procedure is carried out by an automated cycler with instillation and drainage controlled automatically, the treatment sessions may involve four 10 hour sessions per week.

Continuous Ambulatory Peritoneal Dialysis

In 1975, a modification of intermittent peritoneal dialysis was introduced that allowed greater freedom for the patient and the advantages of continuous dialysis. The procedure is called **continuous ambulatory peritoneal dialysis (CAPD)** and involves a permanent indwelling abdominal catheter. A flexible plastic bag containing dialysate is attached by tubing to the catheter, and the bag is raised to allow gravity flow of dialysate into the peritoneal cavity. The bag may then be folded and kept in a pocket. The patient is then free to carry on normal activities for several hours. The bag is positioned below the abdomen to allow the fluid to drain from the peritoneal cavity, and a fresh bag of dialysate is then infused. A typical CAPD schedule involves four exchanges per day, with an approximate dwell time of 4 hours, and 8 hours overnight. The net effect is slow but continuous removal of solutes from the patient's blood.

TABLE 15.6 Preferred methods for the management of acute renal failure associated with various disorders

Peritoneal Dialysis	Hemodialysis
Cerebrovascular accident	Multiple injuries
Cerebral surgery	Abdominal surgery
Head trauma	Severe lung disease
Myocardial infarction	
Acute hemorrhage	

Continuous cyclic peritoneal dialysis (CCPD) is a modification of the CAPD regimen and involves multiple exchanges at night with an automated cycler.

Some advantages of CAPD include (1) continuous removal of wastes, (2) patient freedom, (3) no need for vascular access, and (4) increased clearance of middle size molecules that may play a role in toxicity.

There are circumstances in which one method of dialysis is preferred over the other (table 15.6). For example, risk of hemorrhage with heparin and fluid shifts in the brain during hemodialysis makes peritoneal dialysis the method of choice in treating acute renal failure in cerebral hemorrhage and injury. There is some risk of hypotension and arrhythmias in hemodialysis, and for that reason, peritoneal dialysis is preferred if renal failure occurs after heart surgery or myocardial infarction. There are situations in which hemodialysis is preferred because peritoneal dialysis may not provide adequate clearance of nitrogenous wastes. Peritoneal dialysis is contraindicated by severe lung disease because the instillation of fluid into the peritoneal cavity may interfere with movements of the diaphragm.

COMPLICATIONS

There are potential problems associated with both hemodialysis and peritoneal dialysis. Periodic fistula surgery is necessary in long-term hemodialysis. Infection and thromboses may occur due to the fistula and repeated venipunctures. Patients may experience headache, nausea, vomiting, and hypotension due to the rate of ultrafiltration. Rapid changes in blood potassium ion levels may lead to arrhythmias. Bleeding complications may occur due to the use of heparin. A brain disturbance called **dialysis dementia** can occur after several years of dialysis. The symptoms include speech disturbances, convulsions, and psychoses. Infection around the catheter or infection of the peritoneal lining is a risk in peritoneal dialysis. Moderate protein loss may occur, and the pressure of intraabdominal dialysate may cause a hernia.

Renal transplant is an option for individuals suffering from chronic renal failure. The immunological implications of this procedure are discussed in chapter 28.

SUMMARY

• • •

Acute renal failure is a sudden reduction in glomerular filtration that may be caused by decreased renal perfusion, kidney damage, or obstruction of urine flow. Manifestations include retention of nitrogenous wastes and fluid/electrolyte imbalances. Anemia is a consequence of decreased erythropoietin production by the kidney.

Chronic renal failure is a progressive decline in glomerular filtration accompanied initially by polyuria and electrolyte imbalance. The uremic syndrome, which is an outcome of chronic renal failure, is an array of symptoms that appear as the GFR falls from 20 ml/min to less than 5 ml/min. The symptoms are the result of loss of excretory function of the kidney and include effects on the (1) skin, (2) blood, (3) serous membranes, (4) gastrointestinal tract, (5) neuromuscular system, (6) skeleton, (7) immune system, and (8) fluid/electrolyte balance.

dementia: L. *demens,* out of one's mind

Renal failure is managed by treatment of symptoms, dialysis, and renal transplant. Dialysis involves fluid/electrolyte exchanges between blood and a plasmalike fluid (dialysate) that are separated by a membrane. Hemodialysis requires pumping blood to an artificial kidney. Dialysate is introduced into the peritoneal cavity to allow diffusion between that fluid and capillaries in peritoneal dialysis. Any dialysis procedure has attendant risks including thromboses, infection, and fluid/electrolyte imbalances.

REVIEW QUESTIONS

• • •

1. Define acute and chronic renal failure.
2. List three major categories for causes of acute renal failure.
3. List five possible reasons for decreased renal perfusion.
4. What are three possible reasons for obstruction to urine flow?
5. What effects on blood levels may be predicted in acute renal failure for each of the following? (a) Na^+ (b) K^+ (c) Water (d) Ca^{++} (e) Mg^{++} (f) Urea (g) Creatinine (h) H^+
6. Why do blood levels of creatinine tend to rise slowly?
7. Why do BUN levels tend to rise rather quickly in renal failure?
8. Why is acidosis treated slowly in acute renal failure?
9. In chronic renal failure, why should blood Na^+ levels be watched closely as a low Na^+ diet is implemented?
10. What factors tend to limit hyperkalemia if the GFR is greater than 5 ml/min in chronic renal failure?
11. What is most likely to contribute to increased blood levels of Mg in chronic renal failure?
12. Define the term uremic syndrome.
13. What factors contribute to anemia in the uremic syndrome?
14. What is the evidence of increased bleeding tendencies?
15. What other evidence of the uremic syndrome involves the skin?
16. In what ways does the uremic syndrome involve the gastrointestinal tract?
17. What are some neuromuscular disturbances in the uremic syndrome?
18. How is calcium and the skeletal system involved in the uremic syndrome?
19. List four factors that may be manipulated to control exchange rate in hemodialysis.
20. What controls ultrafiltration in hemodialysis and peritoneal dialysis?

SELECTED READING

• • •

Brown-Cartwright, D. 1989. Continuous ambulatory peritoneal dialysis: An alternative to hemodialysis. PA 89: *Physician Assistants in Primary and Hospital Care* 6:14.

Eisenhauer, F., J. Talartschik, and F. Seheler. 1986. Detection of fluid overload by plasma concentration of human atrial natriuretic peptide (h-ANP) in patients with renal failure. *Klinesche Wochenschrift* 64(Suppl VI):68–72.

Eknoyan, G. 1984. Side effects of hemodialysis. *New England Journal of Medicine* 311:915–17.

Eschbach, J. W. et al. 1987. Correction of the anemia of end-stage renal disease with recombinant erythropoietin. *New England Journal of Medicine* 316:73–78.

Heaton, A. et al. 1986. Evaluation of glycerol as an osmotic agent for continuous ambulatory peritoneal dialysis in end-stage renal failure. *Clinical Science* 70:23–29.

Korkor, A. B. 1987. Reduced binding of [^{3}H] 1,25–dihydroxyvitamin D_3 in the parathyroid glands of patients with renal failure. *New England Journal of Medicine* 316:1573–77.

Lettieri, C. 1980. The history of nephrology and dialysis: The 1970's. *Contemporary Dialysis and Nephrology* 11:35–46.

Nolph, K. D., R. P. Popovich, and J. W. Moncrief. 1978. Theoretical and practical implications of continuous ambulatory peritoneal dialysis. *Nephron* 21:117.

Rodriguez, Soriano, J., B. S. Arant, J. Brodehl, and M. E. Norman. 1986. Fluid and electrolyte imbalances in children with chronic renal failure. *American Journal of Kidney Diseases* 7:268–74.

Runyan, J. D., and H. A. Atterbom. 1990. Exercise and the hemodialysis patient. *Physician Assistant* 14:91–98.

Sherwood, L. M. 1987. Vitamins D, parathyroid hormone, and renal failure. *New England Journal of Medicine* 36:1601–1603.

Slatopolsky, E. et al. 1986. Calcium carbonate as a phosphate binder in patients with chronic renal failure undergoing dialysis. *New England Journal of Medicine* 315:157–61.

Yu, V. L. et al. 1986. *Staphylococcus aureus* nasal carriage and infection in patients on hemodialysis. *New England Journal of Medicine* 315:91–96.

Chapter 16

Practical Applications: Evaluation of Renal Failure

The focus of this unit is renal dysfunction due to injury or disease, and the preceding chapters deal with specific aspects of these topics. The term acute renal failure (chapter 15) implies rapid and frequently reversible loss of kidney function, often manifested by decreased urinary output. In general, acute renal failure is the result of renal hypoperfusion, glomerular/tubular injury, massive infection, or obstruction. Most cases of acute renal failure are associated with surgery or trauma, and inadequate blood flow to the kidney is a common underlying cause.

Chronic renal failure (chapter 15) involves progressive nephron destruction with permanent loss of function and is a major cause of death from kidney disease. The consequences of chronic renal failure include the retention of nitrogenous wastes due to decreased glomerular filtration rate, i.e., azotemia. This may be caused by kidney disease, a decreased blood flow to the kidney, or obstruction to the flow of urine. Uremia is a clinical syndrome associated with chronic renal failure involving metabolic and endocrine dysfunctions due to kidney damage.

The following discussion involves examples of acute and chronic renal failure with emphasis on the significance of laboratory findings.

ACUTE RENAL FAILURE

Acute renal failure sometimes occurs in association with liver disease, which may be a viral hepatitis, a biliary tract obstruction, or a hepatic malignancy. Most often the liver disease is cirrhosis and is usually caused by alcohol abuse (chapter 36). The term hepatorenal syndrome is used for this form of renal failure and, although the precise mechanism for altered renal function is not clear, changes in renal hemodynamics appear to be involved.

HEPATORENAL SYNDROME

The following is an example of hepatorenal syndrome.

> Patient 16.1 was a 69-year-old male, and the diagnosis was cirrhosis of the liver and hepatorenal syndrome. Laboratory findings (normal values in parentheses): BUN 42 mg/dl (7–22), creatinine 3.1 mg/dl (0.6–1.2), alkaline phosphatase 507 U/liter (30–95), blood pH 6.93 (7.35–7.45), HCO_3^- 16.4 meq/liter (22–30), base excess −16 meq/liter (−2.5 to +2.5)

Events preceding renal failure in the hepatorenal syndrome typically include fluid loss due to various causes, such as gastrointestinal hemorrhage, diarrhea, or diuretics. Evidence indicates that the subsequent reduction in glomerular filtration rate is caused by both vasoconstriction and volume loss, with a decrease in renal blood flow. Oliguria is typical in this syndrome. Normally, elevations of BUN and creatinine levels can be correlated with impaired glomerular filtration. There are some additional factors that influence the levels of these nitrogenous substances when liver disease is involved. Protein

metabolism is the source of urea, and in cirrhosis, there is likely to be both decreased dietary protein and impairment of protein synthesis by the liver. Creatinine is synthesized in muscle, so a decrease in muscle mass causes a lower serum creatinine level. The net effect is that the BUN and creatinine levels in the hepatorenal syndrome would be higher if only a reduction in glomerular filtration rate were involved. In the early stages of this syndrome, these values may be normal. This misrepresents the glomerular filtration rate.

This example of hepatorenal syndrome shows elevated BUN and creatinine levels and a high alkaline phosphatase value. The source of the alkaline phosphatase is probably the liver in this case. Acidosis is a characteristic finding in both acute and chronic renal failure and is clearly indicated by a pH of 6.93, HCO_3^- of 16.4 meq/liter and a base excess of −16 meq/liter (chapter 4).

REDUCED RENAL PERFUSION

A sudden reduction in the flow of urine is evidence of acute renal failure. The cause, in general, is reduced blood flow to the kidney, nephron injury, or obstruction to the flow of urine. These mechanisms, which result in diminished kidney function, are identified as (1) prerenal, (2) intrarenal, and (3) postrenal. The following individual was hospitalized and scheduled for surgery when acute renal failure became apparent.

> Patient 16.2 was a 68-year-old female with a long history of hypertension, myocardial infarction, and a recent cerebrovascular accident (stroke). Stenosis of the left carotid artery was identified by angiography. She experienced nausea, anorexia, and vomiting over a 2-week period after this diagnostic procedure.
> Laboratory findings at the time of the angiogram:
> BUN 44 mg/dl (7–22), creatinine 2.7 mg/dl (0.6–1.2).
> Upon second hospital admission two weeks after the angiogram:
> BUN 112 mg/dl (7–22), creatinine 7.0 mg/dl (0.6–1.2), Hct 44% (42–50), Na^+ 127 meq/liter (137–47), K^+ 5.8 meq/liter (3.4–5.0), Cl^- 97 meq/liter (100–10), HCO_3^- 21 meq/liter (22–30)
> Three days later:
> creatinine 8.9 mg/dl (0.6–1.2), Ca 8.6 mg/dl (8.9–10.6), phosphorous 6.6 mg/dl (2.5–4.5), BUN 116 mg/dl (7–21), K^+ 5.9 meq/liter (3.4–5.0), Na^+ 127 meq/liter (137–47)

Surgery for the carotid stenosis was cancelled due to hypertension and evidence of failing kidney function. This patient had a creatinine level of 1.7 mg/dl 2 months prior to hospitalization. The BUN and creatinine values were moderately elevated when the angiogram was performed. Angiography is a diagnostic procedure involving the injection of radiocontrast material into an artery to make visualization by X-ray possible (chapter 38). Radiocontrast agents may injure renal tubules if there is chronic renal insufficiency for any reason.

stenosis: Gk. *stenos*, narrow

This patient's renal function continued to deteriorate with progressive elevations of the BUN and creatinine levels. Urinary output was less than 250 ml in a 12-hour period, and there was no response to the diuretic Lasix.

The probable cause of hyponatremia in this case is the dilutional effect of water retention. The serum potassium level is elevated. Hyperkalemia is a leading cause of death in acute renal failure and is the result of the loss of ability to excrete a normal potassium load. Severe acidosis is not a common feature of early acute renal failure.

Acute renal failure is associated with hypocalcemia and increased blood levels of phosphate, as seen in this patient. During the early stages of failure, there is decreased excretion of phosphate. The consequent hyperphosphatemia leads to a decrease in blood levels of ionized calcium. The mechanism causing hypocalcemia may be the binding and deposit of calcium phosphate in body tissues. Hypocalcemia stimulates parathormone secretion (chapters 3 and 31), which promotes a return of phosphate and calcium balance until the glomerular filtration rate decreases to less than 25% of normal.

The immediate problem with patient 16.2 was to identify the cause of renal failure. She had a long history of hypertension associated with severe peripheral vascular disease. She had suffered a stroke (chapter 38) about a month prior to renal failure. It is likely that there was also underlying renal vascular disease and nephrosclerosis.

A radioisotope scan was done to evaluate blood flow to the kidney. The procedure involves intravenous injection of a radioactive agent which is monitored by a scintillation camera. The results of the test indicated probable renal artery occlusion. The fact that there was continued elevation of the creatinine level two weeks after the angiogram made it unlikely that continued toxic effects of radiocontrast material was responsible. Patient 16.2 was taking digitalis because of the recent myocardial infarction (chapter 23), and there is a strong possibility that the episode of vomiting occurred as the result of digitalis toxicity. It is probable that this identifies the point at which renal arterial occlusion occurred. A renal angiogram was performed to show blood flow to the kidney and to confirm the diagnosis of renal artery occlusion. The patient was put on hemodialysis following the test procedure.

CHRONIC RENAL FAILURE

The most common cause of chronic renal failure is glomerulonephritis followed by diabetic nephropathy, nephrosclerosis, hereditary kidney disease, and chronic pyelonephritis. The following is an example of a patient suffering from chronic renal failure.

> Patient 16.3 was a 70-year-old female and the diagnosis was chronic renal failure secondary to nephrosclerosis or chronic glomerulonephritis; end-stage renal disease.
> Laboratory findings:
> Predialysis lab values: HCO_3^- 8.4 meq/liter (22–30), BUN 298 mg/dl (7–22), creatinine 11.3 mg/dl (0.6–1.2)

NEPHROSCLEROSIS/GLOMERULONEPHRITIS

Patient 16.3 had a long history of hypertension as well as a history of gout. Nephrosclerosis is indicated as a possibility in the diagnosis because of prolonged hypertension. Hypertension may lead to blood vessel injury, including small renal arteries (chapter 14), and cause renal ischemia and, ultimately, death of renal tissue.

Glomerulonephritis was also considered a possible cause, although the medical history of patient 16.3 did not clearly indicate this. Glomerulonephritis may be poststreptococcal, may develop as the result of various systemic diseases, or may be idiopathic (chapter 13). A history of gout is noteworthy in this case. Gout is a condition in which serum urate is elevated, urate salt is deposited around joints, and there may be deposition of monosodium urate crystals in renal interstitial tissue. There may also be uric acid nephrolithiasis. Most individuals with gouty arthritis have some renal dysfunction. With a history of hypertension, nephrosclerosis was deemed likely, although glomerulonephritis was a possibility. Gouty arthritis may have been a contributing factor to renal failure as well.

Metabolic Acidosis

Prior to the initiation of dialysis, patient 16.3 had marked acidosis with a bicarbonate of 8.4 meq/liter. Acidosis is caused by an increase in hydrogen ions or a decrease in bicarbonate and is a manifestation of either acute or chronic renal failure.

Table 16.1 lists events in the kidney that maintain acid-base balance, showing that (1) NH_3 buffers hydrogen ions, so they are excreted as NH_4^+; (2) bicarbonate is reabsorbed as hydrogen ions are secreted; and (3) there is a mechanism for regenerating bicarbonate.

TABLE 16.1 Some events in the nephron that maintain acid-base balance

NH_3 Buffers Hydrogen Ions	
1a	Glutamine $\rightleftharpoons$ NH_3 within renal tubule cell
1b	NH_3 diffuses into tubular lumen
1c	$NH_3 + H^+ \rightleftharpoons NH_4^+$
1d	NH_4^+ excreted in urine
Bicarbonate Is Reabsorbed as Hydrogen Ions Are Secreted	
2a	filtered HCO_3^- reabsorbed in proximal tubule
2b	H^+ secreted in proximal tubule
Mechanism for Regenerating Bicarbonate	
3a	Reaction inside tubule cells: $CO_2 + H_2O \rightleftharpoons H_2CO_3 \rightleftharpoons H^+ + HCO_3^-$
3b	H^+ moves into lumen; HCO_3^- reabsorbed
3c	Above reaction regenerates new HCO_3^-

There is evidence in chronic renal failure that, as nephrons are functionally destroyed, there is increased NH_3 production by the remaining nephrons. At a certain point, nephrons are destroyed at a rate greater than compensatory NH_3 production occurs. Ultimately, NH_3 synthesis falls, and the urinary buffer for hydrogen ions decreases. In addition to this, acid anions produced by metabolism are retained due to a decreased glomerular filtration rate; tubular damage limits hydrogen ion secretion; and to a lesser degree, there is urinary loss of bicarbonate. The overall effect in chronic renal failure is metabolic acidosis, and the bicarbonate value of patient 16.3 indicates this imbalance.

BUN/Creatinine

Patient 16.3 had elevated levels of both BUN and creatinine. These tests are not useful in identifying early renal failure because they remain normal until 50% of renal function has been lost. Initially, serum creatinine levels do not rise as quickly as the BUN. The basis for this observation is that tubular secretion of creatinine increases as glomerular filtration decreases. The BUN/creatinine ratio is high in this case, and this indicates a decreased urine flow. Decreased urine flow does not affect creatinine, but it does allow increased urea reabsorption. A ratio greater than 20:1 is characteristic of dehydration, ureteral obstruction, and decreased renal blood flow.

GLOMERULONEPHRITIS

A second example of chronic renal failure due to glomerulonephritis follows and includes additional laboratory data.

Patient 16.4 was a 45-year-old female with chronic renal failure and a history of glomerulonephritis. Laboratory findings:
Blood pH 7.28 (7.35–7.45), pCO_2 38 mm Hg (35–45), HCO_3^- 18.2 meq/liter (22–30), base excess −7.2 meq/liter (+2.5 to −2.5), anion gap 22 meq/liter (6–16), phosphorous 10.3 mg/dl (2.3–4.2), Ca 7.9 mg/dl (8.6–10.5), Na^+ 126 meq/liter (137–47), alkaline phosphatase 185 U/liter (30–95), hemoglobin 9.2 g/dl (14–18)

Metabolic Acidosis

Metabolic acidosis is confirmed by a bicarbonate of 18.2 meq/liter and a base excess of −7.2 meq/liter. Both of these values indicate a metabolic cause rather than a respiratory origin. A positive base excess value means either metabolic base excess or deficiency of acid, whereas a negative value indicates the reverse, a deficiency of base or excess acid. The pH of 7.28 indicates acidosis. By itself, however, the pH does not show whether the imbalance is of metabolic or of respiratory origin. Hyperventilation is an immediate compensatory response to metabolic acidosis and may lead to a decrease in pCO_2. The pCO_2 in this case is in the normal range.

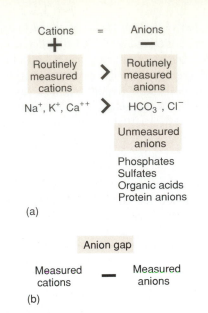

Figure 16.1 The anion gap is the difference between the sum of the measured cations and the sum of the measured anions.

Anion Gap

The anion gap is useful in assessing metabolic acidosis and is based on the following. The sum of cations and anions in the extracellular fluid is always equal, but routine measurements include only two cations and two anions. Sodium and potassium constitute about 95% of the cations, and chloride and bicarbonate represent about 86% of the anions. The difference between these measured cations and anions is called the anion gap. This difference represents unmeasured organic anions, proteins, sulfate, and phosphate. When there is an increase in the anion gap, as there was in this patient, it represents acid anions retained by the body.

The following will clarify the mechanisms involved. Diarrhea causes a loss of bicarbonate, and subsequently, HCl is added to the blood (chapter 5). The result is metabolic acidosis, but lost bicarbonate is equaled by an increase in serum chloride ions. As a result, the sum of the chloride and bicarbonate remains the same, and the anion gap is unaffected. The metabolic acidosis that occurs in chronic renal failure is caused by bicarbonate loss and accumulation of anions in the blood that are not routinely measured. The result is an increased difference between the sum of the cations and the sum of the anions, $(Na^+ + K^+) - (HCO_3^- + Cl^-)$ (figure 16.1). Table 16.2 summarizes the significance of the anion gap.

Phosphorous

In the case of patient 16.4, the serum phosphorous was an elevated 10.3 mg/dl. The phosphorous level usually remains normal until the glomerular filtration rate is about 25 ml/min; and at this point, there is a progressive increase. It is likely that increasing levels of serum phosphorous stimulate parathormone (PTH) production, possibly by causing a transient decrease in the ionized form

TABLE 16.2 Significance of anion gap*

Metabolic Acidosis (Anion Gap Normal)

Hyperchloremic acidosis. Increased blood chloride levels electrically compensate for decreased bicarbonate

Cause	Effect
Renal tubular acidosis (kidney retains H^+)	Increased Cl^-, decreased HCO_3^-
Diarrhea (loss of bicarbonate)	Increased Cl^-, decreased HCO_3^-

Metabolic Acidosis (Anion Gap Abnormal)

Anion gap (>16 meq/liter)

Cause	Unmeasured Anions
Lactic acidosis	Lactate anions
Salicylate poisoning	Salicylate anions
Liver failure	Ketoglutarate anions

Metabolic acidosis may or may not be associated with an accumulation of abnormal, unmeasured anions.

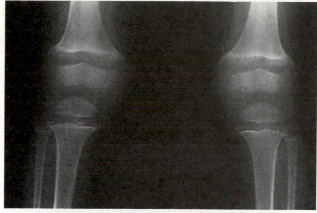

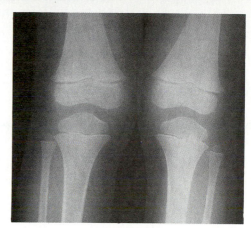

Figure 16.2 (*a*) X-ray of knees of a 2-year-old female with renal osteomalacia. (*b*) Same patient at 4 years. Renal insufficiency, especially in children, is associated with osteomalacia, which is defective mineralization of the organic matrix of bone.

of calcium in the serum. Increased PTH promotes excretion of phosphorous, but it also causes release of calcium from bone and subsequent bone disease. Ultimately, the serum phosphorous level is determined by dietary intake, PTH, and the degree of damage of the proximal tubules.

Calcium

The blood calcium level in patient 16.4 is low, which is common if the glomerular filtration rate is less than 20 ml/min. There are several factors that interact and influence calcium levels. In chronic renal failure, vitamin D metabolism is abnormal. Consequently, a form of vitamin D, 1,25-$(OH)_2D_3$, is decreased. This causes decrease of absorption of calcium from the gastrointestinal tract. Parathormone stimulates reabsorption of calcium by renal tubules and also promotes release of calcium from bone. The tubules develop a resistance to PTH, and the result is usually a decrease in total calcium. Hypercalcemia occasionally occurs and may be due to administration of $CaCO_3$ or other calcium salts, treatment with vitamin D, or severe hyperparathyroidism.

Sodium

The laboratory report of patient 16.4 shows hyponatremia. A decrease in serum sodium ions may be the result of tubular damage, vomiting, or dietary restriction.

Alkaline Phosphatase

Alkaline phosphatase is a family of enzymes produced by the intestine, liver, kidney, and bone. Any of these organs can be the source of an elevated serum level of alkaline phosphatase. Patient history and diagnosis is an indicator of source. This enzyme arises primarily from bone in renal failure. Increases in serum alkaline phosphatase levels in-

dicate the development of skeletal disease. **Renal osteodystrophy** (os″tē-ō-dis′trō-fē) occurs, and this involves bone loss and cystic lesions in response to increased PTH and calcium deficiency. Impaired vitamin D metabolism and decreased calcium leads to a defect in bone mineralization, a condition called **osteomalacia** (figure 16.2). The alkaline phosphatase test is useful in monitoring the administration of vitamin D or calcium compounds for these disorders. This patient has an elevated alkaline phosphatase value, but it should be noted that skeletal disease can exist even if that serum enzyme level is within a normal range.

Anemia

The red blood cell count was not reported, but the hemoglobin value is low and indicative of anemia. Anemia associated with renal insufficiency may be caused by decreased erythropoietin, bone marrow which is unresponsive to erythropoietin, or increased fragility of red blood cells with hemolysis. Table 16.3 summarizes the observations about laboratory data in chronic renal failure.

osteodystrophy: Gk. *osteon*, bone; Gk. *dys*, bad; Gk. *trophe*, nourishment
osteomalacia: Gk. *osteon*, bone; Gk. *malakia*, softness

TABLE 16.3 A summary of expected laboratory data in chronic renal failure with brief explanations

Lab Data	Significance
HCO_3^- decreases; base excess decreases	Metabolic acidosis due to loss of HCO_3^-, accumulation of acids, decreased renal NH_3 production
Creatinine, BUN increases	50% or $>$ renal function lost; decreased GFR
Phosphorous increases	GFR $<$ 25 ml/min; stimulates PTH
Calcium commonly decreases	Factors: decreased vitamin D, decreased reabsorption from GI; PTH causes tubular reabsorption; tubules resistant to PTH
If calcium increases	Ca salts, vitamin D, hyperparathyroidism
Na decreases	Accelerated excretion; dietary restriction
Alkaline phosphatase increases	Demineralization of bone
Anion gap increases	Unmeasured acid anions retained

DIABETIC NEPHROPATHY

The following individual had a long history of diabetes and had not seen a doctor for several years.

Patient 16.5 was a 47-year-old male with 40-year history of diabetes mellitus. Diagnosis: chronic renal failure secondary to diabetic nephropathy, blood glucose 1,300 mg/dl. After hydration and lowering blood sugar, creatinine 4 mg/dl, BUN 60 mg/dl.

This patient had moderate swelling of feet and ankles but no nausea or vomiting. This is an example of the final outcome of kidney involvement with diabetes mellitus (chapter 14). The patient was put on a restricted diet for protein and ultimately was to start dialysis treatments.

RENAL TRANSPLANT REJECTION

Patient 16.6 was admitted to the hospital following acute rejection of a renal transplant. He experienced general malaise, fever, and extreme tenderness in the region of the transplant. He underwent a transplant nephrectomy, and a schedule for hemodialysis was established.

Patient 16.6 was a 40-year-old male. His postdialysis laboratory data is as follows.
WBC 6,000 (4,800–10,800), Hgb 8.8 g/dl (14–18), Hct 15.6% (42–52), Na^+ 141 meq/liter (137–47), K^+ 3.1 meq/liter (3.4–5), Cl 105 meq/liter (100–10), HCO_3^- 18.8 meq/liter (22–30), BUN 82 mg/dl (7–22), creatinine 8.6 mg/dl (0.6–1.2), Ca 6.5 mg/dl (8.6–10.5)

nephrectomy: G. *nephros*, kidney

This patient had undergone parathyroidectomy prior to renal transplant, which is the reason for the low calcium level. Parathyroid secretion of parathormone (PTH) (chapter 3) increases as the glomerular filtration rate falls in renal failure. Hypocalcemia is the major stimulus. Hyperplasia with an increase in parathyroid cell mass may also occur and contribute to PTH hypersecretion. Parathyroidectomy may be indicated in chronic renal failure to prevent elevations of blood calcium levels.

This patient complained of muscle cramps and, in view of his low serum calcium level, was given intravenous calcium to bring the blood level up to 8.5 mg/dl. He was placed on oral calcium supplements.

Anemia is a complication of chronic dialysis; thus, the hematocrit is low in dialysis patients, often in the range of 25–35%. In patients who do not have a kidney, the hematocrit may be 12–20%.

SUMMARY

• • •

Acute renal failure as a part of the hepatorenal syndrome is caused by fluid volume loss and subsequent reduction in GFR. This occurs in association with liver disease, such as cirrhosis. The evidence of acute renal failure is an increase in both BUN and creatinine levels due to the impaired GFR. A decrease in muscle mass associated with liver disease tends to lower creatinine levels. Similarly, impaired protein synthesis leads to a decrease in the BUN. Liver disease moderates the elevations of these nitrogenous substances associated with a decreased GFR. Acute renal failure occurs when there is occlusion of the renal artery. Laboratory findings associated with acute renal failure include (1) hyponatremia due to water retention, (2) hyperkalemia, (3) hypocalcemia, and (4) hyperphosphatemia. The lowered calcium level may be the result of calcium/phosphate binding.

Examples of causes of chronic renal failure included in this chapter are nephrosclerosis, glomerulonephritis, diabetic nephropathy, and renal transplant rejection. A summary of observations related to chronic renal failure are as follows. Metabolic acidosis occurs in renal failure as the result of (1) failure of renal tubular cells to produce NH_3 to buffer hydrogen ions, (2) loss of bicarbonate, and (3) accumulation of acid anions. The laboratory values that show acidosis are blood pH, a measurement of bicarbonate, base excess, and a high anion gap that indicates unmeasured acid anions. The BUN and creatinine levels are general indicators of glomerular filtration rate unless there is liver disease or muscle wasting. A high BUN/creatinine ratio indicates decreased urine flow. Increasing levels of phosphorous stimulate PTH production, and this causes demineralization of bone. Hypocalcemia is common if the GFR is less than 20 ml/min and is due to impaired intestinal absorption and failure of renal tubular reabsorption. Bone is a source of alkaline phosphatase, and blood levels of this enzyme are increased because of bone calcium loss. Anemia is a part of the picture in chronic renal failure and is the result of decreased erythropoietin and increased red blood cell fragility.

REVIEW QUESTIONS

• • •

1. What relationship is there between decreased NH_3 production by renal tubular cells and acidosis?
2. List two other factors that contribute to metabolic acidosis in renal failure.
3. What does a negative base excess value mean?
4. What is the significance of a high BUN/creatinine ratio?
5. Why does creatinine blood level rise slowly in renal failure?
6. What are the interrelationships between phosphorous, PTH, and calcium in chronic renal failure?
7. What two cations and two anions are usually measured in a clinical laboratory?
8. Define the term anion gap.
9. What is the significance of an increase in anion gap?
10. What is the reason for an elevated serum alkaline phosphatase level in chronic renal failure?

SELECTED READING

• • •

Andress, D. L. et al. 1985. Effect of parathyroidectomy on bone aluminum accumulation in chronic renal failure. *New England Journal of Medicine* 312:468–73.

Broadus, A. E. et al. 1984. Evidence for disordered control of 1,25-dihydroxyvitamin D production in absorptive hypercalciuria. *New England Journal of Medicine* 311:73–80.

Gennari, F. J. 1985. Acid-base balance in dialysis patients. *Kidney International* 28:678.

Goldstein, M. B. et al. 1986. The urine anion gap: A clinically useful index of ammonium excretion. *American Journal of Medical Sciences* 292(4):198.

Lindholm, B. 1986. Muscle water and electrolytes in patients undergoing continuous ambulatory peritoneal dialysis. *Acta Medica Scandinavica* 219(3):323–30.

Livio, M. et al. 1986. Conjugated estrogens for the management of bleeding associated with renal failure. *New England Journal of Medicine* 315:731–37.

Mitch, W. E. et al. 1984. The effect of a ketoacid-amino acid supplement to a restricted diet on the progression of chronic renal failure. *New England Journal of Medicine* 311:623–34.

Myers, B. D., and S. M. Moran. 1986. Hemodynamically mediated acute renal failure. *New England Journal of Medicine* 314:97–104.

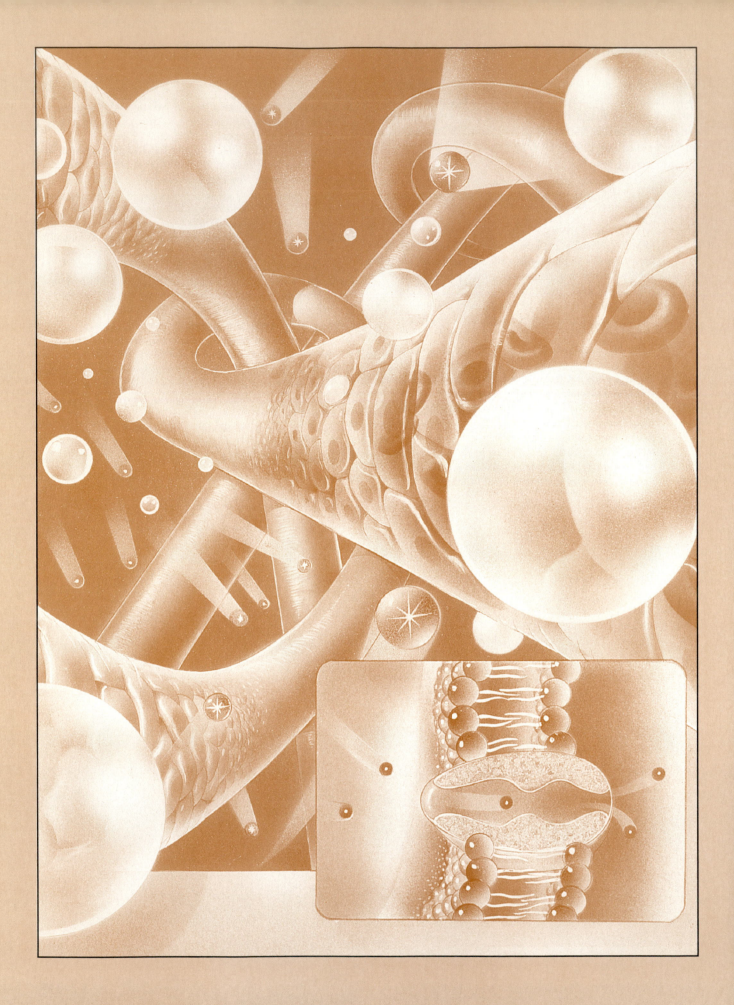

BLOOD DISORDERS: CELLS AND BLEEDING

The introductory portion of chapter 17 combines erythropoiesis, normal red blood cell function, and oxygen delivery to tissues. The related disorders described in the second half of the chapter are selected examples involving inadequate oxygenation and abnormal red blood cell/hemoglobin production.

The introduction in chapter 18 deals with the development and function of normal white blood cells. A summary of reasons for variations in white blood cell numbers precedes the emphasis on malignant proliferation, with reference to three types of leukemias.

Chapter 19 includes a summary of the processes involved in coagulation and fibrinolysis. This summary provides a reference point for the remainder of the chapter. Various bleeding disorders are discussed, including disseminated intravascular coagulopathy.

The purpose of chapter 20 is to discuss hematologic tests important in evaluating blood disorders. Tests for coagulation as well as blood cell tests are included. The last part of chapter 20 involves evaluation of laboratory findings in a case of acute myelogenous leukemia and in a patient with prostatic carcinoma associated with DIC.

Chapter 17

Oxygenation and Red Blood Cell Disorders

The discussion in this chapter centers around red blood cells or erythrocytes. The first section deals with red blood cell formation and the function of oxygen transport, and the remainder of the chapter emphasizes disorders that include abnormalities of cell number and hemoglobin production.

• • •

During fetal development, red blood cell production occurs mainly in the liver, and to some extent, the spleen and lymph nodes. The bone marrow is the exclusive site of erythropoiesis in late gestation and after birth.

• • •

RED BLOOD CELL FORMATION

Mature red blood cells that circulate in the blood develop from unspecialized bone marrow cells called **stem** or **precursor cells.** Stem cells are capable of proliferating to renew the population and differentiating to become specialized in type and function. The least differentiated of the stem cells are called **pluripotent** and may develop into several different blood cell types (figure 17.1). A partially differentiated stem cell described as being **erythropoietin-responsive,** when under the influence of erythropoietin (ĕ-rith″ro-poi′ĕ-tin), undergoes a series of changes to become a mature erythrocyte (figure 17.1). Table 17.1 summarizes the stages of development in the process of red cell formation which is called **erythropoiesis.** Figure 17.2 shows an immature red blood cell called a normoblast that appears in the bone marrow. The size and shape of mature red blood cells are shown in figure 17.3.

• • •

A reticulocyte count is a laboratory test that indicates rate of red blood cell production. The count is expressed as a percentage of total red blood cells, with a normal range in adults between 0.5–1.5%. An increased count shows increased rate of RBC production.

• • •

Erythropoietin, a hormone produced mainly by the kidney, stimulates red blood cell production in response to tissue hypoxia. Possible causes of this oxygen deprivation include hemorrhage, hemolysis, low hemoglobin oxygen content, and decreased cardiac output. Although erythropoietin is the major factor in erythropoiesis, there are other substances listed in table 17.2 essential for the process.

pluripotent: L. *pluris,* several; L. *potens,* able
erythropoiesis: Gk. *erythros,* red; Gk. *poiesis,* production

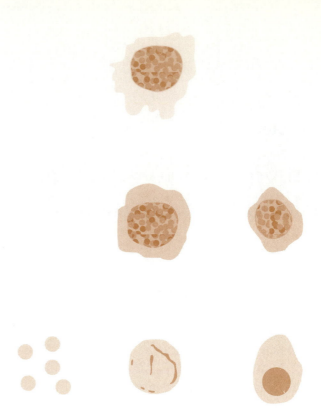

Figure 17.1 The origin and development of red blood cells.

TABLE 17.1 Developmental stages of a mature erythrocyte*

Stage	Description
Proerythroblast	Large round nucleus; diameter 12–19 microns; nongranular cytoplasm stains blue; youngest recognizable precursor of red blood cells
Erythroblast	Decrease in size; rapid hemoglobin synthesis; cytoplasm basophilic (blue), changing to pink mixed with basophilia
Normoblast	Smaller; full complement of hemoglobin; cytoplasm reddish pink
Reticulocyte	No mitosis; period of maturation in bone marrow 2–3 days, 1 day in blood; nucleus absent; may have overall blue appearance with Wright's stain; with special stain, there are granules or filaments due to RNA
Erythrocyte	Nucleus absent; biconcave disk; diameter 6–8 microns

*About 4 days are required for maturation of a proerythroblast into an erythrocyte.

reticulocyte: L. *reticulum,* little net; Gk. *kytos,* cell

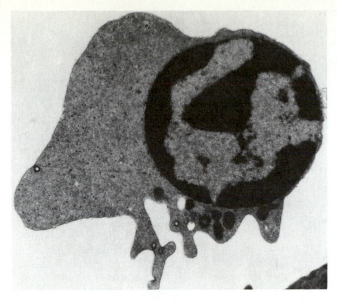

Figure 17.2 Electron micrograph of a normoblast just before extrusion of the nucleus. The cytoplasm reveals mitochondria, a few small vacuoles, a coated pit, and some coated vesicles. The latter are recognizable by their dense outline, a reflection of the coat. The fine dense particles in the cytoplasm are polysomes. ×10,000.
Courtesy of Dorothea Zucker-Franklin, Professor of Medicine at New York University Medical Center.
From M. H. Ross, E. J. Reith, and L. J. Romrell: Histology: A Text and Atlas. *1989 Williams & Wilkins, Baltimore. Reproduced with permission.*

Figure 17.3 The size and shape of a red blood cell.

TABLE 17.2 Factors in red blood cell formation

Factor	Contribution to red blood cell formation
Protein	Important for erythropoietin synthesis
Iron	Incorporated into hemoglobin molecules
Vitamin B_{12}	Required for red blood cell maturation
Folic acid	Required for red blood cell maturation
Vitamin B_6	Required for synthesis of hemoglobin
Cobalt	Incorporated into vitamin B_{12} molecules

Figure 17.4 The hydration of carbon dioxide is catalyzed by the enzyme carbonic anhydrase within red blood cells. Bicarbonate moves outside the cell in exchange for intracellular movement of chloride ions.

• • •

Erythropoietin is a glycoprotein formed mainly by the kidney with 5–10% formed by the liver. Maximum erythropoietin production occurs in response to hypoxia within 24 hours. New red blood cells appear in the circulation 5 days later.

• • •

RED BLOOD CELL FUNCTION

The following paragraphs outline four functions of the red blood cell, three of which are directly related to hemoglobin. Hemoglobin is responsible for (1) oxygen transport, (2) carbon dioxide transport, and (3) buffering hydrogen ions. In addition to this, a reaction between carbon dioxide and water is catalyzed by the enzyme carbonic anhydrase within red blood cells.

FORMATION OF BICARBONATE

The reaction between carbon dioxide and water shown in figure 17.4 occurs in red blood cells in a fraction of a second because of the presence of a large amount of the enzyme

Hemoglobin   **Heme**

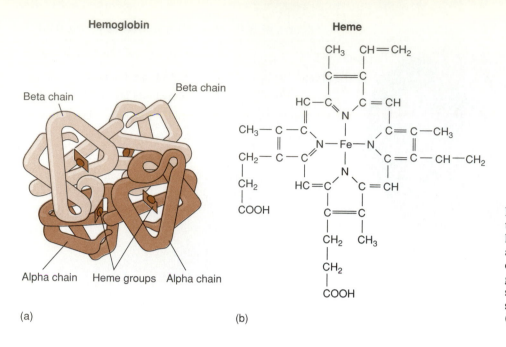

(a)   (b)

Figure 17.5 (*a*) A diagram of the three-dimensional structure of hemoglobin in which the two alpha and two beta polypeptide chains are shown; the four heme groups are represented as flat structures with iron (dark spheres) in the centers.
(*b*) Chemical structure for heme.

carbonic anhydrase (chapter 4). The same reaction occurs in plasma, except the rate is much slower. Bicarbonate, formed intracellularly, diffuses into plasma. Electrical balance is maintained by the diffusion of chloride ions into red blood cells in exchange for bicarbonate. Seventy percent of the carbon dioxide in the plasma is in the form of bicarbonate, a product of this reaction.

HEMOGLOBIN MOLECULE

Hemoglobin is synthesized up through the early reticulocyte stage as the red blood cell matures. The normal adult hemoglobin molecule is made up of four heme groups each with a centrally located iron atom (shown in figure 17.5), and two alpha and two beta polypeptide chains. There are 141 amino acids in each alpha chain and 146 amino acids in each beta chain. Changes in the spatial configuration of the hemoglobin molecule are important in the oxygen and carbon dioxide binding capacity of that molecule. Oxygen combines with the iron atom associated with each heme group and carbon dioxide binds to the terminal amino acids of the polypeptide chains.

• • •

Carbon dioxide is carried in the blood in three forms: (1) bound to hemoglobin (carbaminohemoglobin), about 23%; (2) dissolved in plasma, 7%; and (3) HCO_3^-, 70%. Carbon dioxide dissolved in plasma is expressed in terms of pCO_2 and HCO_3^- is measured as meq/liter (chapter 4).

• • •

TABLE 17.3 The relationship between percent oxyhemoglobin saturation and pO_2 (at pH = 7.40 and temperature = 37°C)

pO_2 (mm Hg)	100	80	61	45	40	36	30	26	23	21	19
Percent oxyhemoglobin	97	95	90	80	75	70	60	50	40	35	30

Arterial blood — Venous blood

Hemoglobin-Oxygen Affinity

The amount of oxygen combined with hemoglobin as compared to the carrying capacity of hemoglobin is expressed as percent saturation (table 17.3). The partial pressure of O_2, expressed in mm of Hg, is an important determining factor in hemoglobin affinity for oxygen. The oxygen dissociation curve in figure 17.6 shows, however, that the relationship between pO_2 and percent saturation of hemoglobin is not a straight-line relationship. There are small increases in hemoglobin saturation between pO_2 values of about 50–100 mm Hg. In contrast to this, there are larger increments of change in hemoglobin saturation when pO_2 is between 20–40 mm Hg. There are four iron binding sites for oxygen on each hemoglobin molecule. Affinity of hemoglobin for oxygen increases after the first oxygen molecule binds. The converse of this is true as well. Affinity decreases when one of the four oxygen molecules dissociates. The steep-slope segment of the oxygen dissociation curve shows the significance of these facts. Changes in hemoglobin saturation, correlated with changes in pO_2, are greater in the pO_2 20–40 mm range. Increasing affinity for oxygen after initial binding makes hemoglobin an efficient oxygen carrier.

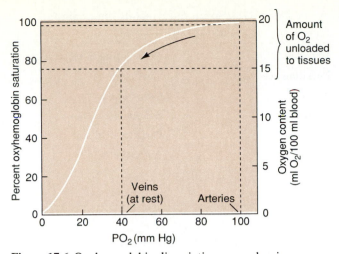

Figure 17.6 Oxyhemoglobin dissociation curve showing oxyhemoglobin saturation at varying pO_2 values. The graph shows that there is about a 25% decrease in percent oxyhemoglobin as the blood passes through the tissue from arteries to veins, resulting in the unloading of approximately 5 ml O_2 per 100 ml to the tissues.

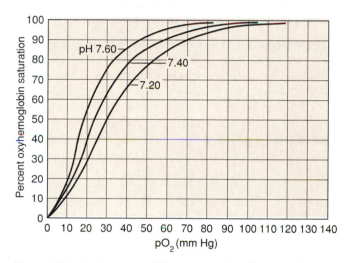

Figure 17.7 An increase in H^+ concentration decreases the affinity of hemoglobin for oxygen at each pO_2 value. The result is a "shift to the right" of the oxyhemoglobin dissociation curve, as shown with a pH of 7.20.

The normal range of values for arterial pO_2 is 75–100 mm Hg, at sea level and breathing room air. These normal values decrease with age and higher altitude. The critical low level for pO_2 is 40 mm Hg or less. In the upper portion of the dissociation curve, a decrease in pO_2 may cause an unremarkable change in hemoglobin oxygen saturation.

Hemoglobin transport of oxygen is influenced by additional factors, i.e., hydrogen ions, carbon dioxide, and **2,3-diphosphoglycerate (2,3-DPG)**. Hydrogen ions and carbon dioxide are products of metabolism, and accumulations of either substance promote the release of oxygen from hemoglobin. This response fulfills the requirement of metabolizing cells for oxygen. The decreased hemoglobin affinity for oxygen represents a shift

to the right in the oxygen dissociation curve (figure 17.7). Hemoglobin affinity for oxygen is influenced also by phosphates, especially 2,3-DPG. This substance, synthesized in red blood cells, binds to hemoglobin and decreases hemoglobin oxygen affinity.

• • •

Blood stored in acid-citrate-dextrose medium undergoes a decrease in 2,3-DPG level during storage. This may be of critical importance in a patient who receives a large volume of this blood because the transfused cells do not unload oxygen. Inosine added to the medium prevents the decrease in 2,3-DPG.

• • •

ACIDOSIS AND OXYGENATION

Since hydrogen ions promote oxygen-hemoglobin dissociation, it follows that acidosis will affect oxygen delivery to tissues. In acute acidosis, there is a shift to the right of the oxygen dissociation curve with increased oxygen delivery to tissues. If the condition continues longer than about 12–36 hours, the acidosis impairs the metabolic pathway from which 2,3-DPG is derived. The 2,3-DPG level is reduced. When 2,3-DPG is deficient, hemoglobin affinity for oxygen is increased; thus, in chronic acidosis, the opposing effects of increased hydrogen ions and decreased 2,3-DPG may result in normal delivery of oxygen to tissues.

• • •

Acidosis induces a decrease in the synthesis of 2,3-DPG, while increased levels of 2,3-DPG levels occur in response to alkalosis. A minimum of 8 hours is required for these changes to occur. Consequently, in chronic acid-base imbalance, the oxyhemoglobin dissociation curve approaches normal due to the 2,3-DPG response. Rapid correction of blood pH in acidosis, for example, would shift the dissociation curve to the left, and the shift would be unopposed by changes in 2,3-DPG for at least 8 hours. The net result is reduced oxygen delivery to tissues.

• • •

The preceding discussion illustrates the fact that there are adaptive mechanisms for meeting oxygen requirements under different physiological conditions. Delivery of oxygen to tissues is maintained, even when there are wide variations in pO_2. Normal oxygen delivery is due to minimal changes in oxyhemoglobin saturation. Increased pCO_2 promotes oxyhemoglobin dissociation and promotes delivery of oxygen to tissues. Decreasing pCO_2, which occurs in capillary blood in the lungs, favors hemoglobin-oxygen affinity and, thus, oxyhemoglobin saturation. Under hypoxic conditions, 2,3-DPG synthesis increases to increase the release of oxygen from hemoglobin.

ERYTHROCYTOSIS

Erythrocytosis or an increase in red blood cell number may be primary, i.e., the direct result of a bone marrow dysfunction, or may be secondary with causes only indirectly related to the bone marrow. Secondary causes of erythrocytosis include pulmonary disease, cigarette smoking, and high-altitude hypoxia. Inadequate oxygenation of tissues is the stimulus for erythropoietin production, which stimulates red blood cell production in the bone marrow. The term **hypoxia** (hī''pok'sē-ah) refers to oxygen deficit with inadequate ventilation, diffusion, or circulation as possible causes. **Hypoxemia** (hī''-pok-sē'mē-ah) indicates a low pO_2 or a low oxygen-hemoglobin saturation in arterial blood.

Pulmonary disease is a frequent cause of secondary erythrocytosis due to hypoxemia. The underlying cause for the hypoxemia may be ventilation/perfusion abnormalities, alveolar hypoventilation, and/or impaired diffusion (chapter 9).

CIGARETTE SMOKING

Cigarette smoking is probably the most common cause of erythrocytosis, the result of an increase in red blood cell number and/or a decrease in plasma volume. The mechanism for the oxygen deficit in smokers involves increased carbon monoxide levels in the blood. Carbon monoxide affinity for hemoglobin is more than 200 times greater than oxygen-hemoglobin affinity. Carbon monoxide not only displaces oxygen from hemoglobin, but also interferes with the dissociation of oxygen-hemoglobin at a cellular level. The reason for a decrease in plasma volume in smokers is not clear.

HIGH-ALTITUDE HYPOXIA

High-altitude hypoxia causes erythrocytosis as a compensatory response to the oxygen deficit. Other compensatory changes include an increased breathing rate, accompanied by a mild respiratory alkalosis (chapter 4). Increased levels of 2,3-DPG in red blood cells is responsible for a shift to the right of the oxygen dissociation curve, which enhances the delivery of oxygen to tissues. These responses, i.e., increases in RBC number, ventilation, and 2,3-DPG, represent a process of acclimatization that defends against hypoxia.

A syndrome called **acute mountain sickness** may accompany high-altitude hypoxia. The following are observations related to this syndrome:

erythrocytosis: Gk. *erythron*, red; Gk. *kytos*, cell; Gk. *osis*, increase
hypoxia: Gk. *hypo*, under; Gk. *oxys*, sharp

TABLE 17.4 Some manifestations of acute mountain sickness

Dizziness	Cheyne-Stokes respiration
Pulsating headache	Confusion
Nausea	Cerebral edema
Vomiting	Pulmonary edema
Insomnia	Peripheral edema
Tachycardia	Antidiuresis
Fever	
Coma	

1. Individuals who make an abrupt ascent above 8,000 ft and remain for at least 2 days may experience acute mountain sickness.
2. Residents of high altitudes who descend below 8,000 ft and then return to a high altitude are subject to acute mountain sickness.
3. Most normal individuals adapt completely within a few weeks, providing that the ascent from sea level to 8,000 ft is gradual.

Individuals affected by acute mountain sickness experience a variety of problems summarized in table 17.4. One problem is **Cheyne-Stokes respiration.** This is periodic breathing in which apnea alternates with rapid breathing. Fluid imbalance in mountain sickness is evidenced by antidiuresis and edema. Pulmonary and cerebral edema, in which the pathogenesis is unclear, may be life-threatening. One theory with regard to pulmonary edema is that hypoxia causes uneven pulmonary vasoconstriction, with increased blood flow in the unconstricted vessels. The result would be increased hydrostatic pressure that, in turn, promotes edema (chapter 2).

Prophylactic use of **acetazolamide** (a diuretic) counters fluid retention. **Dexamethasone,** a synthetic glucocorticoid, reduces cerebral edema. Serious complications of acute mountain sickness are moderated by oxygen therapy, but an imperative course of action is descent to a lower altitude.

POLYCYTHEMIA VERA

Polycythemia (pol''ē-sī-thē'mē-ah) refers to an increase in red blood cell number or an increase in hemoglobin level. Polycythemia vera is a type of primary erythrocytosis in which there is abnormal proliferation of red blood cell precursors that originate from a single stem cell in the bone marrow. Experimental studies indicate that the basic defect is unrelated to erythropoietin metabolism, and the etiology is unclear. There is an increased incidence of acute leukemia (chapter 18) in individuals affected by polycythemia vera. It may be that the original event leading to a mutation affecting the development of the red blood cell line, at times, affects the white blood cell line as well.

TABLE 17.5 Signs and symptoms of polycythemia vera

Ruddy color (plethora), headache, dizziness, vertigo, tinnitus
Itching. Due to histamine release from increased number of
  basophils
Thrombosis. Increased blood viscosity
Hemorrhage. Poor platelet function
Enlarged spleen
Gout. Increased uric acid production from increased
  hematopoiesis
Packed cell volume. 50 to > 70%; hemoglobin 18–24 g/dl
75% of patients have white blood cell count 12,000–25,000/
  mm³
50% of patients have increased platelet count

Manifestations

The evidence of an increase in red blood cell number is an elevated hematocrit or packed cell volume. Blood viscosity increases as packed cell volume goes up, and the result is a slower rate of blood flow. Compensatory responses to decreased rate of flow include an increase of plasma volume and of cardiac output. The overall effect is to maintain normal oxygen delivery to tissues, unless the packed cell volume is greater than 60% and/or there is an associated cardiovascular disorder.

Symptoms of the disorder may develop slowly, or there may be a sudden crisis, such as a cerebrovascular accident (chapter 38), i.e., an occlusion of a cerebral blood vessel. General manifestations may include headache, visual disturbances, dizziness, **vertigo** or a spinning sensation, and **tinnitus** (ti-ni′tus). Patients may have a ruddy complexion and have cyanosis or a blue tinge to nose, ears, and lips. The cyanotic tinge is evidence of inadequate oxygenation. There is an increased work load for the heart and angina (an-ji′nah) (chapter 23) may be a symptom. Occlusion of any blood vessels may occur, and almost any organ system may be involved. Patients commonly experience **pruritus** (proo-ri′tus), and there may be abdominal pain due to ulcers. Table 17.5 summarizes some observations related to this disorder.

• • •

Patients with polycythemia vera may have both clotting and bleeding tendencies. High blood viscosity and an increased platelet count favors clotting, whereas abnormal platelet function (failure to aggregate) leads to a bleeding tendency.

• • •

plethora: Gk. *plethore*, fullness
vertigo: L. *vertere*, to turn
tinnitus: L. *tinnitus*, ringing in ears
pruritus: L. *pruritus*, itching
polycythemia: Gk. *polys*, many; Gk. *kytos*, cell; Gk. *haima*, blood

Management

The objective in the treatment of polycythemia vera is to maintain the packed cell volume at a safe level and to prevent complications for as long as possible. Phlebotomy, i.e., periodic withdrawal of blood, may be sufficient to maintain the packed cell volume below 45–50%. In the case of a very high platelet count, treatment may be instituted to directly suppress bone marrow activity, i.e., chemotherapy or radioactive phosphorous. There are some uncertainties about increased risk of leukemia versus overall improved prognosis when this course of treatment is followed.

Prognosis

The prognosis for the disease is variable. Younger patients and those that initially have minimal symptoms appear to survive longer. Survival times vary from a few months to 20 years. Patients may die of vascular complications, or the polycythemia may convert to anemia with gradual replacement of bone marrow by fibrous connective tissue.

ANEMIAS

Anemia refers to a condition in which there is a deficiency of red blood cells and/or hemoglobin with a consequent reduction in tissue oxygenation. This type of disorder may develop as the result of decreased production of erythropoietin, diminished bone marrow responsiveness to erythropoietin, defective erythropoiesis, or loss of red blood cells by hemolysis or hemorrhage (figure 17.8). Table 17.6 summarizes categories of anemia based on mechanisms by which the disorder develops.

APPEARANCE OF CELLS

A stained blood smear provides evidence with regard to anemia, i.e., the size and shape of red blood cells and an estimate of hemoglobin content (table 17.7). Red blood cells stained with Wright's stain (an alcoholic solution of an acidic and a basic dye) are normally pink because hemoglobin has an affinity for the acidic stain. The biconcave shape is the reason for a clear area in the thin middle region of red blood cells. A large clear area in red blood cells indicates decreased hemoglobin content, and the cells are described as hypochromic. Blue-tinged red blood cells are indicative of immaturity. The condition is called **polychromatophilia** (pol″e-kro″mah-to-fil′e-ah) and represents young red blood cells. With special staining procedures, these cells may be identified as reticulocytes.

phlebotomy: Gk. *phelps*, vein; Gk. *tome*, a cutting
anemia: Gk. *a*, not; Gk. *haima*, blood
polychromatophilia: Gk. *polys*, many; *chroma*, color; *philein*, to love

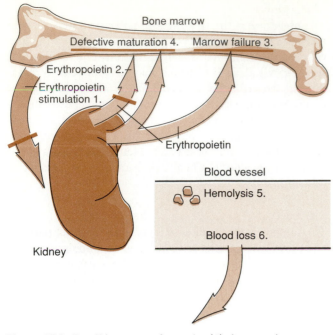

Figure 17.8 Possible causes of anemia: (*1*) decreased production of erythropoietin, (*2*) diminished bone marrow responsiveness to erythropoietin, (*3*) bone marrow failure, (*4*) reduced erythropoiesis due to defective maturation, (*5*) hemolysis, and (*6*) hemorrhage.

TABLE 17.6 Categories of anemia

Inadequate Bone Marrow Response

Decreased marrow stimulation
   Kidney disorders
   Decreased oxygen requirement
      Hypopituitarism, hypothyroidism
Increased oxygen delivery to tissues
   Shift of oxygen dissociation curve
Bone marrow failure
   Radiation, drugs, tumors
   Deficiency of iron or other requirements

Red Blood Cell Loss or Destruction

Blood loss
Hemolysis
   Destruction of cells by spleen or liver
   Inherited abnormalities

Defective Maturation

Folic acid, vitamin B_{12} deficiencies
Abnormal hemoglobin synthesis

There are various abnormalities associated with anemia that may be observed on a blood smear, and these abnormalities depend on the underlying cause of the anemia. For example, disorders involving the kidney may lead to anemia because the kidney is the main source of erythropoietin. Inadequate erythropoietin production causes diminished bone marrow cell proliferation. In this

TABLE 17.7 Terms used to describe red blood cells

Anisocytosis (an-i″so-si-to′sis). Variation in size
Poikilocytosis (poi″ki-lo-si-to′sis). Variation in shape
Hypochromic (hi″po-kro′mik). Decreased color as the result of a decrease in hemoglobin
Normochromic. Normal color, normal amount of hemoglobin
Polychromatophilia. Blue-tinged (young) red blood cells
Normocytic. Red blood cells normal size
Macrocytic. Red blood cells large
Microcytic. Red blood cells small

type of anemia, the blood smear will show red blood cells of normal size and hemoglobin content with no evidence of increased erythropoiesis, i.e., polychromatophilia. In conditions in which there is a decrease in oxygen requirements by tissues, such as hypopituitarism or hypothyroidism (chapter 31), there is no erythropoietin response due to the absence of hypoxia as a stimulus.

Bone marrow damage due to radiation, drugs, or the growth of tumor cells leads to a decrease in red blood cells in spite of increased erythropoietin production. Erythropoietin causes, in the absence of marrow responsiveness, early release of reticulocytes with no increase in the number of reticulocytes. A blood smear shows polychromatophilia in this case, with normocytic and normochromic cells. If tumor cells replace bone marrow cells, nucleated red blood cells may appear in the circulation.

Blood loss or hemolysis of red blood cells normally leads to stimulation of the bone marrow by erythropoietin. The evidence of increased erythropoiesis on a blood smear is polychromatophilia and large red blood cells, i.e., macrocytes.

There are various conditions in which there is a defect in the maturation of red blood cells in the bone marrow. **Nuclear maturation** requires folic acid and vitamin B_{12}. A deficiency of either of these vitamins causes impaired DNA synthesis and interference with cell division. The erythroblasts in the bone marrow become larger than normal, and these abnormally large cells are called **megaloblasts** (meg′ah-lo-blast″). The adult red blood cell is called a macrocyte. The cells frequently are fragile and have irregular shapes. Macrocytes are typical of nuclear maturation defects. Abnormalities of protein synthesis result in defects in **cytoplasmic maturation.** This may be caused by severe iron deficiency or may be inherited as a defect in synthesis of heme or one of the globin chains of hemoglobin. The result is microcytic, hypochromic red blood cells. Table 17.8 summarizes the red blood cell abnormalities associated with the various causes of anemia.

anisocytosis: Gk. *anisos,* unequal; Gk. *kytos,* cell; Gk. *osis,* condition
poikilocytosis: Gk. *poikilos,* diversified
hypochromic: Gk. *hypo,* under; Gk. *chroma,* color

TABLE 17.8 Causes of anemia and associated
red blood cell abnormalities

Decreased marrow stimulation
    Normocytic, normochromic red blood cells; absence of
    polychromatophilic cells
Bone marrow failure
    Marrow damage
        Normocytic, normochromic
        Polychromatophilia (erythropoietin stimulates early
        release of reticulocytes); total reticulocyte count low
    Marrow replaced by tumor
        Nucleated red blood cells in circulation
Red blood cell loss or destruction
    Acute blood loss—normochromic, normocytic red
    blood cells
    Subacute blood loss—modest increase in reticulocytes;
    Hypochromic, microcytic red blood cells
    Chronic blood loss—iron deficiency; hypochromic,
    microcytic red blood cells; anisocytosis
Defective maturation
    Impaired maturation of nucleus—macrocytes; no increase
    in reticulocytes
    Impaired cytoplasmic maturation—microcytic,
    hypochromic red blood cells

TABLE 17.9 Effects of slowly developing,
marked anemia

System	Symptoms
Cardiovascular	Pallor, tachycardia, angina, increased arterial pulsation, reversible cardiac enlargement, edema, ascites (chapter 36), and retinal hemorrhages
Respiratory	Exertional dyspnea
Neuromuscular	Anorexia, nausea, constipation, diarrhea
Urinary tract	Urinary frequency

• • •

About 5 days are required for an increased number of new red blood cells to appear in the circulation in response to blood loss. Blood cell production is limited by the amount of bone marrow and the availability of nutrients, particularly iron. Red blood cell production may increase up to 3 times the normal rate in response to blood loss.

• • •

PHYSIOLOGICAL RESPONSES

Inadequate oxygenation of tissues leads to compensatory responses that include increased oxygen delivery to tissues, changes in tissue perfusion, and increased cardiac output.

Red blood cells adapt to anemia by increasing the synthesis 2,3-DPG. The effect of this response is to shift the dissociation curve to the right and increase oxygen delivery to tissues. The extent of the 2,3-DPG increase is related to the severity of the anemia, and there may be a 40% increase with no change in hemoglobin concentration. The overall effect is to lower venous oxygen content, resulting in a lower oxygen reserve for increased demand, as in the case of exercise.

Redistribution of blood flow occurs in response to anemia. Vasoconstriction of blood vessels of the skin and kidneys shunts blood to more vital organs, such as the heart, brain, and muscle.

In severe cases of anemia, the cardiac output increases with an increase in the volume of blood pumped by the heart each minute. The coronary circulation increases in proportion to the cardiac output to accommodate the increased oxygen requirements of the heart. There may be inadequate oxygenation of the heart muscle and, ultimately, heart failure in those individuals who have severe anemia or in those who have coronary artery disease.

MANIFESTATIONS

The effects of anemia are widespread because of general inadequate oxygenation of tissues. The clinical picture is varied and depends on the rate of onset as well as the severity of the disorder. When there is an acute blood loss, the manifestations are associated with volume depletion, i.e., a falling blood pressure, peripheral vasoconstriction, weak pulse, and tachycardia (chapter 21). A slower onset of anemia is offset by compensatory mechanisms, and there may be either no symptoms or the symptoms may be limited to pallor and fatigue. Marked anemia is characterized by pallor, best observed in mucous membranes, and there may be a low grade fever. Symptoms involving the cardiovascular, respiratory, neuromuscular, and excretory system symptoms are summarized in table 17.9. In chronic anemia, there is vasodilation, possibly the result of tissue hypoxia, and reduced blood viscosity due to a decrease in the number of red blood cells. It is likely that both of these factors contribute to increased return of blood to the heart, which increases the work load on the heart. The effect on the heart in extreme anemia may be **cardiomegaly** (kar″de-o-meg′ah-le) or enlargement of the heart. Increased pressure in pulmonary vessels (chapter 23) is the result of chronic failure of the left ventricle to completely eject the volume of blood entering the cavity. Increased pulmonary pressure leads to pulmonary edema (chapter 2), and peripheral edema may develop. Cardiac failure may be the final outcome.

cardiomegaly: Gk. *kardia,* heart; Gk. *megas,* large

APLASTIC ANEMIA

Aplastic (ah-plas'tik) means inability to form new cells, and the term aplastic anemia refers to an anemia resulting from a decrease in the number of stem cells in the bone marrow. This **bone marrow hypoplasia** leads to decreased numbers of blood cells or **pancytopenia.** There are various causes including drugs, radiation, and viruses, particularly hepatitis viruses. Etiology is unknown in about half the cases of aplastic anemia. There is also a form of this anemia that is inherited and in which pancytopenia develops during childhood.

Diagnosis is based on stained smears of peripheral blood and on examination of bone marrow. The blood smear reveals reduced numbers of red blood cells and platelets and a reduction of all types of white blood cells, with the possible exception of lymphocytes. Bone marrow examination shows loss of cellularity. The symptoms are the result of decreased blood cells. A low platelet count causes bleeding tendencies, such as excessive bruising, retinal hemorrhages, and bleeding of nose or gums. A low white blood cell count leads to increased susceptibility to infection.

The course of the disease depends on the degree of bone marrow damage. Those with severe damage usually die as the result of hemorrhage and/or infection. Treatment involves treating and reducing the risk of infection. Transfusion of red blood cells and platelets supports the patient, so the chances for a spontaneous recovery are enhanced. Another aspect of treatment is the replacement of bone marrow by normal marrow from a suitable donor.

PERNICIOUS ANEMIA

Anemias that develop due to impaired DNA synthesis are characterized by a slow rate of erythropoietic cell division accompanied by normal cytoplasmic maturation. The term **megaloblastic** (meg'ah-lo-blas-tic) **anemia** is used because the red blood cells are larger than normal. Typically, there is excessive destruction of the megaloblastic red blood cell types in the bone marrow. Causes of a megaloblastic type anemia include either a deficiency or malabsorption of vitamin B_{12} and/or folic acid, which are required for DNA synthesis.

Pernicious anemia is the most common cause of vitamin B_{12} deficiency, and the underlying cause is malabsorption (figure 17.9). Vitamin B_{12} is found in meat and animal protein, and the absorption of this vitamin in the ileum requires a protein, called **intrinsic factor,** which is secreted by gastric parietal cells (chapter 34). In pernicious anemia, the gastric mucosa fails to secrete intrinsic factor, and there is evidence to support the idea that there

hypoplasia: Gk. *hypo,* under; Gk. *plasein,* to form
pancytopenia: Gk. *pan,* all; Gk. *kytos,* cell; Gk. *penia,* want
megaloblastic: Gk. *megalos,* greatly; Gk. *blastos,* bud

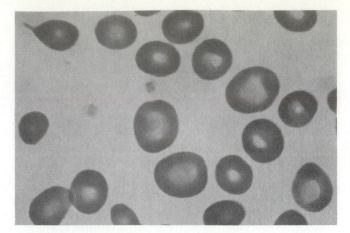

Figure 17.9 Blood smear showing pernicious anemia, characterized by macrocytes and variations in size and shape of red blood cells.
Courtesy of Edwin Reschke.

is, in some cases, an autoimmune reaction (chapter 29) against gastric parietal cells. For example, it has been observed that there is an increased incidence of pernicious anemia associated with diseases believed to have an autoimmune basis, such as Graves' disease, myxedema, and thyroiditis (chapter 31). Most patients with pernicious anemia have antibodies in their blood against gastric parietal cells. Gastritis and gastrectomy lead to pernicious anemia as well.

Vitamin B_{12} is stored in the liver; thus, the manifestations of pernicious anemia develop slowly. There may be weakness, pallor, lightheadedness, anorexia, and diarrhea or constipation. Glossitis, tachycardia, and a slight enlargement of the spleen and liver may occur. Neurologic manifestations include numbness, paresthesias or tingling sensations in the extremities, and a decrease in position and vibratory senses. Loss of coordination and either a decrease or increase in reflexes may occur. There may be irritability and depression as well.

Observations regarding blood cells in the case of pernicious anemia include (1) macrocytes and an increase in mean corpuscular volume (MCV, appendix B), (2) anisocytosis and poikilocytosis, (3) hypersegmented granulocytes, and (4) a decrease in white blood cells and platelets (chapters 18 and 19).

Vitamin B_{12} therapy provides for successful treatment of pernicious anemia.

IRON DEFICIENCY ANEMIA

Iron deficiency anemia is a common type of anemia that most often affects females during their reproductive years and children. Since iron is required for the synthesis of hemoglobin, a deficiency of iron causes a decrease in the

gastrectomy: Gk. *gaster,* stomach
glossitis: Gk. *glossa,* tongue

amount of hemoglobin. Initially, this type of anemia may be normochromic and normocytic, but it develops into a hypochromic, microcytic pattern. The cause of iron deficiency anemia may be (1) nutritional deficiency, (2) faulty iron absorption, (3) increased demand during pregnancy and periods of growth, or (4) excessive loss of iron due to hemorrhage.

IRON METABOLISM

An average adult has about 3.5–5.0 grams of total body iron, ingests 15–20 milligrams per day, and excretes about 1 milligram per day. The body absorbs a small fraction of the ingested iron, approximately 1 milligram per day, so ideally, the amount of iron absorbed is equal to the amount lost from the body. Iron is excreted by way of bile, feces, menstrual blood, and desquamation (des"kwāh-ma'shun) of intestinal epithelial cells.

Iron is reduced from the ferric (Fe^{+3}) state to the ferrous (Fe^{+2}) state in the stomach. Ferrous iron is absorbed from the small intestine, with the rate of absorption regulated by the body's need for iron. Iron is carried in the blood as a protein-iron complex called **transferrin** from which it may be released to any of the cells of the body. Transferrin attaches to receptor sites on cell membranes, moves into the cell to release the iron, and moves back into the blood. The intracellular iron combines with a protein called apoferritin to form ferritin, and excess iron is converted to **hemosiderin.** Bone marrow, liver, and the spleen are major storage areas for ferritin and hemosiderin, and these represent most of the body's iron reserve. The iron may be taken from these reserves and carried by transferrin to erythroblasts in the bone marrow. Mature red blood cells, at the end of their life span, are destroyed with the release of hemoglobin. Hemoglobin is phagocytized by macrophages, and ultimately, the iron from the hemoglobin moves back into the blood to be carried by transferrin (figure 17.10).

Iron is conserved and reused by the body, and there are iron reserves in the form of ferritin and hemosiderin. Iron deficiency anemia develops when there is an insufficient amount of iron in the diet, and there is depletion of the body's store of iron.

THALASSEMIAS

Thalassemias are a group of genetic disorders involving hemoglobin synthesis, and they represent the most common of the inherited hematological disorders. Thalassemia is also called **splenic anemia, Mediterranean anemia,** and **Cooley's anemia.** There are many different genetic defects that lead to either the absence of globin chain synthesis or a reduced rate of α or β globin chain synthesis.

desquamation: L. *desquamare,* to scale off

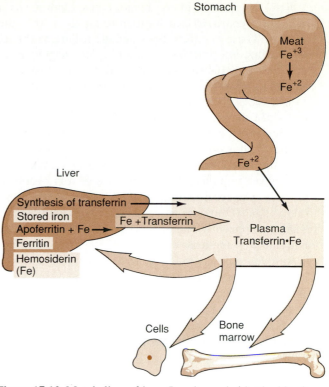

Figure 17.10 Metabolism of iron. Iron is carried in the blood as a protein-iron complex called transferrin. Intracellular iron combines with a protein called apoferritin to form ferritin. Excess iron is converted to hemosiderin. Ferritin and hemosiderin represent most of the body's iron reserve.

TABLE 17.10 Consequences of β-thalassemia

1. Absence or reduction of β chain production

2. β chains precipitate in red cell precursors; interfere with maturation; also interfere with passage through small blood vessels in circulation; thus, there is a decrease in cell survival

3. Increased erythropoietin as a compensatory response; thus, there is an increase in bone marrow and in deformities of skull and long bones

4. Splenomegaly as result of increased destruction of abnormal red blood cells

BETA-THALASSEMIA

Beta-thalassemia, in which there is an absence or reduction of synthesis of the β globin chain, is a common form of the disorder and occurs in every racial group. Regions of high incidence include areas around the Mediterranean and Southeast Asia. Table 17.10 summarizes the consequences of the defect. As noted in table 17.10, although bone deformities occur, they may be prevented by blood transfusions. A consequence of transfusions, however, is the accumulation of iron in the liver, endocrine glands, and the heart.

Beta-thalassemia in the homozygous condition is a severe form characterized by manifestations within the first year. These manifestations include failure to thrive, poor feeding, intermittent fever, and splenomegaly. Regular blood transfusions allow normal development until iron loading effects appear around puberty. In the absence of transfusions, there is progressive splenomegaly, marked anemia, bleeding tendencies, skull deformity, fever, weight loss, sinus infections, and deafness. The homozygous thalassemic child who receives no blood transfusions dies within the first 2 years. If the child is poorly transfused and survives to puberty, he or she dies of iron overload. The child who receives regular transfusions may survive to about age 30 but does not experience normal growth or sexual development. The effects of iron overload include endocrine dysfunctions, such as diabetes, hypoparathyroidism, and adrenal insufficiency (chapters 31 and 32). There is progressive liver and cardiac damage, and heart failure is the usual cause of death.

Individuals who have β-thalassemia in the heterozygous form are referred to as carriers and usually lack symptoms except during periods of stress.

Alpha-Thalassemia

Alpha-thalassemia is an inherited condition in which there is inadequate synthesis or the absence of synthesis of the α globin chains of α hemoglobin. There are various forms of α-thalassemia, and the main classes of this condition along with effects are listed in table 17.11. The term α^o-**thalassemia** is used to identify the genetic defect that leads to the failure of α chain synthesis. In the homozygous state, the result of this defect is fetal or neonatal death. α^+-**thalassemia** causes reduced chain synthesis and, in the homozygous state, causes mild anemia. The heterozygous condition ($\alpha^o\alpha^+$), a combination of the two defects, causes **hemoglobin H** which is hemoglobin with 4 β globin chains. The effects of this abnormal hemoglobin are listed in table 17.12.

The manifestations of hemoglobin H are less severe than those associated with β-thalassemia. The individual usually survives to adulthood and experiences variable degrees of anemia and splenomegaly. Bone abnormalities and growth retardation are unusual. There are episodes of hemolysis because of the shortened life span of the red cells.

Thalassemia Intermedia

Thalassemia intermedia is a term used to identify a group of thalassemias that do not require transfusions but are distinguished from the heterozygous α or β conditions by a more severe anemia. Some of these affected individuals develop marked splenomegaly, bone deformities, and are particularly susceptible to infection.

TABLE 17.11 Consequences of α-thalassemia

1. Deficiency of α globin chains with consequent production of excess β globin chains
2. β globin chains are more soluble than α globin chains; these globin chains do not precipitate in bone marrow
3. Hemoglobin H (4 β globin chains) precipitates in older red cells; the result is a shortened life span
4. Hemoglobin H has high affinity for oxygen, which causes decreased delivery to tissues

TABLE 17.12 Definitions and effects of some types of thalassemia

α^o-thalassemia. Absence of α globin chain production
α^+-thalassemia. Partial reduction of α globin chain production
Homozygous α^o-thalassemia. Causes inability to synthesize fetal or adult hemoglobin; results in prenatal or neonatal death
Heterozygous $\alpha^o\alpha^+$-thalassemia. Results in synthesis of 4 β globin chains; condition called hemoglobin H disease; causes variable degree of anemia and splenomegaly
Homozygous α^+-thalassemia. Causes mild hypochromic anemia

Treatment of Thalassemia

Treatment of thalassemia is symptomatic and may include regular blood transfusions, splenectomy, and administration of desferrioxamine (Desferal), a **chelating (ke'lā-ting) agent.** A chelating agent is a substance that combines with a metal, in this case iron, and promotes its elimination from the body to prevent iron overloading. Prevention is an aspect of management of thalassemia in which carriers of the genes are identified and premarital counseling provided.

Sickle Cell Anemia

Sickle cell anemia is so named because of the sickle shape of variable numbers of red blood cells on a peripheral blood smear (figure 17.11). This is an inherited disorder in which abnormal hemoglobin, called **hemoglobin S,** is produced. The variant hemoglobin differs from normal adult hemoglobin by the substitution of an amino acid, valine for glutamic acid, in the β chain. The affected individual may have the **sickle cell trait** and be a carrier, which is the heterozygous condition. Homozygous individuals have **sickle cell disease.** The disease occurs frequently in black populations, and it also occurs in the Mediterranean region, the Middle East, and India as well.

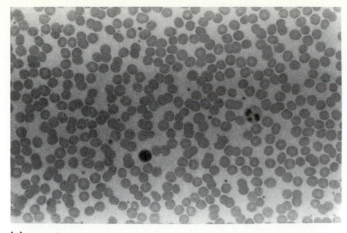

(a)

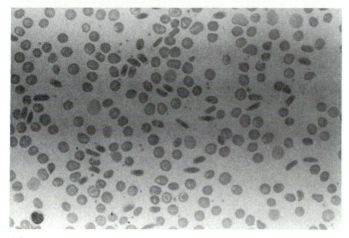

(b)

Figure 17.11 (*a*) Normal blood smear showing red blood cells and two white blood cells (a neutrophil on the right and a lymphocyte on the left). (*b*) A blood smear showing sickled red blood cells and several "target cells" in both the upper and lower portions of the picture. The target cells have a dark center, a clear ring, and a dark outer ring.

MANIFESTATIONS

The effects of hemoglobin S are variable and are the result of decreased solubility of the hemoglobin as it becomes deoxygenated. The hemoglobin molecules assume a rod-like shape arranged in parallel. This, in turn, causes elongation of the normally spherical red blood cell. Individuals with the sickle cell trait only have problems when hypoxic conditions occur, for example, during anesthesia. The manifestations of sickle cell disease are more serious, although there are varying degrees of severity. During infancy, there is some degree of anemia, jaundice, and **hand and foot syndrome.** This involves **dactylitis** which is inflammation of fingers and toes. Typically, there is a lifetime of chronic hemolytic anemia. Growth and development may be normal, or there may be some bone deformities. Throughout life, there are typically periods of crisis involving acute sickling. The symptoms associated with these episodes are summarized in table 17.13.

TABLE 17.13 Symptoms associated with a sickling crisis

Skeletal or abdominal pain
Lung syndrome. Acute dyspnea or pleuritic pain
Hemolytic episodes. Marked decrease in hemoglobin
Enlarged spleen or liver. Result of engorgement by red blood cells; causes marked anemia
Bone marrow aplasia. Defective reproduction of cells

TABLE 17.14 Long-term effects of sickle cell anemia

Bone infarction. Necrosis may lead to deformity of shoulder and hip joints; osteomyelitis
Kidney infarction. In younger children impaired renal function is corrected by transfusion; in later life, renal vascular damage leads to renal failure
Cardiomegaly. Various degrees of enlargement of heart; probably due to chronic anemia
Fibrosis of liver. Caused by repeated infarctions
Eye. Occlusion of retinal vessels causes retinitis proliferans, retinal detachment, and permanent blindness
Spleen. Splenomegaly in early life; repeated infarctions cause decrease in size of spleen

Pain and pulmonary symptoms are the result of occlusion of specific blood vessels, and **bone marrow aplasia** (inactivity) seems to be the result of infection. The most common causes of death in sickle cell anemia are infection and/or sequestering of red blood cells in the liver or spleen. The long-term effects of sickle cell anemia are summarized in table 17.14.

MANAGEMENT

Genetic counseling aimed at prevention is an aspect of management. The sickle cell trait does not require treatment, but sickle cell disease requires medical support. Acute episodes are controlled by intravenous fluid, oxygen, analgesics, and antibiotics to treat infection. In some cases, blood transfusion is required to restore hemoglobin.

SUMMARY

• • •

Red blood cells develop in the bone marrow of adults in response to erythropoietin, which is stimulated by hypoxia. Red blood cells are the site of formation of bicarbonate, which diffuses into plasma. Hemoglobin molecules in red blood cells bind to oxygen, carbon dioxide, and buffer hydrogen ions. Hemoglobin is made up of 4 heme groups each with an iron atom, 2 α polypeptide chains, and 2 β polypeptide chains. Hemoglobin is an efficient carrier of oxygen because affinity for oxygen increases after iron binds to the first oxygen molecule. Increased levels of hydrogen ions, carbon dioxide, and 2,3-DPG promote the dissociation of hemoglobin and oxygen.

Erythrocytosis is an increase in red blood cell number and may be either primary or secondary. Polycythemia vera is a type of primary erythrocytosis in which there is increased hematopoietic activity of the bone marrow, independent of erythropoietin stimulation. The disorder is associated with increased blood viscosity, decreased rate of blood flow, and a risk of occlusion of blood vessels, particularly cerebral vessels. Any organ system may be affected by inadequate oxygenation. Pruritus, headache, and visual disturbances are common symptoms. Management of the disorder involves withdrawal of blood to lower the packed cell volume and, in some cases, suppression of bone marrow activity.

Cigarette smoking and high-altitude hypoxia cause secondary erythrocytosis. Acute mountain sickness may accompany high-altitude hypoxia, particularly if the individual makes an abrupt ascent above 8,000 ft and remains for at least 2 days. Individuals affected by acute mountain sickness experience a variety of problems involving nausea, vomiting, antidiuresis, and edema. Pulmonary and cerebral edema may be life-threatening. Management involves administration of a diuretic, oxygen therapy, and descent to a lower altitude.

Anemia is a deficiency of red blood cells and/or hemoglobin and is associated with reduction in tissue oxygenation. The mechanisms leading to anemia include (1) decreased production of erythropoietin, (2) diminished bone marrow responsiveness to erythropoietin, (3) defective erythropoiesis, or (4) loss of red blood cells by hemolysis or hemorrhage. The appearance of red blood cells on a stained smear is an aid in the diagnosis of anemia. Normocytic, normochromic red blood cells and the absence of polychromatophilic cells is typical in the case of decreased levels of erythropoietin. Bone marrow damage leads to a normocytic, normochromic anemia with polychromatophilia as evidence of early release of reticulocytes due to erythropoietin stimulation. Acute blood loss is associated with a normochromic, normocytic pattern, whereas subacute and chronic blood losses are hypochromic and microcytic. The evidence of defective nuclear maturation is macrocytes in the blood, whereas impaired cytoplasmic maturation causes microcytic, hypochromic red blood cells.

Compensatory responses to inadequate oxygenation of tissues include increased oxygen delivery to tissues, redistribution of blood flow, and increased cardiac output. The effects of anemia are widespread and depend on both the rate of onset and the severity of the disorder. Increased work load on the heart and, ultimately, cardiac failure may be the final outcome.

Aplastic anemia is the result of a decrease in the number of stem cells in the bone marrow. The etiology may be unknown, the condition may be inherited, or it may be caused by various agents such as drugs, radiation,

or viruses. Iron deficiency anemia is relatively common, with children and females most frequently affected. The causes include nutritional deficiency, malabsorption, increased demand, and excessive loss of iron due to hemorrhage.

The thalassemias are caused by the absence or deficiency of α or β polypeptide chains of hemoglobin. Beta-thalassemia is the result of deficiency of β globin and, in the homozygous condition, requires transfusions to sustain life. In the heterozygous state, there are usually no symptoms except in times of stress. There is the absence of α globin in α^o-thalassemia, and the homozygous state causes fetal or neonatal death. In α^+-thalassemia, there is a reduction in α globin synthesis and a homozygous state causes a mild anemia. Heterozygous $\alpha^o\alpha^+$-thalassemia leads to the production of hemoglobin H, which has 4 chains. The manifestations of this form of thalassemia are less severe than in β-thalassemia. Thalassemia intermedia identifies a group of thalassemias that do not require transfusions, but have a more severe anemia than the heterozygous α or β conditions.

Sickle cell anemia is an inherited disorder in which abnormal hemoglobin, with the substitution of one amino acid in the β polypeptide chain, is produced. Sickle cell trait refers to the heterozygous condition, and sickle cell disease is the homozygous state. Typically, the hemoglobin becomes less soluble as it becomes deoxygenated. The molecules assume a rodlike shape that causes the sickle shape of the red blood cells.

REVIEW QUESTIONS

• • •

1. List three substances that promote dissociation of oxygen from hemoglobin.
2. What is the enzyme that catalyzes a reaction between carbon dioxide and water?
3. What does the term chloride shift refer to?
4. How many atoms of iron does a hemoglobin molecule contain?
5. What is the significance of a shift to the right of the oxyhemoglobin dissociation curve?
6. What is the physiological significance of the S-shaped oxyhemoglobin dissociation curve?
7. What is the effect of increased 2,3-DPG synthesis in red blood cells on the oxyhemoglobin dissociation curve?
8. What is life-threatening about acute mountain sickness?
9. What is the basic underlying cause of secondary erythrocytosis?
10. Give an example of a primary erythrocytosis.
11. What is one compensatory response to polycythemia vera?
12. What are two approaches to the management of polycythemia vera?
13. List four basic causes of anemia.

14. Predict the appearance of red blood cells in the case of inadequate erythropoietin synthesis.
15. What is the evidence on a blood smear of increased erythropoietin-stimulation of bone marrow?
16. What kind of maturation defect leads to macrocytes in the blood?
17. What type of red blood cell is characteristic of iron deficiency?
18. List three compensatory responses to hypoxia.
19. Define aplastic anemia.
20. What are four possible causes of iron deficiency anemia?
21. In what form is iron carried in the blood?
22. What are the forms of stored iron?
23. What is the basic defect in the thalassemias?
24. How many polypeptide chains are there in a hemoglobin molecule?
25. Why are there bone deformities in the thalassemias?
26. What is the most severe form of thalassemia?
27. What are the regions of high incidence of beta-thalassemia?
28. What are three modalities of treatment of thalassemia?
29. What is the basic defect in sickle cell anemia?
30. In sickle cell anemia, what is the (a) heterozygous (b) homozygous condition called?

SELECTED READING

• • •

Cain, S. M. et al. 1988. Circulatory adjustments to anemic hypoxia. *Advances in Experimental and Medical Biology* 227:103–15.

Da, W. M. 1988. In vitro studies on the pathogenesis of aplastic anemia in Chinese patients. *Experimental Hematology* 16:336–39.

Embury, S. H. et al. 1987. Rapid prenatal diagnosis of sickle cell anemia by a new method of DNA analysis. *New England Journal of Medicine* 316:656–61.

Erslev, A. 1987. Erythropoietin coming of age. *New England Journal of Medicine* 316:101–3.

Gaston, M. H. et al. 1986. Prophylaxis with oral penicillin in children with sickle cell anemia. *New England Journal of Medicine* 314:1593–99.

Gould, S. A. et al. 1986. Fluosol-Da as a red-cell substitute in acute anemia. *New England Journal of Medicine* 314:1653–56.

Lucarelli, G. et al. 1987. Marrow transplantation in patients with advanced thalassemia. *New England Journal of Medicine* 316:1050–55.

Milledge, J. S. 1987. The ventilatory response to hypoxia: How much is good for a mountaineer? *Postgraduate Medical Journal* 63:169–72.

Miller, B. A. et al. 1987. Molecular analysis of the high-hemoglobin-F phenotype in Saudi Arabian sickle cell anemia. *New England Journal of Medicine* 316:244–49.

Mozzarelli, A., J. Hofrichter, and W. Eaton. 1987. Delay time of hemoglobin polymerization prevents most cells from sickling in vivo. *Science* 237:500–505.

Platt, O. S. et al. 1984. Influence of sickle hemoglobinopathies on growth and development. *New England Journal of Medicine* 311:7–12.

Vetrosky, D. T., H. Sabio, and B. Schmidt. 1987. Diagnosing anemia. *Physician Assistant* 11:24–40.

West, J. B. 1988. High points in the physiology of extreme altitude. *Advances in Experimental and Medical Biology* 227:1–15.

Chapter 18

White Blood Cell Disorders

The preceding chapter deals with the function and disorders of red blood cells. In this chapter, the focus is on white blood cells or **leukocytes,** their normal development and function, and malignant proliferation of these cells.

TYPES OF WHITE BLOOD CELLS

White blood cells are produced in the bone marrow and, as in the case of red blood cells, develop from a pluripotent stem cell. Two major categories of white cells are formed: **granulocytes** and **agranulocytes.** The granulocytes include **neutrophils, eosinophils,** and **basophils. Lymphocytes** and **monocytes** are the agranulocytes. The granulocytes are so named because the cytoplasm of these cells contains many granules that are visible in Wright's stained smears under a light microscope. The cytoplasm of agranulocytes has fewer and smaller granules.

GRANULOCYTES

The granulocytes are named on the basis of the staining characteristics of their granules when the blood smear is stained with Wright's stain. Wright's stain is an alcoholic solution of an acid dye, **eosin,** and **methylene blue** which is basic. Eosinophilic granules are stained by the acidic dye eosin, while basophilic granules take the basic methylene blue. Table 18.1 lists characteristics and functions of granulocytes and figure 18.1 shows two neutrophils.

DEVELOPMENT OF GRANULOCYTES

The granulocytes develop in the bone marrow, and the first distinguishable precursors are myeloblasts and myelocytes. The process of maturation involves reductions in nuclear and cellular volume, condensation of nuclear chromatin, and the appearance of cytoplasmic granules. It is believed that the three types of mature granulocytes follow similar patterns of development, but the stages of differentiation are best documented for neutrophils. The phases of differentiation are outlined, starting with myeloblasts.

    Myeloblasts do not normally appear in the blood and are round, mononuclear cells ranging 11–18 microns in diameter. The nucleus is large and sometimes indented and stained deep purplish blue with Wright's stain. The narrow rim of cytoplasm stains light blue. Granules appear in the cytoplasm in the **promyelocyte** which is the next stage of differentiation. This cell is larger with an eccentric nucleus. The **myelocyte** develops from the promyelocyte and has specific granules that are either neutrophilic, eosinophilic, or basophilic. Further maturation produces the **metamyelocyte** that may be recognized by its smaller size and kidney-shaped nucleus. A **band cell** or **stab cell** is the product of metamyelocyte maturation and is about 13 microns in diameter and has a horseshoe-shaped nucleus.

TABLE 18.1 Comparison of characteristics and functions of the three types of granulocytes

Granulocyte	Characteristics
Neutrophils	Nucleus 2–5 segments; granules lilac or violet-pink; phagocytes; approximately 60% of white blood cells
Eosinophils	Bilobed nucleus; granules red-orange; role in parasitic infections and inflammatory response; 2–5% of white blood cells
Basophils	Nucleus 1–2 lobes; granules dark bluish purple; release mediators that promote inflammation; approximately 0–1% of white blood cells

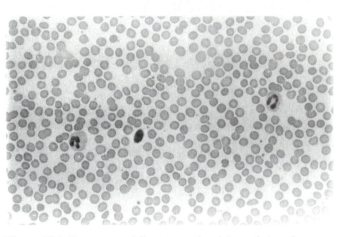

Figure 18.1 Two neutrophils, one on the right and the other on the left.
Courtesy of Dr. M. G. Klein.

These are the youngest neutrophils normally found in blood and make up 1–5% of the total white cell count. The mature neutrophil has a segmented nucleus with 2–5 lobes held together by strands of chromatin. Figure 18.2 shows the developmental forms of granulocytic cells. The maturation process from myelocyte to neutrophil requires about 5 days.

NEUTROPHILS

Neutrophils are also called **polymorphonuclear neutrophils (PMNs)** because of their multilobed nuclei (figure 18.1). Neutrophils constitute about 60% of the white blood cells. Only half the neutrophils in blood are circulating. The other half are sequestered in capillaries, and these neutrophils are described as **marginating** or rolling along the endothelial surfaces of small blood vessels; thus, these sequestered cells are called **marginated cells.**

    The function of neutrophils is to migrate to sites of infection and to phagocytize invading organisms. The marginated neutrophils are immediately available and move to areas of inflammation.

The white blood cell count is a measure of all white blood cells, but, of the neutrophils sampled, only the circulating cells are counted. A shift between the marginated and circulating neutrophils sometimes accounts for changes in the total white cell count. It is likely that a shift of marginated cells to the general circulation is the reason for an increased white cell count during (1) periods of strenuous exercise, (2) emotional states, and (3) subsequent to corticosteroid administration.

EOSINOPHILS

Eosinophils represent from 2–5% of the total number of white blood cells and are distinguished by their large red-orange granules (figure 18.1). Eosinophils are somewhat similar to neutrophils because they exhibit motility and phagocytic capabilities. The precise role that these cells play is not completely clear. There is evidence suggesting that eosinophils (1) regulate inflammatory responses, (2) are involved in hypersensitive reactions, and (3) provide defense against parasitic infestations.

BASOPHILS

Basophils (figure 18.2) are the least common of the five types of white blood cells with a normal value of 0.5% or a range of 0–1%. The function of these cells is uncertain, but it seems to relate to the release of their granular contents. They appear in tissues during hypersensitive states, and they release substances that promote inflammation. It is possible that eosinophils and basophils interact in hypersensitive reactions.

AGRANULOCYTES

Lymphocytes and monocytes are white blood cells classified as agranulocytes. Although this term means an absence of granules, these cells have a few small granules scattered throughout the cytoplasm.

LYMPHOCYTES

Lymphocytes make up about 35% of the white blood cells and represent a varied group of cells. On blood smears, they are described as being small, medium, or large. The small lymphocytes have a diameter of 8–12 microns, a nucleus that occupies most of the cell, and scanty pale blue cytoplasm if stained with Wright's stain (figure 18.3). The larger lymphocytes have more abundant cytoplasm, and all have small numbers of granules scattered in the cytoplasm.

Lymphocytes originate from a pluripotent stem cell, as do all other blood cells, and become either T- or B-lymphocytes. B-lymphocytes differentiate in the bone marrow, whereas precursor cells migrate from bone marrow to the thymus and develop into various specialized types of T-lymphocytes. Both types of cells are important in immune reactions in the body's defense system, and their specific roles will be discussed in chapter 27.

T-lymphocytes are capable of reproducing and renewing the cell population. The B-lymphocyte populations are replenished by marrow stem cells. The following are the recognized stages of development. **Lymphoblasts,** which are difficult to distinguish from myeloblasts, differentiate into prolymphocytes, and these cells develop into the mature lymphocytes.

Only 0.1% of the lymphocytes circulate in the blood. The remainder reside in lymph nodes, spleen, and lymphoid tissue of the gut and pharynx (figure 18.5). Although it is not possible to distinguish T- and B-lymphocytes on Wright's-stained blood smears, 70% of the lymphocytes in the blood are of the T variety. Both types of cells move freely between blood and lymphoid tissue.

MONOCYTES-MACROPHAGES

Monocytes are less common than lymphocytes in the blood and represent 3–7% of the total white blood cells. They originate from stem cells in the bone marrow and are distinguished by their large size, 16–22 microns, and a nucleus that is irregular, folded, or indented (figure 18.4). When circulating in the blood, the monocyte is actually an immature phagocyte that, within a few hours, migrates to tissues to become a fixed or motile macrophage (figure 18.5). Macrophages proliferate in tissues and have relatively long life spans of months to years.

Macrophages are large cells, 20–80 microns, and are remarkably efficient in engulfing and destroying bacteria cellular debris. These cells are also involved in immune responses (chapter 27), interact in the clotting mechanism (chapter 19), and have a regulatory function in the hematopoietic system.

• • •

Newly released neutrophils survive hours to days in the blood, then enter tissues where they die. The monocytes/changed to macrophages survive from months to years. Lymphocytes are generally longer lived and circulate many times between lymphoid tissue and the bloodstream.

• • •

VARIATIONS IN WHITE BLOOD CELL NUMBERS

White blood cells are evaluated on the basis of total count and by a **differential count.** The differential count identifies each type of white blood cell as a percentage of the

phagocyte: Gk. *phagein,* eat; Gk. *kytos,* cell

Hemocytoblast

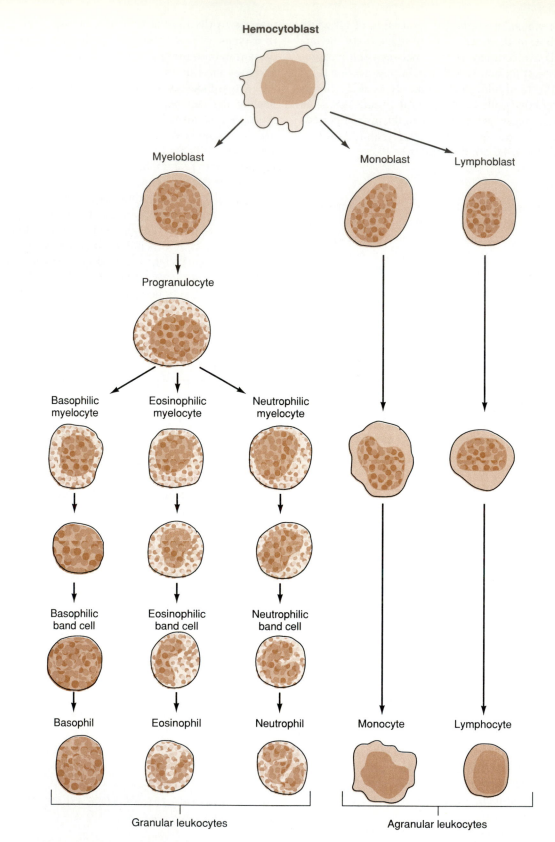

Myeloblast

Monoblast

Lymphoblast

Progranulocyte

Basophilic
myelocyte

Eosinophilic
myelocyte

Neutrophilic
myelocyte

Basophilic
band cell

Eosinophilic
band cell

Neutrophilic
band cell

Basophil

Eosinophil

Neutrophil

Monocyte

Lymphocyte

Granular leukocytes

Agranular leukocytes

Figure 18.2 Undifferentiated stem cells, hemocytoblasts, develop into various types of blood cells. Shown here are granular and agranular white blood cells.

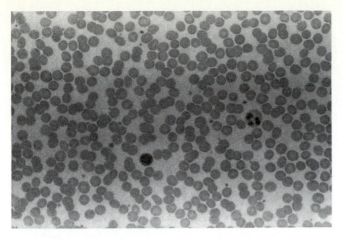

Figure 18.3 A normal blood smear showing a neutrophil on the right and a lymphocyte on the left. *Courtesy of Dr. M. G. Klein.*

Monocytes

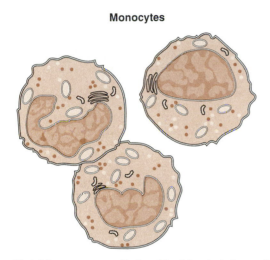

Figure 18.4 Monocytes are distinguished by their large size and a nucleus that is irregular, folded, or indented.

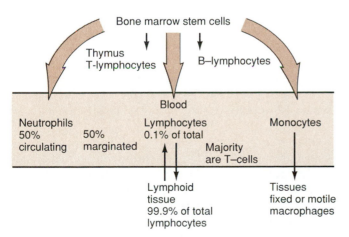

Figure 18.5 Location of the three types of white blood cells.

total. These are relative values, but absolute values for specific types of white blood cells may be reported as well, i.e., number/microliter (μL) of blood.

There are age related variations in both the total and differential white blood cell counts. The normal range in adults for the total count is 5,000–10,000/μL, whereas in neonates the range is 10,000–25,000/μL. Lymphocytes are the predominant white blood cell in children (in neonates, 5,000/μL) and that number declines to an adult normal of 2,000/μL by age 10.

Variations in numbers of lymphocytes and neutrophils are common in association with disease processes. **Neutrophilia** means an increased number of neutrophils and occurs most commonly in response to infection and tissue damage. Drugs may cause neutrophilia by enhancing production or by causing a shift from marginating to circulating blood pools. **Neutropenia,** which is a low number of neutrophils, is usually the result of infiltration of malignant cells into the bone marrow or the consequence of chemotherapy. Neutropenia also occurs with excessive alcohol consumption and in splenomegaly. In the case of an enlarged spleen, the neutropenia is due to cells pooling within the spleen.

Lymphocytosis, an elevated lymphocyte count, is common in viral illnesses and other infections, such as tuberculosis, brucellosis, syphilis, and pertussis. A subnormal lymphocyte count is called **lymphopenia** and may be caused by stress, administration of corticosteroids, or malignancy of lymphoid tissues. Malignant proliferation of white blood cells is discussed in the following section.

LEUKEMIA

Leukemia is a malignant disease characterized by increased production of white blood cells in the bone marrow. These cells, in varying stages of maturity, infiltrate and predominate the bone marrow, appear in peripheral blood, and infiltrate tissue. The first stage is a malignant transformation of one or more hematopoietic (hēm″ah-tō-pōi-e′tik) stem cells with subsequent proliferation of clones of cells. A number of factors have been implicated in this malignant transformation. These factors include (1) viruses, (2) ionizing radiation, (3) chemical exposure, and (4) genetic factors. For example, the risk of developing leukemia is increased in patients receiving long-term chemotherapy for such malignant conditions as ovarian carcinoma, polycythemia vera, and multiple myeloma.

Leukemias are classified on the basis of type and maturity of the predominant cell, severity of symptoms, and the total white blood cell count. **Acute forms of leukemia** are characterized by symptoms that persist for weeks to months, many immature cell forms, and an elevated white

hematopoietic: Gk. *haima,* blood; Gk. *poiesis,* production blood cell count. In untreated **chronic leukemia,** the duration of symptoms is months to years, cells are mostly mature, and the white blood cell count may be either elevated or depressed.

The terms **myelogenous** or **myelocytic** refer to granular leukocytes in the bone marrow. There is an acute and chronic form of myelogenous leukemia in which the predominant cell type belongs to the granulocytic series. There are also acute or chronic, either monocytic or lymphocytic, leukemias. The suffix blast denotes an immature form, so there are **myeloblastic** and **lymphoblastic leukemias.** The more common types of leukemia are discussed in the following paragraphs.

CHRONIC MYELOGENOUS LEUKEMIA

Chronic myelogenous leukemia (CML) accounts for 15% of all leukemias and is basically a disorder in which there is increased production of slightly defective granulocytes (figure 18.6). Clinical manifestations appear years after the malignant transformation occurs in the bone marrow. The median age at the time of diagnosis is 53 years, and there are a few more males than females diagnosed. A distinctive feature is a chromosomal abnormality called a Philadelphia chromosome in 85–90% of the patients. The abnormality involves a translocation of the longer arm of chromosome 22, usually to chromosome 9.

Manifestations
The manifestations of CML are summarized in table 18.2. Splenomegaly or an enlarged spleen is the most common physical finding, and there are general symptoms of malaise.

Laboratory Findings
Laboratory findings listed in table 18.3 indicate the possibility of a somewhat elevated to extremely elevated total white blood cell count in CML. There are immature forms of both red blood cells and cells of the neutrophilic series as well as increased basophils and eosinophils. A fully developed case of CML results in anemia and a white blood cell count of about 200,000/mm³. The anemia is caused by increased numbers of granulocytic cells in the bone marrow and pooling of red blood cells in the spleen. **Leukocyte alkaline phosphatase** activity is absent or diminished in CML. This enzyme is found in neutrophils, and normally, about 20% of the neutrophils show a strong alkaline phosphatase reaction.

Blast Crisis
An increase in malignant cell proliferation occurs, usually 2–6 years after diagnosis of CML, regardless of prior responsiveness to chemotherapy. This is called a blast crisis because of a sudden increase in the immature blast forms of cells. The symptoms become more severe and include

splenomegaly: Gk. *splen*, spleen; Gk. *megas*, large

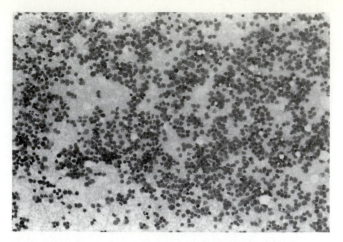

Figure 18.6 Bone marrow in acute myelogenous leukemia showing a predominance of immature white blood cells that belong to the granulocytic series.
Courtesy of Dr. M. G. Klein.

TABLE 18.2 Manifestations of chronic myelogenous leukemia during the chronic phase

Splenomegaly. Most common finding
Hepatomegaly. Often occurs
Malaise, fatigue, weight loss, night sweats
Bone pain
Anemia. Pallor, tachycardia
Lymphadenopathy. Uncommon

TABLE 18.3 Laboratory findings in chronic myelogenous leukemia

Red blood cell count. Decreased
White blood cell count. 20,000–1,000,000/mm³
Platelets. May be extremely elevated, occasionally low
Nucleated red blood cells
Segmented and band forms of neutrophils in blood
Metamyelocytes, myelocytes, myeloblasts in blood
Basophils. Increased
Eosinophils. Increased
Leukocyte alkaline phosphatase. Activity decreased or absent

enlarged lymph nodes or **lymphadenopathy,** hemorrhages, and recurrent infection. The usual course of events is death within 12 weeks.

Management
Chemotherapy is aimed at decreasing the number of malignant cells and relieving symptoms, but it is not a cure. Chemotherapy is not a cure because the malignant cells cannot be completely destroyed without destroying normal hematopoietic cells. Complete destruction of cellular elements in the patient's marrow followed by a bone marrow

hepatomegaly: Gk. *hepar*, liver; Gk. *megas*, large
lymphadenopathy: L. *lympha*, water; Gk. *aden*, gland; Gk. *pathos*, disease

transplant from a suitable donor is an option. Long-term follow-up studies will determine the success of this approach.

Prognosis

The prognosis for long-term survival is poor, with a median survival of 47 months for those treated with chemotherapy. A small percent survive beyond 10 years.

ACUTE LYMPHOBLASTIC LEUKEMIA

Acute lymphoblastic leukemia (ALL), although it occurs in adults, represents over 80% of the acute leukemias in children. There is proliferation of cells in the bone marrow, thymus, and lymph nodes from which mature lymphocytes normally develop. About 80% of the cases represent a proliferation of B-lymphocyte precursors with T-lymphocyte precursors constituting the remainder. There is evidence that implicates genetic aspects, viral causes, and radiation exposure as predisposing factors. There are chromosomal defects in more than 90% of ALL cases.

Manifestations

Manifestations of ALL are summarized in table 18.4 and include general malaise that develops over a period of days or weeks. There is evidence of bleeding problems, bone pain, and enlarged liver and lymph nodes.

Laboratory Findings

Table 18.5 lists some typical laboratory findings in ALL. The white blood cell count at diagnosis varies from low to extremely high. Neutrophils, red blood cells, and platelets are usually decreased. Serum uric acid is elevated after chemotherapy due to uric acid synthesis from nucleoprotein breakdown products.

Management

Three aspects of management of ALL follow:

1. Chemotherapy to induce remission, i.e., to quickly destroy abnormal cells and restore normal hematopoiesis
2. Prophylactic treatment to prevent infiltration of the central nervous system by leukemic cells, involving intraspinal injection of drugs and craniospinal irradiation
3. Maintenance of remission by chemotherapy

Prognosis

In general, there is a more favorable prognosis in children between the ages of 2–10 years as compared to either older or younger patients. Prognosis is somewhat better in females because there is a recurrence of leukemia in the testes of 8–16% of males. Remission occurs in 78% of adult patients after chemotherapy, and the median period of remission is 20 months. The median survival time is 32

TABLE 18.4 Manifestations of acute lymphoblastic leukemia

Pallor, weakness, headache, vomiting, infection
Ecchymoses and purpura. Evidence of hemorrhage
Bone pain. Leukemic infiltration of bone marrow
Leg pain. Infiltration of malignant cells into tissues
Lymphadenopathy. Common
Hepatomegaly. Common
Splenomegaly
Gastrointestinal bleeding. Common; due to thrombocytopenia or chemotherapy
Meningeal or testicular infiltration

TABLE 18.5 Laboratory findings in acute lymphoblastic leukemia

White blood cells. 10% of patients have extremely high counts at diagnosis, 40% either < 10,000/mm³–50,000/mm³
Neutrophils. Decreased
Red blood cells. Decreased
Platelets. Normal to low
Lymphoblasts in peripheral blood
Serum uric acid. Increased after chemotherapy

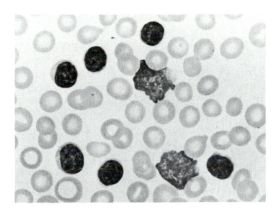

Figure 18.7 Chronic lymphocytic leukemia (CLL) showing small round lymphocytes with dense chromatin clumping. Two smudged (broken) cells are seen, and a larger prolymphocyte is noted. ×1,000.
From S. B. McKenzie: Textbook of Hematology. *1988 Lea & Febiger, Philadelphia. Reproduced with permission.*

months. With chemotherapy, 50–60% of children survive past 5 years, and of those among this group who do not have a relapse, 85% are considered cured.

CHRONIC LYMPHOCYTIC LEUKEMIA

Chronic lymphocytic leukemia (CLL), also called chronic lymphatic leukemia, constitutes about 25% of all leukemias in Western countries (figure 18.7). Most affected individuals are over age 50, and the condition is more common in males. The etiology is unknown, although in up to 50% of the cases there is a chromosome abnormality. The most common of these abnormalities is an extra chromosome 12.

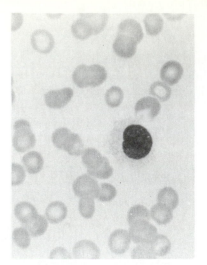

Figure 18.8 Hairy cell leukemia. Large lymphocyte with indented nucleus, diffusely distributed chromatin, small nucleolus, and abundant cytoplasm with peripheral projections or "hairs." This type of leukemia is uncommon, occurs mainly in older males, and is characterized by cells that have fine hairlike projections.
From S. B. McKenzie: Textbook of Hematology. *1988 Lea & Febiger, Philadelphia. Reproduced with permission.*

TABLE 18.6 Manifestations of chronic
lymphocytic leukemia

Reduced exercise tolerance
Chronic fatigue, low-grade fever, anorexia
Lymphadenopathy
Erythematous skin disorder
Splenomegaly
Hepatomegaly
Infections
Diarrhea, intestinal malabsorption

TABLE 18.7 Laboratory findings in chronic
lymphocytic leukemia

Lymphocytes elevated
White blood cell count 5,000–30,000/mm³ or higher
Red blood cells decreased due to autoimmune hemolysis and
    marrow infiltration
Thrombocytopenia due to marrow infiltration
Neutropenia due to marrow infiltration

The disorder results in an absolute increase in the number of normal appearing lymphocytes either of the small or large variety. There is an accumulation of lymphocytes in the blood as well as in the bone marrow, lymph nodes, and spleen. In most cases, the cells belong to the B-lymphocyte series.

Hairy cell leukemia is a variant form of B cell chronic lymphocytic leukemia in which the peripheral blood smear shows a relatively mature lymphoid cell with fine cytoplasmic projections (figure 18.8). The projections can be seen with a light microscope.

Manifestations

Manifestations of CLL are summarized in table 18.6. Common symptoms are those of general malaise, lymphadenopathy, and hepatomegaly. Lymphocytic infiltration of the skin causes **erythematous** skin disorders. Lymphocytes infiltrate other organs, and in the case of the gastrointestinal tract, cause mucosal thickening, ulceration, diarrhea, and intestinal malabsorption.

Laboratory Findings

There is gradual replacement of marrow by lymphocytes, which causes a decrease in red blood cells, platelets, and neutrophils (table 18.7). The anemia is due in part to an immune reaction that causes hemolysis. Neutropenia leads to increased susceptibility to infection, and hemorrhages result from a decreased platelet count. Blood smears show an absolute increase in lymphocytes.

Management

Treatment is symptomatic, and the purpose is to keep the patient comfortable. Modalities of treatment include (1) chemotherapy, (2) local X-ray treatment for enlarged lymph nodes, (3) gamma globulin for recurrent infections, (4) prednisone for autoimmune hemolysis, and (5) packed red blood cell transfusion if therapy for anemia fails.

Prognosis

The survival of individuals affected by chronic lymphocytic leukemia varies considerably. In some cases, the individual dies within a year while others survive 10–15 years. The median survival time in those individuals whose only symptom is lymphocytosis is 150 months. When the symptoms are lymphocytosis and anemia or thrombocytopenia, the median survival time is 19 months.

SUMMARY

• • •

White blood cells may be classified as granulocytes (neutrophils, eosinophils, and basophils) or agranulocytes (lymphocytes and monocytes). Neutrophils are the most abundant of the white blood cells and function as phagocytes at sites of infection. Evidence indicates that eosinophils are involved in the inflammatory response, hypersensitive reactions, and parasitic infestations. Basophils are the least common of the white blood cells and their functions, though somewhat obscure, include the release of substances that promote inflammation.

B- and T-lymphocytes are important in immune responses. T-lymphocytes originate in the bone marrow and differentiate in the thymus whereas B-lymphocytes fully develop in bone marrow. Both types of cells circulate between blood and lymphoid tissue. Monocytes originate in

erythematous: Gk. *erythema,* redness

bone marrow, are transient in the blood, and ultimately reside in tissues where they develop into macrophages. Macrophages have multiple complex functions that include phagocytosis.

Leukemia is a malignant disorder in which there is increased production of white blood cells. The initial event is malignant transformation of a bone marrow stem cell. Factors that may be involved include viruses, ionizing radiation, chemical exposure, and genetic factors. Leukemias are classified on the basis of type and maturity of the predominant cell, severity of symptoms, and the total white blood cell count.

Chronic myelogenous leukemia causes increased production of granulocytes. It is more common in older individuals, and there is an abnormal Philadelphia chromosome in most cases. Usually 2–6 years after diagnosis there is a sudden increase in immature forms of white cells called a blast crisis. The prognosis is poor at this stage.

Acute lymphoblastic leukemia is the common form of acute leukemia in children. In most cases there is a proliferation of B-lymphocyte precursors with lymphoblasts appearing in the blood.

Chronic lymphocytic leukemia usually affects individuals over the age of 50 and represents 25% of all leukemias in Western countries. There is often some kind of chromosomal abnormality associated with the disorder. The disease is characterized by an absolute increase of normal appearing lymphocytes in the blood, bone marrow, lymph nodes, and spleen.

The leukemias are characterized by abnormalities of all of the cellular elements of the blood and bone marrow. Chapter 20 deals further with laboratory findings in a leukemia patient.

SELECTED READING

• • •

Dixon, A., F. Athari, and C. B. Cook. 1989. Removal of leukocytes from donor units. *Laboratory Medicine* 20:685–91.

Evans, J. S. et al. 1986. The influence of diagnostic radiography on the incidence of breast cancer and leukemia. *New England Journal of Medicine* 315:810–15.

James, K. 1988. Immunotyping of lymphomas and leukemias. *Laboratory Medicine* 19:225–27.

Mann, D. L. et al. 1987. HTLV-I-associated B-cell CLL: Indirect role for retrovirus in leukemogenesis. *Science* 236:1103–106.

Marx, J. L. 1987. Leukemia virus linked to nerve disease. *Science* 236:1059–61.

Rinsky, R. A. et al. 1987. Benzene and leukemia. *New England Journal of Medicine* 316:1044–50.

Rivera, G. K. et al. 1986. Intensive retreatment of childhood acute lymphoblastic leukemia in first bone marrow relapse. *New England Journal of Medicine* 315:273–77.

Yunis, J. J. et al. 1984. High-resolution chromosomes as an independent prognostic indicator in adult acute nonlymphocytic leukemia. *New England Journal of Medicine* 311:812–18.

REVIEW QUESTIONS

• • •

1. The myeloblast is a precursor of a (granulocyte, agranulocyte).
2. Which is the more mature cell, myeloblast or myelocyte?
3. What is a major difference in the development of B- and T-lymphocytes?
4. What is the location of most mature lymphocytes?
5. What is the final developmental stage of the monocyte found in the blood?
6. What is the most common cause of neutrophilia?
7. What does the term myelogenous mean?
8. What type of leukemia is common in children and is characterized by a proliferation of B-lymphocyte precursors?
9. An absolute increase in lymphocytes in the blood and elsewhere describes what kind of leukemia?
10. What type of leukemia causes an increased production of granulocytes and is often associated with the Philadelphia chromosome?

Chapter 19

Bleeding Disorders

One type of blood cell, the platelet or **thrombocyte,** has not yet been discussed. Platelets are involved in the blood clotting mechanism, and the broad topic of this chapter is dysfunction of this mechanism. A discussion of platelets and normal coagulation introduces the subject of bleeding disorders. The role of platelets will be considered first.

PLATELETS

Stem cells in the bone marrow develop into precursor cells, **megakaryoblasts.** These in turn enlarge and undergo a series of changes to form **megakaryocytes.** The hormone **thrombopoietin** (throm″bō-pōi-e′tin), probably secreted by the kidney, regulates the rate of megakaryocyte formation and maturation. Megakaryocytes have a maturation time of 5 days and are the largest of the bone marrow cells with a diameter between 50–160 μm. Megakaryocytes fragment and are released into the general circulation as platelets (figure 19.1). These cells are anuclear (ah-nū′kle-ar) or lack a nucleus, have a diameter of 2–4 μm, and a life span of about 9.5 days. Although inactive platelets are thin discs, they undergo changes in shape when stimulated and become irregular or spiny spheres. A normal blood platelet count is in the range of 150,000–350,000/mm³ of blood.

PLATELET AGGREGATION

Vasoconstriction is an immediate response to vascular injury; and within 1–2 minutes, platelets aggregate to form a plug in the area of injury. The first step in clot formation is that platelets become sticky and adhere to subendothelial collagen of the blood vessel wall and flatten to cover the area of injury. Aggregation of platelets in this area is stimulated by various substances, such as adenosine diphosphate (ADP), epinephrine, thrombin, and collagen. Aggregation is most often preceded by a change from a disc shape to a sphere with spicules. Platelets not only undergo a change in shape, become adhesive and aggregate, they also secrete certain substances that promote vasoconstriction, such as epinephrine and serotonin, and ADP that promotes platelet adhesiveness.

OVERVIEW OF COAGULATION AND FIBRINOLYSIS

Table 19.1 summarizes the events that lead to clot formation and, ultimately, to dissolution of that clot. Platelets initially form a plug in response to blood vessel wall injury. Subsequently, inactive substances in the plasma are activated and are involved in a series of reactions. The final product of these reactions is the enzyme thrombin that catalyzes the conversion of fibrinogen to threadlike fibrin.

anuclear: Gk. *a,* not; L. *nucleus,* kernel
thrombocyte: Gk. *thrombos,* lump; Gk. *kytos,* cell
fibrinolysis: L. *fibra,* fiber; Gk. *lysis,* loosing

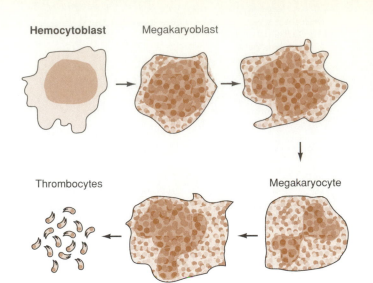

Figure 19.1 The origin and development of platelets or thrombocytes.

TABLE 19.1 Overview of the process of blood coagulation and fibrinolysis

1. a. Injury to blood vessel
   b. Platelets adhere and aggregate
2. a. Blood exposed to vessel wall tissue factor (thromboplastin)
   b. Series of reactions that lead to formation of thrombin
3. a. Thrombin catalyzes conversion of fibrinogen to fibrin
   b. Fibrin strands polymerize to form cross-linked fibrils
4. Platelet peptide growth factors promote healing
5. a. Plasminogen binds to fibrin
   b. Activators catalyze conversion of plasminogen to plasmin
   c. Plasmin splits fibrin and digests clot

Fibrin holds the elements of the clot together until it is dissolved by plasmin. The major aspects of this complex process are discussed in the following paragraphs (figure 19.2).

FIBRIN FORMATION

The processes in fibrin formation are complex. Sequential reactions involving plasma proteins lead to the formation of thrombin and to the formation of fibrin. These plasma proteins circulate in the blood in an inactive form and are activated in response to blood vessel damage. The proteins are called factors and are designated by Roman numerals on the basis of the order in which they were originally identified. The small letter "a" after a Roman numeral indicates that the factor is in an active form.

The chemical reactions that lead to coagulation are designated as either the **extrinsic pathway** or **intrinsic**

intrinsic: L. *intrinsecus,* within

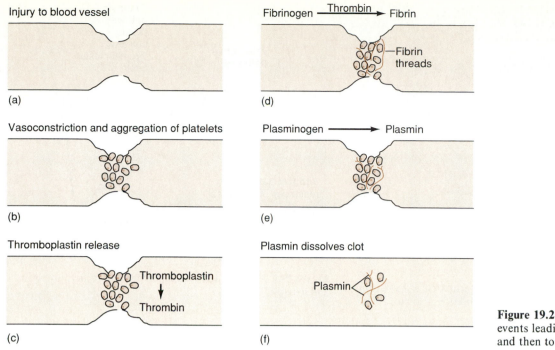

Figure 19.2 An overview of events leading to clot formation and then to clot dissolution.

(a) Injury to blood vessel

(b) Vasoconstriction and aggregation of platelets

(c) Thromboplastin release
Thromboplastin
↓
Thrombin

(d) Fibrinogen —Thrombin→ Fibrin
Fibrin threads

(e) Plasminogen ——→ Plasmin

(f) Plasmin dissolves clot
Plasmin

pathway. The intrinsic pathway is initiated by the contact response of a blood factor to a foreign surface or a nonendothelial surface, such as basement membrane, collagen, or collagen-stimulated platelets. The term intrinsic is used because clotting does not require an additive. The extrinsic pathway is triggered by the release of tissue factor or thromboplastin from damaged tissue.

• • •

The following are Roman numeral designations of clotting factors in which common names are frequently used.
Factor I fibrinogen
Factor II prothrombin
Factor III tissue thromboplastin
Factor IV calcium ion

• • •

INTRINSIC PATHWAY

Table 19.2 provides a summary of events in the intrinsic pathway initiated by the exposure of blood to a number of nonendothelial surfaces. These surfaces include glass, in vitro, or basement membrane and collagen of damaged vessels. Essentially, the intrinsic pathway involves the conversion of **prekallikrein** (pre-kal″li-kre′in) to kallikrein and the conversion of factor XII to the active form XIIa. This triggers the formation of the active forms of **factors XI, IX, X. Kallikrein** provides a positive feedback mechanism (figure 19.3), whereby additional factor XIIa is produced (step 3). Factor Xa is the (1) end product of the intrinsic

TABLE 19.2 Summary of reactions involved in intrinsic pathway

1. Intrinsic pathway initiated by contact with negatively charged surface; blood in contact with subendothelial tissue following vascular injury; reactions occur on surface of platelet membranes

2. Prekallikrein $\xrightarrow[\text{High molecular weight kininogen (HMWK)}]{\text{Factor XII}}$ Kallikrein

3. Factor XII $\xrightarrow[\text{HMWK}]{\text{Kallikrein}}$ Factor XIIa

4. Prekallikrein $\xrightarrow{\text{XIIa}}$ Kallikrein

5. Factor XI $\xrightarrow{\text{Factor XIIa}}$ Factor XIa

6. Factor IX $\xrightarrow{\text{Factor XIa; Ca}^{++}}$ Factor IXa

7. Factor X $\xrightarrow[\text{Platelet phospholipid}]{\text{Factor IXa; factor VIII; Ca}^{++}}$ Factor Xa

pathway, (2) is the product of the extrinsic pathway (as seen in table 19.3), and (3) is involved in the final common pathway leading to fibrin formation.

EXTRINSIC PATHWAY

The extrinsic pathway is initiated by **thromboplastin** also called **tissue factor or factor III.** Thromboplastin, contained in all cell membranes, is released as the result of tissue damage and forms a complex with **factor VII.** Factor VII is activated and subsequently activates factor X. There is also an alternative path by which factor IXa promotes the conversion of factor X to Xa (figure 19.4). Table 19.3 summarizes these events.

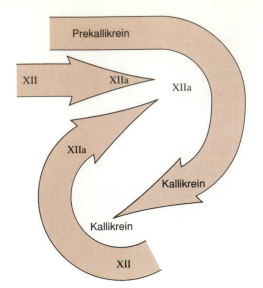

Figure 19.3 Factor XIIa promotes the formation of kallikrein, and in turn, kallikrein promotes the formation of factor XIIa.

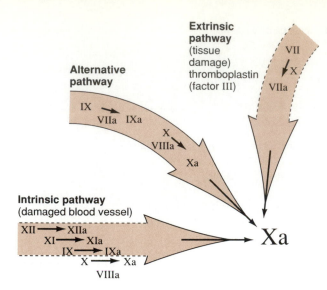

Figure 19.4 Factor Xa is a common product of both intrinsic and extrinsic pathways.

TABLE 19.3 Summary of reactions in the extrinsic pathway leading to formation of factor Xa

1. Initiated by thromboplastin released by damaged tissue
2. Factor VII
   $$\xrightarrow{\text{thromboplastin, phospholipid, Ca}^{++}} \text{Factor VIIa}$$
3. Factor X $\xrightarrow{\text{factor VIIa, thromboplastin}}$ Factor Xa
4. Alternative pathway:
   a. Factor IX
      $$\xrightarrow{\text{factor VIIa-thromboplastin complex}} \text{Factor IXa}$$
   b. Factor X
      $$\xrightarrow{\text{factor IXa, VIII, Ca}^{++}, \text{ platelets}} \text{Factor Xa}$$

• • •

Physiologically, there are interactions between the intrinsic and extrinsic pathways, so both are required for normal hemostasis. Normal function of one pathway does not compensate for a deficiency in the other.

• • •

FINAL COMMON PATHWAY

The final common pathway is shown in table 19.4. The series of reactions is initiated by factor Xa and the end product is fibrin (figure 19.5). **Prothrombin** is converted

hemostasis: Gk. *haima,* blood; Gk. *statikos,* cause to stand

TABLE 19.4 Final common pathway whereby factor Xa leads to the formation of fibrin

1. Factor V $\xrightarrow{\text{Factor Xa (platelet bound)}}$ Factor Va
2. Prothrombin
   $$\xrightarrow{\text{Factors Xa, Va; Ca}^{++}, \text{ platelet factor (PF 3)}} \text{Thrombin}$$
3. Roles of thrombin:
   a. Factor XIII $\xrightarrow{\text{Thrombin, Ca}^{++}}$ Factor XIIIa
   b. Fibrinogen $\xrightarrow{\text{Thrombin, Ca}^{++}, \text{ factor XIIIa}}$ Fibrin

to the enzyme **thrombin,** which catalyzes the change of the soluble protein **fibrinogen** to the insoluble threadlike **fibrin.** Initially, the fibrin molecules are single strands, then they polymerize or combine to form larger molecules, and ultimately form cross-links to stabilize the molecules. The fibrin network covers the area of injury and reinforces the platelet plug. Finally, the clot retracts and becomes more compact.

• • •

Thrombin, the last enzyme in the series of coagulation reactions, has a variety of effects that include cleaving fibrinogen and the activation of factors XIII, V, and VIII as well as protein C. In small amounts, thrombin promotes platelet aggregation, stimulates prostaglandin synthesis in endothelial cells, promotes smooth muscle contraction, and attracts macrophages. Thrombin also cleaves prothrombin, and in sufficient concentration, interferes with its own formation.

• • •

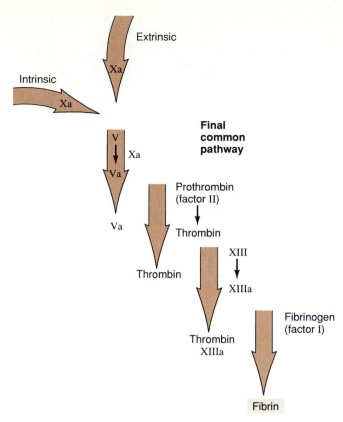

Figure 19.5 The final common pathway of the chemical reactions that lead to coagulation. The reactions are initiated by factor Xa and the final product is fibrin.

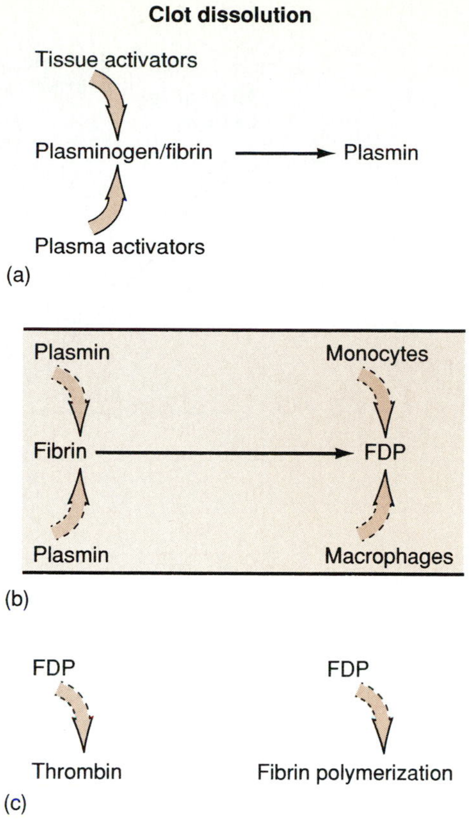

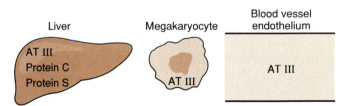

Figure 19.6 The activated form of plasminogen (plasmin) breaks down fibrin. Fibrin degradation products (FDP) interfere with thrombin activity and with polymerization of fibrin.

FIBRINOLYSIS

As healing closes the wound, the clot is slowly dissolved (figure 19.6). A substance called **plasmin** cuts up fibrin threads; thus, the process is called **fibrinolysis. Plasminogen,** which is the inactive form of plasmin, binds to fibrin and is converted to plasmin by various activators. There are both plasma and tissue plasminogen activators. As plasmin digests the clot, fibrin breakdown products block the action of thrombin and interfere with polymerization of fibrin. Monocytes and macrophages aid in the process of clot removal by secreting proteases and ingesting fibrin debris.

• • •

A tissue-type plasminogen activator (t-PA), produced by recombinant DNA technology, is effective in dissolving clots that occlude coronary and pulmonary blood vessels. Streptokinase also promotes dissolution of clots. It is a protein produced by bacteria which is involved in the conversion of plasminogen to plasmin.

• • •

Figure 19.7 Antithrombin III (AT III) and proteins C and S are natural anticoagulants synthesized by the liver. Megakaryocytes and blood vessel endothelium are also sites of synthesis of AT III.

INTRAVASCULAR ANTICOAGULANTS

The clotting mechanism is counterbalanced by natural anticoagulants. There are several of these substances that are either produced by the endothelial lining of blood vessel walls or are activated on the endothelial surface (figure 19.7). The anticoagulant's function is to prevent abnormal clot formation. Major constituents of this intravascular anticoagulant system are (1) **antithrombin III (heparin cofactor),** (2) **protein C,** and (3) **protein S.**

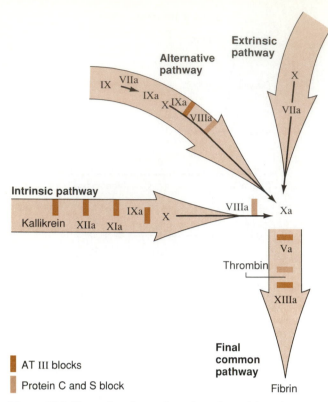

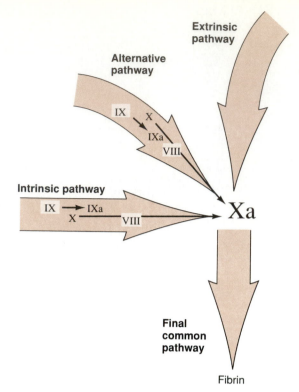

■ AT III blocks

▌ Protein C and S block

Figure 19.8 Natural anticoagulants interfere with various steps in the coagulation pathways.

Figure 19.9 The defect in hemophilia A is a deficiency of factor VIII:C, and there is a deficiency of factor IX in hemophilia B.

Antithrombin III (AT III) is produced by the liver, megakaryocytes, and blood vessel endothelium. It functions as an inhibitor of thrombin, kallikrein, plasmin, and factors XIIa, XIa, Xa, and IXa. Heparin accelerates the action of AT III.

Both protein C and S are synthesized by the liver, require vitamin K for synthesis, and circulate in an inactive form in the blood. Protein C is activated by thrombin and blocks the functions of cofactors Va and VIIIa. Protein S is released by endothelial cells and promotes the binding of protein C to phospholipid surfaces. The presence of protein S enhances inactivation of cofactor Va (figure 19.8).

HEMOPHILIA A

Hemophilia is an inherited bleeding disorder in which there is a deficiency of a clotting factor. **Classic hemophilia,** also called **hemophilia A,** occurs in all ethnic groups and is the most common severe inherited coagulation disorder. It is transmitted as an X-linked recessive disorder and is limited, with few exceptions, to males. Hemophilia A occurs in a small number of females whose fathers are hemophilic and mothers are carriers of the gene. Occasionally, female carriers have a mild bleeding disorder as well.

The bleeding disorder is the result of a deficiency of factor VIII (figure 19.9), and the severity of the disorder is determined by the degree of deficiency. Factor VIII is

actually a complex with two components. One is factor VIII:C, also called antihemophilic factor, that promotes coagulation and is involved in the alternative pathway. The second component is von Willebrand factor (**vWF**) that promotes platelet binding to the subendothelium. The deficiency in hemophilia is that of factor VIII:C.

• • •

There is a high mutation rate for hemophilia A that occurs either during the process of oogenesis or spermatogenesis. As a result, some patients have no apparent affected family members.

• • •

EFFECTS OF HEMOPHILIA A

If the level of factor VIII:C is less than 1% of normal, the affected individual often bleeds in infancy and is afflicted throughout life by serious internal and external hemorrhages. A less severe deficiency may cause occasional episodes of **hemarthroses** or bleeding into joints. Individuals with factor VIII:C levels between 5–25% of normal usually do not have bleeding episodes except after surgery or dental extractions. Individuals typically have a normal bleeding time and usually do not experience prolonged bleeding from small wounds. There is interference with

oogenesis: Gk. *oion*, egg; Gk. *genesis*, production
hemarthrosis: Gk. *haima*, blood; Gk. *arthron*, joint; Gk. *osis*, condition

activation of the intrinsic pathway that leads to fibrin formation. Consequently, there is limited formation of fibrin threads even though a normal platelet plug forms. The usual course of events is that bleeding resumes several hours after injury and tends to occur internally.

Manifestations of the severe form of hemophilia A include the following. Hemarthrosis occurs in which there is effusion of blood into any joint, particularly weight bearing joints, and this causes swelling, inflammation, and pain. **Arthropathy** or **hemophilic joint disease** leads to degeneration of cartilage in the joints and resorption of underlying bone. Bleeding into muscles leads to necrosis of overlying tissue as well as pressure-induced neuropathies and muscular atrophy. Accumulation of blood under the tongue, a **sublingual hematoma,** may pass rapidly into the neck and obstruct the airway. Collections of clots walled off by fibrous connective tissue are called **blood cysts** or pseudotumors. Blood cysts may develop in the hemophilic individual, continue to enlarge due to repeated bleeding, and eventually rupture. These cysts cause pain and are susceptible to infection. Hematuria and gastrointestinal bleeding also occur. Intracranial bleeding accompanied by neurological damage and seizures is the most common cause of death.

MANAGEMENT

Carriers of the defective gene can be identified on the basis of family history and analyses of levels of factor VIII:C. Prenatal diagnosis is possible by testing fetal blood samples at week 12–14 of gestation. Genetic counseling is thus available to individuals at risk.

Treatment of hemophilia involves administration of concentrates of factors VIII:C. There are various types of preparations made from either single-donor-plasma or multiple-donor-pooled plasma. **Cryoprecipitate** (krī″ō-prē-sip′ĭ-tāt) is one such preparation which is the cold-insoluble material that remains after slow thawing of fast frozen plasma. Cryoprecipitate may then be further purified. Some patients develop antibodies against human factor VIII, and it may be necessary to use porcine factor VIII:C for these individuals. A major problem in the treatment of hemophilia is that plasma preparations of factor VIII:C are the vehicles of transmission of both hepatitis and acquired immunodeficiency syndrome (AIDS) viruses.

HEMOPHILIA B

Hemophilia B, also called **Christmas disease,** is a less common form of hemophilia in which there is a deficiency of factor IX (figure 19.9). This disorder is also inherited as an X-linked recessive condition with the same clinical picture as hemophilia A. The similarities between the two

cryoprecipitate: Gk. *kryos,* cold
porcine: L. *porcinus,* hog

types of hemophilia are explained by the fact that factor VIII is the cofactor for activation of factor IX, and the two are interdependent. Fresh frozen plasma is used as replacement therapy for factor IX in mild cases of hemophilia B because this factor is relatively stable in vitro. Plasma concentrate preparations are also used as a source of factor IX.

VON WILLEBRAND'S DISEASE

Von Willebrand's disease (vWD) is a group of bleeding disorders of various types and subtypes. These disorders are inherited and are transmitted as either autosomal dominant or autosomal recessive characteristics. The majority of cases of von Willebrand's disease are heterozygous autosomal dominant. The basic defect is some degree of deficiency of von Willebrand's factor (vWF), a glycoprotein secreted by both the endothelium of blood vessels and megakaryocytes. This factor (1) is carried in the plasma and forms a complex with factor VIII:C to function as a carrier and (2) is stored in platelets and is required for platelet adhesion to subendothelial surfaces.

Affected individuals show a prolonged bleeding time as the result of impaired platelet plug formation. This may be accompanied by decreased factor VIII:C activity with a decrease of fibrin formation. Such individuals experience the following problems. Bleeding problems may appear early in childhood and become less severe as the child matures. The nature of the bleeding includes (1) nosebleeds, (2) bleeding from the mouth, (3) gingival (jin′ji-val) bleeding, (4) bruising, (5) **menorrhagia** (men″o-ra′je-ah) or excessive menstrual flow, and (6) severe bleeding problems following dental extraction or surgery. Treatment for severely affected patients is administration of cryoprecipitate, which contains vWF and factor VIII:C.

DISSEMINATED INTRAVASCULAR COAGULATION

Disseminated intravascular coagulopathy (DIC) is a syndrome in which there is widespread clot formation within small blood vessels that triggers massive fibrinolytic activity. It is thus a disorder involving both excessive clotting and bleeding. DIC is a complication of various disorders including (1) **septicemia** or an infection of the blood, (2) trauma, (3) shock, (4) liver disease, (5) malignancies, (6) renal disease, (7) obstetrical complications, and (8) connective tissue disorders.

gingival: L. *gingivae,* gums
septicemia: Gk. *septikos,* putrefactive; Gk. *haima,* blood

TABLE 19.5 Overview of events in DIC
that cause clotting and bleeding

1. Activation of coagulation factors with generation of fibrin leads to microthrombi
2. Response to step 1 is fibrinolysis; breakdown products have two effects: microthrombi and bleeding
3. Activation of coagulation (step 1) uses up platelets and factors, and this causes microthrombi and bleeding

TABLE 19.6 Initial steps in initiation of DIC*

Step 1. a. Introduction of a substance into circulation that generates fibrin; may be thromboplastin
         b. Increased levels of thrombin
         c. Fibrinogen $\xrightarrow{\text{thrombin}}$ fibrin monomers
         d. Deposition of fibrin plugs in microvasculature

*(Activation of coagulation); events in step 1 of table 19.5

TABLE 19.7 Fibrinolysis in response to fibrin
formation in DIC*

Step 2. a. Plasminogen $\xrightarrow{\text{fibrin monomers, t-PA}}$ plasmin
         b. Fibrinogen $\xrightarrow{\text{thrombin, factor XIIIa}}$ fibrin polymers
         c. Fibrinogen $\xrightarrow{\text{plasmin}}$ fibrinogen degradation products
         d. Microthrombi $\xrightarrow{\text{plasmin}}$ cross-linked fibrin degradation products

*Step 2 of table 19.5

INITIATING EVENT

The initiating event in DIC is the introduction of a **procoagulant** into the circulation, that is, a substance that promotes coagulation and causes the formation of fibrin. This substance may be thromboplastin. Table 19.5 shows an overview of events in DIC that lead both to the formation of small clots (**microthrombi**) and to bleeding.

FIBRINOLYSIS

Table 19.6 lists the reactions that result in widespread deposition of fibrin in small blood vessels. This is the result of activation of clotting factors and the response to this activation is fibrinolysis (table 19.7), which in turn has multiple consequences. The **fibrinogen/fibrin degradation products** (FDP) shown in this table cause the following circulatory problems:

1. Small fragments inhibit platelet function
2. Large fragments promote platelet aggregation
3. Mixtures of soluble fragments increase capillary permeability leading to coagulation outside the vessel (figure 19.10).

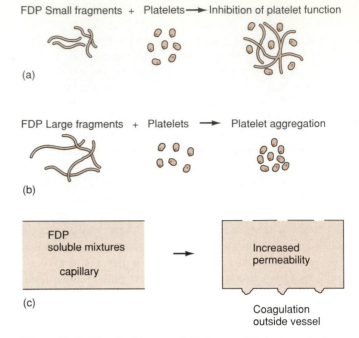

Figure 19.10 The significance of fibrinogen/fibrin degradation products in the blood. (*a*) Inhibition of platelet function, (*b*) promotion of platelet aggregation, and (*c*) increased capillary permeability and coagulation outside of the capillary.

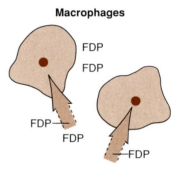

Figure 19.11 Macrophages break down the degradation products of fibrin, but may be overwhelmed by massive production.

TABLE 19.8 Consequences of consumption
of platelets and coagulation factors in DIC*

Step 3. a. Thrombin binds to platelets and causes aggregation
         b. Aggregated platelets are removed by cells of the reticuloendothelial system; the result is bleeding
         c. Coagulation factors are used up; fibrinogen, prothrombin, factors V, VIII, XIII; the result is bleeding
         d. AT III and protein C (oppose thrombin) are used up in process of coagulation resulting in microthrombi

*Step 3 of table 19.5

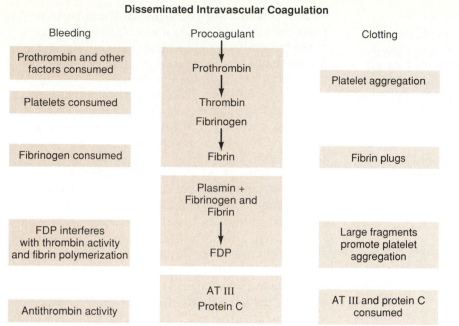

Disseminated Intravascular Coagulation

Bleeding	Procoagulant	Clotting
Prothrombin and other factors consumed	Prothrombin ↓	Platelet aggregation
Platelets consumed	Thrombin Fibrinogen	
Fibrinogen consumed	↓ Fibrin	Fibrin plugs
	Plasmin + Fibrinogen and Fibrin	
FDP interferes with thrombin activity and fibrin polymerization	↓ FDP	Large fragments promote platelet aggregation
Antithrombin activity	AT III Protein C	AT III and protein C consumed

Figure 19.12 Overview of disseminated intravascular coagulation (DIC) in which there is both clotting and bleeding tendencies.

There are specific fragments that cause injury to blood vessel endothelium and probably contribute to adult respiratory distress syndrome (chapter 8) that sometimes occurs during DIC. Macrophages break down these degradation products, but this natural defense mechanism may be overwhelmed by massive production (figure 19.11).

CONSUMPTION OF COAGULATION FACTORS

The overall effect of the fibrinolytic responses is to promote both bleeding and clotting.

Table 19.8 shows additional effects of activation of coagulation. The term **consumption** indicates that certain factors are used up during the coagulation process. Thus, there is consumption of platelets, as shown in this table, as well as several other factors, and this promotes bleeding. There is also consumption of two factors that oppose the action of thrombin, AT III and protein C, and this depletion favors clot formation. The processes involved in DIC are terminated with the total consumption of prothrombin. Figure 19.12 provides an overview of the processes that contribute to both bleeding and clotting in DIC.

MANIFESTATIONS OF DIC

Widespread bleeding occurs in DIC and may include oozing from wounds, bleeding in the skin (purpura), lungs, gastrointestinal tract, or genitourinary tract. There may be massive hemorrhage leading to hypovolemic shock and death. Thrombotic occlusion of small blood vessels may occur in all organs causing coma, ischemia of skin leading to gangrene, acute renal failure, and ulceration of gastrointestinal tract. DIC syndromes are either (1) an acute form that lasts for hours or days, (2) subacute, lasting for days to weeks, or (3) a chronic form continuing from months to years.

MANAGEMENT

Procedures for treating DIC include treatment of the underlying cause and administration of blood and fluids to maintain cardiac output as well as electrolyte balance. Heparin arrests intravascular coagulation, although its role in the treatment of DIC is controversial. Infusion of platelets and cryoprecipitate restores such depleted substances as fibrinogen and factors V and VIII.

• • •

A process called cytapheresis (sīt″ah-fĕ-re′sis) is a technique for separating the cellular elements of blood. An automatic cell separator spins at varying speeds and collects the cells of a particular density. The remainder of the cells are returned to the donor.

• • •

BLEEDING AND VITAMIN K DEFICIENCY

Vitamin K is required for the synthesis of clotting factors VII, IX, X, and prothrombin, as well as the coagulation inhibitors proteins C and S. In addition to dietary sources, vitamin K is synthesized by intestinal organisms. This vitamin, which is fat soluble, requires bile salts for absorption from the intestine and is stored in the liver. Vitamin K deficiency, leading to bleeding tendencies, may develop as the result of poor intestinal absorption, decreased intestinal flora related to antibiotic therapy, or obstruction of bile ducts.

Factor VII is synthesized by the liver in the presence of vitamin K. When this protein is activated by thromboplastin, it initiates the extrinsic coagulation pathway. It has a half-life of 6 hours in the blood, i.e., half the amount present in the circulation is removed in this time; thus, factor VII is the first vitamin K-dependent factor to disappear during vitamin K deficiency.

• • •

THROMBOCYTOPENIC PURPURA

The term **thrombocytopenic** (throm″bō-si″tō-pē′nik) means decreased numbers of platelets, and **purpura** refers to a purple color caused by subcutaneous bleeding. The two terms together refer to a susceptibility to bruising due to low numbers of platelets. One mechanism for removal of platelets from the blood was discussed in which there is increased consumption as in DIC. A second mechanism is an immune reaction involving the coating of platelets by immunoglobulin (chapter 27) with subsequent destruction by phagocytosis. This second form of thrombocytopenia is called **idiopathic thrombocytopenic purpura** or **immune thrombocytopenic purpura** (ITP). Such an immune response may follow the administration of a transfusion or drugs; a disease process, such as a connective tissue disorder; or an infection. Such individuals experience purpura in response to minor trauma with platelet counts between 30,000–50,000/mm³. Spontaneous bleeding and petechiae occur when platelet levels fall to 10,000–30,000. Petechiae are tiny spots of hemorrhage on the surface of the skin. When platelet counts are less than 10,000, there is mucosal and intracranial bleeding.

Recovery from ITP may be spontaneous and require no treatment. Modalities of treatment include (1) intravenous gamma globulin, which blocks phagocytic macrophage receptors, (2) prednisone with mechanism of action uncertain, (3) immunosuppressive drugs, and (4) removal of spleen because it is the major organ responsible for platelet destruction.

SUMMARY

• • •

Response to vascular injury is vasoconstriction, formation of a platelet plug, and ultimately, a stable clot held together by fibrin. A complex series of reactions in which there is activation of clotting factors leads to fibrin formation. One series of reactions, the intrinsic pathway, is initiated by contact with a negatively charged surface and

purpura: L. *purpura*, purple
thrombocytopenic: Gk. *thrombos*, lump; Gk. *kytos*, cell; *penia*, want

results in the formation of factor Xa. Thromboplastin initiates the extrinsic pathway with factor Xa as an end product as well. The final common pathway utilizes factor Xa and results in the formation of thrombin that catalyzes the conversion of fibrinogen to fibrin.

There are also intravascular elements to prevent abnormal clotting. Three substances that are either produced by vascular endothelium or are activated on the endothelial surface are antithrombin III and proteins C and S.

Hemophilia A is an inherited X-linked, recessive bleeding disorder. The basic defect is a deficiency of factor VIII:C, which causes limited formation of fibrin. Typically, bleeding resumes several hours after injury and tends to be internal. There are varying degrees of severity in the disorder, but possible problems include (1) hemarthrosis, (2) pressure-induced neuropathies and muscle atrophy, (3) sublingual hematoma, (4) blood cysts, and (5) gastrointestinal and intracranial bleeding. Cryoprecipitate containing factor VIII:C and porcine factor VIII:C are available for treatment. Hemophilia B is a less common form of hemophilia in which there is a deficiency of factor IX, but the clinical picture is the same as for hemophilia A. A related bleeding disorder is von Willebrand's disease in which there is a deficiency of vWF that normally forms a complex with factor VIII:C. There is impaired platelet plug formation and a prolonged bleeding time.

Disseminated intravascular coagulation is a syndrome in which there is widespread microthrombi formation. There is, in turn, massive fibrinolytic activity, so the overall picture is one of clotting and bleeding. This disorder is associated with a variety of disease processes including septicemia, shock, and malignancies. Treatment involves transfusion, administration of heparin, and infusions of platelets and cryoprecipitate.

Both vitamin K deficiency and decreased numbers of platelets cause bleeding problems. Vitamin K is required for the synthesis of clotting factors VII, IX, X, and prothrombin. A decrease in platelet count is called thrombocytopenia and occurs in DIC. Destruction of platelets also occurs as the result of an immune reaction developing in response to a disease process, infection, or administration of blood or drugs. This is called immune thrombocytopenic purpura and leads to spontaneous bleeding. Modalities of treatment are gamma globulin, prednisone, immunosuppressive drugs, and splenectomy.

REVIEW QUESTIONS

• • •

1. Identify three major steps in the blood coagulation process.
2. What substance degrades fibrin?
3. What is the end product of the intrinsic and extrinsic pathways?
4. What initiates the intrinsic pathway?

5. What initiates the extrinsic pathway?
6. What is the end product of the final common pathway?
7. List two intravascular anticoagulants.
8. What is the defect in hemophilia A? hemophilia B? von Willebrand's disease?
9. List three manifestations of hemophilia A.
10. What is cryoprecipitate?
11. Which of the bleeding disorders causes impaired platelet plug formation?
12. List six disease processes which are sometimes complicated by DIC.
13. In the process of fibrinolysis in DIC, what category of substances cause multiple problems involving both clotting and bleeding?
14. What is the relationship between vitamin K deficiency and bleeding tendencies?
15. List three modalities of treatment for immune thrombocytopenic purpura.

SELECTED READING
• • •

Cines, D. B., A. Tomaski, and S. Tannerbaum. 1987. Immune endothelial-cell injury in heparin-associated thrombocytopenia. *New England Journal of Medicine* 316:581–88.

Clouse, L. H., and P. C. Comp. 1986. The regulation of hemostasis: The protein C system. *New England Journal of Medicine* 314:1298–1304.

Davis, J. M., and K. A. Schwartz. 1989. Bleeding time. *Laboratory Medicine* 20:759–62.

Furie, B. et al. 1988. The molecular basis of blood coagulation. *Cell* 53:505.

Mammen, E. F., and Y. Fujii. 1989. Hypercoagulable states. *Laboratory Medicine* 20:611–15.

Nisen, P. et al. 1986. The molecular basis of severe hemophilia B in a girl. *New England Journal of Medicine* 315:1139–42.

Roberts, G. H. 1988. Thrombotic thrombocytopenic purpura. *Laboratory Medicine* 19:640–44.

Scully, R. E., ed. 1986. Case records of the Massachusetts General Hospital (DIC after dog bite). *New England Journal of Medicine* 315:241–49.

Staub, C. 1990. Complications of massive transfusion. *Physician Assistant* 14:51–63.

Chapter 20

Practical Applications: Blood Disorders

This final chapter of unit IV centers on the laboratory evaluation of blood disorders. The preceding chapters focus both on normal function of the cellular elements of blood and on disorders involving the blood cells. The following paragraphs deal first with specific hematologic tests and then with two examples of blood disorders, i.e., acute myelogenous leukemia and DIC, including interpretation of laboratory tests associated with these two conditions.

This first section describes tests that provide significant information about red and white blood cells as well as about bleeding tendencies.

RED BLOOD CELL TESTS

Red blood cells may be counted and reported as number of cells/volume of blood or, indirectly, may be measured as packed cell volume and reported as a percentage of the volume of blood (figure 20.1). Hemoglobin values indicate hemoglobin content of red blood cells which may also be estimated by the appearance of red blood cells on a stained smear, i.e., normochromic or hypochromic (chapter 17). Variations in size and shape of red cells is also discussed in chapter 17.

Reticulocytes are anucleated young erythrocytes (figure 20.2) that normally constitute 0.5–1.5% of the total number of red cells in the peripheral blood. With special staining techniques, ribonucleic acid in the cell precipitates and appears as blue granules that may form a network. An elevated reticulocyte count shows (1) a functioning hematopoietic system and (2) an increased demand for red blood cells.

Rubricytes, also called normoblasts, are not normally found in peripheral blood and are reported as **nucleated red blood cells.** The presence of the cells on a blood smear indicates increased erythropoiesis. Table 20.1 summarizes these observations regarding red blood cells.

WHITE BLOOD CELL TESTS

White blood cells (WBCs) may be reported in terms of number/volume of blood, which represents a count of all five types of cells. A second method for counting white blood cells is the identification of the white cell types distinguishable on a stained blood smear, called a **differential count.** Each cell type is reported as a percentage of a total of 100 cells identified in a differential count.

A third method for evaluating white blood cell number is a report of **absolute values** for each type of white blood cell, i.e., the number of each type/volume of blood. The absolute number for a specific type of white blood cell may be either counted or calculated as a product of a differential percentage and the total white blood cell count. For example, with a normal white blood cell count of 7,000

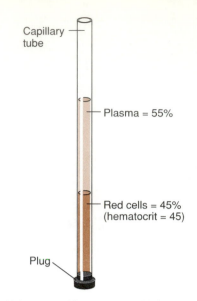

Figure 20.1 When a capillary tube is filled with blood and centrifuged, blood cells are packed in the lower part of the tube. The majority of these cells are red blood cells; hence, this (the hematocrit) is a measure of the percentage of red blood cells.

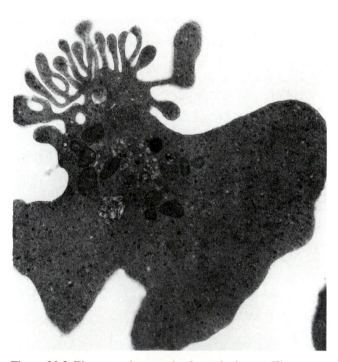

Figure 20.2 Electron micrograph of a reticulocyte. The nucleus is no longer present and the cytoplasm shows the characteristic fimbriated processes that occur just after nuclear extrusion. Mitochondria are still present, as are degradation vacuoles and polysomes. ×16,500.
Courtesy of Dorothea Zucker-Franklin, Professor of Medicine at New York University Medical Center.
From M. H. Ross, E. J. Reith, and L. J. Romrell: Histology: A Text and Atlas. *1989 Williams & Wilkins, Baltimore. Reproduced with permission.*

TABLE 20.1 Laboratory tests that help evaluate red blood cell production

Red Blood Cell Number	Adult Males	Adult Females
Hematocrit (Hct)	40–54%	38–47%
Hemoglobin (Hgb)	12–16 g/dl	13.5–18 g/dl
Reticulocytes	0.5–1.5%	0.5–1.5%
Nucleated red blood cells	Abnormal	Abnormal

Observations about Peripheral Blood Smear (RBCs)

Normocytic; normal size	Poikilocytosis; various shapes
Macrocytic; large	Normochromic; normal color
Microcytic; small	Hypochromic; large clear center
Anisocytosis; various sizes	Polychromatophilia; faintly blue (immature cell)

TABLE 20.2 Normal values for a differential count and normal absolute values

	Percent	Absolute Value
Neutrophils	50–70	1,800–7,000/μl
Bands or stabs	3	0–700/μl
Eosinophils	2–5	0–450/μl
Basophils	0–1	0–200/μl
Lymphocytes	35	1,000–4,800/μl
Monocytes	3–7	0–800/μl

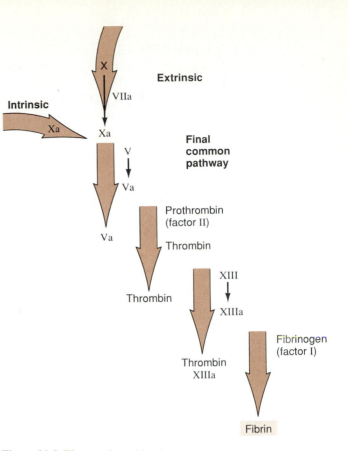

Figure 20.3 The prothrombin time test measures the function of factors VII, X, V, prothrombin, and fibrinogen.

cells/mm³ and 60% neutrophils, a calculated value for the absolute number of neutrophils is the product of 0.6 and 7,000, or 4,200/mm³ (table 20.2). If, however, the white blood cell count is extremely high, it is possible for an increase in absolute numbers to be associated with a decrease in percentage. The reason this may occur is that (1) the differential count involves the identification of 100 cells, an extremely small sample, and (2) the numbers of all white blood cells may be greatly increased. In leukemias, for example, the total white blood cell count may exceed 100,000 with a predominant increase in one cell type. The absolute count for that cell type will be elevated, even though the percentage calculated from a differential count may be low.

COAGULATION TESTS

The coagulant and anticoagulant mechanisms within the vascular system are complex and involve multiple factors (chapter 19). The following are common laboratory tests that help evaluate the status of bleeding tendencies in a patient.

*1 mm³ = 1 ml = 0.001 liter
1μl = 0.000001

QUANTITY OF PLATELETS

Platelets aggregate and form a plug following vascular injury, and this plug is subsequently stabilized by fibrin threads. A decrease in the number of platelets is the most common cause of bleeding, and such a decrease is noted in two ways on laboratory reports. There may be an estimate of normal number from the appearance of a stained blood smear using the terms slight, moderate, or marked decrease. Platelet counts are also performed, and the normal range of values is 150,000–400,000/mm³.*

PROTHROMBIN TIME

The **prothrombin time** or **pro time** measures the function of the extrinsic pathway in coagulation. This pathway is initiated by thromboplastin released by damaged tissue. There is activation of factors VII and X (see table 19.3) with factor Xa contributing in the final pathway to the formation of fibrin. The final common pathway involves (1) activation of factor V, (2) conversion of prothrombin to thrombin, and (3) conversion of fibrinogen to fibrin (table 19.4). Further, calcium ions are required; thus, the prothrombin time test measures the function of factors VII, X, V, prothrombin, and fibrinogen (figure 20.3).

The test is performed by using plasma taken from blood to which a calcium binding anticoagulant has been added (figure 20.4). Thromboplastin and calcium are

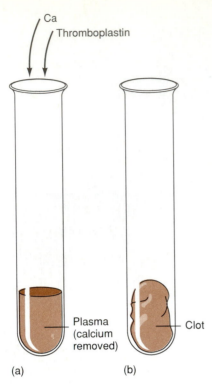

Figure 20.4 The prothrombin time test is a measure of the time required for plasma to clot after the addition of calcium (Ca) and thromboplastin.

TABLE 20.3 The mechanism of action of four therapeutically useful anticoagulants

Aspirin. Interferes with platelet aggregation
Dicumarol and warfarin. Interfere with hepatic synthesis of vitamin K; dependent clotting factors
Heparin. In combination with antithrombin III, inhibit activated factors IX, X, XI, XII and prevent formation of thrombin; heparin inhibits platelet aggregation

• • •

added to the plasma, and the time required for a clot to form is the prothrombin time. Normal values are in the range of 12–14 seconds.

• • •

Anticoagulants are often added to blood drawn for laboratory tests. Anticoagulants used for this purpose include (1) ammonium and potassium oxalate, (2) trisodium citrate, (3) EDTA (ethylenediamine-tetra-acetic acid), and (4) heparin. The first three anticoagulants remove calcium and heparin inhibits thrombin activation.

• • •

The prothrombin time is useful in monitoring the effects of anticoagulants administered to prevent clot formation. Table 20.3 indicates the mechanism of action of

four anticoagulants. The prothrombin time is prolonged by warfarin, dicumarol, and heparin because, as the table shows, the test measures the function of the factors that the anticoagulants inhibit. Prothrombin and factors VII, IX, and X are synthesized in the liver and require vitamin K. Consequently, the prothrombin time is increased when there is liver damage or vitamin K deficiency. Low levels of fibrinogen, which may occur in disseminated intravascular coagulation, result in an increased prothrombin time.

ACTIVATED PARTIAL THROMBOPLASTIN TIME

Thromboplastin is present in all cell membranes and is released as the result of tissue damage (chapter 19). Thromboplastin initiates the extrinsic pathway in coagulation.

A special form of thromboplastin is used in the activated partial thromboplastin test (APTT). The purpose of this test is to identify inhibition or deficiency of factors in the intrinsic and common final pathways in coagulation (tables 19.2 and 19.4, and figure 20.5).

The principles of the test are based on the following facts. Normally, the intrinsic pathway is initiated by contact with a negatively charged surface, and the reactions take place on the surface of platelet membranes. The extrinsic pathway is initiated by thromboplastin.

The basic steps in performing the APTT to identify defects in these two pathways follow:

1. Plasma is activated by the addition of negatively charged particles, such as kaolin, and also by adding thromboplastin. The form of thromboplastin used for the test lacks the ability to compensate for the plasma defect of hemophilia and is called **partial thromboplastin.**
2. Phospholipid is added to substitute for the presence of platelets.
3. Calcium is added.

The activated partial thromboplastin time is the time in seconds required for the plasma to clot, with a normal range of 25–37 seconds (figure 20.6). The APTT test will be prolonged in hemophilia A or B, during heparin therapy, and in cases of defects in the intrinsic or common clotting pathways. These defects include liver disease, vitamin K deficiency, and disseminated intravascular coagulation.

• • •

It is possible to localize coagulation defects to some extent by comparing the results of a pro time and APTT. An example would be a deficiency of factor VIII in which there would be a normal pro time and prolonged APTT.

• • •

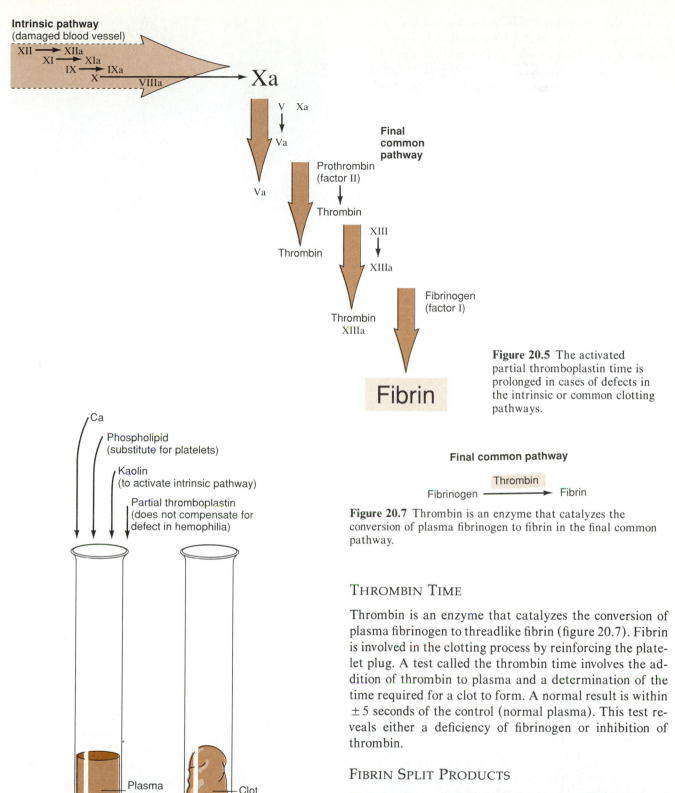

Intrinsic pathway
(damaged blood vessel)

XII ⟶ XIIa
XI ⟶ XIa
IX ⟶ IXa
X
VIIIa

Xa

V Xa

Va

Va

Thrombin

Final common pathway

Prothrombin
(factor II)

Thrombin

XIII

XIIIa

Thrombin
XIIIa

Fibrinogen
(factor I)

Fibrin

Figure 20.5 The activated partial thromboplastin time is prolonged in cases of defects in the intrinsic or common clotting pathways.

Final common pathway

Fibrinogen ⟶ Thrombin ⟶ Fibrin

Figure 20.7 Thrombin is an enzyme that catalyzes the conversion of plasma fibrinogen to fibrin in the final common pathway.

Ca

Phospholipid
(substitute for platelets)

Kaolin
(to activate intrinsic pathway)

Partial thromboplastin
(does not compensate for
defect in hemophilia)

Plasma
(Calcium
removed)

Clot

(a)   (b)

Figure 20.6 The activated partial thromboplastin time test (APTT).

Thrombin Time

Thrombin is an enzyme that catalyzes the conversion of plasma fibrinogen to threadlike fibrin (figure 20.7). Fibrin is involved in the clotting process by reinforcing the platelet plug. A test called the thrombin time involves the addition of thrombin to plasma and a determination of the time required for a clot to form. A normal result is within ± 5 seconds of the control (normal plasma). This test reveals either a deficiency of fibrinogen or inhibition of thrombin.

Fibrin Split Products

Plasmin, also called fibrinolysin, cuts up fibrinogen and fibrin threads and is involved in clot dissolution (chapter 19). Fragments resulting from this lysis are called **fibrin split products** (FSP) or **fibrinogen degradation products** (FDP). Such fragments are normally present in the blood because this mechanism removes clots, but the level of fragments is elevated in thromboembolic conditions. Examples of conditions in which clots occlude small blood vessels are myocardial infarction, deep vein thrombosis, pulmonary embolism, and disseminating intravascular coagulation.

ACUTE MYELOGENOUS LEUKEMIA

In acute myelogenous leukemia (AML), also called acute nonlymphocytic leukemia (figure 20.8), the predominant cell type is the myeloblast of the granulocytic series (chapter 18). The problem is twofold: (1) unregulated proliferation of cells and (2) an arrest in maturation at the blast cell stage. Infiltration of bone marrow by leukemic cells causes failure of normal hematopoiesis; thus, there is anemia, neutropenia, and thrombocytopenia. The consequence of this decrease in normal blood cells is susceptibility to infection and hemorrhage. Cells may infiltrate any tissue including the skin, lymph nodes, and spleen. The treatment of choice is chemotherapy with the goal of destroying malignant cells so that normal hematopoiesis is reestablished. The chemotherapy is not specific for leukemic cells, and the result is destruction of normal cells as well. About 70% of patients with AML have a period of remission after treatment, but only 20% are disease free 5 years or more.

EVALUATION OF LEUKEMIA

The preceding discussion of laboratory tests describes observations that are significant in many disease processes because blood cells are frequently affected. In leukemia, all types of blood cells are affected. The following cases are presented as a basis for further discussion of these tests.

> Patient 20.1 was a 63-year-old male; diagnosis, acute myelogenous leukemia.
> Laboratory findings upon admission (normal values for reporting lab in parentheses): RBC 3 million/mm³ (5.4 ± 0.7), WBC 3,000/mm³ (7,800 ± 3), platelets 10,000/mm³ (130,000–400,000).
>
> One month later: WBC 6,000/mm³, platelets 66,000/mm³. Immediately preceding the second round of chemotherapy: RBC 3.27 million/mm³, WBC 12,940/mm³, platelets 285,000/mm³, hematocrit 35% (47 ± 5); Differential count: bands 9, neutrophils 34, lymphocytes 36, monocytes 7, basophils 1, metamyelocytes 5, myeloblasts 8; RBCs, moderate polychromia, slight hypochromia, 9 nucleated RBC; K⁺ 4.5 meq/liter (3.4–5.0), uric acid 6.2 mg/dl (3.6–8.1), prothrombin time 11.5 sec (9.8–12.6), APTT 26 sec (21–31).

This individual noticed increasing fatigue and shortness of breath. He was hospitalized within a month and, after bone marrow examination, was diagnosed as having acute myelogenous leukemia.

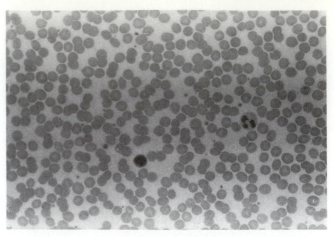

(a)

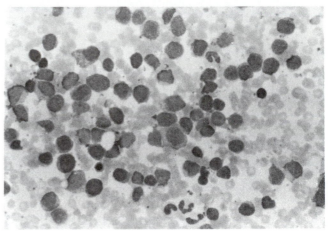

(b)

Figure 20.8 (a) Normal blood smear showing two white blood cells. (b) Acute myelogenous leukemia with many immature white blood cells (myeloblasts). *Courtesy Dr. M. G. Klein.*

The white blood cell count at diagnosis of AML may be normal, increased, or decreased. Over half of the affected individuals do not initially show leukocytosis, and this particular patient was leukopenic. Whether or not the white count is elevated, immature cells, especially blasts, are noted. This patient was both anemic and thrombocytopenic, which is typical of AML at diagnosis. Immediately after chemotherapy, the white blood cell and platelet counts were decreased. Leukopenia results in increased susceptibility to infection, and this individual had some serious problems with infection during this time. Both of these counts showed an increase within a month. Just prior to the next series of treatments, there was evidence of regeneration of bone marrow. Both the white blood cells and platelets showed normal values, although the red blood cell count, hemoglobin, and hematocrit were somewhat low. The differential indicated myeloblasts typical of his leukemia, but the report of polychromia and nucleated red blood cells was evidence of increased erythropoiesis.

BLOOD DISORDERS: CELLS AND BLEEDING

Uric acid is a normal product of nucleic acid metabolism and, although the level is within normal range in patient 20.1, it is frequently elevated after chemotherapy due to increased cell destruction. The lysis of cells releases nucleic acids, and uric acid is an end product of the subsequent degradation. Hyperkalemia may also occur as the result of massive cell destruction, although this did not occur in this patient. The pro time and APTT tests were normal as well.

EVALUATION OF LEUKEMIA/DIC

The following individual was diagnosed with acute myelogenous leukemia and the laboratory values upon hospital admission are indicated. DIC developed subsequent to chemotherapy.

> Patient 20.2 was a 72-year-old male; diagnosis, acute myelogenous leukemia.
> Laboratory findings: RBC 3,694,000/mm³ (5.4 ± 0.7 million), WBC 150,980/mm³ (7,800 ± 3), platelets 25,000/mm³ (130,000–400,000), hematocrit 35.63% (47 ± 5); Differential count: neutrophils 2, lymphocytes 5 unidentified blasts 93.
> After chemotherapy: Fibrinogen 90 mg/dl (200–400), FDP 30 g/ml (10), prothrombin time 25 sec (12–14), APTT 55 sec (25–37).

Patient 20.2 had an extremely elevated white blood cell count with a predominance of blast cells. After chemotherapy, DIC complicated the course of the patient's treatment. DIC may be precipitated by cell breakdown at the beginning of treatment with lysed blast cells releasing procoagulant into the system. This appears to be the case in this patient. DIC may also be precipitated by septicemia when there is an overwhelming infection. The course of events (chapter 19) is (1) procoagulant introduced into the circulation, (2) formation and deposition of fibrin, and (3) lysis of fibrin. The resulting bleeding tendencies are shown in this individual by the prolonged prothrombin time and APTT. Increased fibrinolysis is apparent from the increased level of fibrin/fibrinogen degradation products and decreased fibrinogen.

PROSTATE CARCINOMA/DIC

Patient 20.3 had undergone prostate surgery due to carcinoma. He was admitted to the hospital, complaining of both nausea and bone pain. He had scattered ecchymoses (ek″ĭ-mō′sēz) or bruises on his arms and legs but had not experienced bleeding from his nose, rectum, or bladder.

> Patient 20.3 was an 85-year-old male; diagnosis, prostatic carcinoma with metastasis to bone.
> Laboratory findings upon admission: WBC 5,100/mm³ (4,000–10,700), lymphocytes 8% (20–40), monocytes 10% (2–6), neutrophils 80% (50–60), eosinophils 2% (1–4), basophils 0% (0.5–1), RBC 4.0 million/mm³

ecchymosis: Gk. *ek,* out of; Gk. *chymos,* juice

(4.5–6.0 million), Hgb 11.4 g/dl (13–16), Hct 34% (40–54), platelets 70,000/mm³ (130,000–400,000), pro time 16.5 sec (9.8–12.6), APTT 29 sec (21–31), ALT 14 IU/liter (7–56), GGT 145 IU/liter (7–64), LDH 1685 IU/liter (287–537), PO$_4^{-3}$ 4 mg/dl (2.5–4.5), AP 771 IU/liter (38–126), AST 64 U/liter (8–35), Ca 9.3 mg/dl (8.9–10.6)

The white blood cell count is normal, although there is an increase in the number of neutrophils. The red blood cell count as well as the hemoglobin and hematocrit values are low. The effects of malignancy on blood cells follow.

MALIGNANCY AND BLOOD CELLS

Blood cell abnormalities are frequently associated with neoplasms. These abnormalities are often produced by infiltration of the bone marrow, although infection and the toxic effects of cancer therapies are factors as well. Erythrocytosis, an increase in red blood cell count, occurs in association with tumors and, in more than half these cases, there is an elevated serum level of erythropoietin. Erythrocytosis is associated with an hematocrit of more than 55% in a male or greater than 50% in a female. Anemias occur in cancer patients as well. The underlying mechanisms that lead to anemia include blood loss, chemotherapy or radiation therapy, iron deficiency, and bone marrow invasion. The mechanisms are not always clear, and the term **anemia of chronic malignancy** is sometimes used. Autoimmune hemolytic anemias (AHA), caused by antibody destruction of red blood cells, occur occasionally in association with a tumor, usually in an older individual.

• • •

There may be factors other than erythropoietin contributing to erythrocytosis in cancer patients. Tumors of the adrenal cortex or ovarian tumors may produce prostaglandins that enhance the effects of erythropoietin.

• • •

An increase in the number of granulocytes, neutrophils in particular, may be associated with several types of malignancies. This response is called a **paraneoplastic leukemoid reaction.** There are various possible causes for increased numbers of neutrophils including drugs, infection, metabolic disorders, and emotional stimuli. The underlying mechanism may involve tumor produced stimulating factors. A decrease in the number of granulocytes associated with cancer is usually caused by severe infection, radiation, or chemotherapy.

paraneoplastic: Gk. *para,* beside

. . .

Elevation of both monocyte and neutrophil count may be associated with various types of malignancies including neoplasms of the stomach, pancreas, lung, and brain, as well as Hodgkin's disease.

. . .

MALIGNANCY AND CALCIUM

The calcium level in patient 20.3 is normal, in spite of bone metastasis. Either an increase or a decrease in calcium may occur in malignancy.

Hypercalcemia is common, and there are several ways calcium levels may increase. Possible mechanisms include parathormone (PTH) secretion, ectopic secretion of PTH (chapter 31), tumor production of prostaglandins, and bone metastases. Only about 10% of cancer patients who have hypercalcemia also have bone metastases. The most frequent cause of cancer associated hypercalcemia is the secretion of PTH-like substances by the tumor. Hypocalcemia occurs in cancer patients as well. Approximately 16% of patients with bone metastases have a decrease in serum calcium.

ENZYME LEVELS

The laboratory report shows increased enzyme levels for this patient. The significance of each enzyme is nonspecific, with the possibility of various disease processes causing an elevation.

Alkaline Phosphatase

Alkaline phosphatase (AP) is an enzyme produced mainly in liver and bone. The test for this enzyme is used, for the most part, as one indicator of liver and bone disease. Elevations are frequently associated with obstructive jaundice, cirrhosis, abscesses and cancer of the liver, and bone diseases. Alanine aminotransferase (ALT), also called serum glutamic-pyruvic transaminase (SGPT), is found in high concentrations in the liver and in lower concentrations in the heart, muscle, and kidney. The test for AP enzyme levels is used mainly for the diagnosis of liver disease or to monitor the effects of drugs that have toxic effects on the liver. Increased levels of AP occur in various liver diseases including metastatic liver tumor. The level of AP is elevated in myocardial infarction as well (chapter 25). In addition, there are many drugs that cause elevated levels of AP.

Aspartate Amino Transferase

Aspartate amino transferase (AST) is an enzyme found in tissues with high metabolic activity and is released into the circulation following cell necrosis. It is also called serum glutamic-oxaloacetic transaminase (SGOT). The AST level is proportional to tissue injury, and in the case of severe injury, the blood level rises within hours and re-

mains high for a period of days. Increased levels occur in various conditions including myocardial infarction, liver disease, severe burns, radiation of skeletal muscle, and brain trauma.

Lactate Dehydrogenase

Lactate acid dehydrogenase (LDH) is a widely distributed intracellular enzyme. Tests showing elevated levels of this enzyme indicate cell necrosis with release of the enzyme. The greatest increase of LDH, 2–40 times, occurs in metastasizing cancer and in shock. Increases of 2–4 times are seen in myocardial infarction, hemolytic anemia, and pulmonary infarction. There are slight increases in hepatitis and cirrhosis.

Gamma-Glutamyltransferase

The enzyme gamma-glutamyltransferase (GGT) is found mainly in the liver, kidney, prostate, and spleen. GGT appears to transport amino acids and peptides across cell membranes. GGT is elevated in all forms of liver disease, cancer of the bile duct, barbiturate use, and alcoholism. Hepatotoxic drugs used in the treatment of cancer increase GGT levels as well.

ABNORMALITIES IN CLOTTING CHARACTERISTICS

Table 20.4 is a summary of normal clotting and anticoagulant mechanisms (chapter 19). Cancer patients frequently have problems related to thrombi and hemorrhage. Patient 20.3 had multiple bruises with no other overt signs of bleeding tendencies. His platelet count was low, and the prothrombin time was prolonged. A decrease in numbers of platelets is a common cause of bleeding. The prothrombin time is an important screening test for abnormalities in coagulation. The test measures five coagulation factors including prothrombin and fibrinogen (discussed earlier in this chapter). A prolonged prothrombin time may indicate various conditions including prothrombin deficiency, vitamin K deficiency, liver disease, and DIC.

In general, there may be various complications associated with malignancy that involve coagulation. Ninety-two percent of cancer patients have abnormal results for tests of coagulation factors. The complications are caused by factors, such as (1) treatments resulting in thrombocytopenia, (2) tissue injury caused by either a tumor or therapy, (3) liver disease, (4) anticoagulants in the circulation, (5) vitamin deficiency, and (6) infection. The cause, however, is not always identifiable.

An inflammation of veins associated with thrombosis, i.e., thrombophlebitis, is frequently associated with cancer. Disseminated intravascular coagulopathy (DIC) may occur because of malignancy. DIC may take the form of a chronic condition in which fibrin clot formation predominates. Acute hemorrhage as the result of secondary fibrinolysis is a second possibility, or the abnormality may

TABLE 20.4 Overview of normal clotting
and anticoagulant mechanisms

Clot Formation

Thromboplastin (a tissue factor) initiates clotting
Platelets aggregate
Conversion of prothrombin to thrombin is the result of a series
    of reactions
Fibrinogen converted to fibrin by the enzyme thrombin

Clot Dissolution

Plasminogen binds to fibrin and is converted to plasmin
Plasmin splits fibrin; fibrin/fibrinogen breakdown products

Natural Anticoagulants

Antithrombin III (AT III)
Protein C
Protein S

TABLE 20.5 Overview of events in DIC

Initiating Event

Procoagulant into circulation (may be thromboplastin)

Overview of Process

Widespread clot formation within small blood vessels; triggers
    massive fibrinolytic activity

Coagulation

Uses up platelets and clotting factors (promotes bleeding);
    uses up AT III and protein C that oppose the action of
    thrombin (promotes clot formation); causes both thrombi
    and bleeding; causes widespread deposition of fibrin in
    small blood vessels

Fibrinolysis

Increases formation of fibrin/fibrinogen degradation products;
    small fragments inhibit platelet function; large fragments
    promote platelet aggregation; mixture of soluble
    fragments increases capillary permeability and promotes
    coagulation outside vessel. A decrease in platelets,
    thrombocytopenia, seen in this patient occurs as the result
    of platelet consumption by microthrombi in small blood
    vessels

be recognized only as the result of laboratory tests. DIC may be primarily thrombotic, hemorrhagic, or may be covert.

A sequence of events may occur as follows:

1. Low-grade formation of thrombi, possibly due to the release of thromboplastin from damaged tissue
2. Subsequent accelerated consumption of coagulation factors

3. A compensatory increase in synthesis of fibrinogen, platelets, clotting factors. This represents an overcompensation for intravascular coagulation resulting in an actual elevation of some clotting factors.

An overt hemorrhagic DIC, which is less common, is initiated by intravascular coagulation with deposition of fibrin in small blood vessel walls. The coagulation process is widespread, and it uses up clotting factors and activates massive fibrinolysis. The consumption of clotting factors leads to decreased levels of these factors in the circulation and to hemorrhage. The evidence of fibrinolysis is elevated blood levels of fibrin/fibrinogen split products. These represent fragments resulting from the lysis of fibrin or fibrinogen by plasmin.

Continued monitoring of coagulation tests is required when DIC is suspected.

> Two days after patient 20.3 was admitted to the hospital, there was a problem with frequent voiding in small amounts.
> The laboratory tests showed the following: Platelets 33,000/mm³ (130,000–400,000), pro time 17.1 sec (9.8–12.6), APTT 30 sec (21–31), FSP 256 mcg/ml (0–10), fibrinogen 67 mg/dl (175–400), thrombin time 14.6 sec (11–16), RBC 3.3 million/mm³ (4.5–6.0), Hgb 11.6 gm/dl (13–16)

Patient 20.3 developed an overt DIC with hemorrhage (table 20.5). The hemorrhage in DIC is usually superficial with ecchymoses and petechiae and oozing from the oral mucosa. Oozing from the gastrointestinal and urinary tracts may occur as well. A major hemorrhage, including central nervous system bleeding, may occur.

The elevated level of fibrin/fibrinogen split products is evidence of accelerated fibrinolysis. There are several complications associated with increased FSP. Fibrin split products lead to the following:

1. Inhibition of fibrin monomer polymerization (chapter 19), which interferes with clot formation
2. Binding to thrombin, impairing coagulation
3. Formation of complexes that bind to a growing fibrin clot and weaken it
4. The increase of capillary permeability
5. Both the inhibition of platelet function and increased tendency for platelet aggregation.

These events promote bleeding and increase the prothrombin time. The initiating event in fibrinolysis, as it occurs in DIC, is probably either the release of excessive plasminogen activator into the blood due to endothelial cell injury or plasmin stimulation by fibrin.

The thrombin time in patient 20.3 is within normal limits. This test is designed to show a decrease in fibrinogen. When the thrombin time is prolonged, it indicates interference with the last stage of coagulation at which point fibrinogen is converted to fibrin. There is usually a correlation between this test and the level of FSP and a low fibrinogen level.

A **microangiopathic** (mi″kro-an″je-o-path′ik) **hemolytic anemia** may accompany DIC. Red blood cells are fragmented and lysed by the impact on fibrin strands within small blood vessels. This causes a low red blood cell count as well as a low hemoglobin level, as seen in patient 20.3.

Disorders associated with DIC include adult respiratory distress syndrome, hemolytic uremic syndrome, and oliguric renal insufficiency. The relationship between the events in DIC and these disorders is not clear.

TREATMENT

The management of DIC involves the treatment of such underlying causes as malignancy. There may be other factors, such as volume deficit, acidosis, hypoxemia, or transfusion reactions. The role of heparin in treatment is controversial. Heparin may be used to control the thrombotic complication of DIC.

SUMMARY

• • •

The status of red blood cells is evaluated by direct count, hematocrit or packed cell volume, and hemoglobin. Reticulocytes are young red blood cells, and an increase in number indicates increased erythropoiesis. White blood cells are counted directly as both a total and differential count. Absolute values for the five types of white cells are also reported.

Coagulation tests include (1) platelet counts, (2) prothrombin time, (3) activated partial thromboplastin time, (4) thrombin time, (5) fibrin degradation products, and (6) fibrinogen levels. Prothrombin time measures the function of the extrinsic pathway, while the APTT test is affected by factors in the intrinsic pathway and the final common pathway. A deficiency of fibrinogen or inhibition of thrombin affects the thrombin time test. Increased fibrinolysis causes an elevation of fibrin degradation products and a decrease in fibrinogen.

In acute myelogenous leukemia, the predominant cell type is the myeloblast of the granulocytic series. Infiltration of bone marrow by leukemic cells causes failure of normal hematopoiesis; thus, there is anemia, neutropenia, and thrombocytopenia.

Malignancy may cause erythrocytosis, often due to elevated levels of erythropoietin. Anemias associated with malignancy are often the result of blood loss, radiation or chemotherapy, vitamin and iron deficiency, or bone marrow

microangiopathic: Gk. *mikros,* small; Gk. *aggeion,* vessel; Gk. *pathos,* disease

invasion. Tumor produced stimulating factors cause an elevation in the number of neutrophils, while decreased numbers may be the result of severe infection, radiation, or chemotherapy. Hypercalcemia is common in malignancy, although hypocalcemia occurs as well.

The elevated enzyme levels in patient 20.3 were GGT, LDH, alkaline phosphatase, and AST. Alkaline phosphatase is an indicator of bone disease and, in general, AST levels are proportional to tissue injury. Lactate acid dehydrogenase levels increase with cell necrosis, and GGT increases in response to hepatotoxic drugs, among other factors.

Cancer patients frequently have abnormalities in clotting characteristics, with DIC as a possible sequel to the disease process. The sequence of events in DIC are (1) introduction of a procoagulant, possibly thromboplastin, into the circulation, (2) widespread microthrombi formation with fibrin deposition in small blood vessel walls, (3) consumption of clotting factors, (4) activation of massive fibrinolysis, and (5) hemorrhage. DIC may be revealed only by laboratory tests. The disease process may be predominantly thrombotic, or there may be oozing and hemorrhage. Low-grade coagulation with consumption of clotting factors may result in an overcompensation leading to increased synthesis in fibrinogen, clotting factors, and platelets. When hemorrhagic DIC occurs, there is an increase in fibrin/fibrinogen split products and a decrease in both platelets and fibrinogen. A prolonged prothrombin time indicates a bleeding tendency.

REVIEW QUESTIONS

• • •

1. Why is the hematocrit an indicator of total red blood cells in spite of the fact that all blood cells are spun down and measured?
2. What is the name of non-nucleated, young red blood cells?
3. What is the name of immature neutrophils in the late stage of development?
4. What is the significance of a decreased platelet count?
5. What test is a measure of the intrinsic coagulation pathway?
6. What test is affected by abnormalities in the extrinsic pathway?
7. What test will be prolonged in hemophilia?
8. What is the significance of a prolonged thrombin time?
9. What does an increase in the level of fibrin split products imply?
10. How does DIC affect (a) fibrin split products, (b) pro time, and (c) APTT?
11. What is the predominant cell type in acute myelogenous leukemia?
12. What is a probable cause of erythrocytosis associated with malignancy?

13. What are two possible causes of anemia in malignancy?
14. What are two possible causes of increased numbers of neutrophils in malignancy?
15. A decrease in neutrophils associated with malignancy is likely to be caused by what two factors?
16. What is the most frequent cause of cancer associated hypercalcemia?
17. What enzyme is an indicator of liver and bone disease?
18. What enzyme level is proportional to tissue injury?
19. What is the significance of elevated levels of LDH?
20. Tissue injury results in the release of what procoagulant?
21. What is the initiating event in DIC?
22. What are two consequences of the coagulation that occurs in DIC?
23. What is the significance of increased FSP, increased platelet count, increased fibrinogen?
24. What laboratory test is indicative of fibrinolysis?
25. What are two consequences of an increase in FSP?

SELECTED READING

• • •

Gale, R. P. et al. 1988. Rapid progress in chronic myelogenous leukemia. *Leukemia* 2:321–24.

Grossman, L. et al. 1983. Lactic acidosis in a patient with acute leukemia. *Clinical and Investigative Medicine* 6:85.

Scully, R. E. et al., ed. 1987. Case Records of the Massachusetts General Hospital (leukemia). *New England Journal of Medicine* 316:1259.

Sobol, R. E. et al. 1987. Clinical importance of myeloid antigen expression in adult acute lymphoblastic leukemia. *New England Journal of Medicine* 316:1111.

Unit V

CARDIOVASCULAR SYSTEM

Chapter 21 deals with the concepts of measurement and control of blood pressure. There is a discussion of alterations of blood pressure, i.e., hypertension and circulatory failure. Chapter 22 includes a review of blood vessel wall structure followed by a discussion of vascular disorders. Abnormal vascular tone, injury to arterial walls, and separation of arterial wall layers are the emphasized topics. Disorders of the heart including myocardial infarction, disease of heart valves, etiology of heart failure, and congestive heart failure are discussed in chapter 23. The topic of chapter 24 is congenital heart defects. There is a review of fetal circulation with a discussion of common congenital heart defects and the physiological consequences of each. Chapter 25 integrates topics from preceding units, i.e., the role of the kidney and fluid/electrolyte balance, with volume control in congestive heart failure. Myocardial infarction is discussed in relation to plasma lipid levels and cardiac enzymes, and septic shock is presented with significant laboratory data.

Chapter 21

Blood Pressure Disorders

Evaluation of the status of the cardiovascular system, i.e., the function of the heart and carrying capacity of blood vessels, is central in health and disease. This system determines the adequacy of blood flow to tissues and subsequent oxygen delivery to the cells of the body. The measurement of blood pressure provides a useful clue about the function of the cardiovascular system and the purpose of this chapter is to outline (1) routine measurements and their significance, (2) factors contributing to deviation from normal, and (3) consequences of either an elevation or a fall in blood pressure.

BLOOD PRESSURE

Blood pressure measurement quantifies the force exerted by a volume of blood against a blood vessel wall. The force is expressed as mm of Hg, i.e., pressure equivalent to that required to maintain a column of Hg measured in mm (chapter 2).

SYSTOLIC AND DIASTOLIC PRESSURES

Clinically, **systolic** (sis-tol'ik) pressure is arterial pressure generated during ventricular contraction, while **diastolic** (di"ah-stol'ik) pressure is a lower pressure maintained during ventricular relaxation.

It is possible to measure blood pressure directly by introducing a needle or a catheter into a peripheral artery, although an indirect method utilizing a **sphygmomanometer** (sfig"mo-mah-nom'ē-ter) is used routinely. This involves an inflatable cuff, usually wrapped around the upper arm, and a manometer that measures the pressure in the cuff (figure 21.1). The principle of the procedure is summarized in table 21.1. The cuff is inflated about 30 mm Hg above anticipated systolic pressure and slowly deflated. A stethoscope is applied over the brachial artery (figure 21.2) so that blood flow sounds may be detected. The procedure follows:

1. Cuff inflation occludes the brachial artery and no sound is heard
2. As the cuff is deflated, the first tapping sounds caused by spurts of blood indicate systolic pressure
3. With further deflation, diastolic pressure is indicated when blood flow becomes continuous and the sounds disappear.

The average blood pressure for a normal young adult is systolic/diastolic, 120 mm Hg/80 mm Hg, although there is considerable individual variation. These values tend to increase with age.

systole: Gk. *syn*, together; Gk. *stellein*, to draw
diastole: Gk. *dia*, apart; Gk. *stellein*, to draw
sphygmomanometer: Gk. *sphygmos*, pulse; Gk. *phone*, sound

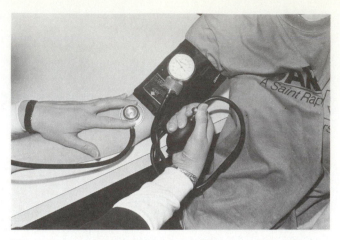

Figure 21.1 A sphygmomanometer cuff is wrapped around the upper arm, and the inflated cuff cuts off the flow of blood in the brachial artery.

TABLE 21.1 The principle of indirectly determining blood pressure by inflating a blood pressure cuff around the upper arm and listening to the blood flow sounds in the brachial artery

Systolic Pressure. Arterial pressure during ventricular contraction

1. a. Cuff inflated to about 30 mm Hg above anticipated systolic pressure
   b. Brachial artery occluded; no blood flow sounds when cuff pressure exceeds systolic pressure
2. a. Cuff slowly deflated
   b. Spurts of blood through brachial artery cause tapping sounds
   c. This occurs when cuff pressure falls just below systolic level; represents systolic pressure

Diastolic Pressure. Arterial pressure during ventricular relaxation

1. a. As cuff is deflated further, more blood flows through brachial artery
   b. Spurts of blood cause louder thuds called Korotkoff sounds
   c. Korotkoff sounds caused by the turbulence of spurts of blood meeting a static column of blood
   d. Korotkoff sounds muffled as cuff pressure approaches diastolic pressure
   e. Sounds disappear when cuff pressure is just below diastolic pressure because blood flow is continuous; when sounds disappear, indicates diastolic pressure

Blood pressure may be expressed as an average or mean of the arterial pressures that occur during a **cardiac cycle,** i.e., the series of events involved in one heartbeat. Assuming a heart rate of 75 beats/min, each cycle requires 0.8 second with ventricular contraction continuing for 0.3 second (see figure 23.1). The mean arterial pressure is not an average of high and low values (120/80) but represents an average of the pressure at every point in the cardiac cycle. The normal mean arterial blood pressure is approximately 93 mm Hg.

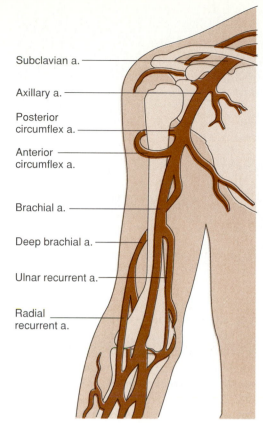

Subclavian a.

Axillary a.

Posterior circumflex a.

Anterior circumflex a.

Brachial a.

Deep brachial a.

Ulnar recurrent a.

Radial recurrent a.

(a)

No sounds
Cuff pressure = 140

First Korotkoff sounds
Cuff pressure = 120
Systolic pressure = 120 mm Hg

Sounds at every systole
Cuff pressure = 100

Last Korotkoff sounds
Cuff pressure = 80
Diastolic pressure = 80 mm Hg

Blood pressure = 120/80

(b)

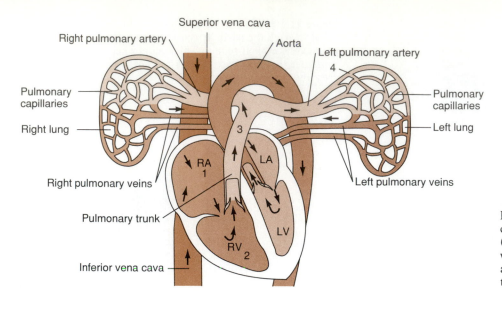

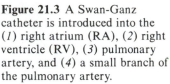

Figure 21.3 A Swan-Ganz catheter is introduced into the (*1*) right atrium (RA), (*2*) right ventricle (RV), (*3*) pulmonary artery, and (*4*) a small branch of the pulmonary artery.

The mean arterial pressure may be approximated with the formula: systolic pressure plus two times diastolic pressure, divided by three.

$$\frac{120 + 2(80)}{3} = \text{approximately 93 mm Hg.}$$

The mean pressure in the aorta is high because blood is rapidly ejected into this large vessel. Beyond the aorta, the arterial circuit subdivides into smaller branches that ultimately become arterioles and capillaries. The blood runs more slowly into these multiple smaller branches, and there is a marked decrease in pressure. Blood pressure at the ends of the arterioles is about 30 mm Hg.

PULMONARY ARTERY WEDGE PRESSURE

It is possible to measure pressures in both the right atrium of the heart and the pulmonary artery by means of a flexible catheter. A **Swan-Ganz catheter,** with a small inflatable balloon at the tip, may be introduced into a peripheral vein in the arm or neck and directed into the right atrium by way of the superior vena cava (figure 21.3). With the balloon inflated, the catheter is then carried by the flow of blood into a small pulmonary artery. The balloon is deflated and the pressure recorded from the tip of the catheter. The balloon is then inflated and the catheter is carried to a smaller branch, where it lodges and obstructs blood flow. The pressure at the obstructed segment of the small artery reflects the pressure of the continuous system beyond that point, i.e., (1) the pulmonary capillary bed

(2) the left atrium, and (3) the left ventricle during diastole. The pressure from the left side of the heart is transmitted to the pulmonary capillaries into the obstructed pulmonary artery to yield the **pulmonary artery wedge pressure,** also called the **pulmonary capillary wedge pressure** (figure 21.4). This reading indicates the left atrial mean pressure as well as the left ventricular diastolic pressure.

The significance of increased pressure in the left heart is that it is communicated to pulmonary capillaries and causes pulmonary edema. A normal average level of pressure is about 8 mm Hg, and a pressure greater than 25 mm Hg indicates pulmonary edema related to cardiac disease. Left ventricular failure is discussed in chapter 23.

FACTORS IN ARTERIAL PRESSURE

The pressure exerted by blood inside arteries is determined by the volume of blood ejected by the left ventricle and the resistance to flow by the vascular system. The factors contributing to both cardiac performance and vascular resistance will be considered in the following paragraphs. Table 21.2 summarizes some terminology related to cardiac function.

CARDIAC OUTPUT

Cardiac output is the volume of blood ejected by the left ventricle per minute. Both heart rate and myocardial contractility are major determinants of cardiac output, and these factors are influenced by a number of factors (table 21.3). The Frank-Starling mechanism determines the cardiac response to changes in venous return to the heart. An increase in the volume of blood returned to the heart causes myocardial muscle fibers to stretch, which in turn stimulates more forceful contractions; thus, the heart responds to accommodate changes in work load.

CARDIOVASCULAR SYSTEM

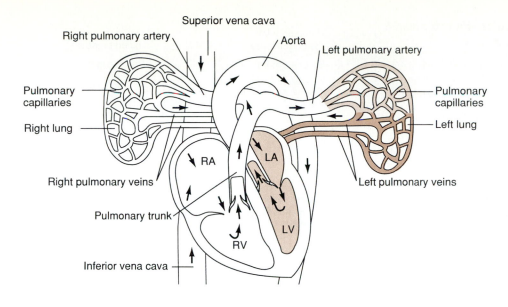

Figure 21.4 The pulmonary capillary wedge pressure is a reflection of the pressure in the shaded area, i.e., left ventricular filling pressure, left atrial pressure, and pulmonary capillary pressure.

TABLE 21.2 Definition of terms related to cardiac function

Frank-Starling law. The heart adapts to changes in the volume of blood returned to the heart; stretching of cardiac muscle fibers, within limits, results in increased force of contraction

Carotid sinus. Slight enlargement located at bifurcation of carotid arteries that contains nerve endings that respond to pressure

Carotid body. Region at bifurcation of carotid arteries that contain chemoreceptors

Aortic body. Chemoreceptors in region of aortic arch

Chemoreceptors. Nerve endings that are stimulated by changes in blood H^+, CO_2, and O_2

Baroreceptors. Nerve endings that are sensitive to changes in pressure

Pressoreceptors. Nerve endings sensitive to stretching

Cardiac output. Volume of blood pumped into aorta by the left ventricle per minute

Stroke volume. Volume of blood pumped by left ventricle during one contraction

• • •

Preload refers to the volume of blood that fills the ventricles during diastole. The total volume of circulating blood influences preload. *Afterload* refers to the resistance to ejection of blood by the left ventricle.

• • •

VASCULAR CONTROL

Arterial blood pressure is influenced by vascular effects as well as cardiac output. Table 21.4 summarizes some aspects of structure of the blood vessels and lists them in the order of flow away from the heart. The larger arteries are thick walled and quite elastic as compared to veins which are thin walled and distensible.

TABLE 21.3 Summary of cardiac control of circulating blood volume

Heart Rate

a. Autonomic nervous control; increased by sympathetic and decreased by parasympathetic activity
b. Hormonal effects; increased by norepinephrine and thyroxine
c. Blood pressure monitored by baroreceptors in aortic arch and carotid sinuses; reflexive decrease in heart rate in response to increased blood pressure
d. Bainbridge reflex. Reflexive increase in heart rate in response to increased volume
e. Cardiac control center in medulla

Myocardial Contractility

a. Frank-Starling mechanism. When myocardial fibers are stretched, there is an increase in contractility
b. Increased peripheral resistance causes heart muscle to adapt without a continued increase in resting fiber length
c. Treppe (staircase) phenomenon. Increased contractility (within limits) in response to increased heart rate
d. Baroreceptor reflex. Increased pressure causes progressive decrease in contractility
e. Hypoxia. Moderate degree stimulates contractility

TABLE 21.4 Differences in structure and function of blood vessels

Arteries. Thick-walled, large arteries have many elastic fibers; less sympathetic innervation than arterioles

Arterioles. Smaller lumen, thick smooth muscle layer; nerve fibers cause constriction; control resistance to blood flow; sympathetic innervation

Metarterioles. Join capillaries, separated by a precapillary sphincter, which opens and closes

Capillaries. Walls are a single layer of endothelial cells; nutrient and gas exchange

Venules. Thin-walled, distensible

Veins. Larger lumen and thinner walls as compared to arteries; sympathetic innervation, especially in smaller veins

TABLE 21.5 Factors in vascular smooth muscle tone and blood flow

Vascular baroreceptors. In carotid sinuses and aortic arch; transmit nerve impulses to vasomotor center; increased pressure inhibits vasoconstriction; result is peripheral vasodilation

Baroreceptors in chambers of heart and pulmonary vessels. Increased pressure inhibits vasoconstriction

Chemoreceptors. In carotid and aortic bodies; primarily influence respiration, minor influence on blood vessels; decreased O_2 causes vasoconstriction

CO_2 and pH. Direct effects on vasomotor center; increased CO_2, increased H^+ cause vasoconstriction

Intrinsic mechanisms. Certain tissues maintain a constant blood flow in spite of changes in pressure

Hormonal effects. Norepinephrine causes vasoconstriction, epinephrine causes both dilation and constriction in different vessels

TABLE 21.6 Distribution of blood in the circulatory system

Systemic circulation. 80–90%
    Veins. 75% of systemic total
    Arteries. 20% of systemic total
    Capillaries. 5% of systemic total
Pulmonary circulation. 10–20%

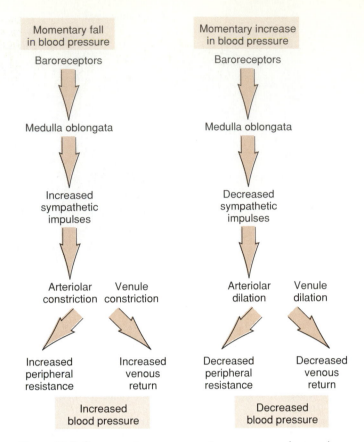

Figure 21.5 Compensatory responses to momentary changes in blood pressure.

The contractile state or tone of vascular smooth muscle is maintained by neural transmissions from the vasomotor center in the medulla. Table 21.5 lists factors that promote vasoconstriction and vasodilation. Sympathetic fibers supply arterioles and smaller veins and, to a lesser extent, supply the larger vessels as well. The effect of arteriolar constriction is to increase resistance to blood flow, and consequently these arterioles are called resistance vessels. The venous side of the circuit is thin walled and distensible and functions as a reservoir for blood and is called the **capacitance system** (table 21.6). The spleen, liver, and large abdominal veins in particular act as blood reservoirs. The distensibility of these vessels allows storage of blood with only small changes in pressure. The effect of vasoconstriction on the venous side of the circuit is an increased venous return to the right atrium. Venous return is determined by vascular tone, and on the arteriolar side, peripheral resistance is maintained by vascular tone.

Baroreceptors are specialized nerve endings that are stimulated by changes in blood pressure (table 21.5). Nerve impulses are transmitted to the vasomotor center in the medulla in response to an increase in blood pressure. This, in turn, causes inhibition of nerve impulse transmission from the vasomotor center to blood vessels. The net effect is vasodilation, with a concomitant decrease in blood pressure due to decreased return of blood to the heart from the venous side. Conversely, low blood pressure

capacitance: L. *capere,* to contain

leads to a decrease in the inhibitory effect of baroreceptors on the vasomotor center, with vasoconstriction and a reflexive increase in blood pressure (figure 21.5).

VASCULAR RESISTANCE AND BLOOD FLOW

There is resistance to the flow of blood through blood vessels because the outer fluid layer in contact with the vessel wall is attracted to and held to the wall. An immediate inner layer forms and slides, against friction, over the stationary layer. Succeeding concentric layers separate and flow over each other. The friction between layers diminishes toward the center, so there is a maximum rate of flow at the center. This is called **laminar flow** and is significant because the size of the vessel determines the degree of resistance. There is increased resistance in a small vessel because the proportion of outer, high-resistance flow is greater as compared to the inner, low-resistance flow.

The relationship between the magnitude of the resistance and the diameter of the blood vessel is expressed by Poiseuille's law (pwah-zuh'yes). This law defines the factors that determine the pressure difference between two ends of a tube. The law states that the pressure difference ($P_1 - P_2$) is directly proportional (1) to the volume per unit time, i.e., flow rate (Q), (2) to the viscosity (v), (3) to the length of the tube (L), and is inversely related to the fourth power of the radius (r).

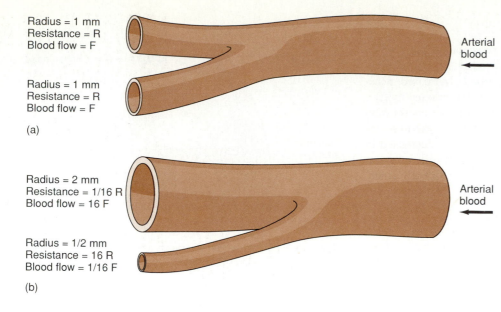

Radius = 1 mm
Resistance = R
Blood flow = F

Radius = 1 mm
Resistance = R
Blood flow = F

(a)

Radius = 2 mm
Resistance = 1/16 R
Blood flow = 16 F

Radius = 1/2 mm
Resistance = 16 R
Blood flow = 1/16 F

(b)

Arterial blood

Arterial blood

Figure 21.6 Relationships between blood flow, vessel radius, and resistance. (*a*) The resistance and blood flow is equally divided between two branches of a vessel. (*b*) A doubling of the radius of one branch and halving of the radius of the other produces a 16-fold increase in blood flow in the former and a 16-fold decrease of blood flow in the latter. The total vascular resistance in (*a*) is equal to that of (*b*).

$$P_1 - P_2 = \frac{Q8Lv}{\pi r^4}$$

The significance of the radius of the tube is easier to visualize if you assume the length and viscosity values are constant, as is the value of π. This equation shows that pressure is greatly influenced by the radius. If the radius is decreased by one-half, then the pressure is increased 16 times to maintain the same flow. It also shows that the rate of flow varies inversely to the 4th power of the radius at a constant pressure. If the radius of a vessel decreases by a factor of 2, then the flow rate is 16 times greater.

The quantity $8Lv \div \pi r^4$ represents resistance and includes those factors that retard flow, i.e., length of tube, viscosity, and radius of vessel. Physiologically, the length and viscosity are usually constant, so resistance is inversely proportional to the 4th power of the radius; thus, a small decrease in radius (constriction) causes a large increase in resistance (figure 21.6).

To summarize, the rate of blood flow is determined by the length of the vessel, viscosity of the blood, the size of the vessel, and the pressure gradient. The mean pressure in the aorta is about 93 mm Hg and the blood flows in the direction of lower arteriolar pressure, which has a mean value of about 30 mm Hg. The significant aspect of this is that the muscular arterioles, by constriction and dilation, maintain a pressure gradient, determine peripheral resistance, and control blood flow in the body.

OVERVIEW OF BLOOD PRESSURE

Blood pressure is the product of cardiac output and the resistance to blood flow. Multiple factors contributing to both resistance and output have been discussed previously and are summarized in table 21.7.

TABLE 21.7 Summary of factors determining blood pressure

Blood Pressure = Cardiac Output × Resistance

Cardiac output is influenced by preload, cardiac contractility, afterload, heart rate
Resistance varies inversely to the 4th power of the radius of the vessel; largely dependent on degree of vasoconstriction

PRELOAD

Preload is defined as the volume of blood that fills the ventricles during diastole. Hemorrhage causes a decrease in preload and a fall in blood pressure. One compensatory response is sympathetic constriction of veins to restore venous return of blood to the heart. Preload is also influenced by the condition of the ventricular walls, either their compliance or stiffness. The ventricles hypertrophy in response to increased peripheral resistance, as occurs in hypertension. This involves an increase in the size and number of myocardial fibers, resulting in an increase in muscle mass and an increase in myocardial stiffness. Ventricular hypertrophy is discussed further in chapters 23 and 24. Preload is closely related to end-diastolic pressure, and this is likely to be higher in a less compliant ventricle.

CARDIAC CONTRACTILITY

The Frank-Starling law of the heart identifies a direct relationship between the stretching of myocardial fibers due to increased volume and a subsequent increased contractility. Within limits, the heart is capable of accommodating the volume of blood returned to it.

AFTERLOAD

Afterload is defined as peripheral resistance against which the left ventricle must pump. Elevated arterial blood pressure represents an increased afterload. Increased resistance to left ventricular ejection of blood is created by (1) a condition in which the aorta is narrowed, or coarctation (ko″ark-ta′shun) of the aorta and (2) aortic valvular stenosis in which the opening into the aorta is decreased in size. Cardiac output may be maintained in spite of an increase in afterload, but the ventricles must work harder.

HEART RATE

Factors that influence heart rate are summarized in table 21.3. Cardiac output is the product of heart rate and the stroke volume (volume ejected during one contraction).

RESISTANCE

Peripheral resistance is determined largely by the degree of vasoconstriction, particularly of the arterioles. The effect of increased resistance is to raise arterial blood pressure.

LONG-TERM CONTROL

The interrelationships of arterial pressure, cardiac output, and peripheral resistance have been discussed in preceding paragraphs ($P = C.O. \times R$). This equation indicates that an increase in cardiac output and/or peripheral resistance causes an elevation of arterial pressure. Baroreceptors and chemoreceptors are sensitive to changes in pressure and blood chemistry and elicit immediate changes in heart rate and vasomotor tone to maintain arterial pressure. The kidney, however, is responsible for long-term control of intravascular volume and arterial pressure. Increased pressure within the upper chambers of the heart inhibits the release of ADH and also inhibits sympathetic renal impulses to promote an increased glomerular filtration rate. Both events promote fluid and sodium loss with a decrease in arterial pressure (chapter 25). The role of the kidney in fluid balance is discussed in chapter 2.

PULSE PRESSURE

The difference between systolic and diastolic pressure is called pulse pressure. The main factors determining that difference are stroke volume and the elasticity or compliance of the arterial system. During ventricular ejection, the aorta stretches to accommodate the sudden increase in volume, which, at that moment, is greater than the peripheral runoff and causes an increase in pressure. In general, when there is an increase in stroke volume, there is a greater increase in systolic pressure and a decrease in diastolic pressure with a resultant wider pulse pressure.

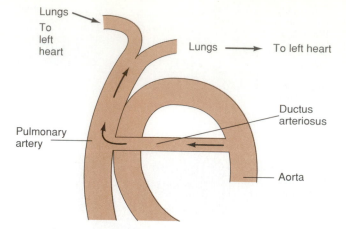

Figure 21.7 Patent ductus arteriosus allows backflow of blood to the lungs. This causes an increased venous return to the left heart resulting in an increased stroke volume.

An example of a condition that causes a widened pulse pressure is patent (pa′tent) ductus arteriosus (chapter 24). The ductus arteriosus is a vessel that connects the pulmonary artery and the aorta during fetal development. Patent ductus arteriosus is a defect in which the passage remains open after birth. High pressure in the aorta causes backflow of blood to the pulmonary artery and to the lungs, which causes an increase in venous return to the left heart (figure 21.7). The overall effect is an increased stroke volume and then an abnormally rapid runoff (by way of the ductus arteriosus). The systolic pressure is elevated, the diastolic pressure decreased, and there is a wider pulse pressure.

An increase in heart rate affects the pulse pressure. As heart rate increases, the stroke volume is diminished and the cardiac output remains unchanged. The decreased stroke volume causes a lower systolic pressure and a narrowed pulse pressure.

The compliance of the arterial system is the second factor that determines pulse pressure. Arteriosclerosis (chapter 22) is a process in which the elastic fibers in arterial walls are replaced by fibrous connective tissue, and consequently, there is a loss of elasticity. This happens both as a pathological condition and as a progressive occurrence in the aging process. There are two consequences of decreased arterial compliance: (1) increased systolic pressure due to stiffness of the aorta and (2) rapid runoff of blood during diastole and a lowered diastolic pressure; hence, there is a widened pulse pressure.

HYPERTENSION

Hypertension refers to an elevated arterial pressure and is categorized as being (1) **essential** or **primary hypertension** with the cause unknown or (2) **secondary hypertension** in which there is an identifiable underlying cause.

patent: L. *patere,* to be open

Table 21.8 Possible causes of secondary hypertension

Oral contraceptives	Renin-secreting tumors
Pregnancy	Aldosteronism
Renal vascular disease	Pheochromocytoma
Chronic renal failure	Cushing's syndrome

Table 21.9 Possible factors operating in primary hypertension

1. Activation of sympathetic nervous system causing constriction of venules
2. Increased circulating norepinephrine
3. Enhanced sympathetic stimulation resulting in increased heart rate, increased contractility; thus, an increase in cardiac output
4. Renin-angiotensin-aldosterone system promote Na^+ and water retention
5. Kallikrein-kinin system promotes local vasodilation and Na^+ excretion; opposes increased blood pressure
6. Prostaglandin E causes local vasodilation and natriuresis; opposes increased blood pressure
7. Abnormal handling of sodium by kidney

More than 95% of the cases of hypertension fall into the former category. Blood pressure values vary with age and, in addition, there are individual variations related to sympathetic activity, state of hydration, and other factors. An adult reading greater than 160/95 mm Hg is considered abnormally high and pressures less than 140/90 are classified as normotensive. Hypertension may be caused by factors that lead to an increase in peripheral resistance or to an increase in cardiac output.

SECONDARY HYPERTENSION

Table 21.8 lists some possible causes of secondary hypertension. Hypertension caused by oral contraceptive use is a common form of secondary hypertension, and the mechanisms by which it occurs are unclear. Pregnancy is associated with hypertension in about 6% of the cases, and the underlying pathophysiology is complex and controversial.

Renovascular hypertension is caused by any lesion that obstructs renal arteries. Atherosclerosis accounts for about two-thirds of the cases of renovascular hypertension. The obstruction leads to increased renin production with subsequent increases in angiotensin and aldosterone that promote sodium/water retention (chapter 2).

Hypertension is frequently associated with chronic renal failure (chapter 15). In most cases, the hypertension parallels sodium and water retention. In some cases there is increased renin production.

Certain types of tumors including juxtaglomerular cell tumors and certain extrarenal neoplasms secrete renin. This leads to an overproduction of angiotensin and aldosterone.

Increased production of aldosterone causes excessive salt and water retention, increased blood volume, and hypertension. Aldosterone-producing adenoma of the adrenal gland is one cause and other causes of aldosteronism are discussed in chapter 32. Pheochromocytoma is a type of tumor that secretes norepinephrine and epinephrine, causing vasoconstriction and hypertensive crises. Cushing's syndrome is caused by excessive levels of glucocorticoids accompanied by salt and water retention. Both pheochromocytoma and Cushing's syndrome are discussed further in chapter 32.

PRIMARY HYPERTENSION

Increased peripheral resistance causes elevation of both systolic and diastolic pressures and an increase in cardiac output has the same effect. The pathogenesis of primary hypertension is not clear, but there are a number of possible contributing factors (table 21.9).

1. In some cases, there may be augmented constriction of venules, which causes an increase in venous return with an increased cardiac output.
2. There are studies that implicate an increase in circulating norepinephrine, which causes vasoconstriction.
3. Evidence indicates that in some cases there is decreased parasympathetic inhibition and increased sympathetic stimulation resulting in an enhanced cardiac output. Stress exacerbates hypertension, and this is indirect evidence of nervous system involvement.
4. When there is sustained elevation of renin levels, the consequent increase in angiotensin and aldosterone promote sodium/fluid retention, expanded blood volume, and elevated arterial pressure.
5. Kallikrein is associated with the membrane of distal tubule cells in the kidney. This is an enzyme that catalyzes the formation of kinins, which promote the excretion of sodium and also cause renal vasodilation. It is possible that the physiological significance of this is that the kinins maintain renal perfusion in spite of the vasoconstricting effects of angiotensin.
6. Prostaglandins are found in almost every tissue including the kidney and blood vessel walls. They are involved in control of blood pressure as well as many other functions. Prostaglandin E exerts antihypertensive effects by causing natriuresis and maintaining renal blood flow.
7. A key aspect in the pathogenesis of hypertension may be a defect in renal excretion of salt and water. It has been shown that in some patients with hypertension, there are alterations in renal blood flow and tubular

function. Normally, the kidney responds to increased pressure by increasing the rate of excretion of sodium and water. If higher pressures are required to bring about this response, this would promote hypertension.

There are also risk factors that, although the mechanisms are not clear, predispose to hypertension. These factors include obesity, increased dietary sodium, family history, alcohol ingestion, smoking, and diabetes mellitus.

MANIFESTATIONS

Uncomplicated hypertension is usually asymptomatic although headache, nocturia, and postural unsteadiness may be noted. Untreated hypertension carries with it a high risk for cardiovascular disease and the complications listed in table 21.10. The incidence of coronary heart disease is clearly increased by hypertension and is manifested by myocardial infarction and sudden death (chapter 23). Hypertension increases afterload, i.e., increases the resistance against which the left ventricle must pump and leads to congestive heart failure (also discussed in chapter 23). Hypertension is the most common cause of renal insufficiency and leads to nephrosclerosis. Nephrosclerosis is hardening of the kidneys and is caused by lesions and necrosis of the renal arteries (chapter 14). There is a gradual reduction in kidney size caused by atrophy and fibrosis, with kidney failure as the final outcome. Finally, hypertension is a leading risk factor for stroke or cerebrovascular accident (CVA). Hypertension causes two types of strokes, those caused by (1) ischemia and death of brain tissue in which blood clots may play a role or (2) hemorrhage into or around the brain (chapter 38).

MANAGEMENT

In general, nondrug therapy is initiated in mild hypertension if the diastolic pressure is not consistently elevated above 94 mm Hg. The measures aimed at blood pressure control include weight loss, moderate dietary sodium restriction, moderate alcohol consumption, and exercise. If the diastolic level is not lowered below 90 mm Hg after several months, drug therapy may be initiated.

TABLE 21.10 A summary of complications of hypertension

1. Acceleration of atherosclerotic process
2. Rupture of blood vessels in retina of the eye
3. Coronary heart disease
4. Hypertrophy of left ventricle and congestive heart failure
5. Nephrosclerosis and renal insufficiency
6. Stroke

CIRCULATORY FAILURE

The preceding section deals with hypertension, whereas the following discussion deals with the other extreme, i.e., low blood pressure or hypotension. Circulatory failure, which is also called shock, refers to an alteration in blood flow resulting in a decrease in blood pressure associated with inadequate delivery of oxygen to tissues. The ultimate result is cell damage and death.

In general, circulatory failure is caused by (1) volume loss, (2) a decrease in cardiac contractility or abnormal rate/rhythm of the heart, (3) increased arterial vasoconstriction that reduces capillary blood flow, (4) impaired capillary/tissue exchange, (5) shunting of blood from arterioles directly to venules, (6) dilation of veins that function as capacitance vessels, and (7) obstruction of blood flow.

COMPENSATORY RESPONSES

Compensatory responses to inadequate perfusion or blood flow include increase in heart rate and peripheral vasoconstriction stimulated by sympathetic nerve impulses and hormonal effects. As the cardiac output decreases, there is redistribution of blood flow. Blood supply to both the heart and brain is protected. Vasodilation of coronary and cerebral vessels occurs in response to rising pCO_2 blood levels and to a decline in pO_2. There is vasoconstriction of vessels supplying the skin, skeletal muscle, viscera, and kidneys. Ischemic injury of vital organs may be the result of prolonged vasoconstriction. Reduced blood flow to the kidney stimulates secretion of renin and, subsequently, angiotensin and aldosterone (chapter 2). Angiotensin is a powerful vasoconstrictor, and aldosterone promotes renal salt and water retention. Increased antidiuretic hormone secretion promotes water retention as well (chapter 31).

• • •

Symptoms of cerebral ischemia, i.e., confusion or disorientation, occur when cerebral blood flow is reduced to less than half. This is likely to be associated with a mean arterial pressure of less than 50 mm Hg.

• • •

SECONDARY EFFECTS

Subsequent events contribute to deterioration in the condition of a patient suffering from shock. Inadequate delivery of oxygen to tissues causes a shift to anaerobic metabolism. Lactic acid is the end product of anaerobic metabolism and leads to metabolic acidosis (chapter 4), which reduces myocardial contractility. Blood lactate levels are an indication of the severity of perfusion deficit.

perfusion: L. *perfusio,* pour over

Increased sympathetic nerve activity causes constriction of postcapillary venules. The result is increased capillary hydrostatic pressure and loss of intravascular fluid to interstitial spaces (chapter 2). This contributes to loss of intravascular volume, regardless of the cause of shock.

MANIFESTATIONS

Characteristic manifestations of shock are hypotension, hyperventilation, and skin that is cold, pale, and clammy. There may be cyanosis, a bluish cast, of the distal extremities. A state of mental confusion is evidence of diminished cerebral perfusion, and decreased renal perfusion leads to oliguria. Prolonged reduction in tissue perfusion causes generalized cellular damage. Acute tubular necrosis may occur (chapter 14), and there may be necrosis of the mucosa of the gastrointestinal tract.

MANAGEMENT

The general principles of management of circulatory failure are to (1) provide for adequate ventilation with oxygen therapy, (2) restore volume with the infusion of fluids, and (3) intervene to improve cardiac performance. These principles are discussed further in the following section that deals with specific causes of shock.

CATEGORIES OF SHOCK

There are four major categories of shock based on hemodynamic effects: (1) hypovolemic shock due to loss of volume, (2) distributive shock caused by alterations in the distribution of blood flow, (3) cardiogenic shock due to decreased contractility or abnormal rate/rhythm of the heart, and (4) obstructive shock caused by obstruction to blood flow. The causes of obstructive shock include obstruction to blood flow in the veins, dissecting aneurysm (chapter 22), and pulmonary embolism (chapter 8). The following discussion focuses on hypovolemic, distributive, and cardiogenic shock.

• • •

Korotkoff sounds are decreased in shock when stroke volume is reduced and arterial resistance is increased. In general, the cuff pressure reading is less than the actual arterial pressure under conditions of decreased blood flow and increased arterial resistance.

• • •

HYPOVOLEMIC SHOCK

Loss of intravascular volume with acute reduction in blood flow is the most common cause of shock. Hemorrhage due to trauma or internal bleeding associated with a disease process may lead to circulatory failure. Excessive fluid losses may occur as the result of vomiting and diarrhea. Diuresis with excessive urinary loss is caused by diabetes mellitus and diabetes insipidus (chapter 2). Increased capillary permeability due to infection or acute allergic reactions causes loss of intravascular fluid to the interstitial compartment (chapter 2) and may lead to circulatory failure.

Compensatory Responses

Plasma water loss has a concentrating effect on the protein fraction and cellular elements of blood. Increased concentration of protein causes an increase in oncotic pressure, which draws extravascular fluid into capillaries (chapter 2). The result is a compensatory restoration of volume. There is an associated effect on hematocrit values, reflecting the concentration of red blood cells. There is an initial increase and a subsequent decrease in the hematocrit.

• • •

There is a compensatory movement of fluid into capillaries in response to increased oncotic pressure following plasma water loss. This compensatory process requires a period of hours and is of little consequence in the case of rapid blood loss.

• • •

The sympathetic autonomic nervous system stimulates arterial vasoconstriction in response to volume loss and reduction in arterial blood pressure. The result is a compensatory increase in peripheral vascular resistance, which maintains mean arterial pressure.

Manifestations

Depending on the extent of the volume deficit, compensatory mechanisms may maintain adequate circulation. Vasoconstriction and increased heart rate may maintain blood pressure and blood flow. If there is blood loss of about 25%, both the cardiac output and systolic pressure fall. Increased sympathetic nerve activity leads to tachycardia, hyperventilation, and vasoconstriction associated with pale, cool skin. Decreased renal blood flow results in oliguria. Apprehension or confusion is evidence of a reduction in cerebral blood flow. The situation becomes life-threatening when compensatory mechanisms are inadequate to maintain cardiac output, blood pressure, and tissue perfusion.

Management

Therapy is directed at restoring the fluid loss and normal cardiac output and correcting metabolic acidosis. Mechanical ventilation may be required to provide adequate oxygenation.

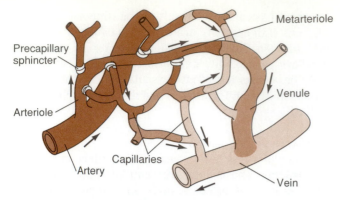

Figure 21.8 Arteriovenous shunting is controlled by metarterioles that provide direct paths between arterioles and venules. Precapillary sphincter muscles regulate the flow of blood through the capillaries.

Distributive Shock

Distributive shock occurs when there is abnormal distribution of blood flow in the absence of alterations in volume, cardiac function, or obstruction to blood flow. The basic dysfunction is altered vascular reactivity. There may be dilation of the veins that constitute the capacitance system, followed by a decrease in venous return to the heart and a decrease in cardiac output. A second alteration is vasodilation of arterioles resulting in lowered peripheral resistance. In the latter case, blood flow may essentially bypass the capillary beds where normally there is blood/tissue exchange of gases and nutrients. This is called **arteriovenous shunting** and is controlled by **metarterioles.** These vessels are an extension of arterioles that form direct channels to the venous side of the system. Capillaries branch from the proximal ends of metarterioles with blood flow into capillaries regulated by contraction or relaxation of bands of smooth muscle, i.e., precapillary sphincters (figure 21.8). Blood flowing directly through the metarterioles does not undergo any oxygen exchange.

In summary, there are two types of alterations associated with distributive shock:

1. There may be normal arterial resistance with vasodilation of venous capacitance system. This is associated with a decreased cardiac output.
2. There may be vasodilation of arterioles with lowered peripheral resistance and arteriovenous shunting. In this case, the cardiac output may be normal or increased, but the blood pressure is low. In both cases, there is inadequate capillary perfusion or blood flow (chapter 25).

The causes of distributive shock include bloodstream infections or septicemia, transection of the spinal cord, and such drug induced alterations as overdose of narcotics, barbiturates, or tranquilizers.

metarterioles: Gk. *meta*, between
septicemia: Gk. *septikos*, putrefactive; Gk. *haima*, blood

The most common cause of distributive shock is septicemia caused by gram-negative bacilli. The mechanisms and clinical manifestations of shock depend on the organism and toxins produced. Gram-positive organisms may be the cause, although gram-negative organisms, such as *Escherichia coli, Proteus, Pseudomonas,* and *Serratia,* are more common causative agents.

• • •

More than one-fourth of the patients who have septicemia caused by gram-negative bacilli develop shock. In many cases, the infections occur during a hospital stay and are sequelae of surgery or invasive procedures.

• • •

Multiple events are involved in the disease process including the following:

1. Bacterial cell wall products activate complement, i.e., a group of serum proteins involved in immune reactions (chapter 27). The result is activation of neutrophils that adhere to each other and to endothelial cells. There is endothelial cell damage and capillary leakiness.
2. Bacterial products activate coagulation with the generation of thrombin and the aggregation of platelets (chapter 19).
3. Bradykinin, a powerful vasodilator, is produced.
4. Endorphins, opiatelike substances, are produced in response to stress. Endorphins relieve pain during stress and, with elevated levels in the blood, cause vasodilation and increased capillary permeability.

Secondary Effects

Complement and white blood cell activation lead to damage of pulmonary capillaries, resulting in pulmonary edema. Adult respiratory distress syndrome (ARDS) may develop with hemorrhage, atelectasis, and hyaline membrane formation (chapter 7). Respiratory failure is a major cause of death and may occur after other alterations are corrected.

Clotting abnormalities occur due to the generation of thrombin and aggregation of platelets. The result is consumption of coagulation factors that may lead to disseminated intravascular coagulation (chapter 19). There is usually no overt bleeding. Microthrombi develop in capillaries and usually disappear as shock is treated.

A decrease in intravascular volume associated with inadequate renal perfusion causes oliguria. Heart failure may occur, possibly due to a substance produced in response to ischemia. Hemorrhage associated with ulcerations of the gastrointestinal tract occurs, and there may be evidence of liver damage.

Manifestations

The manifestations of septicemia-associated shock in the early stages are attributed to arteriovenous shunting, i.e., bypass of capillary exchange vessels. In spite of a normal volume of circulating blood, there is inadequate capillary perfusion. There is vasodilation with an increase in cardiac output and a decrease in arterial resistance. These individuals have warm, dry skin, and both hypotension and oliguria are present. As this stage progresses, the complement-mediated agglutination of neutrophils and capillary damage lead to increased permeability and fluid loss. This represents a second stage in which there is a drop in blood pressure and a decrease in cardiac output in response to intravascular volume loss. There is an increase in peripheral resistance due to vasoconstriction as a compensatory response. Hypotension occurs in spite of increased arterial resistance. There may be a sudden change in mental status with confusion and disorientation, likely due to a decrease in cerebral blood flow. The skin becomes pale, cool, and moist, and there may be peripheral cyanosis. Inadequate perfusion of the kidney and visceral organs lead to oliguria, vomiting, and diarrhea.

Management

The treatment of shock requires restoration of volume by whole blood, plasma, or an appropriate electrolyte solution. Bicarbonate may be added to the solution to treat acidosis. A diuretic, such as furosemide, is a safeguard against pulmonary edema (chapter 2). The administration of oxygen corrects a depressed pO_2. Surgical intervention may be required to remove severe localized areas of infection. The early use of glucocorticoids for brief periods may combat inflammation.

CARDIOGENIC SHOCK

Cardiogenic shock is circulatory failure due to inadequate pumping action of the heart. This may be the result of decreased contractility or abnormal rate and/or rhythm. Myocardial infarction, which is the most common cause of cardiogenic shock, is necrosis of heart muscle due to interruption of blood supply to the myocardium (chapter 23). This sets up a sequence of events that leads to continued deterioration of heart function and a progressive decline in blood pressure in the following manner:

1. Impaired heart function causes a decrease in cardiac output and a concomitant fall in blood pressure.
2. This decrease in cardiac output results in inadequate flow of blood to the heart muscle.
3. Decreased delivery of oxygen causes additional impairment of myocardial function.
4. A decrease in systemic blood flow causes a shift to anaerobic metabolism with lactic acid production and metabolic acidosis. Metabolic acidosis contributes to decreased contractility of the myocardium.

Manifestations

Cardiogenic shock is characterized by marked hypotension, i,e, systolic pressure less than 90 mm Hg, cold clammy skin associated with cyanosis, reduced mental awareness, and oliguria. Further, there is no improvement following the administration of oxygen and pain medication.

Management

Treatment includes the administration of oxygen and adequate control of pain, which reduces oxygen consumption. There is continuous monitoring of both the arterial pressure and the pulmonary capillary wedge pressure as an indication of left ventricular diastolic pressure. Blood volume is controlled, either by infusion of fluids or diuresis, in order to maintain left ventricular pressure at 20 mm Hg. **Vasopressors,** such as dopamine, cause vasoconstriction, raise blood pressure, and are useful in maintaining blood flow to the myocardium.

SUMMARY

• • •

There are four types of blood pressure defined in this chapter. (1) Systolic pressure is arterial pressure generated during ventricular contraction, and (2) diastolic pressure is a lower pressure maintained during ventricular relaxation. (3) The mean arterial pressure represents an average of pressure at every point in the cardiac cycle. (4) Pulmonary wedge pressure is measured in a branch of the pulmonary artery obstructed by an inflated balloon. This reading indicates left atrial mean pressure and left ventricular diastolic pressure. Arterial blood pressure is the product of cardiac output and the resistance to blood flow. The kidney provides long-term blood pressure control by means of water and sodium excretion.

Hypertension refers to elevated arterial pressure and is caused by increased cardiac output or peripheral resistance. Cardiovascular disease and kidney damage are complications of hypertension. Mild hypertension may be controlled by weight loss, sodium restriction, and exercise.

Shock is circulatory failure associated with reduced tissue perfusion and, ultimately, generalized cell damage and death. Shock may be caused by impaired cardiac function, increased arterial constriction, inadequate capillary/tissue exchange, arteriovenous shunting, venous dilation, or obstruction to blood flow. Compensatory responses to shock include increased heart rate and generalized vasoconstriction with preferential flow of blood to heart and brain. Inadequate delivery of oxygen to tissues causes a shift to anaerobic metabolism with lactic acid production and metabolic acidosis. Characteristic manifestations of shock are hypotension, tachycardia, oliguria,

myocardium: Gk. *mys,* muscle; Gk. *kardia,* heart

and skin that is pale, cool, and moist. Therapy is aimed at correcting hypoxemia, restoring volume, and improving cardiac performance.

Hypovolemic shock is caused by overt fluid loss or internal intravascular loss. Vasoconstriction and increased heart rate maintains blood pressure if volume deficit is limited. Therapy is directed at restoration of fluid loss and normal cardiac output. The basic disorder in distributive shock is alteration in blood flow. There may be pooling of blood in the venous capacitance system or reduced arterial resistance with arteriovenous shunting. Septicemia, transection of the spinal cord, and drug induced alterations are possible causes of distributive shock. Septicemia leads to capillary wall damage, generation of thrombin, aggregation of platelets, and vasodilation. Complications of septicemia include adult respiratory distress syndrome and disseminated intravascular coagulopathy. In early stages of septicemia-associated shock, the alteration is arteriovenous shunting associated with a normal volume of circulating blood. The skin is warm and dry at this point. Subsequently, capillary damage leads to fluid loss and the skin becomes pale, cool, and moist. The treatment of septicemia-associated shock includes administering oxygen, restoring volume, treating infection, and possibly using glucocorticoids to combat inflammation early in the infection.

Cardiogenic shock is caused by inadequate pumping of the heart, frequently the result of myocardial infarction. The events tend to be self-perpetuating because decreased myocardial contractility leads to inadequate blood flow and impaired coronary perfusion, which adversely affects contractility of the heart. Treatment involves control of pain and measures to maintain adequate blood flow to the heart.

REVIEW QUESTIONS

• • •

1. What is the lower blood pressure reading called?
2. Which blood pressure reading (systolic or diastolic) represents the pressure maintained during ventricular relaxation?
3. What is the significance of an elevated pulmonary capillary wedge pressure?
4. What is the consequence of increased pressure in the left ventricle?
5. An increased peripheral resistance increases (preload or afterload).
6. How does the heart respond to increased venous return?
7. What controls peripheral resistance?
8. What is the relationship between blood flow and blood pressure?
9. How do veins play a role in determining venous return to the heart?
10. Identify two factors that affect preload.
11. What provides long-term control of blood pressure?
12. What is pulse pressure?
13. What effect does an increased stroke volume have on pulse pressure?
14. How does arteriosclerosis affect blood pressure?
15. What are two basic causes of hypertension?
16. What is the relationship between renin production and blood pressure?
17. What are two major complications of hypertension?
18. List three possible contributing factors to primary hypertension.
19. List three measures that contribute to blood pressure control but do not involve antihypertensive drugs.
20. Define the term shock.
21. Identify two mechanisms whereby shock develops in the presence of normal volume and cardiac function.
22. What are two effects of compensatory sympathetic stimulation due to circulatory failure?
23. Why does metabolic acidosis occur in association with shock?
24. What causes an egress of intravascular fluid to the interstitium in any type of shock?
25. List three characteristic manifestations of shock.
26. What are two possible causes of distributive shock?
27. Why does the hematocrit rise and subsequently fall in hypovolemic shock?
28. What is the most common cause of cardiogenic shock?
29. Why is decreased myocardial contractility self-perpetuating in cardiogenic shock?
30. What laboratory test is a good indicator of tissue perfusion?

SELECTED READING

• • •

Bedoya, L. A., and R. W. Gifford, Jr. 1990. Therapy of mild hypertension. *Physician Assistant* 14:17–42.

Buhler, F. et al. 1988. Platelet membrane and calcium control abnormalities in essential hypertension. *American Journal of Hypertension* 1:42–46.

Holland, O. B. et al. 1988. Metabolic changes with antihypertensive therapy of the salt-sensitive patient. *American Journal of Cardiology* 61:53H–59H.

Hollenberg, N. K. et al. 1988. Angiotensin and the renal circulation in hypertension. *Circulation* 77:159–63.

Kaplan, N. M. 1988. Maximally reducing cardiovascular risk in the treatment of hypertension. *Annals of Internal Medicine* 109:36–40.

Khaw, K., and E. Barrett-Connor. 1987. Dietary potassium and stroke-associated mortality. *New England Journal of Medicine* 316:235–40.

Krieger, D. et al. 1988. Mechanisms in obesity-related hypertension: Role of insulin and catecholamines. *American Journal of Hypertension* 1:84–90.

Lardinois, C. et al. 1988. The effects of antihypertensive agents on serum lipids and lipoproteins. *Archives of Internal Medicine* 148:1280–88.

Lindner, A., M. Kenny, and A. Meacham. 1987. Effects of a circulating factor in patients with essential hypertension on intracellular free calcium in normal platelets. *New England Journal of Medicine* 316:509–13.

Moser, M. 1990. Controversies in management of hypertension: Diuretic use in the 1990s. *Physician Assistant* 14:81–95.

Sites, J. 1990. Treating shock: The clinical and biochemical basis. *Physician Assistant* 12:116–30.

Susman, J. 1988. Orthostatic hypotension. *American Family Physician* 37:115–18.

Weir, M. R. et al. 1988. Physiologic and hemodynamic considerations in blood pressure control while maintaining organ perfusion. *American Journal of Cardiology* 61:60H–66H.

Zemel, M. et al. 1988. Salt sensitivity and systemic hypertension in the elderly. *American Journal of Cardiology* 61:7H–12H.

Chapter 22

Vascular Disorders

The cardiovascular system, consisting of the heart and associated vessels, distributes blood to all body tissues. Blood, in terms of composition, function, and related disorders, was the focus of the preceding unit. The chapters in this unit deal with the structure and function of both the heart and vascular system with an emphasis on disorders related to both. The structure of the blood vessel wall is reviewed followed by a discussion of the following vascular problems: (1) **Raynaud's disease** in which there is vasoconstriction of digital arteries, (2) degeneration of arterial walls caused by **atherosclerosis,** and (3) **dissecting aneurysm,** which involves the splitting of the arterial wall.

BLOOD VESSEL STRUCTURE

In general, three layers constitute blood vessel walls: (1) **tunica intima,** (2) **tunica media,** and (3) **tunica externa** (table 22.1). There are variations in this basic structure among the sizes and types of vessels. Arteries, which carry blood away from the heart, have numerous elastic fibers, and these fibers stretch to accommodate the volume of blood ejected by each cardiac contraction. During diastole or cardiac relaxation, the elastic property of arteries causes the vessels to recoil to maintain blood flow and blood pressure. As table 22.2 indicates, the tunica media of smaller arteries is made up of many smooth muscle cells that contract and control blood flow. Venules and veins have fewer elastic fibers and generally have thinner walls than arteries (figure 22.1). Vasoconstriction, caused by the contraction of vascular smooth muscle, leads to a decrease in the size of the lumen of the vessel and impairs blood flow.

RAYNAUD'S PHENOMENON AND DISEASE

Raynaud's (Rā-nōz′) phenomenon involves vasoconstriction of arteries, arterioles, and perhaps venules of digits, particularly the fingers. The constriction may have various causes including (1) occlusive arterial disease, (2) neurogenic disorders, (3) injury or surgery, (4) drugs, or (5) occupation-related activities, such as playing the piano, typing, or operating a pneumatic hammer.

Raynaud's disease involves a transient decrease in blood flow to the digits due to spasms of arteries and possibly veins. Exposure to cold and in some cases, emotional upsets, are initiating factors. It is characterized by a series of color changes in the digits: (1) pallor due to decreased blood flow, (2) cyanosis as the result of dilated venules and capillaries with stagnant blood, and (3) rubor or redness

vascular: L. *vasculum,* small vessel
diastole: Gk. *diastellein,* dilate
neurogenic: Gk. *neuron,* nerve; Gk. *genesthai,* to be formed
rubor: L. *rubor,* red

TABLE 22.1 Basic structure of blood vessel walls

Tunica intima. Inner lining of squamous epithelial cells called endothelium; connective tissues
Tunica media. Middle layer, usually thick and composed of circular smooth muscle and elastic fibers
Tunica externa (adventitia). Outer layer of connective tissue and longitudinal smooth muscle

TABLE 22.2 Differences in structure of types of blood vessels

Large arteries. Three layers; tunica media is thick with many elastic fibers; stretchable
Smaller arteries. Three layers; tunica media mostly smooth muscle; contracts and causes vasoconstriction
Arterioles. Small lumen; thick tunica media composed of smooth muscle; in small arterioles, the tunica media is reduced and external layer absent
Capillaries. Thin walls; single layer of endothelial cells; no tunica media or tunica externa
Venules. Smallest; have a tunica intima and thin tunica externa; larger venules have a thin tunica media
Veins. Larger lumen and thinner walls than arteries; three layers, the tunica media is thin; the tunica externa is thick with little elastic tissue

caused by **hyperemia** or increased blood flow. Typically, there are sensations of numbness, tingling, and burning with a slight swelling. The symptoms are bilateral and the condition may lead to ulcerations on the tips of the fingers, infection, and even gangrene.

The precise mechanisms causing the vasospasms are uncertain, but may involve abnormal sensitivity of blood vessels to low temperatures or may be caused by nervous stimuli. The disease most often affects females, with onset usually before the age of 40.

Treatment of Raynaud's includes (1) protecting the extremities from cold and injury, (2) drugs to prevent vasoconstriction or to promote dilation, and, in extreme cases, (3) surgery to interrupt sympathetic nerve impulses to blood vessels.

ATHEROSCLEROSIS

Atherosclerosis is a cardiovascular disease in which localized areas of the arterial tunica intima become thickened, with injury to the wall and possible occlusion of the vessel. The disease process involves the formation of a lesion made of cells, connective tissue, and fatty substances. The disease is progressive, beginning in childhood, and is the principal cause of death in adults in Western countries. The following section outlines the nature of the lesion and the disease process, risk factors, and management.

hyperemia: Gk. *hyper,* above; Gk. *haima,* blood
sclerosis: Gk. *sklerosis,* hardening

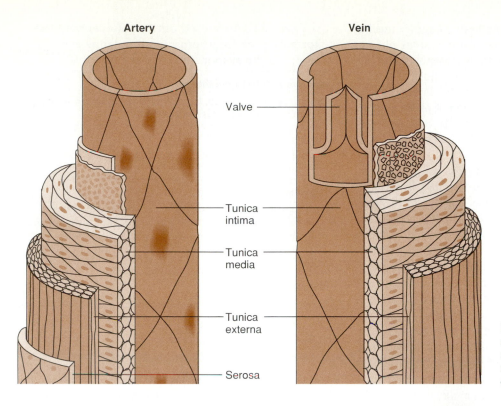

Artery   Vein

Valve

Tunica intima

Tunica media

Tunica externa

Serosa

Figure 22.1 A comparison of the structure of a medium-sized artery and vein showing the relative thickness and composition of the tunicas.

PATHOGENESIS

The nature of the atherosclerotic lesion depends on the stage of formation. The earliest lesion appears as a yellow discoloration in the arterial tunica intima or inner lining and is found in infants and children. At this stage, it is called a **fatty streak** and is made up of lipid laden macrophages and smooth muscle cells (figure 22.2). Evidence suggests that with time a fatty streak at a particular site changes into a more advanced lesion called a **fibrous plaque** or an **atheromatous** (ath″er-o′mah-tus) **plaque.** A fibrous plaque is white and is usually elevated. The elevation consists of cellular debris with, in some cases, cholesterol and calcium crystals covered by an intermingling of (1) smooth muscle cells, (2) macrophages, (3) lymphocytes, and (4) dense connective tissue. The lipid content is variable. Typically, the mound is capped by smooth muscle and connective tissue.

Characteristics of the intimal endothelium provide the following clues about the pathogenesis of fibrous plaque formation.

1. The inner lining of a normal artery consists of a single layer of cells attached to each other by tight junctions. This endothelium provides a smooth surface that functions as a permeability barrier, i.e., controls the movement of molecules into the artery.
2. Endothelial cells grow only in a single layer. This factor limits the response to injury if a damaged area is surrounded by cells incapable of replicating.
3. Endothelial cells have receptors for low-density lipoprotein and for growth factors.

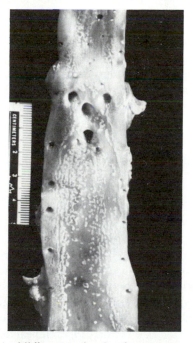

Figure 22.2 A child's aorta showing fatty streaks. *Ritchie, A. C. 1990.* Boyd's Textbook of Pathology. *Lea & Febiger. Reproduced with permission.*

4. The intimal surface may produce substances that both inhibit blood clot formation and promote coagulation.
5. Arterial endothelial cells secrete both growth factors and vasoactive substances that cause vasoconstriction.

vasoactive: L. *vas,* vessel

Vascular Disorders

The **response-to-injury hypothesis** suggests mechanisms that may lead to the formation of fibrous plaques in atherosclerotic disease. This theory assumes that the initiating events are factors that lead to injury of the endothelium. Manifestations of endothelial injury include (1) increased permeability, (2) secretion of procoagulants (chapter 19), (3) increased turnover rate of endothelial cells, (4) secretion of vasoactive substances, and (5) secretion of growth factors.

The series of events leading to fibrous plaque formation, as proposed by the response-to-injury hypothesis, is outlined in table 22.3. In essence, the theory proposes that monocytes migrate between endothelial cells to the subendothelium where they are converted to macrophages (chapter 18). The macrophages accumulate lipids and secrete injurious substances to cause further damage to endothelial cells. These activated macrophages may also secrete growth factors that stimulate the proliferation of smooth muscle cells. If blood flow contributes to endothelial injury, the tight cellular junctions may be disrupted. Platelets may then interact with the subendothelial cells, aggregate, form thrombi, and secrete growth factors as well. This theory suggests that growth factors that stimulate smooth muscle cell proliferation may be secreted by activated endothelial cells and macrophages, as well as by platelets.

. . .

The monoclonal hypothesis proposes that the fibrous plaque lesion of atherosclerosis is derived from the replication of a single smooth muscle cell. It is theorized that the cell is initially transformed by a virus or other factors and subsequently forms a benign neoplasm.

. . .

Risk Factors

Although precise mechanisms for the disease process have not been established, there are risk factors associated with an increased incidence of atherosclerosis. These are (1) cigarette smoking, (2) obesity, (3) family history, (4) diabetes, (5) hypertension, and (6) hypercholesteremia.

Elevated blood levels of cholesterol and associated lipids is a major risk factor. Hypercholesteremia inflicts endothelial injury that leads to accumulation of lipids and proliferation of smooth muscle cells. Plasma lipids, which are the source of fatty deposits in the arterial wall, include **cholesterol, cholesteryl esters, triglycerides,** and **phospholipids.** Since these substances are water insoluble, they are bound to protein carrier molecules called **apoproteins** (ah″po-pro′te-in). Lipid/apoprotein complexes are called **lipoproteins.** There are five major categories of lipoproteins (table 22.4) which are mixtures but are grouped on the basis of size and density. Plasma levels of these lipids

apoprotein: Gk. *apo,* separation or away from

TABLE 22.3 Proposed outline of events in the formation of an atherosclerotic plaque according to the response-to-injury hypothesis

Injury to Arterial Endothelium
Caused by hypertension, hyperlipidemia, cigarette smoking, diabetes, hormone dysfunctions, etc.

Injured Endothelial Cells
Release procoagulants and growth factors (proliferation of smooth muscle cells)

Macrophages
Hyperlipidemia leads to migration of monocytes between endothelial cells and conversion to macrophages; take up lipid; secrete substances which cause further injury; release growth factors; promote proliferation of smooth muscle cells

Platelets
If injury disrupts cell-to-cell attachment, platelets adhere and thrombi form; secrete growth factors; and promote proliferation of smooth muscle cells

are useful in evaluating the risk of atherosclerosis, particularly coronary atherosclerosis. As shown in table 22.5, the LDL fraction is high in cholesterol, and there is evidence that the HDL fraction promotes cholesterol removal from cells. Studies have shown a correlation between elevated total cholesterol levels, high LDL values, and risk for atherosclerosis. A high HDL/LDL cholesterol ratio indicates a decreased risk. A specific apoprotein, apo A-I, is a major protein in HDL and is a predictor of low risk, if present in the blood in high concentrations.

Manifestations

In general, atherosclerotic lesions affect medium and large sized arteries. Typically, the abdominal aorta is involved to a greater degree than the thoracic aorta, and the lesions tend to occur where there are major branches coming off the aorta. The coronary arteries may show extensive involvement. This vascular disease has serious to life-threatening consequences. The lesion may occlude an artery and interrupt blood supply to a vital organ causing an **infarction** (in-fark′shun). The term infarction means necrosis or death of tissue due to inadequate oxygenation. The topic of myocardial infarction is discussed further in chapter 23. The fibrous plaque may progress to a complicated lesion in which fibrosis and calcification cause the vessel to harden. The term **arteriosclerosis** is used to describe this condition (figure 22.3). The force of blood flow against the vessel wall may lead to plaque rupture and ulceration. Such endothelial damage activates platelets, releases thromboplastin, and promotes formation of thrombi. A clot that breaks loose and is carried in the bloodstream is called a **thromboembolism** and may occlude a blood vessel. Finally, atherosclerotic lesions weaken the arterial wall, so there may be dilation or rupture of

thrombus: Gk. *thrombos,* clot

TABLE 22.4 Sources, characteristics, and functions of five categories of lipoproteins

	Source	Characteristics	Function
Chylomicrons	Diet	Largest of lipoproteins; mainly triglycerides	Source of energy or stored
Cholesterol	Diet; 75% is synthesized by liver	Cholesteryl esters transported with LDL to tissues	Cell membrane structure; precursor of steroid hormones and bile salts
VLDL (very low-density lipoprotein)	Synthesized by liver	Mainly endogenous triglycerides	Transport of lipids to sites of use
LDL (low-density lipoprotein)	Synthesized; mainly from VLDL	Major cholesterol carrying fraction	Bound to LDL receptors on cells and metabolized
HDL (high-density lipoprotein)	Synthesized by liver; metabolism of chylomicrons and VLDL	Small particles; relatively rich in protein, cholesterol, and phospholipid	Source of cholesterol for endocrine tissues; may promote cholesterol removal from cells

TABLE 22.5 Cholesterol levels in adults that identify moderate or high-risk for atherosclerosis in different age groups

Age	Moderate Risk Cholesterol (mg/dl)	High Risk Cholesterol (mg/dl)
20–29	200–220	>220
30–39	220–240	>240
over 40	240–260	>260

the vessel, a condition called an **aneurysm** (an'u-rizm) (figure 22.4). A **cerebrovascular accident** (CVA), also called a stroke, is the result of either a thrombus resulting in interruption of blood supply to some part of the brain or rupture of a cerebral vessel (chapter 38).

In summary, there are three major consequences of advanced atherosclerosis: (1) partial or total occlusion of the vessel, (2) formation of microthrombi or small clots, or (3) aneurysm due to a weakened blood vessel wall.

• • •

Lipid content of fibrous plaques and specific risk factors may be associated. Individuals with high blood levels of cholesterol tend to develop high-lipid-content lesions in coronary arteries. On the other hand, individuals who are heavy cigarette smokers frequently develop fibrous plaques that have small amounts of lipid in the femoral arteries.

• • •

TREATMENT

Atherosclerosis is a serious threat to health, particularly if the disease process involves coronary arteries. The approach to treatment involves reducing risk factors by (1) avoiding smoking, (2) losing weight, (3) controlling blood pressure, and (4) controlling blood levels of lipids.

aneurysm: Gk. *aneurysma*, dilatation

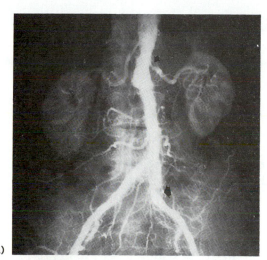

(a)

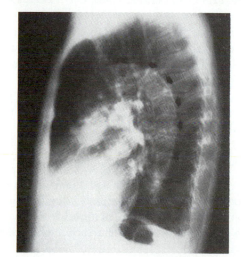

(b)

Figure 22.3 (*a*) Arteriogram of the abdominal aorta and iliac arteries. It is normal on the left with evidence of atherosclerosis on the right. There is stenosis (large arrow). The small arrow, in the area of the patient's kidney, shows stenosis of the origin of the renal artery. (*b*) Lateral chest X-ray that demonstrates a calcified, dilated aortic arch (outlined by arrows).

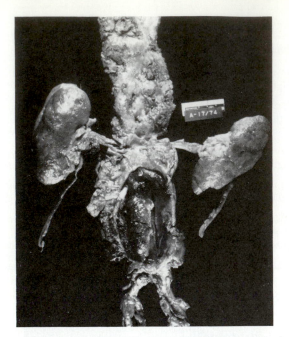

Figure 22.4 An atherosclerotic aneurysm of the abdominal aorta. The aneurysm is below the renal arteries and largely filled with thrombus. The rest of the aorta shows severe atherosclerosis.
Ritchie, A. C. 1990. Boyd's Textbook of Pathology. *Lea & Febiger, Philadelphia. Reproduced with permission.*

Dietary restrictions of cholesterol and saturated fat frequently control triglyceride, total cholesterol, and LDL blood levels. If dietary restraints fail to achieve this goal, drugs may be used to lower cholesterol levels.

DISSECTING ANEURYSM OF AORTA

A dissecting aneurysm is a condition in which an arterial wall is weakened, and there is progressive separation of the middle and outer layers, the tunica media and tunica externa (figure 22.5). In most cases, the site of origin is the ascending aorta or the thoracic aorta just beyond the origin of the left subclavian artery. Dissection that originates in the descending aorta is less common.

PATHOGENESIS

Men are more commonly affected than women and hypertension is the most common associated finding. When dissection of the aorta originates in the descending aorta, it is often associated with severe arteriosclerosis as well as hypertension.

The initial event in the development of a dissecting aneurysm is degeneration of the aortic tunica media. Studies indicate that degenerative changes are the result of injury and repair due to hemodynamic factors, possibly

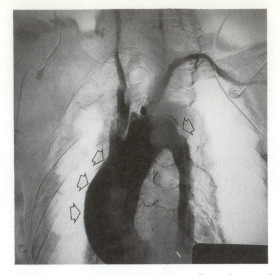

Figure 22.5 An arteriogram that shows a dissecting aortic aneurysm (clear arrows) and total blockage of left carotid artery (solid arrow).

because of pressure on **vasa vasorum,** small vessels that nourish the tunica media. Ultimately, an intimal tear due to the force of ventricular ejection of blood causes blood to enter the tunica media. The pressure of blood inside the vessel wall creates a separation or tear which is self-propagating in both directions. The progress of the dissection is determined by the extent of medial degeneration. If the dissection proceeds toward the heart, blood may fill the sac that surrounds the heart causing a **cardiac tamponade.** The presence of blood around the heart causes compression and interferes with circulating blood filling the chambers of the heart. The result is a decrease in cardiac output, a fall in blood pressure, and ultimately death, if there is no medical intervention. Rupture of the aorta may occur in the thoracic or abdominal cavities. If the dissection proceeds distally, there may be occlusion of major branches of the aorta and subsequent infarction of various organs.

MANIFESTATIONS

Typically, there is sudden onset of severe and persistent pain. The pain is usually in the anterior thorax or posteriorly, between or below the scapulas. Cardiac arrhythmias may occur. As the dissection progresses, the pain migrates downward and abdominal tenderness may be present. Interference with blood supply to the spinal cord may produce paraplegia or paralysis of the lower body. Typically, there is moderate or marked elevation of blood pressure. The clinical picture is variable and changes as the dissection progresses to produce ischemia of various organs.

vasa vasorum: L. *vasa,* vessels; L. *vasorum,* of vessels
tamponade: F. *tampon,* plug
arrhythmia: Gk. *a,* not; Gk. *rythomos,* rhythm

TREATMENT

Treatment involves medical and surgical intervention. The goal of medical intervention is to control blood pressure and decrease the force of ventricular contraction. Surgical procedures are implemented to prevent further dissection, prevent external rupture, and to restore blood flow in occluded vessels. The prognosis in a dissecting aneurysm is poor, particularly if there is a history of hypertension. The dissection may cause death within minutes to hours, and 70% die within 2 weeks. The most common cause of death is rupture into the pericardium.

• • •

Studies in animals have shown that atherosclerotic lesions induced by a high cholesterol diet can regress, following a diet in which cholesterol is restricted. The lesion may become smaller, contain less lipid, and show decreases in connective tissue proteins. It appears that, in some cases, advanced lesions may regress over a period of time in response to lowered cholesterol. There is some evidence that fish oils, high in omega-3-fatty acids, cause a decrease in blood cholesterol levels of individuals with hypercholesteremia.

• • •

SUMMARY

• • •

Raynaud's phenomenon is impaired circulation in the digits that is caused by various factors, such as neurogenic or occlusive vascular disorders, drugs, or trauma. Raynaud's disease is impaired blood flow to the digits precipitated by cold or emotional stress. Mechanisms are unknown, and the condition is treated by protecting extremities from the cold and administering drugs to promote blood flow.

Atherosclerosis is a disease of the arterial tunica intima in which there is a progressive accumulation of cells, connective tissue, lipids, and sometimes calcium. Risk factors include cigarette smoking, obesity, family history, diabetes, hypertension, and hypercholesteremia. The disease process leads to myocardial infarction, thromboembolism, and aneurysm. Treatment includes amelioration of risk factors, with particular attention to lowering blood cholesterol and LDL levels.

Dissecting aneurysm of the aorta is a separation of the layers of the aortic wall due to degeneration of the tunica media. Hypertension is a common finding in affected individuals. The condition is life-threatening with aortic rupture, cardiac tamponade, and infarction of abdominal organs as possible consequences. Control of blood pressure and surgical intervention is required.

REVIEW QUESTIONS

• • •

1. In general, what are the three basic layers that make up a blood vessel wall?
2. In what type of vessel is the middle and outer layer missing?
3. What is present in the walls of large arteries to make them stretchable and capable of rebounding?
4. What is the underlying problem in Raynaud's phenomenon and disease?
5. What are two possible initiating factors in Raynaud's disease?
6. What constitutes an atherosclerotic lesion?
7. List three consequences of atherosclerosis.
8. What are five risk factors for developing atherosclerosis?
9. Which of the lipid fractions measured routinely in clinical laboratories is highest in cholesterol?
10. Define the term atheromatous plaque.
11. High levels of which of the following is a predictor of risk for atherosclerosis? HDL, LDL, VLDL.
12. What is the most common finding associated with dissecting aneurysm?
13. In a dissection of the aorta, what actually causes the tearing after degeneration of the tunica media?
14. What is a cardiac tamponade?
15. List two hazards of an aortic dissecting aneurysm other than cardiac tamponade.

SELECTED READING

• • •

Brown, M. S., and J. L. Goldstein. 1984. How LDL receptors influence cholesterol and atherosclerosis. *Scientific American* 251:58–66.

Cheung, J. Y. et al. 1986. Calcium and ischemic injury. *New England Journal of Medicine* 314:1670–84.

Glagov, S. et al. 1987. Compensatory enlargement of human atherosclerotic coronary arteries. *New England Journal of Medicine* 316:1371–75.

Hulley, S. B. 1988. A national program for lowering high blood cholesterol. *American Journal of Obstetrics and Gynecology* 158:1561–67.

Kistler, J. P., A. H. Ropper, and R. C. Heros. 1984. Therapy of ischemic cerebral vascular disease due to atherothrombosis. *New England Journal of Medicine* 311:27–34.

Lavie, C. J. et al. 1988. Management of lipids in primary and secondary prevention of cardiovascular diseases. *Mayo Clinical Proceedings* 63:605.

Lobo, R. A. 1988. Lipids, clotting factors, and diabetes: Endogenous risk factors for cardiovascular disease. *American Journal of Obstetrics and Gynecology* 158:1584–91.

McPherson, D. D. et al. 1987. Delineation of the extent of coronary atherosclerosis by high-frequency epicardial echocardiography. *New England Journal of Medicine* 316:304–8.

Weiner, B. H. et al. 1986. Inhibition of atherosclerosis by cod-liver oil in a hyperlipidemic swine model. *New England Journal of Medicine* 315:841–46.

Chapter 23

Disorders of the Heart

Cardiovascular disease is the most frequent cause of death in the United States with more than 1 million myocardial infarctions annually and about 700,000 ischemic heart disease deaths per year. In the discussion that follows, ischemic heart disease and myocardial infarction are considered first, followed by disorders involving heart valves; and finally, heart failure is discussed, with emphasis on congestive heart failure. The first section is a brief review of events in the cardiac cycle that introduces abnormalities of cardiac function.

CARDIAC CONTRACTION

Alternating diastole/systole of the heart is synchronized for efficient filling of all chambers and ventricular ejection of adequate volumes of blood. The series of events in a single heartbeat is referred to as the cardiac cycle. Figure 23.1 shows atrial and ventricular contraction during one cycle, assuming that the heart rate is 75 beats/min. The precision of the contraction/relaxation pattern is maintained by the conducting system of the heart (figure 23.2). The conducting system is made up of heart muscle cells specialized for rapid conduction of nerve impulses throughout the myocardium. The sinoatrial (SA) node, located in the right posterior atrial wall, is where nerve impulses originate, and it is called the pacemaker. The subsequent sequence of impulse transmission follows:

1. The atrial walls
2. The **atrioventricular (AV) node** located in the lower interatrial septum
3. Right and left branches down either side of the interventricular septum called the **atrioventricular bundle** or **bundle of His**
4. Terminal fibers called **Purkinje fibers** (pur-kin'jē), which exit from the bundle branches and carry impulses to ventricular myocardium

Ischemia of an area of the myocardium causes the conduction times of the ischemic and nonischemic tissues to be different, which leads to abnormalities of cardiac rhythm. The term ectopic focus is used for conditions and diseases, such as myocardial ischemia or hyperthyroidism that favor the development of pacemakerlike activities in areas that do not normally act as pacemakers. Ischemic heart disease, discussed in the next section, may have life-threatening consequences.

ISCHEMIC HEART DISEASE

Heart muscle requires an abundant blood supply to meet metabolic needs, and this is provided by way of branches from the right and left coronary arteries (figure 23.3). Interruption of blood supply or ischemia leads to death of cardiac muscle called **myocardial infarction.** The terms

ectopic: Gk. *ektopios,* out of place

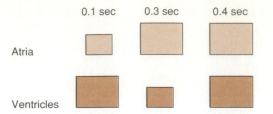

Figure 23.1 Series of events in a cardiac cycle assuming 75 beats/minute. One cycle requires 0.8 second. The upper boxes represent atria and the lower boxes represent ventricles. Systole is signified by small boxes, while the large boxes signify diastole.

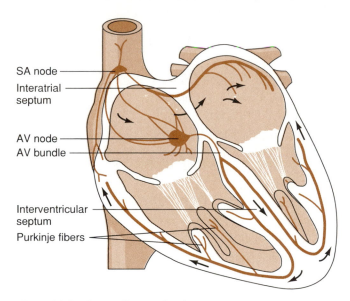

Figure 23.2 The cardiac conduction system.

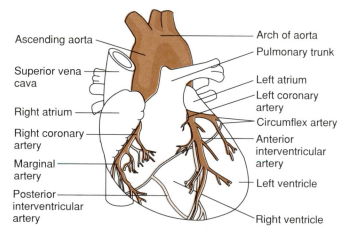

Figure 23.3 The heart muscle is supplied with blood by the right and left coronary arteries that arise from the aorta immediately beyond the aortic semilunar valve. Branches of these arteries serve the atrial and ventricular walls.

used to describe this problem are **coronary heart disease** or **ischemic heart disease.** The clinical manifestations are **angina pectoris,** or chest pain; myocardial infarction; and sudden death.

ANGINA PECTORIS

Angina pectoris (an-ji'nah pek''to-ris) refers to symptoms, usually chest pain, precipitated by exertion and caused by ischemic heart disease. Table 23.1 identifies forms of angina, which by definition is a transient event not associated with clinically detectable myocardial necrosis.

MYOCARDIAL INFARCTION

In contrast to angina, myocardial infarction, due to occlusion of coronary vessels, leads to myocardial necrosis (figure 23.4). Typically, the underlying cause is atherosclerosis with or without thrombosis. Coronary atherosclerosis progresses with age (chapter 22) and is generally less severe in women than in men, up to the age of 50 years. After that point, the difference tends to diminish. Myocardial infarction may be a complication of surgery and, in some cases, is precipitated by exertion.

Manifestations

Typically, the cardinal symptom of myocardial infarction is severe chest pain, although pain is absent in some cases. The pain usually originates in the left chest and frequently radiates to the neck, jaw, shoulders, arms, and fingers. Characteristically, it lasts for more than 30 minutes and often persists for hours. The pain is variously described as being dull, crushing, constricting, burning, or aching.

Other manifestations include a moderate elevation in temperature, usually occurring 24–48 hours after the infarction, with a return to normal within a week. The fever is probably a response to tissue necrosis. Dyspnea, sweating, nausea, and vomiting are common symptoms of myocardial infarction. Although the blood pressure is variable, hypotension and **bradycardia** are common during early stages of an infarction. Pallor, restlessness, and cold extremities are typical, and the patient may feel terror with a sense of impending doom.

Consequences

Life-threatening events subsequent to myocardial infarction include ventricular **tachycardia** and/or fibrillation. Ischemia may cause abnormalities in the conduction of impulses through the AV nodes. A delay or interruption of conduction through the AV node is called atrioventricular block. This leads to independent ventricular contractions. A complete block causes a decrease in cardiac

angina pectoris: L. *angere*, to strangle; L. *pectoralis*, breast
bradycardia: Gk. *bradys*, slow; Gk. *kardia*, heart
tachycardia: Gk. *tachys*, swift

TABLE 23.1 Definitions of forms
of the anginal syndrome

Classic angina. Transient discomfort caused by exertion and relieved by rest or use of nitrates; subsequent myocardial infarction is common
Atypical angina. Similar to classic form, may not be relieved by rest or nitrates
Angina equivalent. Symptom is dyspnea
Variant angina. Onset of symptoms at rest, usually at night, usually no subsequent myocardial infarction
Unstable angina. Prolonged attacks without infarction

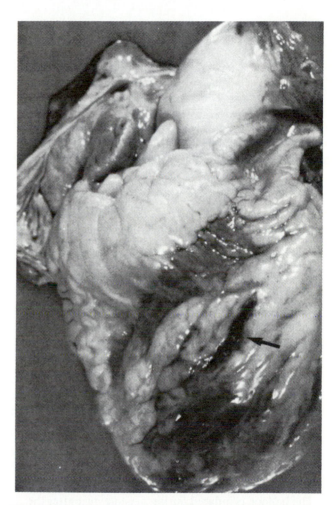

Figure 23.4 The external surface of a heart with a three-day-old infarct. The dark linear tear (arrow) communicates with a rupture of the anterior free wall of the left ventricle. *From R. S. Cottran, V. Kumar, and S. L. Robbins:* Robbins Pathologic Basis of Disease. *1989 W. B. Saunders, Philadelphia. Reproduced with permission.*

output. The heart may fail as a pump and become incapable of supplying the body with an adequate volume of blood. This condition, called **cardiogenic shock,** leads to circulatory failure (chapter 21). A procedure for temporarily stabilizing a patient who has suffered a myocardial infarction is shown in figure 23.5.

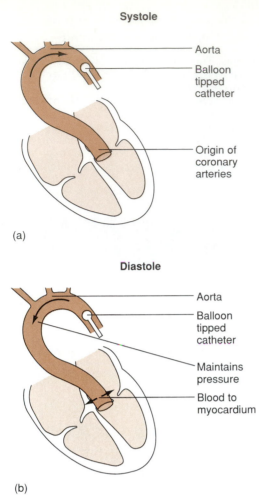

Systole

Aorta

Balloon
tipped
catheter

Origin of
coronary
arteries

(a)

Diastole

Aorta

Balloon
tipped
catheter

Maintains
pressure

Blood to
myocardium

(b)

Figure 23.5 Intra-aortic balloon counterpulsation procedure. (*a*) Balloon deflated during systole. (*b*) Balloon inflated during diastole to decrease runoff of blood. This promotes increased coronary perfusion.

• • •

Intra-aortic balloon counterpulsation is a procedure for temporary hemodynamic stabilization after myocardial infarction associated with cardiogenic shock. A balloon-tipped catheter is inserted into the femoral artery and advanced to the thoracic aorta. The balloon is inflated during diastole, just after the closure of the aortic valve. This decreases runoff of blood during diastole and increases coronary perfusion. The balloon is deflated just before the onset of systole.

• • •

Management

Modalities of treatment of myocardial infarction include (1) measures to alleviate pain and anxiety, (2) administration of oxygen, (3) anticoagulant therapy, (4) therapy to prevent arrhythmias, and (5) control of blood pressure. Streptokinase or tissue plasminogen activator (chapter 19)

may be administered to promote clot dissolution. A patient recovering from myocardial infarction is instructed about dietary restriction of calories, cholesterol, and saturated fat, and the elimination of cigarette smoking. A regular exercise program is instituted.

The management of ischemic heart disease may require coronary artery bypass surgery in which (1) a section of vein is used to form a connection between the aorta and the coronary artery beyond the point of obstruction, or (2) the left internal mammary artery is attached to the coronary artery distal to the obstruction. An alternative procedure is **angioplasty** (an′jē-ō-plas″tē) in which a balloon-tipped catheter is threaded into the coronary artery and the balloon inflated to open the passage of the obstructed segment.

RHEUMATIC HEART DISEASE

Rheumatic fever is an inflammatory condition mainly involving the heart, joints, central nervous system, and skin and is a sequel to a group A β-hemolytic streptococcal infection of the pharynx. The incidence of rheumatic fever is highest in the age group 5–14 years. The disease process is self-limiting except for cardiac involvement, which may result in serious and chronic dysfunction of the heart.

MANIFESTATIONS

Major manifestations of rheumatic fever include (1) arthritis, (2) carditis, (3) chorea, (4) subcutaneous nodules, and (5) erythema marginatum. Arthritic joint pain persisting from 2–4 weeks is a common initial symptom. About half of the affected individuals experience carditis, an inflammation involving the endocardium, myocardium, or pericardium. Symptoms include chest pain, fatigue, cough, and shortness of breath.

Chorea (kō-rē′ah), also called **St. Vitus' dance,** refers to involuntary muscle contractions. This is evidence of central nervous system involvement and is a late manifestation of rheumatic fever. Painless subcutaneous nodules over the spine, scalp, elbows and tendons of hands and feet appear in up to 20% of the cases. **Erythema marginatum,** characterized by a pink raised margin around a central area of normal skin, appears on the trunk and upper extremities.

Minor manifestations that aid in establishing a diagnosis of rheumatic fever include (1) joint pain, (2) fever, (3) leukocytosis, (4) increased sedimentation rate, and (5) prior acute rheumatic fever. Typically, the **sedimentation rate** is elevated in inflammatory diseases and is a

angioplasty: Gk. *aggeion,* vessel; Gk. *plassein,* to form
arthritis: Gk. *arthron,* joint; Gk. *itis,* inflammation
carditis: Gk. *kardia,* heart
chorea: Gk. *choreia,* dance
erythema: Gk. *erythema,* redness

nonspecific test. This test utilizes blood to which anticoagulant has been added and measures the rate at which cells settle in the tube.

The diagnosis of acute rheumatic fever is based on the presence of two major manifestations or a combination of one major and two minor manifestations, plus some evidence of streptococcal infection. A positive throat culture and an elevated **antistreptolysin O titer (ASO titer)** confirm a streptococcal infection. The ASO titer is a measure of the level of antibodies that the body produces against streptolysin O toxin, a poison produced by the organism.

CONSEQUENCES

Table 23.2 indicates the incidence of cardiac valvular damage caused by rheumatic fever. The **mitral** (also called the **bicuspid valve**) is a left interatrioventricular (in"ter-ate"re-oh-ven-trik'u-lur) valve and is most often involved. An autoimmune mechanism (chapter 28) is probable, whereby antibodies against the organism attack cardiac valvular tissue. The mitral valve consists of two cusps anchored to **papillary muscle** by **chordae tendineae** (kor'dē ten'din-ē) as shown in figure 23.6. Rheumatic heart disease causes the cusps to become thick and rigid, and the chordae tendineae fuse and shorten. The overall effect is (1) incomplete closure allowing backflow of blood into the left atrium, called mitral regurgitation or (2) **mitral stenosis,** which is a narrowed valvular opening that obstructs blood flow into the left ventricle. Other heart valves may be affected in a similar way. The burden inflicted on the heart by valvular damage is discussed later in this chapter in connection with heart failure (figure 23.7).

MANAGEMENT

Prompt antibiotic treatment of streptococcal pharyngitis protects against rheumatic fever, provided that the treatment is within 7–8 days of onset of sore throat. In rheumatic fever, it is necessary to prevent recurring attacks by a prolonged course of prophylactic penicillin. Aspirin is given to reduce both fever and inflammation of the joints, and steroids are administered if there is severe carditis. Antibiotics are given in the case of dental procedures as prophylaxis against **subacute bacterial endocarditis.** This refers to a condition in which damaged or congenitally deformed valves are susceptible to infection by organisms, such as *Streptococcus viridans* or *Staphylococcus epidermidis*. These organisms gain entry to the bloodstream during such procedures as tooth extraction and colonize damaged heart valves.

HEART FAILURE

The term heart failure denotes inadequacy of the heart as a pump, i.e., the inability to supply the volume of blood

TABLE 23.2 Incidence of cardiac valvular damage caused by rheumatic fever
Mitral valve 85%
Aortic valve 54%
Tricuspid and pulmonic valves <5%

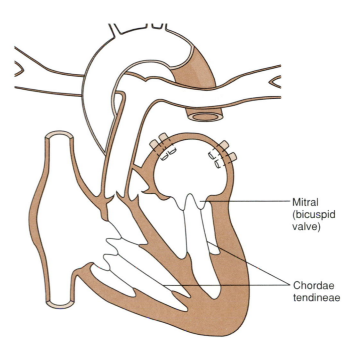

Figure 23.6 The mitral valve and chordae tendineae are frequently involved in rheumatic heart disease.

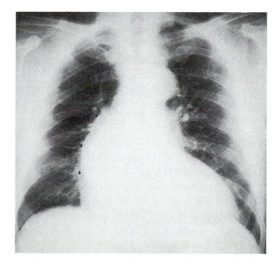

Figure 23.7 A posterior-anterior chest X-ray of a 62-year-old male with a history of rheumatic heart disease, valvular stenosis, and aortic insufficiency. The heart is markedly enlarged, especially on the right border (arrows).

TABLE 23.3 Possible causes or factors in heart failure

Myocardial Disease

Myocardial infarction, arrhythmias
Cardiomyopathies
Myocarditis

Pressure or Volume Overload

Pulmonary embolism
Hypertension
Pulmonary hypertension
Congenital heart disease
Valvular damage

Restriction of Ventricular Filling

Constrictive pericarditis
Cardiac tamponade

Increased Demands

Hyperthyroidism
Anemia

TABLE 23.4 Definition of terms
related to heart disease

Myo = muscle	Itis = inflammation
Cardio = heart	Endo = within
Patho = disease	Infarct = to stuff

Myocardial infarction. Interruption of blood supply to heart
Myocarditis. Inflammation of heart muscle
Cardiomyopathies. Diseases of heart muscle
Endocarditis. Inflammation of valves or inner lining of heart

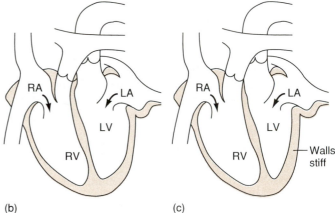

Hypertrophic cardiomyopathy

(a)

Dilated cardiomyopathy   **Restrictive cardiomyopathy**

(b)   (c)

Figure 23.8 Three types of cardiomyopathies.

required to meet the body's needs. A number of mechanisms lead to heart failure and, as summarized in table 23.3, are basically the result of (1) myocardial disease, (2) pressure or volume overload, (3) restriction of ventricular filling, or (4) factors that increase the cardiac work load.

MYOCARDIAL DISEASE

Terms describing problems associated with heart failure are defined in table 23.4. Diseases involving heart muscle that may lead to failure include myocardial infarction. The consequent ischemia causes varying degrees of damage that affects cardiac contractility.

Cardiomyopathies

Three types of cardiomyopathies, i.e., diseases of the heart muscle, lead to heart failure (figure 23.8). **Hypertrophic** (hī″per-trōf′ic) **cardiomyopathy** is a condition in which there is a thickened left ventricular wall and septum, a small left ventricular cavity, and often an obstruction to left ventricular outflow. The cause is unknown and in some

hypertrophic: Gk. *hyper*, beyond; Gk. *trophe*, nourishment
cardiomyopathy: Gk. *kardia*, heart; Gk. *mys*, muscle; Gk. *pathos*, disease

cases appears to be an inherited condition. **Dilated cardiomyopathy** is a heart disease in which the ventricles are enlarged with impaired contractility. The evidence suggests that this condition is caused by a variety of toxic or infectious agents. In some cases, dilated cardiomyopathy may be the result of viral myocarditis. Alcoholism causes dilated cardiomyopathy as well. **Restrictive cardiomyopathy** is a third type of heart muscle disease resulting in impaired ventricular filling. There are various causes that lead to infiltration of ventricular walls with different substances, such as glycogen or fibrous connective tissue. One example is **endomyocardial fibrosis** in which fibrous connective tissue replaces normal heart muscle for no known reason. The infiltration causes the ventricles to become stiff and noncompliant, and this interferes with filling capacity.

Myocarditis

Myocarditis is an inflammation of heart muscle which is sometimes idiopathic or may be caused by an infection. Depending on the area of involvement, this condition may lead to disturbances of heart rate and rhythm and precipitate heart failure.

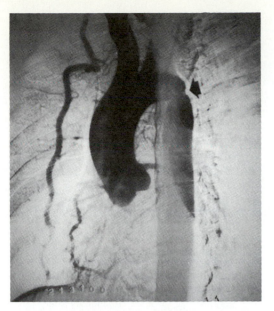

Figure 23.9 An arteriogram of a 26-year-old male showing coarctation of the aorta (arrow). Filling of the distal arch beyond the coarctation is primarily from the left subclavian.

PRESSURE OR VOLUME OVERLOAD

Increased pressure or volume within the chambers of the heart may precipitate or progressively lead to heart failure. Possible causes of heart failure follow:

1. **Pulmonary embolism,** i.e., an occlusion of pulmonary vessels caused by a transported clot or plug. This mechanical obstruction causes increased pulmonary vascular resistance and elevated right ventricular pressure.
2. Hypertension (chapter 21) causes increased systemic resistance that increases the work of the left ventricle.
3. Pulmonary hypertension (chapter 24) develops as the result of proliferation of cells, thickening of walls of pulmonary vessels, and increased resistance to blood flow, which results in right ventricular hypertrophy.
4. The malformations in congenital heart disease (chapter 24) cause pressure and volume overloads in the heart. Coarctation (kō″ark-tā′shun), narrowing of the aorta, impedes blood flow, increases afterload, and causes left ventricular hypertrophy (figure 23.9).
5. Valvular damage allows backflow of blood and creates pressure and volume overloads.

RESTRICTION OF VENTRICULAR FILLING

Constrictive pericarditis is an inflammation of the sac covering the heart, the **pericardium.** Fibrosis following the inflammation causes the sac to become rigid and interferes with ventricular filling during diastole. A second example

coarctation: L. *coarctere*, to press together

TABLE 23.5 Terminology that describes heart failure

Acute. Rapid onset
Chronic. Gradual onset
Low output. Low cardiac output
High output. High cardiac output
Backward failure. Impediment to blood flow; result is increased pressure and volume behind left and/or right ventricles
Forward failure. Inadequate cardiac output
Left-sided failure. Left ventricle dysfunction; blood backs up in left atrium and pulmonary vessels
Right-sided failure. Right ventricle is dysfunctional or overloaded; causes systemic venous congestion

of restriction of inflow of blood to the ventricles is **cardiac tamponade.** This is a situation in which fluid fills the pericardial sac, and it is frequently due to a dissecting aneurysm (chapter 22). The accumulation of pericardial fluid compresses the heart and restricts ventricular filling.

INCREASED DEMANDS

Hyperthyroidism increases the burden on the heart because of an excess of the hormone thyroxine (chapter 31). Hormonal influence increases the force of cardiac contractions resulting in increased oxygen requirements by the heart. Anemia also adds to the work load of the heart. Anemia, due either to a decrease in red blood cells or a decrease in hemoglobin (chapter 17), leads to inadequate oxygen delivery to tissues. Thus, increased cardiac output is required to meet the oxygen needs of the body. This may contribute to the failure of a diseased heart.

CONGESTIVE HEART FAILURE

The term congestive implies excessive filling and refers to increased volume in pulmonary vessels or increased volume on the venous side of the systemic circulation. Causes of heart failure have already been discussed. In many instances, the failure may be characterized as congestive heart failure, a situation in which there is a back up of blood behind one or both ventricles. The following discussion includes the causes, effects, symptoms, and management of congestive heart failure (CHF).

CAUSES

The terminology frequently used to describe heart failure is defined in table 23.5. Heart failure may be associated with either a high or low cardiac output. Low cardiac output is more common, but there are situations in which an increased output occurs and is nevertheless inadequate. Two examples of this, anemia and hyperthyroidism, are listed in table 23.3. The characteristic pattern in heart failure is impaired contractility of the ventricles

tamponade: F. *tampon,* plug

281

TABLE 23.6 Causes of left-sided heart failure

Hypertension. Increased peripheral resistance
Aortic stenosis. Narrow opening into aorta
Coarctation of aorta. Narrow aorta
Mitral regurgitation. Backflow of blood from left ventricle to left atrium

TABLE 23.7 Causes of right-sided heart failure

Pulmonary stenosis. Decreased size of opening into pulmonary artery
Pulmonary hypertension. Elevated pulmonary artery pressure; caused by left ventricular failure and congenital heart defects
Cor pulmonale. Pulmonary hypertension that originates in the lungs; causes right ventricular failure
Atrial septal defect. Opening in septum between the two atria; blood flows from left to right atria

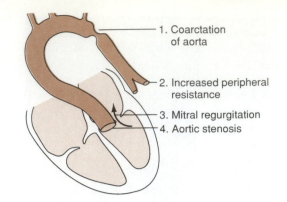

1. Coarctation of aorta
2. Increased peripheral resistance
3. Mitral regurgitation
4. Aortic stenosis

Figure 23.10 Causes of left-sided heart failure.

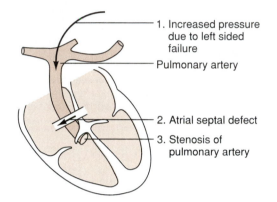

1. Increased pressure due to left sided failure
   Pulmonary artery
2. Atrial septal defect
3. Stenosis of pulmonary artery

Figure 23.11 Causes of right-sided heart failure.

accompanied by a low cardiac output. The focus of this discussion is failure of one or both ventricles, i.e., right- or left-sided failure.

The right and left pumping systems of the heart are continuous and do not behave or respond independently. It is, however, useful to consider the two sides individually.

Left-Sided Failure

Left-sided failure is due most commonly to hypertension in which there is increased peripheral resistance. The left ventricle pumps against this increased afterload and progressively becomes unable to maintain a normal stroke volume (chapter 21). The result is a back up of blood in the left atrium and congestion of the pulmonary vascular system.

Table 23.6 lists other causes of left-sided failure including aortic stenosis (figure 23.10). Stenosis means narrow, and this condition involves a narrowed opening through the aortic valve. The causes include a congenital defect or valvular damage due to rheumatic inflammation. The defect impedes the flow of blood from the left ventricle and, over a period of time, results in hypertrophy of the left ventricle.

Coarctation of the aorta is a congenital abnormality (chapter 24) in which the aorta narrows, typically just distal to the origin of the left subclavian artery. Flow of blood is impeded, leading to hypertension and left ventricular hypertrophy.

Mitral regurgitation refers to a loss of integrity of the mitral valve, which allows backflow of blood from the left ventricle to the left atrium during systole. Rigidity and retraction of valve cusps may be a congenital defect or caused by rheumatic heart disease. The consequence is additional strain on the left atrium and left ventricle, and both chambers dilate and hypertrophy. Ultimately, the left ventricle fails.

stenosis: Gk. *stenos,* narrow

Right-Sided Failure

Defects that lead to right-sided heart failure are listed in table 23.7. These include pulmonary stenosis in which there is a fusion and thickening of valve cusps resulting in a decrease in the size of the opening into the pulmonary artery. The defect may be congenital or the result of rheumatic heart disease. The overall effect is an impediment to the outflow of blood from the right ventricle, which ultimately causes right ventricular hypertrophy (figure 23.11).

Pulmonary hypertension is a possible cause of right-sided heart failure and is discussed in chapter 24 in connection with congenital heart defects. Elevated pulmonary arterial pressure may be the result of increased blood volume and/or pulmonary resistance. Left ventricular failure leads to pulmonary hypertension, and the result is increased resistance to outflow of blood from the right ventricle and ultimate failure of that chamber.

Cor pulmonale (kor pul-mon-a′lē) refers to enlargement of the right ventricle in response to pulmonary hypertension in which left ventricular failure or congenital defects are not involved. In cor pulmonale, the causes of the pulmonary hypertension include lung disease, such as

cor pulmonale: L. *cor,* heart; L. *pulmo,* lung

chronic obstructive pulmonary disease (chapter 7), pulmonary emboli, inadequate chest wall movements, or neuromuscular dysfunction. Factors in the hypertension are (1) obstruction or destruction of pulmonary vessels or (2) hypoxemia, a stimulus for pulmonary vasoconstriction. The pulmonary hypertension in cor pulmonale is distinguished from other types of hypertension discussed in that it is intrinsic within the lungs or the respiratory system. The ultimate outcome is right ventricular failure.

Atrial septal defect refers to an opening in the septum between the two atria and is discussed in chapter 24 as a congenital defect. The opening allows blood to flow from the left to the right atrium. This increases blood flow to the lungs, may cause pulmonary hypertension, and leads to right ventricular hypertrophy.

PHYSIOLOGICAL RESPONSES

Physiological responses to progressive heart failure, i.e., congestion, low cardiac output, and diminished myocardial contractility, are compensatory, albeit deleterious at times. Table 23.8 summarizes these physiological adjustments.

Dilatation

Dilatation (dil-ah-tā′shun) refers to enlargement of a heart chamber in response to a volume overload. The physiological advantage of this response is based on the Frank-Starling law, which states that stretched muscle fibers have increased contractile strength. There is decreased contractile force with excessive stretching, however, and in chronic dilatation, myocardial function is often diminished. The disadvantage of dilatation is that more tension must be generated in the walls of an enlarged heart to create the necessary systolic pressure. According to Laplace's law, the tension required to generate a given pressure increases in proportion to the diameter. An enlarged ventricle requires the expenditure of more energy to eject blood from the chamber.

Hypertrophy

Hypertrophy refers to a thickening of the muscle wall of the heart due to an increase in fiber size or number. This is a characteristic response to high pressure that occurs when there is an impediment to outflow of blood. Within limits, hypertrophy improves the pumping ability of the heart, although ultimately there may be a decrease in myocardial contractile force due to inadequate blood supply compared to the increased muscle mass.

Hormonal Effects

Hormonal responses in heart failure have various effects. There is usually an increase in circulating norepinephrine and epinephrine, which stimulate heart action. There is an increase in renin, angiotensin, and aldosterone. Angiotensin causes vasoconstriction, and aldosterone contributes to the retention of sodium, which occurs in heart

dilatation: L. *dilatatio,* spread out

TABLE 23.8 Physiological adjustments to heart failure and their effects

Compensation for Congestion, Low Cardiac Output, and Diminished Myocardial Contractility

1. Dilatation. Enlargement of heart chambers
   a. Increased fiber length results in increased strength of contraction (within limits)
   b. Increases oxygen requirements; contractile strength decreases with excessive stretching
2. Hypertrophy. Increased myocardial fiber size/number
   a. Commonly results in decreased contractile force
   b. Ischemia due to inadequate blood supply for increased muscle mass
3. Increased circulating norepinephrine and epinephrine
   a. Stimulates heart action
4. Increased renin, angiotensin, aldosterone
   a. Angiotensin causes vasoconstriction
   b. Aldosterone contributes to Na^+ retention
5. Increased circulating vasopressin (antidiuretic hormone)
   a. Contributes to edema
6. Increased 2,3-diphosphoglycerate
   a. Enhances oxygen transport (chapter 17)
7. Pulmonary blood flow directed to upper lobes of lungs
   a. Upper lobes better ventilated, better oxygen exchange

failure. Increased circulating vasopressin (antidiuretic hormone) contributes to edema, although it is uncertain whether this is due to the renal effects of the hormone or whether it is the result of vasoconstriction.

Improved Oxygenation

There are two responses that improve oxygenation. (1) There is an increased level of 2,3-diphosphoglycerate (chapter 17), which promotes oxygen-hemoglobin dissociation. (2) When there is elevated pulmonary venous pressure in heart failure, the flow of blood in the lungs is shunted to the upper lobes. The mechanism is not clear, but the result is that blood is directed to better ventilated lobes with increased oxygenation.

MANIFESTATIONS

Initially, congestive heart failure may predominantly involve one side of the heart, but both sides are affected as the disease process progresses (figure 23.12). Table 23.9 summarizes findings in both left- and right-sided failure.

Left-Sided Failure

Pulmonary effects are prominent in the case of a failing left ventricle. As the left ventricle becomes increasingly unable to eject the appropriate volume of blood with each contraction, blood accumulates in the left atrium and in the pulmonary venous and capillary system. This congestion in the pulmonary vasculature is responsible for increased hydrostatic pressure, which is a force that promotes movement of fluid into the interstitium and alveoli (chapter 2). The consequences are pulmonary edema, lung stiffness, and breathing difficulties. **Dyspnea,** a sense of

dyspnea: Gk. *dys,* hard; Gk. *pnoia,* breath

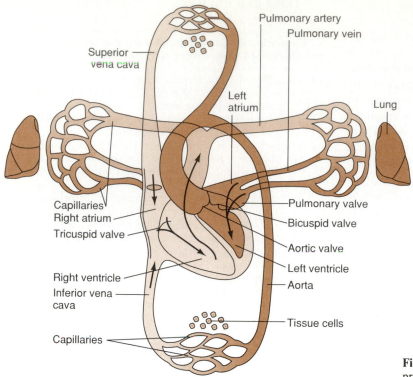

Superior vena cava

Pulmonary artery
Pulmonary vein

Left atrium

Lung

Capillaries
Right atrium
Tricuspid valve

Pulmonary valve
Bicuspid valve
Aortic valve
Left ventricle
Aorta

Right ventricle
Inferior vena cava

Capillaries

Tissue cells

Figure 23.12 Congestive heart failure causes back-pressure behind one or both ventricles.

TABLE 23.9 Manifestations of left- and right-sided heart failure

Left-Sided Failure	Right-Sided Failure
Pulmonary edema	Jugular vein distention
Dyspnea, orthopnea	Peripheral edema
Nonproductive cough	Nocturia
Pleural effusion	Hepatomegaly
Rales, wheezing	
Cardiac asthma	
Cardiomegaly	

breathlessness, and **orthopnea,** a sense of breathlessness while in a supine position, occur. A nonproductive cough is typical, although pink frothy sputum is produced in severe pulmonary edema. Pleural effusion (chapter 6), an excessive accumulation of fluid between the membranous pleura surrounding the lungs, is caused by increased hydrostatic pressure in pulmonary capillaries. Additional pulmonary manifestations are **rales,** difficult breathing with a whistling sound, and a rattling sound with breathing called **wheezing.** The exact mechanisms are not clear, but bronchospasms, referred to as cardiac asthma, occur when pulmonary congestion is severe. Cardiomegaly or cardiac enlargement often occurs in left-sided failure, due mainly to dilatation of the left ventricle.

orthopnea: Gk. *orthos,* straight

Right-Sided Failure

The primary effect of right ventricular failure is increased systemic venous pressure; thus, manifestations of right-sided failure are associated with systemic congestion. Jugular veins in the neck become distended as the result of increased venous pressure. A series of events, initiated by a decrease in blood volume on the arterial side, lead to edema (chapter 2). Decreased arterial blood volume stimulates release of renin by the kidney (chapter 11) with subsequent increase of angiotensin and aldosterone. Aldosterone promotes sodium and water retention. Increased blood volume causes elevation of hydrostatic pressure and outflow of fluid into the interstitium. The resulting edema is often noted first in the lower extremities. Nocturia, which occurs early in congestive heart failure, is the result of redistribution of edematous fluid when the individual lies down. The consequent increase in blood volume leads to an increased glomerular filtration rate and increased urine formation. Hepatomegaly, an enlargement of the liver, often occurs as the result of systemic venous congestion. Venous blood, collected from capillaries of the abdominal viscera by the hepatic portal vein, flows through the sinusoids of the liver, proceeds to hepatic veins, and ultimately is carried by the inferior vena cava to the heart. Venous congestion of both the liver and spleen occurs when there is inadequate pumping action by the right ventricle.

General Manifestations

Cardiac arrhythmias frequently accompany congestive heart failure, and there is an elevation of both systolic and diastolic pressures. Tachycardia is characteristic and is due

TABLE 23.10 Procedures for the diagnosis of heart disease

Chest roentgenogram (X-ray). To determine heart size and shape; configuration of pulmonary artery and aorta; calcium deposits; lung congestion

Electrocardiogram (EKG or ECG). Records electrical activity of heart

Echocardiography. Utilizes ultrasound to visualize internal heart structures

Angiography. Contrast material injected into vessels or heart chambers; visualized by X-ray; may inject radioactive isotope that emits gamma rays which are detected by a gamma camera

Catheterization. Flexible catheter into a blood vessel and into heart to measure chamber pressures and blood gases

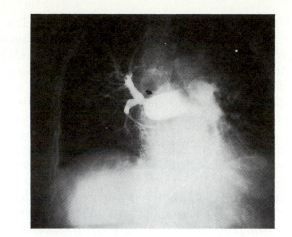

Figure 23.13 A chest X-ray taken during injection of contrast media into right pulmonary artery. Stricture in right pulmonary artery in area of arrow. Artery is dilated proximal to stricture.

in part to increased sympathetic nervous system activity. There are symptoms of inadequate blood flow or perfusion of the skin and skeletal muscle when blood is diverted to the more vital areas, such as the brain and heart. The skin becomes cool, pale, and sometimes cyanotic, a bluish color due to inadequate oxygen. There may be a mild fever because vasodilation is required to radiate heat, and in CHF, norepinephrine and angiotensin II cause peripheral vasoconstriction. As CHF progresses, symptoms of inadequate blood flow to the brain include confusion, insomnia, and depressed levels of consciousness.

Cheyne-Stokes breathing may occur and is characterized by periods of apnea alternating with periods of hyperpnea or rapid breathing (chapter 7). This reaction is probably caused by a prolonged circulation time between the pulmonary venous system and the medullary respiratory centers.

MANAGEMENT

The management of congestive heart failure involves the treatment of contributing or aggravating factors, such as hypertension or valvular disease. In general, direct intervention in the disease process of heart failure includes drug therapy to increase the pumping action of the heart and a program to reduce the work load of the heart, i.e., rest, vasodilators, and treatment of obesity. A third aspect of management is control of sodium and water retention by dietary restrictions and administration of diuretics.

DIAGNOSTIC PROCEDURES

Table 23.10 lists tests which are done to diagnose heart disease. A chest X-ray is useful in showing the size and shape of the heart and is often the first evidence of heart disease. Electrocardiogram tracings show rhythm and electrical activity of the heart. Echocardiography utilizes ultrasound to show internal cardiac structure recorded by an oscilloscope. Ultrasonic beams are sound waves with a frequency, greater than 20,000 cycles per second, that is

not detectable by the human ear. When the beams are directed toward various parts of the heart, they travel at a velocity related to the density and elastic properties of the tissue, and variations in this pattern reveal a picture of the inside of the heart.

Angiography (an″je-og′rah-fe) is a procedure whereby radiopaque or contrast media is injected into a blood vessel in order to visualize vessels and blood flow by means of X-ray (figure 23.13). A radioactive isotope may be injected, and gamma rays are recorded by a gamma camera. Finally, blood pressure and blood gases may be measured by way of a catheter introduced into a blood vessel, such as the subclavian vein, and on into the heart (Swan-Ganz catheter, chapter 21).

SUMMARY

• • •

Ischemic heart disease, usually caused by atherosclerosis, results in angina pectoris and myocardial infarction. Infarction refers to tissue necrosis that typically causes severe chest pain and cardiac arrhythmias and may lead to sudden death. Rheumatic fever is a sequel of streptococcal pharyngitis in which there is a generalized inflammatory reaction, with the possibility of permanent heart valve damage.

In general, heart failure is caused by (1) myocardial disease, (2) volume or pressure overload, (3) restricted ventricular filling, and (4) increased demands. Cardiomyopathies are examples of myocardial disease and these include (1) hypertrophic cardiomyopathy, an idiopathic

angiography: Gk. *aggeion,* vessel

condition in which the walls and septum of the left ventricle are thickened, (2) dilated cardiomyopathy with enlarged ventricles and impaired contractility, and (3) restrictive cardiomyopathy in which ventricular filling is impaired. Evidence suggests that toxic or infectious agents may cause dilated cardiomyopathy. The cause of noncompliant ventricles associated with restrictive cardiomyopathy is infiltration of the heart muscle by glycogen or fibrous connective tissue. Myocarditis, an inflammation of the heart muscle, is a second category of myocardial disease.

Increased pressure or volume within the heart chambers represents an increased work load for the heart and may lead to heart failure. Possible causes of increased work load include pulmonary embolism, hypertension, pulmonary hypertension, congenital heart malformations, and damage to cardiac valves.

Heart failure is caused by restricted ventricular filling, a sequel to an inflammation of the pericardium or the result of an accumulation of pericardial fluid. Hyperthyroidism and anemia add to the work load of the heart and are examples of increased demand that may lead to heart failure.

Congestive heart failure is the failure of one or both ventricles resulting in an accumulation of blood in the lungs and/or the systemic venous system. The body adjusts to the congestion and low cardiac output by (1) cardiac dilatation and hypertrophy, (2) increased sympathetic activity, (3) increased levels of angiotensin, aldosterone, and vasopressin, (4) enhanced oxygen transport, and (5) increased blood flow in upper lobes of the lungs. Modalities of treatment for CHF include drug therapy to increase the pumping action of the heart and a program to reduce the work load of the heart.

Diagnostic procedures to identify the cause of heart disease include chest X-ray, EKG, echocardiography, angiography, and heart catheterization.

REVIEW QUESTIONS
•••

1. What is the most common cause of ischemic heart disease?
2. What is the term for chest pain due to exertion?
3. What are three major consequences of myocardial infarction?
4. List five major manifestations of rheumatic fever.
5. What are four minor manifestations of rheumatic fever?
6. What is the significance of a high antistreptolysin O titer?
7. What are four general categories of problems that lead to heart failure?
8. List three types of cardiomyopathies.
9. How do systemic and pulmonary hypertension contribute to heart failure?
10. What effect does coarctation of the aorta have on the ventricles?
11. What is an example of a cause of volume overload?
12. Identify each of the following as causing either right- or left-sided heart failure: (a) hypertension, (b) mitral regurgitation, (c) pulmonary hypertension, (d) cor pulmonale, (e) atrial septal defect, (f) aortic stenosis.
13. What are three compensatory responses to progressive heart failure?
14. How is water balance affected by left-sided failure?
15. The following are manifestations of (right- or left-) sided failure? (a) jugular vein distention, (b) peripheral edema, (c) nocturia, and (d) hepatomegaly.

SELECTED READING
•••

Barrett-Connor, E., K. Khaw, and S. Yen. 1986. A prospective study of dehydroepiandrosterone sulfate, mortality, and cardiovascular disease. *New England Journal of Medicine* 315:1519–24.

Benowitz, N. L. 1988. Toxicity of nicotine: Implications with regard to nicotine replacement therapy. *Progress in Clinical Biological Research* 261:187–217.

Brachfeld, N. 1989. Acute chest pain: Cardiac or noncardiac? *Physician Assistant* 13:109–18.

Casscells, W. 1986. Special report: Heart transplantation. *New England Journal of Medicine* 315:1365–68.

Felicetta, J. V. et al. 1988. Cardiac involvement in hypertension. *American Journal of Cardiology* 61:67H–72H.

Hegele, R. A. et al. 1986. Apolipoprotein B-gene DNA polymorphisms associated with myocardial infarction. *New England Journal of Medicine* 315:1509–15.

Isner, J. et al. 1986. Acute cardiac events temporally related to cocaine abuse. *New England Journal of Medicine* 315:1438–43.

Kaplan, N. M. 1988. Maximally reducing cardiovascular risk in the treatment of hypertension. *Annals of Internal Medicine* 109:36–40.

Knapp, H. et al. 1986. In vivo indexes of platelet and vascular function during fish-oil administration in patients with atherosclerosis. *New England Journal of Medicine* 314:937–42.

Lobo, R. A. 1988. Lipids, clotting factors, and diabetes: Endogenous risk factors for cardiovascular disease. *American Journal of Obstetrics and Gynecology* 158:1584–91.

Maron, B. et al. 1987. Hypertrophic cardiomyopathy. *New England Journal of Medicine* 316:780–89.

Needleman, P. et al. 1985. Atriopeptins as cardiac hormones. *Hypertension* 7:469–79.

Perry, K. 1988. Indications and complications of percutaneous transluminal coronary angioplasty. *Journal of the American Academy of Physician Assistants* 1:441–47.

Qualey, D. A. 1989. The automatic implantable cardioverter-defibrillator. *Physician Assistant* 13:32–40.

Raine, A. et al. 1986. Atrial natriuretic peptide and atrial pressure in patients with congestive heart failure. *New England Journal of Medicine* 315:533–37.

Reed, J. D., Jr. et al. 1988. Cardiovascular MRI: Current role in patient management. *Radiology Clinical of North America* 26:589–606.

Surawicz, B. 1990. Ventricular arrhythmias in patients with congestive heart failure. *Physician Assistant* 14:111–28.

Tofler, G. H. et al. 1987. Concurrent morning increase in platelet aggregability and the risk of myocardial infarction and sudden cardiac death. *New England Journal of Medicine* 316:1514–18.

Zuza, J. R. 1990. An update on mitral and aortic valvular disease. *Journal of American Academy of Physician Assistants* 13:99–114.

Chapter 24

Congenital Heart Defects

Abnormal embryonic development results in cardiovascular malformations in 1% of all live births. In general, these structural defects cause abnormal patterns of blood flow that have significant physiological consequences. Depending on the severity of the problem, surgical repair is the approach to management.

The following review of fetal circulation provides a basis for the subsequent discussion of congenital defects.

FETAL CIRCULATION

The site of oxygenation of blood determines the differences between fetal circulation and the postnatal pattern of blood flow. Figure 24.1 is a diagram of fetal circulation with a more detailed outline shown in table 24.1. Fetal circulation, for the most part, bypasses both the liver and the lungs. The **ductus venosus** returns blood from the placenta, to the inferior vena cava and to the right atrium. The blood in the right atrium may (1) pass through the interatrial opening, the **foramen ovale** (ōva′lē), to the left atrium or (2) go to the right ventricle, (3) to the pulmonary artery, and (4) either to the lungs or to the aorta by way of the **ductus arteriosus.** Blood returns to the placenta via **umbilical arteries.** Table 24.2 summarizes changes in circulation that occur after birth. There is decreased pulmonary resistance to blood flow combined with constriction or closure of the foramen ovale, the ductus venosus, and the ductus arteriosus after birth. A postnatal pattern of blood flow is established, i.e., to the lungs from the right ventricle via the pulmonary artery and return to left heart from the lungs by way of pulmonary veins (figure 24.2).

Congenital heart defects are the next topics of discussion with emphasis on the physiological effect of each one. Surgical intervention may be required for treatment of these disorders.

PATENT DUCTUS ARTERIOSUS

The ductus arteriosus carries blood from the bifurcation of the pulmonary artery directly to the aorta just distal to the left subclavian artery (figure 24.3). Blood flow to the lungs is not required since fetal blood is oxygenated in the placenta. The term ductus arteriosus indicates that it is a duct that connects two arteries. Normally, this vessel closes functionally within 24 hours of birth due to an increase in arterial oxygen and is anatomically obliterated by intimal proliferation and fibrosis within 2 months. The condition, patent (pā′tent) ductus arteriosus, is the failure of this passage to close after birth.

foramen: L. *foramen,* opening
patent: L. *patere,* to be open

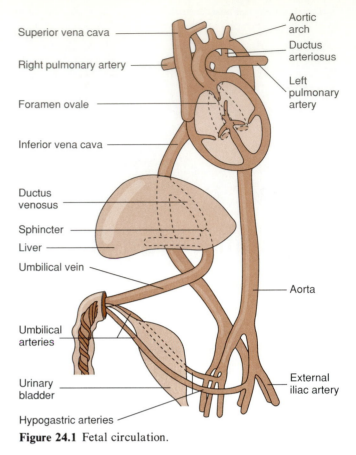

Figure 24.1 Fetal circulation.

TABLE 24.1 Outline of fetal circulation

1. Umbilical vein carries oxygenated blood from placenta
   a. 50% of blood from placenta carried by umbilical vein joins blood in hepatic portal vein; carried by ductus venosus to inferior vena cava
   b. 50% of blood from placenta carried by a branch of the umbilical vein to the liver and then to inferior vena cava
2. Blood from inferior and superior vena cavae to right atrium
3. From right atrium
   a. to left atrium by way of foramen ovale and then to left ventricle and to aorta
   b. to right ventricle and then pulmonary artery; directly to aorta by way of ductus arteriosus
   c. to pulmonary artery and to lungs; then back to left atrium; little blood flow due to high pulmonary resistance
4. Aorta to all parts of body
5. Umbilical arteries (branches of internal iliac arteries) return blood to placenta

OCCURRENCE

Patent ductus arteriosus occurs more commonly in (1) females, (2) in cases of maternal rubella, (3) when there is premature hypoxemia, and (4) in birth at high altitude.

TABLE 24.2 Changes in fetal circulation at birth

Pulmonary Circulation
   a. Fetal pulmonary arteries and arterioles are surrounded by fluid and have relatively thick walls
   b. Inflation of lungs at birth causes pulmonary vessels to be surrounded by air; vasodilation of pulmonary vessels due to increased oxygen; result is decreased pulmonary vascular resistance

Ductus Arteriosus
   a. Closes functionally within 24 hours by constriction, due to increased arterial oxygen content and changes in prostaglandins
   b. Closes anatomically by thrombosis, intimal proliferation, and fibrosis within 2 months

Foramen Ovale
   a. At birth blood flows from left to right atrium and causes flap valve on left side of atrial septum to close
   b. Permanent closure in several months

Ductus Venosus
   a. Constricts after birth and shunts blood to liver

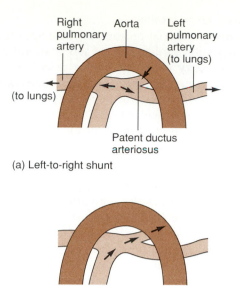

(a) Left-to-right shunt

(b) Right-to-left shunt

Figure 24.3 (*a*) A patent ductus arteriosus after birth. A large defect results in a left-to-right shunt with blood flowing from the aorta to the lungs by way of pulmonary arteries. (*b*) When pulmonary resistance is high, unoxygenated blood goes directly to the aorta from the pulmonary artery.

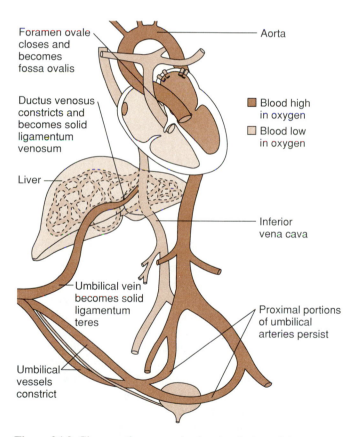

Figure 24.2 Changes that occur in the circulation of the newborn.

TABLE 24.3 Summary of effects of patent ductus arteriosus

Left-to-Right Shunt (blood flows from aorta to lungs)

1. Pulmonary vascular engorgement
   a. Pulmonary hypertension
   b. Pulmonary edema in premature infant
2. Increased volume returned to left heart
   a. Cardiac failure in premature infant
   b. Enlargement of left atrium and ventricle

Right-to-Left Shunt (blood flows from pulmonary artery to aorta; shunt reversal due to increased pulmonary resistance)

1. Lower body cyanosis
   a. Lower body receives inadequately oxygenated blood
   b. Blood flows to upper body proximal to ductus arteriosus; blood oxygenated
2. Aorta somewhat enlarged in older patients

PHYSIOLOGICAL EFFECTS

The consequences of an open ductus arteriosus depend on the size of the defect and on the extent of pulmonary vascular resistance. If the defect is large and there is minimal resistance in pulmonary vessels, there is a **left-to-right**

shunt. Blood flows from the aorta to the pulmonary artery and then to the lungs. The result is increased blood volume in the vasculature of the lungs and, consequently, increased volume returned to the left side of the heart. Pulmonary hypertension develops, and there is an increased work load for the left side of the heart.

Table 24.3 summarizes possible consequences of an open ductus arteriosus, including a shunt in either direction. A **right-to-left shunt** occurs when pulmonary resistance is high. The effect is to direct unoxygenated blood directly to the aorta and to supply the lower body with blood with low oxygen content.

Congenital Heart Defects

Pulmonary Hypertension

An increase in arterial pressure in the lungs occurs with many congenital cardiac defects and is a central feature of the pathology of patent ductus arteriosus. The increased pressure, called **pulmonary hypertension,** is due to increased blood flow and/or pulmonary resistance. Normally, pulmonary vascular resistance falls after birth and there is subsequent thinning of the medial smooth muscle of pulmonary arteries. Two things occur as the result of the shunting of blood to the lungs:

1. The defect interferes with the growth of pulmonary vessels. Normally, the vessels that supply the acini (chapter 6), in particular, increase in size and number throughout childhood, and this increases blood carrying capacity without an increase in pressure. Increased blood volume in pulmonary vessels interferes with development of these intra-acinar (in″trah-uh-sin′ar) vessels and causes increased resistance and pressure.
2. A second factor in the development of pulmonary hypertension is the proliferation of intimal cells and medial thickening of the walls of pulmonary vessels. The causes are unknown, but certain factors have been implicated including increased blood flow, increased pulmonary arterial pressure, polycythemia, and hypoxia. Pulmonary hypertension may cause pulmonary edema in a premature infant (chapter 2).

MANIFESTATIONS

Pulmonary hypertension, as it develops in patent ductus arteriosus, causes a reversal of blood flow, so unoxygenated blood is shunted to the descending aorta. The lower body consequently becomes cyanotic, a bluish discoloration due to low oxygen content of blood. An increase of connective tissue occurs in the toes and is a condition called **clubbing.**

TETRALOGY OF FALLOT

The term tetralogy of Fallot (Falō′) refers to a condition in which there are four abnormalities of the heart (figure 24.4). Two basic defects are (1) incomplete closure of the wall between the two ventricles (ventricular septal defect), usually directly beneath the aortic valve and (2) **pulmonic stenosis,** a narrowing of the pulmonary artery that may include the valves. Two additional abnormalities are (3) aortic override of the ventricular septum, so it receives blood from both ventricles and (4) an enlarged right ventricle due to increased work load. Table 24.4 summarizes other abnormalities that may be associated with tetralogy of Fallot.

intra: L., within

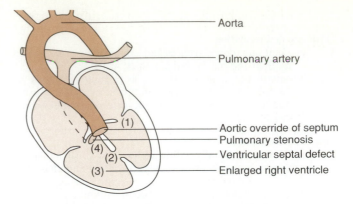

Figure 24.4 Four basic defects in tetralogy of Fallot.

TABLE 24.4 Summary of defects in tetralogy of Fallot

Basic Abnormalities

Ventricular septal defect
Pulmonic stenosis
Aortic override of interventricular septum
Right ventricular hypertrophy

Possible Accompanying Anomalies

Right aortic arch, 25%
Atrial septal defect, 15%
Left pulmonary artery originates from aorta instead of the main pulmonary artery, 3%

OCCURRENCE

Tetralogy of Fallot constitutes 10% of all forms of congenital heart disease and occurs more commonly in males. It is the most common cause of cyanosis after the first year of life.

PHYSIOLOGICAL EFFECTS

The degree of obstruction of outflow of blood from the right ventricle to the lungs determines the manifestations of this congenital abnormality. A severe pulmonic stenosis and a small to moderate ventricular septal defect results in a predominate right-to-left ventricle shunt. The lungs are bypassed and cyanosis is the evidence of inadequate oxygenation. If the pulmonic stenosis is mild to moderate, there may be a minimal right-to-left shunt with **acyanotic tetralogy,** i.e., the absence of cyanosis.

Persistent cyanotic spells lead to metabolic acidosis (chapter 4). Acidosis develops when glucose is metabolized to pyruvic acid and, under anaerobic conditions, is subsequently converted to lactic acid. With adequate oxygenation, pyruvic acid is normally metabolized to carbon dioxide and water.

Inadequate oxygenation stimulates erythropoiesis and causes polycythemia. Cerebral thrombosis is a complication, and the formation of small blood clots in cerebral vessels tends to occur in severely cyanotic children.

tetra: Gk., four

Tetralogy may also be complicated by brain abscesses and an inflammation of the inner heart called **infective endocarditis.**

• • •

Hypercyanotic spells are common complications of tetralogy of Fallot and certain other types of cyanotic heart disease. Typically, there is a marked increase in cyanosis associated with anxiety and increased breathing rate. The cause is a sudden decrease in pulmonary blood flow due to (1) fluctuations in arterial pCO_2 and pH, (2) a decrease in systemic resistance, (3) an increase in pulmonary vascular resistance, or (4) tachycardia associated with right ventricular volume decrease. These episodes may result in convulsions and possibly death. Modalities of treatment include administration of oxygen and placing the patient in a knee-chest position.

• • •

MANIFESTATIONS

Most infants are either cyanotic from birth or develop cyanosis in the first year. This condition causes a bluish color of the skin and mucous membranes and is caused by a decreased oxygen saturation of hemoglobin. Typically, there is retardation of growth and development and a clubbing of the fingers and toes involving a thickening and widening of the terminal phalanges accompanied by convex nails. The underlying cause of clubbing is not clear, but the condition appears to develop in response to arterial hypoxemia. Dyspnea with exertion and resting in a squatting position are characteristic. The squatting behavior increases systemic pressure that helps to alleviate the right ventricular outflow obstruction.

VENTRICULAR SEPTAL DEFECT

A ventricular septal defect is an opening in the septum that separates the right and left ventricles. Small defects frequently close spontaneously during the first year of life, but larger defects cause serious problems.

OCCURRENCE

This is the most common congenital heart abnormality seen in children.

PHYSIOLOGICAL EFFECTS

If pulmonary vascular resistance is normal, the consequence of a ventricular septal defect is a left-to-right shunt that increases right ventricular pressure. The physiological effects are not harmful if the opening is small. In the

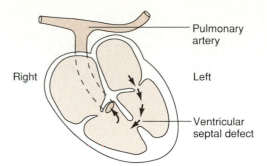

Figure 24.5 Ventricular septal defect causes a volume overload in the right ventricle, which leads to an increased return of blood to the left side via pulmonary veins.

first few weeks of life, elevated pulmonary vascular resistance, which is normal in fetal life, persists and protects against a large left-to-right shunt. The shunt increases with time as pulmonary resistance falls (figure 24.5) and both increased right ventricular pressure and left ventricular volume overload lead to right and left ventricular hypertrophy. Cardiac failure is the ultimate outcome.

MANIFESTATIONS

Large defects are associated with frequent pulmonary infections, growth retardation, and cardiac failure in infancy. Moderate left-to-right shunts cause exercise intolerance and fatigue. Pulmonary hypertension develops over a period of time (discussed in patent ductus arteriosus) and may become irreversible. This causes exertional dyspnea, chest pain, and may reverse the shunt to a right-to-left direction. This in turn leads to cyanosis, clubbing, and polycythemia.

ATRIAL SEPTAL DEFECT

Atrial septal defect is an opening in the interatrial septum usually in the area of the fetal foramen ovale.

OCCURRENCE

This condition is the most common congenital defect seen in adolescents and adults. It occurs more frequently in females in a ratio of 2:1.

PHYSIOLOGICAL EFFECTS

The opening allows blood to flow into the right atrium from the left side. In most cases, the defect is large enough to allow equalization of pressure in both atria, so the magnitude of the left-to-right shunt is determined by the compliance of the right ventricle. In essence, this is the right ventricular resistance to diastolic flow. Characteristically, there is a marked volume overload of the right ventricle (figure 24.6), and over a period of time, this overload leads to fibrosis, a thickening of the ventricular wall, and decreased compliance. There may be some decrease in the

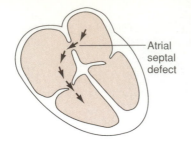

Figure 24.6 Atrial septal defect causes a marked overload in the right ventricle.

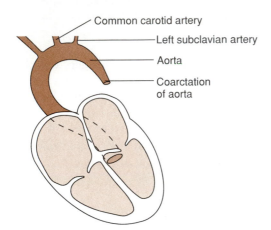

- Common carotid artery
- Left subclavian artery
- Aorta
- Coarctation of aorta

Figure 24.7 Coarctation of the aorta increases the workload of the left ventricle.

magnitude of the left-to-right shunt as the right ventricle becomes less compliant. The overall effect is overload of the right heart and lungs. Pulmonary hypertension develops (as in patent ductus arteriosus) in a small number of cases. If the defect is not surgically repaired it often leads to significant disability and may cause death by age 40.

MANIFESTATIONS

Atrial septal defect is frequently asymptomatic and may not be diagnosed until adulthood. The most frequent complaint is exertional dyspnea, which is probably caused by increased pulmonary blood volume. The incidence of pulmonary hypertension increases with age, and cardiac failure is more common after age 30.

COARCTATION OF AORTA

Coarctation (kō″ark-tā′shun) of the aorta is a congenital narrowing of the aorta typically just distal to the origin of the left subclavian artery (figure 24.7), although the origin of the subclavian artery may be affected as well.

coarctation: L. *coarctare,* to press together

OCCURRENCE

Males are affected by coarctation twice as often as females, and the condition may not be recognized until adulthood.

PHYSIOLOGICAL EFFECTS

The coarctation impedes blood flow and increases the work load (afterload) of the left ventricle (chapter 21). Typically, there is a significant increase in both systolic and diastolic pressures in the arms and a reduced blood pressure in the lower limbs. Only the right arm blood pressure may be increased if the coarctation involves the left subclavian artery. Left ventricular hypertrophy occurs over time, and congestive heart failure (chapter 23) is the ultimate outcome.

MANIFESTATIONS

Individuals affected by coarctation may remain asymptomatic until adulthood. There may be symptoms related to upper body hypertension, such as headache, fatigue, and even cerebral hemorrhage. Symptoms of congestive heart failure (chapter 23) usually occur after the age of 40.

SUMMARY
• • •

Table 24.5 summarizes the effects of the five congenital heart abnormalities discussed in this chapter. Cyanosis may be a symptom of patent ductus arteriosus, tetralogy of Fallot, and ventricular septal defect. In the case of a patent ductus arteriosus, blood flows from the pulmonary artery to the aorta as pulmonary hypertension develops and bypasses the lungs. If there is a severe pulmonic stenosis in tetralogy, the result is bypass of the lungs. Inadequate oxygenation causes metabolic acidosis, polycythemia, cerebral thrombosis, and clubbing of the digits. Ventricular septal defect causes a left-to-right shunt with increased right ventricular pressure and left ventricle volume overload. If pulmonary hypertension develops, the shunt may reverse to a right-to-left direction and cause cyanosis.

Atrial septal defect allows a left-to-right atrium shunt leading to hypertrophy of the right ventricle. Coarctation of the aorta increases the work of the left ventricle, causes upper body hypertension, and leads to congestive heart failure.

TABLE 24.5 Summary of the basic defect and overall effect in five congenital heart abnormalities

Patent ductus arteriosus. Open vessel connects pulmonary artery and aorta; blood flows from aorta to pulmonary artery (left-to-right shunt); if high pulmonary resistance, right-to-left shunt

Tetralogy of Fallot. Ventricular septal defect, pulmonic stenosis, aortic override of ventricular septum, right ventricular hypertrophy; (1) if severe pulmonic stenosis and moderate septal defect, right-to-left shunt with cyanosis, (2) if moderate pulmonic stenosis, then minimal right-to-left shunt and no cyanosis

Ventricular septal defect. Opening in ventricular septum; left-to-right shunt; causes increased right ventricular pressure and increased volume in left ventricle

Atrial septal defect. Opening in atrial septum; left-to-right shunt due to compliance of right ventricle.

Coarctation of aorta. Narrowing of aorta, which increases afterload of the left ventricle

REVIEW QUESTIONS

• • •

1. What fetal vessel returns blood to the inferior vena cava?
2. Where is the ductus arteriosus?
3. What is the foramen ovale?
4. Which direction is blood shunted (right-to-left or left-to-right) in patent ductus arteriosus?
5. Under what conditions is cyanosis a part of the picture in patent ductus arteriosus?
6. What are the four basic abnormalities in tetralogy of Fallot?
7. Which direction is blood shunted in tetralogy of Fallot?
8. What is the abnormal blood flow in ventricular septal defect?
9. Which direction is blood shunted in atrial septal defect?
10. What are the two major effects of coarctation of aorta?

SELECTED READING

• • •

Borman, B. et al. 1987. Using a national register for the epidemiological study of congenital heart defects. *New Zealand Medical Journal* 100:404–6.

Bruyere, H. J. et al. 1987. The causes and underlying developmental mechanisms of congenital cardiovascular malformations: A critical review. *American Journal of Medical Genetics* (Suppl) 3:411–31.

Bush, A. et al. 1988. Correlations of lung morphology, pulmonary vascular resistance, and outcome in children with congenital heart disease. *British Heart Journal* 59:480–85.

Johnson, C. A. et al. 1988. Circulating megakaryocytes and platelet production in children with congenital cardiac defects undergoing cardiac catheterization. *South African Medical Journal* 73:578–80.

Kaplan, S. 1988. The hemodynamic and clinical consequences of common congenital heart defects. *Journal of American Academy of Physician Assistants* 1:454.

Maron, B. J. et al. 1986. Development and progression of left ventricular hypertrophy in children with hypertrophic cardiomyopathy. *New England Journal of Medicine* 315:610–14.

Pietras, R. J. et al. 1988. Large pericardial effusions associated with congenital heart disease: Five- and eight-year follow-up. *American Heart Journal* 115:1334–46.

Chapter 25

Practical Applications: Cardiovascular Disorders

This chapter provides an expanded view of three specific disorders involving the heart and the vascular system, and correlates laboratory findings with the pathophysiology of these conditions. Examples of congestive heart failure, myocardial infarction, and septic shock are included, with a discussion of significant laboratory data.

Congestive Heart Failure

Congestive heart failure causes retention of salt and water resulting in peripheral edema. Normal homeostatic mechanisms are reviewed in the following paragraphs preceding a focus on failure of homeostasis in this and particular disease process. For additional review of fluid and electrolyte balance, see chapters 2, 3, and 11.

Normal Volume Control

Table 25.1 summarizes mechanisms for maintaining osmolality as well as volume of body fluids. Antidiuretic hormone (ADH) maintains osmolality of extracellular fluid by promoting reabsorption of water by the kidney. Increased osmolality, indicated by sodium ion concentration, stimulates the release of ADH by the posterior lobe of the pituitary with subsequent water reabsorption.

Aldosterone and angiotensin are interdependent. The initial stimulus for increased levels of these hormones may be decreased blood pressure, decreased sodium ion concentration, or increased potassium ion levels. The juxtaglomerular apparatus of the kidney releases renin, which catalyzes the production of angiotensin I and II. In turn, angiotensin II promotes the release of aldosterone from the adrenal cortex. The result is that aldosterone favors renal sodium ion reabsorption and consequently, water reabsorption as well. Angiotensin II causes vasoconstriction of renal arterioles, hence a decreased glomerular filtration rate, and decreased urine formation.

Natriuretic (na''tre-u-ret'ik) hormone is a polypeptide that promotes sodium ion excretion, vasodilation, and inhibits aldosterone. There is evidence that this hormone is released by the atria of the heart and that the release, at least in part, is stimulated by increased pressures within the atria. There are higher circulating levels of natriuretic hormone in congestive heart failure, which appears to be an appropriate physiological response to promote sodium ion and water loss.

There are pressure and stretch receptors within atrial walls and within the walls of the pulmonary artery and aorta as well. Increased pressure initiates nerve impulses that are transmitted to the brain and that inhibit the release of ADH. Further, increased pressure inhibits sympathetic renal neural transmission, resulting in an increased glomerular filtration rate with subsequent excretion of sodium ions and water.

TABLE 25.1 A summary of mechanisms for fluid and osmolal homeostasis

Antidiuretic Hormone (Vasopressin)
a. Release in response to increased osmolality of extracellular fluid
b. Promotes renal reabsorption of water

Aldosterone
a. Release in response to angiotensin II
b. Promotes renal reabsorption of Na^+

Angiotensin II
a. Release of renin in response to fall in blood pressure, decreased Na^+, increased K^+; angiotensin II produced, stimulates aldosterone release
b. Aldosterone favors renal retention of Na^+; angiotensin II causes vasoconstriction of renal arterioles, and thus decreased glomerular filtration

Natriuretic Hormone
a. Released by atria in response to increased atrial pressure
b. Favors Na^+ excretion

Atrial and Carotid Baroreceptors
a. Increased pressure initiates nerve impulses to brain; inhibits secretion of ADH
b. Also inhibits sympathetic impulses to kidney; result is increased GFR and increased excretion of Na^+ and water

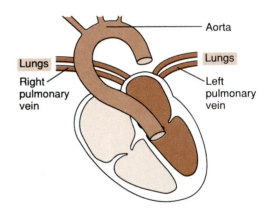

Figure 25.1 Congestive heart failure in which there is congestion behind both ventricles, resulting in pulmonary and systemic venous congestion.

Failure of Homeostasis

Congestive heart failure (CHF) is characterized by a decreased cardiac output and congestion behind both ventricles, i.e., atrial, pulmonary, and systemic venous congestion (figure 25.1). Fluid and salt retention complicates the problem of congestion. As previously discussed, there are physiological responses that would predictably promote fluid loss rather than retention under such circumstances. Although the mechanisms are not clear, there are disturbances in the normal pressure/volume relationships in this disorder that may include the following:

1. Chronic dilatation in CHF may change the sensitivity of atrial neuroreceptors to stretch and interfere with normal neural influence on ADH release and sympathetic control of glomerular filtration rate.
2. There is evidence indicating that the renin-angiotensin-aldosterone system overpowers the natriuretic hormonal influence on fluid and sodium ion excretion.

There are elevated levels of circulating natriuretic hormone in congestive heart failure, which promote loss of water and sodium ions, but opposing effects prevail. A low cardiac output results in diminished blood flow to the kidneys, which leads to a decreased glomerular filtration rate and also elicits a response of renin release.

Patient 25.1 was an 86-year-old male; Diagnosis: congestive heart failure; pleural effusion, cyanosis of extremities; edema of lower extremities.
Laboratory findings: BUN 40 mg/dl (7–22 mg/dl); Na 135 meq/liter (137–47 meq/liter); K 5.6 meq/liter (3.4–5.0 meq/liter); creatinine 2.0 mg/dl (0.6–1.2 mg/dl). (Normal values are indicated in parentheses.)

EDEMA AND ELECTROLYTES

Edema in patient 25.1 is evidence of water retention and increased systemic venous pressure. Congestion in the venous system increases hydrostatic pressure, promoting outflow of fluid into interstitial spaces (chapter 2). In the early stages of heart failure, the cardiac output may only be inadequate in response to exercise, which elicits a sympathetic response for the redistribution of blood flow to meet muscle requirements. There is transient sodium ion/fluid retention to improve cardiac output. As the disease progresses, renal blood flow decreases as the result of redistribution of blood to brain and muscle. Consequently, there is a decrease in glomerular filtration rate and marked sodium ion/fluid reabsorption. If this fluid retention improves cardiac output, sodium ions and water are no longer retained, and sodium ion balance is reestablished, but at a higher volume. In this state, however, the handling of sodium ions is abnormal, so a high dietary intake leads to retention.

In end-stage failure, the cardiac output is inadequate, and even at rest, there is peripheral vasoconstriction, decreased renal blood flow, and marked retention of both sodium ions and water. The concentration of sodium ions reported in patients with congestive heart failure may be in the normal range. Even though there is increased total body sodium ion content, it is in balance with increased volume. The sodium ion value may be lower than normal due to dilution, as it is in patient 25.1.

There is evidence indicating that enhanced reabsorption of sodium ions may not lead to a notable increase in

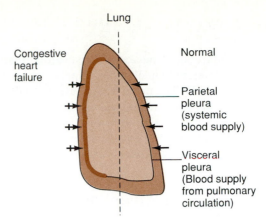

Figure 25.2 Normally, fluid lost from parietal capillaries is reabsorbed by capillaries of visceral pleura. In congestive heart failure, there is increased hydrostatic pressure in capillaries of visceral pleura, which limits reabsorption; thus, fluid accumulates between the two pleura (pleural effusion).

potassium ion secretion. The bulk of sodium ion reabsorption occurs in the proximal tubule, depending on the extent of diminished blood flow to the kidney. When reabsorption occurs in the proximal tubule, there is decreased delivery of sodium ions to the distal tubule where K^+ and Na^+ exchange occurs (chapter 11). Consequently, there may be no increased urinary excretion of potassium ions. Diuretic therapy must also be considered in the evaluation of serum electrolytes.

• • •

A decrease in plasma albumin, which may be caused by renal disease, nutritional deficiency or other disorders, causes a lowered plasma oncotic pressure. This favors edema formation and may limit the effectiveness of diuretic therapy in congestive heart failure.

• • •

PLEURAL EFFUSION

Pleural effusion is noted in patient 25.1. This refers to fluid accumulation between the parietal and visceral pleura surrounding the lungs (chapter 6). Normally, there is fluid loss from systemic capillaries of the parietal pleura balanced by reabsorption by pulmonary capillaries of visceral pleura. In the case of congestive heart failure, there is increased hydrostatic pressure in pulmonary vessels which limits reabsorption; hence, fluid accumulates in the pleural space (figure 25.2).

BUN/CREATININE LEVELS

Both the blood urea nitrogen and creatinine levels for patient 25.1 are elevated and may be correlated with impaired glomerular filtration rate (chapter 16). Further, a BUN/creatinine ratio >20:1 is characteristic of decreased renal blood flow.

MYOCARDIAL INFARCTION

Myocardial infarction is necrosis of myocardial tissue as the result of interruption of blood supply to the heart muscle (figure 25.3). The underlying cause is usually atherosclerosis of the coronary arteries in which there is an accumulation of lipids and proliferation of smooth muscle in the arterial intima. Hypercholesteremia is an important risk factor for atherosclerosis and myocardial infarction (chapter 22).

PLASMA LIPIDS AND CORONARY ARTERY DISEASE

Table 25.2 summarizes guidelines for correlating plasma lipid levels and risk for coronary artery disease. Plasma lipids include cholesterol, triglycerides, and phospholipids transported in plasma as a protein and mixed-lipid complex. These lipoproteins are categorized on the basis of size and density (chapter 22). The low-density lipoprotein (LDL) fraction is the major cholesterol carrying fraction and increased LDL values are associated with high coronary artery disease risk. There is evidence indicating that the high-density lipoprotein (HDL) fraction promotes cholesterol removal from cells and is an indicator of low risk.

Low levels of HDL cholesterol, as well as some elevation of LDL or triglycerides, are frequently observed in patients with coronary artery disease.

Apolipoproteins are protein carriers complexed to plasma lipids (chapter 22). A specific apoprotein, apo A-I, is a major protein in HDL, and studies show that an increased level of apo A-I is a predictor of low risk of coronary artery disease.

Diabetes is associated with a moderate increase in triglyceride levels, a decrease in HDL cholesterol levels, and increased risk for atherosclerosis. The diabetic individual may be unresponsive to the action of insulin or may have a deficiency of insulin (chapter 32). Table 25.3 indicates mechanisms that lead to elevated triglyceride levels in diabetes.

> Patient 25.2 was a 65-year-old female; she had acute chest pain that radiated to her neck and left arm.
> Diagnosis: myocardial infarction.
> Laboratory findings: cholesterol 218 mg/dl (140–250); triglycerides 386 mg/dl (45–232); LDL 106 mg/dl (95–215); HDL 35 mg/dl (35–95)

Patient 25.2 was diagnosed as having suffered a myocardial infarction. Only the triglyceride level is indicative of increased risk for coronary artery disease. The other values are within normal range, and the LDL/HDL ratio of 3 indicates average risk.

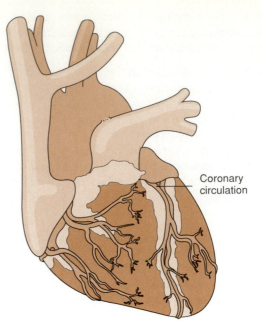

Figure 25.3 Coronary vessels supply the heart muscle with blood.

TABLE 25.2 General guidelines for correlating plasma lipid levels and risk for coronary artery disease

LDL Cholesterol Level

Low Risk	Intermediate Risk	High Risk
<100 mg/dl	100–170 mg/dl	>170 mg/dl

Ratio of LDL/HDL Cholesterol

Low Risk	Moderate Risk	Average Risk	Significant Risk	High Risk
ratio <2	ratio 2–3	ratio 3	ratio 3–5	ratio >5

Triglycerides
Associated with decreased HDL
High levels indicate increased risk
? >200–250 mg/dl borderline range

Apolipoproteins
Apo A-I, higher concentration indicates lower risk

Diabetes Mellitus
Associated with increased triglycerides

CARDIAC ENZYMES

One criterion for diagnosing myocardial infarction is an elevation of serum levels of cardiac enzymes. Injury to heart muscle causes the release of enzymes from myocardial cells (figure 25.4). Table 25.4 summarizes information about three of those enzymes. **Serum glutamic-oxalocetic transaminase (SGOT)** is now called **aspartate aminotransferase (AST)**. An increase of this enzyme is a sensitive test for myocardial infarction, but an increase occurs in other conditions including liver disease, skeletal muscle trauma (such as IM injections), surgery, shock,

TABLE 25.3 Mechanisms leading to elevated triglycerides in diabetes mellitus

Obese diabetics who are insensitive to the action of insulin require increased doses of insulin

1. a. VLDL synthesis is increased secondary to increased insulin
    b. VLDL fraction is mainly endogenous triglyceride

Diabetics who lack insulin

1. a. Lipoprotein lipase promotes breakdown of triglycerides and chylomicrons (tiny fatty particles)
    b. Insulin is required for lipoprotein lipase activity
    c. With insulin deficiency there is decreased lipoprotein lipase activity; the result is accumulation of triglycerides and chylomicrons
2. a. Free fatty acids are required for the synthesis of triglycerides
    b. Insulin causes a decreased release of free fatty acids from fat cells
    c. Insulin deficiency leads to increased release of free fatty acids

and in response to certain drugs. There is a direct relationship between enzyme levels and the degree of heart muscle damage.

Lactate dehydrogenase (LDH) is found in most body tissues and increased levels occur, in addition to myocardial infarction, in liver disease, skeletal muscle trauma (such as IM injections), pulmonary embolism, anemia, leukemia, and others. As table 25.4 indicates, there is a prolonged elevation in myocardial infarction, probably due to the release of the enzyme from inflammatory cells in the area of necrosis.

Creatine kinase (CK) levels in serum is a sensitive test for myocardial infarction. Elevations do not occur in heart failure and liver disease, but occur in skeletal muscle irritation or injury (vigorous exercise, IM injections), diabetes, pulmonary embolism, and others. **Isoenzymes** of creatine kinase are measured to aid in the diagnosis of myocardial infarction. These are forms of the enzyme that have somewhat different catalytic properties. CK MB is an isoenzyme of creatine kinase, and the serum level of CK MB is a sensitive test for myocardial infarction, provided that blood is drawn at the appropriate times. There is an increase within 3–15 hours of the infarction, the peak increase occurs within 12–24 hours, and the duration is 1–3 days or less.

> Patient 25.3 was a 63-year-old male; he had substernal and throat tightness; nausea and vomiting; history of hypertension and diabetes; blood pressure 70/50; diagnosis, myocardial infarction
> Laboratory findings: about 2 hrs. after symptoms—WBC 12,900/ mm³ (4,000–10,700); cholesterol 105 mg/dl (140–250); AST 202 units/liter (13–40); LDH 339 units/liter (87–170); CK 132 units/liter (20–220); 1 hr. later, 328 units/liter

chylomicron: Gk. *chylos,* juice; Gk. *mikros,* small
iso: Gk. *isos,* same

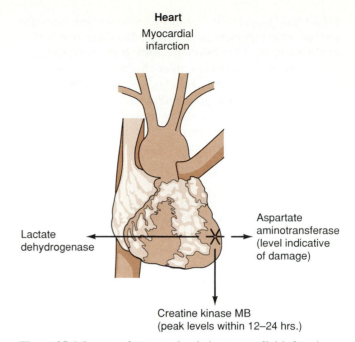

Heart
Myocardial infarction

Lactate dehydrogenase

Aspartate aminotransferase (level indicative of damage)

Creatine kinase MB (peak levels within 12–24 hrs.)

Figure 25.4 Increased enzyme levels in myocardial infarction.

TABLE 25.4 Characteristics and sensitivity of elevations of three enzymes as indicators of myocardial infarction

	Aspartate Aminotransferase (AST)	Lactate Dehydrogenase (LDH)	Creatine Kinase (CK)
Location	Myocardium; liver, skeletal muscle, kidneys, brain	Most tissues	Myocardium; skeletal muscle, brain
High Levels	MI; also in liver disease	MI; others	MI; others
Appearance	8–12 hrs.	24 hrs.	6–15 hrs.
Peak Levels	24–48 hrs.	48–72 hrs.	24–36 hrs.
Duration	3–5 days	7–14 days	1–4 days

Day 2 21 hrs. later—CK 2137 units/liter
12 hrs. later— CK 3029 units/liter
Day 3 24 hrs. later—CK 2225 units/liter

The timing of peak enzyme levels is a clue to the duration of the infarction and the degree of elevation often correlates with the severity of myocardial damage. Marked elevation, 4–5 times normal, indicates a less favorable prognosis. Normal serum enzyme levels during the first 48 hours after the onset of symptoms rules out myocardial infarction. Patient 25.3 showed elevated enzyme levels within a few hours of the onset of symptoms. The creatine kinase level is particularly valuable information, since this enzyme shows an increase in myocardial infarction in 90% of the patients, if the blood is drawn at the right times. It also allows early diagnosis, since the level rises and peaks within the first day or so. Further, this enzyme level is not elevated by liver damage or pulmonary infarction, which sometimes causes confusion in diagnosis. Certainly the timing and the enzyme elevations in patient 25.3 are indicative of myocardial infarction.

LEUKOCYTOSIS

The white blood cell count of this patient is somewhat elevated and this often occurs a few hours after the onset of chest pain in myocardial infarction. The peak is usually 12,000–15,000 cells/mm^3 and usually returns to normal within a week. This increase may be a response to stress and/or to tissue necrosis.

SEPTIC SHOCK

Shock refers to circulatory failure, usually associated with hypotension, resulting in inadequate tissue perfusion and cell injury. Three categories of shock are discussed in chapter 21, i.e., hypovolemic, cardiogenic, and distributive shock. In the case of distributive shock, the most common cause is septicemia. Consequently, the condition is called septic shock. The following is an example of septic shock.

> Patient 25.4 was a 78-year-old female who was admitted to the hospital in a lethargic condition with a respiratory rate of 48 and a pulse of 150. Her systolic pressure was first measured at 90 mm Hg and a short time later at 60 mm Hg.
> Laboratory findings: WBC 8,400/mm^3 (4,000–10,700); Hct 46% (42 ± 5); pO$_2$ 54 mm Hg (75–100); pCO$_2$ 14.7 mm Hg (35–45); pH 7.39 (7.35–7.45); base excess −15.3 (+2.5 to −2.5); Creatinine 5.0 mg/dl (0.6–1.6); BUN 76 mg/dl (10–20)

Patient 25.4 initially felt that she was catching a cold and, over a three day period, complained of discomfort on the right side of the neck. At the time of hospital admission, her breathing was heavy; she was difficult to arouse; and her skin was warm and dry. There was soft tissue swelling and tenderness in the neck region. Her systolic blood pressure increased to 100 mm Hg after the administration of intravenous fluids. A blood culture was taken and the final diagnosis was (1) cellulitis of the neck, (2) septicemia caused by group A *Streptococcus pyogenes*, (3) septic shock, (4) acute renal insufficiency, and (5) acute metabolic acidosis.

CELLULITIS

Cellulitis is an inflammation of the skin and subcutaneous tissues frequently caused by group A streptococcal organisms, as was the case in patient 25.4. The condition is characterized by tenderness, erythema, and fever.

ABNORMALITIES

The manifestations of shock vary, depending on pathogenesis and whether an early or late stage is involved. One pattern of abnormalities in septic shock may be the result of arteriovenous shunting (chapter 21) and include (1) normal cardiac output, (2) a normal or elevated pH, and (3) a decrease in peripheral resistance due to vasodilation. It is likely that in such a case blood is shunted directly from arterioles to venules, bypassing gas exchange in capillaries. The condition is characterized by warm dry skin, hypotension, oliguria, and acidosis due to lactic acid accumulation. Marked metabolic acidosis is a sign of a serious deficiency in tissue perfusion.

In patient 25.4, increased ventilation was a compensatory response to metabolic acidosis; thus, hyperventilation accounts for the low pCO$_2$ (respiratory alkalosis). The normal pH shows overall respiratory compensation. Acidosis occurs as the result of inadequate delivery of oxygen to tissues resulting in anaerobic metabolism with the accumulation of lactic acid. Values for pO$_2$ in septic shock are frequently less than 70 mm Hg, and this was true in patient 25.4.

The white blood cell count was normal, although leukocytosis is common in septic shock. The hematocrit is often initially elevated in shock, due to increased capillary permeability and fluid loss, although the value is normal in this patient.

Oliguria is evidence of inadequate blood flow to the kidneys and renal failure. A BUN/creatinine ratio of 20:1 is a clue that suggests prerenal azotemia (chapter 15).

The condition of patient 25.4 was stabilized after the administration of intravenous fluids, and antibiotic treatment was instituted to treat the suspected septicemia. An indwelling catheter allowed a measurement of the hourly urine output to monitor visceral perfusion.

SUMMARY

• • •

Hormonal influence is a major aspect of volume control and includes ADH, angiotensin, aldosterone, and natriuretic hormone. Increased osmolality is a stimulus for ADH release while release of renin by the kidney occurs in response to decreased blood pressure, decreased sodium ions, or increased potassium ions. The result is an increase in angiotensin/aldosterone, which favor vasoconstriction as well as sodium ion and water retention. Increased atrial pressure stimulates the release of natriuretic hormone with subsequent sodium ion and fluid loss. Atrial and carotid baroreceptors respond to increased pressure by neural inhibition of ADH and inhibition of renal sympathetic impulses to decrease glomerular filtration rate. Congestive heart failure is complicated by fluid and salt retention with disturbances in the normal pressure/volume relationships. Edema is the evidence of water retention and dilutional hyponatremia.

Inadequate cardiac output leads to impaired renal blood flow, a decrease in glomerular filtration rate, and subsequent increase in BUN/creatinine levels.

Increased plasma lipids represent a risk factor for coronary artery disease and myocardial infarction. High levels of the low-density lipoprotein fraction, a high LDL/HDL ratio, and elevated triglyceride levels are indicators of increased risk. Cardiac enzymes are increased as the result of myocardial infarction. These enzymes include aspartate aminotransferase, lactate dehydrogenase, and creatine kinase. The sequence in which these enzymes increase and the duration of the elevation is useful in the diagnosis of myocardial infarction.

Septic shock is circulatory failure due to septicemia, with consequent hypotension and cell injury. Septic shock may cause arteriovenous shunting, and this is associated with decreased peripheral resistance and inadequate capillary/cell gas exchange. The manifestations include warm dry skin, oliguria, and metabolic acidosis. Respiratory alkalosis may occur due to hyperventilation, a compensatory response to metabolic acidosis.

REVIEW QUESTIONS

• • •

1. What hormone promotes sodium ion/fluid loss?
2. What are two hormones that favor fluid retention?
3. What is a normal hormonal response to increased atrial pressure?
4. Why are BUN/creatinine levels increased in CHF?
5. Why are LDL plasma levels significant?
6. What is the significance of HDL plasma levels?
7. What are apolipoproteins?
8. Why is diabetes a risk factor in coronary artery disease?
9. What three enzymes are elevated during myocardial infarction?
10. Which of the three enzymes is the most specific, sensitive test for myocardial infarction?
11. Give an example in which warm dry skin may be associated with shock.
12. In septic shock, what is a possible explanation for hypotension and tissue anoxia?

SELECTED READING

• • •

Cannon, R. et al. 1988. "Microvascular angina" as a cause of chest pain with angiographically normal coronary arteries. *American Journal of Cardiology* 61:1338–43.

Cohn, J. et al. 1984. Plasma norepinephrine as a guide to prognosis in patients with chronic congestive heart failure. *New England Journal of Medicine* 311:819–28.

Colditz, G. et al. 1987. Menopause and the risk of coronary heart disease in women. *New England Journal of Medicine* 316:1105–10.

Kirlin, P. et al. 1988. Spontaneous neurohormonal fluctuation in chronic congestive heart failure. *American Journal of Cardiology* 62:150–51.

LaCroix, A. et al. 1986. Coffee consumption and the incidence of coronary heart disease. *New England Journal of Medicine* 315:978–82.

Laffel, G., and E. Braunwald. 1984. Thrombolytic therapy: A new strategy for the treatment of acute myocardial infarction. *New England Journal of Medicine* 311:710–16.

Lobo, R. 1988. Lipids, clotting factors, and diabetes: Endogenous risk for cardiovascular disease. *American Journal of Obstetrics and Gynecology* 158:1584.

Noma, K. et al. 1988. Evaluation of left ventricular function in an experimental model of congestive heart failure due to combined pressure and volume overload. *Basic Research in Cardiology* 83:58–64.

Packer, M. 1988. Interaction of prostaglandins and angiotensin II in the modulation of renal function in congestive heart failure. *Circulation* 77:164–73.

Sherman, C. et al. 1986. Coronary angioscopy in patients with unstable angina pectoris. *New England Journal of Medicine* 315:913–19.

Unit VI

IMMUNE SYSTEM

The topics of chapter 26 involve general mechanisms of protection and include secretions, mechanical factors, inflammation, phagocytosis, and complement. Chapter 27 deals with specific immune responses and the role of lymphocytes. The focus of chapter 28 is hypersensitive reactions and organ transplant rejection. Chapter 29 deals with immunodeficiencies and autoimmune disorders. The emphasis is on autoimmune disorders characterized by the production of autoantibodies and generally of unknown etiology. Chapter 30 deals with immunologic laboratory tests and includes a discussion of cases of AIDS and lupus erythematosus with related laboratory findings.

Chapter 26

Nonspecific Defense Mechanisms

The body's immune system involves (1) complex mechanisms that protect against invasion and domination by microorganisms, (2) removal of cellular debris, and (3) identification and destruction of abnormal or mutant cells, a process called surveillance. Chapters 26–30 deal with these aspects of the immune system, as well as with immunological disorders. Defense mechanisms include unique responses to infectious agents, products of these agents, certain macromolecules, and tumor cells. These responses lead to the neutralization or destruction of the substances. In contrast, there are aspects of host resistance that are innate and represent a general response to injury and/or infection. These include physical barriers to invasion, secretions and mechanical factors, inflammatory response to injury or infection, and cells/cell products that contribute to removal of offending agents (figure 26.1).

SECRETIONS AND MECHANICAL FACTORS

The body is protected by both mechanical and chemical factors that present a barrier against invading microorganisms. These factors are summarized in table 26.1. Intact skin is a major first-line of defense and in situations in which the skin is injured, such as burns, there is a high risk of infection. There is overgrowth by virulent organisms when normal intestinal flora is diminished by antibiotic therapy, which underlines the importance of the competitive role of intestinal bacteria. The respiratory tract is protected by mucus and the mucociliary escalator (chapter 6), which traps organisms and sweeps them up to the pharynx, where they are either swallowed or expectorated. The urinary tract is protected against infection by the flow of urine. Lysozyme is an enzyme found

immune: L. *immunis*, free

in many types of cells and body fluids, and due to its bactericidal effects, it is a first-line of defense against infection. This enzyme lysis the polysaccharide component of the bacterial cell wall, particularly that of gram positive organisms.

INFLAMMATION

Inflammation is a normal response to tissue damage due to infection, mechanical injury, or harmful immune responses. The inflammatory reaction is a protective mechanism; although under certain circumstances, it may be harmful. Table 26.2 shows a basic outline of the events involved in an inflammatory response. As indicated, the main events in inflammation follow:

1. Hyperemia
2. Edema due to increased vascular permeability
3. Fever induced by bacterial or leukocyte products called pyrogens (pi'ro-jens)
4. Attraction of leukocytes to the injured area
5. Phagocytosis of bacterial cells and debris

In an inflammatory reaction, there usually is localized heat or fever due to increased blood flow and to the release of pyrogens, redness caused by hyperemia, and swelling resulting from increased vascular permeability with egress of fluid into tissue spaces.

TYPES OF INFLAMMATION

The initiating event in an inflammation is either injury or the presence of antigen. An antigen is a substance, often a protein, which is recognized by the body's defense system as being foreign or nonself and elicits antibody production. Bacterial infection is a common cause of inflammation, although extremes of temperature, chemicals, radiation, and mechanical injury may lead to an inflammatory

pyrogen: Gk. *pyros*, fire; Gk. *genes*, producing
hyperemia: Gk. *hyper*, above; Gk. *haima*, blood

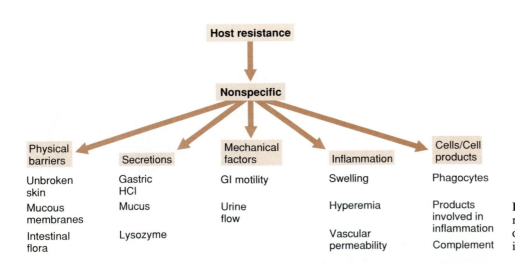

Figure 26.1 Aspects of host resistance that represent general defense mechanisms against injury/disease.

TABLE 26.1 Mechanical factors and secretions that protect against invading organisms

Unbroken Skin

Mechanical barrier
Bactericidal effects of fatty acids on skin surface
Acidic pH of skin

Lining of Gastrointestinal Tract

Acidity of stomach
Rapid movement through the tract
Competition of normal flora of intestine

Lining of Respiratory Tract

Mucus-coating traps organisms
Cilia sweep mucus up to pharynx

Urinary Tract
Flow of urine carries bacteria away

Lysozyme

Bactericidal enzyme

TABLE 26.2 Series of events in an inflammatory response

1. Tissue injury; release of substances to initiate inflammatory response
2. Vasodilation causes increased blood flow and redness
3. Increased permeability of blood vessel wall; plasma leaks into interstitium and causes swelling
4. Granulocytes migrate to area of injury;
    neutrophils phagocytize
    basophils release histamine and serotonin
    eosinophils release antihistamines
5. Fever induced by endotoxins or leukocyte products
6. Phagocytic monocytes migrate to area; lymphocytes in area produce antibodies (chapter 27)

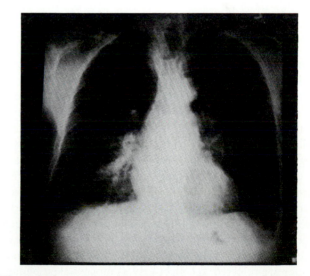

1. Mechanical injury: burns, radiation, infection, chemicals

    Skin

2. Injured cells — Release of substances: initiate inflammatory response

3. Vasodilation ↑Vascular permeability Swelling
    Migration of granulocytes to injury site

4. Neutrophil — Phagocytosis
    Basophil — Histamine and serotonin (enhance inflammation)
    Eosinophil — Antihistamines (control inflammation)

5. Granulocytes or bacterial toxins — Induce fever

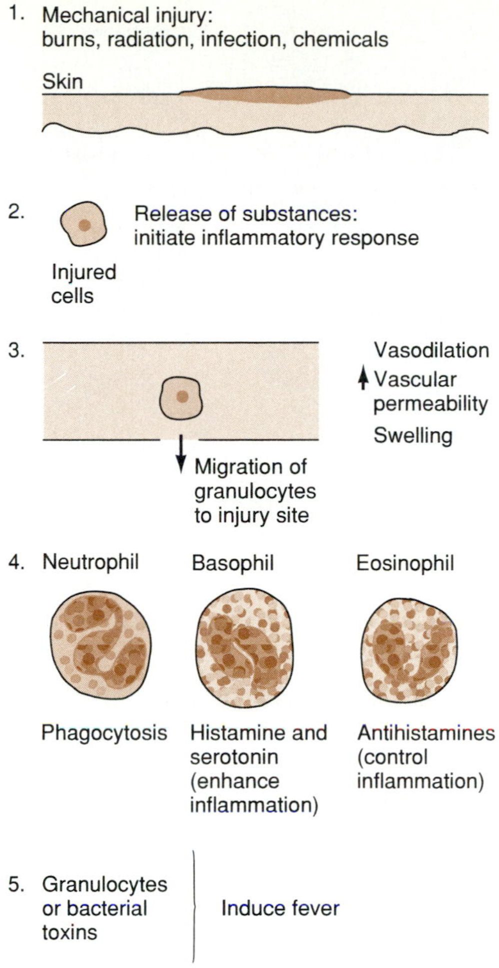

Figure 26.2 The main events in an acute inflammatory response.

---

response as well (figure 26.2). There are different types of inflammatory responses, and the precise nature of the response depends on the nature of the injury and, if antigen is involved, the quantity of antigen and mode of entry into the body (figure 26.3). In some cases, there is an immediate response including vasodilation, edema, bronchoconstriction, and nasal discharge, while in others, there is a slower response that involves mainly swelling (chapter 28).

If the inflammatory response fails to destroy the antigen and promote the complete healing process, a prolonged condition of chronic inflammation may occur. Inflammatory cells remain in the injured area, and fibrous scar tissue may replace normal tissue (figure 26.4). A specific type of chronic inflammation occurs in response to tuberculosis, fungal infections, and to such noninfectious agents as silica. The result is granulomatous (gran″u-lōm′ah-tus) inflammation in which macrophages accumulate in the area, undergo a transformation to form giant

granulomatous: L. *granulum,* small grain; Gk. *oma,* tumor

Figure 26.3 A posterior-anterior chest X-ray of a male with pulmonary interstitial infiltrates following a drug reaction. White patchy areas in the patient's right mid and lower lobe region show the area of infiltration.

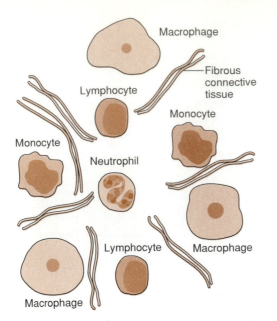

Figure 26.4 In chronic inflammation, lymphocytes and monocytes are more numerous than granulocytes and macrophages accumulate in the area. Fibrosis may occur with the formation of fibrous connective tissue.

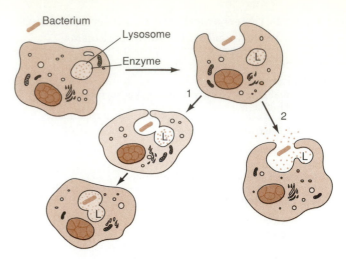

Figure 26.5 Phagocytosis by a neutrophil or macrophage. A phagocytic cell extends its pseudopods around the object to be engulfed, such as a bacterium. Dots represent lysosomal enzymes (L = lysosomes). If the pseudopods fuse to form a complete food vacuole (*1*), lysosomal enzymes are restricted to the organelle formed by the lysosome and food vacuole. If the lysosome fuses with the vacuole before fusion of the pseudopods is complete (*2*), lysosomal enzymes are released into the infected area of tissue.

cells, and are surrounded peripherally by lymphocytes. This reaction may be accompanied by necrosis and scarring.

The magnitude of the inflammatory response may lead to harmful effects, as in the case of fluid accumulations and swelling of the brain in response to trauma. An uncontrolled inflammatory response leads to joint damage in rheumatoid arthritis (chapter 28).

Although inflammation is a protective mechanism, the magnitude and nature of the response varies and may be harmful.

PHAGOCYTOSIS

The first step in phagocytosis is attachment of the antigen to the phagocytic cell, followed by engulfment and digestion of the particle by that cell (figure 26.5). Various cells are capable of phagocytic activity including neutrophils, monocytes, macrophages, and histiocytes. Both neutrophils and monocytes are white blood cells (chapter 18) and migrate from the blood to the site of infection. In the absence of infection, monocytes travel to tissues and differentiate into either fixed or motile macrophages. The fixed macrophages are called histiocytes and are located in the spleen, lymph nodes, and liver; thus, there are several cell types available for the task of phagocytosis. The monocytes circulating in the blood and macrophages in various tissues are referred to collectively as mononuclear phagocytes.

TABLE 26.3 Cells and chemicals involved in inflammation

Cell or Source	Product	Effect
Neutrophils		Phagocytosis
Monocytes		Phagocytosis
Macrophages		Phagocytosis
Histiocytes		Phagocytosis
Microorganisms/ leukocytes	Pyrogens	Fever
Mast cells	Histamine	Vasodilation
Basophils	Histamine	Vasodilation
Platelets	Serotonin	Vasodilation
Eosinophils	Antihistaminics	Control inflammation
Leukocytes	Leukotrienes, prostaglandins	Vascular permeability
Complement	Chemotactic factors	Attract inflammatory cells

CELL PRODUCTS

Table 26.3 includes a partial list of chemicals produced by cells and involved in the inflammatory process. Pyrogens are fever-inducing substances. Some examples of pyrogens include endotoxins, poisonous substances in the cell wall of gram-negative bacteria, as well as leukocyte products. Histamine causes vasodilation and constriction of bronchial smooth muscle and is released by basophils and mast cells. Mast cells are found in connective tissue, and basophils are white blood cells that migrate to the inflammatory site. Platelets release serotonin, which contributes

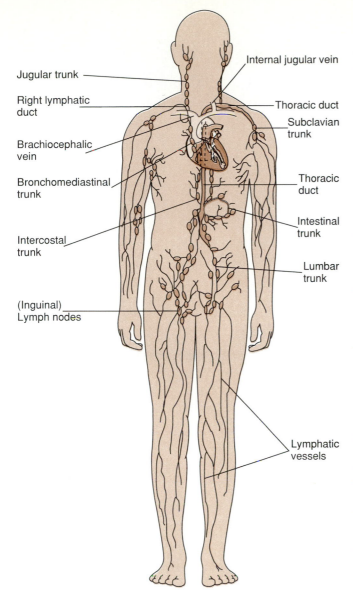

Jugular trunk

Right lymphatic duct

Brachiocephalic vein

Bronchomediastinal trunk

Intercostal trunk

(Inguinal) Lymph nodes

Internal jugular vein

Thoracic duct

Subclavian trunk

Thoracic duct

Intestinal trunk

Lumbar trunk

Lymphatic vessels

Figure 26.6 Lymphatic vessels merge into larger lymphatic trunks, which drain into collecting ducts.

to vasodilation. Eosinophils, also a type of white blood cell, release substances that modify and control the process of inflammation.

Leukocytes, activated by various stimuli, liberate arachidonic acid, which is metabolized to leukotrienes and prostaglandins. These substances have several effects that include increased vascular permeability.

COMPLEMENT

The term complement refers to a group of plasma proteins which, when activated, release factors that enhance phagocytosis and release chemotactic factors as well (chapter 27). Chemotaxis means chemical attraction, and in this case, it refers to the attraction of white blood cells to the site of inflammation.

LYMPHATIC SYSTEM

Cells in the lymphatic system are involved in phagocytosis and in immune responses. This system is composed of a network of vessels and nodes that transport fluid called lymph from the interstitium to the thoracic duct that empties into the left subclavian vein (chapter 2) (figure 26.6). Lymphoid tissue includes tonsils, thymus, and spleen. The involvement of the lymphatic system with immune responses is discussed further in chapter 27.

LOWERED RESISTANCE

The general state of health is an important part of the picture with disease. A number of factors may weaken the resistance of an individual. Injury, poor nutrition, fatigue, and stress contribute to a lowered resistance.

SUMMARY

• • •

The body's first line of defense is intact skin, as well as the lining of the gastrointestinal, respiratory, and urinary tracts. Certain secretions including the enzyme, lysozyme, protect against invading organisms. Inflammation is a protective mechanism and occurs in response to tissue damage. In general, the inflammatory reaction involves vasodilation, increased vascular permeability, and attraction of phagocytes to the area. Chronic inflammation may lead to tissue damage and scarring.

Phagocytosis is accomplished by neutrophils, monocytes, macrophages, and histiocytes. Pyrogens induce fever; histamine and serotonin cause vasodilation; and leukotrienes and prostaglandins lead to increased vascular permeability. Chemotaxis is the result of products of complement activation.

REVIEW QUESTIONS

• • •

1. What are four phagocytic cells?
2. Name a product of basophils involved in inflammation.
3. What causes increased vascular permeability in the inflammatory response?
4. What is the source of serotonin in inflammation?
5. What is one role that complement plays in inflammation?

SELECTED READING

• • •

Appenzeller, T. 1988. Living with yourself: Investigators trace the fate of self-reactive immune cells. *Scientific American* 259:30–32.

Arens, M. et al. 1987. Multiple and persistent viral infections in patient with bare lymphocyte syndrome. *Journal of Infectious Disease* 156:837–41.

Frank, M. 1987. Current concepts: Complement in the pathophysiology of human disease. *New England Journal of Medicine* 316:1525–30.

Syvalahti, E. 1987. Endocrine and immune adaptation in stress. *Annals of Clinical Research* 19:70–77.

Chapter 27

Immune System Function

Chapter 26 provides an overview of general defense mechanisms including phagocytosis and the inflammatory response. The purpose of this chapter is to identify those elements of the body's defenses that recognize specific antigens and that provide both immediate and long-term protection as an immune response.

Tissues, cells, and molecules throughout the body constitute the immune system including (1) the pool of lymphocytes that circulate between blood and lymph, (2) lymph nodes, (3) spleen, (4) thymus, and (5) bone marrow. Lymphocytes, which play a central role in immune responses, are discussed further in this chapter.

LYMPHOCYTES

Lymphocytes are mononuclear white blood cells (figure 27.1) produced in bone marrow and constitute about 35% of the total white blood cell count (chapter 18). Lymphocytes develop from undifferentiated bone marrow cells, and subsequently, some of these cells migrate to lymphoid tissue, i.e., thymus, spleen and lymph nodes (figure 27.2). Special processing occurs in the thymus to transform the lymphocytes into T-(thymus) lymphocytes, cells that have a unique immunologic function. The T cells are produced

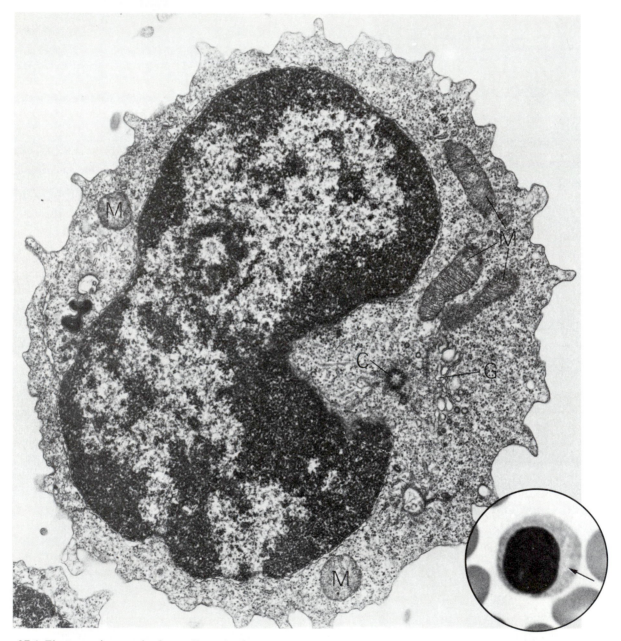

Figure 27.1 Electron micrograph of a medium-sized lymphocyte. The punctate appearance of the cytoplasm is due to the presence of numerous free ribosomes. Several mitochondria (*M*) are evident. The cell center or centrosphere region of the cell (the area of the nuclear indentation) also shows a small Golgi apparatus (*G*) and a centriole (*C*). Inset: Light microscope appearance of the medium-sized lymphocyte from a blood smear. The Golgi-containing centrosphere region is indicated by the arrow.
Electron micrograph courtesy of Dorothea Zucker-Franklin, Professor of Medicine at New York University Medical Center. From M. H. Ross, E. J. Reith, and L. J. Romrell: Histology: A Text and Atlas. 1989 Williams & Wilkins, Baltimore. Reproduced with permission.

314

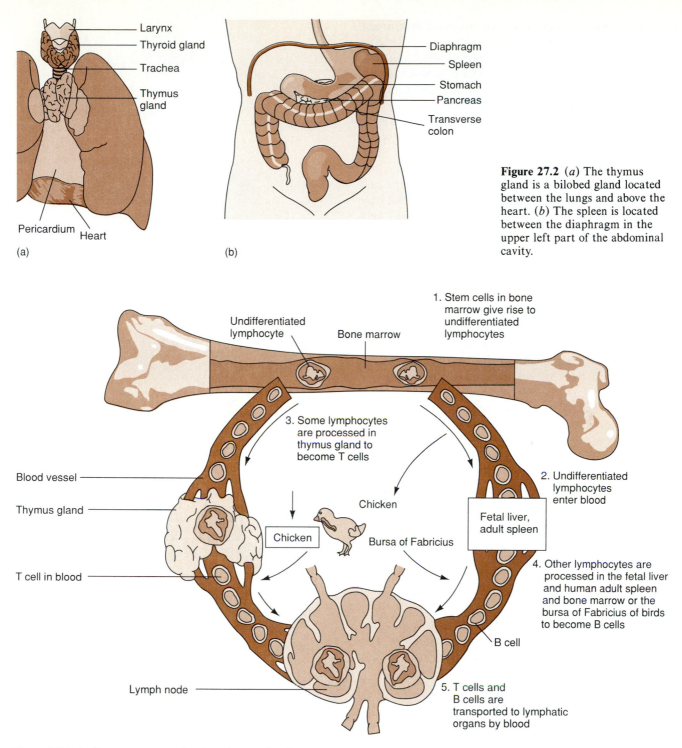

Larynx
Thyroid gland
Trachea
Thymus gland
Pericardium
Heart

Diaphragm
Spleen
Stomach
Pancreas
Transverse colon

(a)

(b)

Figure 27.2 (*a*) The thymus gland is a bilobed gland located between the lungs and above the heart. (*b*) The spleen is located between the diaphragm in the upper left part of the abdominal cavity.

Undifferentiated lymphocyte
Bone marrow

1. Stem cells in bone marrow give rise to undifferentiated lymphocytes

3. Some lymphocytes are processed in thymus gland to become T cells

Chicken

2. Undifferentiated lymphocytes enter blood

Blood vessel

Thymus gland

Chicken

Fetal liver, adult spleen

T cell in blood

Bursa of Fabricius

4. Other lymphocytes are processed in the fetal liver and human adult spleen and bone marrow or the bursa of Fabricius of birds to become B cells

B cell

Lymph node

5. T cells and B cells are transported to lymphatic organs by blood

Figure 27.3 A diagram representing lymphocyte development. Bone marrow releases undifferentiated lymphocytes, which, after processing, become T and B cells.

in utero and early postnatal life and are capable of cell division. Populations of these cells are maintained throughout life. T cells may migrate from the thymus to blood and constitute 70–80% of the circulating lymphocytes. T cells may also go to the spleen and lymph nodes.

The differentiation of a second type of lymphocyte, B-lymphocytes, occurs in the fetal liver and adult bone marrow. These cells go to the blood and to lymphoid tissues as well (figure 27.3). Both B and T cells move freely between blood and lymphoid tissues. Only about 0.1% of the total number of lymphocytes circulate in the blood.

Immune System Function

315

B-lymphocytes in chickens are processed in an organ, an appendage of the cloaca, called the bursa of Fabricius; hence, the use of the letter B to identify these cells. Human beings do not have a bursa, and it is believed that B-lymphocytes are processed in the bone marrow instead.

B-lymphocytes are involved in defending the body against bacterial and some viral infections by producing antibodies in the blood to fight the invasion. This type of immunologic response is called **humoral immunity.** In contrast, T-lymphocytes do not produce antibodies, but provide protection by attacking (1) foreign tissue in the case of organ transplant, (2) cancer cells, and (3) cells that have been invaded by viruses or fungi. This is called **cell-mediated immunity** because there is direct contact between T cells and the target cells.

HUMORAL IMMUNITY

B-lymphocyte antibody production is initiated by exposure of the cells to **antigen** or **immunogen,** i.e., such large foreign molecules as proteins, nucleoproteins, polysaccharides, or glycolipids. Antigens are defined as molecules that stimulate the production of antibodies (proteins), which attach to specific types of antigen. Antibodies are also called **immunoglobulins** and are found either attached to cell surfaces or in blood and other body fluids.

The sequence of events in antibody production is as follows. B-lymphocytes have antibodies on the cell surface, and the antibody molecules have a configuration called a **combining** or **receptor site** that is complementary to a specific antigen.

1. Antigen/antibody attachment on the surface of B-lymphocytes stimulates B cell growth and cell division.
2. The daughter cells mature and become either **plasma cells** or **memory cells.**
3. Memory cells are like the original cell and plasma cells produce antibodies that will attach to the specific type of antigen that initiated the events (figure 27.4).

IMMUNOGLOBULINS

The basic structure of immunoglobulins is two identical heavy (H) and two identical light (L) polypeptide chains in the form of a Y (figure 27.5). As shown in this figure, there is a variable (V) region in the upper portion of the Y that includes both heavy and light chains. Amino acid sequences vary in this region, and this is the location of antigen binding sites called Fab (fragment, antigen binding). There are two Fab sites, one for each pair of H and L chains. The constant (C) domain is a region in which

humoral: L. *humor,* fluid

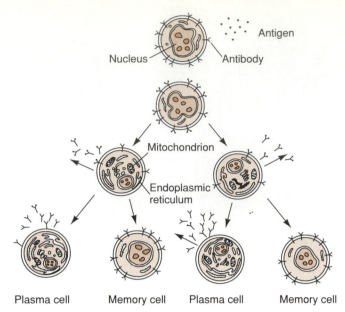

Figure 27.4 The attachment of antigen to antibodies on the B cell surface stimulates cell division and, ultimately, maturation into plasma and memory cells.

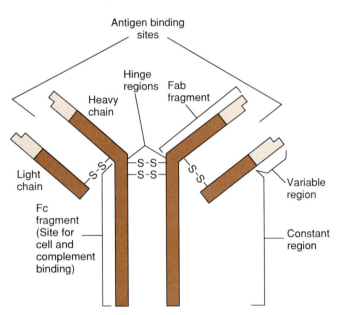

Figure 27.5 An immunoglobin molecule has a basic structure of two identical light chains and two identical heavy chains held together by disulfide bonds.

the amino acid sequences remain the same for immunoglobulins belonging to the same class or subclass. The constant terminal portions of the H chains are called Fc fragments (c for crystallizable), and these are sites for binding complement (chapter 26) and binding to cell surfaces.

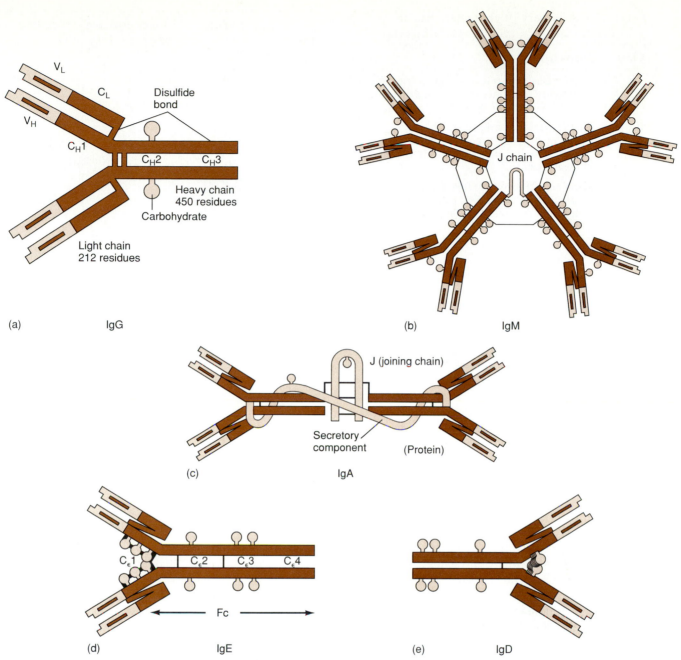

Figure 27.6 The five major categories of immunoglobins. (*a*) The basic structure of human IgG. (*b*) The pentameric structure of human IgM. The circles indicate carbohydrate side chains. (*c*) Dimeric structure of human secretory IgA. (*d*) The monomeric structure of human IgE. (*e*) The monomeric structure of human IgD.

Classes of Immunoglobulins

There are five major categories of immunoglobulins classified on the basis of different types of heavy chains (figure 27.6). These classes of antibodies are both structurally and functionally different. The characteristics of each group are summarized in table 27.1. Immunoglobulin G (IgG) is the most abundant of the immunoglobulins, and it appears in both blood and interstitial fluid. It neutralizes bacterial toxins and attaches to microorganisms to promote phagocytosis.

monomeric: Gk. *monos*, single; Gk. *meros*, part

Immunoglobulin M (IgM) molecules are the largest of the five classes and are produced as a first response to antigen. These immunoglobulins bind to the surface of B-lymphocytes and act as receptors for antigen, may bind to complement to promote clumping or agglutination of antigen, and prepare antigen for phagocytosis.

The major role for immunoglobulin A (IgA) is in protecting the body's external surfaces. IgA is found in secretions of the nose, lungs, genitourinary tract, and gastrointestinal tract, as well as in tears, sweat, and human milk.

TABLE 27.1 Summary of characteristics of the five major classes of immunoglobulins (Ig)

Ig Class	Characteristics
IgG	Highest concentration in serum; crosses placenta; attaches to surface of polymorphonuclear leukocytes and monocytes
IgM	First response to antigen; attaches to B-lymphocytes; largest Ig molecule; found almost exclusively in blood; reacts with complement
IgA	Major role in secretion; found in gastrointestinal tract, respiratory tract, genitourinary system, saliva, tears, human milk; protects surface tissues
IgE	Associated with allergic reaction; main receptor for antigens on mast cells and basophils; may play protective role in parasitic infections
IgD	Minor serum Ig; major surface receptor on B-lymphocytes along with IgM

Immunoglobulin E (IgE) is involved in allergic reactions in which (1) the molecule attaches to the surface of mast cells and basophils and acts as a receptor for antigen, and (2) upon contact with antigen, histamine is released. IgE may also play a protective role in parasitic infections.

The last immunoglobulin, IgD, is present in small amounts in the blood and is a receptor on the surface of B-lymphocytes.

B-LYMPHOCYTES

As described earlier, B-lymphocytes originate from stem cells in the bone marrow where differentiation is initiated. This process of specialization involves the synthesis of immunoglobulin on the cell surface, IgM and IgD, which furnish antigen combining sites. There is a particular type of antigen receptor for each new B-lymphocyte. After acquiring surface immunoglobulin, the B-lymphocyte migrates to lymphoid tissue, such as spleen and lymph nodes, and circulates between the lymphoid tissue and blood. The next phase in development occurs when antigen binds to the Ig receptor. This stimulates proliferation of the B-lymphocytes and induces further differentiation resulting in plasma cells. Plasma cells secrete antibodies or immunoglobulin molecules that have the same receptor site as those Ig molecules on the cell membrane. The Ig molecules may belong to any of the five classes of immunoglobulin. The plasma cells divide to produce daughter cells called clones, and these clones produce identical antibodies. These clone cells, also called memory cells, are long-lived and capable of immediate antibody production when stimulated by the reappearance of antigen.

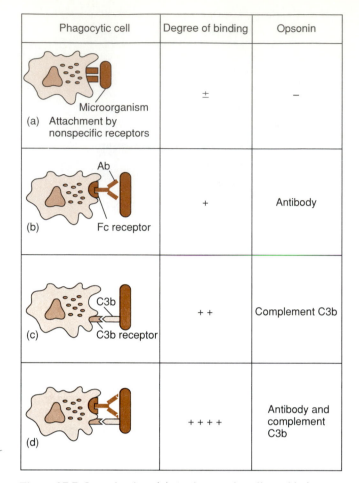

Phagocytic cell	Degree of binding	Opsonin
(a) Attachment by nonspecific receptors — Microorganism	±	−
(b) Ab — Fc receptor	+	Antibody
(c) C3b — C3b receptor	+ +	Complement C3b
(d)	+ + + +	Antibody and complement C3b

Figure 27.7 Opsonization. (*a*) A phagocytic cell can bind directly to an organism via nonspecific receptors. (*b*) This binding ability is enhanced if the organism causes antibodies (Ab) to be produced that act as a bridge to attach the organism to the Fc receptor on the phagocytic cell. (*c*) If complement is activated (C3b), the degree of binding is further enhanced via the C3b receptor. (*d*) If both antibody and complement are involved, binding is greatly enhanced.

Normally, the plasma cells do not produce antibodies against the body's own cells or macromolecules. As B-lymphocytes begin to mature during fetal life, they produce small amounts of antibody. If the antibody binds to antigen at this stage, the source of the antigen is of fetal origin and the immature lymphocyte is killed. The body thus becomes self-tolerant, i.e., it normally does not produce antibodies against antigen belonging to self.

COMPLEMENT

The term complement refers to a group of proteins found in extracellular fluids involved in immune responses. Complement activation occurs when a component of the complement system attaches to an antigen/antibody IgG or IgM complex, with the antibody Fc region as the binding site. Complement is also activated by bacterial endotoxins. Activation involves a cascade of reactions in which a series of complement components are formed.

TABLE 27.2 Summary of events mediated by complement components

Increased vascular permeability
Mast cell release of histamine
Chemotaxis for phagocytes
Antigen/antibody complexes broken into small soluble
 fragments
Promote phagocytosis
Cell lysis

The attachment of a complement protein to an antigen/antibody complex on the membrane of an invading cell is called **complement fixation** and results in injury to the membrane and destruction of that cell. Complement that is not fixed, i.e., is free in fluids, attracts phagocytes to the area. Complement is responsible for **opsonization,** a process of coating the invading cell with subsequent attachment to a phagocyte (figure 27.7).

The role of the activated complement system in an immune response is summarized in table 27.2. In general, complement contributes to the inflammatory response, promotes phagocytosis, and causes cell membrane damage. Complement components indirectly cause increased vascular permeability by stimulating the release of histamine from mast cells and directly promoting permeability as well. The overall effect is that serum antibody and complement accumulates in the area. Chemotactic factors attract phagocytes, and the process of phagocytosis is facilitated further by the coating of the microorganism with antibody/complement complex. Large antigen/antibody complexes tend to cause hypersensitive reactions, and an aspect of complement activity is to break these complexes into small soluble fragments. In addition to these actions, complement is activated by bacterial endotoxins and causes membrane damage that results in bacteriolysis.

OVERVIEW

The role of lymphocytes and plasma cells is summarized in figure 27.4. B-lymphocytes develop from stem cells in the bone marrow, and immunoglobulins (IgM and IgD) are bound to the surface of the cell membrane in the process of differentiation. The cells migrate to lymphoid tissue where antigen attaches to the membrane-bound antibody. The antigen has antigenic determinants, chemical structures that bind to antibody. Antigen thus binds to the lymphocyte that has immunoglobulin with Fab sites complementary to the antigenic determinant. Antigen binding stimulates cells with a particular Fab site to differentiate into plasma cells and to divide to form many cells of the same type. Plasma cells produce an antibody specific for one antigen, and some of these antibodies appear in blood and various tissues, while others attach to the surface of cells. The interaction of antibody with antigen, sometimes in concert with complement, results in

the removal of antigen. Macrophages aid in processing and presenting antigen and secrete factors that influence lymphocytes as well.

Some plasma cells are long-lived and function as memory cells. If the same antigen appears at a later time, the memory cells respond with immediate production of antibody specific for that antigen.

This normal humoral response is the reason for the benefits of immunization. A vaccine is a suspension of live or inactivated bacterial cells, viruses, bacterial capsules, or bacterial toxins called toxoids. These preparations behave as antigen and stimulate the production of specific antibodies to provide protection against infection and provide **active immunity. Passive immunity** is conferred by the injection of preformed antibodies by way of serum, antibodies against bacterial toxin called antitoxin, or **gamma globulin** preparations. Gamma globulin is obtained from pooled human plasma and represents the immunoglobulin fraction of plasma protein. This may be administered prophylactically to protect against such infections as hepatitis A.

CELL-MEDIATED IMMUNITY

Cell-mediated immunity refers to an immune response that does not involve antibody production. Humoral and cell mediated responses are interrelated, although it is convenient to discuss these two aspects of immunologic responses in separate sections.

T-LYMPHOCYTES

Stem cells from the bone marrow migrate to the thymus where differentiation into T-lymphocytes occurs. T-lymphocytes may develop into helper T cells, suppressor T cells, or cytotoxic T cells. The helper and suppressor cells modulate B-lymphocyte production and influence humoral immune responses (figure 27.8).

B cells have surface antibodies for attachment to antigen, whereas T cells lack surface antibodies. Antigen, rather than attaching directly to T cells, must be transported by another cell and presented to the T cell. Both the presenting cell and the T-lymphocyte have surface proteins that are called **histocompatibility antigens (HLA).** There are two classes of HLA antigens, i.e., class I and class II. Class II HLA antigens are produced by the antigen-presenting cells (macrophages), while class I HLA antigens are produced by all cells except for red blood cells.

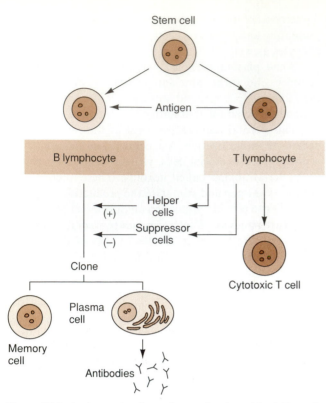

Figure 27.8 Antigen stimulates the production of both B- and T-lymphocytes. B cell production is also influenced by the effects of helper and suppressor T cells.

• • •

All cells, with the exception of red blood cells, have a specific combination of histocompatibility antigens on the cell membrane surface. These antigens cause rejection of an organ transplant. White blood cells are used to determine the tissue type of a prospective transplant recipient, and for this reason, histocompatibility antigen is also called human leukocyte antigen (HLA).

• • •

The events that lead to T cell activation by antigen follow:

1. A macrophage ingests and processes the infecting agent, such as a virus.
2. The macrophage secretes a substance called interleukin-1, and this in turn, stimulates the proliferation of T-lymphocytes.
3. Viral particles become attached to class II HLA antigens on the surface of the macrophage.
4. Receptor proteins on the surface of helper T cells recognize the combination of macrophage class II HLA antigens and viral antigen.

5. T cells are activated and secrete chemical substances called lymphokines (lim'fo-kīns). Interleukin-2 is a lymphokine that stimulates proliferation of both B cells and cytotoxic T cells.
6. Cytotoxic T cells destroy the infected cell (figure 27.9).

ROLE OF CELL-MEDIATED IMMUNITY

The immune response involving T-lymphocytes provides protection against intracellular viral and bacterial infections, causes rejection of organ transplants, and detects certain types of tumors. The role of cell-mediated immunity is discussed further in chapter 28 in relation to delayed hypersensitivity and graft rejection.

SUMMARY
• • •

The humoral immune system protects against antigen by way of plasma cell antibody production. Antibodies are immunoglobulins that have Fab sites which are complementary to specific antigens. Immunoglobulin Fc sites bind to complement and cell surfaces. Complement is a group of proteins found in extracellular fluids that contribute to the inflammatory response, promote phagocytosis, and cause cell membrane damage. Some plasma cells are long-lived and function as memory cells, so a second challenge by antigen elicits an immediate antibody response.

Antigen recognition by B-lymphocytes depends on immunoglobulin on the cell surface. HLA proteins on cell membranes of T-lymphocytes, and macrophages are responsible for antigen recognition and attachment in cell-mediated immunity. Macrophages present the antigen to T cells and ultimately, cytotoxic T cells destroy the invader.

REVIEW QUESTIONS
• • •

1. How do B-lymphocytes recognize antigen?
2. How do T-lymphocytes recognize antigen?
3. Where do T cells differentiate?
4. In what way is complement involved in an immune response?
5. What is the function of helper T cells?

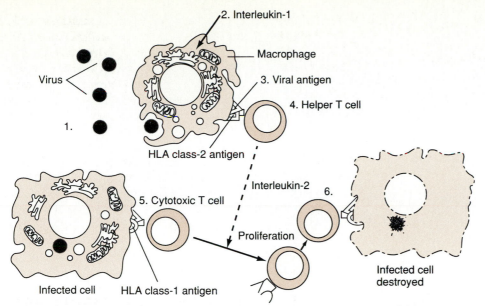

Figure 27.9 A series of events in the cell mediated response to a viral infection. (*1*) Virus is ingested by a macrophage. (*2*) The macrophage secretes interleukin-1, which stimulates the proliferation of T-lymphocytes. (*3*) A processed viral particle (viral antigen) appears on the surface of the macrophage and becomes associated with class-2 HLA antigen. (*4*) A helper T cell is activated as the result of recognizing viral antigen and class-2 HLA antigen. (*5*) Activated helper T cells secrete interleukin-2, which stimulates the proliferation of cytotoxic T cells. (*6*) Cytotoxic T cells destroy cells that are infected by the virus and that have both surface viral antigen and class-1 HLA antigen.

SELECTED READING

• • •

Ada, G., and G. Nossal. 1987. The clonal-selection theory. *Scientific American* 257:62.

Alt, F., K. Blackwell, and G. Yancopoulos. 1987. Development of the primary antibody repertoire. *Science* 238:1079.

deBoer, R. et al. 1987. Immunological discrimination between self and non-self by precursor depletion and memory accumulation. *Journal of Theoretical Biology* 124:343.

Nossal, G. 1987. Current concepts: Immunology. *New England Journal of Medicine* 316:1320.

Weill, J., and C. Reynaud. 1987. The chicken B cell compartment. *Science* 238:1094.

Young, J., and Z. Cohn. 1988. How killer cells kill. *Scientific American* 258:38.

Chapter 28

Detrimental Immune Responses

Although the function of the immune system is that of defense, there are harmful and even life-threatening events that are of immunologic origin. The focus of this chapter is harmful immune responses including (1) anaphylaxis, (2) cytotoxic reactions, (3) immune complex mediated reactions, (4) delayed-type hypersensitivities, and (5) organ transplant rejection.

IMMEDIATE HYPERSENSITIVITY

An immediate hypersensitive reaction is an allergic response to antigen, generally within 30 minutes of exposure. This is called a type I response, and it is caused by IgE antibody. Examples include hay fever, asthma, hives, anaphylaxis, and allergic reactions to insect venom. Table 28.1 lists a variety of antigens that elicit this type of response.

• • •

Breakdown products of penicillin, which become attached to tissue or serum proteins, act as antigen. Allergic reactions are the result of the release of histamine, induced by IgE antibodies. Either hives or anaphylaxis may occur.

• • •

There are varying degrees of severity in immediate hypersensitivity reactions.

ANAPHYLAXIS

Anaphylaxis is a type I reaction that may be life-threatening. Table 28.2 outlines the events in this type of reaction. Interaction between antigen and mast cell- or basophil-bound IgE antibody results in the release of substances that cause the hypersensitive reactions. **Histamine** is released and causes increased vascular permeability, smooth muscle contraction, and vasodilation. **Slow reacting substance** is a leukotriene synthesized in leukocytes that causes bronchospasms and increased vascular permeability. The same response is elicited by **kinins,** which are released from a plasma globulin by kallikrein. Various **chemotactic factors** are released that attract eosinophils and neutrophils. **Platelet aggregating factor** causes both aggregation of platelets and release of **serotonin,** which contributes to inflammation (figure 28.1).

The symptoms of an anaphylactic reaction are mainly due to (1) vasodilation, (2) smooth muscle spasms, and (3) increased capillary permeability. The first symptoms are usually cutaneous with flushed skin and an itching sensation. Respiratory distress, hypotension, and shock are possible secondary results, and the symptoms may last for minutes to hours. An injection of epinephrine, which raises blood pressure and alleviates bronchospasms, is the appropriate immediate treatment.

SRS-A: slow reacting substance of anaphylaxis

TABLE 28.1 Possible causes of immediate hypersensitivity

Penicillin and derivatives	Insulin
Radiographic contrast material	Foods
Animal antiserum	Muscle relaxants
Vaccines	Morphine
Blood	Codeine
Insect venom	Cold temperature

TABLE 28.2 Series of events in anaphylaxis

1. Exposure to antigen
2. B cells differentiate into plasma cells
3. Plasma cells produce IgE that is specific for the antigen
4. IgE antibodies are bound to mast cells or basophils; the result is sensitization of these cells
5. Second exposure to antigen
6. Antigen attaches to sensitized mast cells and basophils
7. Cells release chemicals that cause vasodilation, increased vascular permeability, mucous secretion, and smooth muscle contractions

• • •

Anaphylactoid reactions are similar to anaphylaxis without IgE involvement. Possible causes include radiographic contrast material and plasma expanders, such as gelatin or dextran, and most reactions to aspirin are also anaphylactoid in nature.

• • •

Any route of administration of antigen may cause a hypersensitive reaction, although there must be an initial sensitizing exposure followed by a second dose. Attacks appear to be more severe when the antigen is injected. Localized anaphylaxis may occur, and the symptoms depend on the route of entry of antigen. There may, for example, be upper respiratory symptoms associated with hay fever or asthma or skin eruptions called hives that may be caused by food allergies. Antihistaminic drugs help alleviate the symptoms in such cases.

ANTIBODY-DEPENDENT CELLULAR CYTOTOXICITY

The term used for **type II hypersensitivity** is antibody-dependent cellular cytotoxicity (figure 28.2). This identifies an immune response in which there is interaction between cell-surface antigen and antibodies. The antigen may be a part of the cell membrane or may become associated with it. The result of this interaction is complement activation and ultimate destruction of the cell. The antibodies belong to the IgG or IgM class. The cell destruction is accomplished by various cells including T cells,

anaphylaxis: Gk. *ana*, without; Gk. *phylaxis*, protection

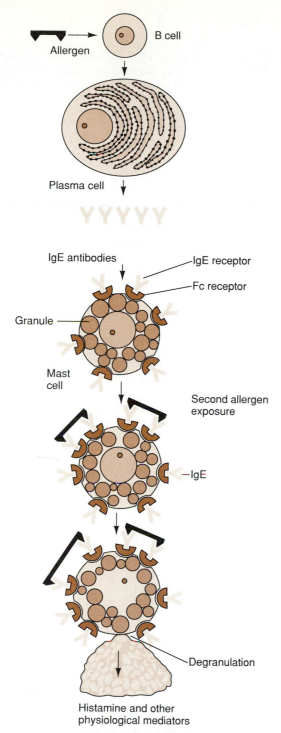

Figure 28.1 Type I (anaphylaxis) hypersensitivity. This type of hypersensitivity occurs when IgE antibodies attach to mast cells. The combination of these antibodies with allergens causes the mast cell to degranulate and produce the physiological mediators that cause the anaphylactic reaction.

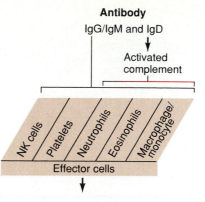

Figure 28.2 Type II (cytotoxic) hypersensitivity. The action of antibody occurs through effector cells or in concert with activated complement, which damages target cell membranes, causing cell destruction.

provide defense against helminths. In some cases, the response is injurious to the host, as in the case of certain autoimmune hemolytic reactions or incompatible blood transfusions.

Goodpasture's syndrome is an example of type II hypersensitivity that, in most cases, is fatal. For no known reason, antibodies develop in the serum against the glomerular basement membrane in the kidney and in the lungs. Tissue damage results and leads to glomerulonephritis, pulmonary hemorrhage, and anemia.

IMMUNE COMPLEX MEDIATED REACTIONS

Immune complex mediated reactions are classified as **type III hypersensitivity** and involve antigen/antibody complexes that precipitate in or near small blood vessels and subsequently activate complement. The result is vascular damage (figure 28.3). This type of reaction is typified by a localized **Arthus reaction** or by **serum sickness,** caused by horse serum antibody preparations. Horse serum antibodies are used to treat conditions, such as rabies and venomous snakebites. Horse antiserum is also used against human lymphocytes to help prevent transplant rejection.

An Arthus reaction occurs at the site of injection of serum in an individual who has had previous exposure to the antigen. Typically, swelling occurs and a hemorrhagic, and ultimately a necrotic, lesion develops. This is the result of deposition of antigen/antibody complex in the skin, activation of complement, and accumulation of neutrophils, which release various substances leading to tissue damage. In serum sickness, immune complex deposition occurs at many sites in capillary networks with possible involvement of cutaneous tissue, heart, kidney, and joints. There may be fever, enlarged lymph nodes, rash, painful joints, vascular inflammation, and nephritis.

neutrophils, and **natural killer (NK)** cells. NK cells are related to, but not the same as, T cells and are sometimes called large granular lymphocytes. These cells provide a first line of cell-mediated defense.

This type of immune response may be important in the destruction of malignant cells, virus-infected cells, and

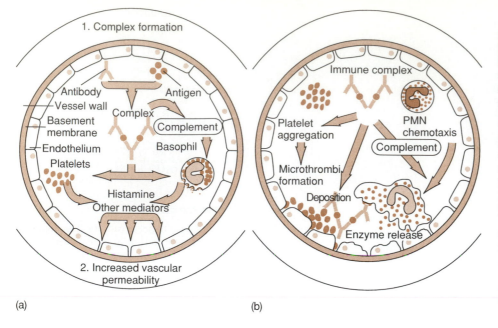

(a)   (b)

Figure 28.3 Type III hypersensitivity. Deposition of immune complex in blood vessel walls. (*a*) Antibody and antigen combine to form an immune complex. This complex acts on complement, which causes basophils and platelets to release histamine and other degranulation mediators. These mediators increase vascular permeability. (*b*) The increased permeability allows the immune complex to be deposited in the blood vessel wall. This then induces platelet aggregation to form microthrombi (blood clots) on the vessel wall. **PMNs** via complement degranulate, causing enzymatic damage to the blood vessel wall.

Most allergic reactions to drugs involve similar mechanisms, and there is evidence of similarities in systemic lupus erythematosus (chapter 29). Group A streptococcal infection may lead to glomerulonephritis (chapter 13), in which it appears that deposition of antibody/bacterial antigen complexes lead to nephron injury.

CELL-MEDIATED IMMUNITY

Cell-mediated immunity (CMI) refers to those immune responses elicited by immune cells rather than by antibodies. One type of CMI response is delayed hypersensitivity (type IV), exemplified by the reaction of an individual to an intradermal injection of tuberculin or purified protein derivative (PPD), a product of *Mycobacterium tuberculosis*. An individual who has had previous exposure to the organism experiences an inflammatory response at the site of injection that peaks within 24–48 hours. Typically, there is erythema and **induration** (hard swelling) of the area.

The inflammatory reaction is initiated by mononuclear cells, mainly monocytes-macrophages and a lesser number of lymphocytes. Monocytes process the antigen and present it to antigen-specific T cells. The T cells secrete soluble factors called **lymphokines.** These include **migration inhibiting factor** and **macrophage activating factor,** which promote the inflammatory process. The intensity of the reaction is limited by **suppressor T cells.**

Contact dermatitis is the result of the mechanism of delayed hypersensitivity involving sensitized T-lymphocytes. Exposure to low molecular weight chemicals from such sources as poison ivy, cosmetics, detergents, and clothing may elicit the skin reaction. The

induration: L. *indurare,* to harden

chemicals combine with tissue proteins to act as antigen, and the skin becomes erythematous with fluid-filled blisters.

It is likely that cell-mediated immunity is involved in those diseases classified as autoimmune in nature (chapter 29).

• • •

The functions of T cells include (1) delayed hypersensitivity, (2) defense against intracellular organisms, (3) organ transplant rejection, (4) destruction of tumor cells, and (5) regulation of antibody production.

• • •

REJECTION OF ORGAN TRANSPLANTS

The transplant of an organ, such as a kidney, may involve a donor that is genetically the same as the recipient, an identical twin. A procedure of this kind is called an **isograft.** A more common procedure is an **allograft** in which the donor and recipient are genetically different.

There are proteins on the surface of the cells of organs that are antigenic and are recognized either as self or foreign by an individual's immune system. These proteins are called **histocompatibility antigen (HLA)** and, if they are nonself, lead to rejection of the organ. Because of this response, it is necessary to match the donor and recipient as closely as possible. The ABO red blood cell types must be

allograft: Gk. *allos,* others

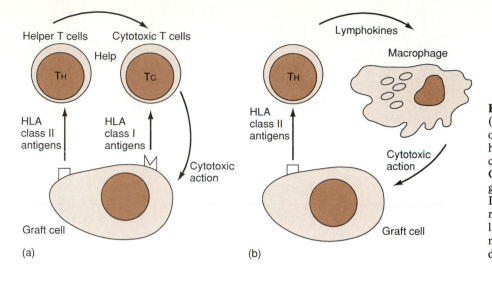

Figure 28.4 Graft cell destruction. (*a*) Foreign HLA class II antigens on the graft cell stimulate host helper T cells to help cytotoxic T cells destroy the target graft cell. Cytotoxic T cells recognize the graft cell via the foreign HLA class I antigens. (*b*) Helper T cells reacting to the graft cell release lymphokines that stimulate macrophages to enter the graft and destroy it via cytotoxic action.

the same, and there must be no antibodies against T cell antigens of the donor. The donor is selected on the basis of the closest HLA match possible.

Rejection of an allograft occurs as B- and T-lymphocytes are first sensitized to the antigenic material and then initiate injury to the graft in various ways (figure 28.4). Helper T cells release lymphokines that activate cytotoxic T cells, macrophages, and antibody producing B cells. This massive immune response requires therapeutic use of anti-inflammatory and immunosuppressive agents to prevent organ rejection.

SUMMARY

• • •

Four types of hypersensitive reactions are summarized in table 28.3. It should be noted that many reactions are not restricted to a particular type and that there may be mixed responses. Examples of type I reactions are hay fever, hives, and anaphylaxis. Type II reactions are typified by incompatible blood transfusions and Goodpasture's syndrome. Serum sickness and allergic reactions to drugs are type III, immune complex mediated reactions. The prototype for cell-mediated immunity, type IV, is delayed hypersensitivity reaction of the tuberculin test, with contact dermatitis and autoimmune disease as other examples.

Organ transplant rejection occurs as the result of an immune response against foreign histocompatibility antigen. Cytotoxic T cells, macrophages, and antibody-secreting B cells are responsible for organ injury.

TABLE 28.3 Types of hypersensitive reactions

Immediate hypersensitivity—type I	IgE antibody, bound to mast cells or basophils; release of chemicals; response within 30 min.
Antibody-dependent cellular cytotoxicity—type II	Cell surface antigen/antibody interaction; complement activation; cell destruction
Immune complex mediated reactions—type III	Antigen/antibody complex precipitates near small blood vessels
Cell-mediated immunity—type IV	Delayed response; mononuclear cells initiate reaction; T cells secrete lymphokines

REVIEW QUESTIONS

• • •

1. List four chemicals that are involved in a type I hypersensitive reaction.
2. What are three physiological events in a type I reaction responsible for symptoms?
3. Damage to blood vessels with various organs involved is characteristic of what type of hypersensitive reaction?
4. What type of hypersensitive reaction is characterized by cell destruction?
5. What type of hypersensitive reaction typically causes a delayed response?

SELECTED READING

• • •

Harjula, A., et al. 1987. Human leukocyte antigen compatibility in heart-lung transplantation. *Journal of Heart Transplant* 6:162–66.

Milstein, C. 1986. From antibody structure to immunological diversification of immune response. *Science* 231:1261–68.

Ramsey, G. et al. 1984. Isohemagglutinins of graft origin after ABO-unmatched liver transplantation. *New England Journal of Medicine* 311:1167–70.

Schifferli, J., Y. Ng, and K. Peters. 1986. The role of complement and its receptor in the elimination of immune complexes. *New England Journal of Medicine* 315:488–95.

Serafin, W., and F. Austen. 1987. Current concepts: Mediators of immediate hypersensitivity reactions. *New England Journal of Medicine* 317:30–34.

Chapter 29

Immune Disorders

The preceding chapters deal with major aspects of immune system function. Lymphocytes play a key role, whereby precursor cells give rise to two lines of development. Lymphocytes differentiate into T and B types under the influence of the thymus or bone marrow, respectively. When stimulated by antigen and regulated by T cells, B cells specialize to produce antibodies. Complement may be activated, and there is a concerted response, including accessory phagocytic cells, to isolate and destroy antigen. Populations of various subsets of T-lymphocytes modulate antibody production, exhibit cytotoxic activity, or produce lymphokines to increase the magnitude of the response.

The fact that there are multiple cellular and molecular events required for normal immune system function indicates there are many possibilities for abnormal or inadequate function of this complex system.

Two categories of disorders are included in this chapter: (1) immunodeficiencies in which there is a decrease or absence of a required factor and (2) autoimmune responses in which there is failure to identify and protect self.

IMMUNODEFICIENCIES

Immunodeficiencies may involve any of the components of the immune system and may be primary, due to genetic factors or failure of embryonic development. Secondary immunodeficiencies may be caused by such disease processes as malignancies and may be transient, depending on treatment of the primary disease. In recent years, an acquired immunodeficiency has become a matter of international concern.

ACQUIRED IMMUNODEFICIENCY SYNDROME (AIDS)

AIDS is a disorder of the immune system caused by a virus. The virus has been variously designated as (1) **lymphadenopathy-associated virus (LAV)**, (2) **human T cell lymphotropic virus type III (HTLV-III)**, and (3) **human immunodeficiency virus (HIV)**. AIDS was first described in the United States in 1981 and was recognized as a new epidemic in 1982 (figure 29.1).

Manifestations
An individual who has AIDS frequently has a history of multiple opportunistic infections caused by viruses (figure 29.2), fungi, and protozoa, as well as bacteria. Table 29.1 includes a partial list of possible infective agents. The symptoms vary, depending on the nature of the infection (figures 29.3 and 29.4). Kaposi's (kap'o-sēz) sarcoma normally occurs primarily in elderly males of Jewish or Italian origin, but it is strikingly common in association with AIDS. It is a neoplasm of vascular endothelial origin and causes purplish-brown lesions on the skin or mucous

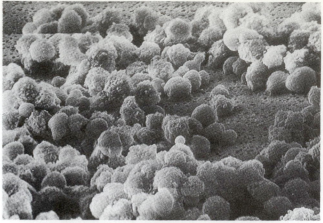

(a)

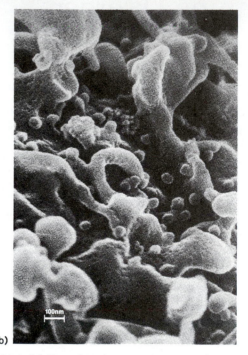

(b)

Figure 29.1 (*a*) Scanning electron micrograph (low magnification) of population of AIDS-infected lymphocytes. (*b*) High magnification of T 4 lymphocyte infected with HTLV–III.

membranes. Patients with AIDS show an increased frequency of B cell lymphomas, squamous carcinoma of the oral cavity, and carcinoma of the rectum as well (chapters 34 and 35). About half of AIDS patients have *Pneumocystis* pneumonia, whereas about one-fourth have Kaposi's sarcoma (chapter 30). The two conditions may also coexist. Lymphadenopathy syndrome (table 29.1) may occur either in the absence of or in association with Kaposi's sarcoma.

Pathogenesis
There are three groups that are at high risk for contracting AIDS: homosexual males, intravenous drug users, and recipients of blood transfusions. The virus has been identified in tissues and body fluids, particularly semen and blood. It is transmitted sexually and by contact with

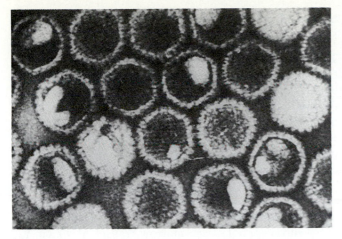

(a)

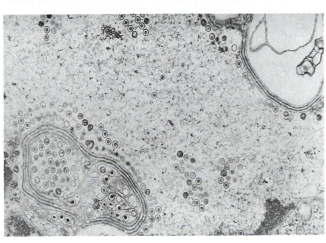

(b)

Figure 29.2 Electron micrograph. (*a*) Herpes simplex.
(*b*) Herpes simplex in nucleus.

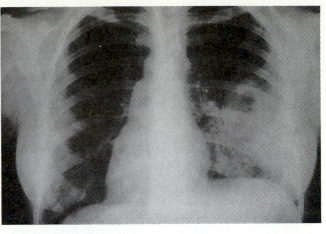

Figure 29.3 Chest X-ray of patient showing nonencapsulated cryptococcosis.

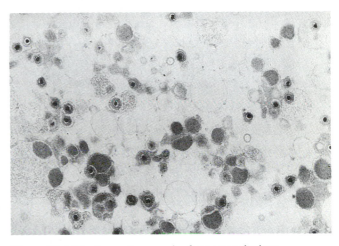

Figure 29.4 Electron micrograph of cytomegalovirus, magnification ×49,200.

TABLE 29.1 Clinical features of the AIDS syndrome

Opportunistic Infections

Pneumocystis carinii. Pneumonia
Candida albicans. Oral thrush or esophagitis
Cryptococcus neoformans. Meningitis
Cytomegalovirus. Fever, disseminated organ involvement
Toxoplasma gondii (protozoan). Encephalitis
Herpes virus infections. Disseminated
Cryptosporidia. Intestinal parasite, diarrhea

Kaposi's Sarcoma

Uncommon endothelial cell tumor

Lymphadenopathy Syndrome

Lymphadenopathy, splenomegaly, fever, weight loss

contaminated blood, including needle-stick injuries. There may be perinatal transmission as well. The highest incidence of infection involves homosexual males.

The disease process is initiated when the virus selectively infects a specific subset of lymphocytes called **T4 lymphocytes,** which are helper cells. These cells normally play a role in activating both humoral and cell-mediated responses. A diminished population of T4 cells leads to defective cell-mediated immunity (chapter 30).

Table 29.2 summarizes immunologic characteristics of AIDS. B cells are stimulated to produce immunoglobulins, and the result is hypergammaglobulinemia, as shown in table 29.2. There is, however, a defective antibody response to antigen introduced by immunization.

Diagnosis

A diagnosis of AIDS is established by a positive test for antibodies to the HIV virus, along with a history of opportunistic infections or Kaposi's sarcoma. Death is likely within 1–2 years for these individuals. Recently infected

perinatal: Gk. *peri,* around; L. *natalis,* birth

Immune Disorders

331

TABLE 29.2 Immunologic characteristics of AIDS

Marked lymphopenia	Lymphokine production abnormal
Reduction in percentage of T cells	Phagocytosis normal
Circulating immune complexes	Complement level normal
Reduced cytotoxic activity	Reduced T4 cells
Diminished antibody response to immunization	Hypergammaglobulinemia

individuals may not test positive for the antibodies, and those who show a positive antibody test may be asymptomatic. There is evidence indicating that the incubation period for the virus is at least 18 months. Those individuals who do not have Kaposi's sarcoma or opportunistic infections, but who exhibit immunologic abnormalities along with fever, weight loss, and lymphadenopathy are diagnosed as having **AIDS related complex (ARC)**. Evidence indicates that not every individual with ARC will develop full-blown AIDS.

Treatment

There is supportive intervention for the complications of AIDS, but there is no effective treatment to combat the HIV virus.

• • •

Individuals who are completely lacking in IgA may have antibodies against IgA. A serious or even fatal anaphylactic reaction may follow a blood transfusion. Administering washed, packed red cells is a safe alternative.

• • •

SELECTIVE IgA DEFICIENCY

Various immunoglobulin deficiencies occur and of these, selective IgA deficiency is the most common, with an incidence in the general population of approximately 1:600. IgA is an antibody fraction found mainly in body secretions (chapter 27). The cause of deficiency is unknown, and both autosomal recessive and autosomal dominant modes of inheritance have been proposed. Individuals with selective IgA deficiency may be asymptomatic because they are capable of making normal amounts of IgG and IgM. Studies indicate that the absence of IgA predisposes to a variety of problems including allergies, sinopulmonary infection, and autoimmune disease. Individuals who are completely lacking in IgA may have antibodies to this immunoglobulin and experience anaphylaxis if transfused with blood containing the IgA.

SEVERE COMBINED IMMUNODEFICIENCY DISEASE

There is an absence of both T and B cell immunity in severe combined immunodeficiency disease (SCID), and this may be the result of failure of differentiation of stem cells or the failure of thymus and lymphoid tissue to develop normally. This is an inherited condition which is transmitted either as an X-linked or autosomal recessive form. Patients may die before diagnosis is made, and the exact incidence of this immune disorder is not known.

There is onset of symptoms by age 6 months, and the affected individual usually succumbs to an opportunistic infection during the first year of life. Treatment consists of a histocompatible bone marrow transplant which, in spite of careful matching, may result in a **graft versus host** reaction. There is onset of symptoms within 7–30 days of the transplant, and the outcome is usually fatal. The exact mechanisms involved are unknown, but the factors in such a reaction are immunocompetent graft cells and host cells which are immunodeficient. Symptoms include rash, diarrhea, jaundice, cardiac irregularities, and pulmonary infiltrate, for which there is no adequate treatment.

• • •

DiGeorge's syndrome is the result of failure in embryogenesis. Facial features in affected infants are abnormal, with low-set ears, low forehead, small lower jaw, and fish-shaped mouth. There is defective development of the thymus gland; thus, there is T cell deficiency. Consequently, recurrent chronic infections occur. Fetal thymus transplants may restore cellular immune function in some patients.

• • •

AUTOIMMUNE DISORDERS

An autoimmune disorder involves an immune response against self, which may be an attack against a specific cell type, chemical constituents, or cellular receptors. Antibodies that are produced against self are called **autoantibodies** and lead to cellular or tissue damage (figure 29.5).

The autoimmune disorders listed in table 29.3 that involve endocrine glands are discussed in unit VII. Goodpasture's syndrome is caused by antibasement membrane antibodies, which lead to glomerulonephritis and pulmonary hemorrhage (chapter 28). Myasthenia gravis is a neuromuscular disorder in which there is voluntary muscle weakness (chapter 40). The basic defect involves receptors for the neurotransmitter acetylcholine (as″ē-til-ko′lēn) at the neuromuscular junction. Antibody attack against these receptors results in impaired nerve impulse transmission.

The remainder of this chapter deals with disorders that have immunologic features and in which, with one exception, there is the production of autoantibodies.

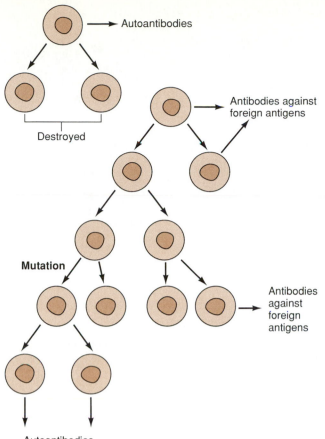

Figure 29.5 A proposed model to explain the increasing frequency of autoimmune diseases with age. Early in life, those lymphocytes that produce autoantibodies may be destroyed. With time, however, some descendants of the remaining lymphocytes may mutate to produce autoantibodies. When this occurs later in life, these cells are not destroyed and may cause autoimmune diseases.

TABLE 29.3 Some autoimmune disorders

Autoimmune hemolytic anemias. Idiopathic; secondary to viral infection or drugs; red blood cell antibodies
Neutropenias, lymphopenias, thrombocytopenias. Blood cell antibodies secondary to various autoimmune disorders
Goodpasture's syndrome. Antibodies against basement membrane
Addison's disease. May be caused by antibodies against the adrenal gland
Juvenile-onset diabetes. Usually have islet cell antibodies
Hashimoto's thyroiditis. Antibodies against thyroid constituents
Myasthenia gravis. Antibodies against acetylcholine receptors
Rheumatoid arthritis. Antibodies against Fc portion of IgG; immune complexes form
Systemic lupus erythematosus. Antinuclear antibodies; various other antibodies

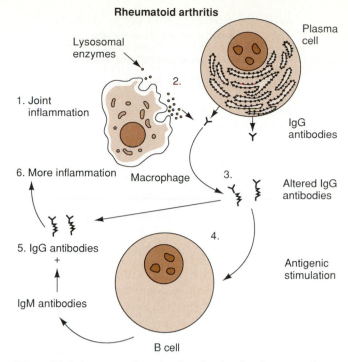

Figure 29.6 A proposed mechanism for the development of rheumatoid arthritis. An initial joint inflammation (*1*) results in the release of lysosomal enzymes (*2*) that cause damage to IgG antibodies (*3*). These damaged IgG antibodies act as antigens (*4*) to stimulate B-lymphocytes, resulting in the production of IgM antibodies antigens. This, in turn, produces further inflammation.

RHEUMATOID ARTHRITIS

Rheumatoid arthritis is a chronic inflammatory disease involving tendon sheaths and connective tissues of joints. Inflammation of peripheral joints, with limited motion and deformity, is characteristic. Autoantibodies are frequently present, and studies implicate the involvement of cellular immunity as well.

The etiology of rheumatoid arthritis is unknown (figure 29.6). Genetic factors are significant in the development of the disorder, and immunologic responses are in some way involved. Rheumatoid factors, antibodies that react with IgG molecules, appear in the circulation of 80% of patients. This disorder affects about 1% of the population and afflicts females 2 to 3 times more often than males.

Manifestations
General symptoms of rheumatoid arthritis include a persistent fever, weight loss, weakness, and diminished endurance. The characteristic finding is inflammation of the peripheral joints. Involvement of the small joints of the hands, wrists, and feet is most common. Frequently, the same joints on both sides of the body are inflamed. The term **synovitis** refers to inflammation of synovial fluid normally present in freely-movable joints. There is excessive fluid and synovial tissue proliferation accompanied by

TABLE 29.4 Effects of rheumatoid arthritis on tissues other than joints

Subcutaneous nodules. Characteristic lesion; various sizes and locations

Lung and pleural lesions. Pleural effusion (chapter 6); interstitial pneumonitis, hypoxemia

Cardiac involvement. Nodules with valvular deformities; pericardial effusions (fluid in pericardium)

Ocular involvement. Nodules on sclera; inflammation of sclera, cornea, and conjunctiva; loss of vision

Laryngeal involvement. Tenderness in hyoid bone or thyroid cartilage; hoarseness

Inflammation of peripheral nerves. Loss of sensation in distal extremities

Lymph node enlargement. Due to inflammatory reaction

pain, limitation of motion, warmth, and redness. In children and young people, these symptoms are often accompanied by a diffuse erythematous rash. This usually occurs at the onset of the disease or during periods of worsening symptoms. Table 29.4 lists the involvement of tissues other than joints. The consequences of rheumatoid arthritis over a period of months or years is damage to cartilage, bone, and structures surrounding the joints, resulting in deformity and impaired function.

Treatment

The modalities of treatment are aimed at controlling the progression of the disease. The course of the disease is variable, with either periods of exacerbations and remissions or a chronic active disease over a period of years. The long-range goal is to maintain maximal function. Certain measures may be taken to make the patient more comfortable, such as a well-balanced diet with weight control, regular rest, the avoidance of physical stresses, heat, supportive splints, and an appropriate exercise program. The purpose of medication is to alleviate pain and to slow the progressive destruction of joints. Salicylates frequently provide adequate control of pain, or other nonsteroidal anti-inflammatory drugs may be given as well. Gold salts appear to slow the progression of erosive lesions. The value of corticosteroid therapy is limited due to serious side effects over prolonged periods of time. Surgical intervention may correct deformities or restore function.

SYSTEMIC LUPUS ERYTHEMATOSUS

Systemic lupus erythematosus (SLE) is an autoimmune disorder of unknown cause in which there may be involvement of any of the systems of the body. A form of the disease that primarily affects the skin is called **discoid lupus erythematosus.** Most affected individuals are diagnosed between ages 13–40 years, although the disorder may occur at any age. Eighty-five percent of the cases are female, and the disorder is three times more common

effusion: L. *effundere*, to pour out

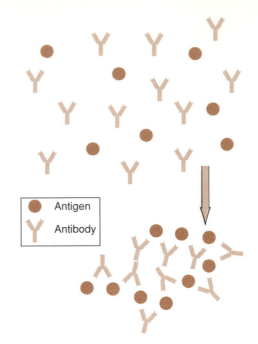

Figure 29.7 Antibodies cross-link antigens forming an antigen/antibody immune complex.

among blacks as compared to whites. There are various precipitating factors including sunlight, infection, stress, and drug therapy. There is some evidence of an hereditary predisposition to the disorder. Various autoantibodies develop in SLE with subsequent inflammatory responses.

Antigen/autoantibody combinations, called immune complexes, circulate and are deposited in tissues throughout the body. An inflammatory response ensues and leads to tissue injury and organ dysfunction (figure 29.7). The course of the disease is variable. It may be acute and quickly result in death or may follow a chronic and intermittent course.

Manifestations

General manifestations include fever, fatigue, weakness, weight loss, and anorexia. The skin, joints, and muscles are commonly involved. Most individuals experience arthritis and arthralgia or joint pain. The arthritis, which is usually transient and does not cause deformity, frequently involves the joints of the knees, ankles, shoulders, elbows, wrists, and hands. There may be sensitivity to light, and about one-third of affected individuals develop an erythematous rash in a butterfly pattern over the nose and cheek area. Ulcerations involving the nasal and oral mucosa, as well as the labia, are common. Ulcers, due to chronic ischemia, may develop on the legs. There may be purpuric (pur-pu'rik) eruptions on the skin due to subcutaneous hemorrhages. The skin may have areas of either hypo- or hyperpigmentation. Hair becomes coarse and thin.

arthralgia: Gk. *arthron*, joint; Gk. *algos*, pain
purpura: L. *purpura*, purple

Cardiovascular involvement may lead to vasculitis, primarily an inflammation of small to medium sized arteries and veins; thus, thrombophlebitis and Raynaud's phenomenon (chapter 22) are frequent manifestations of the disease process. Ulceration and gangrene are possible consequences of interrupted blood supply. Pericarditis and associated chest pain frequently occur, due to an inflammation of the membranous sac that envelops the heart. An inflammation of the heart muscle, myocarditis, may cause tachycardia, cardiomegaly, and cardiac failure. An inflammation of the membrane lining the inner heart and its valves, endocarditis, may occur as well.

Gastrointestinal manifestations include abnormalities of esophageal motility, anorexia, nausea, vomiting, and abdominal pain. Pancreatitis and enlargement of both the liver and spleen may occur. Pleuritis, pneumonia, and atelectasis (chapters 6, 8) occur in approximately half the patients with SLE. Enlarged lymph nodes or lymphadenopathy is also seen in about half the cases. There are various manifestations of involvement of the central nervous system, including double vision, personality changes, psychosis, aseptic meningitis, and seizures. There may be inflammation of nerves accompanied by pain in the extremities.

The kidney is affected in some way in most patients with SLE. The term **lupus nephritis** is used to identify inflammatory changes in the kidney. The effects are variable with mild to moderate involvement or glomerulonephritis and the nephrotic syndrome.

Almost all systems can be involved in SLE, either alone or collectively, and periods of exacerbations interspersed with periods of remission are characteristic. The most frequent cause of death is renal failure or involvement of the central nervous system.

Treatment

The avoidance of nonessential drugs, sunlight, and immunizations are appropriate preventive measures in SLE. Rest and salicylates, which are anti-inflammatory agents, are generally beneficial. Corticosteroids are prescribed when there is severe organ involvement or systemic disease.

SYSTEMIC SCLERODERMA

Systemic scleroderma is also called **progressive systemic sclerosis.** These terms refer to a disorder characterized by a thickening and hardening of the skin due to fibrosis, i.e., increased fibrous connective tissue. There is involvement of small arteries as well. Localized scleroderma involves the skin and underlying muscle exclusively, whereas diffuse or generalized scleroderma also affects internal organs. The cause of scleroderma is unknown, although it

pericarditis: Gk. *peri,* around; Gk. *kardia,* heart
myocarditis: Gk. *mys,* muscle
endocarditis: Gk. *endon,* within

is believed that immunologic factors contribute to tissue injury. The disease occurs most often between ages 35 and 55 and is four times more frequent in females.

Manifestations

In some cases of scleroderma, the skin changes are limited to the fingers, hands, and face and are slowly progressive. The initial symptoms usually include Raynaud's phenomenon (chapter 22) and edema and thickening of the skin of the hands. General manifestations include weakness, weight loss, stiffness, and aching joints. The term **CREST syndrome** is used to identify the following changes:

1. There may be deposition of calcium salts in the skin and subcutaneous tissues, i.e., calcinosis (kal″si-no′sis).
2. Raynaud's phenomenon is the typical initial symptom of scleroderma.
3. Impaired esophageal motility is a major diagnostic sign of the disorder. Anatomical changes in the esophagus include thinning of the mucosa with ulceration and a thickening of the submucosa by the protein, collagen. There is also atrophy of the smooth muscle with replacement by collagen. The walls of small arteries are thickened as well.
4. **Sclerodactyly** (skle″rō-dak′ti-lē) refers to the effects of scleroderma on the fingers. Thick leathery skin may cause both immobilization and ulcers at the ends of digits.
5. **Telangiectasia** may be apparent. This refers to a dilatation of capillaries that causes dark red, wartlike spots that vary in size from small spots up to spots the size of a pea.

Diffuse or generalized scleroderma may be rapidly progressive with involvement of both skin and other tissues and organs. Joint pain due to thickening and fibrosis of tissue surrounding the joints is characteristic. There may be muscle atrophy as well. There may be thickening of intestinal mucosa and replacement of muscle by collagen. Gastrointestinal symptoms include vomiting, abdominal distention, and diarrhea. Constipation and incomplete obstruction are signs of colon involvement.

Manifestations of pulmonary involvement include exertional dyspnea and, less commonly, cough. Alveolar fibrosis leads to impaired diffusion and restrictive lung disease (chapter 8). Sclerosis of pulmonary vessels leads to pulmonary hypertension and right-sided heart failure (chapter 23).

Renal disease may develop, either with or without hypertension. Renal arteries may undergo fibrinoid necrosis, i.e., necrosis of smooth muscle cells with deposition of fibrin (chapter 14). Malignant hypertension may develop and

scleroderma: Gk. *skleros,* hard; Gk. *derma,* skin
sclerodactyly: Gk. *skleros,* hard; Gk. *daktylos,* finger
collagen: Gk. *kolla,* glue; Gk. *genesthai,* to be produced
telangiectasia: Gk. *telos,* end; Gk. *aggeion,* vessel; Gk. *ektasis,* extension

TABLE 29.5 Autoantibodies that may be present in Sjögren's syndrome

Rheumatoid factors. Found in up to 90% of cases
Antigastric parietal cells
Antithyroglobulin
Antimitochondria
Antismooth muscle
Antisalivary duct antigens
SS-B. A nucleoprotein antibody
SS-A. A different nucleoprotein antibody

lead to uremia and death. When there is neurologic involvement, which is rare, abnormalities are attributed to vascular changes and thickening of connective tissue around nerve fibers.

Treatment

Treatment is symptomatic with emphasis on hand care and use of gloves. Management of vasospasms associated with Raynaud's phenomenon includes avoidance of cold, emotional stress, nicotine, and caffeine. Intestinal absorption may be improved by broad-spectrum antibiotic therapy.

SJÖGREN'S SYNDROME

Sjögren's syndrome is an autoimmune disorder characterized by chronic inflammation due to lymphocytic infiltration of tissues (table 29.5). In the primary form called the **sicca complex,** progressive tissue destruction may be limited to the salivary and lacrimal glands. A secondary form of the disease accompanies either rheumatoid arthritis or other connective tissue disorders. Lymphocytic infiltration and inflammation may affect various other body systems in either form of the disease.

The etiology of Sjögren's syndrome is unknown, although various factors appear to be involved, i.e., genetic, immunologic, hormonal, and infectious. Females are affected 10 times more often than males.

Manifestations

Ocular symptoms are the result of atrophy of lacrimal glands and decreased production of tears. There may be sensitivity to light, eye fatigue, diminished visual acuity, and small corneal erosions due to drying. Atrophy of salivary glands leads to a dry mouth with difficulty in swallowing and an increase in dental caries. There may be drying of other mucous membranes including the nose, bronchi, and vagina. There is an increased incidence of respiratory tract infections. Diminished exocrine gland function may cause pancreatitis and atrophy of gastrointestinal mucosa.

Nonglandular tissue involvement is more common in the primary form of Sjögren's syndrome. Raynaud's phenomenon with spasms of digital arterioles (chapter 22) occurs in 20% of patients. Lymphocytic infiltration of the

sicca: L. *siccare,* to dry

lungs may cause pneumonitis leading to dyspnea, or may cause obstruction of the airways. Renal tubular involvement may lead to renal tubular acidosis (chapter 14). Peripheral and cranial neuropathies occur.

Sjögren's syndrome is associated with a 44-fold increase in incidence of lymphoma, i.e., a group of neoplasms of the reticuloendothelial and lymphatic systems.

Treatment

Treatment for Sjögren's syndrome is symptomatic. Artificial tears may be used to alleviate ocular dryness, and increased fluid intake helps to relieve the problem of a dry mouth. Corticosteroid or immunosuppressive therapy is reserved for severe cases in which there is disability or major complications.

• • •

Antinuclear antibodies (ANA) react with antigens of nuclei of host tissue cells and also react with the nuclei of cells from various other animals. ANA are characteristic of autoimmune diseases, such as systemic lupus erythematosus, scleroderma, Sjögren's syndrome, and rheumatoid arthritis. Autoantibody production is usually absent in polyarteritis nodosa.

• • •

POLYARTERITIS NODOSA

Polyarteritis nodosa is a disorder in which there is segmental inflammation and subsequent necrosis of medium to small arteries. The term polyarteritis is used because all except pulmonary arteries may be involved. When the lesion heals, the tunica intima of the vessel proliferates, and there is an accumulation of fibrous connective tissue; thus, the term nodosa refers to nodules.

The etiology of the disorder is unknown, although it is likely that immunologic factors are involved. It occurs more commonly between ages 50 and 70, although it may occur at any age. Two to three times as many males are affected as females.

Manifestations

The manifestations of polyarteritis nodosa depend on the arteries involved and on whether the onset is abrupt or gradual (table 29.6). A biopsy of skin or affected organ, with microscopic examination of the tissue sample, confirms a diagnosis of the disorder.

Treatment

Corticosteroids are effective anti-inflammatory agents and are the treatment of choice in this disorder.

TABLE 29.6 Manifestations of polyarteritis nodosa

Abrupt Onset (features of a multisystem disorder)

Tachycardia	Visceral pain
Fever	Musculoskeletal pain
Weight loss	

Gradual Onset (may present with one system involvement)

Renal. Intermittent proteinuria and hematuria; progressive renal failure

Gastrointestinal. Abdominal pain, diarrhea, vomiting; hepatomegaly, possibly jaundice

Neurologic. Behavioral changes, subarachnoid hemorrhage, sensory and motor nerve involvement

Musculoskeletal. Arthralgia, myalgia

Cardiac. Myocardial infarction, pericarditis, hypertension, and heart failure

Cutaneous. Cutaneous and subcutaneous nodules

SUMMARY

• • •

Immunodeficiencies may involve any of the components of the immune system and may be primary, due to genetic factors, or due to failure of embryonic development. Acquired immunodeficiency syndrome is caused by a virus which infects T4 lymphocytes resulting in impaired cell-mediated immunity. Opportunistic infections caused by viruses, fungi, protozoa, and bacteria are characteristic.

Selective IgA deficiency is the most common of the immunoglobulin deficiencies. This disorder predisposes to allergies, sinopulmonary infection, and autoimmune disease. Severe combined immunodeficiency disease is an inherited condition in which there is an absence of both T and B cell immunity. Autoimmune disorders are the result of an immune response against self. There are many disease processes of autoimmune origin including hemolytic anemias, endocrine dysfunctions, myasthenia gravis, rheumatoid arthritis, and systemic lupus erythematosus.

Rheumatoid arthritis is a chronic inflammatory disorder of the connective tissue of joints and tendons, as well as subcutaneous tissue and other organs. Autoantibodies are present in the circulation of most patients. The characteristic finding is inflammation of peripheral joints and, ultimately, deformity and impaired function. Subcutaneous nodules are characteristic lesions, and there may be involvement of the lungs, heart, and eyes as well.

Systemic lupus erythematosus is an autoimmune disorder of unknown cause in which there may be involvement of any body system. Various autoantibodies develop in SLE with subsequent inflammatory responses. The course of the disease may be acute and quickly lead to death or may follow a chronic and intermittent course.

Systemic scleroderma is characterized by a thickening and hardening of the skin. Small arteries and internal organs may be affected as well. Immunologic factors appear to contribute to tissue injury in this disorder.

Sjögren's syndrome is also an autoimmune disorder characterized by chronic inflammation. Progressive tissue destruction may be limited to the salivary and lacrimal glands or may affect various body systems. Polyarteritis nodosa is a disorder in which there is segmental inflammation and necrosis of medium to small arteries. Immunologic factors appear to be involved. The manifestations vary, depending on the arteries affected.

REVIEW QUESTIONS

• • •

1. What are three names for the virus which causes AIDS?
2. What are three general clinical features of AIDS?
3. What is the basic pathogenesis in AIDS?
4. List five immunologic characteristics of AIDS.
5. Why may an individual with IgA deficiency be asymptomatic?
6. What general problems are often associated with IgA deficiency?
7. What is a major immunologic characteristic of severe combined immunodeficiency disease?
8. Define the term graft versus host reaction.
9. What is an example of an autoimmune reaction involving antibodies against basement membrane?
10. What is the basic defect in myasthenia gravis?
11. What manifestation of rheumatoid arthritis is characteristic in children and young people?
12. What is the most characteristic lesion of rheumatoid arthritis other than inflammation of joints?
13. What treatment appears to slow the progression of erosive joint lesions in rheumatoid arthritis?
14. What is the term for the form of lupus that primarily affects the skin?
15. List four general manifestations of lupus.
16. What are two manifestations of lupus involving the skin or mucosa?
17. In what way(s) may lupus affect the cardiovascular system?
18. List three manifestations of lupus involvement in the gastrointestinal system.
19. What are the consequences of lupus involvement in the lungs?
20. What is the most common cause of death in lupus?
21. What is the name of the disorder in which there is thickening of the skin and walls of small arteries?
22. What is the CREST syndrome?
23. What term refers to the primary form of Sjögren's syndrome?
24. What are two manifestations of the primary form of Sjögren's syndrome?
25. What is the name of the disorder in which there is inflammation and necrosis of medium to small arteries?

Selected Reading

• • •

Barnes, D. 1987. Solo actions of AIDS virus coat. *Science* 237:971–73.

Booth, W. 1987. Combing the earth for cures to cancer, AIDS. *Science* 237:969–70.

Booth, W. 1988. CDC paints a picture of HIV infection in U.S. *Science* 239:253.

Center for Disease Control. 1989. Guidelines of prevention of transmission of human immunodeficiency virus and hepatitis B virus to health-care and public-safety workers. *Laboratory Medicine* 20:783–97.

Eisenbarth, G. 1986. Type I diabetes mellitus: A chronic autoimmune disease. *New England Journal of Medicine* 314:1360–68.

Holmes, V. F., and F. Fernandez. 1989. HIV in women: Current impact and future implications. *Physician Assistant* 13:53–58.

Jochem, A., K. Bork, and M. Loos. 1987. Autoantibody-mediated acquired deficiency of Cl inhibitor. *New England Journal of Medicine* 316:1360–66.

McCarthy, T. A. 1989. Transmission and manifestations of HIV infection in children. *Journal of American Academy of Physician Assistants* 2:325–37.

Mills, M., C. Wofsy, and J. Mills. 1986. The acquired immunodeficiency syndrome: Infection control and public health. *New England Journal of Medicine* 314:931–36.

Morrisett, W. 1988. HIV-antibody counseling and testing. *Physician Assistant* 12:96–102.

Scientific American. 1988. What science knows about AIDS. (entire issue Oct.)

Chapter 30

Practical Applications:
Disorders of the Immune System

This chapter focuses on three immune system disorders: autoimmune hemolytic anemia, lupus erythematosus, and AIDS. Evaluation of laboratory findings associated with each disorder is emphasized. Various laboratory tests identify changes in elements involved in immune responses, and these findings are essential for the diagnosis and treatment of immunologic problems. A discussion of immunologic tests precedes specific examples of immune disorders.

IMMUNOLOGIC TESTS

The following discussion includes selected tests useful in diagnosing immunologic disorders.

AUTOANTIBODIES

Antibodies that develop against the host tissue in autoimmune responses are called autoantibodies. Various autoantibodies have been demonstrated and these include antibodies against (1) the thyroid gland, (2) nucleic acids, (3) lymphocytes, (4) acetylcholine receptors, (5) gastric parietal cells, (6) adrenocortical cells, (7) mitochondria, (8) pancreatic islet cells, and (9) nuclei. A high titer or level of autoantibodies indicates a disease process, but a low titer does not rule out a particular disorder. Autoantibodies are found with increased frequency in the normal population in the elderly. Table 30.1 lists various diseases associated with high titers of autoantibodies.

IMMUNE COMPLEXES

An immune complex is an antigen/antibody combination that (1) varies in size and solubility, and (2) may either be in the general circulation, or deposited in various tissues. Immune complexes have been implicated in many disease processes including autoimmune disease, malignancies, and infectious diseases. There are laboratory tests that demonstrate the presence of such complexes in body fluids; but most are nonspecific tests, since the antigen in the complex is generally unknown. The tests are based on the fact that immunoglobulin properties are changed when the molecules are complexed to antigen. The complexed immunoglobulin has different biological or chemical properties as compared to free immunoglobulin. As a result, there are various tests, such as the test for cryoglobulins (krī″ō-glob′u-lins), which are interpreted as indicators of immune complexes. Tests for immune complexes aid in diagnosis as well as indicate the severity of the disease.

Cryoglobulins

No single test is adequate to determine immune complexes, but the test for cryoglobulins is one indicator. This test identifies altered serum proteins that precipitate at

cryoglobulins: Gk. *kryos*, cold; L. *globulus*, small ball

TABLE 30.1 Diseases associated with high titers of certain autoantibodies

Antibodies Against	Associated Disease
Thyroid	Hashimoto's thyroiditis
Gastric parietal cells	Pernicious anemia
Adrenocortical cells	Addison's disease
Acetylcholine receptors	Myasthenia gravis
Mitochondria	Juvenile diabetes
Nuclei	Systemic lupus erythematosus

low temperature, strikingly at 4°C, and go back into solution as the temperature is raised. Cryoglobulins are associated with such diseases as chronic lymphocytic leukemia, rheumatoid arthritis, poststreptococcal glomerulonephritis, and systemic lupus erythematosus.

COMPLEMENT

Complement is a group of serum proteins, one of which is initially activated by binding to Ig/antigen complex (chapter 27). There are a series of proteolytic reactions resulting in products that promote various aspects of an immune response. These products are proteins recognized by the body as altered and are rapidly removed from the circulation. Measurement of two of these complement products, C_4 and C_3, is commonly done.

Any disease in which IgG or IgM immune complexes are formed, and in which complement is activated, can lead to decreased serum levels of complement. These diseases include systemic lupus erythematosus, vasculitis, and glomerulonephritis.

RHEUMATOID FACTORS

Rheumatoid factors are antibodies that react against antigenic determinants of IgG. The test involves using antigen, usually sensitized red blood cells, that reacts with rheumatoid antibodies. The antigen is added to the patient's serum, and agglutination of cells constitutes a positive reaction. The test results are expressed in terms of the highest dilution of serum in which agglutination is observed. A positive test is found in many diseases in which there is chronic inflammation and immune complex formation. There is increased specificity of the test for rheumatoid arthritis if the titer for rheumatoid factors is high.

ANTIGLOBULIN TEST (COOMBS' TEST)

The purpose of the Coombs' test is to identify immunoglobulin or antibodies that attach to antigen on the surface of red blood cells (figure 30.1). There is a direct test that identifies sensitized red blood cells with antibody attachment. The indirect test shows the presence of antibody circulating in the plasma (table 30.2).

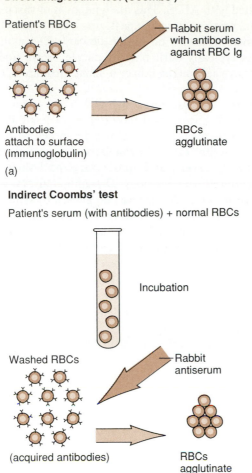

Direct antiglobulin test (Coombs')

Patient's RBCs

Rabbit serum with antibodies against RBC Ig

Antibodies attach to surface (immunoglobulin)

RBCs agglutinate

(a)

Indirect Coombs' test

Patient's serum (with antibodies) + normal RBCs

Incubation

Washed RBCs

Rabbit antiserum

(acquired antibodies)

RBCs agglutinate

(b)

Figure 30.1 (*a*) The direct Coombs' test shows that a patient's red blood cells (RBC) have been coated with antibody. (*b*) The indirect Coombs' test shows that a patient's serum contains antibodies.

TABLE 30.2 Direct and indirect Coombs' tests

Direct Antiglobulin Test (DAT) (Direct Coombs' test)

1. Patient's red blood cells + rabbit antiserum (antibodies)
2. Positive test if red blood cells agglutinate

Indirect Coombs' Test

1. Patient's serum + normal red blood cells (incubation)
2. Red blood cells washed with saline
3. Red blood cells + antiserum
4. Positive test if red blood cells agglutinate (antibodies in the patient's serum sensitized the cells)

The test is called an antiglobulin test because it is necessary to add antibodies against these antibodies (immunoglobulins) to demonstrate their presence.

Red blood cell antibodies develop in various disorders including autoimmune diseases, leukemias, delayed hemolytic transfusion reactions, and they may be drug induced.

• • •

Certain drugs are responsible for a positive DAT. About 15% of patients who are given methyldopa (Aldomet) for 3–6 months or longer show a positive DAT. About 0.8% of these patients have hemolytic anemia.

• • •

AUTOIMMUNE HEMOLYTIC ANEMIA

An autoimmune hemolytic anemia is the result of antibody production against the host's red blood cells with subsequent hemolysis. The mechanisms for induction of the antibody response are unclear. When hemolytic anemia occurs following the administration of drugs it may be (1) direct induction of autoantibodies by the drug, (2) due to adsorption of drug to red blood cells, or (3) due to adsorption of immune complexes to red blood cells.

> Patient 30.1 was a 63-year-old female taking alpha methyldopa 5–6 years for probable benign essential hypertension; urine dark brown; normal values for reporting lab are in parentheses.
> Laboratory findings: Hgb 4.8 g/dl (12–16), WBC 20,000 (4,000–10,700/mm³), reticulocytes 10.4% (0.5–1.5 % of RBC), Coombs' test positive.

The diagnosis in patient 30.1 is Coombs' positive hemolytic anemia associated with Aldomet. The color of the urine is evidence of hemolysis, and the low hemoglobin value indicates anemia. The reticulocyte count is evidence of the body's increased red blood cell production in response to anemia. The white blood cell count is usually elevated in hemolytic anemia. The positive Coombs' test confirms antibodies against red blood cells.

ACQUIRED IMMUNODEFICIENCY SYNDROME

Acquired immunodeficiency syndrome (AIDS) involves impaired T cell immune responses (figure 30.2) that make the individual susceptible to cancer and opportunistic infections (figure 30.3). There is a high incidence of pneumonia caused by *Pneumocystis carinii* associated with AIDS. This organism is a protozoan that usually causes pneumonia either in malnourished infants or in individuals with an inadequate cell-mediated immune response. Adults who develop *Pneumocystis* pneumonia usually belong in one of two categories: (1) those who have undergone chemotherapy for treatment of malignancy or (2) those who have AIDS. Another type of opportunistic infection frequently seen in AIDS is disseminated *Mycobacterium avium-intracellulare*.

Practical Applications: Disorders of the Immune System

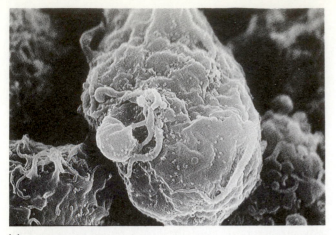

(a)

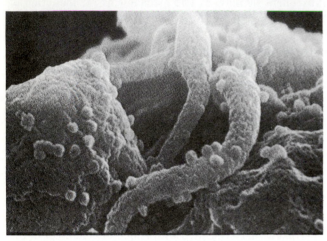

(b)

Figure 30.2 (*a*) Scanning electron micrograph of HTL-III-infected T 4 lymphocytes showing virus budding from the plasma membrane of the lymphocytes. (*b*) Same view, greater magnification.

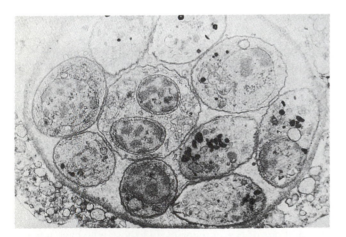

Figure 30.3 *Toxoplasma* cyst (electron micrograph). *Toxoplasma* cysts in tissues are dormant except in immunocompromised patients, as in the case of AIDS. Reactivation can lead to toxoplasmosis that can be fatal.

A type of malignancy called Kaposi's sarcoma is also associated with AIDS. This malignancy is of vascular endothelial origin and, in the past, was usually found in elderly men of Jewish or Italian descent. The malignant lesion is typically dark blue or purple brown and may appear anywhere on the skin or mucous membranes (figure 30.4).

Patient 30.2 was diagnosed with AIDS (figure 30.5), had a typical history, and the disease process followed a characteristic course. He was first diagnosed with AIDS during a hospitalization for *Pneumocystis* pneumonia and subsequently developed Kaposi's sarcoma lesions which were scattered over his body. Disseminated *Mycobacterium avium-intracellulare* was diagnosed by lymph node biopsy as well as by blood culture. Other manifestations of AIDS in patient 30.2 included memory impairment, behavioral disturbances, and tingling and numbness in his hands. Neurologic complications are frequent in AIDS, with AIDS dementia complex as a common primary disorder (chapter 39). This disorder is characterized by cognitive, motor, and behavioral dysfunctions as manifested by this patient.

Patient 30.2 suffered from chronic diarrhea. *Cryptosporidium* may cause diarrhea in AIDS, but frequently, as in this case, there may be persistent diarrhea with no identifiable cause.

> Patient 30.2 was a 25-year-old male, homosexual with a history of hepatitis A and B infections and syphilis; cigarette smoking and alcohol abuse; diagnosed with AIDS 7 months prior to this hospitalization.
> Laboratory findings: alkaline phosphatase 215 U/liter (38–126), AST/SGOT 83 U/liter (8–35), LD/LDH 1,295 U/liter (287–537), gamma-glutamyl transferase (GGT) 265 U/liter (7–64), Ca 8.2 mg/dl (8.9–10.6), RBC 2.9 million/mm³ (4.5–6.0), Hgb 9.8 g/dl (13–16), Hct 28.2% (40–54), WBC 3,200/mm³ (4,000–10,700); Differential count: 57% neutrophils, 9% bands, 18% lymphocytes, 4% monocytes, 10% eosinophils, 2% basophils

Patient 30.2 was admitted to the hospital with chills, a temperature of 103°, and a complaint of episodes of dizziness and shortness of breath. The cause of the fever was not clear, although it is likely that it was related to a relapse of pneumonia or due to some other infectious process.

BLOOD CELLS

Leukopenia, anemia, and idiopathic thrombocytopenia are typical findings in AIDS. Patient 30.2 had a low total white blood cell count with a decrease in the number of lymphocytes. Lymphopenia occurs in about half the patients with Kaposi's sarcoma and in almost all patients with opportunistic infections. A differential count identifies lymphopenia, but does not provide information about subpopulations of lymphocytes. There are two types of tests, not reported for this patient, that indicate quantitative changes in T-lymphocyte subpopulations or show functional capacity of these cells.

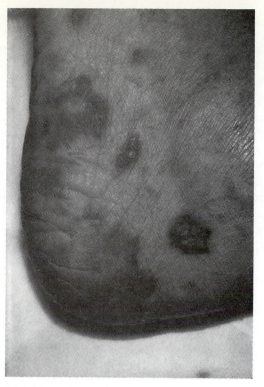

Figure 30.4 Violaceous plaques of Kaposi's sarcoma on the heel and lateral foot.

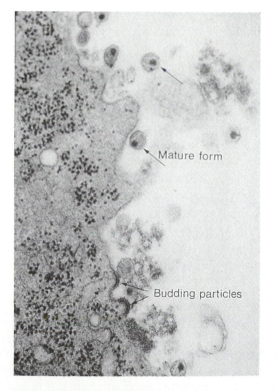

Figure 30.5 HTLV-III/LAV-type virus found in hemophiliac patient who developed AIDS; virus particles range in size from 90–120 nm.

Lymphocyte Subpopulations

The defect in AIDS is selective for the T-lymphocyte. The T4 lymphocytes are decreased in number and are functionally defective, whereas there is little or no change in the number of T8 lymphocytes. A decrease in the T4–T8 lymphocyte ratio in the peripheral blood is an expected finding in AIDS. A normal T4/T8 ratio is 1.5–2.0/1, whereas in AIDS it is less than 1.

Lymphocyte subpopulations may be assessed by means of monoclonal antibody studies. **Monoclonal antibodies** are derived from a single clone of cells and are specific for one type of antigen. These antibodies may be labeled with a fluorescent dye, and after antibody lymphocyte attachment, it is possible to count the particular lymphocyte subpopulation by a fluorescence microscope or a fluorescence-activated cell sorter.

• • •

Subpopulations of T-lymphocytes have different functions. In general, T4 lymphocytes are helper/inducer cells. They help B cells respond to antigen with antibody production and induce an inflammatory response. T8 lymphocytes are suppressor/cytotoxic cells. These lymphocytes limit the scope of immune reactions and are involved in the lysis of antigen bearing target cells.

• • •

Lymphocyte Function

There are tests that assess the functional capabilities of lymphocytes as well. Sensitized T cells normally respond to specific antigen in vitro by undergoing a series of changes that result in cell division. The capacity of T cells to respond to antigen in this way is assessed by measuring deoxyribonucleic acid (DNA) synthesis. Nonspecific stimuli called mitogens (mi'to-jens) cause T cells to undergo cell division in vitro as well. A **lymphocyte mitogen study** is a measure of lymphocyte functional capacity. In AIDS, there is reduced lymphocyte response to mitogens and antigens.

Anemia

The values for hemoglobin, hematocrit, and the red blood cell count are low in patient 30.2. Anemia is associated with chronic conditions including chronic infections, inflammatory disorders, and malignant disease. In chronic disease processes, there is a gradual decrease in red blood cell number, and the low count becomes a constant. The exact mechanism by which this occurs is not clear. Inhibitory factors leading to defective erythropoietin production may be the underlying cause.

in vitro: L. *vitrium,* glass

ENZYME LEVELS

The laboratory report shows elevations of certain enzymes in patient 30.2. Alkaline phosphatase (AP) is found mainly in bone and kidney, and an increase in the level of this enzyme is indicative of liver or bone disease. This patient had a history of hepatitis, and it is likely that there was further involvement of the liver associated with AIDS disease processes (chapter 36). Aspartate amino transferase (AST), also called serum glutamic-oxaloacetic transaminase (SGOT), is present in tissues with high metabolic activity. Increased AST/SGOT levels indicate injury or necrosis of cells. An elevation of alkaline phosphatase is often associated with high levels of AST/SGOT. The lactate dehydrogenase (LD/LDH) value was high in this patient as well. LD/LDH is a widely distributed intracellular enzyme indicative of cellular death when blood levels are elevated. The enzyme gamma-glutamyl transferase (GGT) is found mainly in the liver, kidney, and prostate and is elevated in all forms of liver disease.

DISCOID/SYSTEMIC LUPUS ERYTHEMATOSUS

Systemic lupus erythematosus (SLE) is a multisystem autoimmune disorder of unknown cause (chapter 29). The manifestations are variable, depending on the involvement of specific systems of the body. Nonspecific symptoms include fever, fatigue, weakness, anorexia, and weight loss. Fever is the most common of these symptoms.

Patient 30.3 was admitted to the hospital with a 3 week history of fever, chills, nausea, and vomiting. She had been diagnosed 6 months earlier with discoid or cutaneous lupus, which primarily affects the skin. Extreme sensitivity to sunlight lead to the diagnosis, at which time a marked increase in antinuclear antibodies (ANA) was reported. A skin biopsy was performed with positive results. This test is called a **lupus band test** and involves staining a skin sample for immunoglobulin and complement. The test is also positive in conditions other than lupus, i.e., rheumatoid arthritis, renal disease, and certain dermatologic disorders.

At the time of hospital admission, patient 30.3 complained of coughing, body aches, persistent headaches, and abdominal pain. Examination revealed enlarged inguinal (ing'gwĭ-nal) lymph nodes and areas of hypopigmentation on the skin.

> Patient 30.3 was a 27-year-old female with a history of discoid lupus; 20 pound weight loss during preceding month; fever, chills, weakness, vomiting; abdominal pain; temperature 39°C
> Laboratory findings: Na 133 meq/liter (137–45), alkaline phosphatase 128 U/liter (38–126), AST/SGOT 790 U/liter (8–35), ALT/SGPT 650 U/liter (7–56), LD

inguinal: L. *inguen,* groin

3,345 U/liter (287–537), GGT 284 U/liter (764), WBC 1,400/mm³ (4,000–10,700), RBC 3.9 million (4.2–5.4), Hgb 11.4 g/dl (12–16), Hct 32.1% (37–47), platelets 116,000/mm³ (130,000–400,000) prothrombin time 13.2 sec (9.8–12.6) APTT 33 sec (21–31)

The prothrombin time and activated partial thromboplastin time (APTT) are both prolonged. These are expected results if there is injury to the liver (chapter 20). The elevated enzyme levels may be due to liver damage as well (table 30.3).

The diagnostic problem in patient 30.3 was to identify the cause of the bronchitis and hepatitis and to evaluate the possibility of systemic lupus involvement. The patient's medications prior to hospitalization were prednisone, Plaquenil (hydroxychloroquine sulfate), and aspirin. Plaquenil is an antimalarial drug used to treat lupus (chapter 29). Possible side effects include nausea, vomiting, weakness, and fatigue.

The urinalysis showed nothing to suggest lupus-associated glomerulonephritis, i.e., red blood cells, white blood cells, or protein. She had a low white blood cell count, and the report showed anemia and thrombocytopenia. Leukopenia is characteristic of all phases of activity of systemic lupus erythematosus, and anemia occurs in about 80% of the cases. Antibodies produced in SLE may bind to blood cells and cause these cells to be destroyed.

Antinuclear antibody tests were performed on patient 30.3. Antinuclear antibodies (ANA) are circulating immunoglobulins that react with nuclear antigens. These antibodies are characteristic of autoimmune disorders, and the identification of antibodies for specific nuclear antigens is a useful diagnostic procedure. The tests performed on patient 30.3 were (1) anti-DNA, double strand, (2) anti-Sm, (3) anti-RNP, (4) anti-ENA, (5) anti-SSA, and (6) anti-SSB. The results and significance of these tests are shown in table 30.4.

• • •

Immune disorders may be characterized by an ANA profile that shows a high frequency of specific antinuclear antibodies. Continued monitoring of certain of the antibodies may also be useful in following the patient's response to therapy.

• • •

The complement levels for patient 30.3 were below normal. C3 was reported as 24.4 mg/dl (normal, 76–170) and C4 was 6.7 mg/dl (normal, 12.0–42). A decrease in the serum levels of complement may occur in any disease in which IgG or IgM immune complexes are formed and in which complement is activated and subsequently removed from the circulation. A decrease in complement is seen in SLE as well as other disorders such as vasculitis and glomerulonephritis.

TABLE 30.3 Source and significance of elevated serum enzyme levels

Alanine Aminotransferase (ALT) or Serum Glutamic-Pyruvic Transaminase (SGPT)

Source: High concentration in liver
Significance: Elevated in liver disease

Alkaline Phosphatase (AP)

Source: Mainly in liver and bone
Significance: Elevated in liver and bone disease

Aspartate Amino Transferase (AST) or Serum Glutamic-Oxaloacetic Transaminase (SGOT)

Source: High metabolic tissues
Significance: Elevated in cell injury or necrosis

Gamma-Glutamyl Transferase (GGT)

Source: Mainly in liver, kidney, prostate, and spleen
Significance: Elevated in all forms of liver disease

Lactate Dehydrogenase (LD)

Source: Widely distributed intracellular enzyme
Significance: Elevated in cell necrosis

TABLE 30.4 The significance of antinuclear antibody tests performed on patient 30.3 with possible SLE

Anti-Sm *	Antibody to Smith (Sm) antigen, a nuclear acidic protein; present in 30% of SLE cases; highly specific for SLE
Anti-DNA (double stranded)	Moderate to high titer occurs in 40–60% of cases of SLE
Anti-RNP	Antibody to nuclear ribonucleoprotein; present in low titers in SLE
Anti-ENA	Antibody to extractable nuclear antigen includes RNP and Smith
Anti-SSA	Antibody to a nuclear antigen; characteristic of patients with subacute cutaneous lupus erythematosus
Anti-SSB	Antibody to a nuclear antigen; in 15% of patients with SLE

Boxes indicate positive results in patient 30.3

The diagnosis for patient 30.3 was SLE with lupoid hepatitis, cutaneous involvement, and pulmonary involvement with cough and dyspnea. Treatment included an increased dosage of prednisone with a plan to taper the dosage as the patient improved.

LUPUS/PNEUMONITIS/NEPHRITIS

Patient 30.4 was admitted to the hospital with pancytopenia, diffuse pulmonary infiltrates, and a fever of 39.4°C. He had a history of SLE and lupus nephritis. Pulmonary complications included a history of alveolar hemorrhage secondary to lupus. There had been an episode of *Aspergillus* pneumonia secondary to granulocytopenia due to immunosuppressive therapy as well. Massive hemoptysis (he-mop'ti-sis) from the left lung occurred one month prior to this hospitalization. He was admitted to the hospital for symptoms of shortness of breath, chills, fever, and lower back pain.

Patient 30.4 was a 32-year-old male with a history of lupus nephritis and pulmonary hemorrhage secondary to pulmonary cavity due to *Aspergillus* infection; active lupus with pneumonitis and renal failure.
Laboratory findings: WBC 3,600/mm³ (4,000–10,700), RBC 2.2 million/mm³ (4.7–6.1), Hgb 6.6 g/dl (14–18), Hct 19.6% (42–52), platelets 83,000/mm³ (130,000–400,000), LD 1,327 IU/liter (287–537), BUN 103 mg/dl (7–21), creatinine 5.4 mg/dl (0.6–1.3), direct antiglobulin test (Coombs') negative

A blood transfusion brought the hematocrit up from a value of about 20% to 25%, and the platelet count increased to 162,000/mm³. The white blood count remained relatively the same. The negative direct Coombs' test indicated absence of antibodies attached to the surface of red blood cells. Red blood cell antibodies may develop in various immune disorders including SLE and may cause hemolysis.

A chest X-ray at the time of admission revealed diffuse infiltrates throughout the lungs. Patient 30.4 had a history of pulmonary involvement in the course of his disease. Common manifestations of SLE in the lung include pleurisy and pneumonitis, likely due to the deposition of circulating immune complexes in the alveolar walls. SLE may be associated with pulmonary infections. In this case, there was a history of *Aspergillus* pneumonia. This patient was treated with a high dose cortisone preparation and follow-up chest X-rays showed dramatic improvement. His rapid response to cortisone therapy indicated lupus rather than bacterial or *Aspergillus* pulmonary infection.

Renal function of patient 30.4 was marginal and continued to decline. Laboratory test results follow:

Urinalysis: 3⁺ protein, 5–10 WBC/high power field, 10–20 RBC/high power field
First day: BUN 103 mg/dl (7–21), creatinine 5.4 mg/dl (0.6–1.3), BUN/creatinine ratio 19 (6–22)
Second day: BUN 119 mg/dl, creatinine 6.7 mg/dl
Third day: BUN 130 mg/dl, creatinine 8 mg/dl
Fourth day: BUN 139 mg/dl, creatinine 7.7 mg/dl
Tenth day: BUN 214 mg/dl, creatinine 4.6 mg/dl, BUN/creatinine ratio 47

pancytopenia: Gk. *pan,* all; Gk. *kytos,* cell; Gk. *penia,* want
hemoptysis: Gk. *haima,* blood; Gk. *ptysis,* spitting

Both creatinine and urea appear in the urine by way of glomerular filtration, but urea is partially reabsorbed as well (chapter 11). An increase in these values can be correlated with impaired glomerular filtration rate. In general, a decrease in urine flow does not affect creatinine levels, whereas there is increased blood urea due to urea reabsorption. An increased BUN/creatinine ratio occurs as there is decreased urine flow.

Nephritis occurs in SLE due to deposition of complement/immunoglobulin complexes in the vascular epithelium of glomeruli. Only a few glomeruli may be affected in lupus nephritis, so there is minimal proteinuria and hematuria. There is usually good response to corticosteroids. More extensive involvement of glomeruli is shown by azotemia with protein and both red and white blood cells appearing in the urine. Approximately half of these patients respond to corticosteroids, and those who do not have a remission that progresses to renal failure.

• • •

The immunofluorescent technique for detecting antinuclear antibodies involves (1) a source of nuclei, such as human white blood cells, liver, kidney, or calf thymus, (2) patient's serum, and (3) a fluorescein-labeled antihuman gamma globulin. ANAs in the patient's serum will attach to nuclei and are visible as nuclear fluorescence under an ultraviolet microscope. Patterns of fluorescence are described as homogeneous diffuse, peripheral rim, and speckled.

• • •

OVERVIEW OF LUPUS ERYTHEMATOSUS

Table 30.5 lists the following characteristic laboratory findings in SLE:

1. The cryoglobulin test identifies serum proteins that have been altered to precipitate at cold temperature. A positive test is an indicator of the presence of immune complexes, i.e., antigen/antibody combinations that appear in the circulation and in various tissues in SLE.
2. Complement binds to antigen/antibody complexes, and consequently, low serum complement levels are associated with the formation of immune complexes.
3. Various antinuclear antibodies occur in SLE, including those that react with soluble nucleoprotein, DNA, RNA, and nuclear extracts.

TABLE 30.5 Characteristic laboratory findings in SLE

Cryoglobulins. Evidence of circulating immune complexes
Decreased level of complement. Usually associated with increased severity
Antinuclear antibodies. React against constituents of all nuclei
LE cell phenomenon. Results from action of autoantibodies and phagocytosis of damaged cells
Rheumatoid factors. Associated with chronic inflammation
Positive direct Coombs'. Indicates red blood cell-bound immunoglobulin
Leukopenia. Characteristic of SLE
Thrombocytopenia. Occurs in 15% of cases
Anemia. Mild to moderate degree in most cases
Prolonged prothrombin time. Due to antibodies that are anticoagulants
False-positive test for syphilis. Due to autoantibodies
Hypergammaglobulinemia. Gamma globulins are antibodies

4. The LE cell phenomenon is the result of antinuclear antibodies that react with deoxyribonucleoprotein. The antibodies damage the nucleus of a cell, which is subsequently phagocytized by a neutrophil (figure 30.6). An **LE cell** is a neutrophil filled with an inclusion body or damaged cell. There is a thin rim of cytoplasm, and the neutrophil nucleus is compressed against the cell membrane. The LE cell phenomenon strongly supports a diagnosis of SLE.
5. Rheumatoid factors are autoantibodies associated with diseases in which there is chronic inflammation and immune complex formation. The test for rheumatoid factors is positive in about 35% of patients with SLE.
6. A positive direct Coombs' test indicates autoantibodies against red blood cells. Acute hemolytic anemia with a positive Coombs' test occurs in about 5% of individuals affected by SLE.
7. Leukopenia is characteristic of all phases of SLE.
8. Thrombocytopenia with splenomegaly may occur in SLE.
9. Anemia occurs in three-fourths of the patients with SLE. Bleeding from gastrointestinal lesions exacerbated by a circulating lupus anticoagulant may be the underlying cause. In addition to this, hemolysis may be caused by antibodies that react against red blood cells.
10. There may be a prolonged prothrombin time due to circulating anticoagulants, i.e., antibodies against thrombin or factor VIII (chapter 19).

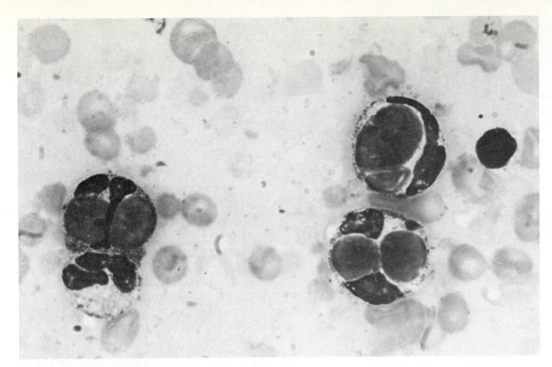

Figure 30.6 LE cells prepared from the blood of a patient with systemic lupus erythematosus, showing the altered nuclei phagocytized by neutrophils.
Courtesy of Dr. J. H. Crookston. From A. C. Ritchie: Boyd's Textbook of Pathology. *1990 Lea & Febiger, Philadelphia. Reproduced with permission.*

11. Autoantibodies are responsible for a false-positive test for syphilis in about 20% of the cases of SLE.
12. Hypergammaglobulinemia occurs in 80% of SLE patients. Gamma globulins are plasma proteins that are mainly antibodies.

SUMMARY

• • •

Autoantibodies and immune complexes develop in autoimmune disorders. A positive cryoglobulin test is an indicator of immune complexes. Tests for various components of the complement system may be diagnostically useful. Any disease in which IgG or IgM immune complexes are formed and in which complement is activated can lead to decreased serum levels of complement. Rheumatoid factors are positive in various diseases, and if there is a high titer, the test is diagnostic for rheumatoid arthritis. The Coombs' test shows the presence of immunoglobulins against red blood cells.

Autoimmune hemolytic anemia is a condition in which there is hemolysis of red blood cells due to autoantibodies. Hemolytic anemia may follow the administration of drugs, and the mechanisms by which antibody induction occurs is unclear.

The acquired immune deficiency syndrome involves impaired T cell immune responses that make the individual susceptible to cancer and opportunistic infection. Leukopenia is characteristic with a decrease in the T4–T8 lymphocyte ratio. Lymphocyte functional capacity, assessed by lymphocyte mitogen studies, is diminished as well. Anemia occurs in AIDS and is generally associated with chronic conditions including chronic infections, inflammatory disorders, and malignant disease. The underlying cause may be impaired erythropoietin production. Enzyme level elevations occur, depending on specific organ involvement.

Systemic lupus erythematosus is a multisystem autoimmune disorder of unknown cause. Identification of antinuclear antibodies is useful in diagnosing and following the course of the disease process. There are specific antinuclear antigen tests. In addition to this, the presence of autoantibodies is confirmed by positive results for the LE cell, rheumatoid factors, direct Coombs', and a false-positive for syphilis. Elevation of serum gamma globulin is a general indication of increased antibody production. A positive test for cryoglobulins is evidence of circulating immune complexes as well. Anemia and leukopenia usually occur in SLE, and there may also be a decreased number of platelets.

1. List five types of autoantibodies.
2. Why is a demonstration of cryoglobulins an indication of immune complexes?
3. Is a positive test for rheumatoid factors diagnostic for rheumatoid arthritis?
4. What is an immune complex?
5. In general, what is the significance of low serum levels of complement components?
6. What is the significance of a positive direct Coombs' test?
7. What is the underlying cause of autoimmune hemolytic anemia?
8. What type of white blood cell is decreased in number in most cases of AIDS?
9. What subpopulation of the leukocyte referred to in question 8 is decreased in AIDS?
10. What is the function of the T8 lymphocyte?
11. What is the probable cause of a prolonged prothrombin time in SLE?
12. What is a probable reason for a low sodium ion level in patient 30.3?
13. Increased serum levels of what two enzymes indicate liver disease?
14. What serum enzyme(s) are elevated in the case of cell necrosis?
15. List three specific antinuclear antibodies.
16. What antinuclear antibodies are highly specific for SLE?
17. What purpose may ANA tests fulfill, other than diagnosis of autoimmune disorders?
18. What is the significance of low levels of C3 and C4?
19. What is discoid lupus erythematosus?
20. In what way(s) may the lungs be affected in SLE?
21. What is the significance of the urinalysis for patient 30.4, i.e., 3+ protein, white blood cells, red blood cells?
22. What does a rising blood creatinine level imply?
23. What causes nephritis in SLE?
24. Define an LE cell.
25. What are two possible causes of anemia in SLE?

SELECTED READING

• • •

Clements, D. et al. 1988. Autoimmune haemolytic anaemia complicating ulcerative colitis. *British Journal of Hospital Medicine* 40:72.

Jondeau, G. et al. 1988. Autoimmune neutropenia after renal transplantation. *Transplantation* 46:589–91.

Olsen, R. et al. 1987. Serological and virological evidence of human T-lymphotropic virus in systemic lupus erythematosus. *Medical Microbiology and Immunology* 176:53–64.

Stone, M. 1988. Cutaneous necrosis at sites of transfusion: Cold agglutinin disease (letter). *Journal of American Academy of Dermatology* 19:356–67.

Unit VII
ENDOCRINE SYSTEM

The general theme of this unit is hormonal imbalance with a focus on disorders associated with both hypersecretion and diminished function of endocrine glands. Chapter 31 deals with the pituitary, thyroid, and parathyroid glands. The first section is concerned with the interrelationships of the pituitary gland and the hypothalamus. Pituitary hormonal imbalance may affect all pituitary hormones or, more commonly, may involve a clinically significant increase or decrease of a single hormone. The discussion centers around growth hormone imbalance, general pituitary hypofunction, prolactin excess, and deficiency of vasopressin. The discussion of thyroid function disorders includes the consequences of both hyper- and hypofunction of the thyroid as well as thyroid malignancy. The last part of chapter 31 deals with parathormone, which is secreted by the parathyroid glands, and the effects of hormonal imbalance.

Chapter 32 deals with endocrine function of the adrenal gland and the pancreas. The main topics related to the adrenal gland are glucocorticoid and aldosterone imbalance. The last section is concerned with disorders of the endocrine pancreas, i.e., diabetes mellitus and hyperosmolar hyperglycemic nonketosis.

The purpose of chapter 33 is to evaluate uncontrolled endocrine pancreas dysfunction. Examples of ketoacidosis and hyperosmolar hyperglycemic nonketosis are included. The discussion emphasizes laboratory findings associated with these conditions and the pathophysiology of consequent electrolyte and acid-base imbalances. The significance of laboratory findings related to these two disorders is the theme of chapter 33.

Chapter 31

Pituitary, Thyroid, Parathyroid Disorders

The focus of this chapter is dysfunction of three endocrine glands. Endocrine glands are ductless glands with secretory cells that empty hormonal products into extracellular spaces and, subsequently, into the bloodstream (figure 31.1). Hormones are chemicals released by a cell and which act on another cell. Hormones are not only produced by endocrine glands, but also by various other cells and tissues, such as neurons or the epithelium of the gastrointestinal tract. This chapter is limited to hormonal disorders related to the pituitary, thyroid, and parathyroid glands.

PITUITARY AND HYPOTHALAMUS

The pituitary gland occupies a protected position at the base of the brain, partially covered by the bony sella turcica (figure 31.2). The gland has two distinct lobes, i.e., the posterior lobe, an enlargement of the pituitary stalk, and the anterior lobe (figure 31.3). Six hormones are synthesized and secreted by the anterior pituitary, and two hormones, although synthesized by the hypothalamus, are stored and released by the posterior pituitary (table 31.1).

There is a relationship between the hypothalamic area of the brain and the pituitary gland. Nerve cell bodies located in the supraoptic and paraventricular nuclei of the hypothalamus (figure 31.3) have axons that extend to the posterior lobe of the pituitary. These nerve cell bodies synthesize antidiuretic hormone and oxytocin, both of which are carried to the posterior lobe of the pituitary.

The anterior pituitary is indirectly connected to the hypothalamus by way of capillaries and portal veins. Axons in the hypothalamus secrete hormones that enter capillary blood, which carries the hormones to the anterior pituitary. These hypothalamic hormones act on cells of the anterior pituitary and regulate the secretion of pituitary hormones (figure 31.4). Table 31.2 lists hypothalamic hormones and their functions.

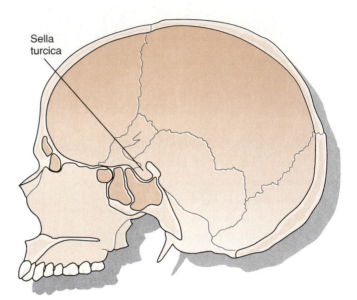

Figure 31.2 The sella turcica houses the pituitary gland.

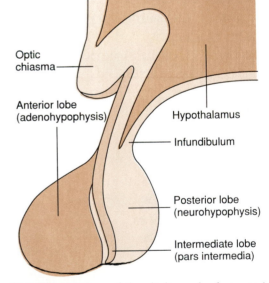

Figure 31.3 The structure of the pituitary gland as seen in the sagittal view.

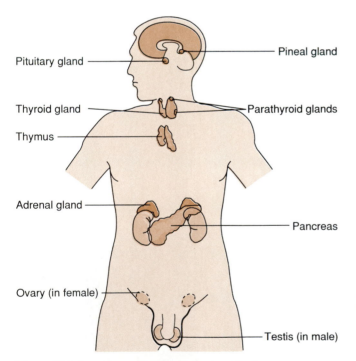

Figure 31.1 Locations of major endocrine glands.

ENDOCRINE SYSTEM

TABLE 31.1 Pituitary hormones and their functions

Anterior Pituitary (Adenohypophysis)

Thyroid-stimulating hormone (TSH). Thyroid hormone
  synthesis and secretion
Adrenocorticotrophic hormone (ACTH). Glucocorticoid
  synthesis and secretion
Growth hormone (GH). Growth, carbohydrate and fat
  metabolism
Luteinizing hormone (LH). Gonad control; female, ovulation,
  progesterone and estrogen production; male, androgen
  production
Follicle-stimulating hormone (FSH). Gonad control; female,
  follicle development, estrogen synthesis; male,
  spermatogenesis
Prolactin. Milk synthesis

Posterior Pituitary (Neurohypophysis)

Antidiuretic hormone (ADH). Renal water retention
Oxytocin. Release of milk during lactation

TABLE 31.2 Names and functions of hormones produced by the hypothalamus that control anterior pituitary function

Thyrotropin releasing hormone (TRH). Stimulates secretion
  of TSH
Corticotropin releasing factor (CRF). Stimulates secretion of
  ACTH
Growth hormone releasing hormone (GRH). Stimulates
  secretion of growth hormone
Luteinizing hormone releasing hormone (LH-RH). Stimulates
  LH secretion, and to lesser extent, stimulates FSH
  secretion
Somatostatin. Inhibits GH and TSH secretion

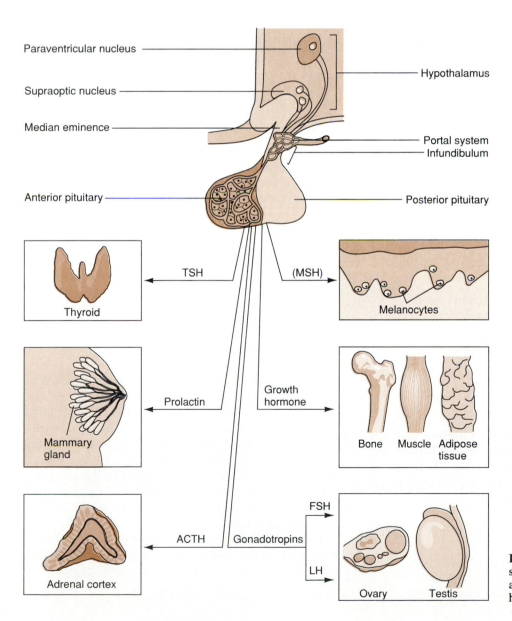

Figure 31.4 The hormones secreted by the anterior pituitary and the target organs for those hormones.

Pituitary Hormone Imbalance

Imbalance of pituitary hormonal secretion may involve one or more hormones and may be due to either a pituitary disorder or to a dysfunction of the hypothalamus.

Anterior Pituitary Hyperfunction

Pituitary adenomas are usually not malignant tumors, but they frequently secrete pituitary hormones. Tumors that produce TSH, LH, or FSH are rare. Excessive secretion of ACTH by the pituitary stimulates the adrenal cortex and may lead to Cushing's disease (chapter 32).

Growth Hormone Excess

The most common cause of growth hormone excess is hypersecretion by a pituitary tumor. Elevated levels of growth hormone during childhood cause an increase in linear growth and, ultimately, extreme height. The result of hypersecretion of growth hormone in an adult is a condition called acromegaly. Acromegaly is characterized by enlargement of the head, hands, and feet, with increased growth of connective tissue, muscle, skin, and internal organs (figure 31.5). Table 31.3 summarizes effects of excessive amounts of growth hormone. There is evidence indicating that growth hormone activity is the result of the formation of a substance in the blood called somatomedin (so″mah-to-me′din), and it is somatomedin that mediates the effects of growth hormone. Both growth hormone and somatomedin levels are elevated in acromegaly.

Acromegaly may progress over a period of years without being recognized. Diagnosis of the disease depends on the following observations:

1. Oral administration of 100 g of glucose causes growth hormone levels to fall to less than 2 mg/ml, while those levels are not less than 5 mg/ml in acromegaly.
2. Administration of TRH stimulates growth hormone increase after 30 minutes in acromegaly, but there is no growth hormone stimulation in normal individuals.

Surgery or radiation therapy may be effective therapy for pituitary tumor hypersecretion of growth hormone.

Prolactin Excess

The hypothalamus inhibits prolactin secretion, and it is likely that the inhibiting agent is dopamine. Tumors that affect the hypothalamus may reduce this inhibition and lead to prolactin secretion. Pituitary adenomas may be the source of prolactin secretion as well.

acromegaly: Gk. *akros*, extremity; Gk. *megalo,* large

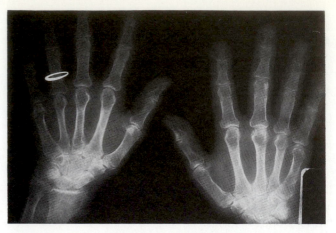

Figure 31.5 X-ray of hands of 74-year-old female with acromegaly.

Table 31.3 Effects of excessive amounts of growth hormone

Stimulation of growth. Linear growth; growth of connective tissue, muscle, skin, and viscera
Increased protein synthesis
Antagonism to action of insulin. Results in elevated blood sugar level
Diabetes or glucose intolerance
Increased breakdown of fat (lipolysis)
Enlarged head, feet, and hands
Enlarged facial sinuses. Deep, resonant voice
Enlarged mandible
Overgrowth of bone and cartilage in spine and knees
Thickened, coarse skin

• • •

Certain drugs, such as phenothiazine, tricyclic antidepressants, and methyldopa may cause excessive secretion of prolactin. Other possible causes include stress, administration of estrogen, and hypothyroidism.

• • •

Manifestations of elevated levels of prolactin include both galactorrhea (gah-lak″to-re′ah) and effects on the reproductive system. The term galactorrhea refers to milk production that normally occurs only after giving birth. Excessive secretion of prolactin can occur in either males or females at any age, but galactorrhea occurs in association with this condition only if estrogen is also available to help promote milk production. High levels of prolactin may lead to cessation of menses in females or interference with testicular function in males.

Plasma levels of prolactin may be determined to diagnose a suspected case of excessive prolactin secretion, i.e., hyperprolactinemia. A prolactin-secreting pituitary

galactorrhea: Gk. *gala,* milk; Gk. *rhoia,* flow

tumor may be treated surgically, although there is a significant recurrence rate. Administration of the drug bromocriptine suppresses prolactin secretion and may be used to successfully treat hyperprolactinemia.

ANTERIOR PITUITARY HYPOFUNCTION

Suppressed anterior pituitary function may lead to a single hormonal deficiency or a deficiency of multiple hormones. The underlying cause may be either a pituitary disorder or a dysfunction of the hypothalamus. Pressure of a growing tumor or a condition that leads to ischemic necrosis is a possible cause of pituitary hypofunction.

Sheehan's Syndrome

Sheehan's syndrome, also called postpartum hypopituitarism, (hi''po-pi-tu'i-tah-rizm'') develops subsequent to childbirth and is characterized by hyposecretion of pituitary hormones. The condition develops subsequent to moderate to severe blood loss and hypotension accompanying childbirth. Ischemic injury of the pituitary is a probable cause. The pituitary gland increases 40–60% in size during pregnancy and is unusually susceptible to damage in the case of inadequate blood supply.

Sheehan's syndrome causes failure to lactate postpartum, amenorrhea (ah-men''o-re'ah), evidence of hypothyroidism, and possibly adrenal insufficiency (chapter 32). The symptoms may develop over a period of years, but more commonly, symptoms appear soon after childbirth. The syndrome is treated by hormone replacement therapy.

• • •

If gonadotropin (LH and FSH) deficiency develops during adulthood, the result is amenorrhea and infertility in females. In males, there is reduced sperm production and possibly impotence. Gonadotropin deficiency in children interferes with sexual development during puberty.

• • •

Other Causes of Hypofunction

Adrenocorticotrophic hormone stimulates the secretion of glucocorticoids (cortisol) by the adrenal cortex (chapter 32). Decreasing blood levels of glucocorticoids leads to increased ACTH secretion, whereas high glucocorticoid levels inhibit ACTH secretion. Cortisol is used as an anti-inflammatory agent for the treatment of disorders, such as arthritis or severe asthma. Prolonged use of cortisol suppresses ACTH secretion; thus, therapeutic use of cortisol is a possible cause of pituitary hypofunction in terms of ACTH secretion.

Pituitary adenoma may cause hypofunction as well as hyperfunction of the gland. In an earlier part of this

postpartum: L. *post*, after; L. *partus*, bringing forth

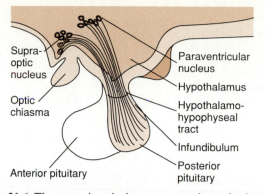

Figure 31.6 The posterior pituitary, or neurohypophysis, stores and secretes hormones (vasopressin and oxytocin) produced in neuron cell bodies within the supraoptic and paraventricular nuclei of the hypothalamus. These hormones are transported to the posterior pituitary by nerve fibers of the hypothalamo-hypophyseal tract.

chapter, pituitary adenoma was identified as a possible source of excessive levels of growth hormone. This tumor may also cause destruction of pituitary tissue or may interfere with pituitary-hypothalamic communication. Pituitary hypofunction is the result.

Growth Hormone Deficiency

Most cases of growth hormone deficiency are due to disorders of either the hypothalamus or the pituitary, trauma at birth, or asphyxia. Growth hormone deficiency may also be hereditary, inherited as an autosomal recessive characteristic. In some instances, low somatomedin levels are associated with normal growth hormone production. The result of growth hormone or somatomedin deficiency is pituitary dwarfism. If the condition is untreated, body proportions are normal, but short stature is the outcome. No treatment is required for growth hormone deficiency in an adult. In a child, growth hormone administration promotes normal growth.

• • •

Growth hormone levels are variable. There is increased growth hormone secretion during sleep and exercise and in response to both hypoglycemia and stress. Growth hormone levels are lower after the ingestion of glucose, and a reversible growth hormone deficiency occurs in emotionally deprived children.

• • •

POSTERIOR PITUITARY HORMONE IMBALANCE

The posterior pituitary secretes two hormones, vasopressin (antidiuretic hormone) and oxytocin (figure 31.6). Both hormones are synthesized in the supraoptic and paraventricular nuclei of the hypothalamus and are transported down axons to the posterior lobe of the pituitary

TABLE 31.4 Factors that stimulate
vasopressin secretion

Increased osmolality	Angiotensin II
Decreased blood pressure	Morphine
Decreased blood volume	Nicotine
Nausea	Alcohol
Acute insulin-induced hypoglycemia	Chemotherapy
Acute hypoxia	Stress
Acute hypercapnia	

for storage and secretion. Oxytocin promotes the secretion of milk during lactation. Vasopressin increases renal reabsorption of water and is a powerful vasoconstrictor as well.

Vasopressin Secretion

Osmolality is the most important factor controlling the secretion of vasopressin. Special neurons called osmoreceptors are located in the hypothalamus and are sensitive to changes in osmolality. A 1% increase in osmolality stimulates an increase of vasopressin secretion. Changes in blood volume or blood pressure affect plasma vasopressin levels as well. An acute fall in blood pressure causes increased secretion of vasopressin. If there is severe hypotension, there is a large vasopressin response. Table 31.4 summarizes various factors that stimulate the release of vasopressin.

• • •

Not all solutes are equally effective in evoking the secretion of vasopressin. Sodium ions and associated anions are the most effective stimulus. Intravenous infusion of sugars, such as mannitol and sucrose, is an effective stimulus as well. An increase in osmolality due to urea or glucose has little effect in a normal individual, although hyperglycemia in a person with insulin-deficient diabetes mellitus stimulates vasopressin secretion.

• • •

Diabetes Insipidus

Diabetes insipidus is classified as either neurogenic or nephrogenic. Neurogenic diabetes insipidus is characterized by a deficiency of vasopressin, while the nephrogenic form of the disorder is caused by renal unresponsiveness to the hormone.

Neurogenic diabetes insipidus may be caused by meningeal inflammation, head trauma, tumors, or the condition may be idiopathic. Rarely, the disorder is inherited as an autosomal dominant trait. The nephrogenic form of the disease may be a complication of drug therapy. The disorder may be inherited as an X-linked trait, although this is uncommon. The normal effect of vasopressin is to promote reabsorption of water by the distal and collecting tubules of the kidney. Consequently, either

hormonal deficiency or renal unresponsiveness leads to an increased urinary output of 3–20 liters per day. Thus, polyuria accompanied by thirst is a major manifestation of diabetes insipidus.

The vasopressin test is useful in distinguishing between the neurogenic and nephrogenic forms of diabetes insipidus. Vasopressin (Pitressin) is administered parenterally, and changes in urine osmolality (chapter 5) are monitored for a 2 hour period. An increase of urine osmolality of at least 50% is indicative of the neurogenic form of the disorder, while a lesser increase implicates the nephrogenic form.

It is possible to treat diabetes insipidus by adequate hydration, providing the vasopressin level is only partially depressed and the urinary output does not exceed 2–6 liters per day. Vasopressin preparations are available either for injection or for use as a nasal spray. Desmopressin is a synthetic form of vasopressin that may be administered by the nasal route and has the advantage of increased potency and longer duration of action (8–20 hours) compared to vasopressin. These agents are useful in treating the neurogenic form of the disease. Chlorpropamide is a hypoglycemic agent effective in reducing urinary output in patients in which there is partial depression of vasopressin secretion. Clofibrate, used to treat hyperlipidemia, also has an antidiuretic effect in partial neurogenic diabetes insipidus. Thiazide diuretics decrease the volume of urine in both neurogenic and nephrogenic diabetes insipidus.

THYROID GLAND

The thyroid gland lies on either side of the trachea along the lateral margins of the thyroid cartilage. The two lobes are connected by a band of tissue called the isthmus which is in the area of the cricoid (kri'koid) cartilage (figure 31.7). Two pairs of parathyroid glands are located on or beneath the posterior surface of the lobes of the thyroid.

FUNCTION OF THYROID GLAND

The function of the thyroid gland is to incorporate iodine into thyronine to produce the active hormones, thyroxine and triiodothyronine (tri''i-o''do-thi'ro-nēn). These hormones are designated T_4 and T_3 respectively, on the basis of the number of iodine atoms in the molecule (figure 31.8).

THYROID HORMONES

The thyroid gland secretes thyroxine (T_4) and lesser amounts of T_3 daily (figure 31.8). For the most part, these hormones are carried in the plasma in a protein-bound form, and it is only the small protein free fraction that is physiologically active. Thyroxine is converted to T_3, and about 70% of the circulating T_3 is the result of that conversion. It is likely that most, if not all, of the metabolic activity of thyroid hormones is due to T_3. The physiological effects of thyroid hormones are summarized in table 31.5.

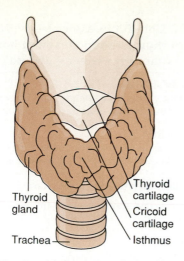

Figure 31.7 The thyroid gland. Its relationship to the larynx and trachea.

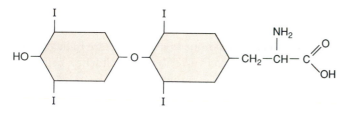

Thyroxine, or tetraiodothyronine (T₄)

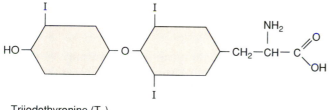

Triiodothyronine (T₃)

Figure 31.8 The thyroid hormones: thyroxine (T₄) and triiodothyronine (T₃) are secreted in a ratio of 9:1.

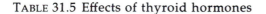

TABLE 31.5 Effects of thyroid hormones

Lower serum cholesterol
Stimulate growth
Required for central nervous system development
Promote protein synthesis
Reduce sensitivity to insulin
Accelerate insulin degradation

CONTROL OF THYROID FUNCTION

Thyroid-stimulating hormone is a major factor in regulating both the structure and function of the thyroid gland. Synthesis and secretion of thyroid hormones is controlled by a negative feedback mechanism as shown in figure 31.9.

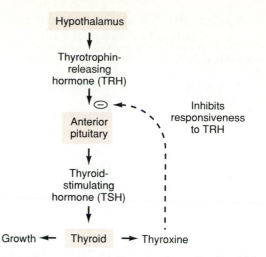

Figure 31.9 The secretion of thyroxine from the thyroid is stimulated by the thyroid-stimulating hormone (TSH) from the anterior pituitary. The secretion of TSH is stimulated by the thyrotropin releasing hormone (TRH) secreted from the hypothalamus into the hypothalamo-hypophyseal portal system. This stimulation is balanced by the negative feedback inhibition of thyroxine, which decreases the responsiveness of the anterior pituitary to stimulation by TRH.

Increasing blood levels of thyroid hormones inhibit the secretion of thyroid stimulating hormone by the pituitary gland. In addition, there is overall control of pituitary TSH by the hypothalamus. The availability of iodine is an additional requirement for thyroid hormone synthesis.

THYROID HYPERFUNCTION

There are various causes for excessive levels of circulating thyroid hormones, several of which will be discussed.

Graves' Disease

The most common cause of hyperthyroidism is Graves' disease, also called toxic diffuse goiter or autoimmune-hyperthyroidism. This disease is classified as an autoimmune disorder in which IgG antibodies stimulate thyroid secretion of hormones. Originally, the term long-acting thyroid stimulator (LATS) was used, but **thyroid-stimulating immunoglobulin (TSI)** is now used to refer to these antibodies. Although the cause of this disorder is uncertain, there may be an inherited defect involving suppressor T-lymphocytes (chapter 27).

The incidence of Graves' disease in the United States is estimated to be 0.4% of the population. It is most common between the ages of 30–50 years and occurs in females in a ratio of 7–10:1. Heredity is a factor, and emotional stress frequently precedes the appearance of symptoms.

The manifestations of hyperthyroidism caused by Graves' disease are listed in table 31.6. The term thyrotoxicosis is used for the effects of excessive amounts of circulating thyroid hormones. Ocular changes are apparent in about 70% of individuals with Graves' disease.

TABLE 31.6 Manifestations of hyperthyroidism caused by Graves' disease

Exophthalmos	Fatigue
Goiter	Weight loss
Pretibial myxedema	Increased appetite
Clubbing of fingertips	Thirst
Elevated TSI levels	Muscular weakness
Low TSH levels	Mild polyuria
Nervousness	Frequent bowel movements
Sweating	Insomnia
Heat intolerance	Dyspnea
Tachycardia	Disturbed menstrual function

TABLE 31.7 Conditions associated with elevated levels of thyroid hormones (thyrotoxicosis) other than Graves' disease

Toxic Multinodular Goiter
Thyroid hormone excess mild compared to Graves'; develops from a nontoxic goiter; after age 50; no exophthalmos

Subacute Thyroiditis
Mild thyrotoxicosis in early stages due to hormone leakage; etiology unknown; may develop hypothyroidism

Excess Thyroid-Stimulating Hormone (Rare)
Pituitary tumor or unresponsiveness to thyroid hormone inhibition

Toxic Adenoma
Long history of lump in neck; usually ages 30–50; hyperfunction of thyroid with no increase in TSH

These include irritation, lacrimation, blurred vision, and periorbital edema. There may be retraction of the eyelids, resulting in a staring appearance. Exophthalmos means protrusion of the eyeballs, and this response may be observed in Graves' disease. The thyroid gland is often enlarged at least 2 to 3 times normal size, forming a goiter. A small number of patients have pretibial myxedema, a swelling of the tissue over the tibia, that probably is the result of an immunologic reaction.

Modalities of treatment for Graves' disease include thyroidectomy, the use of antithyroid drugs, and radioactive iodine to reduce the synthesis of thyroid hormones.

Thyroid Storm

The term thyroid storm, also called thyrotoxic crisis, refers to when fever, marked tachycardia, and other symptoms of hyperthyroidism suddenly become life-threatening. Although the mechanism is uncertain, this is a complication that occurs when Graves' disease is not treated or is incompletely controlled. Infection, trauma, and surgical procedures are common precipitating factors. Management of this crisis requires therapy to inhibit the synthesis and release of thyroid hormone as well as general supportive measures. Although there is an approximate mortality rate of 20%, response to treatment usually occurs within 1–2 days.

Other Causes of Thyrotoxicosis

There are various conditions other than Graves' disease that lead to elevated levels of thyroid hormones (table 31.7). The term toxic goiter refers to an enlarged thyroid that secretes excessive thyroid hormones. Toxic multinodular goiter is a condition in which, for unknown reasons, there are several hyperfunctioning nodules. Inflammation of the thyroid may occur, and this causes various types of thyrotoxicosis. In subacute thyroiditis, there is an early phase of thyrotoxicosis due to leakage of thyroid hormone from the gland. Thyroid adenomas typically exhibit hyperthyroid function. These tumors are benign, noninvasive neoplasms that do not metastasize or spread to distant tissues.

Thyroid Carcinoma

The functional units of the thyroid gland are follicles made up of a single layer of cuboidal epithelial cells surrounding a colloid-filled lumen (figure 31.10). Malignant neoplasms of the thyroid are usually of epithelial origin and consequently are called carcinomas. Three types of thyroid carcinomas that originate in the follicles are listed in table 31.8.

Thyroid Hypofunction

Inadequate thyroid hormone production may be due to disorders of the thyroid, the pituitary, or occasionally the hypothalamus. Table 31.9 lists some possible causes of hypothyroidism. Iodine deficiency leads to inadequate synthesis of thyroid hormones and causes elevated levels of TSH. Thyroid-stimulating hormone causes enlargement of the thyroid gland, a condition called endemic goiter (figure 31.11). The incidence of endemic goiter is high in regions of the world, such as the Alps, Himalayas, Andes, and formerly the Great Lakes area of the United States, where iodide content of the soil is low. A preventive measure is the use of iodized salt in the diet.

Hashimoto's disease, an autoimmune thyroiditis, is the most common cause of hypothyroidism associated with goiter. There are circulating antibodies against the thyroid gland, and there is faulty synthesis of hormones associated with an elevated TSH level. Goiter is the most obvious feature of this disorder.

Hypothyroidism may occur after the surgical or medical treatment of Graves' disease. It is likely that there is an autoimmune response that results in functional destruction of the thyroid.

exophthalmos: Gk. L. *ex*, out; Gk. *ophthalmos*, eye

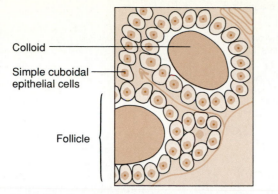

Figure 31.10 The functional units of the thyroid gland are follicles made up of a single layer of cuboidal epithelial cells surrounding a colloid-filled lumen.

TABLE 31.8 Malignant neoplasms of the thyroid

Papillary carcinoma. Most common of the thyroid carcinomas; slow growing; more frequent in children and young adults; tends to invade lymph nodes
Follicular carcinoma. Mimics normal thyroid tissue; metastasizes; most occur after age 40
Anaplastic carcinoma. Least common; most occur after age 50; rapid, painful enlargement of thyroid; invasive; usually incurable

TABLE 31.9 Possible causes of hypothyroidism

Thyroid malignancy	Radioactive iodine therapy
Iodine deficiency	Cellular resistance to thyroid hormone
Autoimmune thyroiditis	
Prolonged ingestion of large doses of iodide	Subtotal thyroidectomy
	Sequel of treatment of Graves' disease

• • •

Chronic administration of large doses of iodine causes goiter and hypothyroidism in certain susceptible individuals. This occurs in a small percentage of patients with chronic respiratory disease who are given potassium iodide as an expectorant.

• • •

Thyroid hormones are essential for growth and development during embryonic life. Inadequate hormonal levels during prenatal development causes cretinism. The child is mentally retarded and exhibits typical physical characteristics including short stature, short limbs, broad nose, puffy eyelids, thick tongue, and husky voice (figure 31.8). Thyroid hormone therapy begun soon after birth may allow normal mental and physical development.

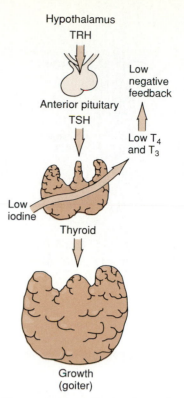

Figure 31.11 The mechanism of goiter formation in iodine deficiency. Low negative feedback inhibition results in excessive TSH secretion, which stimulates abnormal growth of the thyroid.

TABLE 31.10 Manifestations of hypothyroidism

Fatigue	Enlarged heart
Cold intolerance	Muscle aches
Weight gain	Decreased sweat
Coarse, dry hair	Constipation
Puffiness of face and body	Hoarseness
Dyspnea	Deafness
Difficulty in concentrating	Enlarged tongue
Periorbital edema	Coma in severe cases

Effects of Hypothyroidism

In general, hypothyroidism causes decreased metabolic activity with multiple consequences (table 31.10). The term **myxedema** is used for severe hypothyroidism in which nonpitting edema causes puffiness, particularly around the eyes, the hands, and feet, as well as enlargement of the tongue. Severe cases of hypothyroidism lead to coma and death. Therapy consists of thyroid hormone replacement.

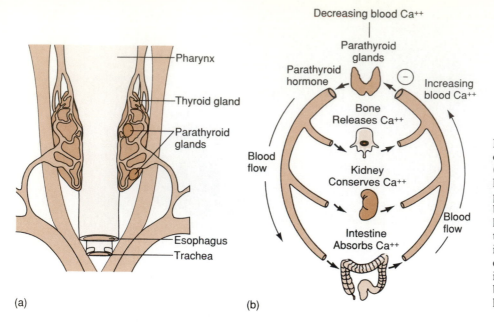

Figure 31.12 (a) A posterior view of the parathyroid glands. (b) Actions of parathyroid hormone. An increased level of parathyroid hormone causes the bones to release calcium, the kidneys to conserve calcium loss through the urine, and the intestines to increase absorption of calcium. Negative feedback of increased calcium levels in the blood inhibits the secretion of this hormone.

PARATHYROID GLANDS

The parathyroid glands secrete (figure 31.12) parathyroid hormone (PTH), which promotes (1) bone resorption, (2) renal calcium reabsorption, (3) intestinal calcium absorption, and (4) decreased renal reabsorption of phosphate. The overall effect of PTH is to increase serum levels of calcium. Calcium metabolism is discussed in chapters 2 and 11.

HYPERPARATHYROIDISM

Hyperparathyroidism causes increased calcium and lowered phosphate levels in the blood. Adenoma of the parathyroid, which is a glandular growth, and an increase in the size of the gland (hyperplasia) are causes of primary hyperparathyroidism. The parathyroids are stimulated to secrete PTH when serum calcium levels are decreased, and consequently, disorders associated with lowered calcium levels lead to secondary hyperparathyroidism. These disorders include chronic renal disease, renal tubular acidosis, and vitamin D deficiency.

Primary hyperparathyroidism usually occurs between ages 30 and 70 and affects females more often. Elevated serum calcium levels may lead to kidney stone formation, and if the condition continues over a period of time, there may be interstitial renal calcium deposits or nephrocalcinosis (nef″ro-kal″si-no′sis). High levels of PTH may lead to bone lesions. Table 31.11 summarizes the manifestations of primary hyperparathyroidism. Surgery is required to correct the condition.

TABLE 31.11 Manifestations of hyperparathyroidism

Renal effects. Nephrolithiasis, renal colic
Skeletal effects. Increased turnover of Ca in bone; <15% of patients have bone lesions
Neurological effects. Malaise, fatigue, irritability, psychosis
Muscular effects. Muscle weakness
Gastrointestinal effects. Nausea, vomiting, constipation, increased incidence of ulcer

HYPOPARATHYROIDISM

Decreased secretion of PTH occurs in hypoparathyroidism and results in hypocalcemia and hyperphosphatemia. This disorder may be idiopathic or may develop after surgical removal of the parathyroid glands. Decreased serum calcium levels increase neuromuscular excitability, with symptoms ranging from tingling and numbness to muscle cramps and convulsions. Hypoparathyroidism is treated with vitamin D and calcium supplements.

SUMMARY

• • •

The anterior lobe of the pituitary secretes hormones that influence reproductive functions, stimulate growth, promote glucocorticoid synthesis by the adrenal cortex, and stimulate thyroid hormone production. The posterior lobe of the pituitary secretes ADH and vasopressin, both of which are produced in the hypothalamus. The hypothalamus exerts control over the anterior pituitary by way of hormones that stimulate the release of pituitary hormones. The hypothalamic hormone somatostatin is an exception because it inhibits the secretion of TSH and GH.

Pituitary tumor is the most common cause of growth hormone excess. Hypersecretion of growth hormone during childhood causes extreme height. Acromegaly is the result of growth hormone excess in an adult and causes increased growth of connective tissue, muscle, skin, and internal organs. The hypothalamus inhibits prolactin secretion and tumors involving the hypothalamus may lead to prolactin secretion. High levels of prolactin may lead to cessation of menses in females or interference with testicular function in males.

Sheehan's syndrome is characterized by hyposecretion of pituitary hormones that follows severe blood loss associated with childbirth. Amenorrhea, hypothyroidism, and adrenal insufficiency are manifestations of the syndrome. Therapeutic use of cortisol leads to decreased ACTH secretion. Growth hormone deficiency may be caused by lesions of the hypothalamus or pituitary, asphyxia, or it may be an inherited condition. Pituitary dwarfism is the result when the deficiency occurs in childhood and may be treated by administration of growth hormone.

Vasopressin secretion by the posterior lobe of the pituitary is stimulated by increased osmolality of extracellular fluids and leads to reabsorption of water by the kidney. Diabetes insipidus is a condition in which there is either vasopressin deficiency or renal unresponsiveness to the hormone. The result is increased urinary output, which may be treated either by adequate hydration or by drug therapy that reduces urinary output.

Thyroid development and function are stimulated by TSH secreted by the anterior pituitary. Thyroid hormones stimulate growth, promote protein synthesis, and increase metabolic activity. Graves' disease is caused by hyperthyroidism and is characterized by goiter, exophthalmos, and circulating antibodies against the thyroid. Iodine deficiency is one cause of hypothyroidism manifested by goiter and symptoms of a lowered metabolism. Hyperactivity of the parathyroid glands results in hypercalcemia and hypophosphatemia, whereas calcium and phosphate levels are reversed in hypoparathyroidism.

Review Questions

• • •

1. What two pituitary hormones are actually synthesized in the hypothalamus?
2. List six hormones synthesized by the anterior lobe of the pituitary.
3. What hypothalamic hormone inhibits GH and TSH secretion by the anterior pituitary?
4. List hypothalamic hormones that stimulate the secretion of anterior pituitary hormones.

5. What is the most common cause of growth hormone excess?
6. What is the condition called when there is excess growth hormone in an adult?
7. Growth hormone activity is probably due to the formation of what substance in the blood?
8. What is the meaning of the term galactorrhea, and what is a possible cause?
9. What is another term for postpartum hypopituitarism?
10. What is a probable cause of postpartum hypopituitarism?
11. List four possible causes of growth hormone deficiency.
12. What are two causes of pituitary dwarfism?
13. What are two physiological effects of vasopressin?
14. Define neurogenic diabetes insipidus.
15. Define nephrogenic diabetes insipidus.
16. Which of the two thyroid hormones is the most physiologically active?
17. What are three effects of thyroid hormones?
18. Goiter is a manifestation of (hyperthyroidism, hypothyroidism).
19. What condition, other than Graves' disease, causes thyrotoxicosis?
20. What is the effect of inadequate thyroid hormones on a developing fetus?
21. What are three possible causes of hypothyroidism?
22. Define the term myxedema.
23. By what means does PTH increase serum calcium levels?
24. What is a cause of primary hyperparathyroidism?
25. What general effect is the result of hypoparathyroidism?

Selected Reading

• • •

Barkai, G. et al. 1988. In utero thyroxine therapy for the induction of fetal lung maturity; long-term effects. *Journal of Perinatal Medicine* 16:145.

Holm, L., H. Blomgren, and T. Lowhagen. 1985. Cancer risks in patients with chronic lymphocytic thyroiditis. *New England Journal of Medicine* 312:601.

Phillips, D. et al. 1988. Iodine in milk and the incidence of thyrotoxicosis in England. *Clinical Endocrinology* 28:61.

Valenta, L. et al. 1985. Pituitary dwarfism in a patient with circulating abnormal growth hormone polymers. *New England Journal of Medicine* 312:214.

Chapter 32

Adrenal and Pancreatic Endocrine Disorders

This chapter deals with both hormonal overproduction and insufficiency of the adrenal cortex as well as a tumorous condition of the adrenal medulla. The focus in relation to the pancreas is diabetes mellitus and the action of insulin. A condition related to diabetes called hyperglycemic, hyperosmolar nonketosis is the last topic of this chapter.

ADRENAL GLANDS

The adrenal glands are located on the upper surface of each kidney and are made up of an outer cortex and an inner medulla (figure 32.1). The adrenal cortex secretes steroid hormones including the glucocorticoids, mineralocorticoids, and sex steroids. Cholesterol is the precursor of the steroid hormones (figure 32.2). The adrenal medulla produces catecholamines (kat″ĕ-kōl′ah-mēns), epinephrine and norepinephrine.

GLUCOCORTICOIDS

Cortisol is a glucocorticoid that is secreted by the adrenal cortex in physiologically significant amounts. This hormone is essential for life and is involved with metabolism as well as various other processes (table 32.1).

Glucocorticoid secretion is stimulated by the pituitary hormone ACTH, which is under the control of corticotropin releasing hormone produced in the hypothalamus. There is a negative feedback loop whereby increasing blood levels of glucocorticoids suppress ACTH release by means of a direct effect on the anterior pituitary, as well as by inhibition of CRH (figure 32.3). Stress is a stimulus for the release of cortisol, and an increased blood level of the hormone occurs in response to surgery, fever, infection, hypoglycemia, and psychosis (figure 32.4).

• • •

The lipolytic and hyperglycemic actions of cortisol become apparent when cortisol secretion is stimulated by stress. Hyperglycemia and lipolysis occur in response to glucagon, epinephrine, and growth hormone. Cortisol secretion promotes and amplifies these effects.

• • •

Glucocorticoid preparations are useful in the treatment of various disorders and are particularly used as anti-inflammatory and immunosuppressive agents (table 32.2).

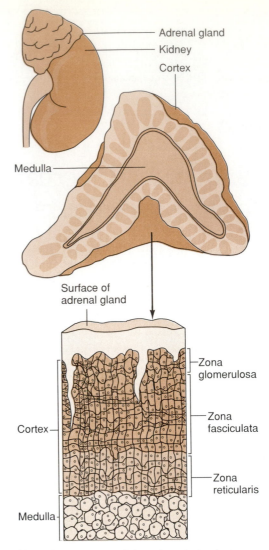

Figure 32.1 The structure of the adrenal gland showing the three zones of the adrenal cortex.

• • •

Cortisol is therapeutically useful for the treatment of severe or life-threatening symptoms of tissue injury. On the other hand, prolonged administration of cortisol increases susceptibility to infection and may prevent wound healing. The choice of glucocorticoid preparation for therapeutic use is determined by various factors. A lower sodium retaining potency is preferable when long-term therapy is required. Prednisone and dexamethasone exert a lower sodium retaining effect than either cortisone or cortisol. Those preparations, such as dexamethasone, which have a relatively long plasma half-life, have greater anti-inflammatory potency.

• • •

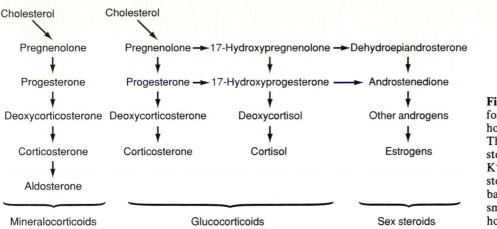

Zona glomerulosa

Cholesterol
→ Pregnenolone
→ Progesterone
→ Deoxycorticosterone
→ Corticosterone
→ Aldosterone

Mineralocorticoids

Zona fasciculata and zona reticularis

Cholesterol
→ Pregnenolone → 17-Hydroxypregnenolone → Dehydroepiandrosterone
→ Progesterone → 17-Hydroxyprogesterone → Androstenedione
→ Deoxycorticosterone → Deoxycortisol → Other androgens
→ Corticosterone → Cortisol → Estrogens

Glucocorticoids   Sex steroids

Figure 32.2 Simplified pathways for the synthesis of steroid hormones in the adrenal cortex. The adrenal cortex produces steroids that regulate Na^+ and K^+ balance (mineralocorticoids), steroids that regulate glucose balance (glucocorticoids), and small amounts of sex steroid hormones.

TABLE 32.1 Physiological effects of cortisol

Mobilizes muscle protein for gluconeogenesis
    Accelerates protein catabolism
    Inhibits protein synthesis
Stimulates conversion of amino acids to glucose and then to glycogen
Required for increased protein degradation during fasting
Amplifies the breakdown of glycogen stores in response to glucagon and epinephrine
Decreases tissue sensitivity to insulin
Increases appetite and caloric intake
Stimulates lipogenesis in certain areas
Required for fat catabolism stimulated by epinephrine and growth hormone
Inhibits bone formation and increases rate of bone resorption
Inhibits collagen formation
Contributes to maintenance of blood pressure
    Permits constriction of arterioles in response to catecholamines
    Decreases vascular permeability
    Enhances renal reabsorption of Na^+
Increases glomerular filtration rate
Influences response to trauma and infection
    Inhibits normal responses of capillary dilation and trapping of white blood cells at the site
    Reduces release of proteolytic enzymes
    May decrease phagocytic activity of neutrophils
    Inhibits cell-mediated immunity and may decrease the function of helper T-lymphocytes
Influences emotional function
    In excess, may elevate or depress mood

TABLE 32.2 Some disorders for which glucocorticoids are therapeutically useful

Severe asthma	Organ transplant rejection
Rheumatoid arthritis	Autoimmune hemolytic
Systemic lupus	anemia
erythematosus	Brain edema
Ulcerative colitis	Hypercalcemia due to
Acute rheumatic fever	neoplasm

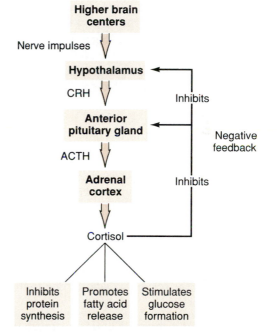

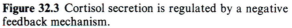

Figure 32.3 Cortisol secretion is regulated by a negative feedback mechanism.

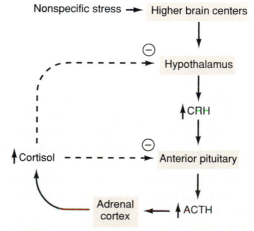

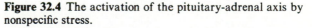

Figure 32.4 The activation of the pituitary-adrenal axis by nonspecific stress.

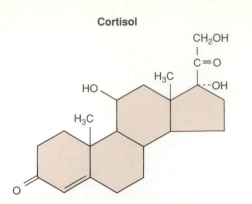

Cortisol

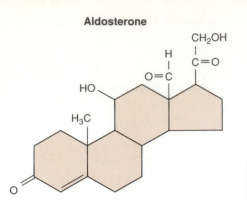

Aldosterone

Figure 32.5 Cortisol and aldosterone are steroids with similar molecular structure.

ALDOSTERONE

Aldosterone is the major mineralocorticoid secreted by the adrenal cortex (figure 32.5). This hormone stimulates the renal reabsorption of sodium from distal tubules, which results in the passive reabsorption of water. Aldosterone also promotes potassium and hydrogen ion secretion into the urine. Aldosterone maintains extracellular fluid volume as well as sodium and potassium ion balance. Stimuli for the secretion of aldosterone include a decrease in renal arterial blood flow and pressure and a decrease in sodium ion concentration.

A series of events occur in response to sodium ion depletion or hypovolemia:

1. The juxtaglomerular cells of the kidney secrete the enzyme renin (chapter 11). Renin promotes the conversion of angiotensinogen in the blood to angiotensin I.
2. Angiotensin I is cleaved to produce angiotensin II.
3. Angiotensin II (a potent vasoconstrictor) stimulates the release of aldosterone from the adrenal cortex.

In addition to this mode of hormonal regulation, an increase of potassium ions in the blood directly stimulates secretion of aldosterone by the adrenal cortex. Aldosterone secretion is inhibited when sodium, potassium, or fluid balance is restored.

• • •

The relationship between Na^+ reabsorption and H^+/K^+ secretion in response to aldosterone is dependent upon the delivery of sodium ions to the distal tubules. It is probable that K^+ secretion occurs as the result of an electronegative condition that develops in the glomerular filtrate when sodium ions are reabsorbed. Aldosterone increases K^+ excretion when there is a high sodium intake. Aldosterone does not, however, increase K^+ excretion when there is a deficiency of sodium.

• • •

TABLE 32.3 Sex steroids secreted by the adrenal cortex

Androgens
Dehydroepiandrosterone sulfate (DHEA-S)
Dehydroepiandrosterone (DHEA)
Androstenedione
Testosterone (small amounts)

Estrogens
Estradiol
Estrone

SEX STEROIDS

The adrenal cortex secretes sex steroids, androgens and estrogens, which contribute to maintaining secondary sexual characteristics. Androgens are male sex hormones, and the main adrenal products with androgenic activity are listed in table 32.3. These adrenal products are relatively insignificant in the male because the potent male sex hormone, testosterone, is produced in the testes. In females, androgens sustain pubic and axillary hair. Increased levels of prolactin appear to stimulate androgens, and ACTH is a stimulus as well. Small amounts of estrogens, the female sex hormones, are produced by the adrenals (table 32.3). In addition, the adrenal androgens may be converted to estrogen in subcutaneous fat and other tissues; thus, the adrenal is a source of estrogens in postmenopausal women.

CUSHING'S SYNDROME

Cushing's syndrome is the result of excessive levels of glucocorticoids, either due to hypersecretion by the adrenal cortex or due to prolonged administration for therapeutic purposes.

Etiology

The most common cause of Cushing's syndrome is iatrogenic (ī-at″rō-jen′ik), or the result of therapeutic use of glucocorticoids over a period of time. Cushing's syndrome may be primary, which implies hypersecretion of cortisol

iatrogenic: Gk. *iatro*, physician; Gk. *genesthai*, to be produced

by adrenocortical tumors. Both adenomas and carcinomas may be cortisol-secreting and occur more commonly in females. Carcinomas may secrete a mixture of other steroids as well.

Secondary forms of Cushing's syndrome are caused by ACTH, the source of which may be a pituitary adenoma or some other malignant tissue. Pituitary tumor is the most common endogenous cause of Cushing's syndrome, with a higher incidence in females between the ages of 25–40 years. High levels of ACTH may be due to increased CRH secretions by the hypothalamus as well.

Manifestations

Table 32.4 lists manifestations of cortisol excess. Cushing's syndrome is characterized by muscle weakness, obesity, a round puffy face called moon face, fat deposits above the clavicles, and a buffalo hump formed by a dorsal fat pad. The legs and arms are typically thin as compared to the body. Most individuals with Cushing's syndrome have hypertension. A majority of female patients experience hirsutism (her'soot-izm), male distribution of facial and truncal hair accompanied by menstrual dysfunction. This is the result of weak androgenic activity of cortisol itself and an excess of androgens converted to testosterone in the circulation. Virilism, a masculinizing effect, may occur in females and leads to male pattern baldness, enlargement of the clitoris, and deepening of the voice. Adrenal adenoma or carcinoma may cause feminizing effects in men including gynecomastia (jin″ĕ-ko-mas′te-ah), loss of body and facial hair, and decreased sexual potency. Excess glucocorticoids cause thinning of subcutaneous connective tissue, resulting in red to purple striae or stretch marks on the abdomen, thighs, buttocks, and breasts. An increase in both ACTH and lipotropin stimulates the production of melanin, which causes hyperpigmentation of the skin. Laboratory findings associated with cortisol excess include an increase in red blood cells, white blood cells, blood calcium, and blood sugar level. Hypokalemia and alkalosis may occur with prolonged glucocorticoid excess (table 32.5).

Diagnostic Tests

Increased production of cortisol associated with Cushing's syndrome is confirmed by the measurement of increased urinary excretion of either cortisol or cortisol metabolites, measured in the urine as 17-hydroxycorticosteroids (17-OHCS).

The dexamethasone suppression test is a means of evaluating the glucocorticoid-ACTH negative feedback mechanism whereby increased levels of glucocorticoids inhibit pituitary secretion of ACTH. Dexamethasone is a glucocorticoid preparation used both for therapeutic purposes and to differentiate the etiology of Cushing's syndrome. The suppression test involves the administration of low and/or high doses of dexamethasone. The effect of dexamethasone is determined by measuring urinary levels of cortisol or 17-OHCS. In a normal individual, low doses

TABLE 32.4 Manifestations of cortisol excess with probable causes

Weakness due to catabolism of protein
Obesity and redistribution of fat resulting in puffy face and fat deposits in supraclavicular and retrocervical areas; truncal and abdominal fat
Thin arms and legs due to protein catabolism
Hypertension due to increased mineralocorticoid activity of cortisol leading to Na⁺ and water retention
Hirsutism, menstrual irregularities, and acne in females due to excess androgens
Striae caused by thinning of subcutaneous connective tissue
Osteoporosis resulting from increased bone resorption
Hyperglycemia or carbohydrate intolerance due to gluconeogenesis
Increased susceptibility to infection
Other manifestations. Depression, psychosis, emotional lability, bruising tendency, gastric ulcers, edema, hyperpigmentation

TABLE 32.5 Laboratory findings associated with cortisol excess

Increased red blood cell count. Primarily due to androgen secretion
Increased white blood cell count. Increased neutrophils; decreased lymphocytes, monocytes, and eosinophils
Increased blood calcium level. Due to increased bone resorption
Hyperglycemia. Due to gluconeogenesis and insulin resistance
Hypokalemia and alkalosis. Occur with prolonged glucocorticoid excess; cortisol has mild mineralocorticoid activity; promotes Na⁺ retention and K⁺/H⁺ excretion

of dexamethasone suppress ACTH and thus also suppress cortisol. The following are expected results of the dexamethasone suppression test:

1. In Cushing's syndrome, there is failure to suppress ACTH and cortisol in response to low doses of dexamethasone.
2. Higher doses of dexamethasone suppress ACTH if a pituitary tumor is the cause of Cushing's syndrome.
3. There is no response to high doses of dexamethasone in the case of an adrenal tumor.
4. There is no change in cortisol secretion regardless of the dexamethasone dose if the ACTH secretion is ectopic, i.e., the source is nonpituitary.

Plasma levels of ACTH are indicative of the etiology of Cushing's syndrome as well. Plasma ACTH levels are elevated if the source is ectopic. In contrast to this, an adrenal tumor is associated with low or undetectable plasma levels of ACTH. There is a normal or somewhat elevated ACTH level in pituitary Cushing's syndrome.

ectopic: Gk. *ektopos,* away from a place

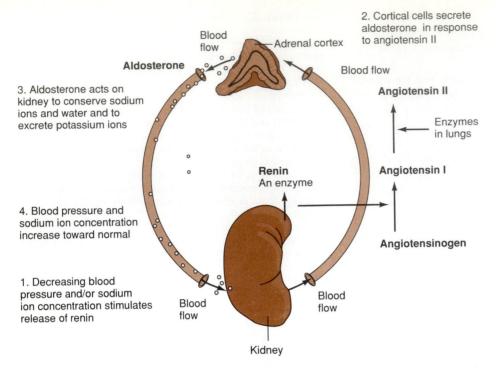

Blood flow — Adrenal cortex

2. Cortical cells secrete aldosterone in response to angiotensin II

Aldosterone

Blood flow

3. Aldosterone acts on kidney to conserve sodium ions and water and to excrete potassium ions

Angiotensin II

Enzymes in lungs

Angiotensin I

Renin
An enzyme

4. Blood pressure and sodium ion concentration increase toward normal

Angiotensinogen

1. Decreasing blood pressure and/or sodium ion concentration stimulates release of renin

Blood flow

Blood flow

Kidney

Figure 32.6 Aldosterone causes an increase in blood volume and blood pressure by promoting the conservation of sodium ions and water.

Treatment

The treatment for a pituitary lesion is surgical removal of the tumor. Alternative treatments are radiation therapy or the administration of ACTH-blocking drugs, such as bromocriptine. A benign nonpituitary tumor that secretes ACTH may be surgically removed, whereas hypercortisolism associated with a metastatic malignant tumor may be best controlled by an enzymatic inhibitor, such as metyrapone. Adrenalectomy is indicated as medical intervention for adrenocortical tumors.

ALDOSTERONISM

The term aldosteronism refers to an excessive secretion of aldosterone. Normal stimuli for aldosterone secretion from the adrenal cortex include (1) decreased sodium ion concentration, (2) angiotensin II, (3) increased potassium ion concentration, and (4) ACTH in minimal amounts (figure 32.6).

Etiology

Excessive amounts of aldosterone may be the result of a disorder involving the adrenal cortex, i.e., aldosteronism may be primary. Hyperplasia, or increased growth of the adrenals, is one possible cause of primary aldosteronism and is an idiopathic condition. Adrenal adenoma may also lead to hypersecretion of aldosterone.

Secondary aldosteronism is associated with high levels of renin. The stimulus for renin production may be unknown or may be in response to a decrease in renal blood flow. Arterial hypovolemia is a probable stimulus for renin, (and aldosterone secretion) in disorders in which edema is a prominent feature. Disorders that lead to secondary aldosteronism include the nephrotic syndrome (chapter

TABLE 32.6 Underlying causes of the manifestations of primary aldosteronism

Diastolic hypertension. Related to sodium reabsorption and increased extracellular volume
Polyuria. Impaired renal concentrating ability
Hypernatremia. Sodium retention and water loss due to polyuria
Hypokalemia. Renal reabsorption of Na^+ and excretion of K^+
Alkalosis. Both K^+ and H^+ excretion in exchange for Na^+ reabsorption; increased HCO_3^- reabsorption; K^+ lost from cells and H^+ migrate into cells
Edema absent. In chronic condition, there is escape from sodium retaining effects of aldosterone

13), congestive heart failure (chapter 23), and cirrhosis (chapter 36). Other causes of secondary aldosteronism include (1) renal artery stenosis, (2) renin secreting tumor of the kidney, (3) pregnancy, and (4) oral contraceptives.

Manifestations

Manifestations of primary aldosteronism (table 32.6) are the result of (1) aldosterone-stimulated sodium reabsorption, (2) passive water reabsorption, (3) excretion of potassium and hydrogen ions in exchange for sodium retention, and (4) reabsorption of bicarbonate due to hydrogen ion secretion (chapter 11). Diastolic hypertension is characteristic of the condition and is accompanied by hypernatremia, metabolic alkalosis, and hypokalemia. Muscle weakness and transient paralysis may occur because of potassium depletion. Typically, there is an absence of edema. Edema occurs initially because of sodium retention, but this stops after a few days and reabsorption of water is minimized.

A syndrome similar to aldosteronism occurs as the result of ingestion of large amounts of licorice. The chemical in licorice that causes sodium retention is glycyrrhizinic (glis″ĭ-ri′zin-ic) acid.

• • •

Diagnostic Tests

Primary aldosteronism may be suspected in a patient who has hypertension, hypokalemia, and suppressed renin activity. The possibility of secondary causes must be eliminated, and certain tests may be performed to aid in a differential diagnosis. Aldosterone levels in a 24-hour urine sample are indicative of aldosterone production in the body. The urinary aldosterone test is done under conditions of high sodium intake (120 meq or greater). If salt intake is lower, renin is stimulated, and the result is an increase in aldosterone. An elevated urinary aldosterone level associated with high sodium intake is indicative of primary aldosteronism.

The saline infusion test is a method of evaluating the normal suppression of aldosterone. The procedure involves the infusion of 2 liters of isotonic saline over a 4-hour period while the patient is lying down. The principle of the test is that saline suppresses renin and decreases the plasma aldosterone level. In normal individuals, an infusion of saline causes the plasma aldosterone level to fall below the normal range of 5–15 ng per deciliter. A patient who has an aldosterone secreting adenoma has decreased plasma renin activity. Saline infusion has little effect when plasma renin activity is already suppressed. In the case of an adrenal adenoma, the test usually shows a plasma aldosterone concentration of 10 ng/dl or more. The saline infusion test eliminates primary aldosteronism if the plasma aldosterone concentration is less than 6 ng/dl following the administration of saline.

Plasma renin activity is evaluated by the ability of the plasma to generate angiotensin I from renin substrate. Patients with primary aldosteronism have decreased plasma renin activity even when the test is performed subsequent to dietary restriction of salt. There is a high level of renin activity in secondary aldosteronism.

Treatment

Primary aldosteronism may be successfully treated by dietary restriction of sodium and the administration of spironolactone, an aldosterone antagonist (chapter 11). Antihypertensive agents may be required to control blood pressure in the case of idiopathic aldosteronism. Amiloride is another option for medical therapy. This drug, acting independently of aldosterone, decreases both potassium excretion and sodium retention and lowers blood pressure. Surgical removal is the treatment of choice for adrenal adenoma.

ADDISON'S DISEASE

Addison's disease is a disorder in which there is hypofunction of the adrenal glands resulting in manifestations of deficiency of adrenocortical hormones.

Etiology

The most common cause of Addison's disease is autoimmune destruction of the adrenal gland. Other causes include tuberculosis involving the adrenals, malignancy, and hemorrhage into the gland due to anticoagulant therapy or abdominal trauma. The autoimmune form of the disease occurs more frequently in females, and the incidence of all forms is predominantly during young adult to middle years. Secondary causes of glucocorticoid insufficiency include suppression of ACTH secretion due to prolonged cortisone therapy or ACTH deficiency due to a pituitary or hypothalamic disorder.

• • •

The adrenal cortex is made up of three layers: the outer zona glomerulosa which is the source of aldosterone, the middle glucocorticoid-secreting zona fasciculata, and the sex steroid producing inner layer called the zona reticularis. A diffuse autoimmune reaction involving the adrenals usually affects aldosterone and cortisol production equally. The pattern of adrenal destruction in tuberculosis or fungal infections involves the glucocorticoid-secreting layer before the outer layer is affected.

• • •

Manifestations

In primary adrenal insufficiency, there are symptoms of glucocorticoid and mineralocorticoid deficiency. Hypoadrenalism due to secondary causes (ACTH deficiency) leads to symptoms of only glucocorticoid deficiency. Table 32.7 summarizes manifestations of both glucocorticoid and aldosterone insufficiency due to hypoadrenalism. There are general symptoms of fatigue and weakness accompanied by gastrointestinal symptoms. Glucocorticoids normally stimulate gluconeogenesis, and when there is glucocorticoid deficiency, hypoglycemia occurs. Decreased tolerance to stress is characteristic of glucocorticoid insufficiency. Increased skin pigmentation typically occurs over pressure areas, such as elbows and knees, and over skin folds. This is either due to ACTH or to a high level of circulating beta-lipotropin. Adrenal insufficiency in females leads to loss of axillary hair and amenorrhea due to a deficiency of sex steroids.

Aldosterone normally promotes sodium and water retention and excretion of potassium. Aldosterone deficiency leads to sodium loss, volume depletion, and hypotension. In acute hypoadrenalism, fluid and electrolyte

TABLE 32.7 Manifestations of hypoadrenalism

Fatigue	Sodium loss
Weakness	Volume depletion
Weight loss (anorexia)	Hyperkalemia
Nausea, vomiting	Hypotension
Alternating constipation and diarrhea	
Hypoglycemia	
Impaired tolerance to stress	
Increased skin pigmentation	

loss occurs by way of both the kidneys and the gastrointestinal tract. There is total body sodium depletion, although plasma sodium levels may be low or the values may be in the normal range because of loss of volume. In the chronic condition, there is a mild hyponatremia, hyperkalemia, and acidosis due to aldosterone deficiency. Volume depletion and decreased glomerular filtration rate lead to moderate increases in blood urea nitrogen and creatinine levels. Hypercalcemia is observed and may be related to cortisol deficiency. It is believed that cortisol normally has an antagonistic effect on vitamin D activity, which plays a role in calcium metabolism (chapter 3). Calcification of the pinna (outer ear) is an unusual finding in hypoadrenalism.

Acute adrenal insufficiency, also called Addisonian crisis, is a condition in which there is a sudden onset of life-threatening symptoms in a patient who has chronic adrenal insufficiency. The crisis may be caused by the stress of trauma, surgery, or febrile processes. Manifestations include abdominal pain, hypoglycemia, nausea, vomiting with volume depletion, hypotension, and shock.

Sudden withdrawal from prolonged glucocorticoid therapy results in adrenal insufficiency, although this may not be apparent except under conditions of stress. The reason that this occurs is that elevated levels of glucocorticoids suppress the secretion of hypothalamic and pituitary hormones that normally stimulate the adrenal cortex. A safe regimen for glucocorticoid withdrawal involves gradually tapering the daily dose and then following an alternate day schedule until the drug is discontinued.

Diagnostic Tests

Typically in primary hypoadrenalism, there are low levels of glucocorticoids in the plasma and urine and elevated amounts of plasma ACTH. The responsiveness of the adrenal cortex to ACTH may be evaluated by the ACTH stimulation tests. The screening test involves a single intramuscular injection of cosyntropin (Cortrosyn), whereas the prolonged test involves intravenous infusion of cosyntropin for 8 hours on 2 or 3 successive days. The response in a normal individual is increased levels of plasma cortisol or urinary 17-hydroxycorticosteroids. If such an increase does not occur, it demonstrates primary adrenal insufficiency. In the case of hypopituitarism, there will be a daily increase in urinary 17-hydroxycorticosteroids in response to the long ACTH stimulation test.

Treatment

The treatment for adrenal insufficiency is glucocorticoid and mineralocorticoid replacement therapy, resulting in an excellent prognosis for a normal life. Androgen replacement is seldom necessary. Adrenal crisis must be treated promptly by intravenous infusion of a saline/dextrose solution and glucocorticoid injection.

PHEOCHROMOCYTOMA

Pheochromocytomas (fe-o-kro″mo-si-to′mah) are chromaffin cell tumors that usually arise in the adrenal medulla. The adrenal medulla normally secretes catecholamines, and most pheochromocytomas produce both epinephrine and norepinephrine. These tumors may develop in or close to sympathetic ganglia as well. Hypertension is the most common manifestation, and it may be sustained, periodic, or may result in a hypertensive crisis. Manifestation of a crisis includes headache, profuse sweating, apprehension, nausea, and vomiting. Although pheochromocytoma accounts for only 0.1% of all cases of hypertension, it is significant both because it can be fatal and because it is correctable by surgery.

PANCREAS

The pancreas secretes an enzyme-containing alkaline fluid into the duodenum and thus has an exocrine function. The pancreatic islets of Langerhans exhibit an endocrine function, with two main cell types that produce hormones (figure 32.7). Alpha cells secrete glucagon, which has an overall effect of increasing blood glucose levels, whereas beta cells produce insulin. Insulin, stimulated by increasing blood glucose levels, acts to lower those levels (figure 32.8). Additional effects of insulin are discussed in chapter 33. Insulin binds to receptor sites on the cell membrane and promotes cellular uptake of glucose with subsequent metabolism of the sugar. The action of insulin is particularly important on liver, muscle, and adipose cells.

DIABETES MELLITUS

Diabetes mellitus is an insulin-related disorder of glucose metabolism in which there is both hyperglycemia and disturbances of lipid metabolism accompanied by various side effects. The incidence of diabetes in the United States is approximately 2.3% of the population with a higher incidence among both females and nonwhites.

Classification

Diabetes is classified as (1) insulin-dependent diabetes mellitus (IDDM), (2) noninsulin-dependent diabetes mellitus (NIDDM), (3) secondary to other disorders, and (4) gestational diabetes mellitus (GDM). In addition,

pheochromocytoma: Gk. *phaios*, gray; Gk. *chroma*, color; Gk. *kytos*, cell; Gk. *oma*, tumor

ENDOCRINE SYSTEM

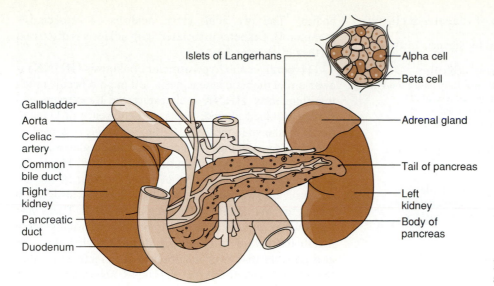

Figure 32.7 The pancreas and the associated islets of Langerhans.

Metabolism

Absorption of meal (↑ Glucose)	Fasting (↓ Glucose)
↑ Insulin	↓ Insulin
↓ Glucagon	↑ Glucagon
↑ Insulin/ glucagon ratio	↓ Insulin/ glucagon ratio
Formation of glycogen, fat, and protein	Hydrolysis of glycogen, fat, and protein + Gluconeogenesis and ketogenesis
Blood	**Blood**
↓ Glucose	↑ Glucose
↓ Amino acids	↑ Amino acids
↓ Fatty acids	↑ Fatty acids
↓ Ketone bodies	↑ Ketone bodies

Figure 32.8 The inverse relationship between insulin and glucagon secretion during the absorption of a meal and during fasting. Changes in the insulin: glucagon ratio tilts metabolism toward anabolism during the absorption of food and toward catabolism during fasting.

there are various abnormalities related to glucose tolerance (chapter 33). Table 32.8 identifies characteristics of IDDM and NIDDM.

Long-Term Effects

In general, there are prominent vascular and metabolic complications associated with diabetes mellitus. The precise mechanisms are not clear, but possible causes of abnormalities and damage include (1) high glucose

TABLE 32.8 A comparison of two main categories of diabetes mellitus

Insulin-Dependent Diabetes Mellitus (Type I) (Juvenile Onset)

Average age of onset 12, abrupt onset of symptoms
Cause unknown; tends to be familial
Insulin deficiency
Islet cell antibodies; associated with other autoimmune
    disorders
Viruses may play a role

Noninsulin-Dependent Diabetes Mellitus (Type II) (Adult Onset)

Rarely detect islet cell antibodies
Cellular resistance to insulin (may have normal or high insulin
    levels)
Abnormal beta cell response to stimulation
Genetic factors
Obesity may play a role

levels, (2) increased intracellular levels of sorbitol, and (3) increased tendency for platelets to aggregate. Table 32.9 summarizes the complications of diabetes. Diabetic retinopathy is a leading cause of blindness in adults less than 65 years of age. There is an increased incidence of cardiovascular disease, and diabetic nephropathy leads to end-stage renal failure. There are various manifestations of diabetic neuropathy (table 32.9), including retention of residual urine due to abnormal relaxation of the bladder. This is called atony and is a contributing factor in urinary tract infections. The combined effect of nerve damage and ischemia lead to skin ulcers, particularly on the feet.

Treatment

The main goal in the management of diabetes mellitus is to stabilize blood glucose levels within a normal range. IDDM requires daily injections of insulin or a constant

atony: Gk. *a*, not; Gk. *tonos*, tone

Table 32.9 Long-term effects of diabetes mellitus

Injury of capillaries of the retina lead to blindness (diabetic retinopathy)

Increased formation of sorbitol (a sugar alcohol); accumulation of intracellular sorbitol may cause damage

Atherosclerosis occurs at a younger age (accelerated)

Increased blood levels of triglycerides and cholesterol

Coronary artery disease; cerebral vascular disease; lightheadedness, faintness

Glomerular lesions, pyelonephritis (diabetic nephropathy); gradual decline of renal function

Urinary tract infections

Injury to nerve fibers (diabetic neuropathy); diminished sensation in legs, feet, hands, arms; impotence; pain, discomfort in calf and thigh muscles and buttocks

Decreased sweat, heat intolerance

Skin lesions, especially of the feet

Gastrointestinal symptoms, variability of gastric emptying; diarrhea

Table 32.10 Factors that contribute to HHNK

Dehydration

Diuretics
Volume loss; vomiting, diarrhea, infection
Impaired ability to ingest fluids
Hyperglycemia causes osmotic diuresis

Hyperglycemia

NIDDM (may have no history of diabetes)
Drugs that impair glucose tolerance
Ingestion of large amounts of sugar-containing beverages
Intravenous hypertonic glucose solutions
Drugs that affect insulin secretion (chlorpromazine)
Decreased glomerular filtration rate due to volume depletion

infusion of insulin delivered by a portable infusion device. Dietary regulations include (1) regularity of meals and snacks, (2) restriction of refined carbohydrates, and (3) limited caloric intake to maintain ideal body weight. Moderate exercise contributes to better control of diabetes, although there is risk of hypoglycemia unless it is properly timed. Exercise, weight reduction, and diet is sufficient to control blood sugar levels in some cases of non-insulin-dependent diabetes. Patients with NIDDM may require both dietary restriction and oral administration of drugs that lower blood sugar levels, while others may require insulin for blood sugar control.

Hyperosmolar Hyperglycemic Nonketosis (HHNK)

In uncontrolled diabetes, glucose is unavailable as an energy source and the body metabolizes fat as an alternative source of energy. The breakdown of fat as the sole source of energy leads to the accumulation of acetoacetic acid, beta-hydroxybutyric acid, and acetone (ketone bodies). The two acids cause acidosis or ketoacidosis (chapter 4). Diabetes associated with acidosis is discussed further in chapter 33.

Hyperglycemic hyperosmolar nonketosis (HHNK) is a variant of diabetic ketoacidosis, although it occurs much less frequently. HHNK is characterized by extreme elevations of blood glucose levels in the absence of ketoacidosis. As shown in table 32.10, the overall picture involves dehydration combined with increased blood glucose levels with each condition contributing to the other. Hyperglycemia (sometimes in excess of 1,000 mg/dl), in the absence of ketoacidosis, implies that there is insufficient insulin to maintain normal blood glucose levels. There may be, however, adequate insulin to promote limited glucose metabolism, which prevents significant ketoacidosis. It is also possible that liver metabolism is defective and that this inhibits ketone synthesis. HHNK occurs most often in individuals past age 50 and frequently is a complication of such disorders as stroke, myocardial infarction, pneumonia, or burns.

There is extreme dehydration accompanied by orthostatic hypotension, accompanied by evidence of such central nervous system impairment as confusion, seizures, or coma. The mortality rate may be as high as 50%.

Infusion of fluids to correct the volume deficit is an urgent aspect of treatment, and insulin is required to control elevated blood sugar levels.

SUMMARY

• • •

The adrenal cortex secretes glucocorticoids, mineralocorticoids, and androgens, whereas the adrenal medulla releases catecholamines. Cortisol is a glucocorticoid essential for life and is secreted by the adrenal cortex. The physiological effects of cortisol include (1) protein catabolism, (2) gluconeogenesis, (3) decreased insulin sensitivity, (4) localized lipogenesis, (5) anti-inflammatory reaction, and (6) influence on emotional function. Stress and the pituitary hormone ACTH stimulate the release of cortisol. Glucocorticoids are therapeutically useful as anti-inflammatory and immunosuppressive agents. Aldosterone is the major mineralocorticoid secreted by the adrenal cortex. This hormone promotes the reabsorption of sodium and water, as well as the secretion of potassium and hydrogen ions. A decrease in renal arterial blood flow or a decrease in sodium ion concentration stimulate the renal release of the enzyme renin. Renin promotes the formation of angiotensin I, and subsequently, angiotensin II is formed. Angiotensin II stimulates the release of aldosterone. In addition to the secretion of glucocorticoids and mineralocorticoids, the adrenal cortex secretes sex steroids, which contribute to maintaining secondary sexual characteristics.

Cushing's syndrome is the result of excessive levels of glucocorticoids, either due to hypersecretion or to prolonged therapeutic use. The most common cause is iatrogenic and other causes include adrenocortical tumors, ACTH-secreting pituitary tumors, and increased CRH secretion by the hypothalamus. Notable manifestations of cortisol excess are obesity and redistribution of fat, muscle wasting, hypertension, osteoporosis, striae, and increased susceptibility to infection. Increased production of cortisol is confirmed by the measurement of urinary excretion of cortisol or its metabolites (17-hydroxycorticosteroids). The dexamethasone suppression test is used to differentiate the etiology of Cushing's syndrome, and plasma levels of ACTH are indicative of etiology as well.

Aldosteronism refers to an excessive level of aldosterone. The condition may be primary, due to adrenal hyperplasia or adenoma, or it may be secondary and caused by high levels of renin. Diastolic hypertension is characteristic of the aldosteronism and is accompanied by hypernatremia, metabolic alkalosis, and hypokalemia. An elevated urinary aldosterone level under conditions of high sodium intake is indicative of primary aldosteronism. The saline infusion test is a method of evaluating whether or not renin, and thus aldosterone production, can be suppressed by sodium. This test eliminates primary aldosteronism if, following the administration of saline, the plasma aldosterone concentration is less than 6 ng/dl. The test for plasma renin activity distinguishes primary aldosteronism, associated with decreased renin activity, from secondary aldosteronism in which there is a high level of renin activity.

Addison's disease is hypoadrenalism caused by injury or destruction of the adrenal gland or by ACTH suppression. The disorder results in manifestations of both glucocorticoid and mineralocorticoid deficiency. These manifestations include decreased stress tolerance, hypoglycemia, weight loss, sodium and volume depletion, hyperkalemia, and hyperpigmentation. An individual with chronic adrenal insufficiency may experience a sudden life-threatening acute condition called Addisonian crisis. Primary hypoadrenalism is associated with low levels of glucocorticoids in the plasma and urine and elevated plasma ACTH. The ACTH stimulation tests aid in evaluating the responsiveness of the adrenal cortex to the influence of ACTH.

Pheochromocytoma is a chromaffin cell tumor that often involves the adrenal medulla. It is characterized by secretion of catecholamines, which causes hypertension.

Insulin deficiency or cellular resistance to insulin causes diabetes mellitus. The overall picture is that of impaired glucose metabolism, which may require injections of insulin or may be controlled by exercise and diet. Predominant long-term effects are vascular and metabolic complications. Hyperglycemic hyperosmolar nonketosis is a variant of diabetes in which there is dehydration and extreme hyperglycemia in the absence of ketoacidosis.

REVIEW QUESTIONS

• • •

1. List six physiological effects of cortisol.
2. What two hormones are involved in stimulating the secretion of glucocorticoids?
3. What is a nonhormonal stimulus for the release of cortisol?
4. What are the physiological effects of aldosterone?
5. Outline the events leading to aldosterone secretion in response to either a decrease of sodium ions or hypovolemia.
6. List three possible causes of Cushing's syndrome.
7. Identify four manifestations of Cushing's syndrome.
8. What electrolyte imbalance(s) may occur with prolonged glucocorticoid excess?
9. What are the expected results of the dexamethasone suppression test in a normal individual?
10. What do the following results of the dexamethasone test indicate?
    a. Low doses of dexamethasone, no ACTH or cortisol suppression
    b. ACTH suppression in response to high doses of dexamethasone
    c. No response to high doses of dexamethasone
    d. No change in cortisol secretion with either low or high doses of dexamethasone
11. Would you predict plasma ACTH levels to be low or high in the case of an adrenal tumor?
12. What are two possible causes of secondary aldosteronism?
13. List four possible causes of secondary aldosteronism.
14. What electrolyte imbalance(s) occur in primary aldosteronism?
15. Why is edema absent in chronic aldosteronism?
16. What is the significance of an elevated urinary aldosterone level associated with a high sodium intake?
17. What is the principle of the saline infusion test?
18. What are the results of the saline infusion test if the plasma renin activity is low prior to the test?
19. Compare renin activity in primary and secondary aldosteronism.
20. What is the most common cause of Addison's disease?
21. What causes symptoms of glucocorticoid deficiency in the case of secondary hypoadrenalism?
22. What is the most common manifestation of pheochromocytoma?
23. What is the role of insulin in glucose metabolism?
24. What is a metabolic effect of diabetes other than carbohydrate intolerance?
25. What is a major feature that distinguishes HHNK from diabetes mellitus?

Ansari, A. 1988. Severe postural hypotension, dizziness, and recurrent syncope in a young woman. *Physician Assistant* 12:59.

Cerami, A., H. Vlassara, and M. Brownlee. 1987. Glucose and aging. *Scientific American* 256:90.

Evan-Wong, L., R. Davidson, and J. Stowers. 1985. Alterations in erythrocytes in hyperosmolar diabetic decompensation: A pathophysiological basis for impaired blood flow and for an improved design of fluid therapy. *Diabetologia* 28:739.

Gold, P. et al. 1986. Responses to corticotropin-releasing hormone in the hypercortisolism of depression and Cushing's disease. *New England Journal of Medicine* 314:1329.

Gold, P. et al. 1986. Abnormal hypothalamic-pituitary-adrenal function in anorexia nervosa. *New England Journal of Medicine* 314:1335.

Kolata, G. 1987. Clinical promise with new hormones. *Science* 236:517.

Martin, D. 1986. Type II diabetes: Insulin versus oral agents. *New England Journal of Medicine* 314:1314.

Niijima, A., and N. Mei. 1987. Glucose sensors in viscera and control of blood glucose level. *News in Physiological Sciences* 2:164–67.

Wood, F., and E. Bierman. 1986. Is diet the cornerstone in management of diabetes? *New England Journal of Medicine* 315:1224.

Chapter 33

Practical Applications: Diabetes Mellitus and HHNK

This chapter focuses on a common disorder of the endocrine pancreas, diabetes mellitus and its complications. The perspective is evaluation and management related to laboratory findings. Fluid, electrolyte, and acid-base imbalances are considered in terms of the pathophysiology of the disease processes. Diagnostic tests that evaluate the control of blood sugar levels are discussed first.

DIAGNOSTIC TESTS

The diagnosis of diabetes is clear when there are symptoms, such as thirst, polyuria, and weight loss, as well as an elevated fasting blood glucose level. In some cases of IDDM and in many cases of NIDDM, the fasting glucose level may be borderline, which makes the diagnosis less certain. A normal fasting plasma glucose level is <115 mg/dl, and a normal blood glucose level is 15% lower. In order to be diagnostic of diabetes, the glucose level in two or more tests should be >140 mg/dl in plasma or >120 mg/dl for venous blood.

ORAL GLUCOSE TOLERANCE TEST

In borderline cases, the oral glucose tolerance test (OGTT) aids in the diagnosis. The purpose of the OGTT is to test the patient's ability to metabolize a loading dose of glucose. An adult patient in a fasting state, ingests 75 g of glucose dissolved in water. Blood glucose levels are determined at 30 minute intervals for at least 2 hours. The test indicates diabetes when it is performed on more than one occasion and both the 2 hour and one other sample are >200 mg/dl for plasma or >180 mg/dl for venous blood. The test is indicative of impaired glucose tolerance if the fasting glucose level is between normal and diabetic values (plasma >140 mg/dl, blood >120 mg/dl) and if the 2 hour sample in the OGTT is between normal and diabetic values.

GLYCOHEMOGLOBIN

Glucose combines with hemoglobin slowly and continuously during the life span of the red blood cell. A small percentage of the hemoglobin molecules become glycosylated and are called glycohemoglobin. The amount of glycohemoglobin formed depends on the blood glucose level and is proportional to the average glucose levels during the 2 to 3 months preceding measurement. Hemoglobin A_{1c} is the most abundant of these hemoglobins although there are two other components, hemoglobin A_{1a} and A_{1b}. Normal values (A_{1a}, A_{1b}, A_{1c}) vary with the reporting laboratory, but are in the range of 4–8%. The glycohemoglobin test is unaffected by recent food or exercise and indicates long-term blood glucose control.

CONSEQUENCES OF HORMONAL INFLUENCE

Insulin promotes the cellular uptake of glucose and the subsequent metabolism of this sugar. The formation of glycogen, a starch, and the control of protein synthesis in muscle are aspects of insulin activity as well. Insulin stimulates potassium uptake by muscle and liver cells and is important in maintaining the distribution of this ion. Insulin is a potent inhibitor of lipolysis, the breakdown of fat in fat cells.

Glucagon, a hormone produced by pancreatic alpha cells (figure 33.1), essentially exerts effects that oppose insulin (figure 33.2). It promotes an increase in blood glucose levels by stimulating glycogenolysis, the breakdown

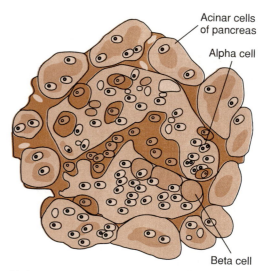

Figure 33.1 A normal pancreatic islet is made up of alpha and beta cells.

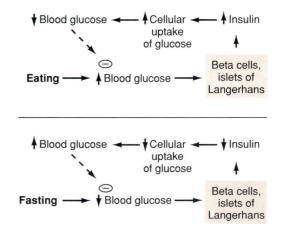

Figure 33.2 The negative feedback control of insulin secretion by changes in the blood glucose concentration. Dashed arrows and negative signs indicate that negative feedback loops compensate the initial changes in blood glucose concentrations produced by eating or fasting.

of glycogen to glucose, and by gluconeogenesis, a synthesis of glucose from amino acids. In addition, glucagon promotes lipolysis.

EFFECTS OF INSULIN DEFICIENCY

Insulin deficiency (1) blocks glucose utilization, (2) enhances the breakdown of protein in muscle, (3) activates lipolysis, and (4) leads to an increase in glucagon (figure 33.3). The breakdown of protein results in an increase of amino acids which may be converted to glucose by way of gluconeogenesis. Lipolysis releases free fatty acids which are metabolized to ketones. When glucagon is unopposed by insulin, there is increased gluconeogenesis, glycogenolysis, and ketone body formation. The overall effect of insulin deficiency is hyperglycemia and acidosis due to the ketone bodies, acetoacetic acid, and beta-hydroxybutyric acid.

KETOACIDOSIS

Ketoacidosis is a metabolic acidosis resulting from increased utilization of fat in response to inadequate glucose metabolism. Free fatty acids are products of fat metabolism that, in part, are converted to acetone, acetoacetic acid, and beta-hydroxybutyric acid. Metabolic acidosis is caused mainly by these two acids and, to a lesser extent, by free fatty acids and lactic acid (figure 33.4).

Ketoacidosis may be precipitated by inadequate insulin, stress, or by illness. Hormones released during stress include epinephrine, norepinephrine, glucocorticoids, and growth hormone. These oppose the action of insulin, stimulate glucagon release, and exert a hyperglycemic effect.

DIABETIC COMA VS. INSULIN SHOCK

Both ketoacidosis accompanied by hyperglycemia and hypoglycemia induced by an overdose of insulin may result in coma. The two are distinguished on the basis of differences in both clinical and laboratory findings. Osmotic diuresis due to elevated blood glucose levels leads to excessive fluid loss in ketoacidosis. Ketoacidosis is characterized by the following:

1. Evidence of volume depletion, dry skin and mouth
2. Hypotension
3. Weak, rapid pulse

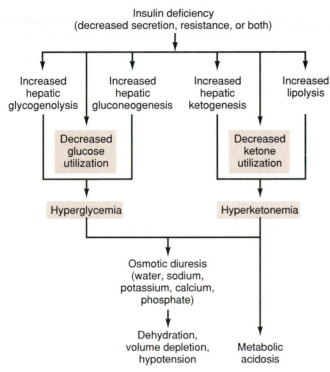

Figure 33.3 The sequence of events by which an insulin deficiency may lead to coma and death.

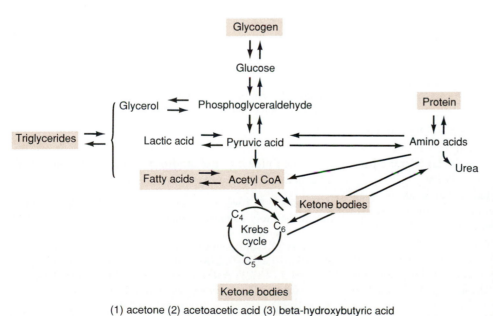

(1) acetone (2) acetoacetic acid (3) beta-hydroxybutyric acid

Figure 33.4 Simplified metabolic pathways showing how glycogen, fat, and protein can be interconverted. When glucose cannot be utilized, the metabolism of fat increases. The result is production of ketone bodies and metabolic acidosis.

4. Kussmaul respiration due to acidosis (chapter 9)
5. Odor of acetone on breath
6. Glycosuria
7. Ketonuria
8. Hyperglycemia

Insulin is administered and steps are taken to restore fluid/electrolyte losses.

In contrast, there is no evidence of volume depletion in insulin shock. The skin is moist, the mouth is not dry, and blood pressure is normal. The blood glucose level is low so that appropriate treatment is the administration of glucose.

OVERVIEW OF LABORATORY FINDINGS

Laboratory findings in ketoacidosis are summarized in table 33.1. There is usually an increase in blood lipid levels in the form of very low-density lipoproteins (VLDL), triglycerides, and a less marked increase in cholesterol. Insulin deficiency leads to (1) increased fat catabolism and release of free fatty acids, (2) the conversion of free fatty acids to either ketones or triglycerides, and (3) an increase in the VLDL fraction due to triglycerides. Elevation of the triglyceride level correlates with hyperglycemia; thus, it correlates with poor control of diabetes.

The occurrence of metabolic acidosis has been discussed previously. A low pH reflects this condition and is accompanied by a wide anion gap (chapter 16). The anion gap is the difference between the sum of the measured cations ($Na^+ + K^+$) and the sum of the measured anions ($Cl^- + HCO_3^-$) (figure 33.5). Since the numerical value for serum potassium is small, this anion gap can be approximated by subtracting the sum of chloride and bicarbonate concentration from the value for sodium ions. Since the sum of the cations is equal to the sum of the anions, the anion gap represents unmeasured anions. The wide anion gap in ketoacidosis is the result of an accumulation of unmeasured anions, i.e., beta-hydroxybutyrate and acetoacetate.

Elevated, low, or normal levels of sodium ions may coexist with hyperglycemia (figure 33.6). Osmotic diuresis may result in a greater loss of water as compared to sodium ions and lead to hypernatremia. The volume of water ingested is a factor in serum sodium. There is a dilutional factor due to shift of intracellular water in response to increased extracellular osmolality.

Hyperkalemia is usually noted in ketoacidosis, even though total body stores of potassium are depleted by urinary losses (figure 33.7). The elevated serum level of potassium is due to a shift of intracellular potassium to extracellular spaces in response to acidosis. Normally, the efflux of potassium is opposed by insulin. When insulin is administered, the movement of these ions is reversed to an intracellular position.

TABLE 33.1 Laboratory findings and their contributing factors in ketoacidosis

1. Blood glucose levels usually > 300 mg/dl
2. Serum osmolality almost always elevated when consciousness impaired; due to increased glucose
3. High osmolality leads to osmotic diuresis and volume depletion
4. Osmotic diuresis causes loss of Na^+, K^+, HCO_3^-, and Cl^-
5. If osmotic diuresis is prolonged, the result is decreased glomerular filtration rate (GFR)
6. Blood urea nitrogen (BUN) increased; due to increased protein catabolism and/or decreased GFR related to volume depletion
7. Increased very low-density lipoproteins (VLDL), increased triglycerides, less marked increase in cholesterol due to decreased insulin
8. $HCO_3^- < 7.25$ meq/liter; used up by buffering; urinary losses
9. pH < 7.25 due to acetoacetic and beta-hydroxybutyric acids
10. Wide anion gap
11. Na^+ variable; low due to urinary loss or due to dilution
12. Hyperkalemia

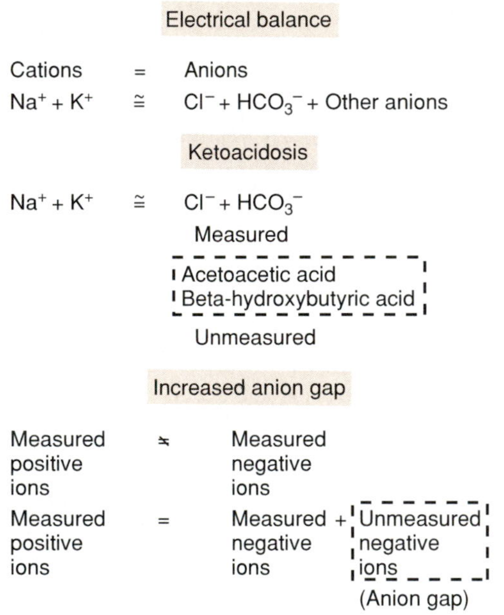

Electrical balance

Cations = Anions
$Na^+ + K^+$ ≅ $Cl^- + HCO_3^-$ + Other anions

Ketoacidosis

$Na^+ + K^+$ ≅ $Cl^- + HCO_3^-$
Measured

Acetoacetic acid
Beta-hydroxybutyric acid

Unmeasured

Increased anion gap

Measured positive ions ≠ Measured negative ions

Measured positive ions = Measured negative ions + Unmeasured negative ions
(Anion gap)

Figure 33.5 An increased anion gap is indicative of an accumulation of unmeasured anions.

The following case studies are three examples of diabetes mellitus, two of which involve ketoacidosis.

Patient 33.1 was a 48-year-old male; IDDM, ketoacidosis. Laboratory findings (normal values in parentheses): BUN 17 mg/dl (7–22), Na 149 meq/liter (137–47), K 4 meq/liter (3.4–5.0), glucose 380 mg/dl (70–110), anion gap 20 meq/liter (6–16), osmolality 313 mOsm/kg (274–300), pH 7.04 (7.35–7.45), pCO$_2$ 11 mm Hg (35–45), base excess −25 meq/liter (−2.5 to +2.5)

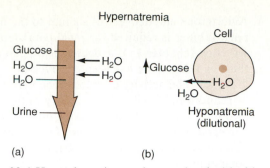

Hypernatremia

(a)   (b)

Figure 33.6 Hyperglycemia may be associated with either (a) an apparent increase in sodium ions due to excessive loss of water or (b) an apparent decrease in sodium ions as the result of a dilutional effect.

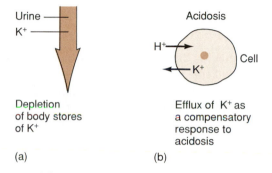

(a)   (b)

Figure 33.7 (a) Ketoacidosis is associated with depletion of body stores of potassium. (b) Hyperkalemia may exist due to a compensatory efflux of potassium ions out of cells.

In patient 33.1, laboratory tests show hypernatremia, probably the result of excessive diuresis. The anion gap shows an increase in unmeasured anions, in this case ketoacids. The increased blood sugar level contributes to elevated serum osmolality. The blood pH shows acidosis and the base excess, due to ketoacids, confirms metabolic acidosis. The low pCO_2 is evidence of increased respiration, a compensatory response to acidosis.

Patient 33.2 was a 25-year-old male; IDDM, ketoacidosis. Laboratory findings: BUN 26 mg/dl (7–22), Na 127 meq/liter (137–47), K 6.1 meq/liter (3.4–5.0), glucose 1,149 mg/dl (70–110), anion gap 35 meq/liter (6–16), osmolality 318 mOsm/kg (274–300), pH 6.88 (7.35–7.45), pCO_2 12 mm Hg (35–45), HCO_3^- 2.3 meq/liter (12–30), base excess −29 meq/liter (+2.5 to −2.5)

In patient 33.2 the BUN is elevated, probably due to protein catabolism. In contrast to patient 33.1, there is hyponatremia that, considering the extreme elevation of blood glucose, is likely dilutional. The pH is close to the point that is incompatible with life. The low pCO_2 is evidence of a compensatory response to acidosis, and bicarbonate has been used up in buffering hydrogen ions. The extreme base excess value mirrors the level of accumulated ketoacids.

Patient 33.3 was a 56-year-old male; NIDDM. Laboratory findings: Hgb A_{1c} 9.6% (5–8), glucose 242 mg/dl (70–110), anion gap 11 meq/liter (6–16), pH 7.33 (7.35–7.45), base excess +0.9 meq/liter (+2.5 to −2.5)

The laboratory tests for patient 33.3 do not indicate a problem with acidosis. The glucose level is elevated and the glycohemoglobin value, Hgb A_{1c}, indicates that the blood glucose has not been completely controlled for the past 2–3 months.

COMPREHENSIVE ANALYSIS OF KETOACIDOSIS

Ketoacidosis occurs less often in undiagnosed diabetics than in individuals who have been diagnosed and are receiving insulin. There are various factors that may precipitate this crisis in a diabetic, i.e., inadequate insulin or illnesses, such as gastrointestinal disorders, pancreatitis, myocardial infarction, or infections. In addition, an excess of hormones normally antagonistic to insulin effects may lead to ketoacidosis. Insulin antagonists include glucagon, epinephrine, cortisol, and growth hormone. Acromegaly, pregnancy, the stress of surgery, and hyperactivity of the adrenal cortex may be precipitating factors. Certain medications and emotional stress may contribute as well.

The clinical presentation may include low blood pressure, postural hypotension, and tachycardia. This is the result of osmotic diuresis associated with persistent hyperglycemia that leads to an increase in urinary output and a decrease in extracellular fluid volume. The fluid loss causes reduced skin turgor and dry mucous membranes. Anorexia, nausea, vomiting, and abdominal pain frequently occur in ketoacidosis. The cause of the abdominal pain is not clear, but it may be related to increased levels of triglycerides. Acidosis causes Kussmaul respiration, an increase in both rate and depth of respiration, as a compensatory response. The fruity odor of acetone is typically on the breath. The affected individual may or may not be comatose. A degree of drowsiness is common while less than 10% of patients are in a coma. There is not a good correlation between mental alertness and blood pH or glucose levels. There is a closer relationship between osmolality of the blood and the state of consciousness.

Patient 33.4 was a 59-year-old female who had a family history of diabetes mellitus. She had been diagnosed as IDDM years earlier and received daily injections of insulin. She was admitted to the hospital complaining of flu like symptoms, i.e., abdominal pain, vomiting, and headache, which had persisted for 3 days. The patient was alert and cooperative.
Laboratory findings (upon admission): Na^+ 146 meq/liter (137–45), K^+ 4.3 meq/liter (3.4–5.0), Cl^- 111 meq/liter (100–10), pH 7.06 (7.35–7.45), pCO_2 23.1 mm Hg (35–45), anion gap 32.6 meq/liter (6–16), HCO_3^- 6.7 meq/liter (22–30), base excess −21.4 meq/liter (−2.5 to

+2.5), WBC 13,800/mm³ (5,000–10,000), RBC 4.6 million/mm³ (4.8), creatinine 1.2 mg/dl (0.6–1.2), BUN/creatinine ratio 18 (6 to 22), glucose 628 mg/dl (70–110 mg/dl)

The diagnosis upon admission was diabetic ketoacidosis. The blood pH is low, and both the base excess and the low bicarbonate values indicate metabolic acidosis. The medical history and remarkably elevated blood glucose level of patient 33.4 confirms ketoacidosis. Possible precipitating factors were not identified.

POTASSIUM LEVELS

The serum potassium level in patient 33.4 is within normal limits. There are various factors that determine the blood level of this ion:

1. Glucose-induced hyperkalemia may occur. Elevated glucose level causes an increase in serum osmolality, which induces a shift of water and potassium out of cells. If insulin is available, it enhances an opposing effect, the cellular uptake of potassium. The lack of insulin predisposes the diabetic to hyperkalemia.
2. The overall relationship between acidosis and hyperkalemia is influenced by both the cause and duration of the acidosis. In general, hydrogen ions diffuse into cells and are buffered intracellularly, which promotes the efflux of potassium ions. The extent of potassium efflux, however, is determined by the source of the hydrogen ions. In the case of ketoacidosis, one of the chemicals responsible for the acidosis is beta-hydroxybutyric acid. The acid dissociates to release hydrogen cations and butyrate anions. Hydrogen ions may be accompanied by a butyrate anion to maintain electrical neutrality as intracellular diffusion occurs. This diffusion limits the extent of potassium efflux and the degree of hyperkalemia. Acidosis due to acids other than beta-hydroxybutyric acid may be associated with higher serum-potassium levels.
3. The duration of the acidosis is an additional factor which affects potassium homeostasis. As time passes and more hydrogen ions move into cells, the serum-potassium level tends to rise.
4. Acute metabolic acidosis seems to displace potassium from renal epithelial cells. The effect is to reduce urinary excretion of potassium. With time, however, there is enhanced potassium excretion.
5. Renal insufficiency is a common complication of long-standing diabetes and an increased serum-potassium level is associated with this problem.
6. Aldosterone and delivery of sodium to the distal renal tubules is required for renal potassium ion secretion (chapter 11). A decrease in renin and, thus, in aldosterone has been observed in diabetics. The activity of aldosterone may also be reduced in diabetes due to injury to the renal medulla. This results in collecting duct unresponsiveness to aldosterone. The overall effect is sodium wasting and retention of potassium.
7. In poorly controlled diabetes, cell necrosis may contribute to hyperkalemia.

In diabetic ketoacidosis, the initial serum level of potassium may be elevated. The total body potassium is usually depleted by renal losses or vomiting.

SODIUM LEVELS

Characteristically, there is loss of total body sodium associated with ketoacidosis. Laboratory determination of serum sodium reflects the ratio of sodium and extracellular fluid, and the report may indicate normal, high, or low values. As in the case of potassium, there are various contributing factors:

1. Hyperglycemia induces osmotic diuresis, which causes sodium wasting as well as loss of water and potassium.
2. Renal disease is commonly associated with diabetes. There is, for example, a higher incidence of pyelonephritis among diabetics. This may lead to necrosis of renal papilla with subsequent urinary sodium loss due to the injury.
3. A decrease of aldosterone activity, discussed in connection with potassium balance, leads to sodium loss.
4. Elevated glucose levels osmotically draw water from cells into extracellular fluid. This has a dilutional effect and leads to a laboratory determination of low sodium values.
5. Osmotic diuresis causes a decrease in extracellular fluid volume. Antidiuretic hormone is released in response to volume loss and reduces the ability to excrete ingested water. This represents another possible dilutional factor.
6. Insulin promotes renal tubular reabsorption of sodium and sodium loss may be the direct result of insulin deficiency.
7. If water loss exceeds water ingestion, the patient may have hypernatremia.

In summary, sodium loss occurs in ketoacidosis, although laboratory determinations may not indicate this. The sodium value is slightly above normal in this patient. Hypernatremia implies that water ingestion is less than water loss.

ACID-BASE BALANCE

The increased anion gap in patient 33.4 is due to an accumulation of unmeasured anions, i.e., acetoacetate and beta-hydroxybutyrate. The low bicarbonate indicates that bicarbonate is being used up in buffering the acid load. In metabolic acidosis, approximately 60% of the excess hydrogen ions are buffered intracellularly, while about 40% are buffered in the extracellular compartment. The extent of decrease of bicarbonate provides an estimate of this buffering.

OTHER LABORATORY FINDINGS

The BUN/creatinine ratio reported for patient 33.4 is within the normal range. The significance of an elevated BUN/creatinine ratio follows. Extracellular volume loss leads to a decrease in glomerular filtration rate, resulting in an increase in both BUN and creatinine values. Under conditions of low urine flow, there is increased reabsorption of urea, resulting in an increased blood level of urea. The overall result is a greater increase in the BUN as compared to creatinine. These mechanisms were apparently not operative in this patient.

The plasma osmolality, primarily determined by sodium and the high glucose level, is elevated in this patient. Values greater than 300 mOsm/liter are not normal, and a value above 320 mOsm indicates a significant hyperosmolar condition.

TREATMENT

Intravenous saline with potassium chloride and insulin were administered to patient 33.4. Fluid replacement is vital because the volume depletion may be life-threatening. In spite of normal plasma potassium values, there is potassium depletion in ketoacidosis that requires replacement. If the initial laboratory report indicates hypokalemia, it represents a severe depletion with a high risk for cardiac arrhythmias or respiratory paralysis. After therapy has been initiated, the serum potassium decreases due to potassium entry into cells and dilution because of volume replacement. Potassium is administered to prevent hypokalemia.

Electrolytes, blood gases, and blood glucose levels were determined every 2 hours in patient 33.4. The day following admission the laboratory results were as follows:

pH 7.24 (7.35–7.45), pCO_2 10.7 mm Hg (35–45), HCO_3^-, 4.6 meq/liter (22–30), base excess −19.3 meq/liter (−2.5 to +2.5)

At this point, bicarbonate was added to the intravenous fluid. In general, bicarbonate therapy is considered when the bicarbonate level is less than 5–10 meq/liter. The patient's ability to regenerate bicarbonate in a short period of time is one consideration in the decision to treat metabolic acidosis with bicarbonate. Bicarbonate therapy may not be required if the bicarbonate level is greater than 10 meq/liter and if respiratory acidosis is not a compli-

cation. Regeneration of bicarbonate occurs in the following way. Hydrogen ions are buffered by bicarbonate with the formation of carbonic acid, i.e., $H^+ + HCO_3^- \rightleftharpoons H_2CO_3$. As ketoacidosis is corrected, hydrogen ions combine with the anions acetoacetate and beta-hydroxybutyrate. The resulting acids are then metabolized to neutral products. The removal of hydrogen ions causes the following reaction to occur, $H_2CO_3 \rightleftharpoons H^+ + HCO_3^-$, with the formation of bicarbonate ions. These events occur rather quickly as the metabolic disorder is corrected.

Caution with regard to bicarbonate therapy is related to the consequent intracellular movement of potassium ions (chapter 3). Correction of the potassium deficit in ketoacidosis is combined with bicarbonate therapy to avoid hypokalemia.

Insulin dosage was adjusted for this patient to maintain blood glucose levels at about 250 mg/dl for the first 48 hours, with the sugar levels monitored every 2 hours. The purpose of this approach is to minimize the risk of cerebral edema. It is theorized that unidentified substances, ideogenic osmoles, accumulate within brain cells as a protective mechanism against water loss due to hyperosmolar conditions. As hyperosmolarity is corrected, there is a tendency for an osmotic shift of water into brain cells due to these intracellular substances (chapter 38).

• • •

Thrombosis is a potential complication of HHNK. A diminished extracellular fluid volume, increased platelet aggregation, and increased fibrinogen levels are contributing factors. Heparin, an anticoagulant, is used to treat this complication.

• • •

HYPEROSMOLAR HYPERGLYCEMIC NONKETOSIS

Hyperosmolar hyperglycemic nonketosis (HHNK) is an acute emergency which is characterized by extreme elevations of blood glucose, increased osmolality, and the absence of ketones (chapter 32). The patient is usually past age 50 with noninsulin-dependent diabetes accompanied by cardiovascular, renal, or other disease. The patient often is taking a variety of medications, as in the case of patient 33.5.

Patient 33.5 was a 41-year-old male with IDDM who received insulin twice a day. He was admitted to the hospital after becoming lethargic and vomiting once. The diagnosis was HHNK.
Laboratory findings: blood glucose 1,401 mg/dl (70–110), Na^+ 116 meq/liter (137–47), K^+ 6 meq/liter (3.4–5.0), Cl^- 84 meq/dl (100–110), HCO_3^- 15 meq/liter (22–30), BUN 90 mg/dl (7–22), serum osmolality 346.1 mOsm/kg (280–300), anion gap 23 meq/liter (6–16)

Patient 33.5 had a history of hypertension for 3 years and had experienced a right hemispheric stroke 3 months prior to this hospital admission. He also had a past history of ulcer disease, urinary tract infections, and probable pulmonary infections. He was taking, in addition to insulin, medication for hypertension, gastroesophageal reflux, and ulcer.

Hyperosmolar hyperglycemic nonketosis is characterized by profound hyperglycemia, dehydration, and little or no acidosis. This individual was mildly lethargic, but responsive and showed signs of dehydration. He had experienced increased frequency of urination over the last 24 hours, poor skin turgor, and dry mucosal membranes.

Laboratory Findings

The blood sugar of patient 33.5 was extremely elevated which is the basis for dehydration. Increased glucose in the extracellular compartment causes both a shift of water out of cells and an osmotic diuresis. The diuresis is associated with urinary loss of sodium and potassium as well. The loss of potassium associated with this disorder is due to both osmotic diuresis and to aldosterone release induced by dehydration (chapter 32). Electrolyte concentrations associated with this disorder frequently show elevated or normal sodium and potassium concentrations, normal chloride, and mildly reduced bicarbonate concentrations. Various factors that contribute to blood levels of electrolytes were previously discussed. In patient 33.5, the potassium value is high and sodium and chloride values are low. The anion gap of 23 meq/liter is somewhat elevated. Patient 33.4, with ketoacidosis, has a greater anion gap associated with an elevated, but lower blood glucose, compared to patient 33.5. This indicates, in the latter patient, limited ketone production, i.e., unmeasured anions acetoacetate and β-hydroxybutyrate.

The reason for limited ketone production is unclear. One possibility is that secretion of a small amount of insulin inhibits the production of free fatty acids which are the source of ketones, while the insulin is inadequate to maintain normal glucose utilization (chapter 32).

The fact that there is limited ketoacidosis probably allows for a more prolonged course and a greater degree of dehydration. The elevated BUN is indicative of impaired renal blood flow secondary to dehydration and a resulting decrease in glomerular filtration rate. There is increased reabsorption of urea when urine flow is diminished.

Possible Complication

Lactic acidosis is a possible serious complication of either diabetic ketoacidosis or HHNK. A marked decrease in extracellular fluid volume leads to poor oxygen delivery to tissues. Tissue hypoxia results in anaerobic glycolysis, i.e., the metabolism of glucose to pyruvic and lactic acids. The overall effect is the (1) accumulation of lactic acid molecules, (2) buffering by bicarbonate ions, and (3) a decrease in blood bicarbonate levels. The term bicarbonate consumption is used to describe these events. This implies that a hydrogen ion combines with bicarbonate to form carbonic acid. The carbonic acid, in turn, breaks down to carbon dioxide and water, and the carbon dioxide is eliminated in the process of respiration (chapter 4).

Lactic acid and pyruvic acid also increase the anion gap because the lactate and pyruvate ions are unmeasured anions. Lowered bicarbonate values and an increased anion gap may be indicative of lactic acidosis.

Treatment

The following principles determined the course of treatment of the hyperosmolar condition of patient 33.5. The first priority is volume replacement with fluids that are relatively hypotonic (chapters 2 and 5). This corrects the problem of increased osmolality and reestablishes urine flow. Potassium salts are added to intravenous fluids to correct total body potassium depletion. Insulin is administered to reduce blood sugar levels. There are frequent measurements of glucose, electrolytes, blood pressure, and urine volume to monitor the course of the patient's recovery.

Summary

· · ·

An elevated blood glucose level accompanied by symptoms, such as thirst, polyuria, and weight loss, are diagnostic for diabetes mellitus. The oral glucose tolerance test confirms diagnosis for those cases in which the fasting blood glucose level is borderline. The glycohemoglobin test indicates the average blood glucose levels over a 2–3 month period and is useful in monitoring insulin dosage and patient compliance.

Insulin promotes cellular uptake of glucose as well as stimulating the formation of glycogen, protein synthesis, and cellular uptake of potassium. Glucagon opposes the action of insulin by stimulating glycogenolysis, gluconeogenesis, and lipolysis. Ketoacidosis is the consequence of increased fat utilization and may be precipitated by stress, illness, or inadequate insulin. Coma may occur due to hyperglycemia associated with ketoacidosis as well as hypoglycemia resulting from insulin overdose. Ketoacidosis is characterized by symptoms of volume depletion as well as hyperglycemia and metabolic acidosis.

Laboratory findings in ketoacidosis include (1) an increase in triglycerides, (2) evidence of metabolic acidosis, (3) hyperglycemia, (4) osmotic diuresis, and (5) variable effects on potassium and sodium ions.

The blood level of potassium ions is influenced by a number of factors. (1) Hyperglycemia induces a shift of water and potassium out of cells. (2) Insulin is antagonistic to cellular loss of potassium. Thus, insulin deficiency predisposes to hyperkalemia. (3) Acidosis promotes exchange of intracellular potassium and extracellular hydrogen ions. Potassium ion efflux is limited if hydrogen ions are accompanied by an anion. (4) Prolonged acidosis favors hyperkalemia. (5) Metabolic acidosis initially reduces urinary excretion of potassium and later enhances the elimination of potassium. (6) Decreased aldosterone activity promotes renal retention of potassium. (7) In poorly controlled diabetes, cell necrosis may contribute to hyperkalemia. In ketoacidosis, there is total body depletion of potassium regardless of reported blood values.

Sodium loss occurs in ketoacidosis as well. Factors that influence blood levels of sodium include osmotic dilution due to hyperglycemia, renal loss of sodium associated with osmotic diuresis, and ingestion of water balanced by renal fluid loss.

The anion gap is increased in ketoacidosis due to the anions acetoacetate and β-hydroxybutyrate and bicarbonate is used up in buffering hydrogen ions. Blood urea nitrogen levels increase under conditions of volume depletion due to a decrease in glomerular filtration rate and increased reabsorption of urea.

Treatment of ketoacidosis includes intravenous fluids with potassium and insulin. Bicarbonate may be administered if the bicarbonate level is less than 5–10 meq/liter. Blood glucose levels may be maintained at about 250 mg/dl for the first 48 hours to minimize the risk of cerebral edema.

Hyperosmolar hyperglycemic nonketosis is characterized by extreme elevations of blood glucose, increased osmolality, and the absence or near absence of ketones. There is marked dehydration and little or no acidosis. Lactic acidosis due to tissue hypoxia is a possible complication of either diabetic ketoacidosis or HHNK. The approach to treatment involves volume replacement with fluids that are relatively hypotonic, insulin to reduce blood sugar levels, and correction of potassium depletion.

REVIEW QUESTIONS

• • •

1. What is a definitive test to diagnose borderline hyperglycemia?
2. What does a value for hemoglobin A_{1c} identify?
3. Volume depletion is a major event in (insulin shock or diabetic coma).
4. What is the effect on triglycerides in ketoacidosis?
5. What is the expected effect of ketoacidosis on the anion gap?
6. What is the reason for osmotic diuresis in ketoacidosis?
7. What are the three ketone bodies produced in ketoacidosis?

8. What is the significance of a wide anion gap?
9. How does hyperglycemia influence serum-potassium levels?
10. What is the relationship between acidosis and potassium homeostasis?
11. Why may there be differences in the degree of hyperkalemia comparing ketoacidosis and other types of metabolic acidosis?
12. What is the normal effect of insulin on potassium?
13. What is the expected effect of diabetes on aldosterone activity?
14. What is the effect of diabetic ketoacidosis on total body potassium?
15. How does hyperglycemia affect serum-sodium levels?
16. What is the significance of hypernatremia in ketoacidosis?
17. What laboratory value provides an estimate of hydrogen ion buffering?
18. Why do blood levels of urea increase in ketoacidosis?
19. What are the main determinants of plasma osmolality in ketoacidosis?
20. What are three steps in treating ketoacidosis?
21. Why does serum potassium decrease as acidosis is corrected?
22. How is bicarbonate regenerated in the body as acidosis is corrected?
23. What is a danger associated with bicarbonate therapy?
24. What are two possible complications of diabetic ketoacidosis or HHNK?
25. What is a major difference between diabetic ketoacidosis and HHNK?

SELECTED READING

• • •

Keller, R., and J. Wolfsdorf. 1987. Isolated growth hormone deficiency after cerebral edema complicating diabetic ketoacidosis. *New England Journal of Medicine* 316: 857–59.

Kemmer, F. et al. 1986. Psychological stress and metabolic control in patients with type I diabetes mellitus. *New England Journal of Medicine* 314:1078–84.

Walden, C. et al. 1984. Sex differences in the effect of diabetes mellitus on lipoprotein triglyceride and cholesterol concentrations. *New England Journal of Medicine* 311:953–58.

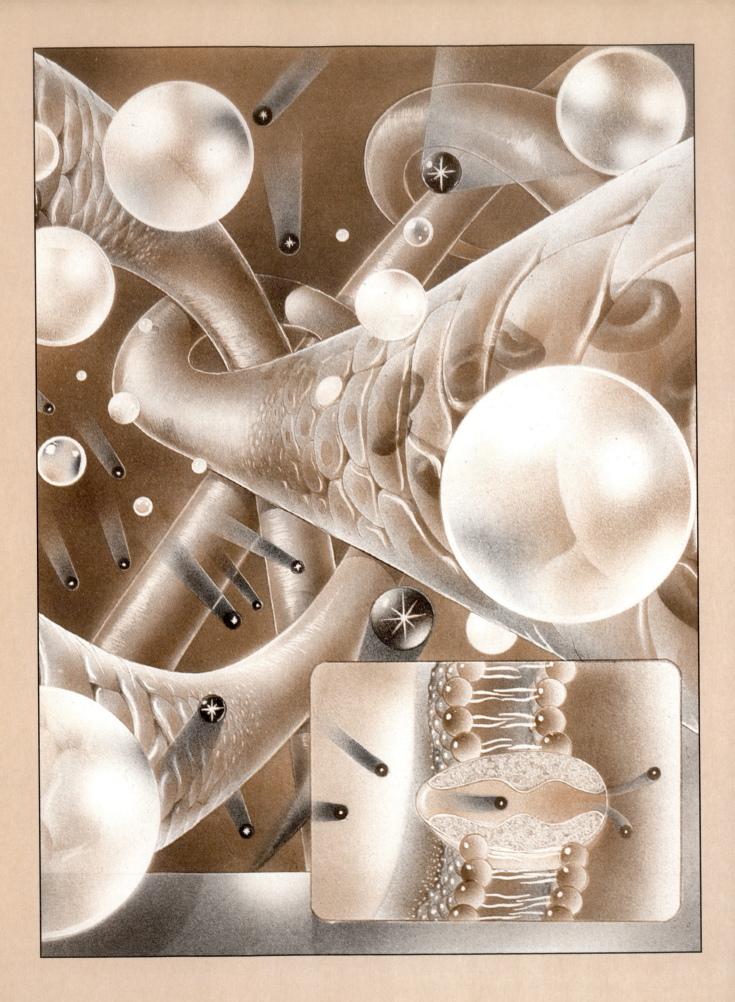

GASTROINTESTINAL SYSTEM

The introduction of chapter 34 deals with normal anatomy and function of the upper gastrointestinal tract. This is followed by a discussion of swallowing difficulties, inflammation, ulcers, and gastric carcinoma. The focus of chapter 35 is on the lower gastrointestinal tract. Various types of intestinal disorders are discussed in this chapter, i.e., diverticula, mucosal injury, defective absorption, inflammation, enzyme deficiency, absence of motility, and malignancy.

Chapter 36 includes a brief overview of anatomy of the liver, associated bile ducts, and blood vessels, as well as bilirubin metabolism. This introduces the general topic of jaundice with emphasis on hepatitis and cirrhosis. The main focus is on the liver with selected disorders of the gallbladder and pancreas included.

Liver function tests are discussed in the first part of chapter 37. Laboratory test results are correlated with disorders involving both the liver and pancreas.

Chapter 34

Disorders of the Upper Gastrointestinal Tract

This chapter includes a brief review of normal anatomy and function of the upper gastrointestinal tract. Esophageal disorders include motility dysfunction and inflammation, due to both reflux of gastric contents and also associated with AIDS. The stomach and duodenum are the focus of the remainder of the chapter, i.e., ulceration of the mucosal lining, complications of ulcer surgery, and gastric carcinoma.

ANATOMY AND FUNCTION

The esophagus carries food down through an opening in the diaphragm, the **hiatus** (hi-a′tus), into the stomach. The opening between the esophagus and stomach is somewhat to the left of the midline and is controlled by a ring of muscle, the **lower esophageal sphincter.** The stomach lies in the upper abdomen on the left side and distally joins the duodenum to the right of the midline. The **pyloric sphincter** controls the opening into the duodenum.

Figure 34.1 shows the following regions of the stomach:

1. The **cardiac region** in the area of the junction of the esophagus
2. The **fundic region** which bulges upward against the diaphragm
3. The largest portion, which is in the middle and is called the **body**
4. The **antrum,** which tapers to the right
5. The narrowed distal portion called the **pyloric canal**

Figure 34.2 is an X-ray picture of the stomach.

Table 34.1 lists the major gastric secretions including mucus, hydrochloric acid, the proteolytic enzyme **pepsin,** and the hormone **gastrin.** Figure 34.3 identifies the cells and the anatomy of gastric glands in the fundus.

ACHALASIA

The term achalasia (ak″ah-la′ze-ah) means inability to relax. This is a motility disorder of the esophagus associated with **dysphagia** (dis-fa′je-ah), difficulty in swallowing. This condition is associated with an absence of esophageal contractions and failure of relaxation of the lower esophageal sphincter. The etiology of the disorder is uncertain, although there is evidence of denervation of smooth muscle. The dysphagia tends to become progressive and is accompanied by weight loss, and occasionally, vomiting. Symptomatic relief is provided by **pneumatic dilatation** of the lower esophageal sphincter. This involves introducing a bag into the sphincter and then enlarging

achalasia: Gk. *a,* not; Gk. *chalasis,* relaxation
dysphagia: Gk. *dys,* hard; Gk. *phagein,* to eat

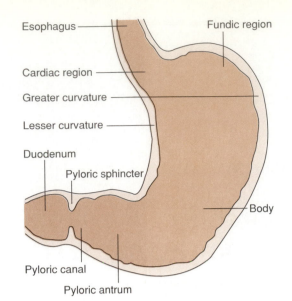

Figure 34.1 Major regions of the stomach.

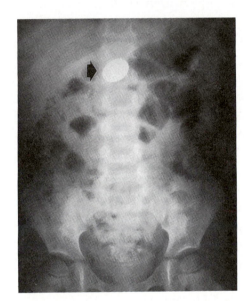

Figure 34.2 An X-ray of the abdomen of a 3-year-old female. The round opaque object over the spine (arrow) is a penny in the stomach.

TABLE 34.1 Gastric secretions and their sources

Goblet cells. Secrete mucus
Mucous glands in cardiac and pyloric regions. Secrete mucus
Glands in fundus and body
    Chief cells secrete pepsin
    Parietal cells secrete HCl
Pyloric region. Produces the hormone gastrin which stimulates gastric secretion

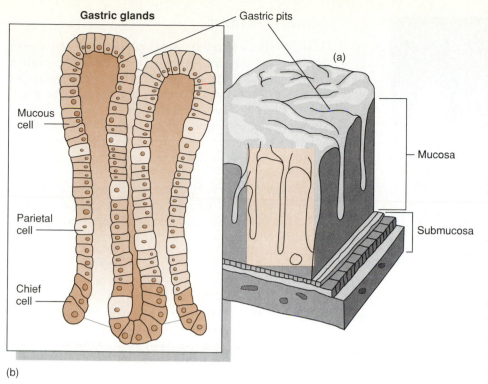

Gastric glands

Gastric pits

(a)

Mucous cell

Parietal cell

Chief cell

Mucosa

Submucosa

(b)

Figure 34.3 Gastric pits and gastric glands of the mucosa. (*a*) Gastric pits are the openings of the gastric glands. (*b*) Gastric glands consist of mucous cells, chief cells, and parietal cells, each of which produces a specific secretion.

the opening by inflating the bag. An alternative procedure is **surgical myotomy** (mi-ot′o-me) in which the circular muscle of the sphincter is incised.

REFLUX ESOPHAGITIS

The term **gastroesophageal** (gas″tro-e-sof″ah-je′al) **reflux** refers to the passage of gastric contents from the stomach upward into the esophagus. The lower esophageal sphincter normally prevents backflow of gastric contents, although everyone experiences reflux on occasion. Reflux esophagitis, also called **gastroesophageal reflux disease,** is a disorder in which the rate of reflux is usually greater than in normal individuals and causes clinical symptoms or histologic changes. It is generally a chronic condition characterized by recurrent symptoms. There may be erosion of the epithelial lining of the esophagus or hyperplasia, an increase in cell number.

ETIOLOGY

It is most likely that reflux esophagitis is caused by a combination of factors including the following:

1. An inadequate antireflux mechanism may be the result of either lower esophageal sphincter relaxation or abdominal muscle contractions leading to increased intra-abdominal pressure.

2. Increased gastric fluid volume may be a contributing factor and may be the result of delayed gastric emptying.
3. Pepsin, a proteolytic enzyme in gastric contents, induces esophageal injury by attacking epithelial protein.
4. Inadequate esophageal clearance may contribute to the problem. The normal response to gastroesophageal reflux is primary peristalsis, propulsive muscular contractions elicited by swallowing. A delay in esophageal acid clearance occurs if peristalsis is abnormal or if salivation is decreased.
5. A natural resistance to acidic injury is provided by a surface layer of mucus with bicarbonate ions, a layer of epithelial cells impermeable to hydrogen ions, and a deeper layer of epithelial cells that have a transport mechanism to maintain normal pH. Hydrogen ions and such agents as pepsin and bile acids must overwhelm this protective barrier in order to cause injury.

Most patients with severe esophagitis have a sliding hiatus hernia. This is a condition in which the stomach pushes up into the thorax through the opening in the diaphragm called the hiatus (figures 34.4 and 34.5). It appears that the hernia does not cause reflux esophagitis because the majority of individuals with a sliding hiatus hernia do not have the reflux disease.

reflux: L. *re*, back; L. *fluere*, to flow
hyperplasia: Gk. *hyper*, beyond; Gk. *plasis*, molding

Figure 34.4 An X-ray of an adult female in a lateral position (spine to the right). This film was taken after the ingestion of barium and demonstrates a large paraesophageal hernia with rotation of the stomach 180° up. The area above the line of the arrows is the herniated region.

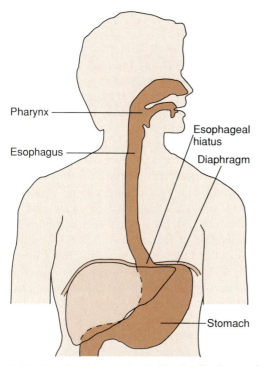

Figure 34.5 The hiatus is an opening in the diaphragm by which the esophagus enters the abdominal cavity.

MANIFESTATIONS

Manifestations of reflux esophagitis include retrosternal burning or chest pain, regurgitation, increased salivation, and dysphagia. The individual may experience a sensation of a lump in the throat, possibly due to increased esophageal sphincter pressure. Hemorrhage usually does not occur. Reflux and regurgitation may result in aspiration of gastric contents into the air passages. Respiratory symptoms may include morning hoarseness, asthma, and pneumonia.

Severe cases may be complicated by inflammation, fibrosis, and loss of compliance of the esophageal wall. Stricture and possibly occlusion may occur over a period of months.

• • •

Barrett's esophagus is a condition in which the esophagus is lined with columnar epithelial cells rather than the normal squamous epithelium. The esophagus is lined entirely with columnar cells during fetal development, and this condition may persist after birth. Most cases are caused by an epithelial regenerative process associated with reflux esophagitis. This histologic change increases the risk of esophageal adenocarcinoma.

• • •

DIAGNOSTIC TESTS

A presumptive diagnosis of reflux esophagitis may be made in an individual who has a typical history. Diagnostic tests may be required for those patients with an atypical history or for those who do not respond to treatment. An instrument may be introduced into the esophagus for direct visualization of the lining of the esophagus. This procedure is called **esophageal endoscopy.** The diagnosis of Barrett's esophagus requires microscopic examination of esophageal tissue, which may be excised during endoscopic examination.

TREATMENT

Table 34.2 indicates measures directed toward either alleviating symptoms or minimizing reflux associated with reflux esophagitis. The broad goals of therapy, as shown in table 34.2, are to (1) decrease reflux, (2) neutralize gastric contents, (3) promote esophageal clearance, and (4) protect the esophageal mucosa.

retrosternal: L. *retro,* behind

TABLE 34.2 Modalities of treatment of reflux esophagitis

Decrease Gastric Volume

1. Avoid large meals
2. Avoid snacks before bed
3. Restrict dietary fat which slows gastric emptying
4. Drugs to promote gastric emptying

Decrease Acidity of Gastric Fluid

1. Antacids
2. Drugs to suppress acid secretion

Improve Esophageal Clearance

1. Raise the head of the bed
2. Increase salivation. Chew gum or suck on a lozenge
3. Drugs that promote motility

Maintain Esophageal Sphincter Pressure

1. Restrict dietary fat, chocolate, alcohol, coffee
2. Avoid smoking
3. Drugs that promote motility also increase sphincter pressure
4. Rate of sphincter relaxation is decreased when gastric volume is low

Protect Esophageal Mucus

1. Sucralfate seems to form a protective barrier

ESOPHAGITIS ASSOCIATED WITH AIDS

Patients with autoimmune deficiency syndrome frequently experience both dysphagia and odynophagia (od″ĭ-no-fa′je-ah) due to esophagitis. The majority of these patients have a yeast infection, *Candida*, which may be associated with cytomegalovirus and herpes virus. AIDS patients may also have reflux esophagitis. Drug therapy for the treatment of esophageal infection results in symptomatic improvement, although it does not necessarily prolong survival for the AIDS patient.

GASTRITIS

Gastritis is an inflammation of the mucosal lining of the stomach. It may be acute, with abrupt onset of symptoms, or have a chronic course that is either asymptomatic or is characterized by prolonged symptoms.

ACUTE GASTRITIS

Acute gastritis may be due to chemical irritants, mechanical or thermal injury, or may be caused by bacteria. Ingestion of alcohol or salicylate are two common causes. Symptoms include **epigastric** pressure (in the area over the stomach), headache, nausea, and vomiting. If the ingested substance is poisonous or is a strong acid or alkali,

odynophagia: Gk. *odyne,* pain

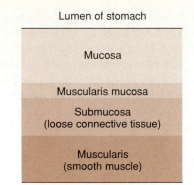

Figure 34.6 The main layers of the stomach wall.

immediate dilution or neutralization is indicated. Gastric **lavage** (lah-vahzh′), washing out stomach contents, may be required. Gastritis due to a bacterial agent may be treated with antibiotics.

CHRONIC GASTRITIS

Chronic gastritis is diagnosed on the basis of the microscopic appearance of the gastric mucosa. **Superficial gastritis** results in inflammation of the luminal surface of the mucosa with infiltration by lymphocytes and plasma cells. **Atrophic** (ah-trof′ik) **gastritis** leads to thinning of the entire mucosal layer, atrophy of glandular structure, and a decrease in chief and parietal cells.

If the upper portion of the stomach is involved, the most likely causes are alcohol, salicylates, anti-inflammatory agents, or autoimmune diseases. In some cases, it is possible to demonstrate the presence of antibodies against parietal cells. Reflux of bile and pancreatic enzymes from the duodenum may be the cause if the lower third of the stomach is involved. Chronic gastritis is common among alcoholics and heavy smokers. Manifestations include anorexia, food intolerance, nausea, and vomiting. Long-term treatment of chronic gastritis involves antacids, a diet restricted to well cooked foods, and the avoidance of caffeine, spices, alcohol, and cigarettes.

There are at least three possible complications associated with gastritis. Normally, parietal cells produce a glycoprotein called **intrinsic factor,** a substance that promotes absorption of vitamin B_{12}. Since vitamin B_{12} is required for maturation of red blood cells, chronic gastritis may be associated with anemia due to vitamin B_{12} deficiency. Chronic gastritis predisposes to gastric ulcers, and there is suggestive evidence that it may predispose to gastric carcinoma.

GASTRIC ULCER

Figure 34.6 shows the main layers of the stomach. The **mucosa** lines the lumen of the stomach and consists of an epithelial layer of columnar cells, a middle layer of connective tissue with blood vessels and glands, and a thin outer layer of smooth muscle called the **muscularis mucosa.** A gastric ulcer is a break in the mucosa, which may be superficial or deep enough to extend through the muscularis mucosa. The most common site of gastric ulcers is on the lesser curvature or the antrum (figure 34.1). Ulcers of the upper gastrointestinal tract are sometimes called **peptic ulcers,** a term which indicates the involvement of the enzyme pepsin in the pathogenesis of the lesion. Approximately 4 million people in the United States have gastric ulcers, with a peak incidence between 55–65 years. About 3,000 individuals die of gastric ulcer disease each year.

ETIOLOGY

There are various contributing factors in ulcer formation including the (1) proteolytic enzyme pepsin, (2) hydrochloric acid, and (3) lowered mucosal resistance, which normally protects against mechanical and chemical injury. Ulcers that occur on the lesser curvature in the body of the stomach appear to be associated with both hyposecretion of acid and impaired mucosal resistance. There is usually acid hypersecretion in the case of ulcers that develop in the region of the antrum.

Table 34.3 identifies factors involved in the **gastric mucosal barrier.** There are various reasons for impairment of this physiological barrier including (1) salicylates, which increase mucosal permeability to acid, (2) inhibition of mucosal prostaglandin production that normally protects epithelial cells, (3) decreased blood flow, and (4) reflux of bile acids and pancreatic enzymes from the duodenum.

• • •

The ultimate source of mucosal injury is hydrogen ions that penetrate the surface mucus and cross cell membranes. The result is destruction of cells and injury to capillaries and venules.

• • •

MANIFESTATIONS

A major symptom of gastric ulcer is dull, aching, or burning pain. The pain may be abdominal or may radiate to the back or the sternum. Other symptoms include nausea, vomiting, weakness, weight loss, and gastrointestinal bleeding. Bleeding leads to varying degrees of anemia. Hypokalemia and metabolic alkalosis occur, depending on the frequency and duration of vomiting. Chapter 5 describes the basis for these observations. If the ulcer leads to a perforation of the stomach wall, the pain is both acute and constant with **peritonitis** as a subsequent complication.

peritonitis: Gk. *peritonaion,* membrane containing lower viscera; Gk. *itis,* inflammation

TABLE 34.3 Factors that constitute the gastric mucosal barrier that protects against mucosal injury

Thin Layer of Mucus Adheres to Surface of Mucosa

a. Mechanical protection
b. HCO_3^- is secreted by surface epithelium
   HCO_3^- diffuses into and is trapped by mucus
   HCO_3^- neutralizes acid

Surface Epithelial Cells of Mucosa

a. Mechanical barrier
b. Prostaglandins protect surface of cells

Adequate Blood Supply to Mucosa

a. Nutrition for cells
b. Blood flow removes accumulations of H^+

DIAGNOSTIC TESTS

Measurement of the secretion of acid is not useful in the diagnosis of gastric ulcer because the test results are normal for the majority of ulcer patients. Most ulcers can be detected by X-rays following the ingestion of a suspension of barium sulfate. Endoscopy, a procedure involving the introduction of an instrument into the stomach for direct visualization, is an aid in differentiating a benign from a malignant ulcer.

TREATMENT

The goals of medical therapy of gastric ulcer are to (1) neutralize gastric acid, (2) prevent acid secretion, and (3) to promote local healing. The avoidance of injurious substances is important, and these substances include alcohol, anti-inflammatory agents, caffeine, and cigarettes. Diets that do not cause pain are planned on an individual basis.

• • •

Histamine plays a role in stimulating gastric acid secretion. Drugs that compete with histamine for gastric histamine receptors (H_2 receptors), inhibit acid secretion. Cimetidine and ranitidine are both H_2 antagonists. Sucralfate promotes ulcer healing by forming a protein-drug complex over the ulcer.

• • •

A bleeding ulcer may be treated with laser or thermal techniques to stop the bleeding. Surgery is indicated when there is failure to heal, perforation, or recurrent bleeding.

DUODENAL ULCER

The first segment of the duodenum, called the duodenal bulb, is the common site for duodenal ulcers. The incidence of duodenal ulcer is approximately equal in both

Figure 34.7 Gastric erosions and small acute ulcers scattered in the mucosa of the stomach.
From A. C. Ritchie: Boyd's Textbook of Pathology. *1990 Lea & Febiger, Philadelphia. Reproduced with permission.*

sexes and is four times more common than gastric ulcer. As in the case of gastric ulcer, the pathogenesis involves injury inflicted by stress, acid, and pepsin to a mucosa with inadequate protective mechanisms. Defense against such injury includes mucus, adequate blood supply to the mucosa, active cell turnover for repair, and production of prostaglandins that protect cells.

Epigastric pain, which is relieved by food or antacids, is characteristic. The symptoms are usually intermittent. Gastrointestinal bleeding is a possible complication. Sucralfate for cytoprotection, H₂ blockers, and antacids are effective in promoting healing.

GASTRIC MUCOSAL STRESS EROSIONS

Gastric mucosal stress erosions or **stress ulcers** represent an entity separate from chronic gastric and duodenal ulcers (figure 34.7). Various conditions of trauma and illness predispose to the development of these gastric lesions, which are typically multiple, shallow, and well demarcated. There is no intense inflammatory reaction, such as occurs with chronic peptic ulcer, and the erosion usually does not extend down through the submucosa.

ETIOLOGY

As indicated in table 34.4, gastric mucosal erosions are associated with trauma, chronic ingestion of drugs, and serious illness. This condition associated with severe burns is called **Curling's ulcer,** named after the individual who described it in 1842. Ulcers occur more frequently in patients with burns of 35% or more of the body. In some burn patients, both gastric and duodenal ulcers are present. Trauma or surgery involving the central nervous system leads to mucosal lesions in the esophagus, stomach, or

TABLE 34.4 Conditions associated with gastric mucosal stress erosions

Severe trauma	Extensive burns
Sepsis	Central nervous system disease
Chronic illnesses	Cerebral or spinal cord injuries
Respiratory failure	Renal failure
Hypotension	Jaundice

duodenum. Unlike other conditions associated with stress induced ulcers, central nervous system trauma is characterized by increased gastric acid secretion. The ulcers tend to be deep and susceptible to perforation. Other conditions that predispose to the development of gastric mucosal ulceration include respiratory failure, renal failure, hypotension, and jaundice.

Drug-induced acute gastric mucosal ulcers are indistinguishable from the stress-induced type. Nonsteroidal anti-inflammatory agents and ethyl alcohol are possible causes. Aspirin, of the anti-inflammatory agents, appears to have a greater propensity for causing mucosal irritation as compared to indomethacin, phenylbutazone, and ibuprofen.

MANIFESTATIONS

Acute gastric mucosal ulceration usually does not cause epigastric pain. Bleeding from multiple shallow ulcers may occur. A small percent of affected individuals will develop life-threatening hemorrhage. It has been estimated that 10–20% of patients receiving intensive care will have evidence of upper gastrointestinal bleeding, if there are no preventive measures. There is a greater risk of bleeding in patients who are acutely ill.

Disorders of the Upper Gastrointestinal Tract

Gastric bleeding may result in hematemesis (hem''ah-tem'ĕ-sis), which is the appearance of blood in vomitus. Melena (mĕ-lē'nah), black stools due to blood, may occur as well.

DIAGNOSTIC TESTS

Endoscopy is a procedure in which an instrument is introduced into the intestinal tract for the purpose of direct visualization of the mucosa. Photographs or videos may also be taken. An upper gastrointestinal endoscopy is useful in identifying the source of bleeding.

TREATMENT

There is an 80% mortality rate associated with hemorrhage from acute gastric mucosal ulceration. Consequently, prophylactic measures are an important aspect of managing this condition. Antacids, such as Mylanta or Maalox, are effective in the prevention of gastrointestinal bleeding in both critically ill and in burn patients. Histamine H_2 receptor antagonists may be useful as adjunctive therapy. Sucralfate, a basic aluminum salt of sucrose octasulfate, is an effective prophylactic agent as well.

• • •

There is an unresolved question relating to more neutral gastric contents. It is theorized that increased gastric pH allows greater bacterial growth and that this may contribute to a greater frequency of pneumonia in critically ill patients on ventilators. The pneumonia may be the result of chronic aspiration of gastric contents. Sucralfate, administered prophylactically to prevent bleeding, may be preferred over antacids in such cases because it does not increase gastric pH.

• • •

The initial step in treating gastric bleeding is to correct the underlying problem, which may be sepsis or hypovolemia, for example. Methods for controlling gastric bleeding include (1) intra-arterial infusion of vasopressin, a vasoconstrictor; (2) occlusion of the bleeding vessel with a clot, a procedure called transcatheter embolization; and (3) endoscopically guided laser beam therapy. Surgical intervention may be required in extreme cases.

hematemesis: Gk. *haema*, blood; Gk. *emesis*, vomiting
melena: Gk. *melas*, black

TABLE 34.5 Definition of terms that refer to surgical procedures in the treatment of peptic ulcer

Truncal vagotomy. All nerve fibers below the esophagus are cut
Selective vagotomy. All nerve fibers to the stomach are cut
Gastrectomy. Surgical removal of stomach
Subtotal gastric resection. Most of stomach removed
Pyloroplasty. Surgical procedure involving the pylorus that forms a large opening into duodenum
Gastroenterostomy. Formation of a communication between the stomach and either the duodenum or jejunum

• • •

Stress ulcers can be produced in rats by placing them in a confined space. The restrained animals develop ulcers within 2–24 hours when the temperature is 4°C. The erosions are superficial and usually heal within a week. Rats also develop stress ulcers in response to forced exertion and traumatic shock.

• • •

ULCER SURGERY

Surgical intervention for the treatment of ulcers may be required if (1) the symptoms are debilitating, (2) there is perforation of the gastric wall, (3) there is obstruction, or (4) there is severe hemorrhage. Table 34.5 lists various surgical procedures for the treatment of peptic ulcer.

In principle, these surgical procedures are designed to decrease gastric secretion, remove injured tissue, and promote gastric emptying. Fibers of the vagus nerve, which are parasympathetic, innervate the esophagus, stomach, and to a lesser extent, the small intestine. The vagus stimulates the gastric secretion of pepsin and acid; thus, vagotomy results in decreased secretion. Truncal or selective vagotomy leads to gastric obstruction in 20% of patients. For this reason, **pyloroplasty** (pi-lo'ro-plas''te) or **gastroenterostomy** (gas''tro-en-ter-os'to-me) is performed with vagotomy to promote gastric emptying of liquids.

COMPLICATIONS

Ulcer operations lead to an increased rate of gastric emptying. There are symptoms following ingestion of food, **postcibal** (pōst-si'bal) symptoms, which may be explained on the basis of rapid movement of food into the intestine. The most common observation is early satiety (sah-ti'e-ty), a sensation of being full, and/or postcibal vomiting accompanied by pain. Chronic diarrhea, usually within an hour or two of eating, may be a sequel to ulcer

pyloroplasty: Gk. *pyloros*, gatekeeper; Gk. *plassein*, to form
gastroenterostomy: Gk. *gaster*, belly; Gk. *enteron*, intestine; Gk. *stoma*, mouth
postcibal: L. *post*, after; L. *cibus*, food
satiety: L. *satiatus*, to fill

surgery. Patients may also experience a "dumping syndrome." This involves a combination of postcibal alimentary symptoms, i.e., satiety, nausea, vomiting, cramps, pain, and/or diarrhea, associated with sweating, weakness, and palpitations.

Many patients lose weight and maintain that loss after ulcer surgery. Limited food intake due to satiety may be the most important factor in poor nutrition. Ineffective digestion resulting in malabsorption of fat and other nutrients is another factor.

Anemia develops after subtotal gastrectomy, with iron deficiency as the most important cause. Ulcer surgery leads to bone disease with bone pain, loss of bone mineral content, and increased susceptibility to fracture. Bone demineralization may be caused by malabsorption of calcium or vitamin D.

There is evidence to support the view that ulcer surgery is associated with an increased risk for gastric cancer. It also appears that ulcer disease, with or without surgery, is associated with an increased prevalence of gastric cancer.

• • •

There is a pattern of gastrointestinal motility during fasting called **migrating myoelectric complexes.** It involves the appearance of contractions in the antrum about every two hours. The contractions increase in rate and strength for about 15 minutes and then slowly migrate down the small intestine for the next two hours. The cycle is then repeated. The gastrointestinal tract is thus cleared of debris. Bezoars (be'zōrs), particles of indigestible residue, tend to form after ulcer surgery, due to the absence of a normal migrating myoelectric complex. Bezoars may be asymptomatic or may cause a partial obstruction.

• • •

GASTRIC CARCINOMA

Most gastric malignancies are **adenocarcinoma,** which involves mucous cells. The term **early gastric cancer** is used to identify the disease when the depth of invasion is limited to the mucosa or submucosa. Advanced gastric cancer involves the muscular layer of the stomach and is usually associated with **metastasis,** spread to other parts of the body.

The death rate for gastric carcinoma has decreased in the United States during the past 40 years. The mortality rate currently is less than 7 per 100,000, with the incidence involving males about twice that of females. Japan has the highest gastric cancer mortality rate, 63.1 for males and 30.3 for females, per 100,000 population.

TABLE 34.6 Possible factors in gastric cancer

Studies suggest an increased association between gastric cancer and the following:

Salted fish and meat	Nitrates
Smoked foods	Nitrites
Starch	Pickled vegetables

Studies suggest an inverse relationship between gastric cancer and the following:

Refrigeration	Whole milk
Citrus fruits	Fresh vegetables

Possible precancerous conditions:

Chronic gastritis	Atrophic gastritis of pernicious anemia
Gastric ulcer	Achlorhydria (decreased HCl secretion)

ETIOLOGY

The etiology of gastric cancer is unknown. Table 34.6 shows a list of factors that are possible elements in the disease process. Studies have consistently shown a correlation between gastric cancer and increased ingestion of salt. Nitrates and nitrites may be contributing factors. Refrigeration, which inhibits the conversion of nitrate to nitrite, combined with the use of less nitrate as a preservative, may have contributed to a decline in gastric cancer in the United States.

MANIFESTATIONS

Early gastric cancer is frequently asymptomatic. In advanced gastric cancer, nausea, a feeling of fullness, weight loss, and abdominal pain may occur. If the cancer causes obstruction, dysphagia or vomiting is the result. There may be gastrointestinal bleeding, and in about 20% of cases, a palpable abdominal mass. Distant metastasis may involve the liver, lymph nodes, lungs, colon, and other tissues.

DIAGNOSTIC TESTS

Barium X-rays are useful in differentiating benign from malignant ulcers. Endoscopic excision of tissue from the ulcer margin for microscopic examination is a method of evaluating malignancy.

TREATMENT

Treatment involves surgical resection, the removal of the involved segment. Radiation and chemotherapy may be beneficial. Prognosis is more favorable in middle-aged patients with a limited depth of invasion, no lymph node involvement, and no metastasis. The average survival time is about 11 months after diagnosis in untreated cases. Approximately 10% of all patients treated for gastric carcinoma survive 5 years.

SUMMARY

• • •

Achalasia is a motility disorder of the esophagus of unknown etiology, which causes difficulty in swallowing. Pneumatic dilatation or myotomy of the lower esophageal sphincter provides symptomatic relief. Reflux esophagitis is caused by backflow of gastric contents into the esophagus. Manifestations include chest pain, regurgitation, increased salivation, and dysphagia. Inflammation, fibrosis, and stricture may occur in severe cases. Treatment involves measures directed toward alleviating symptoms or minimizing reflux. AIDS patients frequently have esophagitis, usually due to an infection caused by *Candida*.

Gastritis is an inflammation of the gastric mucosa, which may be either acute or chronic. The condition is caused by ingested irritants, bacterial toxins, and mechanical or thermal injury. Chronic gastritis predisposes to ulcers and possibly to gastric carcinoma. Ulceration of the gastric mucosa is the result of impaired resistance and/or an increase in injurious substances, which allow hydrogen ions to diffuse into mucosal cells. The gastric mucosal barrier normally provides protection against injury. The elements of this physiological barrier are (1) surface mucus, (2) HCO_3^-, (3) mucosal epithelial cells, (4) prostaglandins, and (5) adequate blood supply. Treatment involves measures to neutralize gastric acid, prevent acid secretion, and to promote local healing. Duodenal ulcer involving the first segment of the duodenum is more common than gastric ulcer. Pathogenesis is similar to that of gastric ulcer. Epigastric pain is usually relieved by food or antacids.

Gastric mucosal stress erosions or stress ulcers are associated with trauma, chronic ingestion of drugs, and serious illness. Epigastric pain is absent in most cases, although bleeding from multiple shallow ulcers may occur. Modalities of prevention and treatment include antacids, H_2 receptor antagonists, and sucralfate. Methods for controlling gastric bleeding involve intra-arterial infusion of vasopressin, transcatheter embolization, or laser beam therapy.

Surgical procedures for the treatment of peptic ulcer may involve resection or vagotomy and result in a decrease in gastric secretion or in improved gastric emptying. Complications of ulcer surgery include early satiety, postcibal vomiting and pain, chronic diarrhea, weight loss, anemia, and increased risk of gastric cancer.

The etiology of gastric cancer is unknown, although certain substances, such as salt, nitrates, nitrites, and smoked foods, have been implicated by suggestive evidence. Early gastric cancer may be asymptomatic, but in more advanced stages, there may be nausea, weight loss, abdominal pain, bleeding, and possibly a palpable mass. Surgical resection, radiation therapy, and chemotherapy are modalities of treatment. The overall 5-year survival rate for patients with gastric cancer is approximately 10%.

REVIEW QUESTIONS

• • •

1. What is the meaning of the term achalasia?
2. What are four underlying mechanisms that may be contributing factors in reflux esophagitis?
3. List three characteristic symptoms of reflux esophagitis.
4. What is a possible complication of severe reflux esophagitis?
5. Define Barrett's esophagus.
6. What are three broad goals of treatment of reflux esophagitis?
7. What is the most common infective agent causing esophagitis in AIDS?
8. What are the most likely causes of chronic gastritis if the upper portion of the stomach is involved?
9. What are two substances normally secreted by mucosal cells that are potentially injurious to gastric mucosa?
10. List five factors which protect the gastric mucosa against injury.
11. What is the ultimate cause of a gastric ulcer?
12. List three substances which are ingested that are probable factors in gastric carcinoma.
13. Which is most common in the United States, gastric or duodenal ulcer?
14. What two general types of conditions cause stress ulcers?
15. In what ways are stress ulcers different from chronic gastric and duodenal ulcers?
16. Under what condition is increased gastric acid secretion associated with stress ulcers?
17. Why are prophylactic measures against stress ulcers important in acutely ill patients?
18. List three types of treatment for the prevention of stress ulcers.
19. What are two modalities of treatment of gastric bleeding?
20. What is the difference between truncal and selective vagotomy?
21. What is the meaning of the term subtotal gastric resection?
22. List four complications of ulcer surgery.
23. List three substances which are ingested that are probable factors in gastric carcinoma.
24. What factors are most important in the prognosis of gastric cancer?
25. What is the treatment of choice in gastric cancer?

Brasitus, T., and E. Foster. 1988. Peptic ulcer update: Approaches to treatment. *Physician Assistant* 9:71–86.

Collins, R., and M. Langman. 1985. Treatment with histamine H_2 antagonists in acute upper gastrointestinal hemorrhage. *New England Journal of Medicine* 313: 660–65.

Krejs, G. et al. 1987. Laser photocoagulation for the treatment of acute peptic ulcer bleeding. *New England Journal of Medicine* 316:1618–21.

Narain, S., D. B. Smith, and A. L. Goldman. 1989. A 53-year-old man with nausea, vomiting, diarrhea, and weakness. *Physician Assistant* 13:54–66.

Sievert, W. et al. 1988. Gastrointestinal emergencies in the acquired immunodeficiency syndrome. *Gastroenterology Clinics of North America* 17:409–18.

Steiner, J. F. 1990. Peptic ulcer disease: Treatment and therapeutic controversies. *Journal of Academy of Physician Assistants* 3:64–70.

Windoffer, E. 1988. Current concepts in gastroesophageal reflux disease. *Journal of American Academy of Physician Assistants* 1:183–88.

Wood, S. 1988. Toward the expedient diagnosis and treatment of esophageal carcinoma. *Journal of American Academy of Physician Assistants* 1:344–50.

Chapter 35

Disorders of the Lower Gastrointestinal Tract

This chapter includes representative examples of various types of intestinal disorders. These include disorders involving (1) outpouchings of the intestinal wall, (2) injury to the intestinal mucosa, (3) absorption of nutrients, (4) inflammatory reactions, (5) digestive enzyme deficiency, (6) absence of motility, and (7) malignancy.

DIVERTICULOSIS OF THE COLON

A diverticulum is an outpouching or herniation of intestinal mucosa (table 35.1). The most common site is the sigmoid colon (figure 35.1) although these may occur in other parts of the gastrointestinal tract as well. The walls of a true diverticulum contain all layers of the intestine. Most herniations are pseudodiverticula (soo''do-di''ver-tik'u-lah), outpouchings of the mucosa and submucosa through the muscular coats of the colon. These two layers, the mucosa and submucosa, push up between bundles of circular muscle fibers where nutrient arteries pass through the submucosa. The pockets may be 1 cm or more in diameter (figure 35.2).

Diverticulosis is common in developed Western countries with an estimated incidence of 10%. There is a correlation between incidence and increasing age. The condition is uncommon before the age of 40, but there is an increased incidence with each decade thereafter. Figure 35.3 shows a diverticulum of the terminal ileum.

ETIOLOGY

The factors that contribute to the formation of diverticula are believed to be (1) increased pressure within the lumen of the colon and (2) areas of weakness in the intestinal wall. There is evidence which suggests that there is increased distensibility of the inner colon wall, possibly due to deposition of elastin in the longitudinal muscles. Pressure develops within the colon as the result of contraction of rings of circular muscle accompanied by longitudinal muscle contractions. **Flatus** or retention of gas may contribute to intraluminal pressure as well.

The law of Young and Laplace states that there is a proportional relationship between intraluminal pressure and wall tension due to muscle contractions, whereas there is an inverse relationship between pressure and the radius of the intestine. Thus, a decrease in the radius leads to an increase in pressure or an increase in radius is associated with a decrease in pressure. These principles apply to events in the colon, and to the sigmoid colon in particular, because this is the narrowest portion of the colon. It has been shown that dietary changes leading to increased stool mass results in a decrease in motility. In accordance with

diverticulosis: L. *diverticulum*, a byway; Gk. *osis*, condition
herniation: L. *hernia*, rupture
flatus: L. *flatus*, a blowing

TABLE 35.1 Basic layers of the gastrointestinal wall, listed in order from the outside to the inner lining of the lumen

Tunica serosa. Outer layer of connective tissue covered with a serous membrane
Tunica muscularis. Layer of muscle consisting of an outer longitudinal layer and an inner layer with a circular arrangement
Tunica submucosa. Layer of connective tissue with blood and lymph vessels, nerves, and glands
Tunica mucosa. Inner lining of digestive tract; made up of a layer of epithelial cells, connective tissue with blood vessels, lymph nodes, glands, and a layer of smooth muscle

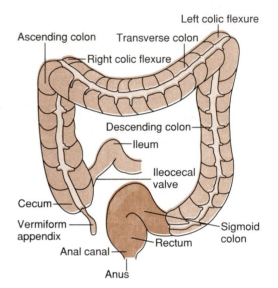

Figure 35.1 The major parts of the large intestine.

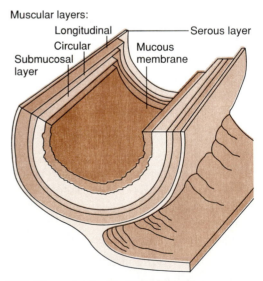

Figure 35.2 The major layers of the intestine.

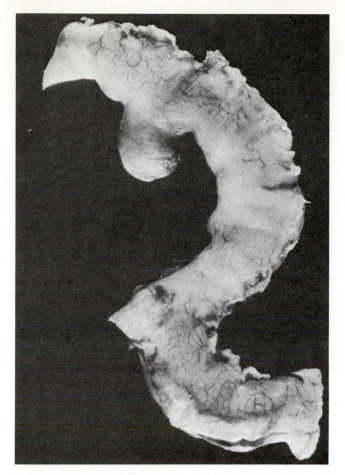

Figure 35.3 Meckel's diverticulum. Derived from an abnormal persistence of the omphalomesenteric duct, this diverticulum is most often located in the terminal ileum.
From R. S. Cottran, V. Kumar, and S. L. Robbins: Robbins Pathologic Basis of Disease. *1989 W. B. Saunders, Philadelphia. Reproduced with permission.*

the law of Young and Laplace, (1) decreased wall tension leads to decreased intraluminal pressure, and (2) increased radius contributes to a decrease in intraluminal pressure.

MANIFESTATIONS

There are no symptoms in most cases of uncomplicated diverticulosis. When the disorder is symptomatic, there may be years of periods of remissions interspersed with days or weeks of relapse. The most common symptom is pain, lasting for hours to days, and usually located in the left abdominal quadrant. The individual may also experience constipation, diarrhea, and flatulence. Painless rectal bleeding is a possible symptom. Severe blood loss occurs in 3–5% of those individuals suffering from diverticulosis.

DIAGNOSTIC TESTS

Administration of an enema containing a barium sulfate suspension followed by X-ray studies will reveal diverticula, as well as shortening or narrowing of the distal colon.

An X-ray film after evacuation of the barium may show barium filled diverticula. **Sigmoidoscopy** and **colonoscopy** are procedures whereby an instrument is introduced into the colon for the purpose of direct visualization of the mucosa of the large bowel. The orifices of diverticula may be visible with this kind of procedure.

TREATMENT

The approach to the treatment of uncomplicated diverticulosis involves measures to both relieve symptoms and to avoid complications. Increased dietary intake of vegetable fiber leads to an increase of stool mass, lower intraluminal pressure, and relief of pain. Symptomatic relief may be achieved by adding increasing amounts of wheat bran to the diet until a recommended level is achieved over a period of 4–6 weeks. Other dietary sources of fiber include whole wheat bread, various breakfast cereals, potatoes, leafy vegetables, bananas, and apples. The use of hydrophilic colloid laxatives, containing agar or methylcellulose, also increase stool bulk and water content. This type of laxative may be tolerated better than wheat bran during the early course of treatment.

Glucagon acts as a smooth muscle relaxant and an intravenous injection of this substance provides prompt relief of symptoms of diverticulosis. Analgesics for the relief of pain must be selected with care. Meperidine may be used, whereas morphine raises the intraluminal pressure in the sigmoid colon and is contraindicated.

• • •

There are two mechanisms by which dietary fiber causes an increase in stool mass. Undigested fiber in the large intestine absorbs water and contributes to increased fecal mass. The major factor is that most fiber is degraded by flora in the colon. The degradation products, i.e., propionate, butyrate, and other anions, create an osmotic gradient. An increase in water content of the stool is the result.

• • •

DIVERTICULITIS

Diverticulitis is a complication of diverticulosis and involves inflammation of the herniations. The inflammatory process may be localized or may spread to surrounding areas. Approximately 10–20% of the individuals affected by diverticulosis will at some time develop diverticulitis.

orifice: L. *orificium,* opening

The probability of the occurrence of an inflammatory complication of diverticulosis is increased if (1) there are many diverticula, (2) if the condition develops at a young age, (3) if a large portion of the colon is involved, or (4) if the condition has existed for a number of years.

• • •

Etiology

The inflammatory process is initiated by inspissated (in-spis'āt-ed) fecal material in the outpouching. The localized inflammation progresses to necrosis, rupture, and release of fecal material. The result is inflammation, usually localized, around the diverticulum. Typically, a small abscess develops with subsequent deposition of fibrous connective tissue. Repeated fibrotic reactions may cause obstruction of the colon. Large abscesses may erode into the lumen of the colon or may rupture into surrounding structures. The result may be the development of a **fistula** between the colon and such structures as bladder, vagina, or ureter. The rupture of a diverticulum usually does not lead to widespread peritonitis.

Manifestations

Acute abdominal and back pain and fever are notable symptoms of diverticulitis. There may also be chills, anorexia, nausea, and vomiting. Either diarrhea or constipation is a common feature. If the bladder is involved, dysuria (dis-u're-ah) and frequency of urination is a part of the picture. An elevated white blood cell count with a marked increase in neutrophils is associated with these symptoms.

Diagnostic Tests

Radiologic studies involving both supine and erect X-ray films of the abdomen aid in evaluating the degree of intestinal obstruction. Barium enema studies are useful in diverticulitis and may show barium outside a diverticulum or demonstrate a fistula leading from the colon.

• • •

The most common type of fistula is the result of a rupture and extension of diverticular abscess between the sigmoid colon and the posterior bladder wall. This is called a colovesical fistula and leads to recurrent urinary tract infections and fecaluria.

• • •

inspissated: L. *inspissatus,* thick
fibrosis: L. *fibra,* fiber; Gk. *osis,* condition
fistula: L. *fistula,* tube
dysuria: Gk. *dys,* difficult or painful; Gk. *ouron,* urine
vesicle: L. *vesica,* bladder

Treatment

A patient suffering from diverticulitis is provided bed rest, given nothing by mouth, and administered antibiotics for infection. Intravenous fluids are instituted to maintain caloric intake as well as fluid and electrolyte balance. Surgery is indicated in acute cases in which (1) peritonitis occurs, (2) there is an abscess in spite of antibiotic treatment, (3) there is bowel obstruction, or (4) a severe urinary tract infection persists due to a colovesical fistula.

TROPICAL SPRUE

Tropical sprue is defined as a disorder of unknown etiology in which there are structural abnormalities of the intestinal mucosa accompanied by malabsorption and nutritional deficiencies. It is a chronic condition acquired in the tropics and may be successfully treated with folic acid and tetracycline.

Although it can occur as an epidemic outbreak, tropical sprue typically involves isolated cases. Adults are more frequently affected than children. It occurs in native residents of the tropics as well as in visitors to the region. Most often the condition develops after two or more years residence, although it may occur within a week.

• • •

It appears that tropical sprue occurs only in individuals who have lived in or visited the tropics. Symptoms may not occur for months or years after the individual has moved to a temperate climate.

• • •

Etiology

The causative factor or factors in tropical sprue have not been identified. There is evidence supporting the idea that the underlying cause is a chronic infection of the small intestine and that this infection involves the production of toxins by noninvasive strains of coliform bacteria. Continued exposure to toxins may be responsible for the progressive nature of the intestinal abnormalities. The initiating event in the development of tropical sprue may be an overgrowth of coliforms in the jejunum due to infectious diarrhea. If this occurs, the reason for failure to subsequently eliminate these coliforms from the small intestine is not clear. There is evidence to suggest that dietary intake of long-chain unsaturated fatty acids, i.e., linoleic acid, in some way changes the intestinal flora, so there is decreased resistance to pathogenic colonization. Finally, since administration of folic acid leads to improvement of sprue, folic acid deficiency may be a contributing factor. It has been observed that folic acid deficiency combined with excessive ethyl alcohol intake causes abnormal changes in the intestinal mucosa. There

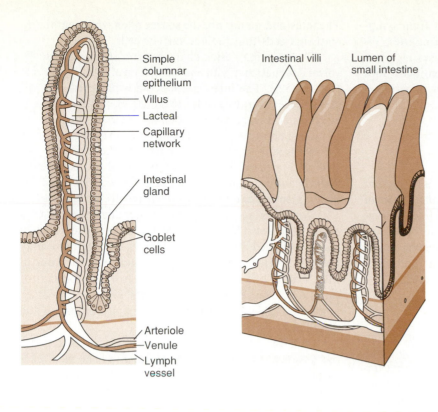

Simple
columnar
epithelium

Villus

Lacteal

Capillary
network

Intestinal
gland

Goblet
cells

Arteriole

Venule

Lymph
vessel

Intestinal villi

Lumen of
small intestine

Figure 35.4 Villi, tiny fingerlike projections, extend from the mucosa into the lumen of the small intestine. These projections increase surface area and absorb nutrients.

is the possibility that the intestinal lesion of sprue is caused by (1) inadequate dietary intake of folic acid and (2) the production of ethyl alcohol by pathogenic coliforms in the small intestine.

MANIFESTATIONS

The onset of the disease is usually acute, with diarrhea, malaise, fever, and weakness. The symptoms become less severe after about a week, and the condition is characterized by chronic diarrhea, abdominal cramps, and excessive gas. Subsequently, there may be intolerance to both milk and alcohol. A condition called **steatorrhea,** or fatty stools, occurs in the majority of cases.

After several weeks structural changes occur, first in the jejunum and later in the entire small intestine. The villi (figure 35.4) become thickened and fused, there are epithelial cell changes, and there is infiltration by inflammatory cells. Enzyme activity is reduced, and there is an abnormal pattern of lipid accumulation.

Structural abnormalities lead to jejunal malabsorption within several months. This causes vitamin deficiencies including deficiency of folic acid, anorexia, and weight loss. Over a period of months, an inflammation of the tongue called **glossitis** occurs and is the result of deficiency of both folic acid and vitamin B_{12}. Anemia develops and edema, due to hypoalbuminemia, may become apparent (chapter 2). Other symptoms of nutritional deficiencies include night blindness, emaciation, and stomatitis.

steatorrhea: Gk. *stear,* fat; Gk. *rhoia,* flow
glossitis: Gk. *glossa,* tongue; Gk. *itis,* inflammation
stomatitis: Gk. *stoma,* mouth

TABLE 35.2 The effects of treatment of tropical sprue by tetracycline and folic acid

Tetracycline Therapy

1. Structure of villi improves
2. Enzyme activity increases
3. Absorption improves
4. Improved appetite, reversal of glossitis
5. Steatorrhea improves
6. Improvement of anemia is delayed

Folic Acid Therapy

1. Structure of villi improves
2. Enzyme activity increases
3. Malabsorption of vitamin B_{12} usually persists
4. Improved appetite, reversal of glossitis
5. Steatorrhea usually persists
6. Remission of anemia

DIAGNOSTIC TESTS

The **xylose tolerance test** provides a measure of intestinal absorptive capacity. Xylose is a sugar excreted unchanged in the urine. The test involves the administration of xylose dissolved in water to a patient who is in a fasting state. Urine is collected over a 5-hour period and tested for xylose. Xylose concentration in the urine is decreased when there is malabsorption in the small intestine.

TREATMENT

Tetracycline and folic acid are effective in treating tropical sprue (table 35.2). Combining folic acid and either tetracycline or a sulfonamide provides relief from symptoms and leads to healing of intestinal abnormalities. Vitamin B$_{12}$ may be administered in those cases in which there is a deficiency. Treatment should be continued until normal intestinal function is restored, possibly a period of months.

CELIAC SPRUE

Celiac sprue is a disorder in which there is a gluten-induced intestinal lesion that leads to malabsorption of nutrients in the affected segment of the intestine. **Gluten-sensitive enteropathy** is another term for the condition. Gluten is a water insoluble protein found in such grains as wheat, barley, and rye. The disease affects both children and adults and occurs throughout the world.

ETIOLOGY

Gluten is the agent that causes injury to the intestinal mucosa in celiac sprue, although the mechanisms by which the injury occurs are not clear. There is evidence that implicates immune mechanisms in the disease process. In addition to this, genetic factors play a role. The fact that there is increased incidence of the disease in relatives of affected individuals indicates that inheritance is a factor, and genetic studies further support this idea.

* * *

A study done in 1950 showed there was a decreased incidence of celiac sprue in children living in Holland during World War II. During this period, there was limited availability of grain products. After the war there was an increased incidence of celiac sprue that coincided with a more plentiful supply of cereal grains. Subsequently, it was shown that wheat, barley, or rye flour, introduced into a normal segment of intestine of an individual with treated celiac sprue, causes cramps and diarrhea within a few hours. Rice or corn flour elicits no symptoms under the same circumstances.

* * *

MANIFESTATIONS

The symptoms of celiac sprue are due to intestinal malabsorption caused by structural changes in the intestinal mucosa. The inner surface of the intestine normally has a thick covering of tiny projections called villi (vil'li). These projections are made up of a single layer of epithelial cells interspersed with goblet cells (figure 35.4). The epithelial cells have an absorptive function, whereas the less numerous goblet cells secrete mucus. There are many intestinal glands (crypts of Lieberkühn) at the bases of the villi.

enteropathy: Gk. *enteron*, intestine; Gk. *pathos*, disease

The intestinal glands are the source of enzyme-containing epithelial cells that replace surface cells.

The intestinal lesion characteristic of celiac sprue involves destruction of villi and elongation of intestinal crypts (figure 35.5). The intestinal mucosa is flattened, the relatively few absorptive cells are damaged, and the mucosal enzymes necessary for digestion are altered. This pattern of destruction and injury may involve the entire length of the intestine or may be more marked in the proximal portion with decreasing severity in the distal segments. Thus, manifestations of the disease vary considerably, depending on the extent of intestinal involvement.

Symptoms may first appear in later adult years or may occur in infants. If evidence of the disease occurs early and is untreated throughout childhood, the symptoms may diminish during the teenage years and reappear later in life. Most of the symptoms are due to malabsorption of fat, water, electrolytes, carbohydrates, proteins, and other nutrients. Table 35.3 lists possible manifestations of the disorder.

Impaired calcium absorption may lead to osteopenic (os″te-o-pen′ik) bone disease resulting in bone pain. Hyperparathyroidism in response to low calcium levels promotes bone resorption and contributes to osteopenia. In addition, low serum calcium levels may cause tingling or crawling sensations called paresthesias, muscle cramps, or possibly even tetany.

Patients with a severe form of sprue experience nasal, gastrointestinal, vaginal, or renal bleeding. The underlying cause of bleeding tendencies is impaired absorption of vitamin K which is required for hepatic synthesis of clotting factors (chapter 19). This bleeding, as well as impaired iron or folic acid absorption, are factors in the development of anemia.

Rarely, in severe cases of celiac sprue, there may be lesions of the nervous system that cause muscle weakness, sensory loss, paresthesias, and ataxia or muscle incoordination. The cause is uncertain.

* * *

There are several probable mechanisms causing diarrhea in celiac sprue. (1) There is an increased osmotic load delivered to the large intestine. (2) Bacterial action on fat in the large intestine causes production of hydroxy fatty acids. These substances are potent cathartics. (3) In this disease, water and electrolytes are secreted into the lumen of the small intestine. (4) Bile salts in the colon, normally absorbed in the ileum, may have a cathartic effect.

* * *

osteopenia: Gk. *osteon*, bone; Gk. *penia*, poverty
paresthesias: Gk. *para*, beyond; Gk. *aisthesis*, perception
cathartic: Gk. *kathartikos*, to purify
ataxia: Gk. *ataxia*, disorder

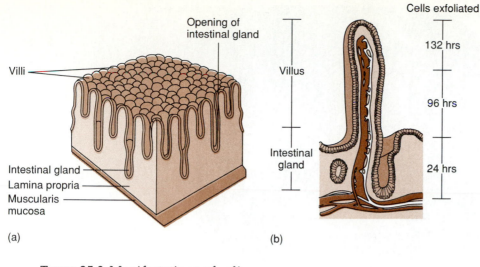

Villi

Opening of intestinal gland

Villus

Intestinal gland

Intestinal gland
Lamina propria
Muscularis mucosa

(a)

Cells exfoliated

132 hrs

96 hrs

24 hrs

(b)

Figure 35.5 Intestinal villi and intestinal glands, or crypts of Lieberkühn are shown in (*a*). The crypts serve as sites for production of new epithelial cells. The time required for migration of these new cells to the tip of the villi is shown in (*b*). Epithelial cells are exfoliated from the tips of the villi.

TABLE 35.3 Manifestations of celiac sprue

Diarrhea	Anemia
Constipation may occur	Dehydration
Flatulence	Hypokalemia
Malodorous, oily stools (steatorrhea)	Abdominal distention
Weight loss	Bleeding tendencies
Osteopenic bone disease	Night blindness

• • •

There may be marked electrolyte depletion if the diarrhea associated with celiac sprue is severe. Low blood levels of sodium, potassium, chloride, and bicarbonate may occur. Bicarbonate loss in the stool can cause metabolic acidosis.

• • •

DIAGNOSTIC TESTS

Malabsorption of fat causes steatorrhea. The stool is tan or grayish, greasy, and malodorous. A qualitative procedure, a useful screening test for fat, involves staining a stool suspension with Sudan III or IV after the addition of acetic acid. Orange or red fat globules may be observed under the microscope.

The xylose absorption test is useful in diagnosing malabsorption. Xylose is normally absorbed in the proximal portion of the small intestine, which corresponds to the region most severely affected in celiac sprue. Xylose is administered orally, and the amount eliminated in the urine is measured. Xylose excretion is suppressed in celiac sprue.

A barium meal followed by X-ray films of the small intestine will reveal dilation of some of the loops of the small intestine, as well as a distortion of the mucosal pattern in cases of celiac sprue.

The most valuable diagnostic test is biopsy of the small intestine. An instrument is introduced by way of a biopsy tube, it is positioned and visualized under a fluoroscope, and mucosal samples are excised from the duodenaljejunal (du″o-de′nal-jĕ-joo′nal) junction. Stained sections of tissue are examined under the microscope.

TREATMENT

Treatment for celiac sprue is dietary restriction of gluten-containing cereal grains, i.e., wheat, barley, rye, and oats. The diet should be well balanced and contain normal amounts of fat, protein, and carbohydrates. Dietary restriction must be continued throughout life. The intestinal mucosa rapidly recovers normal structure and function in the absence of gluten (figure 35.6). Primary and secondary manifestations disappear. In the absence of treatment, severe celiac sprue leads to malnutrition and may be fatal. Table 35.4 compares celiac and tropical sprue.

• • •

Wheat flour, in particular, is found in many processed foods including ice cream, salad dressing, canned soups, candy bars, peanut butter, wieners, mustard, and catsup.

• • •

INTESTINAL INVOLVEMENT IN AIDS

Diarrhea is the most common gastrointestinal symptom in AIDS and, in most cases, there is at least one identifiable pathogen. Protozoa are the most frequent diarrheal pathogens. *Cryptosporidium* is a protozoan that causes severe diarrhea, profound weight loss, and malnutrition. There is a remarkable increase in the incidence of recurrent *Salmonella* infections in AIDS patients. The

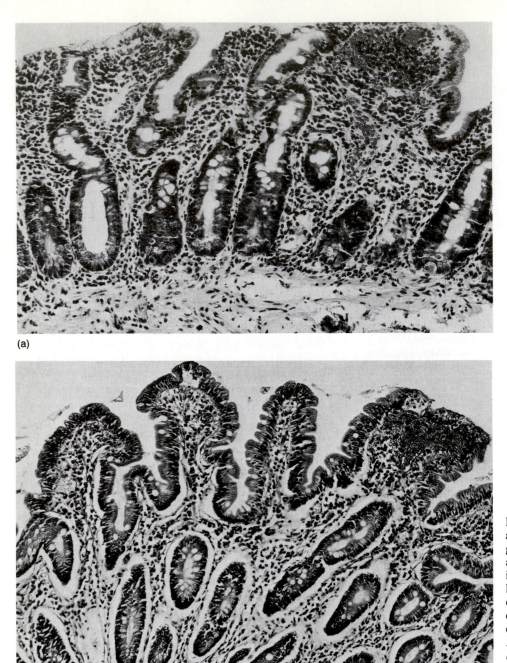

(a)

(b)

Figure 35.6 (*a*) Celiac disease—a jejunal peroral biopsy of gluten enteropathy with atrophy and blunting of villi and an inflammatory infiltrate in the lamina propria (thin layer of connective tissue). (*b*) Celiac disease—same patient after five days on a gluten-free diet. *From R. S. Cottran, V. Kumar, and S. L. Robbins:* Robbins Pathologic Basis of Disease. *1989 W. B. Saunders, Philadelphia. Reproduced with permission.*

TABLE 35.4 A comparison of celiac and tropical sprue

Celiac Sprue	Tropical Sprue
Gluten intolerance	Cause unknown
Atrophy of villi	Villi usually present
Diarrhea	Diarrhea
Steatorrhea	Steatorrhea
Treated by dietary restriction of gluten	Treated with folic acid and antibiotics

symptom is watery, nonbloody diarrhea associated with an increased probability of bacteremia, a bacterial invasion of blood. Cytomegalovirus (CMV) is also a pathogen associated with AIDS. The virus commonly affects the colon causing ulceration, colitis, and watery diarrhea.

The term **idiopathic AIDS enteropathy** is used for those cases in which the patient has diarrhea without evidence of enteric infections. It may be that these patients either have a noninfectious cause of diarrhea or that the

enteric: Gk. *enteron,* intestine

pathogen is unidentified. Many viral and possibly proto-zoan pathogens exist that are not detectable by current diagnostic methods.

Prolonged antibiotic therapy may be required for the treatment of enteric pathogens, such as *Salmonella, Shigella,* and *Camphylobacter.* Antibiotics are inef-fective against *Cryptosporidium* and viruses. Symptom-atic treatment, i.e., antidiarrheal drugs and intravenous fluids, is required when the infection is not treatable.

CROHN'S DISEASE (REGIONAL ENTERITIS)

Crohn's disease involves an inflammation of the alimen-tary tract (figure 35.7). A common site of inflammation is the ileum, with or without involvement of the jejunum and the colon. It is estimated that the incidence of this disorder is approximately 2 cases per 100,000 population. The etiology is unknown, although inheritance seems to be a factor. Symptoms may be mild or severe, and typi-cally, there are intermittent periods of remission. Symp-toms include abdominal pain in the right lower quadrant, diarrhea, low-grade fever, and arthralgia. Table 35.5 sum-marizes some possible complications of Crohn's disease.

Treatment of Crohn's disease is symptomatic. Cor-ticosteroids are useful as anti-inflammatory agents, and antibiotics treat the infection of abscesses and fistulas. A high calorie diet with adequate protein is recommended. If there is intestinal obstruction, there should be restric-tion of raw fruits and vegetables. Surgical removal of a segment of the small bowel may be required.

ULCERATIVE COLITIS

Ulcerative colitis is a chronic inflammation of the mucosa of the large intestine (figure 35.8). This term is used when the inflammation extends to an area proximal to the rectum. The inflammation is classified as ulcerative **proc-titis** when it is restricted to the rectum. The etiology is unknown. In some cases, however, there is a family history that suggests genetic factors. Infectious agents or autoim-mune mechanisms are possible, but not established as causes. The incidence of the disease in the United States is approximately 2–7/100,000 population. It occurs more commonly in young people, but older age groups are not exempt. It affects women more frequently than men.

Manifestations of the disease include abdominal cramps, fever, and a bloody diarrhea, which may be mild or profuse. As the disease progresses, there may be an-orexia and weight loss. In severe and chronic disease, there is low serum albumin due to protein loss. Diarrhea leads to electrolyte imbalance as described in chapter 5. Other manifestations are arthritis in the knees or wrists and a

alimentary: L. *alimentarius,* nourishment

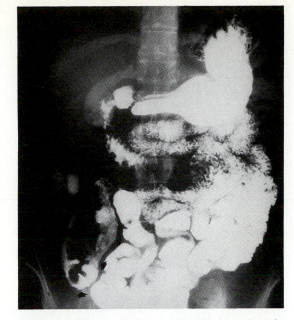

Figure 35.7 An X-ray of abdomen after the ingestion of a barium meal. The stomach and small intestine are filled with barium. There is a change in the pattern (arrows) in the distal portion of the ileum due to Crohn's disease.

TABLE 35.5 Complications of Crohn's disease

Fistula. Abnormal canal leading from an abscess
Mechanical intestinal obstruction
Malabsorption leading to nutritional deficiencies
Bowel stricture. A narrowing of a segment of the intestine
Bowel perforations that develop slowly (small openings)
Increased risk of colon cancer

lesion of the skin called **erythema nodosum.** This lesion usually appears on the legs as a red painful nodule. Ul-cerative colitis may be complicated by massive hemor-rhage, hemorrhoids, and **anal fissures** or cracks.

Treatment for a mild form of the disease is dietary restriction of fiber and antidiarrheal agents to control the diarrhea. More severe cases require cortisone as an anti-inflammatory agent and possibly surgical removal of a segment of the intestine. The disease follows a chronic course, and there is an approximate 10% chance of car-cinoma of the colon.

• • •

It may be difficult to distinguish between ulcerative colitis and Crohn's disease. In general, there is diar-rhea and no rectal bleeding in Crohn's disease. In ulcerative colitis, there is bloody diarrhea.

• • •

fissure: L. *fissura,* cleft

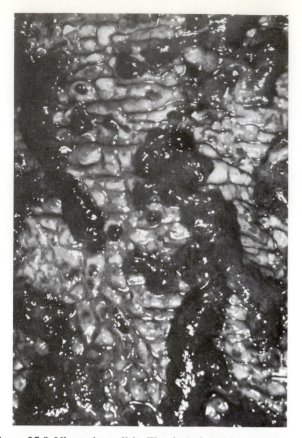

Figure 35.8 Ulcerative colitis. The dark, irregular pattern comprises ulcerations that have in many instances coalesced, leaving virtual islands of residual, paler mucosa. A tendency toward pseudopolyp formation is already evident. *From R. S. Cottran, V. Kumar, and S. L. Robbins:* Robbins Pathologic Basis of Disease. *1989 W. B. Saunders, Philadelphia. Reproduced with permission.*

LACTASE DEFICIENCY

Lactose is a disaccharide made up of glucose and galactose which is found in milk. The enzyme lactase is required to cleave the sugars in order for absorption to occur. Approximately two-thirds of the world's population develops a lactase deficiency after weaning. The group that maintains high lactase levels throughout life are generally of European descent. Lactase is found in the microvilli on the surface of the intestinal lining, and secondary lactase deficiency develops when there is injury to the intestinal mucosa.

Fermentation of undigested lactose in the colon leads to production of both gas and acids, which are absorbed. Large amounts of lactose result in the presence of unfermented lactose in the colon. This osmotically draws water into the intestine and causes diarrhea. Dietary restriction of lactose solves the problem.

PARALYTIC ILEUS (ADYNAMIC ILEUS)

Wavelike contractions of the small intestine called peristalsis cause chyme to move toward the anus. Para-

ileus: Gk. *eileos,* intestinal obstruction

TABLE 35.6 Some factors or agents of pathogenesis of colorectal cancer

High dietary fat	Genetic predisposition
Fecapentaenes	Risk after age 40
Charbroiled meat and fish	Ulcerative colitis

TABLE 35.7 Definition of terms used in association with colorectal tumor

Neoplasm. Abnormal growth or tumor
Polyp. Lesion that develops on the luminal surface of gastrointestinal tract and protrudes into lumen
Pedunculated polyp. On a stalk
Sessile polyp. Flat and without a stalk
Adenoma. A type of polyp; benign tumor that is glandularlike, either in origin or cell arrangement
Adenocarcinoma of large intestine. Malignant adenoma; most common type of cancer in the large intestine
Squamous cell carcinoma of the large intestine. Second most common type of malignant tumor of the large intestine
Villous adenoma. Spongy growths larger than adenomatous polyps; high association with malignant transformation

lytic ileus is an impairment of peristalsis that may cause intestinal obstruction. There are various causes including (1) surgery, (2) peritonitis, (3) vertebral fractures, (4) anticholinergic drugs, and (5) electrolyte imbalances. The symptoms are continuous abdominal pain, vomiting, constipation, and abdominal distention. Treatment involves restriction of oral intake or gastrointestinal suction in prolonged cases.

COLORECTAL CANCER

Colorectal cancer, cancer of the colon and rectum, is the second most common cancer in the United States, excluding skin cancer. Worldwide, it is the third most common cancer in males and the fourth most common in females. The incidence is highest in North America, Australia, and New Zealand. Approximately 6% of the people in the United States will develop colorectal cancer.

ETIOLOGY

The causes of colorectal cancer are not known, but there is circumstantial and experimental evidence implicating various factors and agents for a role in pathogenesis (table 35.6). Studies suggest that high amounts of fat in the diet predispose to colorectal cancer. Increased fiber intake is associated with a low incidence of the disease, and it is theorized that increased stool bulk dilutes cancer causing substances in the colon and promotes their removal by decreasing transit time. Studies in experimental animals show that cellulose decreases the level of bacterial enzymes in the colon, and it is theorized that this may decrease the activation of carcinogens. In many cases, genetic predisposition plays a role. A group of compounds synthesized by bacteria in the colon, fecapentaenes (fē-kuh-pēn'tuh-enes), may play a role as carcinogens.

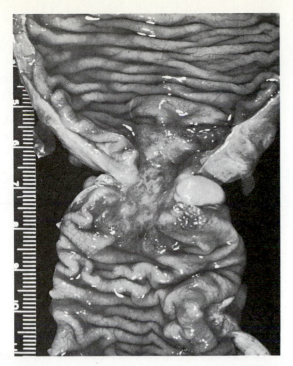

Figure 35.9 Carcinoma of left colon that has completely encircled the lumen. Dilation of proximal bowel lumen is evident.
From R. S. Cottran, V. Kumar, and S. L. Robbins: Robbins Pathologic Basis of Disease. *1989 W. B. Saunders, Philadelphia. Reproduced with permission.*

Polyps of the colon (definitions, table 35.7) have a significant potential for becoming malignant. There is a strong correlation between adenomas, a type of polyp, and adenocarcinoma of the colon. The probability of cancer increases with increased size of the adenoma. Polyps are frequently asymptomatic, but they may cause bleeding and should be removed because of their association with colon cancer.

MANIFESTATIONS

Adenocarcinoma of the colon and rectum may grow for as long as 5 years without causing symptoms. As the cancer develops, the growth invades the muscularis mucosa as well as blood and lymphatic vessels (figure 35.9). Although the liver is the most common site of metastasis, the cancer may spread to other locations, such as distant lymph nodes, lungs, and vertebrae. General symptoms include malaise, anorexia, and weight loss. Specific symptoms are changes in bowel habits, bleeding, an abdominal mass, or signs of liver involvement.

SCREENING AND DIAGNOSTIC TESTS

A test for fecal occult blood, blood not grossly visible, is a means of screening asymptomatic individuals for colorectal cancer. Colorectal cancers and adenomas bleed intermittently. Sampling more than one stool specimen increases the probability of a positive test in the presence of a malignancy.

Proctosigmoidoscopy (prok″to-sig″moi-dos′ko-pe) is a procedure that allows examination of both the rectum and sigmoid colon. A rigid proctosigmoidoscope is a tubular instrument that has a distal light and magnifying lens for viewing. Flexible instruments allow visualization of a greater length of the bowel and are preferred for examining individuals with rectal bleeding or for screening asymptomatic individuals.

A barium enema is a diagnostic procedure for evaluating various disorders including diverticulitis, polyps, and malignancy. Barium is instilled through a rectal tube, while the flow into the large intestine is observed by means of a fluoroscope, and X-rays are taken.

Colonoscopy involves visual examination of the mucosal lining of the entire length of large intestine. An instrument is inserted through the anus to the ileocecal valve that separates the ileum and the large intestine. Surveillance for malignancy or rectal bleeding are reasons for performing a colonoscopy.

Serum **carcinoembryonic antigen** (CEA) is a nonspecific test useful in follow-up of patients with colorectal cancer. This glycoprotein was initially isolated from adenocarcinoma and from fetal gastrointestinal tissue. Serum CEA levels are elevated in most cases of malignancy involving the colon, pancreas, stomach, lung, and breast. The CEA test is not reliable as a screening test for early diagnosis of colorectal cancer, but it provides valuable information on the efficacy of treatment and on possible recurrence of the disease.

• • •

Nonmalignant conditions in which serum CEA may be elevated include cigarette smoking, inflammatory bowel disease, hepatitis, alcoholic cirrhosis, and chronic pulmonary disease.

• • •

TREATMENT

The treatment of choice in most cases of colorectal cancer is surgical removal of the neoplasm. There is a 42% overall 10-year survival rate after surgery for colorectal cancer.

• • •

Villous adenomas occasionally secrete large amounts of mucus and electrolytes, and this may lead to hypokalemia and hypoalbuminemia.

• • •

proctosigmoidoscopy: Gk. *proktos*, anus; Gk. *sigmoeides*, crescent shaped

Diverticulosis is a condition in which there are outpouchings of the intestinal mucosa. It is common in Western developed countries with an increased incidence after the age of 40. Contributing factors to the formation of diverticula are both increased pressure within the lumen of the colon and areas of weakness in the intestinal wall. Uncomplicated diverticulosis may be asymptomatic, or there may be left abdominal pain. Treatment involves measures to both relieve symptoms and to avoid complications. Diverticulitis is an inflammation of diverticula. Symptoms are abdominal and back pain and fever. Treatment involves bed rest, administration of antibiotics, and restriction of ingested food.

Tropical sprue is a chronic condition of unknown etiology acquired in the tropics. It is associated with abnormalities of the intestinal mucosa involving thickening and fusion of villi. Malabsorption leads to vitamin and other nutritional deficiencies. The xylose tolerance test provides a measure of intestinal absorptive capacity. Characteristic symptoms include diarrhea, malaise, fever, and steatorrhea. Tetracycline and folic acid are effective in treating tropical sprue.

Celiac sprue involves a gluten-induced lesion that leads to intestinal malabsorption. The intestinal lesion involves destruction of villi and elongation of intestinal crypts. Possible symptoms include diarrhea, osteopenic bone disease, and anemia. Treatment for celiac sprue is dietary restriction of gluten-containing cereal grains.

Diarrhea is a common symptom in AIDS and is usually associated with at least one identifiable pathogen. Crohn's disease is an inflammation of any part of the alimentary tract with frequent involvement of the ileum. The etiology is unknown and symptoms include abdominal pain and diarrhea. The disorder is treated symptomatically. Ulcerative colitis is a chronic inflammation of the mucosa of the large intestine due to unknown factors. Bloody diarrhea, abdominal cramps, and fever are characteristic of the disease. Modalities of treatment include dietary restriction of fiber, antidiarrheal agents, anti-inflammatory agents, and surgical resection.

Lactase deficiency leads to diarrhea, which can be controlled by dietary restriction of lactose. Impaired peristalsis called paralytic ileus is often caused by surgery and leads to intestinal obstruction.

Colorectal cancer is a common type of malignancy. High dietary fat may be a contributing factor, and genetic predisposition also may play a role. Over a period of time, adenocarcinoma invades the muscularis mucosa and lymphatic vessels and metastasizes to distant sites. Symptoms include anorexia, weight loss, bleeding, and a change in bowel habits. The treatment of choice is surgical removal of the neoplasm.

1. What are the four basic layers of the gastrointestinal tract, listed in order and starting from the luminal side?
2. In general, what are the relationships between both wall tension and radius and intraluminal pressure in the large intestine?
3. Is diverticulosis asymptomatic?
4. What is the most common symptom of diverticulosis?
5. What is the main approach to treatment of diverticulosis?
6. What is the difference between diverticulosis and diverticulitis?
7. List three complications of diverticulitis.
8. What are two major manifestations of diverticulitis?
9. What is a possible cause of tropical sprue?
10. What is the underlying cause of manifestations of tropical sprue?
11. Why is xylose used in a test for sprue?
12. What is an effective treatment for tropical sprue?
13. In what way may bones be affected by celiac sprue?
14. Why may night blindness occur in celiac sprue?
15. What is the main cause of anemia in celiac sprue?
16. Intestinal villi are destroyed in which type of sprue?
17. What type of sprue is characterized by gluten intolerance?
18. Bloody diarrhea is characteristic for (Crohn's disease or ulcerative colitis).
19. What food intolerance is caused by lactase deficiency?
20. What is a major cause of paralytic ileus?
21. What is a major dietary factor in colorectal cancer?
22. List two dietary substances that appear to protect against colorectal cancer.
23. Define the term adenoma.
24. Define the term adenocarcinoma.
25. What is the most common site of metastasis for colorectal adenocarcinoma?

Borcich, A., and D. P. Kotler. 1990. Combating chronic diarrhea in AIDS patients. *Physician Assistant* 14:101–14.

Clouse, R., and D. Alpers. 1988. Irritable bowel syndrome: A systematic approach to diagnosis and management. *Physician Assistant* 12:43–53.

Collins, R., M. Feldman, and J. Fordtran. 1987. Colon cancer, dysplasia, and surveillance in patients with ulcerative colitis. *New England Journal of Medicine* 316:1654–58.

Isenberg, J. et al. 1987. Impaired proximal duodenal mucosal bicarbonate secretion in patients with duodenal ulcer. *New England Journal of Medicine* 316:374–78.

Ogle, B. 1989. Colorectal cancer: Reducing mortality. *Physician Assistant* 13:58.

Sirlin, S., K. Benkov, and N. LeLeiko. 1988. Inflammatory bowel disease. *Physician Assistant* 12:24–36.

Tornberg, S. et al. 1986. Risks of cancer of the colon and rectum in relation to serum cholesterol and beta-lipoprotein. *New England Journal of Medicine* 315:1629–38.

Chapter 36

Disorders of Liver, Gallbladder, and Pancreas

Bile, produced by the liver, and pancreatic juices enter the duodenum to promote digestion. The gallbladder is a sac for the storage and concentration of bile. The liver not only produces bile, but has many complex metabolic functions as well. This chapter emphasizes inflammatory and degenerative diseases of the liver and includes selected disorders of the gallbladder and pancreas. The following review of liver and gallbladder anatomy introduces a discussion of the effects of inflammation of the liver.

ANATOMY OF LIVER AND GALLBLADDER

The liver is the largest organ of the body. It lies directly beneath the diaphragm on the right side and extends into the left side. The gallbladder is a pear-shaped organ that stores and concentrates bile from the liver. It lies against the posterior surface of the liver and usually projects a short distance beyond the inferior margin of the liver (figure 36.1).

HEPATIC DUCTS

Bile, a viscous greenish fluid produced by the liver, is transported by way of ducts both to and from the gallbladder as shown in figure 36.2. Right and left hepatic ducts emerge from the undersurface of the liver and join to form the common hepatic duct. The cystic duct provides a channel between the gallbladder and the common hepatic duct. The word cyst means bladder, thus the term cystic duct. Beyond the junction of the cystic and common hepatic ducts, the common bile duct opens into the duodenum.

MICROSCOPIC ANATOMY OF THE LIVER

The liver is made up of many lobules, hexagonal cylinders approximately 2 mm high and 1 mm in diameter. There is a branch of the hepatic vein in the center of each lobule

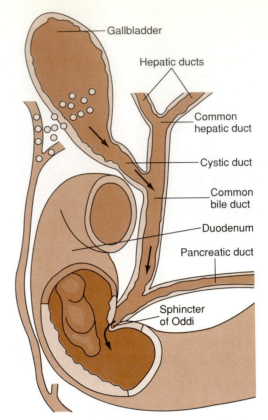

Figure 36.2 The flow of bile from the liver by way of hepatic ducts to the gallbladder and duodenum.

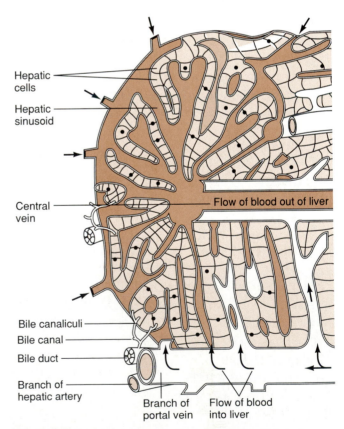

Figure 36.3 A cross section of a hepatic lobule which is the functional unit of the liver.

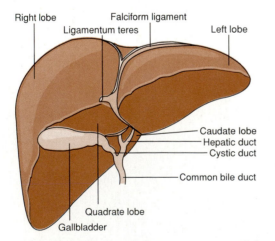

Figure 36.1 An anterior view of the gross structure of the liver.

with hepatic cells arranged around the vein. There are branches of the hepatic artery, portal vein and hepatic bile duct around the outside of each lobule. Tiny channels called bile canaliculi (kan-"ah-lik'u-li) form networks around each cell. These features of hepatic organization are shown in figures 36.3 and 36.4.

HEPATIC BLOOD SUPPLY

Capillary networks associated with the stomach, intestine, pancreas, and spleen join to form the portal vein which enters the liver and ultimately provides branches around each liver lobule (figure 36.5). From these branches, blood flows through hepatic sinusoids to each central vein. Central veins join to form the right and left hepatic veins,

Figure 36.4 (*a*) Light micrograph of portion of liver lobule showing radial organization of hepatocytes and sinusoids around the central vein. CV, central vein; H, hepatocyte; S, sinusoid; arrows, points of entry of sinusoids into central vein; bile canaliculus, ×300. (*b*) and (*c*) Matched pair of light and electron micrographs of serial sections taken through portal triad of monkey liver lobule. BD, bile duct; H, hepatocyte; HA, hepatic artery; PV, portal vein; S, sinusoid. (*b*) ×600; (*c*) ×1,000.

From D. T. Moran and J. C. Rowley: Visual Histology. 1988 Lea & Febiger, Philadelphia. Reproduced with permission.

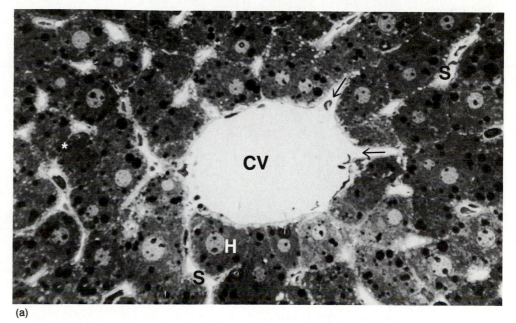

(a)

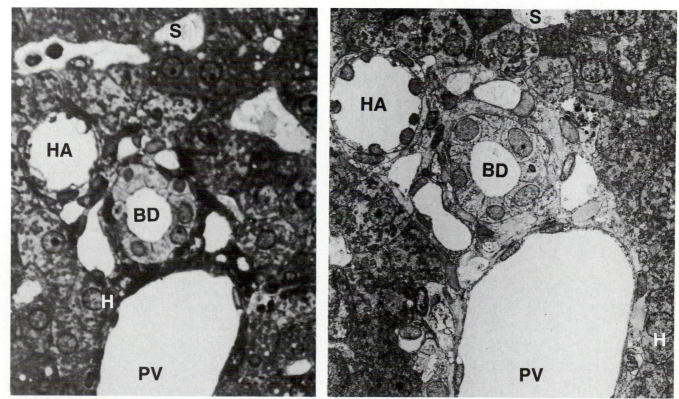

(b)

(c)

Disorders of Liver, Gallbladder, and Pancreas

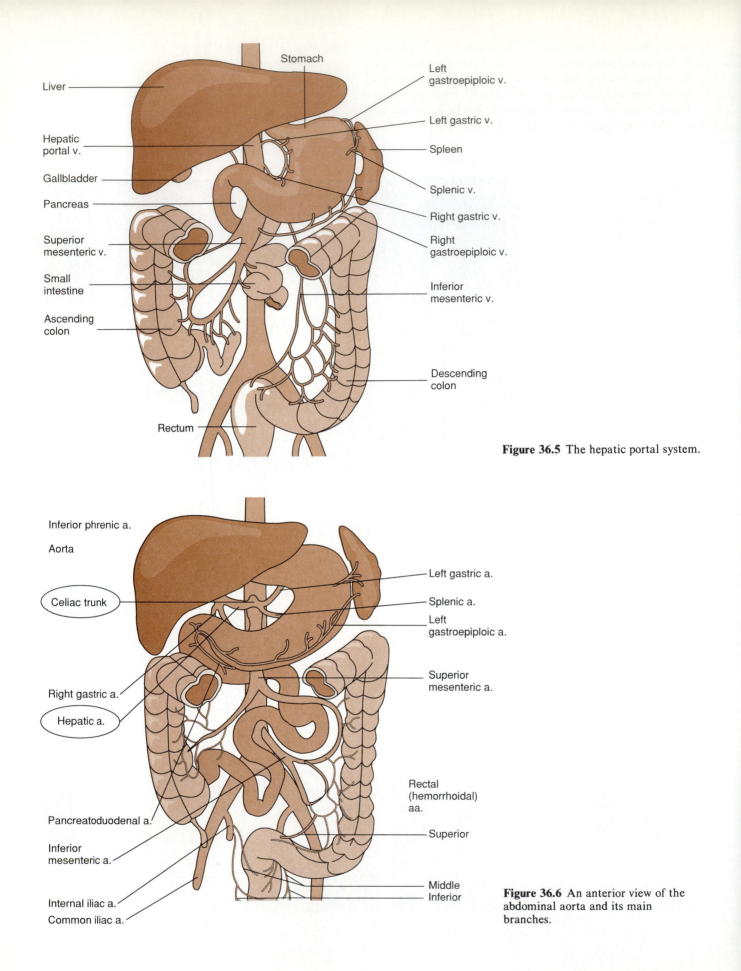

Liver

Stomach

Left gastroepiploic v.

Hepatic portal v.

Left gastric v.

Gallbladder

Spleen

Pancreas

Splenic v.

Superior mesenteric v.

Right gastric v.

Small intestine

Right gastroepiploic v.

Ascending colon

Inferior mesenteric v.

Descending colon

Rectum

Figure 36.5 The hepatic portal system.

Inferior phrenic a.

Aorta

Left gastric a.

Celiac trunk

Splenic a.

Left gastroepiploic a.

Right gastric a.

Superior mesenteric a.

Hepatic a.

Rectal (hemorrhoidal) aa.

Pancreatoduodenal a.

Superior

Inferior mesenteric a.

Middle

Internal iliac a.

Inferior

Common iliac a.

Figure 36.6 An anterior view of the abdominal aorta and its main branches.

which empty into the inferior vena cava. The celiac artery, which is a branch of the abdominal aorta, gives rise to the hepatic artery (figure 36.6). The hepatic artery goes to the liver and provides branches around each liver lobule. These branches empty into hepatic sinusoids to mix with venous blood and flow into a central vein.

SIGNIFICANT OBSERVATIONS

There is particular significance to two aspects of the anatomical features just described. (1) Bile is formed within liver cells and easily enters the surrounding canaliculi to progress to branches of the hepatic bile duct. (2) Hepatic sinusoids are the equivalent of a capillary network, so the portal vein is uniquely situated between two capillary systems. This means that obstruction to blood flow within the liver results in stasis of flow and increased pressure in the portal vein due to visceral capillary resistance to backflow. There are serious consequences associated with increased portal vein pressure, which will be discussed later in this chapter.

BILE FORMATION

Bile is produced by liver cells and has three main constituents:

1. Bile salts that emulsify fats in the small intestine

2. Cholesterol from dietary sources and also synthesized by the liver
3. Bilirubin, the main pigment in bile

The accumulation of bilirubin in body tissues causes jaundice and is significant in many liver diseases.

BILIRUBIN METABOLISM

Bilirubin is the main pigment in bile and is orange-red. The major source of bilirubin is hemoglobin released from aging red blood cells. The events involved in the metabolism and transport of bilirubin are listed in table 36.1. Heme is first converted to biliverdin, a green pigment, and this, in turn, is metabolized to bilirubin (figure 36.7). These reactions occur in the spleen, liver, and bone marrow. Bilirubin circulates in the plasma and is taken up by hepatic cells where it is converted to a conjugated form. Conjugated bilirubin passes into the bile and ultimately is converted to urobilinogens and excreted (figure 36.8). Table 36.2 summarizes the characteristics of biliverdin and bilirubin.

• • •

The changing color of a bruise is visible evidence of the reactions by which heme is converted into bilirubin.

• • •

TABLE 36.1 Steps in the metabolism
and transport of bilirubin

1. Hemoglobin released from red blood cells	8. Conjugated bilirubin passes into bile in canaliculi
2. Hemoglobin broken down to heme + globin (protein)	9. Bile collects in ducts and moves to the gallbladder and into the intestine
3. Heme converted to biliverdin	10. In the large intestine, bilirubin is converted to a series of products, collectively called urobilinogens
4. Biliverdin converted to unconjugated bilirubin	
5. Unconjugated bilirubin released into plasma; bound to albumin	11. Approximately 80% of urobilinogens excreted
6. Uptake of unconjugated bilirubin by hepatic cells	12. Urobilinogen also reabsorbed; re-excreted in bile or appears in urine
7. Bilirubin conjugated to glucuronic acid	

Heme

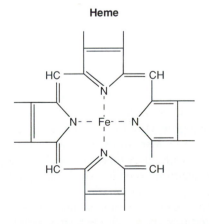

Biliverdin($C_{33}H_{34}O_6N_4$)

Bilirubin($C_{33}H_{36}O_6N_4$)

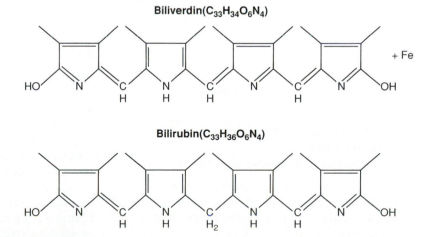

Figure 36.7 When a hemoglobin molecule is decomposed, the heme portions are broken down into iron (Fe) and biliverdin. Most of the biliverdin is then converted to bilirubin.

Disorders of Liver, Gallbladder, and Pancreas

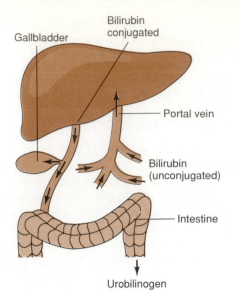

Figure 36.8 Unconjugated bilirubin is carried to the liver where the process of conjugation occurs inside of liver cells. Conjugated bilirubin appears in bile and goes to the intestine. In the large intestine, bilirubin is converted to urobilinogen, which for the most part, is eliminated from the body.

TABLE 36.2 A summary of characteristics of biliverdin, conjugated bilirubin, and unconjugated bilirubin

Biliverdin. Green pigment; not normally in plasma; water soluble; nontoxic

Unconjugated bilirubin. Orange-red pigment; lipid soluble; readily diffuses across cell membranes; circulates in plasma; bound to albumin

Conjugated bilirubin. Formed in liver cells; water soluble; only the conjugated form appears in urine; can reflux into plasma; accounts for less than 5% of total bilirubin in the blood

JAUNDICE

Jaundice refers to an abnormal yellow color of the sclera of the eyes, mucous membranes, and skin caused by an increased level of bilirubin in the blood. The condition may be preceded or accompanied by dark urine.

CAUSES

Jaundice may be categorized on the basis of three major causes:

1. Abnormal metabolism of bilirubin
2. Hepatocellular disease
3. Obstruction to outflow of bilirubin (obstructive)

Overproduction of bilirubin due to hemolysis is an example of abnormal metabolism. Massive destruction of red blood cells may occur as the result of an incompatible blood transfusion. Jaundice occurs in sickle cell anemia and is due to hemolysis of fragile red blood cells.

Liver diseases accompanied by jaundice include hepatitis and cirrhosis (sir-ro′sis). Liver damage may result

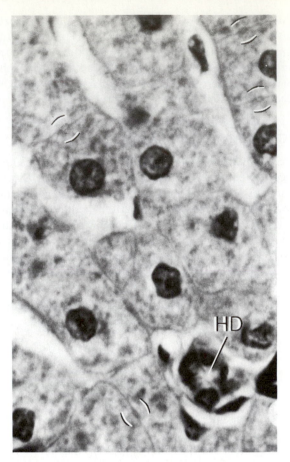

Figure 36.9 Light micrograph of liver cells and a small bile duct (HD). This duct is still surrounded by hepatocytes and is called a canal of Hering. The brackets mark cross-sectioned bile canaliculi. ×1,240.
From M. H. Ross, E. J. Reith, and L. J. Romrell: Histology: A Text and Atlas. *1989 Williams & Wilkins, Baltimore. Reproduced with permission.*

in impaired uptake of bilirubin by liver cells, impaired conjugation, and impaired transport into bile canaliculi (figure 36.9).

Obstruction to bile flow may be intrahepatic, i.e., due to liver damage (figure 36.10). Inflammation or gallstones associated either with bile ducts or the gallbladder may obstruct flow of bile as well (figure 36.11).

KERNICTERUS

Bilirubin has a particular affinity for elastic connective tissue; but in severe jaundice in adults, most body tissues except the central nervous system become stained. In infants, if there is an extreme elevation of bilirubin (20 mg/dl or greater) the gray matter of the central nervous system, as well as other tissues, may take up bilirubin and may cause death. The condition is called kernicterus and is treated by phototherapy, i.e., illumination of the infant by a strong white or blue light. This causes unconjugated bilirubin to be converted to a water-soluble form which is rapidly excreted. The discussion that follows deals with hepatitis, cirrhosis, and cholelithiasis, all of which may lead to jaundice.

kernicterus: Gk. *ikteros,* jaundice

420

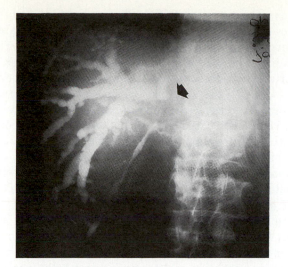

Figure 36.10 This is an X-ray of a percutaneous transhepatic cholangiogram. A long needle is inserted through the right lateral aspect of the abdomen through the length of the liver. Contrast material is injected as the needle is slowly withdrawn. The hepatic system is filled with contrast media. This film demonstrates obstruction, with severely dilated ducts.

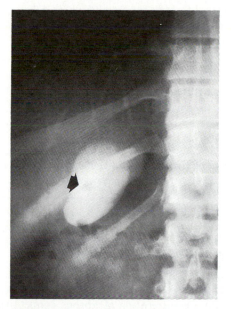

Figure 36.11 This is an X-ray of a 32-year-old female following the ingestion of contrast media. A gallstone is in the area of the arrow.

HEPATITIS

Hepatitis is an inflammation of the liver, and it may be viral or toxic, i.e., caused by various drugs.

ACUTE VIRAL HEPATITIS

The causes of acute viral hepatitis include hepatitis A virus, B virus, and hepatitis C (formerly non-A, non-B) virus. Characteristics of these viral infections are summarized in table 36.3. There is also a hepatitis D virus,

TABLE 36.3 A summary of characteristics of hepatitis type A, hepatitis type B, and hepatitis type C

Type A	Type B	Type C
Fecal-oral transmission	Blood infectious	Blood infectious
Mainly children, young adults	Adults	Adults
Acute, self-limited	Acute, self-limited or chronic	Acute, self-limited or chronic
No chronic carriers	Chronic carriers	Chronic carriers

TABLE 36.4 Manifestations of acute viral hepatitis

Malaise	Chills	Muscle and joint pain
Anorexia	Fever	Jaundice
Nausea	Sore throat	Abdominal pain
Vomiting	Cough	Dark urine
Fatigue	Runny nose	Clay-colored stools

which does not cause disease by itself but requires another agent, such as a hepatitis B virus antigen to invade cells.

Manifestations and Course of Disease

Manifestations of acute viral hepatitis are listed in table 36.4. Hepatomegaly and splenomegaly may be observed, and jaundice is common. The incubation period for hepatitis A is an average of 30 days, while the incubation period for hepatitis B and hepatitis C averages 60–90 days and 50 days, respectively. The patients with hepatitis A recover within 1–2 months, and most recover from types B and C within 3–4 months. A small percentage of cases of hepatitis B progress to chronic hepatitis, while 50% of the cases of hepatitis C progress to chronic hepatitis. About 20% of the patients with chronic hepatitis C develop cirrhosis, which is associated with increased risk of carcinoma of the liver.

Hepatitis Virus Antigens/Antibodies

Table 36.5 lists the antigens and antibodies associated with the hepatitis viruses. There are serological tests to identify the types of viral hepatitis. The test for HCV infection detects antibodies against the virus. Because HCV antibodies may not appear in the blood for 6–12 months, a negative test does not rule out hepatitis C infection. The test also shows false-positive results in screening blood donors for hepatitis C.

Treatment

There is no specific treatment that will cure viral hepatitis. Therapy is symptomatic and includes bed rest and rehydration. Chronic hepatitis caused by hepatitis C virus may be treated by subcutaneous injections of interferon alfa-2b, three times a week for 6 months. The treatment results in improvement in 50% of the patients. When

TABLE 36.5 Antigens and antibodies
of the hepatitis viruses

Hepatitis A Virus (HAV)

Antigen
Hepatitis A virus = hepatitis A antigen (HAAg). Found in
    liver, bile, stools, and blood during late incubation period

Antibodies
IgM type anti-HA appears in the serum immediately
IgG type anti-HA appears during convalescence and persists
    indefinitely; provides permanent immunity

Hepatitis B Virus (HBV)

Antigens (3)
Hepatitis B surface antigen (HBsAg)
Hepatitis B core antigen (HBcAg)
e antigen (HBeAg). Appears early in the disease process;
    found only in HBsAg-positive serum; persists in serum in
    chronic hepatitis

Antibodies
Anti-HBs; anti-HBc; anti-HBe
Anti-HBs appears in serum after recovery and provides
    immunity;
Chronic HBsAg carriers are reservoirs for infection

Hepatitis D Virus (HDV)

Antigen
Hepatitis D virus is the hepatitis D antigen (HDAg). Infection
    only occurs with HBV; found in nuclei of liver cells and
    occasionally in serum

Antibodies
Anti-HD

Hepatitis C Virus (HCV)

Antigen
Hepatitis C virus is the hepatitis C antigen (HCAg)

Antibodies
Anti-HC; may take 6–12 months to appear

treatment is discontinued, improvement is maintained in only half the patients. Side effects of the drug include chills, fever, headache, and muscle aches. There may also be fatigue, anorexia, and difficulty in concentrating. Less frequently, bone marrow suppression and psychological changes occur.

• • •

The hepatitis A virus is toxic to liver cells. Hepatitis B virus is not toxic but elicits a cell-mediated immune attack against liver cell membranes. A healthy carrier state is the result of failure of an immune response to HBV. An inadequate response can result in chronic hepatitis.

• • •

Prevention

Type A hepatitis is transmitted almost exclusively by the fecal-oral route, i.e., by fecally contaminated food and water, ingestion of raw shellfish, or inadequate personal hygiene. Hepatitis type B is transmitted by blood, and the modes of transmission include blood transfusion, shared needles by drug users, and accidental needle sticks. The virus has been identified in all body fluids, but there is uncertainty about the infectivity of these fluids. Infection may occur as the result of sexual contact. Maternal transfer of the virus to a child during delivery occurs as well. Hepatitis C virus causes most of the cases of post-transfusion (non-A, non-B) hepatitis and is a major cause of hepatitis among intravenous drug users. HCV is the cause of some cases of acute hepatitis not associated with parenteral transmission. Heterosexual transmission of HCV has not been established with certainty.

Immune serum globulin (gamma globulin), the antibody fraction obtained from pooled human plasma, may be administered prophylactically to individuals at risk of exposure to hepatitis A. Vaccines against both hepatitis A and B are available, which provide active immunity against the diseases. Recipients of blood transfusions are protected from viral hepatitis by screening tests of blood donors.

CHRONIC ACTIVE HEPATITIS

Chronic active hepatitis is a prolonged inflammation of the liver of at least 6 months duration. The condition may be asymptomatic, mild, or severe and may be idiopathic or initiated by viruses or drugs. It may lead to liver failure or cirrhosis. Patients with the severe idiopathic disease may be treated with cortisone, but there is no established therapy for the viral induced disease.

• • •

Any form of chronic liver disease seems to carry the increased risk of hepatic carcinoma.

• • •

CIRRHOSIS

Cirrhosis is defined as a condition in which necrosis of the liver leads to a proliferation of fibrous connective tissue, a process called fibrosis (figure 36.12). This condition is accompanied by the formation of regenerative nodules and progressive intrahepatic obstruction. Cirrhosis may be the end result of alcohol abuse, sometimes called **Laennec's cirrhosis,** which is the most common type. Early hepatic changes in alcoholism involve fat accumulations due to improper metabolism of fatty acids and increased synthesis of triglycerides.

Postnecrotic cirrhosis may be a sequel to hepatitis B, or hepatitis C, or to drug induced hepatitis. The liver ultimately becomes shrunken and composed of nodules. **Biliary cirrhosis** is the outcome of prolonged obstruction to bile flow and **cardiac cirrhosis** is due to right-sided congestive heart failure. In cardiac cirrhosis, liver damage

cirrhosis: Gk. *kirrhos,* tawny, brownish yellow; Gk. *osis,* condition

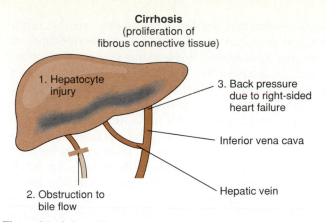

Cirrhosis
(proliferation of fibrous connective tissue)

1. Hepatocyte injury

2. Obstruction to bile flow

3. Back pressure due to right-sided heart failure

Inferior vena cava

Hepatic vein

Figure 36.12 Possible causes of cirrhosis.

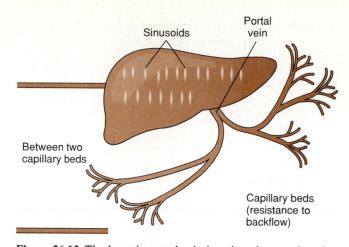

Sinusoids

Portal vein

Between two capillary beds

Capillary beds (resistance to backflow)

Figure 36.13 The hepatic portal vein is unique inasmuch as it is between two capillary beds. If there is obstruction in the liver, the result is increased portal pressure and back-pressure in the capillary beds behind it.

results from elevated venous pressures involving the inferior vena cava and hepatic veins, leading ultimately to congestion of the liver. The consequences of cirrhosis are discussed in the following paragraphs.

• • •

Wilson's disease is a rare inherited disorder in which copper accumulates first in the liver and then in other organs. Symptoms vary with organ involvement, but hepatitis and cirrhosis frequently occur.

• • •

PORTAL HYPERTENSION

The portal system, shown in figure 36.13, includes (1) capillaries of the stomach, spleen, and intestines, (2) the portal vein formed by the superior and inferior mesenteric veins and the splenic vein, and (3) the sinusoids of the liver. The term portal hypertension refers to increased hydrostatic pressure within the portal vein. Liver disease, associated with obstruction to blood flow through the liver, is the most common cause of portal hypertension. Congestive heart failure and portal vein thrombosis are also possible causes.

• • •

There are several factors in increased hepatic resistance to blood flow in cirrhosis. Regenerative nodules compress vessels, there is damage to sinusoids, and branches of the portal vein are invaded by fibrous tissue.

• • •

Ascites

Ascites (ah-si'tēz) refers to an accumulation of fluid in the peritoneal cavity that may be imperceptible or massive (figure 36.14). The mechanisms for this accumulation are basically those discussed in relation to fluid imbalance in chapter 2. Increased resistance to blood flow in the liver leads to (1) increased hydrostatic pressure, (2) vascular

Figure 36.14 Portal hypertension leads to ascites, i.e., an accumulation of fluid in the peritoneal cavity.

fluid loss, (3) increased rate of hepatic lymph formation, and (4) escape of lymph into the peritoneal cavity. In addition, there is increased permeability of the sinusoidal endothelium, so the fluid lost is rich in protein. The protein in lymph/ascites fluid causes an elevated oncotic pressure, which draws out intravascular fluid. Although the mechanisms are poorly understood, renal retention of sodium and water is an additional factor.

Bacterial peritonitis occurs in approximately 10% of patients with ascites. The infecting organisms are usually enteric, and there are probably several factors that lead to the passage of bacteria through the intestinal wall. An edematous bowel wall may play a role, or the failure of a damaged liver to remove bacteria from the blood may be a factor. A second complication of ascites is umbilical hernias due to intra-abdominal pressure. This involves protrusion of abdominal contents through the umbilical ring. Other complications of ascites include respiratory distress, abdominal pain, gastroesophageal reflux, and inguinal (ing'gwĭ-nal) hernia in the groin region.

Paracentesis (par"ah-sen-te'sis) is a procedure by which a needle is used to puncture the abdominal wall to remove ascites fluid. This is a temporary measure to relieve pain or respiratory distress. The restriction of sodium and administration of diuretics are major aspects of treating ascites.

Disorders of Liver, Gallbladder, and Pancreas

TABLE 36.6 Factors in formation, complications, and treatment of cirrhotic ascites

Factors in Formation

Portal hypertension. Caused by resistance to flow in central veins and branches of hepatic vein; result is increased hydrostatic pressure

Endothelium of sinusoids is highly permeable

Renal retention of sodium and water

Complications

Bacterial peritonitis

Umbilical hernia

Pressure against diaphragm

Treatment

Sodium restriction

Diuretics

Paracentesis

Surgical procedure to direct venous blood from the liver to the inferior vena cava

---

Table 36.6 summarizes the etiology, complications, and treatment of cirrhotic ascites. Other possible causes of ascites include neoplasms, congestive heart failure, and tuberculosis.

Esophageal Varices

The term varices (vār′i-sēz) refers to dilated blood vessels. Increased pressure within the portal vein causes congestion in blood vessels that supply the portal vein. As a result, there is a tendency for both esophageal and gastric varices to develop and rupture. This is commonly manifested by hematemesis, vomiting blood. There is an associated mortality rate of 30–60%, and immediate restoration of volume is imperative in a bleeding episode. Bleeding may be controlled by vasopressin, a potent vasoconstrictor, or by balloon tamponade. The latter involves inserting a balloon-tipped tube and inflating the balloon either in the esophageal or gastric region.

Hypersplenism

Portal hypertension causes congestion of blood in the spleen with subsequent hyperactivity in the destruction of blood cells. The result is decreased numbers of red blood cells, white blood cells, and platelets.

Other Effects of Portal Hypertension

The stasis of blood flow in portal hypertension leads to dilation of abdominal veins with a ring of varicosities around the umbilicus. The term used to describe this condition is caput medusae (kap′ut medu′sae), which means the head of Medusa. Dilation of rectal veins causes internal hemorrhoids to develop.

HEPATIC ENCEPHALOPATHY

Hepatic encephalopathy (en-sef″ah-lop′ah-the) is a disorder of cerebral and neuromuscular function associated with liver disease. The manifestations of the disorder include restlessness, euphoria, irritability, apathy, confusion, disorientation, somnolence, coma, incoordination, and impaired handwriting. Toxic substances accumulate and cause encephalopathy. Blood ammonia levels are frequently elevated in cirrhosis. Ammonia is a product of protein metabolism and normally is converted to urea by the liver. Other metabolic abnormalities include increased levels of glutamine, fatty acids, and mercaptans. Mercaptans are produced in the intestinal tract and are normally removed by the liver. The sweetish, musty odor on the breath of encephalopathic patients is probably due to mercaptans. This characteristic odor is called fetor hepaticus or hepatic fetor. It is possible that encephalopathy is the cumulative effect of accumulated toxins associated with hypoxia, hypovolemia, and other abnormalities. Dietary restriction of protein is essential in the treatment of hepatic encephalopathy.

• • •

Lactulose (cephulac) is a nonabsorbable synthetic disaccharide that acidifies colon contents and causes retention of ammonium ions. The result is decreased ammonia absorption.

• • •

ABNORMAL SEX HORMONE METABOLISM

Liver cell damage results in abnormal metabolism of male and female sex hormones. Manifestations of these abnormalities include testicular atrophy, amenorrhea, and the development of breasts, gynecomastia (jin″ē-ko-mas′te-ah), in males. It is probable that sex hormone abnormalities are also responsible for loss of axillary and pubic hair, palmar erythema, spider nevi, and **telangiectases.** Telangiectases are small red tumors of the skin formed by dilated capillaries. In cirrhosis, these tumors appear on exposed areas of the skin. A spider nevus is a type of telangiectases consisting of a central red dot with vessels radiating around it. It is usually found on the upper half of the body. Palmar erythema is a mottled redness of the palms of the hands.

Table 36.7 provides an overview of the effects of portal hypertension and hepatocellular failure in cirrhosis.

PROGNOSIS

The 5-year survival rate in advanced cirrhosis is about 35%. Hematemesis, jaundice, and ascites indicate a poor prognosis.

HEPATOMEGALY AND AIDS

Hepatomegaly is a frequent finding in patients who have AIDS (chapter 29). There may or may not be jaundice, although one or more liver function tests are usually ab-

telangiectasis, pl. ses: Gk. *telos,* end; Gk. *angeion,* vessel; Gk. *ektasis,* extension

TABLE 36.7 Overview of the effects of portal hypertension and hepatocellular failure in cirrhosis

Portal Hypertension

Ascites	Hypersplenism
Esophageal varices	Internal hemorrhoids
Caput medusae	Edema

Hepatocellular Failure

Failure to synthesize clotting factors. Nosebleeds, gingival bleeding

Hepatic encephalopathy. Disturbed mental and neuromuscular functions

Abnormal sex hormone metabolism. Testicular atrophy, gynecomastia, amenorrhea, loss of axillary and pubic hair, spider nevi, palmar erythema

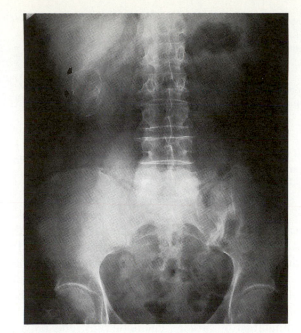

Figure 36.15 An X-ray of an abdomen. The clear area (arrow) is a porcelain gallbladder. Deposition of calcium may be a sequel to acute cholecystitis, and this is called porcelain gallbladder.

normal. There are various causes of jaundice and/or hepatomegaly in AIDS including (1) *Mycobacterium avium-intracellulare*; (2) *Cryptococcus*; (3) Kaposi's sarcoma, a malignancy associated with AIDS (chapter 29); (4) cytomegalovirus; (5) *Histoplasma,* an opportunistic pathogen associated with disseminated fungal disease; (6) sulfonamides used to treat *Pneumocystis* infections; and (7) chronic active hepatitis in children. An unusual form of chronic active hepatitis has been identified in children with AIDS or ARC. It is uncertain whether this is caused by a new type of infection associated with AIDS or is the direct result of HIV infection.

There has been special interest in a possible relationship between the hepatitis B virus (HBV) and AIDS. The majority of individuals with AIDS also have serologic evidence of past or present hepatitis B infection. It is unusual, however, for symptomatic HBV liver disease to be associated with AIDS. The fact that there is a low incidence of chronic liver disease in AIDS patients with HBV hepatitis suggests the possibility that an inadequate immune response may prevent T-lymphocyte-mediated liver injury (chapter 28). The role of HBV in AIDS remains an unanswered question.

CHOLELITHIASIS

Cholelithiasis (ko″le-li-thi′ah-sis) refers to the formation of gallstones (figure 36.15). The mechanism of formation is not clear, but there appears to be at least three predisposing factors: (1) stasis of bile flow, (2) a high cholesterol level in bile, and (3) infection (cholecystitis). Either cholesterol or a calcium salt of bilirubin are the predominant constituents of these stones. The stones may form in the gallbladder or the bile passages. The incidence is higher in females and patients with diabetes mellitus, and there is an increasing incidence in older age groups. In the United States, 10% of the men and 20% of the women between ages 55–65 have gallstones. The obstruction to bile flow may lead to obstructive jaundice and cirrhosis.

ACUTE PANCREATITIS

Acute pancreatitis is a disorder of unknown etiology in which pancreatic proteolytic enzymes are released to attack pancreatic tissue (figure 36.16). It is characterized by epigastric pain of abrupt onset. The pain is dull, often agonizing, and may radiate to the back. Manifestations include nausea, vomiting, and weakness, and a mild jaundice is common. The patient often has a history of alcohol abuse or biliary tract disease. Other predisposing factors include abdominal trauma, drugs, viral infections, and hyperlipidemia.

The pancreas becomes edematous with trapped fluid, and fluid may accumulate in the peritoneal cavity as well. Fluid trapped in a third space (chapter 2) is a fluid loss, and the volume depletion may be profound, resulting in shock. Pancreatic abscess may occur, and acute respiratory distress syndrome (chapter 8) or acute renal failure are possible complications.

Treatment of this condition includes withholding food and water by mouth, bed rest, medication for pain, and volume replacement. The mortality rate is high and recurrences are common.

PANCREATIC CARCINOMA

Pancreatic carcinoma is the second most common neoplasm of the digestive tract and the fourth most common cause of cancer deaths. Males are affected twice as often

Disorders of Liver, Gallbladder, and Pancreas

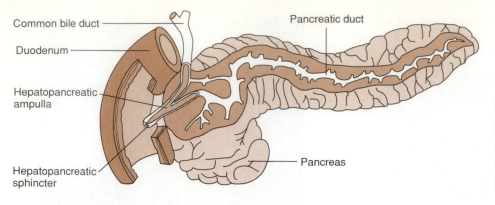

Common bile duct
Duodenum
Hepatopancreatic ampulla
Hepatopancreatic sphincter
Pancreatic duct
Pancreas

Figure 36.16 The pancreas stores and secretes proteolytic enzymes: trypsin, chymotrypsin, and carboxypeptidase. Normally, these enzymes are in an inactive form and are activated by other enzymes in the small intestine.

as females, and the incidence is higher in the elderly. Cigarette smoking is associated with increased risk. Presenting symptoms are nonspecific and include abdominal pain, weight loss, and vomiting. Serum alkaline phosphatase levels (chapter 37) are frequently 5–10 times above normal. Carcinoma of the pancreas is usually treated by a combination of surgery, chemotherapy, and radiation therapy. The prognosis is poor.

SUMMARY

• • •

Jaundice is a common manifestation of liver disease because bilirubin metabolism is a liver function. Obstruction to flow of bile and hemolysis also causes jaundice.

Hepatitis is an inflammation of the liver that may be caused by drugs or viruses. The causes of viral hepatitis include hepatitis A virus, B virus, and hepatitis C virus. Type A is transmitted by fecal-oral means, and hepatitis B virus is transmitted by blood, sexual contact, and by perinatal transfer. Type C hepatitis virus is transmitted by blood and causes some cases of acute hepatitis that are not associated with parenteral transmission. Cirrhosis may be a sequel to viral hepatitis or follow either toxic hepatitis or bile flow obstruction. Cirrhosis is a condition in which necrosis leads to fibrosis accompanied by the formation of regenerative nodules. Three major manifestations of cirrhosis are ascites, esophageal varices, and hepatic encephalopathy. Hepatomegaly associated with abnormal liver function tests is a frequent finding in AIDS. The causes include various infectious agents and Kaposi's sarcoma.

Cholelithiasis or gallstones occur more frequently in females and in individuals with diabetes mellitus. These stones may obstruct bile flow, cause obstructive jaundice, and lead to cirrhosis. Pancreatitis is a condition in which pancreatic enzymes attack pancreatic tissue. Predisposing factors include biliary tract disease, alcohol abuse, hyperlipidemia, and abdominal trauma. There is a high mortality rate associated with this condition. Pancreatic carcinoma is the fourth most common cause of cancer deaths and has a poor prognosis.

REVIEW QUESTIONS

• • •

1. Which of the hepatitis viruses is typically transmitted via blood?
2. What hepatitis virus is most commonly transmitted by blood transfusions?
3. Define the term cirrhosis.
4. Why is the portal vein uniquely susceptible to developing hypertension?
5. What are three consequences of portal hypertension?
6. What are three major effects of hepatocellular failure in cirrhosis?
7. What are three predisposing factors for the formation of gallstones?
8. Describe the basic disease process in pancreatitis.
9. List three predisposing factors in pancreatitis.
10. What are the most common presenting symptoms in pancreatic carcinoma?

SELECTED READING

• • •

Cello, J. et al. 1987. Endoscopic sclerotherapy versus portacaval shunt in patients with severe cirrhosis and acute variceal hemorrhage. *New England Journal of Medicine* 316:11–14.

Eriksson, S., J. Carlson, and R. Velez. 1986. Risk of cirrhosis and primary liver cancer in alpha-antitrypsin deficiency. *New England Journal of Medicine* 314:736–39.

Hoofnagle, J. et al. 1986. Treatment of chronic non-A, non-B hepatitis with recombinant human alpha interferon. *New England Journal of Medicine* 315:1575–78.

Lipsig, L., and R. Davis. 1987. Diagnosis and management of acute calculous cholecystitis. *PA 87 Physician Assistants in Primary and Hospital Care* 4:307–310.

Moody, F. 1988. Pancreatitis as a medical emergency. *Gastroenterology Clinics of North America* 17:433–43.

Seeff, L. et al. 1987. A serologic follow-up of the 1942 epidemic of post-vaccination hepatitis in the United States army. *New England Journal of Medicine* 316:965–70.

Sollinger, H. 1988. Experience with simultaneous pancreas-kidney transplantation. *Annals of Surgery* 208:475–83.

Chapter 37

Practical Applications: Disorders of Liver and Pancreas

The introduction in this chapter deals with liver function tests and the remainder of the chapter is concerned with the interpretation of laboratory test values related to specific liver and pancreatic disorders.

LIVER FUNCTION TESTS

Normal liver function involves many chemical reactions. Liver function tests measure some of these reactions and are useful in identifying hepatic cell injury. A number of tests that aid in evaluating liver function are discussed in the following paragraphs.

BILIRUBIN

The heme portion of the hemoglobin molecule is the source of most of the bilirubin produced by the body (chapter 36). The measurement of bilirubin in the blood is a significant indicator of liver function because liver cells remove the unconjugated form from the blood. Further, bilirubin is converted to a conjugated form within those cells and is subsequently secreted into bile. A laboratory report of total bilirubin represents the sum of conjugated and unconjugated forms. A direct bilirubin, which may be reported as a direct van den Bergh reaction, is a measure of conjugated bilirubin in the serum. An indirect van den Bergh reaction indicates unconjugated serum bilirubin and that fraction is the difference between the total and the direct bilirubin. The normal total bilirubin is less than 1.0–1.5 mg/dl, and approximately 85% of the total is normally unconjugated bilirubin.

An increase in unconjugated bilirubin alone is indicative of a hemolytic condition resulting in release of excessive amounts of hemoglobin. Liver disease or obstruction to flow of bile leads to elevated levels of conjugated bilirubin. Liver disease usually causes an increase in both conjugated and unconjugated forms of bilirubin. When liver cells are damaged, the uptake as well as conjugation of bilirubin is likely to be affected. Liver disease often results in obstruction to flow of bile, and when this occurs, conjugated bilirubin appears in the blood. Bilirubin in the urine is the conjugated form because unconjugated bilirubin is albumin-bound and is not filtered by the glomerulus. Bilirubin in the urine indicates elevated levels of serum conjugated bilirubin.

SERUM PROTEINS

Table 37.1 indicates sources of serum albumin and globulin and lists conditions that lead to variations in serum levels. Albumin has a relatively long half-life of 15–20 days. This represents the time required for half of the total albumin to be removed from the blood. There is little change in serum albumin values in acute liver disease due to this long half-life. Albumin levels are decreased in chronic liver disease. The reason for this decrease is either

TABLE 37.1 Sources of albumin and globulin and reasons for changes in serum levels

Serum albumin

Synthesized by liver cells
Albumin decreased in malnutrition, inflammatory bowel disease, nephrotic syndrome, chronic liver disease

Serum globulins

Gamma globulins produced by B-lymphocytes
Alpha and beta globulins produced by liver cells
Gamma globulins increased in chronic infections, inflammatory disorders, chronic liver disease

diminished synthesis or loss from the vascular compartment in ascites fluid.

Gamma globulins are increased in chronic liver disease, such as cirrhosis. The mechanisms for the increase are not completely clear, but they may be related to either the failure of the liver to sequester antigens or the release of antigenic material from injured liver cells. Increased antigen in the circulation stimulates antibody production with an increase in the gamma globulin fraction. As table 37.1 indicates, tests for serum proteins are not specific for liver disease but are useful in the overall evaluation of liver function.

ENZYMES

Some liver enzymes are found in the blood in low concentrations and may be elevated due to liver cell injury or obstruction to flow of bile.

• • •

Serum glutamic-oxaloacetic transaminase (SGOT) is named on the basis of the products of the reaction it catalyzes. It is also called aspartate transaminase (AST), so named because of the amino donor. Serum glutamic-pyruvic transaminase (SGPT) is similarly called alanine transaminase (ALT).

• • •

The aminotransferases, also called transaminases, are liver enzymes that may be indicators of liver disease. Tests for serum levels of these enzymes are useful for screening purposes even though they are nonspecific and may be elevated in various conditions. Table 37.2 summarizes some aspects of the significance of serum aspartate transaminase (AST) and alanine transaminase (ALT). In general, values exceeding 10 times normal usually indicate hepatic or biliary disease. When there is massive necrosis of liver cells, however, serum aminotransferase levels may fall due to a decline in hepatic enzyme activity. In most disease processes the ALT level is equal to or greater than AST with the exception of alcoholic hepatitis. An AST/ALT ratio greater than 2:1 is an indicator of probable alcohol liver disease.

TABLE 37.2 Significance of elevations of serum aminotransferases

Normal values. AST 13–49 U/liter; ALT 10–50 U/liter
Moderate to high values: Musculoskeletal diseases including trauma and intramuscular injections; acute pancreatitis; acute heart failure; any type of liver cell injury; obstructive jaundice rarely >500; cirrhosis usually <300; acute alcoholic hepatitis, <200–300
Extreme elevations. >100 U/liter almost exclusively in extensive hepatocellular injury due to drug and viral hepatitis; also in sudden obstruction due to passage of gallstone

Alkaline phosphatase is a group of enzymes found in various tissue including the liver, bone, kidney, and leukocytes. Serum alkaline phosphatase activity is usually increased in bile duct obstruction and with hepatic lesions due to increased enzyme synthesis. There may be an increase with other conditions, such as bone disorders, pregnancy, normal growth, and malignancies. As a general rule, alkaline phosphatase activity which is less than three times a normal value is indicative of any type of liver disease and fourfold elevations occur with obstruction to bile flow, liver cancer, and conditions in which there is rapid bone turnover.

PROTHROMBIN TIME

The liver synthesizes almost all of the blood clotting factors; thus, bleeding tendencies are associated with liver disease. The process of blood coagulation involves a complex series of reactions in which ultimately, prothrombin is converted to the enzyme thrombin and in turn thrombin catalyzes the conversion of fibrinogen to fibrin (chapter 19). The prothrombin time test measures the rate at which prothrombin is converted to thrombin. The results of the test depend on plasma concentration of prothrombin, fibrinogen, and other clotting factors, all of which are synthesized by the liver. Prothrombin time is expressed as seconds required for clotting and is compared to a normal control. A prolonged prothrombin time occurs in liver disease as the result of decreased synthesis of clotting factors. Since vitamin K is required for the synthesis of several of these factors, vitamin K deficiency also leads to an abnormal prothrombin time test. Vitamin K deficiency may be caused by a malabsorption syndrome associated with bile flow obstruction. Intestinal bacteria synthesize vitamin K and occasionally antibiotic suppression of normal intestinal flora leads to vitamin K deficiency.

• • •

In the absence of vitamin K deficiency, a prothrombin time prolonged more than 5 seconds indicates extensive liver necrosis.

• • •

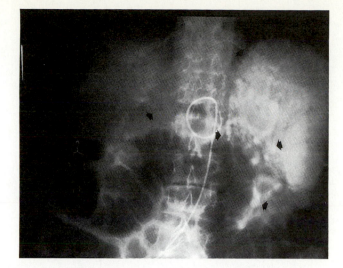

Figure 37.1 This X-ray is a delayed venous filling film showing varices (arrows). The contrast media was injected into a "pigtail" catheter and positioned in the abdominal aorta. Late filming is done to allow the contrast to get to the venous stage.

LIVER DISORDERS

The next section is devoted to examples of liver disorders with emphasis on interpretation of test results associated with each disease process.

CIRRHOSIS

Patient 37.1 was admitted to a hospital with a diagnosis of cirrhosis which was confirmed by physical examination and the following blood tests.

Patient 37.1 was a 58-year-old male with a history of alcoholism and with a diagnosis of cirrhosis.
Laboratory findings: (normal values in parentheses) total bilirubin 3.9 gm/dl (0.5–1.5), direct bilirubin 2.5 mg/dl (0–0.5), ammonia 123 g/dl (80–110), hematocrit 27.7% (42–52), pro time 16.4 seconds (control 11.4), Na+ 131 meq/liter (137–47)

Patient 37.1 had a 30-year history of alcoholism. He was in jail for the 7 months preceding hospitalization and had not ingested any alcohol during this time. Upon hospital admission he was semicomatose, slightly jaundiced, and was bleeding from the nose and ears. Ascites, peripheral edema, and hepatomegaly were noted. He was admitted to the hospital after vomiting blood and having bright red stools. Subsequent X-ray studies showed esophageal varices. Figure 37.1 is an example of X-ray evidence of varices.

The laboratory findings show that total and conjugated bilirubin values are elevated, which correlates with the obvious jaundice of this individual. Ammonia, produced by protein metabolism, is normally converted to urea by the liver. In patient 37.1 the ammonia level is high, which is a probable factor in hepatic encephalopathy and,

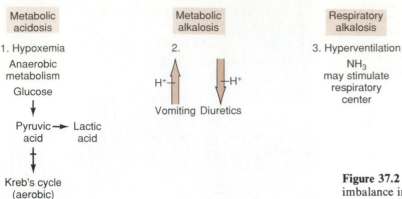

Acid-base imbalance in chronic liver disease

Metabolic acidosis	Metabolic alkalosis	Respiratory alkalosis
1. Hypoxemia	2.	3. Hyperventilation
Anaerobic metabolism		NH₃ may stimulate respiratory center

Figure 37.2 Various factors may be involved in acid-base imbalance in the course of chronic liver disease.

at least in part, accounts for his semicomatose condition. The low hematocrit is evidence of blood loss, and the prolonged prothrombin time is further evidence of liver damage. Alkaline phosphatase and AST values are in the normal range. The sodium ion value is low and may be explained on the basis of the dilutional effect of water retention. In cirrhosis, there is impaired salt and water excretion by the kidney, so that sodium retention is the rule. Water retention occurs as well, as evidenced in this patient by edema, and accounts for the apparent low value for sodium.

> Patient 37.2 was a 68-year-old male with alcoholic cirrhosis. There was neither ascites nor edema. Hepatic fetor was apparent, and there were multiple bruises on his body.
> Laboratory findings: pH 7.43 (7.35–7.45), pCO_2 36 mm Hg (35–45), base excess 0.4 meq/liter (+2.5 to −2.5), pO_2 67 mm Hg (75–80), K 7.2 meq/liter (3.4–5.0), BUN 77 mg/dl (7–22), LDH 256 U/liter (87–270), AST/SGOT 93 U/liter (13–40), ALT/SGPT 69 U/liter (10–50), 233 U/liter (30–95), total bilirubin 28 mg/dl (0.1–1.0), direct bilirubin 22.7, albumin 2.2 g/dl (3.5–5.0), red blood cells 4.1 million/mm³ (4.7–6.1), white blood cells 4,000/mm³ (4,800–10,800).

A mild degree of hypoxemia is common in chronic liver disease, and the pO_2 is below normal in patient 37.2. A correlation between ascites and areas of atelectasis with poorly ventilated regions of the lung is a possible reason (chapter 9). There is a decrease in oxygen-hemoglobin affinity in cirrhosis. Consequently, there is a shift of the oxygen-hemoglobin dissociation curve to the right (chapter 17) which may lead to a decrease in the percentage of oxygen-hemoglobin saturation. The reason for this change in affinity is an increase of 2,3-diphosphoglycerate in red blood cells.

Laboratory tests for patient 37.2 show a normal pH and both respiratory and metabolic acid-base balance. Metabolic acidosis may occur in chronic liver disease due to anaerobic metabolism and lactic acid accumulation

TABLE 37.3 Typical blood gases and laboratory findings related to acid-base balance in cirrhosis

Arterial pO_2 decrease	Marked decrease in ascites; perhaps due to atelectasis
Arterial pCO_2 decrease	Hyperventilation; cause uncertain; ammonia may stimulate respiratory center
Oxygen-Hgb saturation decrease	Decreased oxygen affinity due to increased 2,3-diphosphoglycerate in red blood cells
Metabolic acidosis	Anaerobic metabolism causes lactic acid increase
Metabolic alkalosis	Vomiting, diuretics
Respiratory alkalosis	Hyperventilation; renal compensation may lead to decreased HCO_3^-

(figure 37.2). Metabolic alkalosis may develop when there is vomiting (chapter 5). Metabolic alkalosis may also occur in those individuals with ascites who are treated with diuretics. Respiratory acid-base imbalance occurs in liver disease as well. Hyperventilation is associated with both chronic and acute liver disorders, and the result is respiratory alkalosis. The cause is uncertain, although it has been suggested that accumulations of ammonia may directly stimulate the respiratory center. If there is prolonged hyperventilation, there will be a compensatory response by the kidney with increased excretion of bicarbonate. Table 37.3 is a summary of typical blood gases and laboratory findings related to acid-base balance in cirrhosis.

The elevated BUN in the case of patient 37.2 may be the result of hepatorenal syndrome. This is a renal dysfunction of unknown cause associated with liver disease. The syndrome occurs most often in acute alcoholic hepatitis. This patient has a high potassium level, which may be due to renal failure as well. Hypokalemia, rather than elevated potassium levels, is often seen in liver disease. Diuretics, vomiting, or diarrhea may be underlying causes.

TABLE 37.4 Blood chemistry and hematological findings in cirrhosis

BUN increase	Hepatorenal syndrome
Hyperkalemia	Renal failure
Hypokalemia	Commonly occurs; diuretics, vomiting, diarrhea; affected by acidosis/alkalosis (chapter 3)
Hyponatremia	Renal sodium and water retention; dilutional effect
Albumin decrease	Decreased synthesis by liver cells or loss in ascites fluid; little change in acute liver disease due to long half-life
Liver enzymes increase	Values that exceed a tenfold increase usually indicate hepatic or biliary disease
Anemia	Alcohol suppression of erythropoiesis, folate deficiency, hemolysis, blood loss
White blood cell count	Increase due to infection or decrease due to hypersplenism
Clotting factors decrease	Decreased hepatic synthesis
Bilirubin increase	Decreased uptake of unconjugated bilirubin; decreased secretion of conjugated bilirubin into bile

Hepatocellular injury is reflected in patient 37.2 by the elevated enzyme and bilirubin values. The albumin level is low, due to decreased hepatocellular synthesis. When ascites occurs, there may be loss of albumin in this extracellular fluid.

Both the red blood cell and white blood cell counts are low in patient 37.2. Anemia in heavy drinkers is due to (1) direct suppression of erythropoiesis by alcohol, (2) folate deficiency (chapter 17), (3) gastrointestinal blood loss, and (4) hemolysis. The white blood cell count may be elevated in cirrhosis due to infection or may be low as the result of hypersplenism. Table 37.4 summarizes expected results of blood chemistry and hematological tests in cirrhosis.

CHRONIC ACTIVE HEPATITIS

Patient 37.3 was admitted to the hospital with a fractured hip and a diagnosis of chronic active hepatitis.

Patient 37.3 was a 74-year-old female.
Laboratory findings: Hgb 7.3 g/dl (12–16), Hct 22.3% (37–47), total bilirubin 1.2 mg/dl (0.1–1.0), alkaline phosphatase 195 U/liter (30–95), AST/SGOT 99 U/liter (13–40), ALT/SGPT 159 U/liter (10–50), LDH 473 U/liter (87–170)

Patient 37.3 had elevated liver function tests 3 years prior to this hospital admission. An infectious cause was not established at that time. Liver biopsy three months before admission confirmed the diagnosis of chronic active hepatitis. This disorder is the most serious form of chronic hepatitis because it may lead to cirrhosis and ultimately to liver failure. About 20% of the cases are associated with hepatitis B infection, although in many cases, the etiology is unknown. Anemia, as indicated in this patient, occurs in about half the cases and a decrease in platelets is observed in about 25% of affected individuals. Bilirubin levels are elevated in most cases when there is severe inflammatory activity. The total bilirubin value is greater than 3 mg/dl in about half of these cases. There are variations in the degree of elevations of liver enzymes.

INFECTIOUS HEPATITIS

Patient 37.4 was admitted to the hospital with a diagnosis of infectious hepatitis and pulmonary edema. His temperature was 39.3°C, and he complained of myalgia, nausea, vomiting, and diarrhea.

Patient 37.4 was a 20-year-old male.
Laboratory findings: pH 7.27 (7.35–7.45), pCO_2 32.8 mm Hg (35–45), HCO_3^- 15.1 meq/liter (22–30), pO_2 84 mm Hg (75–80), cholesterol 68 mg/dl (140–300), total bilirubin 13.2 mg/dl (0.1–1.0), albumin 2.3 g/dl (3.5–5.0), AST/SGOT 1,731 U/liter (13–40), LD/LDH 14,276 U/liter (87–170), thrombin time 120 sec (11–16), prothrombin 12.8 sec (9.8–12.6), APTT 37 sec (21–31) fibrin degradation products 256 mcg/ml (10)

Patient 37.4 was critically ill. The low pCO_2, associated with respiratory alkalosis, is an expected result of pulmonary edema. Bicarbonate loss occurs in diarrhea and causes metabolic acidosis (chapter 5). The predominant acid-base imbalance is reflected in the low pH, which indicates acidosis. The bilirubin and AST values are indicative of a liver disorder.

Patient 37.4 experienced oozing from the oral mucosa, and the laboratory tests confirm DIC as a complication of this acute illness. The prolonged thrombin time is indicative of decreased fibrinogen. The high value for fibrin degradation products is evidence of accelerated fibrinolysis (chapter 20).

HEPATIC FAILURE

Patient 37.5 had a prolonged hospitalization with multiple problems and, when there was evidence of hepatic failure as a complication, a consulting physician was asked to evaluate her case. Laboratory data are as follows.

Patient 37.5 was a 60-year-old female with a diagnosis of hepatic failure.
Laboratory findings: platelets 43,000/mm³ (130,000–400,000), hemoglobin 8.8 g/dl (14–18), total bilirubin 13.5 mg/dl (0.1–1.0), alkaline phosphatase 760 U/liter (30–95), AST 43 U/liter (13–40), ALT 7 U/liter (10–50)

Patient 37.5 had a complex history that included breast cancer, chronic cholecystitis and subsequent cholecystectomy, a seizure disorder, a prolonged infection, and finally hepatic failure as evidenced by jaundice, ascites, and laboratory tests. Identifying the cause of the hepatic failure was a difficult diagnostic problem. Probable causes were hepatic ischemia due to multiple transfusions, infection, surgeries, and drug toxicity. A low-grade disseminated intravascular coagulopathy was ongoing and evidenced by a low hemoglobin level and low platelet count (chapter 19). This is a likely contributing factor to increased bilirubin. Dilantin and phenobarbital were two of the drugs she was receiving, and these drugs can elevate the alkaline phosphatase level.

PANCREATIC DISORDERS

The last part of this chapter deals with disorders of the pancreas with a focus on the correlation of test results and the disease process.

• • •

Acute pancreatitis occurs in approximately 8% of cases of hyperparathyroidism. It also occurs in cases of hypercalcemia due to vitamin D poisoning and metastatic breast cancer. The chronic type of pancreatitis occurs in hyperparathyroidism but is not associated with other hypercalcemic disorders. It is possible that elevated calcium levels lead to pancreatitis by activating trypsinogen.

• • •

ACUTE PANCREATITIS

Most cases of acute pancreatitis are associated with excessive alcohol ingestion or biliary tract disease. There are other causes including hypercalcemia, viral infections, vasculitis, and abdominal trauma. Drugs, such as prednisone and thiazides, are related to acute pancreatitis as well. Pancreatitis may occur in patients with elevated triglyceride levels who also drink alcohol excessively, or in those who combine modest alcohol use with excessive lipid ingestion. The history of patient 37.6 did not reveal a possible cause for his episode of acute pancreatitis.

Patient 37.6 was a 50-year-old male diagnosed with acute pancreatitis.
Laboratory findings: pH 7.42 (7.35–7.45), pCO$_2$ 26 mm Hg (35–45), HCO$_3^-$ 16.7 meq/liter (22–30), pO$_2$ 90 mm Hg (75–80) Na 137 meq/liter (137–47), K 3.3 meq/liter (3.4–5.0), calcium 9.1 mg/dl (8.6–10.5), BUN 21 mg/dl (7–22), glucose 120 mg/dl (70–110), anion gap 15 meq/liter (6–16), albumin 3.4 g/dl (3.50–5.0), amylase 978 U/liter (44–128), triglycerides 110 mg/dl (45–232)

The laboratory tests show that the blood glucose level is somewhat elevated in this case. Hyperglycemia often occurs in pancreatitis and may be due to increased glucagon and/or decreased insulin.

The calcium level of patient 37.6 is within the normal range. The majority of patients with acute pancreatitis have a low serum calcium level, although severe hypocalcemia is unusual. Calcium levels less than 7 mg/dl are associated with a poor prognosis in acute hemorrhagic pancreatitis. The reason for hypocalcemia is uncertain, but may be due to a reaction between calcium and fatty products to form soaps in areas of fat necrosis within the pancreas. A large number of patients have low parathormone (PTH) levels (chapter 31), and this may be a factor in maintaining hypocalcemia. It is possible that circulating proteolytic enzymes from the pancreas are responsible for defective PTH secretion.

The serum amylase level of patient 37.6 is markedly elevated. Amylase is an enzyme produced by the pancreas, salivary glands, and some malignant tumors. The lungs and fallopian tubes produce lesser amounts of the enzyme. The amylase produced by both the salivary glands and the pancreas enters the gastrointestinal tract. In pancreatitis, amylase appears in the pancreatic lymph and blood supply. Serum amylase levels in acute pancreatitis are elevated within 24 hours in 90% of the cases.

• • •

Serum amylase levels are elevated, not only in pancreatitis, but in other disorders as well. An increase in amylase may occur in association with such conditions as perforated peptic ulcer, bowel infarction, lung disease, salivary gland disease, malignant tumors, and diabetic ketoacidosis.

• • •

Patient 37.7 was a 70-year-old male with a diagnosis of acute pancreatitis.
Laboratory findings: glucose 305 mg/dl (70–110), calcium 7.2 mg/dl (8.6–10.5), albumin 2.6 g/dl (3.5–5.0), amylase 828 U/liter (44–128), LDH 409 U/liter (87–170)

Patient 37.7 has an elevated glucose level and a low level of both calcium and albumin. There is a correlation between the calcium and albumin (chapter 3). Approximately half of the total serum calcium is bound to albumin, with the remainder in the physiologically active, ionized form. If there is fluid loss associated with pancreatitis, by way of plasma accumulations either within the pancreas or the peritoneal cavity, the result is loss of albumin. Hypoalbuminemia causes a decrease in total calcium, although there may be no reduction in the ionized form of calcium. Adding 0.8 mg to the total calcium value for each g/dl that the albumin is decreased corrects for the effect of hypoalbuminemia. In patient 37.7, the albumin level is approximately 1 g/dl below normal; thus, the calculated total calcium level is 7.2 plus 0.8 or approximately 8.0 mg/dl.

The amylase value for patient 37.7 is high. The extent of elevation of this enzyme does not, however, correlate with the severity of the pancreatitis. If the elevation extends beyond 1 week, it is usually indicative of fairly severe pancreatitis.

The lactate dehydrogenase (LDH) level is high. This is an intracellular enzyme widely distributed in the body (chapter 25). An increase in the LDH value is usually the result of cell necrosis with escape of the enzyme from cells.

> Patient 37.8 was an 80-year-old female diagnosed with hypertension, diabetes, and acute pancreatitis.
> Laboratory findings: amylase 3,948 U/liter (44–128), Hct 42% (37–47), glucose 231 mg/dl (70–110), LDH 176 U/liter (87–170)

The elevation of the amylase is extreme in patient 37.8. The LDH level indicates cell necrosis. The hematocrit is normal, although in some cases it may be elevated because of the concentrating effect of massive loss of plasma into the peritoneal space. In such a case, intravascular volume depletion may lead to prerenal azotemia, i.e., increased creatinine and BUN levels (chapter 16).

Table 37.5 summarizes expected laboratory findings in acute pancreatitis.

CARCINOMA OF PANCREAS

Carcinoma is the most common neoplasm of the pancreas. The majority of cases involve the head of the pancreas and may involve the ampulla of Vater, the common bile duct, and the duodenum. The common bile duct is frequently compressed. Consequently, the liver function tests are those of obstructive jaundice. Patient 37.9 was admitted to the hospital with a mass at the head of the pancreas with biliary obstruction.

> Patient 37.9 was a 75-year-old male with a diagnosis of probable carcinoma.
> Laboratory findings: Na+ 139 meq/liter (137–47), K+ 4.3 meq/liter (3.4–5.0), Cl− 104 meq/liter (100–10), BUN 18 mg/dl (7–22), creatinine 1.0 mg/dl (0.6–1.2), anion gap 11 meq/liter (6–16), glucose 291 mg/dl (70–110), white blood cells 6,400/mm³ (4,800–10,800), red blood cells 5,230,000/mm³ (4.7–6.1 million), Hgb 15.4 g/dl (14–18), alkaline phosphatase 890 U/liter (30–95), AST/SGOT 80 U/liter (13–40), amylase 510 U/liter (44–128), pro time 20 sec (9.8–12.6), total bilirubin 2 mg/dl (0.1–1.0)

Patient 37.9 had a 15-year history of diabetes and was having difficulty keeping his blood glucose levels within normal range. Many of the laboratory test results are in a normal range. The red blood cell count is normal, although anemia may develop in pancreatic carcinoma for several reasons, i.e., nutritional deficiency, blood loss by way of the gastrointestinal tract, or there may be the anemia of chronic disease (chapter 20). Patient 37.9 has

TABLE 37.5 Probable reasons for laboratory findings in acute pancreatitis

Hyperglycemia	Increased glucagon or decreased insulin
Amylase increased	Source is pancreas; enzyme enters pancreatic lymph and blood supply
LDH increased	Cell necrosis
Hematocrit increased	Hemoconcentration due to fluid loss
BUN increased	Prerenal azotemia due to intravascular volume depletion
Albumin decreased	Loss in pancreatic ascites
Calcium decreased	Ca plus fatty products from fat necrosis form soaps; decreased PTH (proteolytic enzymes may cause defective PTH secretion); loss of albumin causes decreased protein bound Ca

an elevated serum amylase, which occurs if there is an inflammation of the pancreas associated with the malignancy. An alkaline phosphatase value greater than 5–10 times normal, as it is in this patient, is indicative of common bile duct obstruction or metastatic involvement of the liver. The prolonged prothrombin time in this case is likely due to obstruction of intrahepatic bile flow. The AST/SGOT enzyme level increases when the bile duct obstruction is prolonged. The total bilirubin of patient 37.9 is above normal. In the case of bile duct obstruction, there is a continued rise of bilirubin, and the conjugated form predominates.

SUMMARY

• • •

The source of bilirubin, for the most part, is hemoglobin. The unconjugated form is taken up by liver cells, converted to the conjugated form, and secreted into bile. Unconjugated bilirubin makes up approximately 85% of the total serum bilirubin. Hemolysis with a release of excess hemoglobin leads to an elevation of unconjugated bilirubin in the blood. Liver cell damage usually affects the processes of uptake and conjugation and causes an increase of both forms of serum bilirubin. Obstruction to bile flow causes increased levels of conjugated bilirubin to appear in blood.

Serum albumin is synthesized by liver cells along with both alpha and beta globulins. Gamma globulins are produced by B-lymphocytes. Chronic liver disease leads to decreased serum albumin and gamma globulins are increased, possibly due to antigenic stimulus. The liver enzymes aspartate transaminase (AST) and alanine transaminase (ALT) may be elevated in liver disease as well as in various other conditions. Extreme elevations are due almost exclusively to massive hepatocellular injury due to drug and viral hepatitis. Alkaline phosphatase is found in various tissues including the liver. In general, alkaline phosphatase activity less than 3 times above normal is due to liver disease, while elevations 4 times above normal occur with obstruction to bile flow, liver cancer, and conditions in which there is rapid bone turnover.

Blood clotting factors are synthesized by the liver. Prothrombin is one of these factors, and the prothrombin time test measures the rate at which prothrombin is converted to thrombin. A prolonged prothrombin time occurs in liver disease because of decreased synthesis of clotting factors.

Expected laboratory findings in cirrhosis include (1) decreased pO_2, pCO_2, and oxygen-hemoglobin saturation; (2) acid-base imbalances including metabolic acidosis or alkalosis and respiratory alkalosis; (3) increased BUN due to the hepatorenal syndrome; (4) hypokalemia and hyponatremia; (5) decreased albumin; (6) increased levels of liver enzymes; (7) anemia and variable white blood cell counts; (8) decreased clotting factors; and (9) increased bilirubin levels.

Hyperglycemia and elevated serum amylase levels are associated with acute pancreatitis. A decrease in calcium and albumin may occur when there is pancreatic ascites, and hemoconcentration with an increased hematocrit may be observed as well. Intravascular volume depletion leads to prerenal azotemia. Pancreatic carcinoma may cause obstruction of the common bile duct resulting in elevated alkaline phosphatase levels.

REVIEW QUESTIONS

• • •

1. Which form of bilirubin is normally predominant in the blood?
2. What is the most likely cause of a high indirect bilirubin test?
3. What normal hepatocellular functions involve bilirubin?
4. What occurs in liver disease that leads to an elevation of serum conjugated bilirubin?
5. What form of bilirubin is found exclusively in urine?
6. Chronic liver disease generally has what effect on albumin/globulin values?
7. Blood levels of what two liver enzymes tend to be elevated in liver disease?
8. What is the most likely cause of extreme elevations of the enzymes elevated in liver disease?
9. What is the most likely cause of a fourfold increase in alkaline phosphatase?
10. What is the possible significance of a prolonged prothrombin time?
11. What is a likely reason for a low pO_2 in cirrhosis?
12. What is the reason for a low pCO_2 in cirrhosis?
13. What is a possible cause of metabolic acidosis in cirrhosis?
14. What may cause metabolic alkalosis in cirrhosis?
15. What respiratory acid-base imbalance may occur in cirrhosis?
16. What occurs in cirrhosis that may lead to decreased oxygen-hemoglobin saturation?
17. What differences in albumin levels may be observed in acute liver disease as compared to a chronic condition?
18. The elevation of what enzyme level is most important in diagnosing acute pancreatitis?
19. In what way are calcium and albumin levels interrelated?
20. Why may there be an increased BUN in pancreatitis?

Kaplan, M. et al. 1986. A prospective trial of colchicine for primary biliary cirrhosis. *New England Journal of Medicine* 315:144–54.

Kaplan, M. 1987. Primary biliary cirrhosis. *New England Journal of Medicine* 316:521–28.

Nicholls, K. et al. 1986. Sodium excretion in advanced cirrhosis: Effect of expansion of central blood volume and suppression of plasma aldosterone. *Hepatology* 6:235–38.

Scully, R., E. Mark, and B. McNeeley. 1984. Case records of Massachusetts general hospital (non-A, non-B hepatitis). *New England Journal of Medicine* 311:904–11.

NERVOUS SYSTEM

Chapter 38 deals with the flow of cerebrospinal fluid, fluid accumulation in the cerebral hemispheres, and alterations in cerebral blood flow. The first topic is obstruction to flow of cerebrospinal fluid associated with hydrocephalus or water on the brain. There is discussion of various causes of brain edema, and the subject of arteriovenous malformations is included. The last part of chapter 38 deals with interruption of cerebral blood flow with stroke as the outcome.

Chapter 39 focuses on cerebral disorders. The first two topics are epilepsy and intracranial tumors followed by examples of brain injury. Degenerative disorders included in chapter 39 are Alzheimer's disease, Creutzfeldt-Jakob disease, and AIDS dementia. Chapter 40 includes selected examples of sensory and motor disorders. The remainder of chapter 40 is devoted to movement disorders, categorized on the basis of the nature of the defect. Chapter 41 focuses on laboratory data and diagnostic tests related to examples of seizure, transient ischemic attack, and stroke.

Chapter 38

Cerebral Fluid and Vascular Disorders

This chapter focuses on alterations in the flow of cerebrospinal fluid, accumulation of fluid in the brain, and problems related to cerebral blood flow. The topic of cerebrospinal fluid is considered first.

CEREBROSPINAL FLUID

Extracellular fluid (chapter 2) includes plasma, lymph, interstitial fluid, and cerebrospinal fluid. Cerebrospinal fluid (CSF) circulates in and around both the brain and spinal cord and acts as a cushion for the central nervous system. The brain and spinal cord are covered by meninges (mě-nin'jēz) or membranes: the outer **dura mater,** the middle **arachnoid mater,** and a delicate inner layer that adheres to the brain surface called the **pia mater** (pi'ah ma'ter) (figure 38.1). The subarachnoid space between the arachnoid mater and pia mater is filled with cerebrospinal fluid. The source of that fluid is the vascular tissue of the choroid plexuses that project into the ventricles of the brain (figure 38.2).

The fluid circulates from the two lateral ventricles to the third ventricle, in the region of the midbrain, and to the fourth ventricle in the medulla. The fluid flows into subarachnoid spaces by way of the foramina of Luschka and the foramen of Magendie. Cerebrospinal fluid circulates through the spinal canal, down the posterior subarachnoid space of the spinal cord, and continues up the anterior surface of the spinal cord. The fluid is returned to the blood from the subarachnoid space of the brain through arachnoid villi or granulations. These are projections of arachnoid tissue that extend into the blood-filled venous sinuses (figure 38.2). The dura mater is made up

dura mater: L. hard mother
arachnoid: Gk. *arachnes,* spider; Gk. *eidos,* form
pia: L. gentle

of a double membrane around the brain and provides a channel for venous sinuses, which return blood from the brain, mainly into the internal jugular vein. Cerebrospinal fluid forms at the rate of about 500 ml/day resulting in an average renewal of that fluid about four times each day.

HYDROCEPHALUS

Hydrocephalus (hi-dro-sef'ah-lus) means water on the brain and involves both an increased volume of cerebrospinal fluid and dilation of cerebral ventricles (figure 38.3).

ETIOLOGY

Obstructive hydrocephalus is the most common form of this disorder and is due to a blockage within the ventricles or in the subarachnoid spaces. Various sites of obstruction follow:

1. Foramen of Monroe
2. Third ventricle
3. Aqueduct of Sylvius
4. Fourth ventricle
5. Foramina of Luschka, two foramina in lateral walls of fourth ventricle
6. A medial opening in fourth ventricle called the foramen of Magendie
7. Subarachnoid spaces at the base of the brain or over the cerebral hemispheres

Table 38.1 summarizes the possible causes of obstructive hydrocephalus. The condition may be congenital, may be due to hemorrhage or inflammation, or may be caused by a mass lesion. In the case of hemorrhage, obstruction to CSF flow occurs when a clot forms within the ventricles or when fibrosis follows an inflammation of the arachnoid

congenital: L. *congenitus,* born together with

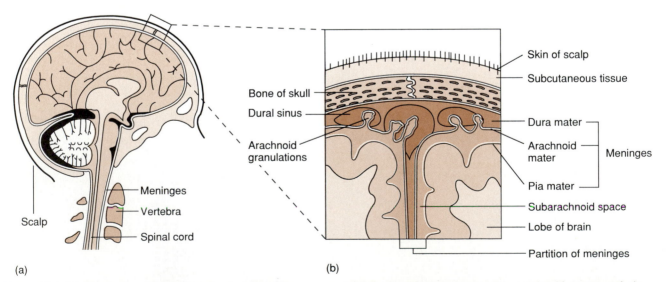

(a)   (b)

Figure 38.1 (*a*) The brain and spinal cord are enclosed by bone and by membranes called meninges. (*b*) The meninges include three layers: dura mater, arachnoid mater, and pia mater.

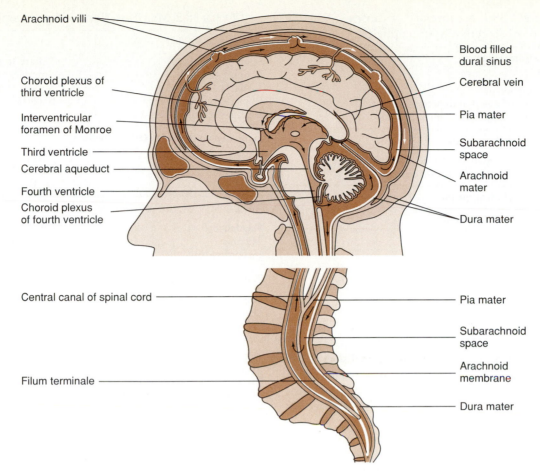

Arachnoid villi

Choroid plexus of third ventricle

Interventricular foramen of Monroe

Third ventricle

Cerebral aqueduct

Fourth ventricle

Choroid plexus of fourth ventricle

Blood filled dural sinus

Cerebral vein

Pia mater

Subarachnoid space

Arachnoid mater

Dura mater

Central canal of spinal cord

Filum terminale

Pia mater

Subarachnoid space

Arachnoid membrane

Dura mater

Figure 38.2 The flow of cerebrospinal fluid. Cerebrospinal fluid is secreted by choroid plexuses in the ventricular walls. The fluid circulates through the ventricles and central canal, enters the subarachnoid space, and is reabsorbed into the blood of the dural sinuses through the arachnoid villi.

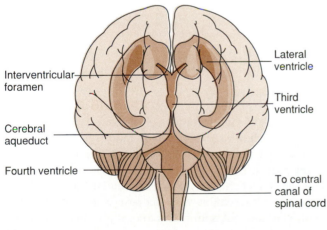

Interventricular foramen

Cerebral aqueduct

Fourth ventricle

Lateral ventricle

Third ventricle

To central canal of spinal cord

(a)

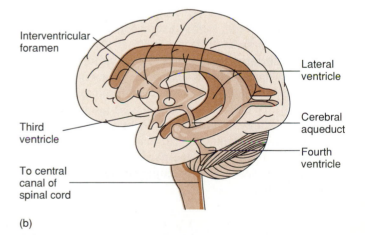

Interventricular foramen

Third ventricle

To central canal of spinal cord

Lateral ventricle

Cerebral aqueduct

Fourth ventricle

(b)

Figure 38.3 (*a*) Anterior view of the ventricles within the cerebral hemispheres and brain stem. (*b*) Lateral view of the ventricles of the brain.

TABLE 38.1 A summary of causes of obstructive hydrocephalus

Congenital Malformations

Genetic
Intrauterine infection
Intracranial hemorrhage due to prematurity
Intracranial hemorrhage due to birth trauma

Posthemorrhagic Hydrocephalus

Hemorrhage within ventricles or medulla of adults
Subarachnoid bleeding due to trauma or aneurysm

Postinflammatory Hydrocephalus

Arachnoiditis caused by meningitis

Mass Lesions

Intracranial neoplasms
Vascular abnormalities

TABLE 38.2 Manifestations of hydrocephalus

Infants

Skull enlargement	Failure to thrive
Exophthalmos	Prominent skull veins
Visual loss	Muscle wasting
Impaired motor development	Seizures
Impaired intellectual development	

Hydrocephalus Due to Otitis Media

Fever	Eardrum perforation
Lethargy	Paralysis, cranial nerve VI

Adults

Headache	Dementia
Malaise	Altered consciousness
Incoordination	

Normal Pressure Hydrocephalus

Insidious onset	Ataxia of gait
Dementia	Urinary incontinence

mater. There is some evidence indicating that packed red blood cells in the arachnoid villi impair reabsorption of CSF following hemorrhage. Inflammation associated with meningitis may lead to hydrocephalus. Intracranial neoplasm or vascular abnormalities may cause obstructive hydrocephalus as well.

An increased rate of secretion of CSF occurs in the presence of choroid plexus papilloma, a neoplasm of choroid plexus epithelium. Oversecretion of CSF leads to hydrocephalus.

Impaired absorption of CSF by arachnoid villi causes hydrocephalus and may occur in the case of chronic middle ear infection, or as the result of mastoiditis. Thrombosis associated with these infections may impair cerebral venous drainage and lead to hydrocephalus.

Normal pressure hydrocephalus is a syndrome that may be associated with a mass lesion or follow various conditions including subarachnoid hemorrhage, head trauma, or meningitis. Normal cerebrospinal fluid pressure, measured during a lumbar puncture, is observed, although there may be intermittent increases in intracranial pressure prior to diagnosis. The cause of the syndrome is unclear. There may be some degree of occlusion of the subarachnoid space and/or impaired absorption in the arachnoid villi.

• • •

Hydrocephalus may be caused by the congenital absence of arachnoid villi, which causes impaired CSF absorption. Hydrocephalus also occurs when CSF protein exceeds 500 mg/dl, as in the case of spinal cord tumor. In this situation, the protein may interfere with CSF absorption.

• • •

papilloma: L. *pappilla*, nipple; Gk. *oma*, tumor

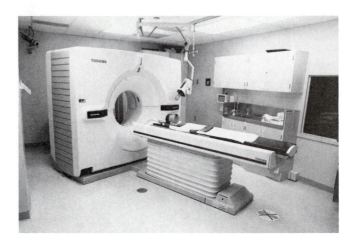

Figure 38.4 A CAT scanner.

MANIFESTATIONS

A summary of manifestations of different types of hydrocephalus is shown in table 38.2. In infants, if the condition is untreated, there is retardation of motor and intellectual development, and there may be scalp necrosis with CSF leakage, infection, and death. In adults, the manifestations of hydrocephalus vary because enlargement of the ventricles does not uniformly cause injury to surrounding white and gray matter.

DIAGNOSTIC TESTS

Measurements of the skull and skull X-rays aid in following the course of hydrocephalus. Computerized tomography and nuclear magnetic resonance imaging are the most useful procedures for diagnosing all types of hydrocephalus.

dementia: L. *dementia*, madness

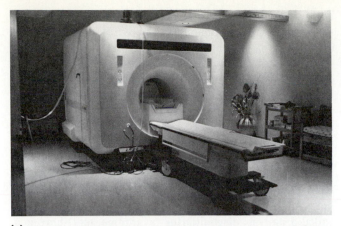

(a)

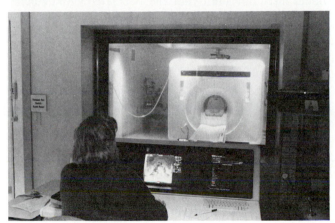

(b)

Figure 38.5 (*a*) NMRI unit. (*b*) Control console.

Various terms have been used for computerized tomography (CT) including computerized axial tomography (CAT) (figure 38.4) and computerized transmission tomography (CTT). These terms refer to a technique for X-ray imaging of thin serial sections of tissue. A source of X-rays rotates around the head. A narrow beam of X-rays passes through the head and is detected by a sensing device. A computer processes the information and records differences in density of areas of each thin section. A picture is built up from this information and can be printed on X-ray film. The procedure is virtually without risk and can be repeated to follow a patient's progress subsequent to such events as intracranial hemorrhage, the occlusion of cerebral blood supply, or dilation of ventricles in hydrocephalus.

The basis for nuclear magnetic resonance imaging (NMRI) is the fact that nuclei of certain elements in the body behave like magnets and will interact with an external magnetic field. Hydrogen is the most abundant atom in the body and is strongly magnetic. The patient is placed in the center of the NMRI unit (figure 38.5), which is a long tube, and is exposed to a magnetic field about 10,000 times that of the earth's magnetic field. A small population of hydrogen nuclei align with this field. A series of magnetic pulses of a certain frequency tumble the nuclei into a new position into a state of higher energy. When the nuclei return to a lower energy state, they emit that energy, and the energy is recorded. This information is processed and converted into an image. NMRI has no apparent adverse side effect and is useful for diagnosing various conditions, including spinal cord tumors and hydrocephalus.

• • •

The absorption and radiation of energy by hydrogen nuclei is an example of resonance. The term nuclear magnetic resonance is used because it also involves magnetic interaction with a nucleus.

• • •

Lumbar puncture is a possible diagnostic procedure in suspected hydrocephalus. It involves insertion of a needle between the third and fourth vertebrae into the subarachnoid space (figure 38.6). This procedure allows examination of CSF and measurement of CSF pressure. CSF pressure may reflect increased ventricular pressure, which usually develops in hydrocephalus.

TREATMENT

Medical treatment involving the use of diuretics, such as acetazolamide or furosemide, is not effective as long-term therapy for hydrocephalus. Surgical procedures to shunt fluid away from ventricles are effective in alleviating intracerebral pressure.

BRAIN EDEMA

Increased water and sodium content of the brain causes an increase in brain volume or brain edema. There are various causes including (1) brain tumor, (2) head injury, (3) stroke, (4) cerebral infections, (5) hypoxia, (6) complications of dialysis and diabetic ketoacidosis, and (7) obstructive hydrocephalus.

CLASSIFICATION

There are three forms of brain edema classified on the basis of pathogenesis: vasogenic, cellular or cytotoxic, and interstitial. Vasogenic edema is the result of increased permeability of endothelial cells of brain capillaries. Cellular brain edema is caused by swelling of neurons, glial cells, and endothelial cells. Interstitial edema occurs when there is obstruction to CSF flow, i.e., obstructive hydrocephalus.

ETIOLOGY

Disorders leading to increased capillary permeability, brain cell swelling, and interstitial edema are listed in table 38.3. There are two possible mechanisms that cause cellular brain edema. The swelling may be due to either hyponatremia or to an accumulation of excessive solutes in the

vasogenic: L. *vas*, vessel

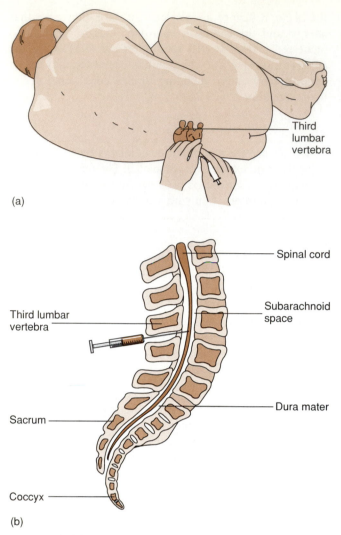

(a)

(b)

Figure 38.6 (*a*) A lumbar puncture is performed by inserting a fine needle between the third and fourth lumbar vertebrae and (*b*) withdrawing a sample of cerebrospinal fluid from the subarachnoid space.

TABLE 38.3 Disorders associated with three categories of brain edema

Vasogenic Brain Edema	
Brain tumor	Infarction
Abscess	Trauma
Hemorrhage	
Cellular Brain Edema	
Dilutional hyponatremia	Disequilibrium syndromes
Acute sodium depletion	Acute hypoxia
Interstitial Brain Edema	
Obstructive hydrocephalus	

brain cells. (1) An acute decrease in extracellular fluid osmolality occurs with inappropriate secretion of ADH accompanied by a dilutional hyponatremia. A decreased osmolality also occurs with acute sodium depletion. Hypoosmolality causes water to enter brain cells (chapter 2). (2) Accumulation of intracellular solutes in the brain is a protective mechanism against shrinking in the presence of hyperosmolar conditions which occur, for example, in uremia and diabetic ketoacidosis.

During rapid dialysis the movement of solutes from plasma to dialysate may be faster than solute diffusion from cells into plasma. The result in the brain is an osmotic gradient that draws fluid into brain cells. This is called a **dialysis disequilibrium syndrome,** and there is a similar disequilibrium syndrome associated with diabetic ketoacidosis. In addition to the disequilibrium syndromes, acute hypoxia causes cellular edema, which is osmotically determined by increased brain cell solute.

MANIFESTATIONS

Manifestations of cerebral edema include **myoclonus** or irregular muscle contractions, seizures, disturbances of consciousness, stupor, and coma.

DIAGNOSTIC TESTS

Magnetic resonance imaging and CT demonstrate increased brain water and volume in vasogenic edema. The electrical activity of the brain is slowed in both vasogenic and cellular brain edema. Brain electrical activity may be recorded by means of an electroencephalogram (EEG). The test involves the placement of electrodes on the scalp to detect electrical changes by means of a recorder and pens, which trace the pattern of activity picked up by each electrode. The EEG is often normal in interstitial brain edema.

TREATMENT

Treating the underlying cause, such as intracranial infection or mass lesion, is an important aspect of the management of cerebral edema. Intravenous fluids should contain salt, i.e., a normal saline solution or 5% glucose in saline, in order to maintain plasma osmolality. Glucocorticoids are useful for the treatment of vasogenic edema, but not for the treatment of edema associated with hypoxia or ischemia. Administration of a hypertonic solution of mannitol may alleviate intracranial pressure by establishing an osmotic gradient.

ARTERIOVENOUS MALFORMATIONS

Arteriovenous malformations (AVMs) are congenital defects involving a tangle of abnormal arteries and veins (figure 38.7). The lesions may be located in the brain, which is more common, or may involve the dura mater.

myoclonus: Gk. *mys,* muscle; Gk. *klonos,* confused motion

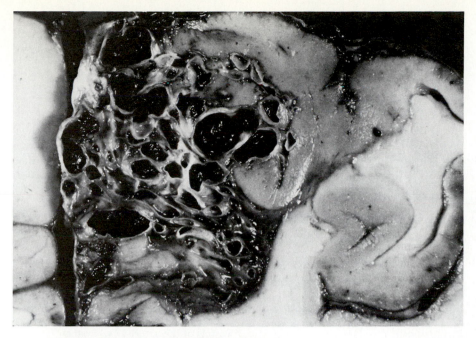

Figure 38.7 A cross section of an arteriovenous malformation in the brain showing the labyrinth of intertwining blood vessels making up the lesion. The brain adjacent to the malformation is infarcted because the blood flow through the malformation was so great that it stole blood from the nearby normal tissue.
Courtesy of Dr. J. H. N. Deck. From A. C. Ritchie: Boyd's Textbook of Pathology. *1990 Lea & Febiger, Philadelphia. Reproduced with permission.*

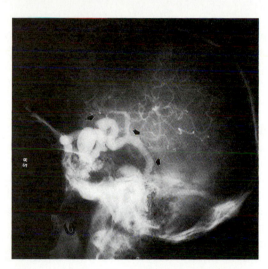

Figure 38.8 Lateral skull X-ray of a male. Contrast was injected in the right carotid and right temporal fossa. An arteriovenous malformation is outlined by arrows and an aneurysm (large oval structure) is anterior to the sella turcica.

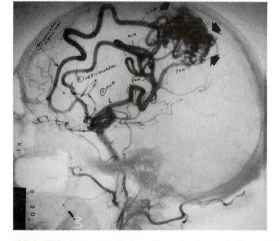

Figure 38.9 This is a subtraction film of a lateral view of the skull taken during a cerebral arteriogram. The dark convoluted mass in the area of the arrows demonstrates an arteriovenous malformation.

The malformations usually do not grow, although there is a tendency for collateral or secondary circulation to develop. The lesions are frequently associated with cerebral hemorrhage. In the case of AVMs limited to the dura mater, there may be subarachnoid hemorrhage and subdural hematoma. Over a period of many years patients experience headaches, seizures, and progressive neurologic abnormality. The final outcome may be disability and death. Injection of contrast medium into an artery allows visualization of an AVM, which appears as tortuous dilated vessels (figures 38.8 and 38.9). The lesions are treated surgically.

STROKE

The term stroke or cerebrovascular accident (CVA) refers to injury to the brain due to (1) rupture of blood vessels and hemorrhage in the brain or (2) occlusion of blood vessels supplying the brain. In spite of the fact that there has been a decline in the incidence of stroke in the last 20 years, it ranks third after heart disease and cancer as a cause of death in the United States. The following discussion of stroke is divided into two parts: (1) cerebrovascular occlusion and (2) intracranial hemorrhage.

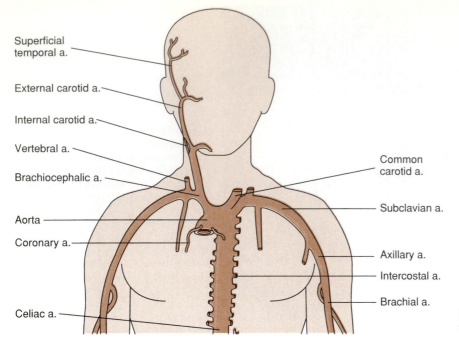

Superficial temporal a.

External carotid a.

Internal carotid a.

Vertebral a.

Brachiocephalic a.

Aorta

Coronary a.

Celiac a.

Common carotid a.

Subclavian a.

Axillary a.

Intercostal a.

Brachial a.

Figure 38.10 Major upper body vessels of the arterial (a.) system.

CEREBROVASCULAR OCCLUSION

The aortic arch gives rise to the brachiocephalic artery on the right side, which bifurcates to form the right common carotid and subclavian arteries. The left common carotid originates from the aortic arch distal to the brachiocephalic artery (figure 38.10). Both common carotids bifurcate to form internal and external carotids. The external carotid arteries supply most of the head and neck, whereas the internal carotid arteries supply the brain.

A blood vessel may be occluded by an **embolus** or a **thrombus.** An embolus is a transported clot or plug that may be composed of cholesterol crystals, platelet-fibrin fragments, air, tissue cells, fat, or clumps of bacteria. A thrombus is a blood clot. In most cases, cerebrovascular disease is associated with atherosclerosis and hypertension. Atherosclerosis (chapter 22) causes various abnormalities in the carotid arteries which lead to disturbances in blood flow to the brain. Fragmentation of an atherosclerotic plaque results in emboli that may lodge in branches of cerebral arteries. Atherosclerotic plaques cause stenosis or occlusion of the carotids. The atherosclerotic disease process may cause the carotid to elongate and buckle and interfere with blood flow. Atherosclerosis of coronary blood vessels leads to myocardial infarction with a potential for small blood clots and cerebral emboli. Cardiac arrhythmias, such as atrial fibrillation, predispose to clot formation and possibly cerebral emboli. The disease processes involved in atherosclerosis may affect the carotid arteries and blood flow to the brain or cause cerebral ischemia due to emboli.

embolism: Gk. *embolisma,* to throw in
thrombus: Gk. *thrombos,* lump

Other causes of emboli include cardiac valvular disease, endocarditis, and cardiac catheterization. An air embolus may be introduced during head and neck surgery, and fracture of the long bones may result in a fat embolus.

Transient Ischemic Attacks

The term transient ischemic attack refers to a brief episode in which there is interruption of cerebral blood flow and temporary cerebral dysfunction. The event usually lasts minutes and may occasionally last up to 24 hours. The effects of a transient ischemic attack are temporary. The term **progressing stroke** refers to an episode in which there is increasing dysfunction, which is evidence of brain cell death or infarct. When there is prolonged and permanent dysfunction, it is called a **completed stroke.**

The manifestations of this kind of event vary with the location and severity of the brain injury. Table 38.4 lists possible effects of a transient ischemic attack.

There is a 4–10 times increase in risk for stroke in a patient who has suffered a TIA, with the greatest danger in the period immediately following the episode. Approximately one-third of the individuals who have experienced a TIA develop a completed stroke. Arteriography makes it possible to visualize the site of occlusion or malformation. If there is evidence of carotid artery disease, vascular reconstructive surgery is an option. Medical treatment for TIA involves administration of anticoagulants, such as heparin or coumarin or medications that decrease platelet adhesiveness, i.e., salicylates and dipyridamole (di″pi-rid′ah-mōl). The purpose of these drugs is to prevent thromboembolism.

TABLE 38.4 Manifestations of a transient
ischemic attack

Motor Deficits

Hemiplegia. Paralysis of one side
Dysarthria. Impaired articulation
Ataxia. Uncoordinated muscle action

Sensory Deficits

Touch
Vision
Balance

Integrative Deficits

Memory
Recall
Judgment
Language skills
Perceptual disturbances
Disorders of voluntary movement

Strokes Caused by Occlusion

Brain cell death or infarction is the result of occlusion of vessels providing cerebral blood supply. Temporary occlusion causes TIA, and a persistent occlusion leads to ischemic stroke. The event is called a **minor ischemic stroke** if there is improvement within 24 hours, with no residual neurologic abnormality. Another term for this event is **reversible ischemic neurologic deficit** (RIND). The consequences of interrupted cerebral blood flow are determined by the site of occlusion, the region of the brain affected, and the development of **collateral channels** or alternate pathways for blood flow. The following discussion deals with effects of occlusion of the internal carotid and small arteries that penetrate brain tissue.

Atherosclerosis is the most common cause of obstruction of the common or internal carotid artery. The internal carotid gives rise to the **ophthalmic artery** that supplies blood to the retina of the eye. The **anterior cerebral artery** and the **middle cerebral artery** are terminal branches of the internal carotid that supply the brain with blood. There may be no symptoms of occlusion of the internal carotid if only one side is affected and if there is adequate collateral circulation. Neurologic abnormalities may be motor, sensory, or cognitive in nature. These abnormalities include (1) hemiparesis, a partial paralysis on one side, (2) paresthesias, tingling or burning sensations of the skin, (3) monocular visual problems, (4) facial weakness, (5) aphasia, impaired ability to use words, (6) dysarthria or stammering, and (7) headache. Headache, which occurs on the side of the occlusion, is a less common feature and is due to dilation of collateral channels. Occasionally, the patient experiences head noise due to turbulence of blood flow. The cerebral infarction may lead to massive edema, brain compression, and death.

hemiparesis: Gk. *hemi*, half; Gk. *paresis*, paralysis
aphasia: Gk. *a*, not; Gk. *phasis*, speech
dysarthria: Gk. *dys*, hard or bad; Gk. *arthron*, joint

It is believed that hypertension affects small arteries that penetrate the brain and causes small cerebral infarcts called **lacunae** (lah-ku′ne). The areas of infarction are usually less than 10 mm across. The term **lacunar state** refers to a condition in which there are multiple lacunae. The site of infarction is usually the basal ganglia, which are regions of gray matter within cerebral tissue, the pons, or the thalamus. White matter of the cerebral hemispheres or the cerebellum may be involved as well. Lacunae nearly always occur in association with prolonged hypertension.

There is usually evidence of mild neurologic deficit associated with lacunae and this may include sensations of numbness, heaviness, or heat and muscle weakness. There is rapid recovery from each episode, but multiple attacks may lead to permanent neurologic abnormalities including dementia.

Prevention is a major aspect of management of ischemic cerebrovascular disease. Prevention involves treating hypertension and elevated cholesterol levels and eliminating cigarette smoking. Modalities of treatment of cerebral embolism include carbon dioxide inhalation, which may be helpful as a vasodilator during the acute phase of embolism when spasms occur at the site of the embolus. Anticoagulants may be given, provided that there is no evidence of bleeding. Embolic infarctions may cause convulsions that can be prevented by the administration of diphenylhydantoin (di-fen″il-hi-dan′to-in). Patients who have experienced a complete infarction have varying degrees of neurologic deficit and require rehabilitation.

• • •

Arterial occlusion usually causes cellular edema in response to acute hypoxia. This condition develops over a period of minutes to hours and, within days, is followed by vasogenic edema. These events are described by the term ischemic brain edema. There is a delay in demonstrating this by CT, since cell defects of the blood vessel wall require time to develop.

• • •

INTRACRANIAL HEMORRHAGE

A stroke may be caused by spontaneous intracranial hemorrhage. The term **intracerebral hemorrhage** is used if the bleeding involves the brain tissue or if there is bleeding into the subarachnoid space between the arachnoid and pia maters, the term **subarachnoid hemorrhage** is used. Hypertension is a leading cause of intracranial hemorrhage. Aneurysms of cerebral vessels, weakened dilated segments, make such a vessel susceptible to rupture. Other possible causes of brain hemorrhage include tumors, leukemia, anticoagulants, and disseminated intravascular coagulopathy. About 10–20% of cerebrovascular accidents involve intracranial hemorrhage. The incidence in

lacune: L. *lacuna*, hole

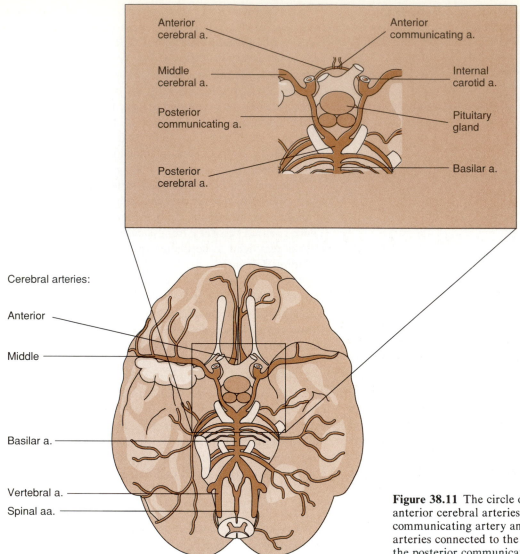

Cerebral arteries:

Anterior

Middle

Basilar a.

Vertebral a.

Spinal aa.

Anterior
cerebral a.

Middle
cerebral a.

Posterior
communicating a.

Posterior
cerebral a.

Anterior
communicating a.

Internal
carotid a.

Pituitary
gland

Basilar a.

Figure 38.11 The circle of Willis is formed by the anterior cerebral arteries connected by the anterior communicating artery and by the posterior cerebral arteries connected to the internal carotid arteries by the posterior communicating arteries.

the United States is higher in women than in men and higher in blacks and Asians than in whites. There has been a decline in frequency of hemorrhagic stroke in the last 20 years, probably due to better control of hypertension. In spite of this, intracranial hemorrhage accounts for 10–20% of all sudden deaths.

Aneurysm

A **saccular aneurysm** is a dilation of an artery due to a defect in the arterial wall. The arterial wall is made up of (1) the inner tunica intima, (2) an internal elastic lamina, (3) a middle layer, the tunica media, (4) an external elastic lamina, and (5) the outer tunica externa (chapter 22). Aneurysms arise in the arterial elastic lamina, which are made up of elastic fibers, and may involve the tunica media. The term **berry aneurysm** is used for small saccular aneurysms.

aneurysm: Gk. *aneurysma*, dilatation
lamina: L. *lamina*, a thin layer

It is uncertain whether or not these defects may be congenital. The role of atherosclerosis is uncertain as well. Hypertension may be a contributing factor in aneurysm formation and rupture. There is increased risk of aneurysm rupture resulting in subarachnoid hemorrhage associated with ingestion of alcohol, smoking cigarettes, and oral contraceptive use after age 35.

Saccular aneurysms may lie within the subarachnoid space or may be located within the brain. Most occur at the circle of Willis or its major branches. The vessels that form the circle of Willis, which surrounds the pituitary infundibulum, are shown in figure 38.11.

The factors that determine the manifestations of intracranial aneurysm are (1) compression of cranial nerves, (2) compression of brain tissue, (3) the presence of a clot in the aneurysm, and (4) rupture and bleeding. There may be evidence of pressure on cranial nerves, or there may be seizures or evidence of involvement of peripheral nerves if

saccular: L. *sacculus*, little sac

there is compression of brain tissue. A blood clot may be the source of emboli causing a TIA or infarction. Many patients with aneurysms have no symptoms unless there is rupture.

The first symptom of aneurysmal rupture is usually sudden severe headache. This may occur after physical exertion, but it frequently happens during normal activity or sleep. Some individuals remain alert, while others experience confusion, delirium, or coma. Massive hemorrhage may lead to death.

• • •

There may be days or weeks of warning signals preceding the rupture of an aneurysm. Suspicious symptoms include the abrupt onset of prolonged headache, stiff neck, vomiting, and visual disturbances.

• • •

Examination of cerebral spinal fluid reveals red blood cells, **xanthochromia** or a yellow color, increased protein, and low glucose levels. CSF pressure is usually elevated. The xanthochromia and the appearance of red blood cells may be delayed for a period of hours, and both signs disappear in about 2 weeks. Computed tomography may show hematoma, although initial CT scans may show no abnormality. Cerebral arteriography, involving the injection of contrast medium into an artery, provides visualization of blood vessels by X-ray and reveals the presence of an aneurysm in the majority of cases.

Saccular aneurysms are treated surgically. If surgery is delayed, the patient is kept on strict bed rest and sedated. Analgesics and laxatives may be administered, and drugs that control coughing and vomiting may be indicated as well. Fluid restriction aids in reducing cerebral edema and intracranial pressure.

• • •

Antifibrinolytic agents may delay the lysis of a clot inside an aneurysm and prevent a second rupture. Epsilon aminocaproic acid and tranexamic acid are two such agents. There is, however, a higher incidence of vasospasm associated with the use of antifibrinolytic agents.

• • •

CEREBRAL INFARCTION IN YOUNG ADULTS

Stroke is less common in the younger age groups. The incidence of stroke is approximately 3.5/100,000 population in those individuals less than age 35. There are about

xanthochromia: Gk. *xanthos*, yellow; Gk. *chroma*, color

TABLE 38.5 Probable mechanisms whereby drugs, alcohol, and cigarette smoking cause stroke

Methamphetamines. Necrotizing vasculitis leads to hemorrhages
Lysergic acid diethylamide (LSD). Arteriospasms lead to infarction
Heroin. Allergic vascular hypersensitivity leads to infarction
Cocaine. Vasoconstriction leads to subarachnoid hemorrhage and cerebral infarction
Ethyl alcohol. Changes in blood pressure, platelets, cerebral blood flow
Intravenous drug use. Embolism
Cigarette smoking. Vasoconstriction, platelet aggregation, increased fibrinogen and blood viscosity, transient increase in blood pressure

TABLE 38.6 Etiology of strokes in young patients

Atherosclerosis. Hypertension, diabetes, hyperlipidemia
Cardiogenic emboli. Rheumatic heart disease, myocardial infarction, mitral valve prolapse
Thrombosis. Pregnancy, oral contraceptives, blood disorders, diabetes
Subarachnoid hemorrhage. Aneurysm, AVM
Parenchymal hemorrhage. Hypertension, blood disorders

40 cases/100,000 between ages 35 and 44, with increasing incidence thereafter. Males and females are affected equally up to age 45, but stroke victims are predominantly male after this age.

The stroke risk in young adults is increased by drug or alcohol use and by cigarette smoking. Table 38.5 lists possible causes of stroke associated with these risk factors.

There are other causes of stroke in young adults (table 38.6). Atherosclerosis is a possible factor in stroke in younger patients. Hypertension, diabetes, or hyperlipidemia may lead to atherosclerotic disease of cerebral blood vessels and cause infarction. The most common cause of strokes in younger individuals is thrombosis and embolus. Emboli originating in the heart, **cardiogenic emboli,** may occur in association with such conditions as rheumatic heart disease, myocardial infarction, and mitral valve prolapse. In most cases, cardiogenic emboli are the result of rheumatic heart disease. The mechanisms by which mitral valve prolapse causes cerebral ischemia is uncertain. Thrombosis of venous sinuses or veins of the cerebral cortex may occlude the vessels and lead to infarction or hemorrhage. Venous thrombosis occurs in association with pregnancy, oral contraceptive use, blood disorders, and diabetes mellitus. Hemorrhagic stroke may be the result of aneurysms, AVMs, hypertension, or blood disorders.

prolapse: L. *prolapsus*, falling

Oral contraceptive use increases the risk of thrombotic stroke 5–9 times and doubles the risk of hemorrhagic stroke. Cigarette smoking, hypertension, and an older age further increase the risk. Pregnancy carries with it a higher risk of ischemic stroke than oral contraceptive use. This is due to a decrease in fibrinolysins and an increase in fibrinogen during pregnancy, which promotes clotting.

...

SUMMARY

...

Hydrocephalus refers to an increased volume of CSF with dilation of cerebral ventricles. The etiology may be obstruction to CSF flow, increased rate of CSF secretion by choroid plexuses, or impaired absorption of CSF by arachnoid villi. Computerized tomography and nuclear magnetic resonance imaging are the most useful methods for diagnosing hydrocephalus. Lumbar puncture allows examination of CSF and measurement of CSF pressure. There are surgical procedures to shunt fluid away from the ventricles to relieve intracerebral pressure.

Brain edema caused by increased permeability of brain capillaries is called vasogenic edema, which may be associated with brain tumor, abscess, hemorrhage, infarction, and trauma. Cellular brain edema is caused by swelling of all cell types found in the brain. The swelling may be due to hyponatremia or to increased solutes in the brain cells. Interstitial brain edema is caused by obstructive hydrocephalus. Manifestations of cerebral edema include seizures, stupor, and coma. Plasma osmolality must be maintained during treatment. Glucocorticoids are useful in treating vasogenic edema. Administration of a hypertonic solution of mannitol osmotically alleviates intracranial pressure.

An arteriovenous malformation is a tangle of arteries and veins that represents a congenital defect. This type of lesion is frequently associated with cerebral and subarachnoid hemorrhage and may be treated surgically. Stroke refers to brain injury due to vascular rupture and hemorrhage or to occlusion of cerebral blood flow. Occlusion is often the result of atherosclerosis. A blood vessel may be occluded by an embolus or a thrombus. A transient ischemic attack refers to a brief interruption of cerebral blood flow. A completed stroke is an event in which there is permanent injury and brain cell death called an infarction. Resulting neurologic abnormalities may be motor, sensory, or cognitive. Lacunae are small infarcts within brain tissue, usually associated with mild neurologic deficit followed by rapid recovery. An aneurysm is a weakened segment of an arterial wall resulting in a dilatation. Saccular aneurysms may be in the subarachnoid space or may be within the brain. Aneurysmal rupture may result in massive hemorrhage and may result in death. Stroke is less common in younger age groups, with increased risk associated with drug or alcohol use or with cigarette smoking. Reducing risk factors is an important aspect of management of stroke.

REVIEW QUESTIONS

...

1. What is the source of CSF?
2. Where does CSF circulate after it leaves the ventricles?
3. How is CSF returned to the general circulation?
4. What are four general categories of causes of obstructive hydrocephalus?
5. What are two mechanisms leading to hydrocephalus other than obstruction?
6. What are the most useful procedures for diagnosing hydrocephalus?
7. List three forms of brain edema, classified on the basis of pathogenesis.
8. What are two mechanisms by which cellular brain edema occurs?
9. Why, in the treatment of cerebral edema, should intravenous fluids contain salt?
10. What are two types of abnormalities involving cerebral vessels that may lead to hemorrhage?
11. What are two types of events that lead to stroke?
12. List four types of emboli which may occlude cerebral vessels.
13. Distinguish between a transient ischemic attack and a completed stroke.
14. What two categories of drugs are useful in the treatment of TIA?
15. What are two important factors that determine the consequences of interrupted cerebral blood flow?
16. Why is vision likely to be affected if there is occlusion of the internal carotid?
17. Define the terms hemiparesis, aphasia, dysarthria, and paresthesia.
18. Define the term lacuna.
19. What is a saccular aneurysm?
20. What is the first symptom of aneurysmal rupture?
21. What does CSF examination reveal in the case of intracranial bleeding?
22. What diagnostic procedure is useful in identifying an aneurysm?
23. What factors related to life-style increase risk of stroke in young adults?
24. What are two disorders that may lead to cardiogenic emboli?
25. What are two possible factors associated with venous thrombosis?

Buchanan, J. F. 1989. Cocaine intoxication: Presentation and management of medical complications. *Physician Assistant* 13(11):87–103.

Cohen, S. N. 1990. An approach to the management of cerebrovascular diseases. *Journal of American Academy of Physician Assistants* 3:11–30.

del Zoppo, G. J. 1989. Antiplatelet agents in arterial vascular disease. *Laboratory Medicine* 20(10):673–81.

Grasch, A. L., and H. E. Roberts. 1989. Carotid bruit: Clinical significance, implications, diagnosis, and management. *Journal of American Academy of Physician Assistants* 2:447–59.

Harrington, M. 1988. Cerebrospinal fluid protein analysis in diseases of the nervous system. *Journal of Chromatography* 429:345–58.

Kunkel, R. S. 1990. Mixed headache. *Physician Assistant* 14:94–102.

Pearl, M. 1989. Nontrauma emergencies: Respiratory, central nervous system, and pelvic. *Physician Assistant* 13:25–41.

Staijch, J. M. 1989. Common neurologic disorders part 1: Vascular syndromes and epilepsy. *Physician Assistant* 13(11):13–30.

Chapter 39

Cerebral Function Disorders

The preceding chapter focuses on disorders involving cerebrospinal fluid and cerebral blood flow and the consequences related to cerebral function. The emphasis of this chapter is cerebral function as it relates to other disorders, such as seizures, tumors, injury, and degenerative processes.

Cerebral disorders may be the result of a localized lesion, or there may be diffuse involvement of brain tissue. The term **focal deficit** refers to the manifestations of a localized lesion. Symptoms attributable to a lesion involving a specific region of the brain may include (1) hemiplegia (hem"e-ple'je-ah), a paralysis of one side of the body; (2) aphasia or impairment of the ability to use words; and (3) hemianopia (hem"e-ah-no'pe-ah), blindness in half the field of vision. In contrast to this, there may be diffuse or generalized involvement that can cause delirium, dementia, or seizures. Delirium leads to impaired consciousness, usually due to systemic disease or drug intoxication. Degeneration or structural disease of the brain causes dementia, which results in progressive intellectual deterioration characterized by impairment of memory, orientation, ability to learn, and visual perception.

The following discussion deals with the topic of seizures, with a description of the classification and manifestations of epileptic seizures.

EPILEPSY

Epilepsy means seizure, and the term is used to describe a cluster of symptoms related to sudden, temporary hyperactivity of brain cells. This burst of nerve impulses may be restricted to a small area of the cerebrum or may spread to involve a larger region.

ETIOLOGY AND INCIDENCE

The basic defect in epilepsy is excessive excitability of neurons in the cerebral cortex. There are four main categories of possible causes:

1. The cause may be metabolic and be the result of factors, such as electrolyte or acid-base imbalance, hypoglycemia, hypoxia, drug intoxication, or drug withdrawal.
2. Lesions of the cortex due to tumors, infarction, contusion, or meningitis may lead to seizures.
3. The cause may be a congenital defect that is developmental or the result of injury at birth.
4. The most common types of seizures are idiopathic with no apparent lesion.

It is estimated that 1% of the population of the United States have epilepsy. The incidence is highest in children less than 5 years of age, and there is an increased incidence after age 50.

hemiplegia: Gk. *hemi,* half; Gk. *plege,* stroke
hemianopia: Gk. *hemi,* half; Gk. *a,* not; Gk. *opsis,* vision

TABLE 39.1 Symptoms of types of partial seizures

Simple Partial Seizures

Partial motor attacks. Involve motor activity of head, face, limbs
Autonomic attacks. Thirst or desire to micturate associated with other manifestations
Somatosensory attack. Paresthesia, flashing lights, buzzing sound, sensations of taste and smell
Psychic seizures. Fear, depression, visual and auditory distortions

Complex Partial Seizures

Start as simple partial seizure and progress to impaired consciousness or start with impaired consciousness

CLASSIFICATION

Epileptic seizures are categorized as being either partial or generalized. A partial seizure is localized on one side of the cerebral hemispheres and may or may not spread to the other side. In generalized seizures, brain dysfunction is bilateral.

Partial Seizures

The symptoms of partial seizures are summarized in table 39.1. Partial seizures are categorized as either simple or complex. A simple partial seizure does not involve impaired consciousness. In the complex type, there is unresponsiveness or a degree of altered awareness.

The somatosensory and psychic attack symptomatic of a simple partial seizure may also be the first event in either a complex partial or generalized seizure. The term **aura** is used for sensory distortions that precede a complex or generalized seizure. Complex partial seizures are characterized by impaired consciousness during which the individual is unresponsive, either initially or subsequent to an aura. During the unresponsive state, the individual may perform simple acts, such as chewing, swallowing, looking at objects, or walking around. This usually lasts for several minutes and is followed by loss of memory of the events. The term **automatisms** is used for the behavior during this unresponsive state.

Generalized Seizures

Generalized seizures are characterized by brain dysfunction on both sides of the cerebral hemispheres. Characteristics of two types of generalized seizures are listed in table 39.2. The **absence seizure** is a common pattern that occurs almost exclusively in children. It lasts for a period of a few seconds during which the individual suddenly stops and stares and sometimes blinks repeatedly. There is quick recovery with complete or partial amnesia of the incident.

The manifestations of a **tonoclonic seizure** are more dramatic than those of an absence seizure. The individual falls and the tonic phase involves stiffening of muscles. This phase usually lasts less than a minute. The tonic phase is followed by the clonic phase with involuntary muscle jerks

TABLE 39.2 Characteristics of generalized seizures

Absence seizures (petit mal). Abrupt onset, absence of aura, interruption of consciousness for a period of seconds, few to hundreds of incidents per day, immobility, simple automatic movements

Tonoclonic seizures (grand mal). Complete loss of consciousness, body stiffens, jerking of extremities, incontinence, frothing at mouth, confused recovery period and amnesia

for a period of several minutes. There may be incontinence and increased salivation with frothing at the mouth. The individual feels confusion during the recovery period and has no memory of the seizure.

• • •

Hypoglycemia may cause partial seizures. Other metabolic disturbances are more likely to lead to generalized seizures. There is greater danger of seizures in response to rapid metabolic changes as compared to gradual change.

• • •

TREATMENT

It is possible to completely control seizures with anticonvulsant drugs in over half the patients. These drugs include phenytoin, phenobarbital, carbamazepine, and valproic acid. Anticonvulsant therapy reduces the frequency of seizures in most patients. **Status epilepticus** is a special case requiring immediate medical attention. This condition is characterized by repeated generalized convulsions without recovery of consciousness. The consequent hypoxia, hypoglycemia, and acidosis are life-threatening. There must be drug therapy for the seizures, adequate ventilation, administration of bicarbonate to treat acidosis, and glucose to correct hypoglycemia.

INTRACRANIAL TUMORS

Brain tumor is the second most common cause of death associated with the nervous system. Most brain tumors occur after age 45. Tumors of the central nervous system are, however, the second leading cause of cancer deaths in individuals under 15 years and are the third leading cause of death in the age group 15–34 years. Most brain tumors are the result of metastasis from other sites such as breast, lung, kidney, gastrointestinal tract, and melanoma. Primary brain tumors may develop from a common stem cell prior to differentiation or may develop from various differentiated cell types found in the brain.

TABLE 39.3 Symptoms often associated with specific regions of the brain

Frontal lobe tumors. Personality changes, headache, mild slowing of hand movement, mood elevation, loss of initiative, difficulty in speaking or understanding language (dysphasia)

Temporal lobe tumors. Slight changes in perception, impairment of recent memory, auditory hallucinations, aggressive behavior

Parietal lobe tumors. Affect sensory more than motor function, severe sensory loss

Occipital lobe tumors. Aberrations of color, size, or location of objects

Brain stem tumors (medulla oblongata and pons). Hearing loss, facial weakness and pain or numbness, gait ataxia (foot raised and brought down flat)

Cerebellum. Headache, nausea, vomiting

MANIFESTATIONS

In general, symptoms produced by a brain tumor are either the result of increased intracranial pressure or by a focal or a localized disturbance due to infiltration of brain tissue by the tumor. The manifestations are determined by the size, location, and rate of tumor growth.

The cranium represents a fixed space for the brain, blood, and cerebrospinal fluid. When neoplasms growing within that space become larger than three centimeters in diameter, the result is increased intracranial pressure. Blood vessels associated with a growing tumor have gaps that allow the escape of plasma. The result is cerebral edema, which increases volume further.

Headache is common as an initial symptom of brain tumor, and the onset and fluctuation of pain seem to correlate with changes in intracranial pressure. Gastrointestinal symptoms, such as loss of appetite, nausea, and vomiting are common. There may be changes in personality, mood, mental capacity, and ability to concentrate. Slowed psychomotor activity and a tendency to sleep more are possible symptoms.

Focal manifestations of infiltrating tumors include motor or sensory seizures as well as generalized seizures. Although there may be marked variations, certain symptoms help provide clues to which region of the brain is involved. Table 39.3 lists symptoms which may be associated with the different lobes of the brain (figure 39.1). Metastatic brain tumors are more likely than primary tumors to produce acute symptoms that develop over a short period of time.

• • •

Frontal and temporal tumors usually cause headache behind the eyes or in the frontal or temporal regions.

• • •

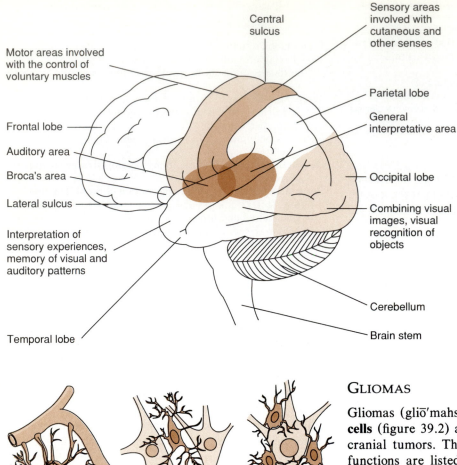

Motor areas involved with the control of voluntary muscles

Central sulcus

Sensory areas involved with cutaneous and other senses

Frontal lobe

Parietal lobe

General interpretative area

Auditory area

Broca's area

Occipital lobe

Lateral sulcus

Combining visual images, visual recognition of objects

Interpretation of sensory experiences, memory of visual and auditory patterns

Temporal lobe

Cerebellum

Brain stem

Figure 39.1 The lobes of the left cerebral hemisphere showing the principal motor and sensory areas of the cerebral cortex. Dotted lines show the division of the lobes.

Figure 39.2 Various types of neuroglial cells that connect, support, and nourish neurons. (*a*) An astrocyte in contact with a capillary of a blood vessel serving the brain; (*b*) a microglia with processes extending to two nerve cell bodies; (*c*) oligodendrocytes near a nerve cell body.

TABLE 39.4 Characteristics of neuroglial cells found in the central nervous system

Astrocytes. Cells with star-shaped bodies and numerous processes that provide structural support
Oligodendrocytes (*oligo,* small). Small with fewer processes than astrocytes; processes form myelin sheath around axons
Microglia. Smallest of neuroglial cells; phagocytes
Ependymal cells. Line the ventricles of the brain

GLIOMAS

Gliomas (glīo′mahs) arise from primitive forms of **glial cells** (figure 39.2) and make up nearly half of all intracranial tumors. The types of neuroglial cells and their functions are listed in table 39.4. Three types of neoplasms that arise from neuroglial cells are discussed.

Astrocytoma

Malignant astrocytoma constitutes three-fourths of glial tumors in adults. The terms **glioblastoma** or **grade III astrocytoma** are also used for this type of neoplasm. It is an invasive, rapidly growing tumor with the highest incidence occurring between ages 40–70 years. It commonly occurs in the cerebral white matter, the frontal, parietal, and temporal lobes. With aggressive radiation, chemotherapy, or surgery, the median survival time is 62 weeks. **Astrocytomas, grades I and II,** are slower growing and infiltrate the white matter of both brain and spinal cord. Cerebellar astrocytomas are more common in children. This type of tumor may develop over a period of years. Treatment involves total surgical removal if possible. There is a median life expectancy of 67 months with brain tumors of this type.

glia: Gk. *glia,* glue

•••

The astrocytes are believed to contribute to the blood-brain barrier because many of their processes are in contact with blood vessels in the brain. It has been observed that many substances pass into capillaries in the brain slowly.

•••

Glioblastoma Multiforme

Glioblastoma multiforme is a neoplasm that arises from the differentiation of mature glial cells, most of which are astrocytes. This type of tumor is fast growing with the highest incidence between ages 40–70 years. The term multiforme is used because there is typically a variation in the appearance and consistency of the tumor. The disease progresses rapidly, and the manifestations are those of both increased intracranial pressure and localized disturbances. Surgery followed by radiation therapy offers the most hope. The prognosis is poor with median survival time of 9–10 months.

MENINGIOMA

There are three layers of connective tissue surrounding the central nervous system that collectively are called meninges. The outer layer is called the dura mater, the middle layer is the arachnoid mater, and the pia mater is the innermost layer (chapter 38). A meningioma is a neoplasm that arises from arachnoidal fibroblasts. This type of tumor occurs mainly in adults with a peak incidence at about age 45. It occurs more commonly in women. The growth is slow, but eventually the tumor becomes large, causes edema, compresses the cerebrum, and leads to death. The tumor may invade the skull. The recurrence after 5 years following surgical removal is about 10%. There is a higher probability of surgical treatment of meningiomas than in any other brain tumor.

CEREBRAL INJURY

Cerebral dysfunction due to seizures and tumors have been discussed in the preceding sections. The following is a discussion of two disorders associated with injury to the brain, i.e., minimal brain dysfunction and cerebral palsy.

MINIMAL BRAIN DYSFUNCTION

Minimal brain dysfunction (MBD) is characterized by various learning disabilities or abnormal behavior patterns, such as a tendency to be easily distracted, impulsiveness, hyperactivity, and short attention span. It is estimated that 5–10% of school-age children are in some way affected by MBD. The disorder occurs more often in males, in a ratio of 4:1.

TABLE 39.5 Manifestations of minimal brain dysfunction syndrome

Abnormal Behavior

Short attention span, inability to ignore trivial stimuli or to sit still, excessive running, abnormal concentration on a single object, emotional lability, aggressiveness, rage, emotional immaturity

Learning Disabilities

Difficulties with abstract ideas, reading, writing, spelling, calculations

Speech Disorders

Poor articulation, misuse of words

Impaired Coordination

Clumsiness, poor fine and gross motor coordination

Etiology

The etiology of MBD is uncertain. There is evidence indicating that brain damage may cause the disorder. There are similarities in behavior patterns, for example, in individuals who have experienced severe head injury and those who have recovered from encephalitis. There is increased risk of the disorder in premature infants or in those who suffer from anoxia at birth. Other possible causes include chronic lead poisoning and alcohol ingestion during pregnancy. There may be genetic factors as well. There is some evidence that depletion of dopamine, a CNS neurotransmitter, may be a factor in MBD.

Manifestations

Various manifestations of the minimal brain dysfunction syndrome are listed in table 39.5. In general, there are behavior problems, learning disabilities, and problems with speech and coordination. Hyperactivity is common, although some affected children are hypoactive. New or stressful situations exacerbate hyperactive tendencies. The most common cognitive dysfunction is dyslexia, a misperception of written words. There may be language difficulties and problems with activities requiring fine and gross motor coordination. Few of the manifestations may be evident in some cases of MBD.

Treatment

Medication may be required to improve behavior. Psychostimulants have an unexpected calming effect on hyperactive children, and this type of drug is frequently used, although the mechanism of action is not clear. Medications that may be given include dextroamphetamine, methylphenidate, and pemoline. A program for treating MBD includes family counseling and proper school placement, as well as medication.

TABLE 39.6 Definition of terms used to describe manifestations of cerebral palsy

Spastic. Increased muscle tone; sudden muscle contraction
Paresis. Partial paralysis
Hemiparesis. Weakness on one side
Tetraparesis. Weakness involving all four limbs
Diplegia (diparesis). Paralysis on both sides of the body
Athetosis. Slow continual change of position of fingers, toes, hands, and feet
Dyskinesia. Abnormal involuntary movements
Ataxia. Uncoordinated body movements
Dysarthria. Difficulty in enunciation
Dysphagia. Difficulty in swallowing

CEREBRAL PALSY

Cerebral palsy refers to various nonprogressive motor disorders associated with dysfunction of motor areas of the brain. The cause is prenatal injury, injury at birth, or trauma early in life.

Etiology

There are various prenatal events or factors believed to cause cerebral palsy, i.e., abnormal implantation of ovum, toxins, metabolic disease, or maternal disease. The most common factor associated with cerebral palsy is a birth weight less than 2,500 g. Complications of childbirth that are associated with hypoxia or intracranial bleeding may cause cerebral palsy. After birth, the most likely causes are infections and trauma.

Manifestations

Various terms which describe manifestations of cerebral palsy are defined in table 39.6. There are three categories of this disorder: spastic, dyskinetic, and ataxic.

The **spastic type** is characterized by a form of paralysis associated with rigidity due to continuous muscle contractions. When the limbs on only one side are affected, it is called **spastic hemiparesis.** Posture and walk are affected by the condition. There may be impaired sensory-cortical function which affects perception of size, shape, and texture. The term **tetraparesis** is used when all four limbs are equally affected. Spastic tetraparesis is often associated with dysarthria, dysphagia, and drooling. **Spastic diplegia** (di-ple'je-ah) refers to the condition when the legs and feet are more affected than the arms and hands.

Dyskinetic cerebral palsy involves abnormal involuntary movements, especially **athetosis,** that result in hypertrophy of the constantly moving muscles. Athetosis refers to a continual slow change in position of hands, feet, and other parts of the body. The movement subsides in sleep.

spastic: Gk. *spastikos*, drawing in
dyskinetic: Gk. *dys*, hard or bad; Gk. *kinesis*, motion
athetosis: Gk. *athetos*, without position
paresis: Gk. *paresis*, paralysis
diplegia: Gk. *di*, two; Gk. *plege*, a stroke

Ataxic cerebral palsy is characterized by uncoordinated movements. Asymmetric kicking and backward head thrust are manifestations of cerebral palsy in an infant.

• • •

The manifestations of cerebral palsy may change, even though the disorder is not progressive. Weight gain may cause a patient to become nonambulatory, and failure to use muscles and joints affects mobility. Uncontrolled seizures may lead to loss of mental ability.

• • •

Treatment

Although there is no cure for cerebral palsy, early treatment usually results in improvement. Treatment is individualized with the purpose of improving function, controlling seizures, and providing emotional support. Modalities of treatment include speech therapy, as well as physical and occupational therapy. On occasion, surgery may improve a particular function.

• • •

Motor function is impaired in cerebral palsy, and there may also be impairment of intellectual development, vision, and hearing. Cerebral palsy is often associated with mental retardation, epilepsy, and minimal brain dysfunction.

• • •

DEGENERATIVE DISORDERS

The last part of this chapter focuses on disorders in which there are degenerative changes in the brain. In the cases of AIDS dementia and Creutzfeldt-Jakob disease an infectious agent is involved, and the etiology is unknown in the case of Alzheimer's disease.

ALZHEIMER'S DISEASE

Alzheimer's disease is a degenerative disorder of the brain which causes dementia, i.e., impaired intellectual function. This disorder and **senile** (se'nil) **dementia** are now considered the same disease. Males and females appear to be equally affected, with the usual age of onset between 65–70 years. The disease process may become apparent as early as age 40. It is estimated that there are 700,000 individuals in the United States who are severely affected by the disease.

senile: L. *senilis*, pertaining to old age

Figure 39.3 shows a microscopic view of a normal cerebrum. In Alzheimer's disease, there is diffuse atrophy of the cerebral cortex, which is the outer layer of gray matter. Microscopically, there is loss of neurons in the cortex. There may also be a demyelination, a loss of the fatty myelin covering of axons that make up the subcortical white matter. There are two characteristic microscopic observations: **senile plaques** and **neurofibrillar** (nu″ro-fi-bril′ar) **tangles.** Senile plaques are made of a core of a type of protein called amyloid. Degenerating axon nerve endings surround this core. Degenerating nerve endings contain filaments that accumulate to form tangles within the body of a swollen neuron. These constitute neurofibrillar tangles which may be scattered throughout the cerebral cortex.

Etiology

In some cases, there is evidence that the disease is associated with an autosomal dominant trait, although there is usually no apparent pattern of inheritance. The possibility of chromosomal changes in Alzheimer's disease has been suggested because older individuals with **Down's syndrome** experience changes that mimic Alzheimer's disease. Down's syndrome is a condition in which there is an extra number 21 chromosome; thus, the condition is called **trisomy 21.** Down's syndrome results in mental retardation. No consistent chromosomal abnormality has been demonstrated in Alzheimer's disease.

The most consistent biochemical change identified in Alzheimer's disease is a reduction in **choline acetyltransferase,** an enzyme involved in the synthesis of acetylcholine. Acetylcholine is a neurotransmitter released by cholinergic neurons (figure 39.4). There is evidence that the degree of impaired mental function in life is approximately proportional to the loss of choline acetyltransferase, which is determined at autopsy.

Manifestations

Alzheimer's disease is characterized by a progressive dementia. Manifestations early in the course of the disease include (1) memory loss, (2) slowed thought processes, (3) difficulty in comprehending words, oral or written, (4) speech disturbances, (5) depression, (6) agitation, and (7) restlessness. As the disease progresses there are changes in reflexes, general weakness, and a shuffling gait. Ultimately, there are seizures and loss of bowel and bladder function. A vegetative state precedes death. The rate of mental and physical deterioration may vary, and there may be periods in which there is little change. The usual course of the disease is 4–10 years.

Treatment

Treatment involves supportive care. No drugs have been shown to slow the course of the disease or to result in improvement.

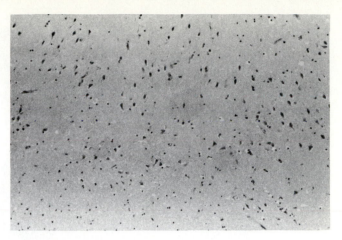

Figure 39.3 Microscopic view of normal cerebrum. *Courtesy of Dr. M. G. Klein.*

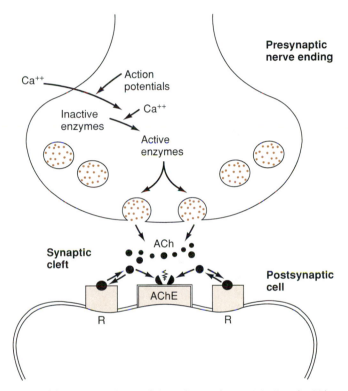

Figure 39.4 Mechanisms of the release of acetylcholine (ACh) from presynaptic nerve endings and the binding of ACh to receptor proteins (R) in the postsynaptic membrane. Acetylcholine that combines with acetylcholinesterase (AChE) in the postsynaptic membrane is hydrolyzed and thus inactivated.

CREUTZFELDT-JAKOB DISEASE

Creutzfeldt-Jakob disease is a rapidly progressive dementia that is uniformly fatal. It is a disease of middle age with an average age of onset of 52 years. The incidence is approximately one per million each year, and the distribution is worldwide. The disease process involves degeneration and atrophy of the brain and spinal cord. The

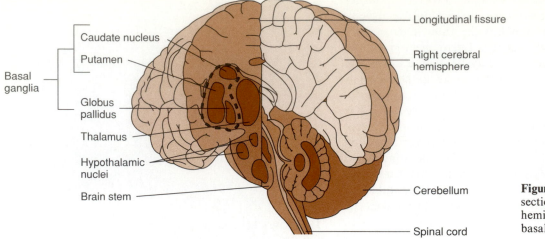

Basal ganglia
- Caudate nucleus
- Putamen
- Globus pallidus

Thalamus

Hypothalamic nuclei

Brain stem

Longitudinal fissure

Right cerebral hemisphere

Cerebellum

Spinal cord

Figure 39.5 A frontal section of the left cerebral hemisphere reveals the basal ganglia.

cortex and the basal ganglia of the cerebral hemispheres are predominantly involved (figure 39.5). Microscopically, there is loss of supporting cells, the astrocytes. It is characterized by an absence of inflammation.

Etiology

The etiologic agent for Creutzfeldt-Jakob disease is a virus, which does not elicit an antibody response. The mode of transmission is uncertain, and it is possible that genetically determined susceptibility is a factor. An isolated case of transmission by a corneal transplant has been reported. There have been two reports of transmission by formalin-sterilized electrodes used during craniotomy.

• • •

The etiologic agent of Creutzfeldt-Jakob disease is extremely resistant to formalin and other routine methods of decontamination. Consequently, there are special problems in handling both tissue taken at autopsy and surgical and other instruments. Current recommendations include incineration of disposable equipment and either autoclaving instruments for 1 hour at 121°C or immersing in 0.5% sodium hypochlorite for 1 hour. An iodine solution, such as Betadine, hexachlorophene, or Lysol, which are phenolic compounds; 0.5% sodium hypochlorite; or a 1:3000 potassium permanganate solution may be used to cleanse a wound resulting in percutaneous exposure.

• • •

Manifestations

There are gradual changes in personality, recent memory, and intellectual abilities. There may be vague symptoms of unusual behavior, fatigue, impaired judgment, dizziness, and headache. There may be an impaired ability to use words, a condition called aphasia, and visual misperceptions may occur. Disorders of movement include reflex changes, weakness and stiffness of limbs, muscle atrophy, tremors, rigidity, and **myoclonus.** Myoclonus involves convulsive muscle contractions that may be triggered by auditory and various other stimuli. Seizures may occur and there may be signs due to involvement of any part of the brain or spinal cord. Ultimately, a vegetative state develops and death ensues.

Treatment

There is no treatment for Creutzfeldt-Jakob disease, and death usually occurs within 9 months of onset of symptoms.

• • •

In the past, growth hormone was prepared from collections of human pituitary glands taken at autopsy. This form of growth hormone was administered to approximately 10,000 individuals in the U.S. prior to 1985, when it was discontinued. Several young adults who were treated with this growth hormone have developed Creutzfeldt-Jakob disease, 4–21 years after therapy.

• • •

AIDS Dementia

Acquired immunodeficiency syndrome (AIDS) is caused by the human immunodeficiency virus (HIV). The virus is transmitted by sexual contact and by parenteral routes (chapter 29). There is generalized involvement of body systems as the viral infection is established. Gastrointestinal complications of AIDS are discussed in unit VIII. **AIDS dementia complex** is a common complication of AIDS involving the central nervous system. Other neurologic complications include opportunistic infections and neoplasms.

craniotomy: Gk. *kranion*, skull; Gk. *tomos*, cutting
percutaneous: L. *per*, through; L. *cutis*, skin

myoclonus: Gk. *mys*, muscle; Gk. *klonos*, violent motion
aphasia: Gk. *a*, not; Gk. *phasis*, speech

TABLE 39.7 Manifestations of AIDS dementia complex

Gait unsteadiness	Paranoia
Leg weakness	Hallucinations
Loss of coordination	Urinary incontinence
Handwriting changes	Fecal incontinence
Apathy	Tremor
Agitation	Myoclonus
Confusion	Agitated psychosis

It is estimated that 90% of AIDS patients ultimately have neurologic involvement. It is probable that the AIDS virus infects the brain soon after exposure, although neurologic symptoms usually occur after the disease is diagnosed.

Manifestations

Meningitis may occur, during which time HIV appears in the cerebrospinal fluid. The AIDS dementia complex develops after this and is associated with various cognitive, motor, and behavior disorders. Various manifestations are listed in table 39.7.

In addition to the AIDS dementia complex, affected individuals may experience several types of peripheral neuropathies. The most common type develops late in the course of the disease and causes sensory loss and motor weakness accompanied by painful **dysesthesia** (dis″es-the′ze-ah), or impaired sense of touch.

The course of the disease varies, so deterioration may progress slowly over a period of months, up to two or more years, or there may be abrupt impairment with rapid deterioration.

Treatment

Treatment of the AIDS dementia complex is supportive and involves the control of symptoms.

OTHER CAUSES OF DEMENTIA

Alzheimer's disease accounts for more than half the cases of dementia, while vascular disease is responsible for approximately 25% of the cases. Dementia caused by multiple strokes is called **multi-infarct dementia** and may be the result of thrombosis, emboli, or hemorrhages. The degree of dementia is determined by the amount of brain tissue destroyed by infarction. Impairment is apparent when 50–100 g or more of cerebral tissue is destroyed. Diffuse mental impairment is usually associated with diabetic or hypertensive vascular disease.

Other causes of dementia include brain tumors, subdural hematomas, thiamine deficiency usually noted in alcoholics, hypothyroidism, and chronic intoxication due to prolonged administration of drugs.

dysesthesia: Gk. *dys*, hard or bad; Gk. *aisthesis*, sensation

SUMMARY

• • •

The term epilepsy refers to symptoms resulting from temporary neuronal hyperactivity in the brain. For the most part, the cause is unknown and results in either partial or generalized seizures. A partial seizure is initiated on one side of the brain and may or may not spread to the other side. Simple partial seizures do not lead to impaired consciousness, whereas complex seizures result in some degree of altered awareness. Generalized seizures are characterized by brain dysfunction on both sides of the cerebral hemispheres. Anticonvulsant drugs are used to control or reduce the frequency of seizures.

Intracranial tumors may be the result of metastasis from other sites or may develop from different cell types in the brain. Symptoms of a brain tumor are the result of either increased intracranial pressure or to tumor infiltration of the brain. Gliomas are common brain tumors that arise from neuroglial cells. Astrocytoma is the most common type of glioma. A meningioma is a neoplasm that arises from one of the meninges, the arachnoid mater.

Examples of cerebral dysfunctions due to brain injury are minimal brain dysfunction and cerebral palsy. Minimal brain dysfunction is characterized by various learning disabilities or abnormal behavior patterns. The etiology is uncertain, but there is evidence indicating that brain damage may be the cause. Hyperactivity and behavior problems are common, with dyslexia the most common cognitive dysfunction. Paradoxically, psychostimulants are useful in the management of MBD. Cerebral palsy refers to various nonprogressive motor disorders caused by prenatal injury, injury at birth, or trauma early in life. There is a spastic type characterized by a form of paralysis associated with rigidity. Dyskinetic cerebral palsy involves abnormal involuntary movements and uncoordinated movements characterize ataxic cerebral palsy. There is no cure for cerebral palsy, although it is possible to improve function.

The degenerative disorders discussed in this chapter include Alzheimer's disease, Creutzfeldt-Jakob disease, and AIDS dementia. The etiology of Alzheimer's disease is unknown and is characterized by mental and physical deterioration ending in death. Creutzfeldt-Jakob disease is a rapidly progressive dementia caused by a submicroscopic infectious agent. The disease is uniformly fatal. AIDS dementia complex is a common complication of AIDS involving various cognitive, motor, and behavior disorders.

1. What is a major difference between partial and generalized seizures?
2. Distinguish between a simple and a complex partial seizure.
3. What are the meanings of the terms tonic and clonic?
4. What type of glial cell is most frequently involved in neoplasms?
5. Which of the three types of brain tumors discussed in this chapter is most treatable by surgery?
6. List two possible causes or risk factors in MBD.
7. What are three major categories of problems that an MBD child may have?
8. What are three types of events that may cause cerebral palsy?
9. What are three categories of cerebral palsy?
10. Define the terms (a) spastic, (b) paresis, and (c) athetosis.
11. Define the terms (a) dyskinesia and (b) ataxia.
12. What is the term that means difficulty in enunciation?
13. What is the term for difficulty in swallowing?
14. Identify two characteristic microscopic findings in Alzheimer's disease.
15. What biochemical change has been identified in Alzheimer's disease?
16. What period of time is usually involved between onset of symptoms and death in Alzheimer's disease?
17. What is the cause of Creutzfeldt-Jakob disease?
18. How long does a patient with Creutzfeldt-Jakob disease usually survive after onset of symptoms?
19. List six manifestations of AIDS dementia complex.
20. What is the meaning of the term myoclonus?

Berkovic, S. F. et al. 1986. Progressive myoclonus epilepsies: Specific causes and diagnosis. *New England Journal of Medicine* 315:296–304.

Borson, S. et al. 1989. Impaired sympathetic nervous system response to cognitive effort in early Alzheimer's disease. *Journal of Gerontology* 44:M8–12.

Dichter, M., and G. Ayala. 1987. Cellular mechanisms of epilepsy: A status report. *Science* 237:157–63.

Franceschi, M. et al. 1988. Neuroendocrinological function in Alzheimer's disease. *Neuroendocrinology* 48:367–70.

Green, R. et al. 1988. Epilepsy surgery in children. *Journal of Child Neurology* 3:155–56.

Kety, S. S. 1979. Disorders of the human brain. *Scientific American* 241:202–14.

Lecso, P. 1989. Murder-suicide in Alzheimer's disease. *Journal of American Geriatric Society* 37:167–78.

Pontecorvo, M. et al. 1988. Age-related cognitive as assessed with an automated repeated measures memory task: Implications for the possible role of acetylcholine and norepinephrine in memory dysfunction. *Neurobiology Aging* 9:617–25.

Rasool, G. G., and D. J. Selkoe. 1985. Sharing of specific antigens by degenerating neurons in Pick's disease and Alzheimer's disease. *New England Journal of Medicine* 312:700–705.

Ribak, C. E. 1987. Epilepsy hypothesis. *Science* 238:1292.

St. George-Hyslop, P. H. et al. 1987. Absence of duplication of chromosome 21 genes in familial and sporadic Alzheimer's disease. *Science* 238:664–69.

Chapter 40

Sensory and Motor Disorders

The topics of this chapter include pain of the head and face, and a sensory disorder involving the inner ear. Disorders of movement are discussed and include types that are due to (1) motor neuron abnormalities, (2) degenerative muscle changes, (3) defective neuromuscular transmission, (4) motor pathway abnormalities, and (5) demyelination.

HEAD PAIN

Head pain is a frequent complaint and cause of disablement. Although the brain itself is not sensitive to pain, there are both intracranial and external structures that are pain sensitive. External causes of head pain include (1) contraction of scalp and neck muscles, (2) inflammation of sinuses, (3) increased intraocular pressure, (4) inflammation involving the eyes, and (5) inflammation or dilation of blood vessels. Dilation of intracranial blood vessels occurs in response to severe hypertension, hypoxia, fever, and hyperthyroidism. Meningitis and subarachnoid hemorrhage cause headache. Brain tumors, hydrocephalus, and abscess cause headache if pain sensitive structures are distorted. The most common types of headaches, by far, are either tension-muscle contraction or migraine headaches.

MIGRAINE HEADACHE

Migraine headache represents a group of disorders of unknown etiology characterized by a pattern of repeated vascular type headaches. The underlying cause of pain seems to be dilation of a branch or branches of the carotid artery, which stimulates arterial nerve endings.

The pathogenesis may involve the release of a chemical substance that causes a lowered pain threshold. Suspected chemicals include serotonin, histamine, bradykinins, and prostaglandins. The pain usually pulsates and involves one side of the head.

Migraine affects approximately 15% of the population, with a 3:2 female to male ratio.

The **classic migraine** is characterized by neurologic manifestations preceding the attack. These include visual manifestations, such as bright flashing lights and tingling or burning sensations of the skin of the hand or mouth. The latter cutaneous sensations are called **paresthesias.** The neurologic disturbances last about 30 minutes and are followed by a throbbing unilateral headache frequently accompanied by nausea and vomiting. There may also be an intolerance for noise and light. The attack may persist for hours to a day or more.

More individuals suffer from **common migraine** as compared to the classic type just described. Some people experience both types. The neurologic manifestations preceding common migraine include depression, hyperactivity, or increased appetite. The headaches are frequently unilateral and are accompanied by nausea, vomiting, and light sensitivity. Most affected individuals have a family history of migraine.

Both common and classic migraine may be due, for unknown reasons, to painful dilation of arteries supplying the scalp. It appears that there are various precipitating factors including (1) stress, (2) too much sleep, (3) menstruation, (4) bright lights, (5) alcohol, (6) specific foods, such as chocolate, nuts, and cheese, and (7) oral contraceptive use. There is some evidence that suggests oral contraceptive use not only increases frequency of migraine, but also increases the risk of permanent neurologic damage, i.e., partial paralysis on one side or loss of vision on one side.

• • •

Cerebral blood flow in classic and common migraine is different. In classic migraine, there is decreased cerebral blood flow associated with the neurologic symptoms preceding the attack of head pain. There may be cerebral vasodilation and increased cerebral blood flow during the attack of classic migraine. Common migraine appears not to be associated with changes in cerebral blood flow.

• • •

Control of precipitating factors is an important aspect of management of migraine. Drugs may be given prophylactically, i.e., aspirin with or without codeine, ergotamine, antiemetics, or beta blockers, such as propranolol.

TENSION HEADACHE

Tension or muscle contraction headache is the most common type of headache and the cause is uncertain. In some cases, neck or scalp muscle spasm may be a factor. In other cases, the basis for the problem may be stress or other unrecognized factors. Tension headaches tend to involve both sides of the head and tend to be both dull and prolonged. They may be either mild and occasional or extremely severe and occur on a daily basis. An individual may experience intermittent common migraine and tension headaches. Analgesics, and in the case of chronic tension headaches, antidepressant agents are useful for treatment.

CLUSTER HEADACHE

Cluster headaches, which are less common than migraines, involve several episodes a day, continuing for weeks or months. There are intermittent periods of relief with a repeating cycle of headaches. This pattern of headaches usually affects males and typically begins between ages 30–50. The pain is agonizing and nonthrobbing and frequently awakes the individual from sleep. Each episode usually lasts for less than one hour. One eye is often red and watering, the lid may droop, and the nostril on the same side may be congested or running. The cause of cluster headache is unknown. It may involve edema, inflammation, or spasm of the internal carotid artery. The preferred treatment is prophylactic, using methysergide, lithium carbonate, or prednisone.

• • •

Postlumbar puncture headache may occur hours to days after withdrawal of cerebrospinal fluid. This is caused by leakage of CSF resulting in intracranial hypotension. The headache develops when the individual is in an upright position. The pain involves the head, cervical area, and shoulders. The problem may persist for a few days to weeks.

• • •

OTHER SENSORY DISORDERS

One category of sensory disorders is **neuralgia** (nu-ral'je-ah), which refers to severe periodic pain along the course of a nerve. The most common type of neuralgia is trigeminal neuralgia.

TRIGEMINAL NEURALGIA

The trigeminal nerve is the fifth cranial nerve and has three divisions: the **ophthalmic, maxillary,** and **mandibular nerves** (figure 40.1). The trigeminal nerve is a mixed nerve with motor neurons to the muscles of mastication. Branches of this nerve constitute the major sensory nerves of the face.

Trigeminal neuralgia, also called **tic douloureux** (tik doo-loo-roo'), is characterized by burning pain on one side of the face. It involves one or more branches of the trigeminal nerve. The etiology is unknown, although in certain instances there may be pressure on the nerve due to blood vessels or a tumor. The disorder is uncommon before age 35 and occurs slightly more frequently in females.

The pain is sudden, searing, and recurring and may persist for 1–15 minutes or more. The frequency of the attacks varies from multiple daily events to an occasional episode. The eye on the affected side may water. An attack may be set off by touching a particular part of the face, often the cheek, lip, or nose. Movement of facial muscles or the act of chewing may initiate an attack. An affected individual sometimes refuses to eat in order to avoid an episode of pain. The pain is limited to the branches of the trigeminal nerve, usually involving the second or third division. Motor and sensory function of the nerve remains normal. Characteristically, periods of remission occur, but rarely on a permanent basis.

Analgesics, other than opiates, do not relieve the severe pain of this neuralgia. Anticonvulsants, such as phenytoin or carbamazepine, usually provide relief. Other

neuralgia: Gk. *neuron*, nerve; Gk. *algos*, pain
tic douloureux: F. *tic*, twitching; F. *douloureux*, painful

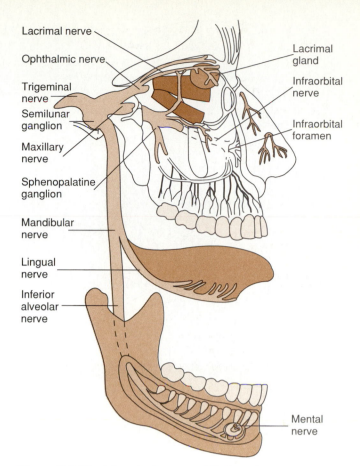

Figure 40.1 The trigeminal nerve and its branches.

measures sometimes employed include (1) alcohol injection of the nerve, (2) partial section of the nerve, and (3) percutaneous thermal destruction of the nerve.

• • •

Tumors of the gasserian ganglion of the trigeminal nerve (figure 40.1) produce facial pain that is different from that of trigeminal neuralgia. The pain usually caused by tumor is both prolonged and steady.

• • •

BELL'S PALSY

Bell's palsy is a disorder involving paralysis of the seventh cranial nerve. Cranial nerve VII, also called the **facial nerve,** provides numerous branches to the muscles of the face and scalp and neurons to lacrimal, mucous, and salivary glands (figure 40.2). There are sensory neurons from the taste buds of the tongue as well. The cause of Bell's palsy is believed to be swelling of the nerve. The paralysis

lacrimal: L. *lacrima*, a tear

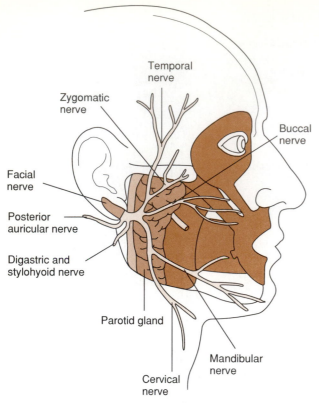

Temporal
nerve

Zygomatic
nerve

Buccal
nerve

Facial
nerve

Posterior
auricular nerve

Digastric and
stylohyoid nerve

Parotid gland

Mandibular
nerve

Cervical
nerve

Figure 40.2 Facial nerve and its branches.

frequently occurs after exposure to cold. The disorder may occur at any age, although it is somewhat more frequent in the age group of 30–50 years.

The first manifestation of Bell's palsy is a feeling of stiffness in the face usually not accompanied by pain. Complete paralysis of the nerve leads to an absence of voluntary facial movement and sagging of muscles of the lower half of the face. An attempt to smile results in a distortion of muscles with the lower muscles pulled to the opposite side. An attempt to close the eye on the affected side results in the eyeball rolling upward and slightly inward. Whether or not lacrimation is affected depends on the location of the lesion. The same is true for salivary secretion and loss of the sense of taste. Weakness of upper and lower halves of the face may occur when there is partial injury to the facial nerve. Partial or complete recovery usually occurs in Bell's palsy. There may be continued disturbances of facial muscle movements and lacrimal secretions. If spontaneous recovery does not occur, surgical procedures may be helpful.

Meniere's Syndrome

Meniere's (men″e-ārz′) syndrome is characterized by acute attacks of **vertigo,** which involves a perception of spinning objects or a sensation of giddiness. This is associated with **tinnitus,** a buzzing sound, and hearing loss.

vertigo: L. *vertigo,* whirling around
tinnitus: L. *tinnitus,* a ringing

Attacks last from minutes to hours and symptoms may include nausea, vomiting, and profuse perspiration. The disorder is a chronic disease and the attacks usually become increasingly frequent with time. The symptoms may stop spontaneously after many years.

The cause of Meniere's syndrome is unknown. It has been observed that there is dilation of the endolymphatic labyrinth of the inner ear (figure 40.2) in affected individuals. In most patients, the condition develops between ages 50–70 years, and men are affected two to three times more often than women. The estimated incidence is 1–2 per 10,000. The acute episodes may not require treatment.

Motor Disorders

The motor disorders discussed in the first part of this next section are diseases in which the primary problem is weakness and wasting of muscle. These disorders fall into two categories: disease processes involving motor neurons and pathologic changes in muscle. An example of a motor neuron disease is considered first.

Amyotrophic Lateral Sclerosis

Amyotrophic (ah-mi″o-trof′ik) lateral sclerosis (ALS) is a disease in which there is loss of motor neurons resulting in muscle weakness, and it ultimately ends in death. The incidence of the disease is estimated to be two in 100,000 population, and it most often affects the age group of 50–70 years. The cause is unknown and usually leads to death within 2–6 years.

There is a familial type of ALS that is believed to be inherited in approximately 10% of the cases. There is usually an autosomal dominant pattern of inheritance. The sporadic form is the most common and affects males in a ratio of 2:1. The disease process is the same in both familial and sporadic form, but there are some differences in spinal cord pathways affected by degenerative changes. There is a high incidence of ALS among the Chamorros on Guam. It is called the Guamanian form of ALS, and there is suggestive evidence that the cycad nut contains the causative agent.

Disease Process

There are motor neurons concentrated in the cerebral cortex in the region called the precentral gyrus and in the anterior horn of the gray matter of the spinal cord (figure 40.3). In addition, there are motor nerve cell bodies located in the brain stem associated with cranial nerves. Motor nerve fibers, which carry impulses from the brain down the spinal cord to lower motor neurons, make up pyramidal and extrapyramidal tracts (figure 40.4). In general, extrapyramidal tracts are involved in movements of groups of muscles that affect posture and balance. The

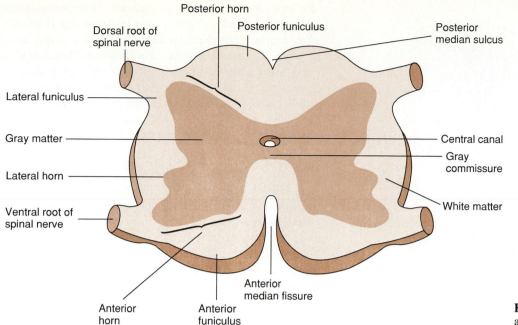

Posterior horn

Dorsal root of spinal nerve

Posterior funiculus

Posterior median sulcus

Lateral funiculus

Gray matter

Central canal

Gray commissure

Lateral horn

White matter

Ventral root of spinal nerve

Anterior median fissure

Anterior horn

Anterior funiculus

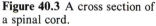
Figure 40.3 A cross section of a spinal cord.

pyramidal tracts affect fine skeletal muscle movements. The disease process in amyotrophic lateral sclerosis causes abnormalities in these motor neurons and their axons. The changes are most pronounced in the region of the cervical spinal cord and lower cranial nerves, although both upper and lower motor neurons are affected. There is loss of neurons and a concomitant increase in the number of astrocytes (chapter 39). Morphologic changes in the remaining motor neurons include shrinkage of both cytoplasm and nucleus. In addition, there is loss of myelin, a fatty insulating material, from pyramidal tract fibers.

Manifestations

The patient remains alert and aware through the course of the disease. The primary manifestation of ALS is weakness, initially involving muscles of the mouth, throat, legs, and arms. Any skeletal muscle may be affected, although cardiac and smooth muscle are not affected, and eye muscles are never involved. There may be facial weakness, as well as dysarthria and dysphagia. **Fasciculations** (fah-sik″u-la′shuns), simultaneous contractions of groups of muscle fibers, and muscle cramps occur. Reflexes may be active or depressed. There are no sensory complaints or pain, although there may be dysuria when abdominal muscles are weak. Affected muscles atrophy and, ultimately, muscles of respiration become involved. The cough mechanism is impaired, and the patient usually dies of pulmonary infection.

Treatment

There is no specific treatment for this disease.

amyotrophic: Gk. *a*, not; Gk. *myos*, muscle; Gk. *tropha*, to nourish
sclerosis: Gk. *sklerosis*, hardening
fasciculation: L. *fasciculus*, small bundle

MUSCULAR DYSTROPHIES

The term muscular dystrophy refers to a group of inherited muscle disorders in which there is muscle degeneration and symptoms due to progressive weakness. A discussion of two common types of muscular dystrophy follows.

Duchenne Muscular Dystrophy

Duchenne muscular dystrophy is nearly always limited to boys and is inherited as an X-linked disorder. The incidence in the United States is estimated to be 1.9–4 per 100,000 population. There are two main aspects of the disease, i.e., muscle weakness and shortening of muscles called **contractures.** The onset of symptoms is usually before age 3, and symptoms include difficulty in walking, toe-walking, and difficulty in getting up from a chair. Over a period of time, there are postural changes with a forward curvature of the spine called **lordosis.** The abdomen protrudes, while the upper trunk and head are tilted back. There is both muscle wasting and infiltration of muscle fibers by fat and connective tissue. Most patients are confined to a wheelchair by age 12. There is no effective drug therapy for the condition.

Myotonic Muscular Dystrophy

Myotonic muscular dystrophy is characterized by the failure of muscles to relax after a forceful contraction. This disorder is inherited as an autosomal dominant characteristic with an estimated incidence of 5 per 100,000. Myotonia may affect all muscles, but symptoms are usu-

dystrophy: Gk. *dys*, bad; Gk. *trophe*, nourishment
lordosis: Gk. *lordosis*, curvature of spine
myotonic: Gk. *mys*, muscle; Gk. *tonos*, tone

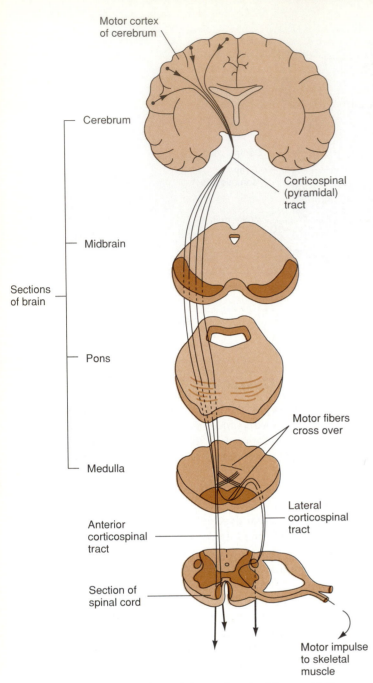

Figure 40.4 Motor fibers of the corticospinal (pyramidal) tract begin in the cerebral cortex, cross over in the medulla, and descend in the spinal cord. There they synapse with neurons that have fibers that lead to spinal nerves supplying skeletal muscles.

ally limited to the hands. In addition, there is weakness and wasting of muscles that may be disabling. It usually affects muscles of the face and neck first, and this leads to problems in lifting and turning the head, as well as swallowing and speech difficulties. Other problems include arrhythmias, cataracts, and abdominal cramps. This disorder is slowly progressive, so the individual may or may not be incapacitated. In some cases, the disease process results in confinement to bed or a wheelchair.

NEUROMUSCULAR TRANSMISSION

Muscle contraction occurs in response to nerve impulse transmission. Nerve endings lie in invaginations of the muscle membrane. This region of close contact between axon and muscle fiber is called the neuromuscular junction (figure 40.5). The nerve ending releases a neurotransmitter, **acetylcholine,** (as″ĕ-til-ko′lēn), which then attaches to receptors on the muscle fiber membrane. The result is transfer of nerve impulses to the muscle and subsequent muscle contraction. The following discussion deals with a disorder resulting from a defect in neuromuscular transmission.

Myasthenia Gravis

Myasthenia gravis (mi″as-the′ne-ah gra′vis) is characterized by fluctuating weakness that varies, even in the course of a single day. The disease process is related to antibodies against acetylcholine receptors and subsequent destruction of these receptors. Myasthenia gravis is thus an autoimmune disease of unknown etiology. A persistent viral infection may be the underlying cause. The thymus gland is almost always abnormal and this may be the site of antibody-forming cells.

The disease is most common between ages 20–40 years. The incidence in the United States is 3 per 100,000 population and is three times more common in females under age 40.

In many cases, ocular muscle weakness occurs first, resulting in a drooping of the eyelid and double vision. Weakness of muscles of the face, throat, extremities, and neck also occurs. Respiratory muscle weakness may cause difficulty in breathing. Fluctuations in weakness are typical. The weakness is made worse by exercise, emotional stress, and other illnesses.

Most patients do not die of myasthenia gravis, although a crisis involving respiratory muscle weakness may require assisted ventilation. Treatment includes anticholinesterase medication. The enzyme **cholinesterase** normally destroys acetylcholine after its release from nerve endings. Anticholinesterase drugs interfere with this reaction and prolong the effect of acetylcholine. Other modalities of treatment include surgical removal of the thymus gland, glucocorticoids, and immunosuppressant drugs.

myasthenia: Gk. *mys,* muscle; Gk. *astheneia,* weakness

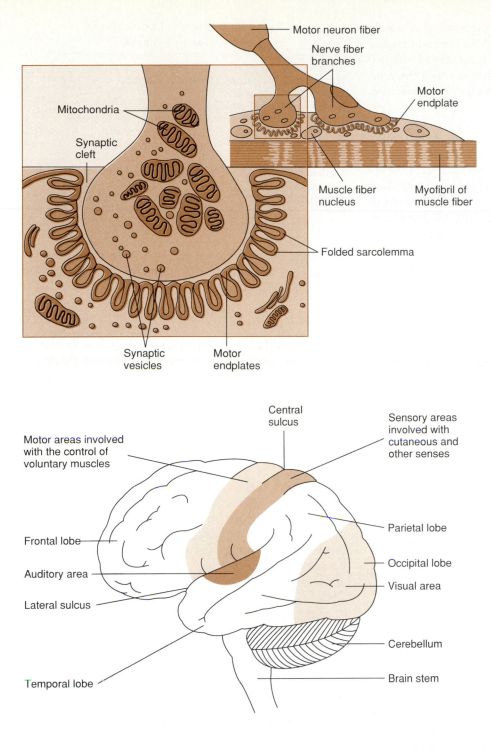

Figure 40.5 A neuromuscular junction includes the end of a motor neuron and the motor end plate of a muscle fiber. The neurotransmitter is stored in vesicles at the end of the nerve fiber.

Figure 40.6 Some motor and sensory areas of the cerebral cortex.

Autoantibodies as well as acetylcholine receptor antibodies are often present in the blood of myasthenia gravis patients. These include antithyroid, antinuclear, and antiparietal cell antibodies.

EXTRAPYRAMIDAL DISORDERS

Neurons concentrated in the precentral gyrus of the cerebral cortex are **pyramidal** neurons and are involved in somatic motor function (figure 40.6). Motor neurons located elsewhere, i.e., the premotor area of the cerebral cortex and the **basal ganglia,** constitute the **extrapyramidal** system. Basal ganglia are islands of gray matter, essentially aggregations of nerve cell bodies, within the cerebral hemispheres and brain stem. It appears that the neurons of the basal ganglia have some inhibitory effects which are balanced by stimulatory effects of pyramidal neurons. The extrapyramidal system is involved in muscle movements required for walking and for the maintenance of posture. The following discussion is concerned with two disorders of the extrapyramidal system in which there are abnormal involuntary movements, changes in muscle tone, and postural instability.

Parkinson's Disease

Parkinson's disease is also called **shaking palsy** or **paralysis agitans.** This disease involves abnormalities of the extrapyramidal system that cause involuntary tremor, slow movements, and postural instability. The incidence of this disease is approximately 90–100 per 100,000 population.

The cause of Parkinson's disease is not identifiable in most cases. In a small group of patients, there is a high incidence of the disease within families, although studies of twins indicate that inheritance is not involved. In a few instances, parkinson-like symptoms develop after carbon monoxide poisoning, encephalitis, or the administration of such drugs as alpha-methyldopa or reserpine.

There is a marked deficiency of the neurotransmitter **dopamine** in the basal ganglia of patients with Parkinson's disease. There also is a loss of neurons in the extrapyramidal system, degenerative changes in the cortex of the cerebral hemispheres, and depigmentation of the **substantia nigra.** The substantia nigra is an area of extrapyramidal gray matter in the brain stem. It is so named because of dark melanin pigment in the region.

Table 40.1 lists major manifestations of Parkinson's disease. The first symptoms of the disease are usually tremor and a loss of facial expression with infrequent eye blinking. The tremor may ultimately become disabling. The progress of the disease is usually slow. Initially, the patient may complain of aching pains in the limbs, neck, and back. A change in handwriting may be observed. With time, the individual experiences a sense of resistance to movement. There is difficulty in getting started walking and walking around obstacles. There is failure to swing the arms when walking, and the gait is typically a shuffle with small steps. Typically, postural changes occur due to an increase in muscle tone. There is flexion of the neck, back, knees, and ankles. Frequent falls are the result of the absence of postural reflexes. The affected individual tends to drool due to loss of automatic swallowing. Constipation and delayed emptying of the bladder are additional problems. The voice lacks volume and after talking for a period of time, may be reduced to a whisper. Loss of intellect may occur, and the changes may be mild and may be due to depression.

Parkinson's disease shortens life expectancy. Treatment is symptomatic and is aimed at providing sustained and improved mobility. Dopamine does not cross the blood-brain barrier. **Levodopa** (L-dihydroxyphenylalanine) is effective in replacing dopamine in the brain because it is transported to the brain and is subsequently converted to

TABLE 40.1 Major manifestations
of Parkinson's disease

Tremor
Hypokinesia (decreased muscle movements)
Loss of facial expression
Bradykinesia (slow movements) and muscle rigidity
Difficulty in walking; flexion of all joints; loss of speech volume
Postural instability
Intellectual decline
Decline in interest; poor short-term memory; anxiety and confusion;
Depression

dopamine. This drug combined with a **decarboxylase inhibitor** alleviates much of the disability of the disease for a prolonged period.

• • •

Decarboxylase inhibitors prevent the conversion of levodopa to dopamine in the gastrointestinal tract. Consequently, there is a larger amount of levodopa available for transport to the brain; thus, decarboxylase inhibitors make it possible to reduce the dose of levodopa.

• • •

Huntington's Disease

Huntington's disease is an hereditary disorder in which there is loss of neurons in the basal ganglia, as well as in the cerebral cortex. Biochemical abnormalities accompany these structural changes. There is a decrease in chemical substances that are neurotransmitters in the basal ganglia, i.e., **gamma-aminobutyric acid** and acetylcholine.

The disease is a form of chorea (ko-re'ah), which involves involuntary movements of muscles of the extremities and the face. There is an autosomal dominant pattern of inheritance. A child of an affected parent has a 50% chance of having the disease. The incidence in the United States is in the range of 4–8 per 100,000 population.

The onset of symptoms usually occurs between the ages of 35–40, although there is a wide variation in this. Table 40.2 summarizes typical manifestations of Huntington's disease grouped on the basis of (1) involuntary movements, (2) emotional disturbances, and (3) intellectual decline.

There is no cure for Huntington's disease, and the average time between diagnosis and death is about 14 years. It is usually necessary to commit the individual to a mental institution. Treatment is symptomatic with drug therapy for involuntary movements.

TABLE 40.2 Manifestations of Huntington's disease

Chorea. Clumsiness, facial grimacing, jerking movements, exaggerated leg and arm movements during walking, speech and swallowing difficulties

Intellectual decline. Apathy, memory loss, loss of intellectual capacity

Emotional disturbances. Irritability, depression, paranoia, violent behavior

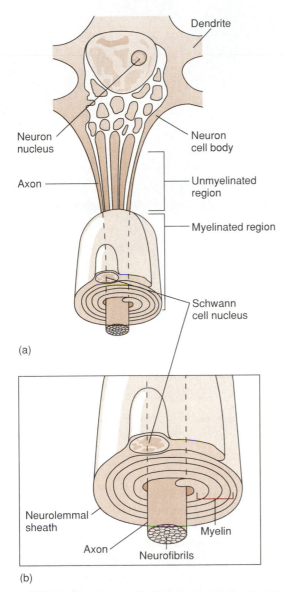

(a)

(b)

Figure 40.7 (*a*) The portion of a Schwann cell that is tightly wound around an axon forms the myelin sheath. (*b*) The cytoplasm and nucleus of the Schwann cell, remaining on the outside, form the neurolemmal sheath. The region of a nerve fiber that possesses myelin is said to be myelinated.

chorea: Gk. *choreia,* dance

DEMYELINATING DISORDERS

Myelin is a fatty covering on the surface of axons formed by Schwann cells in the peripheral nervous system and, in the central nervous system, is formed by **oligodendrocytes** (ol″ĭ-go-den′dro sīts) (figures 40.7 and 40.8). Oligodendrocytes are supporting or **neuroglial** cells that send out processes that wrap around axons in the central nervous system.

Multiple Sclerosis

Multiple sclerosis is a demyelinating disease of the central nervous system. In multiple sclerosis, there is disintegration of the axonal myelin and degeneration of oligodendrocytes. Later, there is proliferation of another type of neuroglial cell, **astrocytes,** with the formation of scars. The name multiple sclerosis indicates there are multiple axonal lesions with scarring. Sections of brain taken at autopsy show grayish areas in the white matter of the cerebral hemispheres. These lesions of multiple sclerosis are also called plaques. Similar lesions are found in the brain stem, cerebellum, and spinal cord.

Multiple sclerosis is mainly a disease of young adults with age of onset in most cases between 20–40 years. The disease affects women more often than men in a ratio of 1.5–1. Multiple sclerosis is almost nonexistent at the equator, but there is increased incidence with latitude, both north and south. The number of cases in the northern United States and Europe is about 60 per 100,000 population.

The cause of multiple sclerosis is unknown. There is some evidence that a viral infection with a long incubation period may be the initiating event in the disease. An altered immune response, possibly controlled by genetic predisposition, may be involved. Environmental factors, such as emotional stress or dietary changes, may allow an autoimmune attack on myelin.

Most individuals with multiple sclerosis experience intermittent periods of acute symptoms followed by a period of remission. Characteristically, there are multiple symptoms over a period of decades, although in a few cases death occurs within a few months. Some common manifestations are muscle weakness, ocular disturbances, urinary problems, and mental disturbances. Table 40.3 defines some terms that describe other manifestations of the disease. Ocular disturbances include blurred vision and diplopia. Weakness of the limbs is the most common sign of the disease, and facial muscle weakness is also common. Urinary problems include incontinence and frequency of urination. Numbness and tingling may involve the face, trunk, or limbs. Euphoria, depression, and memory disturbances frequently occur.

The course of multiple sclerosis is unpredictable. There may be a pattern of (1) mild symptoms and complete remission with no disability, (2) frequent acute symptoms followed by less complete remission and some

sclerosis: Gk. *sclerosis,* hardening

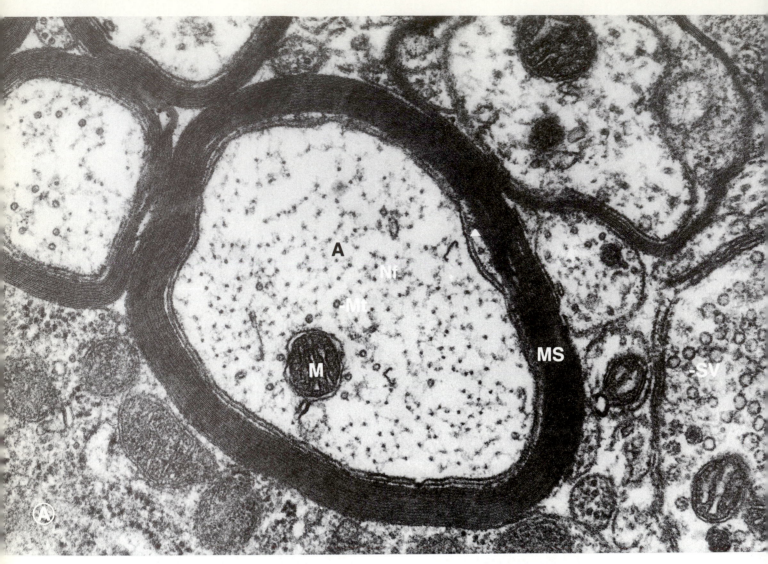

Figure 40.8 Electron micrograph of a cross section taken through a small myelinated axon in the spinal cord. A, axon; MS, myelin sheath; M, mitochondrion. ×68,000.

From D. J. Moran and J. C. Rowley: Visual Histology. *1988 Lea & Febiger, Philadelphia. Reproduced with permission.*

TABLE 40.3 Definition of terms that describe some manifestations of multiple sclerosis

Paresthesias. Burning or tingling sensation of the skin
Gait ataxia. Manner of walking in which the foot is raised high and brought down flat and suddenly
Dysarthria. Stammering
Diplopia. Double vision
Dysphagia. Swallowing difficulties
Nystagmus. Abnormal movements of eyeball
Spasticity. Increased muscle tension
Spastic paraplegia. Paralysis of legs with increased muscle tone

disability, (3) fewer remissions and increasing disability, or (4) steady progressive deterioration.

Treatment of the disease is symptomatic. Management of urinary problems is essential to prevent infections and stone formation. Skin care to prevent decubitus ulcers is important when there is partial paralysis of the lower extremities. Physical therapy may be helpful in maintaining mobility.

Guillain-Barré Syndrome

Guillain-Barré syndrome is a demyelinating neuropathy involving both peripheral and cranial nerves. The cause is unknown, although it is frequently preceded by a viral infection. The disorder occurs in approximately one out of 100,000 population. It affects males and females equally at any age. The disease process involves demyelination of segments of peripheral and cranial nerves.

The symptoms of the syndrome are frequently preceded by surgery, immunization, or viral infection. The intervening period may be several days to 3 weeks. Early manifestations include weakness of limbs accompanied by paresthesias. The majority of affected individuals have (1) facial diplegia, a paralysis on both sides, (2) dysphagia or swallowing difficulties, and (3) dysarthria, or stammering. Reflexes are impaired and sensory impairment is variable. There may be loss of perception of vibration, pain, and temperature. Muscle tenderness and sensitivity to pressure occurs. Symptoms may continue for several weeks and, in most cases, a period of months is required for recovery.

Plasmapheresis (plaz″mah-fĕ-re′sis) is a possible modality of treatment in severe cases of Guillain-Barré syndrome. The procedure involves removal of blood from the patient, separating the cellular and plasma fractions, and reintroducing the cellular component. Mechanically assisted ventilation may be required in some cases.

• • •

When untreated, about one-third of individuals affected by Guillain-Barré syndrome experience permanent weakness of distal muscles or a slight facial paralysis. A smaller percentage of patients have a partial recovery followed by a relapse. In about 2% of the cases, there is full recovery and then a recurrence of the disease.

• • •

SUMMARY

• • •

Head pain is due to both intracranial and external structures. The cause may be muscle contraction, inflammation, dilation of blood vessels, or intracranial pressure that distorts pain sensitive structures. Migraine headache is a group of disorders of unknown etiology. The underlying cause may be dilation of branches of the carotid artery. Classic migraine is preceded by dramatic neurologic manifestations, whereas manifestations preceding common migraine are more subtle. Control of precipitating factors is an important aspect of management of migraine. Tension headache is the most common type of headache involving various contributing factors. Tension headache usually involves both sides of the head and tends to be dull and prolonged. Cluster headaches are short episodes of head pain, one following another, for weeks or months. Males are affected more frequently than females and the cause is unknown.

Other sensory disorders include trigeminal neuralgia and Bell's palsy, both of which involve cranial nerves. Trigeminal neuralgia is characterized by stabbing intense pain, usually along the course of the second or third division of the trigeminal nerve. Bell's palsy is a disorder of the seventh cranial nerve, also called the facial nerve. It leads to unilateral facial paralysis and may affect lacrimation, salivation, and sense of taste. Meniere's syndrome is a sensory disorder, as well. The cause is unknown, but there is involvement of the inner ear. The manifestations of Meniere's syndrome include vertigo, tinnitus, and vomiting.

The first motor disorder considered in this chapter, amyotrophic lateral sclerosis, is due to loss of motor neurons. The cause is unknown, and the primary manifestation is muscle weakness that leads to death in 2–6 years. The muscular dystrophies are a group of inherited muscle disorders in which there is muscle degeneration and progressive weakness. Duchenne muscular dystrophy is essentially limited to boys and is transmitted as an X-linked disorder. It is characterized by muscle weakness, contractures, and lordosis. Most patients are confined to a wheelchair by age 12. Myotonic muscular dystrophy is inherited as an autosomal dominant disease. Myotonia is usually limited to the hands, but there is weakness and wasting of other muscles. The course is variable, so some affected individuals are not incapacitated.

Myasthenia gravis is an autoimmune disease in which antibodies destroy acetylcholine receptors. The result is weakness of muscles of eyelids, face, neck, extremities, and respiration. Parkinson's disease involves abnormalities of the extrapyramidal system that cause involuntary tremor, slow movements, and postural instability. Huntington's disease is an inherited, autosomal dominant disorder in which there is loss of neurons in the cortex and basal ganglia. The major manifestations are involuntary movements, emotional disturbances, and intellectual decline. Multiple sclerosis is a demyelinating disease of the central nervous system in which there is muscle weakness, ocular disturbances, urinary problems, and mental disturbances. Guillain-Barré syndrome is also a demyelinating neuropathy involving both peripheral and cranial nerves. The cause is unknown, although it is frequently preceded by a viral infection. The manifestations include facial paralysis and swallowing difficulties. Recovery usually requires a period of months.

REVIEW QUESTIONS

• • •

1. In comparing classic and common migraine, how are preceding neurologic manifestations different?
2. List four possible precipitating factors for migraine.
3. Describe the phenomenon of cluster headache.
4. Define the terms (a) vertigo, (b) tinnitus, (c) contracture, (d) lordosis, (e) myotonic.
5. In what disease discussed in this chapter is there an autoimmune reaction against acetylcholine receptors?
6. List three main manifestations of Meniere's syndrome.
7. What muscle disorder is associated with muscle weakness, contractures, and lordosis in males?

8. What condition is characterized by demyel[in]. nerves in the central nervous system?
9. What are three major manifestations of myotonic muscular dystrophy?
10. What does the term basal ganglia refer to?
11. What disease is associated with a decrease of dopamine in the basal ganglia?
12. What are three major manifestations of Huntington's disease?
13. Why is levodopa used in the treatment of Parkinson's disease rather than dopamine?
14. What symptom characterizes trigeminal neuralgia?
15. What often triggers an attack in trigeminal neuralgia?
16. What nerve is involved in Bell's palsy?
17. What is a characteristic symptom of Bell's palsy?
18. What disease process is involved in amyotrophic lateral sclerosis?
19. What is the primary manifestation of ALS?
20. What is the underlying disease process in Guillain-Barré syndrome?

SELECTED READING

• • •

Eldridge, R. et al. 1984. Hereditary adult-onset leukodystrophy simulating chronic progressive multiple sclerosis. *New England Journal of Medicine* 311:948–52.

Hawkes, C. et al. 1989. Motoneuron disease: A disorder secondary to solvent exposure. *Lancet* 1:73–76.

Husain, F. et al. 1989. Concurrence of limb girdle muscular dystrophy and myasthenia gravis. *Archives of Neurology* 46:101–2.

Madrazo, I. et al. 1987. Open microsurgical autograft of adrenal medulla to the right caudate nucleus in two patients with intractable Parkinson's disease. *New England Journal of Medicine* 316:831–34.

Rowland, L. 1984. Looking for the cause of amyotrophic lateral sclerosis. *New England Journal of Medicine* 311:979–82.

Speer, M. et al. 1988. Prenatal diagnosis using deletion studies in Duchenne muscular dystrophy. *Prenatal Diagnosis* 8:427–37.

Taft, J. M. 1990. Respiratory management of amyotrophic lateral sclerosis. *Journal of American Academy of Physician Assistants* 3:131–36.

Tarnopolsky, R. 1990. When the patient's world goes 'round. *Journal of American Academy of Physician Assistants* 3:117–29.

Weavil, P., and G. Harkness. 1989. Reye's syndrome update. *Physician Assistant* 13: 15–23.

Chapter 41

Practical Applications: Central Nervous System Disorders

The purpose of this final chapter in the unit is to consider selected examples of disorders involving the central nervous system. Examples of seizure and stroke are included, with emphasis on laboratory findings and diagnostic tests related to each disorder. The first topic involves an adult female who suffered a grand mal seizure for the first time.

SEIZURE

Patient 41.1 was observed to be shaking, with teeth clenched and eyes rolled back. The episode lasted 3–5 minutes and was followed by some periodic breathing and an interval of confusion.

> Patient 41.1 was a 55-year-old female hospitalized due to first grand mal seizure.
> Laboratory findings: (Normal values in parentheses) Na$^+$ 110 meq/liter (137–47), K$^+$ 2.6 meq/liter (3.4–5.0), chloride 81 meq/liter (100–10), HCO$_3^-$ 16.5 meq/liter (22–30) BUN 4 mg/dl (7–22), creatinine 0.4 mg/dl (0.6–1.2), glucose 96 mg/dl (70–110), Hgb 12.4 g/dl (12–16), Hct 35.9% (37–47), platelets 235,000/mm³ (130,000–400,000), urine, specific gravity 1.002, thyroxine 8.4 μg/dl (5–12.5)

A CT scan of the head (chapter 39) failed to reveal lesions or hemorrhages as possible cause of the seizure. Patient 41.1 had a history of drinking 3–4 gallons of water each day and marked hyponatremia was noted at the time of hospital admission. Hyponatremia is a common electrolyte imbalance and dilution is the most common cause (chapter 3).

There are three main categories of causes of a low serum sodium level based on the amount of total body water, i.e., circumstances in which there is (1) fluid loss replaced by hypotonic fluid, (2) near normal extracellular fluid volume associated with inappropriate secretion of ADH or drugs that limit renal excretion of water, and (3) increased extracellular fluid volume due to various conditions, such as cirrhosis, congestive heart failure, renal failure, or acute water intoxication.

Hemoglobin and hematocrit values in the case of patient 41.1 exclude major blood loss and the BUN/creatinine levels show adequate renal excretion of nitrogenous wastes. The low urine specific gravity is evidence that the renal excretion of water is maximal.

The thyroxine level of patient 41.1 is in the normal range. Evaluation of thyroid function is pertinent in this case because there may be impaired excretion of water associated with hypothyroidism. Psychological evaluation indicated that this was not a case of psychogenic polydipsia (chapter 2), and the underlying cause was not determined.

The consequences of compulsive water drinking are hyponatremia associated with an increase in blood volume. The low levels of sodium, potassium, and chloride in patient 41.1 are likely due to dilution. The manifestations of water intoxication are the result of hyponatremia and may include vomiting, seizures, and coma. Lactic acidosis may be induced by seizure activity, resulting from anaerobic glycolysis in muscles.

Patient 41.1 was treated with an infusion of hypertonic saline, fluid restriction, intravenous and oral potassium, and given Dilantin to prevent additional seizures.

TRANSIENT ISCHEMIC ATTACK/ARTERITIS

A transient ischemic attack (TIA) is a brief episode of ischemia that does not lead to infarction (chapter 39). The event usually lasts a few minutes and never more than an hour. The most frequent cause is stenosis or occlusion of major arteries, rather than an interrupted blood supply in intracranial arterial branches. Occlusion of the carotid artery or its branches may lead to (1) transient monocular blindness in which there is painless temporary impairment of vision or (2) a transient hemisphere attack. In the case of the latter, the affected cerebral region is supplied by a terminal branch of the internal carotid, the middle cerebral artery. The result is motor and sensory deficits, usually involving fingers, hand, and forearm. Language and behavior may be affected as well. Stenosis of the vertebral or basilar arteries affects blood supply to the pons and cerebellum. The manifestations of a TIA involving the vertebrobasilar arteries include double vision, numbness around the mouth, incoordination, and speech impairment. The diagnosis for patient 41.2 was transient ischemic attack.

Patient 41.2 had a history of asthma, which was treated with theophylline and periodic steroids. He was admitted to the hospital for an overdose of theophylline. The toxic effects of theophylline include vomiting, arrhythmias, seizures, shock, and death. This patient did not have arrhythmias or seizures, but experienced persistent vomiting resulting in hypokalemia (chapter 5). The theophylline blood level was significantly lowered after 10 hours of dialysis. He received potassium supplementation to correct the hypokalemia.

> Patient 41.2 was a 65-year-old male diagnosed with a theophylline overdose with subsequent transient ischemic attack.
> Laboratory findings: hematocrit 41.5% (42–52), pro time 12 sec (9.8–12.6), APTT 22 seconds (21–31), Na$^+$ 145 meq/liter (137–45), K$^+$ 2.6 meq/liter (3.4–5.0), BUN 9 mg/dl (7–21), total bilirubin 0.5 mg/dl (0.1–1.3), LDH 593 U/liter (287–537)

There was an episode of slurred speech the following day with right facial weakness and numbness on the left side. Twenty-four hours later, there was a similar episode. He experienced transient double vision and a persistent dull generalized headache that started to throb if he sat up. Subsequent to this event, patient 41.2 experienced some blurred vision and trouble with color vision. There was some clumsiness associated with the left hand. At this point, there was no obvious explanation for a stroke, except for the possibility of theophylline induced cerebral vascular vasoconstriction.

Subsequent diagnostic tests include CT head scan, echocardiogram, magnetic resonance study, and cerebral arteriogram.

COMPUTED TOMOGRAPHY HEAD SCAN

Plain X-ray films or radiographs of the head and neck are of little value in diagnosing stroke, although the films can show calcified arteries or arteriovenous malformations. Computed tomography (CT), with or without contrast media, provides significant information about tissue damage resulting from stroke. A beam of X-rays pass through the head as the X-ray source rotates around the patient's head. A computer reconstructs an image based on X-ray beam intensity differences due to differences in tissue density. The term **X-ray attenuation** is used to identify a diminished X-ray beam intensity.

CT scans allow the following:

1. Differentiation of white and gray matter
2. Differentiation of thalamus and basal ganglia
3. Visualization of major arteries after contrast infusion
4. Identification of hematomas as high-density regions during the first week
5. Identification of infarction as a low attenuation lesion.

The major limitation is in the area of the posterior fossa, the bony structures that house the pons, medulla, and cerebellum. Artifacts due to attenuation of the X-ray beam by thick bony structures may obscure a diagnosis of infarction.

In patient 41.2, the CT scan demonstrated a low-density area, i.e., an area of diminished attenuation, in the left cerebellum and a questionable abnormality in the right cerebellum. The CT scan was performed both with and without contrast media. The contrast media, after intravenous injection, enhances differences of tissue density.

ECHOCARDIOGRAM

An echocardiogram was obtained to evaluate the heart and the flow of blood through the heart of patient 41.2. Echocardiography provides an image of the heart by means of a beam of ultrasound waves directed toward the heart. Tissue density determines the reflection of the ultrasonic beam, and a recording of the reflected sound waves results in a cardiac image. An echocardiogram on patient 41.2 indicated the absence of cardiac valve abnormality and a normal-sized left ventricle.

MAGNETIC RESONANCE IMAGING

Magnetic resonance imaging (MRI) provided additional information in regard to the condition of patient 41.2. This is a valuable diagnostic procedure for both hemorrhagic

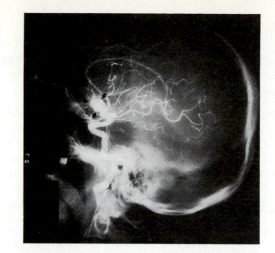

Figure 41.1 This is a lateral view of the skull taken during the injection of contrast during a cerebral arteriogram. There is a large cerebral aneurysm outlined by the arrows.

and ischemic stroke. This imaging technique is based on radio wave excitation of hydrogen nuclei in body tissue. Excitation occurs within a magnetic field, absorbed energy is released, and images are reconstructed from signals obtained from these events. When the absorbed energy is completely released, the tissue is in a relaxed state and tissue-specific relaxation constants can be measured. On this basis, images are identified as (1) T_1-dependent, i.e., the spinal fluid has decreased signal intensity, and fat has increased signal as compared to the brain; (2) T_2-dependent, i.e., the cerebrospinal fluid has an increased signal compared to the brain; and (3) a balanced image with comparable signals from brain and spinal fluid.

An MRI in patient 41.2 showed two areas of increased signal in the cerebellum in both T_1 and T_2 images. This observation is suggestive of subacute hemorrhage, possibly areas of infarction or petechial hemorrhage.

CEREBRAL ARTERIOGRAM

The purpose of a cerebral arteriogram is to visualize the blood vessels and evaluate the blood supply to different parts of the brain (figure 41.1). The procedure is useful in identifying arteritis or arterial occlusions and stenosis due to atherosclerotic disease.

The test involves injection of a radiopaque contrast agent into an artery supplying the brain, which is then followed by serial X-rays of the head and neck. The results of the test in patient 41.2 showed dramatic segmental narrowing of cerebellar arteries. A tentative diagnosis of arteritis was based on the results of the arteriogram. Arteritis is an inflammation of arteries due to unknown causes, which leads to swelling and stenosis of the vessels. The final evaluation of this case was TIA with arteritis as the probable cause.

SUBARACHNOID HEMORRHAGE

Patient 41.3 was admitted to the hospital after collapsing at home. The tentative diagnosis upon admission was cerebrovascular accident.

> Patient 41.3 was a 67-year-old female who collapsed and lost consciousness at home. At admission, the patient was confused, complained of severe headache and neck pain, and vomiting, and was acutely ill. She had no history of drug/alcohol use. White blood cells, 7,000/mm³ (4,000–10,700), red blood cells 4,500,000/mm³ (4.2–5.5), hematocrit 39.4% (37–47), platelets 224,000/mm³ (130,000–400,000), prothrombin time 10.2 sec (9.8–12.6), fibrinogen 214 mg/dl (200–400), Na⁺ 133 meq/liter (137–47), K⁺ 3.5 meq/liter (3.4–5.0), pH 7.34 (7.35–7.45), pCO₂ 47 mm (35–45)

PRESENTING SYMPTOMS AND HISTORY

Presenting symptoms in the case of patient 41.3 are characteristic of subarachnoid hemorrhage (SAH), although the diagnosis was not established until tests were completed. More than 75% of patients with SAH experience sudden, severe headache at the onset, and this may be accompanied by neck pain. Convulsions did not occur. In the initial evaluation, it is necessary to consider various possible reasons for headache including meningitis, flu, migraine headache, alcohol or drug use. The following tests were done to establish a diagnosis.

HEMATOLOGY

Intracranial hemorrhage (ICH) may be the result of various factors, such as trauma, hypertension, aneurysm, anticoagulants, or vascular malformation. It can be a complication of cerebral blood vessel thrombosis. Other conditions that may be responsible for ICH are leukemia and disseminated intravascular coagulopathy. There is no indication of an underlying problem with coagulation in patient 41.3 who has normal values for platelets, prothrombin time, and fibrinogen.

DIAGNOSTIC TESTS

Patients with a suspected subarachnoid hemorrhage are evaluated by a computed tomography (CT) study and cerebrospinal spinal fluid (CSF) examination. A CT scan will demonstrate the presence of blood in subarachnoid spaces or the presence of a localized clot. A clot suggests the location of a ruptured aneurysm and is best visualized if the scan is done within 48 hours of the initial event.

The cerebrospinal fluid of patient 41.3 showed elevated pressure and the presence of blood. The CSF examination and a CT scan confirmed a diagnosis of a ruptured saccular aneurysm. A saccular aneurysm, also called a berry aneurysm (figure 41.2), is the most common cause of subarachnoid hemorrhage and is the result of a localized weakness or the absence of the arterial media (chapter 22). Rupture of a saccular aneurysm usually

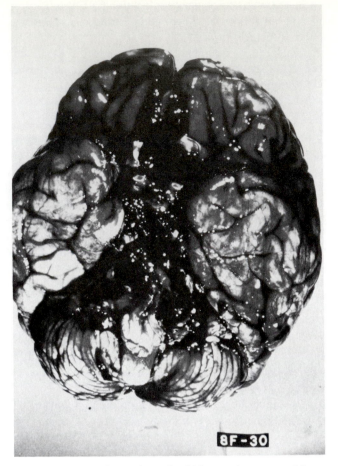

Figure 41.2 Extensive, subarachnoid hemorrhage caused by the rupture of a berry aneurysm filling the subarachnoid cisterns at the base of the brain with dark blood. *Courtesy of Dr. J. H. N. Deck. From A. C. Ritchie:* Boyd's Textbook of Pathology. *1990 Lea & Febiger, Philadelphia. Reproduced with permission.*

occurs at a bifurcation of large arteries at the base of the brain. A less common type of aneurysm is a mycotic aneurysm in which there is a weakness of a vessel wall due to infection. This type of aneurysm usually ruptures into the subarachnoid space over the cortical surface of the cerebral hemispheres. Arteriography (cerebral angiography) was subsequently performed to demonstrate the size and origin of the rupture in preparation for surgery. Figure 41.3 shows an example of an X-ray view of a large aneurysm.

FLUID AND ELECTROLYTE DISTURBANCES

The laboratory results for patient 41.3 are used as an example of SAH in this chapter and show some abnormalities in blood gas and electrolyte values. The following paragraphs deal with the underlying causes of such imbalances that occur in SAH.

Diabetes insipidus or the syndrome of inappropriate secretion of antidiuretic hormone (ADH) may develop

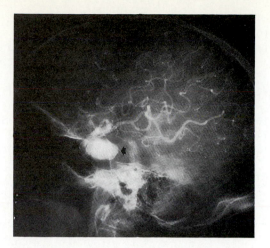

Figure 41.3 This is a lateral skull X-ray where contrast has been injected into the internal carotid artery. A large aneurysm is in the area of the arrows.

(chapter 2). ADH with water retention may lead to a dilutional hyponatremia resulting in convulsions and coma. It is possible to have, rather than water retention, a net loss of sodium and water and hyponatremia.

A blockage of flow of CSF through the ventricles and the subarachnoid space leads to hydrocephalus in about 20% of SAH patients. Cerebral edema is also seen in some patients immediately after SAH.

BLOOD GAS VALUES

Hypoventilation in patients suffering from SAH leads to hypercapnia and respiratory acidosis. Oxygen saturation, pH, and pCO_2 are determined immediately to assess the need for supplemental oxygen and assisted respiratory support.

OTHER COMPLICATIONS

Various other complications may occur with SAH. Patient 41.3 suffered from deep vein thrombosis of the right leg, pulmonary emboli, and anterolateral infarction. Vasospasm of cerebral arteries, subsequent to SAH, can lead to cerebral ischemia and infarction. Other possible complications include cardiac arrhythmias, hypertension, and stress-induced ulcer with gastrointestinal bleeding. Primary causes of death are hydrocephalus, cerebral edema, increased intracranial pressure, and convulsions.

CEREBROVASCULAR ACCIDENT/ HYPERTENSION

Hypertension, cardiac disease, and diabetes are major risk factors for a cerebrovascular accident. Patient 41.4 had a history of hypertension, alcohol abuse, cigarette smoking, and angina.

Patient 41.4 was a 50-year-old male admitted to the hospital with a 2-day history of sudden onset of slurred speech and difficulty in swallowing. He exhibited left tongue deviation and right facial droop.
Laboratory findings: Blood tests all within normal range.

The patient denied having experienced chest pain, and serum enzyme levels that would be indicative of myocardial infarction were in the normal range. The ALT/SGPT level may be elevated due to alcoholism in about 50% of the cases. Creatine kinase may also be elevated if there is alcohol related myopathy, i.e., muscle dysfunction. These enzyme levels were in the normal range for patient 41.4.

A CT head scan showed no abnormal areas of increased or decreased attenuation. There was no indication of any masses, intracranial hemorrhage, or abnormalities involving the ventricles. Magnetic resonance imaging showed normal ventricles and brain stem. There were multiple areas of increased signal in the deep white matter of the brain. This observation is suggestive of ischemic changes, possibly secondary to hypertension. The diagnosis for this patient was cerebrovascular accident associated with hypertension.

SUMMARY

• • •

Hyponatremia may lead to seizure. Low serum sodium levels occur when fluid loss is replaced by hypotonic fluid, when there is inappropriate secretion of ADH, or when there is impaired renal excretion of water. Increased extracellular fluid volume is also the result of cirrhosis, congestive heart failure, or acute water intoxication.

A transient ischemic attack is a brief episode of ischemia that does not lead to infarction. The following tests are useful in evaluating a possible TIA. Computed tomography provides significant information about tissue damage resulting from TIA or stroke. A CT scan allows identification of hematoma as a high-density region and infarction as a low-density lesion. The test also makes it possible to visualize arteries after injection of contrast media. An echocardiogram is a diagnostic test by which an image of the heart is obtained by recording the reflection of a beam of ultrasound waves. Magnetic resonance imaging is a valuable diagnostic procedure for both hemorrhagic and ischemic stroke. A cerebral arteriogram involves injection of a contrast agent into an artery in order to visualize occlusion or stenosis. All of these tests were used in diagnosing the case of TIA included in this chapter.

The most common presenting symptom of subarachnoid hemorrhage is headache. Blood tests are done to identify possible underlying coagulation disorders. Diagnostic tests that establish a diagnosis of SAH are the CT scan and cerebrospinal fluid examination showing red blood cells. The most common cause of SAH is rupture of a saccular aneurysm. A mycotic aneurysm is a less common cause. Fluid imbalances are associated with SAH and include diabetes insipidus, syndrome of inappropriate secretion of ADH, cerebral edema, and hydrocephalus. Dilutional hyponatremia may occur. Hypoventilation may lead to hypoxemia and respiratory acidosis. Other complications include gastrointestinal bleeding, cardiac arrhythmias, and deep vein thrombosis.

An example of a cerebrovascular accident associated with hypertension is included. Diagnosis was established in this case by a CT scan and by MRI.

REVIEW QUESTIONS

• • •

1. What is the most common cause of hyponatremia?
2. List three circumstances in which hyponatremia occurs.
3. In the example of water intoxication in this chapter, why is a report of the thyroxine level important?
4. What is the difference between TIA and a stroke?
5. List three types of information obtained by a CT scan.
6. What is another valuable diagnostic procedure for evaluation of hemorrhagic and ischemic stroke (other than CT scan)?
7. What procedure is useful in identifying stenosis of cerebral arteries?
8. What is the principle of magnetic resonance imaging?

9. How can arteritis lead to impaired blood flow?
10. What is an echocardiogram?
11. Why are laboratory results from such tests as a platelet count and pro time important in suspected SAH?
12. What is the most important observation in CSF examination in cases of SAH?
13. Of what value is a CT scan in the diagnosis of SAH?
14. What is the purpose of arteriography in SAH?
15. Why do hyponatremia, hypoxemia, and acidosis occur in SAH?

SELECTED READING

• • •

Luyendijk, W. et al. 1988. Hereditary cerebral hemorrhage caused by cortical amyloid angiopathy. *Journal of Neurological Sciences* 85:267.

Mercado, A. et al. 1989. Cocaine, pregnancy, and postpartum intracerebral hemorrhage. *Obstetrics and Gynecology* 73:467.

Nalls, G. et al. 1988. Subcortical cerebral hemorrhages associated with cocaine abuse: CT and MR findings. *Journal of Computer Assisted Tomography* 13:1.

Volpe, J. 1989. Intraventricular hemorrhage in the premature infant-current concepts. Part I. *Annals of Neurology* 25:3.

Appendix A

Answers to Review Questions

1. Polar
2. Hydrophilic
3. Hydrophobic
4. Lipid soluble substances
5. Ions and polar molecules
6. All of them
7. The flow of Na^+, which is diffusion, from high to low concentrations
8. It appears that membrane protein molecules form hydrophilic channels
9. Conformational changes in the carrier protein
10. The breakdown of ATP

CHAPTER 2

1. Plasma, lymph, interstitial fluid, cerebrospinal fluid
2. Cell membranes of endothelial cells that make up the capillary wall
3. Cell membranes
4. Na^+
5. Protein
6. Into
7. Loss
8. Lymphatic pump
9. 6.3 mm out of capillary
10. Into cells
11. a. Pulmonary capillaries are leaky to protein
    b. Narrow interstitial spaces
12. Increased
13. Pitting edema
14. Cellular hydration
15. Water retention

16. Decreased ADH
17. Water is drawn from cells
18. Lesser
19. Osmotic diuresis
20. The percentage of body weight that is water decreases with age; impaired sense of thirst

CHAPTER 3

1. Dehydration
2. Blood-brain barrier
3. Accumulation of idiogenic osmoles
4. Lethargy, coma, hyperirritability, convulsions
5. Rapid infusion of fluid can cause cerebral edema
6. Into cells
7. Volume depletion stimulates thirst; increased ingestion of water may cause hyponatremia
8. SIADH, psychogenic polydipsia, diuretics
9. False
10. K^+ move into cells in exchange for H^+, which move out of cells
11. Muscle weakness, paralysis, cardiac arrhythmias
12. It is an aldosterone antagonist; aldosterone favors Na^+ retention and elimination of K^+
13. Calcium gluconate solution, sodium chloride solution, sodium bicarbonate solution
14. K^+/H^+ are eliminated as Na^+ are retained by kidney; if H^+ unavailable, H^+ are eliminated
15. Promotes bone resorption; increases intestinal absorption; promotes kidney tubule Ca^{++} reabsorption
16. Increases
17. Ca^{++}, protein-bound calcium, calcium bound to substances other than protein
18. Ca^{++}
19. Nerve impulse transmission, muscle contraction
20. Lethargy, depression, insomnia, decreased memory
21. Muscle weakness and pain; nausea and vomiting, constipation, cardiac arrhythmias
22. Excess vitamin D, hyperparathyroidism, immobilization, malignancy, milk-alkali syndrome
23. Increase nerve impulse transmission, abnormal muscle contractions
24. Numbness and tingling, muscle spasms, tetany
25. Enlargement of heart, heart failure

CHAPTER 4

1. H_2CO_3, lactic, pyruvic
2. $\dfrac{[H^+][A^-]}{[HA]}$
   For a strong acid, the numerator is larger
3. Increase in H^+ concentration

4. Decreased breathing rate or impaired diffusion of CO_2 into air sacs
5. CO_2 is eliminated, shift to left, decreased H^+
6. $H_2PO_4^{-2}$ acid, HPO_4^{-2} base
7. Alkalosis
8. Metabolism, $CO_2 + H_2O$, food
9. Occurs in plasma, but most quickly in red blood cells
10. a. $CO_2 + H_2O \rightleftharpoons H_2CO_3 \rightleftharpoons H^+ + HCO_3^-$ in red blood cells
    b. Above reaction in parietal cells of gastric mucosa, HCO_3^- diffuses into blood
11. Hypercapnia due to hypoventilation
12. Hyperventilation
13. Lesion of respiratory center, airway obstruction, collapse of lung
14. Respiratory acidosis is caused by increased CO_2; metabolic acidosis is nonrespiratory
15. Buffer systems
16. Renal control of H^+/HCO_3^- excretion
17. HCO_3^-
18. Protein and biphosphate buffer system
19. Hypoventilation
20. Metabolic alkalosis

CHAPTER 5

1. Loss of H^+ with addition of HCO_3^- to the blood; contraction alkalosis due to decreased extracellular volume with no change in HCO_3^-
2. Reduced dietary intake, aldosterone promotes urinary K^+ loss, alkalosis promotes K^+ cellular uptake
3. Low, chloride ions are lost in vomitus
4. Decrease in osmolality due to water retention
5. Isotonic
6. Thirst, ADH and aldosterone secretion
7. Accompanied by HCO_3^- or Cl^-; exchanged for K^+ or H^+
8. Hypovolemia, total body loss of Na^+, hypochloremia, renal elimination of K^+, alkalosis
9. Both acidosis and alkalosis
10. In secretory diarrhea, there may be a greater loss of HCO_3^- as compared to Cl^-

CHAPTER 6

1. Bronchioles are smaller, lack cartilage
2. Lobules
3. Respiratory bronchioles are microscopic; alveoli open into lumen
4. Surrounded by alveolar sacs
5. Basic functional unit of lung; supplied by a terminal bronchiole (includes respiratory bronchioles and air sacs)
6. Goblet cells, mucous glands

7. Alveolar wall (epithelium, basement membrane, connective tissue); capillary wall (endothelium, basement membrane)
8. Type II cells; decreases surface tension, prevents alveolar collapse
9. Ingestion of foreign particles
10. In allergic reactions release histamine
11. Remove interstitial fluid, prevent pulmonary edema
12. Smoking, chronic inflammation, excessive drying
13. Bronchioles
14. Decreased surface area for gas exchange
15. Macrophages; causes breakdown of elastic fibers
16. The arrangement of the protein fibers elastin and collagen
17. Decreased expandability; increased effort to breathe
18. Parietal pleura, branches of systemic arteries; visceral pleura, branches from pulmonary circulation
19. Edema, fibrosis, bronchiolar obstruction
20. Hydrostatic pressure in capillaries, protein in pleural fluid, subatmospheric intrapleural pressure

CHAPTER 7

1. a. Emphysema
    b. Asthma
    c. Chronic bronchitis
2. Cigarette smoking, air pollution
3. Hypertrophy of mucous glands, increased number of goblet cells, hypertrophy of smooth muscle
4. Excessive mucus, bronchoconstriction, hypertrophy of smooth muscle
5. Macrophages, leukocytes
6. Destroy elastic fibers in alveolar walls and around bronchioles
7. Synthesized in the liver
8. Inactivates elastase
9. Smoking causes an increased number of macrophages and leukocytes (sources of elastase)
10. Dyspnea
11. Bronchospasms
12. a. Intrinsic
     b. Extrinsic
     c. Intrinsic
     d. Extrinsic
13. Mast cells, basophils
14. Mast cells, basophils
15. Histamine causes bronchoconstriction and increases capillary permeability; chemotactic factors attract leukocytes

16. Sets the stage for a continued inflammatory reaction
17. Difficult expiration with wheezing; sometimes cough with mucus production
18. Increased thickness of alveolar basement membrane; hypertrophy of mucous glands; increased number of goblet cells; hyperplasia of smooth muscle; infiltration of eosinophils
19. Bronchoconstriction, edema of mucous membranes, viscous mucus plugs, hyperplasia of smooth muscle
20. Hypoxemia, acidosis

CHAPTER 8

1. Deficiency of surfactant
2. Cellular debris and protein fibers; inner lining of alveoli
3. Inadequate inflation of alveoli, hyaline membrane, edema, decreased compliance
4. Preceding acute illness or trauma, latent period of 24–72 hours; dyspnea, tachypnea, hypoxemia, pulmonary edema
5. Interstitial fibrosis
6. Necrotic tissue that is not cheeselike, and it is granulated (has red granules); noncaseating epithelioid granulomas
7. Blood flow stasis, damage to blood vessel wall, increased tendency for blood to clot
8. Necrosis of lung tissue due to inadequate blood supply
9. No gas exchange, loss of surfactant, alveolar collapse, pulmonary edema
10. Vasodilation and increased permeability
11. Inactive cough mechanism, anesthesia may impair respiratory defenses
12. Viscid exocrine gland secretion, increased Na^+, K^+, Cl^- in sweat
13. Malignant. Rapid growth, necrosis, invasive, metastasizes, almost never encapsulated
14. Squamous carcinoma, adenocarcinoma, small cell carcinoma, large cell carcinoma
15. May be asymptomatic, cough, dyspnea, hemoptysis, stridor
16. Chest X-ray, cytological examination of sputum, bronchoscopy
17. Cuboidal and columnar cells that form glandular structures
18. Carcinoma originating in epithelial tissue, sarcoma from undifferentiated connective tissue, leukemias from white blood cells, lymphomas from lymphoid tissues
19. T1 = noninvasive tumor, 3 cm or less, NO, MO = no lymph node involvement and no metastasis, respectively
20. Any tumor with varying lymph node involvement; metastasis

CHAPTER 9

1. Alveolar pO_2 = 100 mm; blood pO_2 = 40 mm; oxygen diffuses into blood
2. Oxygen must be removed from plasma to form oxyhemoglobin for diffusion into blood to continue
3. Percent saturation of hemoglobin changes little over a wide range of pO_2
4. Usually, because increased ventilation in undamaged lung areas causes a proportional increase in the elimination of CO_2
5. Inadequate airflow
6. pO_2 decreases, pCO_2 increases
7. Respiratory muscle weakness, depression of respiratory center, atelectasis, deep sleep, severe obstructive/restrictive pulmonary disease
8. Cardiac arrhythmias and pulmonary hypertension
9. Constriction of airways in the affected area, loss of surfactant, atelectasis
10. Hypoxemia, hypocapnia
11. Respiratory alkalosis due to an increased breathing rate
12. Thickened respiratory membrane, decreased surface area for gas exchange
13. Carbon dioxide diffusion occurs in spite of an alveolar, capillary block
14. Bronchospasms, thick mucus secretions, edema and inflammation
15. Hypoventilation, uneven ventilation, or impaired diffusion
16. Hypoxemia, hypercapnia
17. Restriction of thoracic movements
18. Increased production of lactic acid from pyruvic acid (a product of anaerobic metabolism)
19. Retention of bicarbonate
20. Decreased respiratory drive, airway obstruction

CHAPTER 10

1. Prostaglandins
2. Implantation of endometrial tissue outside the uterus
3. Ectopic pregnancy
4. Endometrial hyperplasia
5. Enlarged ovaries, hyperplasia of ovarian stroma, layer of follicles beneath capsule of ovary, multiple cysts, high LH/FSH ratio, absence of ovulation
6. Infertility, amenorrhea
7. Abnormal increase in the number of cells
8. Majority of epithelial cells of cervix are abnormal in size and shape
9. All cells from the basement membrane to the surface have the appearance of cancer cells
10. Multiple sexual partners at an early age, smoking, venereal infections

11. Stimulation of the endometrium by either endogenous or exogenous estrogens unopposed by progesterone
12. Germ cells, surface epithelial cells, ovarian stromal cells, hilum cells
13. Galactorrhea
14. Fibrosis, cysts, epithelial hyperplasia, adenosis
15. Infiltrating duct carcinoma
16. Late first pregnancy, Western culture, family history, exposure to ionizing radiation, prolonged postmenopausal estrogens, benign breast disease with atypical epithelial hyperplasia
17. More hopeful
18. Remove estrogens as a source of stimulation of breast cancer growth
19. Balanitis, posthitis
20. Unretractable prepuce; prepuce cannot be pulled to cover glans penis
21. Prolonged erection due to failure of drainage of venous blood
22. Peyronie's disease
23. Cryptorchidism
24. Epididymitis
25. Orchitis
26. Testicular torsion
27. Hesitancy to initiate voiding, straining to void, dribbling after micturition, urinary retention, nocturia, dysuria
28. Seminoma, embryonal carcinoma
29. Teratoma
30. Acid phosphatase

CHAPTER 11

1. Medial area of kidney where ureter and renal vein exit and where renal artery enters
2. Tubules and blood vessels
3. Compartments that receive urine; major calyces form the renal pelvis
4. Glomerulus, peritubular capillaries
5. Glomerular hydrostatic pressure
6. Plasma oncotic pressure, filtrate hydrostatic pressure
7. Reabsorption
8. Descending. Permeable to H_2O; ascending impermeable to H_2O
9. Proximal convoluted tubule
10. Proximal, distal convoluted tubule; collecting duct
11. Juxtamedullary nephrons
12. Reabsorption of Na^+, Cl^- from ascending limb; water not reabsorbed from ascending limb; urea reabsorption from collecting ducts; water reabsorbed from descending limb, which concentrates the filtrate

13. Vasa recta, which has a loop configuration like the loop of Henle; in the descending segment there is net efflux of H_2O and influx of solute; process reverses in ascending segment, and water is carried away
14. Stimulates
15. Juxtaglomerular cells
16. Reaction in renal tubular cells, $CO_2 + H_2O \rightleftharpoons H_2CO_3 \rightleftharpoons H^+ + HCO_3^-$
17. Secretion of H^+, reabsorption of HCO_3^-
18. a. $H^+ + HCO_3^- \rightleftharpoons H_2CO_3 \rightleftharpoons CO_2 + H_2O$
    b. $H^+ + HPO_4^{-2} \; H_2PO_4^-$
    c. $H^+ + NH_3 \rightleftharpoons NH_4^+$
19. Reabsorption of Cl^- or HCO_3^-; secretion of K^+ or H^+
20. Formation of NH_3 from amino acids and subsequent elimination of NH_4^+ in urine
21. Protein metabolism in the liver
22. Filtration, reabsorption
23. Androgens, growth hormone (anabolic); corticosteroids, thyroxine (catabolic)
24. Fluid imbalance, protein catabolism, renal dysfunction
25. Low urine flow causes increased solute concentration that favors urea reabsorption
26. Elevated, increased time for reabsorption
27. Chemical reaction in muscle involving creatine
28. Dehydration, ureteral obstruction, decreased renal blood flow
29. 120 ml plasma required per min to supply the amount of creatinine at the rate excreted in the urine; it is glomerular filtration rate
30. Nuclear magnetic resonance scan
31. Substance is not reabsorbed, creates an osmotic gradient, draws water into the tubule
32. Potassium loss leads to intracellular hydrogen ion movements in exchange for potassium
33. Acidosis because there is inhibition of H^+ secretion
34. Inhibition of reabsorption of Na^+ and Cl^-
35. Spironolactone
36. Because of increased Na^+ presented to distal tubule where K^+ is secreted; causes increased K^+ secretion
37. Cl^- excreted, HCO_3^- reabsorbed; result is alkalosis
38. Collecting duct
39. Loop diuretics inhibit sodium reabsorption in the ascending limb, and the sodium content of the filtrate overwhelms the reabsorptive capacity of the distal tubule and collecting duct
40. Alkalosis

CHAPTER 12

1. Bacteria, pyuria, red blood cells
2. Vesicoureteral reflux
3. Chills and fever, nausea, vomiting, diarrhea, costovertebral tenderness
4. Dilatation of renal pelvis due to back-pressure of urine; obstruction
5. Increased pressure in renal pelvis and tubules; medullary ischemia; decreased GFR; loss of tubular functions; hypertension; secondary infection; stones; renal failure
6. Procedure involving shock waves or ultrasound to cause kidney stone disintegration
7. Control urine pH, increase fluid intake, dietary control
8. Calcium oxalate, $Ca_3(PO_4)_2$, $MgNH_4PO_4$
9. Prostate gland encircles the urethra
10. Failure of reabsorption in the proximal tubule

CHAPTER 13

1. Antigen/antibody complexes and complement
2. Release of proteolytic enzymes and subsequent activation of macrophages and platelets
3. Deposition of complement
4. Release of proteolytic enzymes and factors that promote cellular proliferation
5. Increase glomerular permeability by releasing serotonin and histamine; promote cellular proliferation; activate clotting system; secrete cationic proteins, which neutralize a charge barrier and allow protein to cross
6. Formation of thrombi and promotion of crescent-shaped accumulations of cells within or around glomeruli
7. Negatively charged molecules which are constituents of the glomerular basement membrane; prevent crossing of albumin which is anionic
8. Pyoderma or pharyngitis caused by group A beta-hemolytic Streptococci
9. Macroscopic hematuria, edema, oliguria
10. Protein, white blood cells, red blood cells, casts, high specific gravity
11. Proliferation of epithelial cells in the form of a crescent, which fill Bowman's space and are within and around the glomerulus
12. Characterized by glomerular disease, antiglomerular basement membrane antibodies, pulmonary hemorrhage
13. These antibodies react both with the basement membrane of lung capillaries as well as the glomeruli
14. >3.5 g protein/day in urine; hypoalbuminemia; edema; hyperlipidemia; lipiduria

CHAPTER 14

1. Autosomal dominant
2. Flank pain, hematuria, hypertension
3. 30–50 years

4. Autosomal recessive
5. Impaired H^+ secretion
6. Hypokalemia
7. Causes hypercalciuria
8. Impaired K^+/H^+ secretion
9. Distal tubular cell defect, decreased aldosterone, renal disease, urinary tract obstruction
10. Impaired HCO_3^- reabsorption
11. Defective renal response to ADH
12. Renal cell carcinoma (renal adenocarcinoma)
13. Hemolytic anemia, thrombi, acute renal failure
14. Viral infection or bacterial gastroenteritis
15. Childbirth, oral contraceptives
16. Proteinuria
17. Thickening of glomerular basement membrane, hyaline thickening of afferent and efferent arterioles, increase in mesangial matrix
18. Fibrinoid necrosis of afferent arterioles
19. Skin, lungs, heart, kidney, gastrointestinal tract
20. Females, 30–50 years

CHAPTER 15

1. Acute failure. Sudden reduction of GFR; chronic failure. Progressive reduction of GFR, 20 ml/min to less than 5 ml/min
2. Prerenal, renal, postrenal
3. Hemorrhage, vomiting, burns, diarrhea, vasodilation, congestive heart failure
4. Tumors, kidney stones, hypertrophy of prostate
5. a. Hyponatremia due to water retention
   b. Hyperkalemia
   c. Edema
   d. Hypocalcemia due to decreased vitamin D metabolite production and skeletal resistance to PTH
   e. Hypermagnesemia
   f. Elevated BUN
   g. Elevated creatinine
   h. Acidosis
6. Creatinine is secreted to some extent; this secretion increases as filtration decreases
7. Urea is both filtered and reabsorbed; reabsorption increases as renal blood flow decreases
8. Acidosis increases ionized form of calcium, which is the physiologically active form; sudden reduction may cause tetany
9. Sodium excretion increases as GFR declines; this may continue even with a low sodium diet
10. Increased tubular secretion, increased fecal loss
11. Mg containing antacids
12. Array of symptoms that result as GFR falls below 20 ml/min
13. Decreased erythropoietin, increased red blood cell fragility

14. Ecchymoses
15. Grayish bronze color, pruritis, uremic frost
16. Stomatitis, erosive gastritis, uremic colitis, anorexia, nausea, vomiting
17. Emotional lability, insomnia, coma, convulsions, paresthesia, hypalgesia
18. Growth retardation; bone demineralization; calcium phosphate deposits in arteries, joints, and organs
19. Pore size and area of membrane, composition of dialysate, hydrostatic pressure, rate of blood and dialysate flow
20. a. Hydrostatic pressure
    b. Osmotic gradient created by glucose

CHAPTER 16

1. Ammonia is an important buffer in urine; combines with H^+ to form NH_4^+
2. Loss of HCO_3^- and retention of metabolic acid anions due to decreased GFR; tubular damage limits H^+ secretion
3. Excess acid or decreased base
4. Urea is reabsorbed, creatinine is not; if BUN is high, implication is decreased urine flow and additional time for reabsorption
5. As GFR decreases, rate of creatinine secretion increases
6. Blood phosphorous increases, stimulates PTH secretion; PTH causes calcium bone loss, favors excretion of phosphorous, and reabsorption of calcium
7. Na^+, K^+, Cl^-, HCO_3^-
8. Difference between the sum of the cations in answer to question 7 and the sum of the anions
9. Implies accumulation of unmeasured metabolic acid anions
10. Bone is a source of alkaline phosphatase; breakdown of bone

CHAPTER 17

1. H^+, CO_2, 2,3-DPG
2. Carbonic anhydrase
3. Chloride ions move into red blood cells as bicarbonate moves out
4. Four
5. There is increased oxygen delivery to tissues
6. The oxyhemoglobin saturation does not change significantly over a wide range of pO_2 values
7. Shift to right of dissociation curve; enhances oxygen delivery to tissues
8. Pulmonary and cerebral edema
9. Inadequate oxygenation
10. Polycythemia vera
11. Increased plasma volume and increased cardiac output

12. Phlebotomy; bone marrow suppression
13. Decrease in marrow stimulation, diminished bone marrow responsiveness to erythropoietin, bone marrow failure, loss of red blood cells by hemorrhage or hemolysis
14. Normochromic, normocytic
15. Macrocytes and polychromatophilia
16. Nuclear maturation defect; vitamin B_{12} and folic acid deficiency
17. Hypochromic, microcytic
18. Increased synthesis of 2,3-DPG, redistribution of blood flow to vital organs, increased cardiac output
19. Decrease in stem cells causing decreased red blood cell production
20. Nutritional deficiency, malabsorption, increased demand, excessive loss
21. Protein-iron complex called transferrin
22. Ferritin, hemosiderin
23. Absence or reduction in synthesis of either α or β globin chains
24. Two alpha chains, two beta chains
25. Increase in bone marrow due to stimulation of erythropoietin, which is a compensatory response
26. α^o-thalassemia; failure of α globin chain synthesis; result is fetal or neonatal death
27. Mediterranean, Southeast Asia
28. Blood transfusions, splenectomy, administration of a chelating agent
29. One amino acid substitution in β globin chain
30. a. Sickle cell trait
    b. Sickle cell disease

Chapter 18

1. Granulocyte
2. Myelocyte
3. B-lymphocytes differentiate in bone marrow; T-lymphocytes differentiate in the thymus
4. Lymphoid tissue
5. Migrates to tissues to become either a fixed or motile macrophage
6. Infection
7. Granular leukocytes in the bone marrow
8. Acute lymphoblastic leukemia
9. Chronic lymphocytic leukemia
10. Chronic myelogenous leukemia

Chapter 19

1. a. Platelets form a plug
   b. Formation of thrombin
   c. Formation of fibrin
2. Plasmin
3. Factor Xa

4. Contact of blood with surfaces, such as glass or collagen of damaged blood vessels
5. Thromboplastin
6. Fibrin
7. AT III, protein C
8. Deficiency of factor VIII:C; deficiency of factor IX; deficiency of vWF
9. Hemarthrosis, arthropathy, blood cysts, pressure-induced neuropathies and muscular atrophy, intracranial and gastrointestinal bleeding
10. Cold-insoluble material left after slow thawing of frozen plasma; contains factor VIII:C
11. von Willebrand's disease
12. Connective tissue disorders, septicemia, trauma, shock, obstetrical complications, malignancies, liver disease, renal disease
13. Fibrinogen/fibrin degradation products
14. Vitamin K is required for the synthesis of clotting factors VII, IX, X, and prothrombin
15. Gamma globulin, prednisone, immunosuppressive drugs, splenectomy

Chapter 20

1. Vast majority of cells are red blood cells
2. Reticulocytes
3. Bands or stabs
4. Result is bleeding tendency
5. APTT
6. Pro time
7. APTT
8. There is either a deficiency of fibrinogen or inhibition of thrombin
9. Increased fibrinolysis
10. a. Increased
    b. Prolonged
    c. Prolonged
11. Myeloblast
12. Increased erythropoietin
13. Blood loss, iron deficiency, bone marrow invasion, radiation or chemotherapy
14. Drugs, infection, metabolic disorders, emotional stimuli
15. Infection, therapy
16. Secretion of PTH-like substances by the tumor
17. Alkaline phosphatase
18. AST
19. Cell necrosis with release of the enzyme
20. Thromboplastin
21. Introduction of a procoagulant into the circulation
22. Consumption of clotting factors, fibrinolysis
23. Low-grade coagulation followed by overcompensation (increased synthesis of clotting factors)
24. Increased FSP

25. Inhibition of fibrin monomer polymerization, interference with thrombin, weaken fibrin clots, inhibit platelet function, increase platelet aggregation

Chapter 21

1. Diastolic
2. Diastolic
3. Indicates increased pulmonary capillary pressure, as well as increased left atrial and ventricular pressures; pulmonary edema may occur
4. Increased pulmonary capillary pressure and pulmonary edema
5. Afterload
6. More forceful contractions
7. Constriction of arterioles
8. Blood flows down a pressure gradient, i.e., in the direction of lower pressure
9. Veins are distensible, act as reservoirs; when constricted they increase venous return
10. Diastolic volume, compliance of ventricular walls
11. Kidney
12. Difference between systolic and diastolic pressures
13. Wider pulse pressure
14. Increased systolic pressure due to stiffness of the aorta; rapid runoff of blood during diastole and a lowered diastolic pressure
15. Increased peripheral resistance and cardiac output
16. Renin leads to production of angiotensin and aldosterone; salt and water retention occurs; increased blood volume causes hypertension
17. Cardiovascular disease and renal failure
18. Augmented constriction of venules, vasoconstriction due to norepinephrine, increased cardiac output due to sympathetic stimulation, elevated renin levels, defect in renal excretion of sodium and water
19. Weight loss, dietary sodium restriction, limited alcohol intake, exercise
20. Reduced blood flow that leads to a fall in blood pressure and inadequate tissue perfusion
21. Arterial vasoconstriction, arteriovenous shunting, dilation of veins
22. Increased heart rate, vasoconstriction
23. Inadequate delivery of oxygen to tissues causes a shift to anaerobic metabolism with lactic acid production
24. Sympathetic nerve activity causes constriction of postcapillary venules; increased capillary hydrostatic pressure causes intravascular fluid loss

25. Hyperventilation, hypotension, tachycardia, pale cool skin, cyanosis, confusion, oliguria
26. Septicemia, drugs, transection of spinal cord
27. Due to initial intravascular loss; there is then compensatory restoration of fluid due to increased oncotic pressure
28. Myocardial infarction
29. Decreased cardiac contractility leads to:
    a. Decreased cardiac output
    b. Decreased coronary perfusion
    c. Inadequate perfusion further impairs myocardial function
30. Lactate level

Chapter 22

1. Tunica intima, tunica media, tunica externa
2. Capillaries
3. Elastic fibers
4. Vasoconstriction of digital arteries and arterioles
5. Emotional stress, exposure to cold
6. Smooth muscle cells, accumulations of intracellular and extracellular lipids, macrophages, lymphocytes, and connective tissue; possibly calcium
7. Myocardial infarction, thromboembolism, aneurysm
8. Cigarette smoking, obesity, family history, diabetes, hypertension, hypercholesteremia
9. Low-density lipid (LDL) fraction
10. Elevation consisting of cellular debris, cholesterol, calcium; covered by smooth muscle, lymphocytes, macrophages, and connective tissue
11. LDL
12. Hypertension
13. A tear in the tunica intima allows blood to enter the subendothelium
14. Blood in the pericardial sac compresses the heart
15. Rupture of aorta

Chapter 23

1. Atherosclerosis
2. Angina pectoris
3. Myocardial necrosis, atrioventricular block, cardiogenic shock, death
4. Arthritis, carditis, chorea, subcutaneous nodules, erythema marginatum
5. Joint pain, fever, leukocytosis, increased sedimentation rate
6. It indicates a high level of antibodies against streptolysin O produced by streptococcal organism

7. Myocardial disease, pressure or volume overload, restriction of ventricular filling, increased demands on the heart
8. Hypertrophic, dilated, and restrictive cardiomyopathy
9. Systemic hypertension causes increased resistance and increased work for the left ventricle; pulmonary hypertension increases the work of the right ventricle
10. Left ventricular hypertrophy
11. Valve damage that allows backflow of blood, i.e., mitral regurgitation
12. a. Left
    b. Left
    c. Right
    d. Right
    e. Right
    f. Left
13. Cardiac dilatation and hypertrophy; increased levels of norepinephrine, epinephrine, angiotensin, aldosterone, vasopressin, increased 2,3-DPG; pulmonary blood flow directed to upper lobes
14. Pulmonary edema
15. Right-sided failure

Chapter 24

1. Ductus venosus
2. Connects the pulmonary artery and the aorta
3. Opening in atrial septum
4. Left-to-right shunt if pulmonary vascular resistance is low
5. If high pulmonary resistance develops and causes a right-to-left shunt
6. a. Ventral septal defect
    b. Pulmonic stenosis
    c. Aortic override of ventricular septum
    d. Right ventricular hypertrophy
7. If severe pulmonic stenosis, right-to-left shunt; if moderate pulmonic stenosis, then minimal right-to-left shunt
8. From left to right atrium
9. Left to right atrium
10. Upper body hypertension and increased left ventricular work load

Chapter 25

1. Natriuretic hormone
2. ADH, aldosterone
3. Natriuretic hormone
4. Due to decreased renal blood flow and impaired glomerular filtration rate
5. LDL is a major cholesterol carrying fraction and is a risk factor in coronary artery disease

6. HDL promotes removal of cholesterol from cells and is an indicator of low risk for coronary artery disease
7. Protein carriers complexed to plasma lipids
8. There is an association between diabetes and increased triglyceride levels
9. Aspartate aminotransferase, lactate dehydrogenase, creatine kinase
10. Creatine kinase
11. Septic shock with arteriovenous shunting
12. Vasodilation and arteriovenous shunting

Chapter 26

1. Neutrophils, monocytes, macrophages, histiocytes
2. Histamine
3. Leukotrienes, prostaglandins
4. Platelets
5. Chemotaxis, enhances phagocytosis

Chapter 27

1. Fab sites of surface immunoglobulin combine with specific antigen
2. Surface class II MHC-encoded protein recognizes specific macrophage-antigen complexes
3. Thymus
4. Promotes inflammation, phagocytosis, and cell membrane damage
5. Some are memory cells; some promote antibody production by plasma cells

Chapter 28

1. Histamine, slow-reacting substance, kinin, platelet aggregating factor, serotonin, chemotactic factors
2. Increased vascular permeability, vasodilation, smooth muscle contractions
3. Immune complex-mediated reaction, type III
4. Antibody-dependent cellular cytotoxicity, type II
5. Cell-mediated immunity, type IV

Chapter 29

1. Lymphadenopathy-associated virus (LAV), human T cell lymphotropic virus type III (HTLV-III), human immunodeficiency virus (HIV)
2. Opportunistic infections, Kaposi's sarcoma, lymphadenopathy syndrome
3. Viral infection of T4 lymphocytes (helper cells); these cells play a role in activating both humoral and cell-mediated responses

4. Lymphopenia, decreased T cells, circulating immune complexes, reduced cytotoxic activity, decreased antibody response to immunization, abnormal lymphokine production
5. They make normal amounts of IgG and IgM
6. Allergies, sinopulmonary infection, autoimmune disease
7. Absence of T and B cell immunity
8. Reaction between immunocompetent graft cells, such as bone marrow transplant, and immunodeficient host cells
9. Goodpasture's syndrome
10. Antibodies against acetylcholine receptors at the neuromuscular junction
11. Diffuse erythematous rash
12. Subcutaneous nodules
13. Gold salts
14. Discoid lupus erythematosus
15. Fever, fatigue, weakness, weight loss, anorexia
16. Light sensitivity, butterfly rash, ulcerations of nasal and oral mucosa and labia, ulcers on legs, purpuric skin eruptions
17. Vasculitis leads to thrombophlebitis and Raynaud's disease; pericarditis, myocarditis, endocarditis
18. Decreased esophageal motility; anorexia, nausea, vomiting, abdominal pain; pancreatitis, hepatomegaly, splenomegaly
19. Pleuritis, pneumonia, atelectasis
20. Renal failure and CNS involvement
21. Systemic scleroderma
22. Calcinosis, Raynaud's disease, esophageal dysmotility, sclerodactyly, telangiectasia
23. Sicca complex
24. Decreased production of tears, dry mouth
25. Polyarteritis nodosa

Chapter 30

1. Against the thyroid, nucleic acids, lymphocytes, acetylcholine receptors, gastric parietal cells, adrenocortical cells, mitochondria, pancreatic islet cells, nuclei
2. Immunoglobulin properties change when bound to antigen; in some cases, they precipitate out in the cold
3. Only if there is a high titer
4. Antigen/antibody combination
5. IgG or IgM immune complex formation leads to complement activation and decreased serum levels
6. Indicates immunoglobulin attachment to red blood cells
7. Antibodies that react with red blood cells
8. Lymphocytes
9. T4 lymphocytes

10. Suppressor cytotoxic cells; limit scope of immune reactions and are involved in lysis of antigen bearing cells
11. Circulating anticoagulants which are antibodies against factor VIII or against thrombin
12. Vomiting (see chapter 5)
13. ALT/SGPT, alkaline phosphatase, GGT
14. AST/SGOT, LD/LDH
15. Anti-DNA, anti-Sm, anti-RNP, anti-ENA, anti-SSA, anti-SSB
16. Antibody to Smith (Sm) antigen
17. May be useful in following a patient's response to therapy
18. Indicates immune complex formation with the activation and removal of complement
19. Primarily involves the skin
20. Deposition of immune complexes in the alveolar walls; susceptibility to pulmonary infections
21. Possible lupus nephritis
22. Decrease in GFR
23. Deposition of immune complexes in vascular epithelium of glomeruli
24. A neutrophil containing a phagocytized damaged cell; the neutrophil is filled with this inclusion body; nucleus of neutrophil is compressed against the cell membrane
25. Bleeding from gastrointestinal lesion; may be made worse by lupus anticoagulants; antibodies against red blood cells may cause hemolysis

Chapter 31

1. Oxytocin, vasopressin (ADH)
2. TSH, ACTH, GH, LH, FSH, prolactin
3. Somatostatin
4. TRH, CRF, GRH, LH-RH
5. Pituitary tumor
6. Acromegaly
7. Somatomedin
8. Abnormal milk production, prolactin-secreting tumor
9. Sheehan's syndrome
10. Ischemic injury of pituitary
11. Lesion of hypothalamus, lesion of pituitary, asphyxia, trauma at birth, inheritance
12. Deficiency of either GH or somatomedin
13. Renal reabsorption of water, vasoconstriction
14. Deficiency of vasopressin
15. Renal unresponsiveness to vasopressin
16. T_3
17. Lower serum cholesterol, stimulate growth, promote protein synthesis, reduce sensitivity to insulin
18. May occur in either
19. Thyroiditis, toxic adenoma
20. Mental retardation, abnormal physical development (cretinism)

21. Malignancy, iodine, deficiency, autoimmune thyroiditis, radioactive iodine therapy, thyroidectomy
22. Severe hypothyroidism with puffiness of face, hands, and feet
23. Promotes bone resorption, renal calcium reabsorption, intestinal calcium absorption
24. Parathyroid adenoma or hyperplasia
25. Increased neuromuscular excitability

CHAPTER 32

1. Protein catabolism, gluconeogenesis, localized lipogenesis, increased bone resorption, decreased insulin sensitivity, a factor in maintaining blood pressure, inhibition of cell-mediated immunity, influence on mood
2. ACTH, CRH
3. Stress
4. Sodium and water reabsorption; potassium and hydrogen ion secretion
5. Juxtaglomerular cells secrete renin; renin promotes conversion of angiotensinogen to angiotensin I; production of angiotensin II; angiotensin II stimulates aldosterone secretion
6. Adrenal tumors, iatrogenic, ACTH-secreting pituitary adenoma, hypersecretion of CRH by hypothalamus
7. Truncal and facial obesity, supraclavicular fat deposits, buffalo hump, muscle wasting causes thin arms and legs, striae, osteoporosis, increased susceptibility to infection
8. Hypercalcemia, hypokalemia, alkalosis
9. Low doses of dexamethasone suppress ACTH, and thus cortisol
10. a. Cushing's disease
    b. Pituitary tumor
    c. Adrenal tumor
    d. Ectopic ACTH secretion
11. Low or undetectable plasma ACTH level
12. Hyperplasia, adrenal adenoma
13. Nephrotic syndrome, congestive heart failure, cirrhosis, renal artery stenosis, renin secreting renal tumor, pregnancy, oral contraceptives
14. Hypernatremia, metabolic alkalosis, hypokalemia
15. In the chronic condition, there is an escape from the sodium (and water) retaining effects of aldosterone
16. Indicative of primary aldosteronism
17. Saline suppresses renin and decreases plasma aldosterone in normal individuals
18. Saline infusion has little effect on aldosterone
19. Primary. Decreased plasma renin activity; Secondary. High level of plasma renin activity
20. Autoimmune destruction of the adrenal gland
21. ACTH deficiency

22. Sustained or periodic hypertension
23. Promotes cellular uptake of glucose
24. Increased levels of triglycerides and cholesterol
25. Absence of ketoacidosis

CHAPTER 33

1. Oral glucose tolerance test
2. Percent of glycohemoglobin is an indication of average blood glucose level over 2–3 month period
3. Diabetic coma
4. Elevated
5. Increase
6. Increased osmolality (hyperglycemia)
7. Acetone, acetoacetic acid, beta-hydroxybutyric acid
8. There is an accumulation of anions normally not measured
9. Osmotic shift of water and potassium out of cells; osmotic diuresis promotes renal loss of potassium
10. Acidosis causes efflux of intracellular potassium; acute metabolic acidosis reduces urinary potassium loss; prolonged acidosis enhances potassium excretion
11. In ketoacidosis, the efflux of potassium ions may be limited by the fact that β-hydroxybutyrate anion may accompany the intracellular movement of hydrogen ions. This maintains electrical neutrality and does not require potassium ion efflux.
12. Enhances cellular uptake of potassium
13. Decreased aldosterone has been observed; aldosterone activity may be reduced due to renal injury
14. Depletion
15. Draws water out of cells and has a dilutional effect; causes osmotic diuresis and promotes renal excretion of sodium
16. It implies that the ingestion of water is less than renal loss
17. Bicarbonate
18. Decreased GFR and, due to low urine flow, increased reabsorption of urea
19. Glucose and sodium
20. Intravenous fluids, potassium, insulin
21. Dilution due to volume replacement and intracellular movement of potassium ions
22. By the reaction $H_2CO_3 \rightleftharpoons H^+ + HCO_3^-$ as hydrogen ions combine with ketoacid anions
23. Hypokalemia due to intracellular movement of potassium ions
24. Cerebral edema, lactic acidosis, thrombosis
25. Little or no acidosis in HHNK; greater degree of hyperglycemia in HHNK

CHAPTER 34

1. Literally means inability to relax; motility disorder of esophagus
2. Inadequate antireflux mechanism, increased gastric volume, injury due to pepsin, inadequate esophageal clearance
3. Retrosternal burning or chest pain, regurgitation, increased salivation, dysphagia
4. Stricture or occlusion
5. A condition in which the esophagus is lined with columnar epithelial cells
6. Decrease reflux, neutralize gastric contents, promote esophageal clearance, protect esophageal mucosa
7. *Candida*
8. Alcohol, salicylates, anti-inflammatory agents, autoimmune disease
9. Hydrochloric acid, pepsin
10. Mucus, HCO_3^-, epithelial cells, prostaglandins, adequate blood flow
11. Diffusion of HCl into mucosal epithelial cells
12. Neutralize gastric acid, prevent acid secretion, promote local healing
13. Duodenal ulcer
14. Trauma and severe illness
15. Stress ulcers are usually multiple and shallow, and there is no intense inflammatory reaction
16. Associated with central nervous system trauma
17. To avoid risk of gastrointestinal bleeding
18. Antacids, H_2 receptor antagonists, sucralfate
19. Intra-arterial infusion of vasopressin, transcatheter embolization, laser beam therapy
20. Truncal vagotomy, all fibers below the esophagus are cut; selective vagotomy, all fibers to the stomach are cut
21. Most of the stomach is removed
22. Early satiety, postcibal vomiting and pain, chronic diarrhea, weight loss, anemia
23. Salt, nitrates, nitrites, smoked foods, pickled vegetables, starch
24. Invasiveness, lymph node involvement, metastasis
25. Surgical resection

CHAPTER 35

1. Mucosa, submucosa, muscularis, serosa
2. a. Increased wall tension correlates with increased intraluminal pressure
   b. Decreased radius correlates with increased intraluminal pressure
3. Usually
4. Pain in left abdominal quadrant
5. Measures to increase stool mass
6. Diverticulosis is an outpouching of intestinal wall; diverticulitis is an inflammation which may cause rupture of diverticula

7. Rupture, abscess, fibrosis and obstruction, fistula
8. Acute abdominal and back pain, fever
9. Toxin produced by bacteria
10. Malabsorption
11. Xylose is normally absorbed in the intestine and appears unchanged in the urine; test for intestinal absorptive capacity
12. Tetracycline and folic acid
13. Weakening of bones due to malabsorption of calcium
14. Vitamin A deficiency
15. Impaired iron or folic acid absorption
16. Celiac sprue
17. Celiac sprue
18. Ulcerative colitis
19. Milk intolerance
20. Surgery
21. High fat intake
22. Fiber, calcium
23. Benign tumor that is glandularlike
24. Malignant adenoma
25. Liver

CHAPTER 36

1. Hepatitis B and C
2. Hepatitis C
3. Condition in which necrosis of liver leads to proliferation of fibrous connective tissue accompanied by formation of regenerative nodules
4. Because it is between two capillary beds
5. Ascites, esophageal varices, hypersplenism, caput medusae, internal hemorrhoids
6. Deficiency of clotting factors, hepatic encephalopathy, abnormal sex hormone metabolism
7. Stasis of bile flow, high level of cholesterol in bile, infection
8. Digestion of pancreatic tissue by pancreatic enzymes
9. Alcohol abuse, biliary tract disease, abdominal trauma, hyperlipidemia, drugs, viral infections
10. Abdominal pain, weight loss, vomiting

CHAPTER 37

1. Unconjugated
2. Indirect equals unconjugated bilirubin; hemolysis
3. Cellular uptake, conjugation, secretion
4. Obstruction to flow of bile
5. Conjugated bilirubin
6. Decreased albumin, increased gamma globulin
7. AST, ALT
8. Extensive hepatocellular injury due to drug and viral hepatitis

9. Obstruction to flow of bile, liver cancer, rapid bone turnover
10. Decreased synthesis of clotting factors by the liver due to hepatocellular damage
11. Atelectasis causing areas of hypoventilation, particularly if there is ascites
12. Hyperventilation, possibly due to direct stimulation of respiratory center by ammonia
13. Anaerobic metabolism with an accumulation of lactic acid
14. Vomiting, diuretics
15. Respiratory alkalosis due to hyperventilation
16. Increased 2,3-diphosphoglycerate in red blood cells causes decreased oxygen affinity
17. Half-life of albumin is 15–20 days; decreased hepatic synthesis may not be apparent for that period of time
18. Amylase
19. About half of calcium is bound to albumin; loss of protein-bound calcium may not affect the physiologically active ionized form
20. Massive loss of plasma leads to intravascular volume depletion; thus, prerenal azotemia may occur

CHAPTER 38

1. Choroid plexuses of the ventricles of the brain
2. Flows into subarachnoid spaces surrounding spinal cord and brain
3. By way of arachnoid villi which project into blood filled venous sinuses
4. Congenital malformations, posthemorrhagic, postinflammatory, mass lesions
5. Oversecretion of CSF, impaired absorption by arachnoid villi
6. Computerized tomography, magnetic resonance imaging
7. Vasogenic, cellular, interstitial
8. Hyponatremia or accumulation of solutes in brain cells
9. To maintain plasma osmolality to avoid exacerbation of cerebral edema
10. AVMs, aneurysms
11. Cerebrovascular occlusion, intracranial hemorrhage
12. Cholesterol crystals, platelet-fibrin fragments, air, tissue cells, fat, clumps of bacteria, thromboembolus (blood clot)
13. TIA can last for minutes or up to 24 hours, and there is no permanent damage; completed stroke results in permanent neurologic deficit
14. Anticoagulants and medications decrease platelet adhesiveness
15. Site of occlusion, area of the brain that is affected, and development of collateral channels
16. The internal carotid gives rise to the ophthalmic artery, which supplies the retina

17. Hemiparesis, a partial paralysis on one side; aphasia, impaired ability to associate words and ideas; dysarthria, stammering; paresthesia, tingling or burning sensation of the skin
18. Area of small infarction within brain tissue
19. Dilatation due to a defect in an arterial wall
20. Sudden, severe headache
21. Red blood cells, xanthochromia, increased protein
22. Arteriography
23. Drug or alcohol use, cigarette smoking
24. Myocardial infarction, mitral valve prolapse
25. Pregnancy, oral contraceptive use, blood disorders, diabetes mellitus

CHAPTER 39

1. A partial seizure is initiated on one side; a generalized seizure involves both sides
2. A complex type involves impaired consciousness
3. Tonic muscle stiffening; clonic jerking movements
4. Astrocyte
5. Meningioma
6. Anoxia at birth, chronic lead poisoning, alcohol ingestion during pregnancy, possible genetic factors
7. Behavior, learning, speech, coordination
8. Prenatal events, complications of childbirth, infections, trauma
9. Spastic, dyskinetic, ataxic
10. a. Sudden muscle contractions
    b. Partial paralysis
    c. Slow continual change of position of toes, fingers, hands, and feet
11. a. Abnormal involuntary movements
    b. Uncoordinated body movements
12. Dysarthria
13. Dysphagia
14. Senile plaques and neurofibrillary tangles
15. Reduced choline acetyltransferase
16. 4–10 years
17. A virus with a long incubation period
18. About 9 months
19. Gait unsteadiness, loss of coordination, handwriting changes, agitation, confusion, paranoia, urinary and fecal incontinence, psychosis
20. Involuntary, irregular muscle contractions

CHAPTER 40

1. a. Classic. Dramatic visual disturbances, paresthesias
   b. Common. More subtle, depression, hyperactivity, increased appetite
2. Stress, too much sleep, menstruation, bright lights, alcohol, specific foods, oral contraceptives

3. Short episodes of agonizing pain, frequently repeated
4. a. Perception of spinning objects
   b. Ringing or buzzing in the ears
   c. Shortening of muscle
   d. Postural abnormality, abdomen forward with upper torso and head back
   e. Muscle spasms
5. Myasthenia gravis
6. Vertigo, tinnitus, hearing loss
7. Duchenne muscular dystrophy
8. Multiple sclerosis, Guillain-Barré syndrome
9. Myotonia, muscle weakness and wasting, arrhythmias, cataracts
10. Islands of gray matter within the cerebral hemispheres; caudate nucleus, putamen, pallidus globus
11. Parkinson's disease
12. Involuntary movements, emotional disturbances, intellectual decline
13. Dopamine does not cross the blood-brain barrier
14. Unilateral, stabbing pain along the course of the second or third divisions of the trigeminal nerve
15. Chewing or touching the cheek, lip, or nose
16. Cranial nerve VII (facial nerve)
17. Partial facial paralysis
18. Loss of and abnormalities of motor neurons
19. Muscle weakness
20. Demyelination of both cranial and peripheral nerves

CHAPTER 41

1. Dilution
2. Fluid loss replaced by hypotonic fluid, inappropriate secretion of ADH, impaired renal excretion of water, cirrhosis, congestive heart failure, renal failure, water intoxication
3. There may be impaired excretion of water associated with hypothyroidism
4. TIA lasts a few minutes up to an hour, with no infarction; a stroke is prolonged and causes infarction
5. Differentiation of white and gray matter; differentiation of thalamus and basal ganglia; visualization of major arteries with a contrast agent; hematomas show as high-density regions; infarction shows as low-density region
6. Magnetic resonance imaging
7. Cerebral arteriogram
8. Radio wave excitation of hydrogen nuclei in a magnetic field; absorbed energy is released, and images are reconstructed from these signals
9. Because of swelling associated with the inflammation
10. Ultrasound waves are directed toward the heart; a recording of the reflected sound waves results in an image of the heart
11. To identify underlying coagulation disorders
12. Blood in CSF
13. Will demonstrate blood in subarachnoid space or the presence of a clot
14. To show size and origin of rupture prior to surgery
15. Hyponatremia is dilutional; hypoventilation causes hypoxemia and respiratory acidosis

Appendix B

Normal Laboratory Values

	Conventional	SI (Système Internationale d'unités)
Amylase	4–25 units/ml	4–25 arb. unit
Bilirubin	Direct: up to 0.4 mg/100 ml. Total: up to 1.0 mg/100 ml	Up to 7 μmol/liter Up to 17 μmol/liter
Calcium	8.5–10.5 mg/100 ml (slightly higher in children)	2.1–2.6 mmol/liter
Carbon dioxide content	24–30 meq/liter	24–30 mmol/liter
Chloride	100–106 meq/liter	100–106 mmol/liter
CK isoenzymes	5% MB or less	
Creatine kinase (CK)	Female: 10–79 U/liter Male: 17–148 U/liter	167–1317 nmol · sec^{-1}/liter 283–2467 nmol · sec^{-1}/liter
Creatinine	0.6–1.5 mg/100 ml	53–133 μmol/liter
Glucose	Fasting: 70–110 mg/100 ml	3.9–5.6 mmol/liter
Iron	50–150 μg/100 ml (higher in males)	9.0–26.9 μmol/liter
Lactic acid	0.6–1.8 meq/liter	0.6–1.8 mmol/liter
Lactic dehydrogenase	45–90 U/liter	750–1,500 nmol · sec^{-1}/liter
Lipase	2 units or less	Up to 2 arb. unit
Lipids 　Cholesterol 　Triglycerides	 120–220 mg/100 ml 40–150 mg/100 ml	 3.10–5.69 mmol/liter 0.4–1.5 g/liter
Osmolality	280–96 mOsm/kg water	280–96 mmol/kg
Oxygen saturation (arterial)	96–100%	0.96–1.00
pCO$_2$	35–45 mm Hg	4.7–6.0 kPa
pH	7.35–7.45	Same
pO$_2$	75–100 mm Hg (dependent on age) while breathing room air; above 500 mm Hg while on 100% O$_2$	10.0–13.3 kPa
Phosphatase (alkaline)	13–39 U/liter; infants and adolescents up to 104 U/liter	217–650 nmol · sec^{-1}/liter; up to 1.26 μmol sec^{-1}/liter
Phosphorus (inorganic)	3.0–4.5 mg/100 ml (infants in first year up to 6.0 mg/100 ml)	1.0–1.5 mmol/liter
Potassium	3.5–5.0 meq/liter	3.5–5.0 mmol/liter
Protein: Total	6.0–8.4 g/100 ml	60–84 g/liter
Albumin	3.5–5.0 g/100 ml	35–50 g/liter
Globulin	2.3–3.5 g/100 ml	23–35 g/liter
Sodium	135–45 meq/liter	135–45 mmol/liter
Transaminase, SGOT (aspartate aminotransferase)	7–27 U/liter	117–450 nmol · sec^{-1}/liter
Transaminase, SGPT (alanine aminotransferase)	1–21 U/liter	17–350 nmol · sec^{-1}/liter
Urea nitrogen (BUN)	8–25 mg/100 ml	2.9–8.9 mmol/liter
Uric acid	3.0–7.0 mg/100 ml	0.18–0.42 mmol/liter

Hormones

Adrenocorticotropin (ACTH)	15–70 pg/ml	3.3–15.4 pmol/liter
Growth hormone	Below 5 ng/ml	<233 pmol/liter
	Children: Over 10 ng/ml	>465 pmol/liter
	Male: Below 5ng/ml	<233 pmol/liter
	Female: Up to 30 ng/ml	0–1395 pmol/liter
	Male: Below 5 ng/ml	<233 pmol/liter
	Female: Below 5 ng/ml	<233 pmol/liter
Insulin	6–26 μU/ml	43–187 pmol/liter
	Below 20 μU/ml	<144 pmol/liter
Parathyroid hormone	<25 pg/ml	<2.94 pmol/liter
Renin activity	Supine:	
	1.1 $\pm$ 0.8 ng/ml/hr	0.9 $\pm$ 0.6 nmol/liter/hr
	Upright:	
	1.9 $\pm$ 1.7 ng/ml/hr	1.5 $\pm$ 1.3 nmol/liter/hr
Thyroid-stimulating hormone (TSH)	0.5–5.0 μU/ml	0.5–5.0 arb. unit
Total triiodothyronine (T_3)	75–195 ng/100 ml	1.16–3.00 nmol/liter
Total thyroxine by RIA (T_4)	4–12 μg/100 ml	52–154 nmol/liter
Free thyroxine index (FT_4I)	1–4	

Hematologic values

Coagulation factors:		
Factor I (fibrinogen)	0.15–0.35 g/100 ml	4.0–10.0 μmol/liter
Factor II (prothrombin)	60–140%	0.60–1.40
Factor VIII (antihemophilic globulin)	50–200%	0.50–2.0
Prothrombin time	<2 sec deviation from control	<2 sec deviation from control
Partial thromboplastin time (activated)	25–38 sec	25–38 sec
Fibrinogen split products	Negative reaction at >1:4 dilution	0 (at 1:4 dilution)
Thrombin time	Control $\pm$ 5 sec	Control $\pm$ 5 sec
"Complete" blood count:		
Hematocrit	Male: 45–52%	Male: 0.45–0.52
	Female: 37–48%	Female: 0.37–0.48
Hemoglobin	Male: 13–18 g/100 ml	Male: 8.1–11.2 mmol/liter
	Female: 12–16 g/100 ml	Female: 7.4–9.9 mmol/liter
Leukocyte count	4300–10,800/mm³	4.3–10.8 $\times$ 10^9/liter
Erythrocyte count	4.2–5.9 million/mm³	4.2–5.9 $\times$ 10^{12}/liter
Mean corpuscular volume (MCV)	86–98 μm³/cell	86–98 fl
Mean corpuscular hemoglobin (MCH)	27–32 pg/RBC	1.7–2.0 pg/cell
Erythrocyte sedimentation rate	Male: 1–13 mm/hr	Male: 1–13 mm/hr
	Female: 1–20 mm/hr	Female: 1–20 mm/hr
Ferritin (serum)		
Iron deficiency	0–12 ng/ml	0–4.8 nmol/liter
	13–20 Borderline	5.2–8 nmol/liter Borderline
Iron excess	>400 ng/liter	>160 nmol/liter
Platelet count	150,000–350,000/mm³	150–350 $\times$ 10^9/liter
Reticulocyte	0.5–2.5% red blood cells	0.005–0.025

Miscellaneous values

Rheumatoid factor	<60 IU/ml
Antinuclear antibodies	Negative at a 1:8 dilution of serum
Anti-DNA antibodies	Negative at a 1:10 dilution of serum
Antibodies to Sm and RNP (ENA)	None detected
Antibodies to SS-A (Ro) and SS-B (La)	None detected

From The New England Journal of Medicine, *314: 39–49, 2 January 1986. Copyright © 1986 The New England Journal of Medicine, Waltham, MA. Reprinted by permission.*

References

Chapter 1

Kutchai, H. C. 1983. Cellular membranes and transmembrane transport of solutes and water, p. 13. *In* R. M. Berne and M. N. Levy (eds.), Physiology. C. V. Mosby Co., St. Louis.

Stein, W. D., and W. R. Lieb. 1986. Transport and diffusion across cell membranes, p. 7, 12, 16, 23–24, 33, 69, 102, 337–38, 351, 363, 524, 611. W. B. Saunders, Philadelphia.

Stryer, L. 1981. Biochemistry, 2nd ed., p. 206, 215, 227, 862, 866, 869. W. H. Freeman & Co., San Francisco.

Vick, R. L. 1984. Contemporary medical physiology, p. 3, 6. Addison-Wesley, Menlo Park, Cal.

Chapter 2

Alvis, R., M. Geheb, and M. Cox. 1985. Hypo- and hyperosmolar states: Diagnostic approaches, p. 186. *In* A. I. Arieff and R. A. DeFronzo (eds.), Fluid, electrolyte, and acid-base disorders. Churchill Livingstone, New York.

Andreoli, T. E. 1985. Disorders of fluid volume, electrolyte, and acid-base balance, p. 515–16, 519, 525, 527, 529. *In* J. B. Wyngaarden and L. H. Smith (eds.), Cecil textbook of medicine, 17th ed. W. B. Saunders, Philadelphia.

Andreoli, T. E. 1985. The posterior pituitary, p. 1268, 1270. *In* J. B. Wyngaarden and L. H. Smith (eds.), Cecil textbook of medicine, 17th ed. W. B. Saunders, Philadelphia.

Arieff, A. I., and R. W. Schmidt. 1980. Fluid and electrolyte disorders and the central nervous system, p. 1438–39, 1451. *In* M. H. Maxwell and C. R. Kleeman (eds.), Clinical disorders of fluid and electrolyte metabolism, 3rd ed. McGraw-Hill, New York.

Cox, M., M. Geheb, and I. Singer. 1985. Disorders of thirst and renal water excretion, p. 162. *In* A. I. Arieff and R. A. DeFronzo (eds.), Fluid, electrolyte, acid-base disorders. Churchill Livingstone, New York.

Goldberger, E. 1975. A primer of water, electrolyte, and acid-base syndromes, 5th ed., p. 27, 35, 48, 71–73. Lea & Febiger, Philadelphia.

Guyton, A. C. 1986. Textbook of medical physiology, p. 276, 349, 476, 478, 959. W. B. Saunders, Philadelphia.

Knochel, J. P. 1980. Clinical physiology of heat exposure, p. 1524. *In* M. H. Maxwell and C. R. Kleeman (eds.), Disorders of fluid and electrolyte metabolism, 3rd ed. McGraw-Hill, New York.

Koppel, J. D., and M. J. Blumenkrantz. 1980. Total parenteral nutrition and parenteral fluid therapy, p. 463–64. *In* M. H. Maxwell and C. R. Kleeman (eds.), Clinical disorders of fluid and electrolyte metabolism, 3rd ed. McGraw-Hill, New York.

Pestana, C. 1979. Fluids and electrolytes in the surgical patient, p. 76. Williams & Wilkins, Baltimore.

Rose, B. D. 1977. Clinical physiology of acid-base and electrolyte disorders, p. 102, 273, 412, 439. McGraw-Hill, New York.

Tilelli, J. A., and J. P. Ophoven. 1986. Hyponatremic seizures as a presenting symptom of child abuse. *Forensic Science International* 30:213.

Weitzman, R., and C. R. Kleeman. 1980. Water metabolism and the neurohypophyseal hormones, p. 536, 607, 623. *In* M. H. Maxwell and C. R. Kleeman (eds.), Clinical disorders of fluid and electrolyte metabolism, 3rd ed. McGraw-Hill, New York.

Williams, T. D. et al. 1985. Vasopressin and oxytocin responses to acute and chronic osmotic stimuli in man. *Journal of Endocrinology* 108(1):163.

Chapter 3

Agus, Z. S., and S. Goldfarb. 1985. Calcium metabolism: Normal and abnormal, p. 511, 520, 522–23, 535–56, 540, 543, 545–56, 552. *In* A. I. Arieff and R. H. DeFronzo (eds.), Fluid, electrolyte, and acid-base disorders. Churchill Livingstone, New York.

Andreoli, T. E. 1989. Disorders of fluid volume, electrolyte, and acid-base balance, p. 529. *In* J. B. Wyngaarden and L. H. Smith (eds.), Cecil textbook of medicine, 17th ed. W. B. Saunders, Philadelphia.

Arieff, A. I. 1985. Effects of water, acid-base, and electrolyte disorders on the central nervous system, p. 999, 1015. *In* A. I. Arieff and R. A. DeFronzo (eds.), Fluid, electrolyte, and acid-base disorders. Churchill Livingstone, New York.

Arnaud, C. D. 1985. Mineral and bone homeostasis, p. 1416. *In* J. B. Wyngaarden and L. H. Smith (eds.), Cecil textbook of medicine, 17th ed. W. B. Saunders, Philadelphia.

Dollery, C. T. 1985. Arterial hypertension, p. 280. *In* J. B. Wyngaarden and L. H. Smith (eds.), Cecil textbook of medicine, 17th ed. W. B. Saunders, Philadelphia.

Epstein, F. H. 1980. Signs and symptoms of electrolyte disorders, p. 500, 511, 513, 515, 517–18. *In* M. H. Maxwell and C. R. Kleeman (eds.), Clinical disorders of fluid and electrolyte metabolism, 3rd ed. McGraw-Hill, New York.

Goldberger, E. 1975. A primer of water, electrolyte and acid-base syndromes, 5th ed., p. 42–43, 343–44, 380–81, 425, 432, 442–44, 452. Lea & Febiger, Philadelphia.

Guyton, A. C. 1986. Textbook of medical physiology, p. 1055–56, 1062. W. B. Saunders, Philadelphia.

Liddle, G. W. 1979. Regulation of adrenal steroid secretion, p. 2147. *In* P. B. Beeson, W. McDermott, and J. B. Wyngaarden (eds.), Cecil textbook of medicine, 15th ed. W. B. Saunders, Philadelphia.

Lyles, K. W. 1989. Osteoporosis, p. 2601–6. *In* Wm. N. Kelley (ed.), Textbook of internal medicine. J. B. Lippincott Co., Philadelphia.

Mudge, G. H. 1985. Agents affecting volume and composition of body fluids, p. 874. *In* A. G. Gilman et al. (eds.), Goodman and Gilman's pharmacological basis of therapeutics, 7th ed. Macmillan, New York.

Papper, S. (ed.). 1982. Sodium, its biological significance, p. 255–56, 259–60, 266, 269–73. CRC Press, Boca Raton, Fla.

Parfitt, A. M., and M. Kleerekoper. 1980. Clinical disorders of calcium, phosphorous, and magnesium metabolism, p. 120–21, 948, 983, 1033–35, 1038. *In* M. H. Maxwell and C. R. Kleeman (eds.), Clinical disorders of fluid and electrolyte metabolism, 3rd ed. McGraw-Hill, New York.

Parfitt, A. M., and M. Kleerekoper. 1980. The divalent homeostatic system-physiology and metabolism of calcium, phosphorous, magnesium, and bone, p. 359–66, 369. *In* M. H. Maxwell and C. R. Kleeman (eds.), Clinical disorders of fluid and electrolyte metabolism, 3rd ed. McGraw-Hill, New York.

Pestana, C. 1979. Fluids and electrolytes in the surgical patient, p. 108. Williams & Wilkins, Baltimore.

Rose, B. D. 1977. Clinical physiology of acid-base and electrolyte disorders, p. 212–13, 385, 389, 396, 411, 419, 424–25, 465, 483, 486, 490, 493. McGraw-Hill, New York.

Rowland, L. P. 1979. Familial periodic paralysis, p. 921. *In* P. B. Beeson, W. McDermott, and J. B. Wyngaarden (eds.), Cecil textbook of medicine, 15th ed. W. B. Saunders, Philadelphia.

Schwartz, W. B. 1979. Disorders of fluid, electrolyte, and acid-base balance, p. 1959. *In* P. B. Beeson, W. McDermott, and J. B. Wyngaarden (eds.). Cecil textbook of medicine, 15th ed. W. B. Saunders, Philadelphia.

Smith, J. D., M. J. Bia, and R. DeFronzo. 1985. Clinical disorders of potassium metabolism, p. 424, 447, 449, 454–6, 461, 463, 465, 481, 483. *In* A. I. Arieff and R. A. DeFronzo (eds.), Fluid, electrolyte, and acid-base disorders. Churchill Livingstone, New York.

Stryer, L. 1981. Biochemistry, 2nd ed., p. 478. W. H. Freeman, San Francisco.

Williams, M. E. et al. 1985. Catecholamine modulation of rapid potassium shifts during exercise. *New England Journal of Medicine* 312:823–27.

Chapter 4

Andreoli, T. E. 1985. Disorders of fluid volume, electrolyte, and acid-base balance, p. 541–43. *In* J. B. Wyngaarden and L. H. Smith (eds.), Cecil textbook of medicine, 17th ed. W. B. Saunders, Philadelphia.

Cohen, J. J., and J. P. Kassirer. 1980. Acid-base metabolism, p. 216–17, 223, 225. *In* M. H. Maxwell and C. R. Kleeman (eds.), Clinical disorders of fluid and electrolyte metabolism, 3rd ed. McGraw-Hill, New York.

Guyton, A. C. 1986. Textbook of medical physiology, p. 504–47. W. B. Saunders, Philadelphia.

Levinsky, N. G. 1987. Acidosis and alkalosis, p. 211–13. *In* E. Braunwald et al. (eds.), Harrison's principles of internal medicine, 11th ed. McGraw-Hill, New York.

Schultze, R. G., and A. R. Nissenson. 1980. Potassium: physiology and pathophysiology, p. 131. *In* M. H. Maxwell and C. R. Kleeman (eds.), Clinical disorders of fluid and electrolyte metabolism. McGraw-Hill, New York.

Scully, R. E. 1986. Case records of the Massachusetts General Hospital. *New England Journal of Medicine* 314:39.

Tortora, G. J. 1988. Introduction to the human body, p. 376. Harper & Row, New York.

Chapter 5

Alvis, R., M. Geheb, and M. Cox. 1985. Hypo- and hyperosmolar states: Diagnostic approaches, p. 186–87, 205, 211. *In* A. I. Arieff and R. A. DeFronzo (eds.), Fluid, electrolyte, and acid-base disorders. Churchill Livingstone, New York.

Brody, M. J., and F. M. Abboud. 1976. Tissue perfusion, p. 29. *In* E. D. Frohlich (ed.), Pathophysiology: Altered regulatory mechanisms in disease, 2nd ed. J. B. Lippincott Co., Philadelphia.

Dobbins, J. 1985. Gastrointestinal disorders, p. 837–40, 842, 844. *In* A. I. Arieff and R. A. DeFronzo (eds.), Fluid, electrolyte, and acid-base disorders. Churchill Livingstone, New York.

Goldberger, E. 1975. A primer of water, electrolyte and acid-base syndromes, 5th ed., p. 18–19, 23, 26, 28, 366–67, 369, 380. Lea & Febiger, Philadelphia.

Guyton, A. C. 1986. Textbook of medical physiology, p. 872, 877, 1001. W. B. Saunders, Philadelphia.

Norins, R. G. et al. 1985. Metabolic acid-base disorders: Pathophysiology, classification, and treatment, p. 270, 277, 319–20, 335, 339–40, 344, 354. *In* A. I. Arieff and R. A. DeFronzo (eds.), Fluid, electrolyte acid-base disorders. Churchill Livingstone, New York.

Papper, S. (ed.). 1982. Sodium, its biological significance, p. 220, 243. CRC Press, Boca Raton, Fla.

Pestana, C. 1979. Fluids and electrolytes in the surgical patient, p. 63–65, 161–62. Williams & Wilkins, Baltimore.

Rose, B. D. 1977. Clinical physiology of acid-base and electrolyte disorders, p. 162, 229, 306, 308, 315–16, 379, 489, 509, 520. McGraw-Hill, New York.

Sebastian, A. et al. 1985. Acid-base and electrolyte disorders associated with adrenal disease, p. 894. *In* A. I. Arieff and R. A. DeFronzo (eds.), Fluid, electrolyte, and acid-base disorders. Churchill Livingstone, New York.

Siegel, N. J., and W. E. Lattansi. 1985. Fluid and electrolyte therapy in children, p. 1216–17. *In* A. I. Arieff and R. A. DeFronzo (eds.), Fluid, electrolyte, and acid-base disorders. Churchill Livingstone, New York.

Smith, J. D., M. J. Bia, and R. A. DeFronzo. 1985. Clinical disorders of potassium metabolism, p. 468. *In* A. I. Arieff and R. A. DeFronzo (eds.), Fluid, electrolyte, and acid-base disorders. Churchill Livingstone, New York.

Chapter 6

Brody, J. S. 1985. Diseases of the pleura, mediastinum, diaphragm, and chest wall, p. 447–48. *In* J. B. Wyngaarden and L. H. Smith (eds.), Cecil textbook of medicine, 17th ed. W. B. Saunders, Philadelphia.

Cherniak, N. S., M. D. Altose, and S. G. Kelsen. 1983. Respiratory system mechanics, p. 635, 738. *In* R. M. Berne and M. N. Levy (eds.), Physiology. C. V. Mosby Co., St. Louis.

Crofton, J., and A. Douglas. 1981. Respiratory diseases, p. 16–27, 306–10, 353–54. Blackwell Scientific, London.

Crouch, J. E. 1982. Essential human anatomy: A text atlas, p. 417. Lea & Febiger, Philadelphia.

Ganong, W. F. 1977. Review of medical physiology, p. 506. Lange Medical Publishers, Los Altos, Cal.

Guyton, A. C. 1986. Textbook of medical physiology, p. 468, 482. W. B. Saunders, Philadelphia.

Hirschmann, J. V., and J. F. Murray. 1987. Pneumonia and lung abscess, p. 1082. *In* E. Braunwald et al. (eds.), Harrison's principles of internal medicine, 11th ed. McGraw-Hill, New York.

Ingram, R. H. 1987. Diseases of the pleura mediastinum, and diaphragm, p. 1123. *In* E. Braunwald et al. (eds.), Harrison's principles of internal medicine, 11th ed. McGraw-Hill, New York.

Macleod, J., C. Edwards, and I. Bouchier. 1984. Davidson's principles and practice of medicine, p. 198–99. Churchill Livingstone, Edinburgh.

Netter, F. H. 1980. The Ciba collection of medical illustrations, vol. 7, p. 30, 253–54, 274. Ciba Pharmaceutical Company, Summit, N.J.

Robbins, S. L., R. S. Cottran, and V. Kumar. 1984. Pathologic basis of disease, p. 705. W. B. Saunders, Philadelphia.

Schlossberg, L., and G. D. Zuidema. 1977. The Johns Hopkins atlas of human functional anatomy, p. 78. Johns Hopkins University Press, Baltimore.

Spence, A. P. 1986. Basic human anatomy, p. 504–11. Benjamin-Cummings, Menlo Park, Cal.

Tortora, G. J. 1988. Introduction to the human body, p. 10, 71, 368–72, 525. Harper & Row, New York.

Vick, R. L. 1984. Contemporary medical physiology, p. 383. Addison-Wesley, Menlo Park, Cal.

West, J. B. 1987. Disturbances of respiratory function, p. 1049, 1053. *In* E. Braunwald et al. (eds.), Harrison's principles of internal medicine, 11th ed. McGraw-Hill, New York.

Chapter 7

Cherniak, N. S., M. D. Altose, and S. G. Kelsen. 1983. Respiratory system mechanics, p. 645. *In* R. M. Berne and M. N. Levy (eds.), Physiology. C. V. Mosby Co., St. Louis.

Daniele, R. P. 1985. Asthma, p. 394–95. *In* J. B. Wyngaarden and L. H. Smith (eds.), Cecil textbook of medicine, 17th ed. W. B. Saunders, Philadelphia.

Nadel, J. A. 1988. Obstructive diseases: General principles and diagnostic approach, p. 989–91. *In* J. F. Murray and J. A. Nadel (eds.), Textbook of respiratory medicine. W. B. Saunders, Philadelphia.

Robbins, S. L., R. S. Cottran, and V. Kumar, 1984. Pathologic basis of disease, p. 569–70, 718, 723–24, 729, 732. W. B. Saunders, Philadelphia.

Snider, G. L. 1988. Chronic bronchitis and emphysema, p. 1069–78. *In* J. F. Murray and J. A. Nadel (eds.), Textbook of respiratory medicine. W. B. Saunders, Philadelphia.

Thurlbeck, W. M. 1976. Chronic airflow obstruction in lung disease, p. 80–86, 298–99, 306–7. W. B. Saunders, Philadelphia.

Woolcock, A. J. 1988. Asthma, p. 1039–55. *In* J. F. Murray and J. A. Nadel (eds.), Textbook of respiratory medicine. W. B. Saunders, Philadelphia.

Chapter 8

Black, L. F. 1985. Neoplasms of the lung, p. 441–44. *In* J. B. Wyngaarden and L. H. Smith (eds.), Cecil textbook of medicine, 17th ed. W. B. Saunders, Philadelphia.

Carr, D. T., and P. Y. Holoye. 1988. Bronchogenic carcinoma, p. 1177–81, 1187. *In* J. F. Murray and J. M. Nadel (eds.), Textbook of respiratory medicine. W. B. Saunders, Philadelphia.

Colten, H. R. 1985. Cystic fibrosis, p. 1086. *In* E. Braunwald et al. (eds.), Harrison's principles of internal medicine, 11th ed. McGraw-Hill, New York.

Croften, J., and A. Douglas, 1981. Respiratory diseases, p. 197–98, 551–59, 640–41, 649, 714. Blackwell Scientific, London.

Crystal, R. G. 1987. Sarcoidosis, p. 1445–57. *In* E. Braunwald et al. (eds.), Harrison's principles of internal medicine, 11th ed. McGraw-Hill, New York.

Flenley, D. C. 1981. Respiratory medicine, p. 72, 86, 128–89, 204–5. Bailliere Tindall, London.

Geraint, J. D., and P. R. Studdy. 1982. Color atlas of respiratory diseases, p. 220–21. Year Book Medical Publishers, Chicago.

Macleod, J., C. Edwards, and I. Bouchier. 1984. Davidson's principles and practices of medicine, p. 260, 312. Churchill Livingstone, Edinburgh.

Minna, J. D. et al. 1989. Cancer of the lung, p. 592, 598, 600. *In* V. T. DeVita, S. Hellman, and S. A. Rosenberg (eds.), Cancer: Principles and practice of oncology, 3rd ed. J. B. Lippincott Co., Philadelphia.

Robbins, S. L., R. S. Cottran, and V. Kumar. 1984. Pathologic basis of disease, p. 215–23, 495, 717, 735. W. B. Saunders, Philadelphia.

CHAPTER 9

Carr, D. T., and P. Y. Holoye. 1988. Bronchogenic carcinoma, p. 1181, 1184. *In* J. F. Murray and J. A. Nadel (eds.), Textbook of respiratory medicine. W. B. Saunders, Philadelphia.

Cherniak, N. S., M. D. Altose, and S. G. Kelsen. 1983. Gas exchange and gas transport, p. 688. *In* R. M. Berne and M. N. Levy (eds.), Physiology. C. V. Mosby Co., St. Louis.

Cherniak, R. M., and L. Cherniak. 1983. Respiration in health and disease, p. 332. W. B. Saunders, Philadelphia.

Cohen, J. J., and J. P. Kassirer. 1980. Acid-base metabolism, p. 210. *In* M. H. Maxwell and C. R. Kleeman (eds.), Clinical disorders of fluid and electrolyte metabolism, 3rd ed. McGraw-Hill, New York.

Comroe, J. H. et al. 1962. The lung: Clinical and pulmonary function tests, p. 213–14, 221–24, 232–34, 238–39, 246, 248. Year Book Medical Publishers, Chicago.

Crofton, J., and A. Douglas. 1981. Respiratory diseases, p. 409. Blackwell Scientific, London.

Cummings, G., and S. J. Semple. 1980. Disorders of the respiratory system, p. 233, 434–42. Blackwell Scientific Publishers, London.

Ganong, W. F. 1977. Review of medical physiology, p. 205, 530. Lange Medical Publishers, Los Altos, Cal.

Green, J. F. 1977. Mechanical concepts in cardiovascular and pulmonary physiology, p. 149. Lea & Febiger, Philadelphia.

Hopewell, P. C. 1985. Critical care medicine, p. 472. *In* J. B. Wyngaarden and L. H. Smith, Jr. (eds.), Cecil textbook of medicine, 17th ed. W. B. Saunders, Philadelphia.

Ingram, R. H. 1987. Chronic bronchitis, emphysema, and airways obstruction, p. 1092. *In* E. Braunwald et al. (eds.), Harrison's principles of internal medicine, 11th ed. McGraw-Hill, New York.

Moser, K. M. 1987. Pulmonary thromboembolism, p. 1106. *In* E. Braunwald et al. (eds.), Harrison's principles of internal medicine, 11th ed. McGraw-Hill, New York.

Murray, J. F. 1986. The normal lung, p. 189–91. W. B. Saunders, Philadelphia.

Robbins, S. L., R. S. Cottran, and V. Kumar (eds.), 1989. Pathologic basis of disease, 4th ed., p. 761. W. B. Saunders, Philadelphia.

Saunders, N. A., and C. E. Sullivan. 1984. Sleep and breathing, p. 365–71, 383–85. Marcel Dekker Inc., New York.

Scully, R. E. 1986. Case records of the Massachusetts General Hospital. *New England Journal of Medicine* 314:39.

West, J. B. 1987. Disturbances of respiratory function, p. 1050–52, 1106, 1108. *In* E. Braunwald et al. (eds.), Harrison's principles of internal medicine, 11th ed. McGraw-Hill, New York.

CHAPTER 10

Babayan, R. K. 1987. Genitourinary neoplasms, p. 1999, 2000. *In* J. Noble (ed.), Textbook of general medicine and primary care. Little, Brown & Co., Boston.

Babayan, R. K. 1987. Scrotal masses, p. 1991–93. *In* J. Noble (ed.), Textbook of general medicine and primary care. Little, Brown & Co., Boston.

Brendler, C. B. 1989. Benign disorders of the prostate, p. 2600. *In* Wm. N. Kelley (ed.), Textbook of internal medicine. J. B. Lippincott Co., Philadelphia.

Copeland, L. J. 1990. Cancer of the cervix, p. 767, 769–70. *In* N. G. Kase, A. B. Weingold and C. M. Gershenson (eds.), Principles and practice of clinical gynecology, 2nd ed. Churchill Livingstone, New York.

Crowley, W. 1990. Disorders of the ovary, p. 2230. *In* J. H. Stein (ed.), Internal medicine, 3rd ed. Little, Brown & Co., Boston.

Gershenson, D. M. 1990. Cancer of the ovary, p. 868–89. *In* N. G. Kase, A. B. Weingold, and D. M. Gershenson (eds.), Principles and practice of clinical gynecology, 2nd ed. Churchill Livingstone, New York.

Goldstein, I. 1987. Disorders of the penis, p. 1967, 1969. *In* J. Noble (ed.), Textbook of general medicine and primary care. Little, Brown & Co., Boston.

Kissane, J. M. (ed.). 1990. Anderson's pathology, p. 1657–64, 1670–91. C. V. Mosby Co., St. Louis.

Loughlin, K. R., and W. F. Whitmore III. 1987. Benign and malignant enlargement of the prostate, p. 563, 566–67, 569–70, 575, 578. *In* Wm. T. Branch, Jr. (ed.), Office practice of medicine, 2nd ed. W. B. Saunders, Philadelphia.

Loughlin, K. R., and W. F. Whitmore III. 1987. Prostatitis, epididymitis, and testicular enlargement, p. 556–58, 561. *In* Wm. T. Branch, Jr. (ed.), Office practice of medicine, 2nd ed. W. B. Saunders, Philadelphia.

Muggia, F. M. 1989. Approach to the management of gynecologic cancers, p. 1161–62, 1266–67. *In* Wm. N. Kelley (ed.), Textbook of internal medicine. J. B. Lippincott Co., Philadelphia.

Osborne, C. K. 1990. Breast cancer, p. 1156–62. *In* J. H. Stein (ed.), Internal Medicine, 3rd ed. Little, Brown & Co., Boston.

Page, D. L. 1990. Disorders of the breast, p. 1138. *In* J. H. Stein (ed.), Internal medicine, 3rd ed. Little, Brown & Co., Boston.

Polk, B. F., and Wm. T. Branch, Jr. 1987. Sexually transmitted diseases, p. 516. *In* Wm. T. Branch, Jr. (ed.), Office practice of medicine, 2nd ed. W. B. Saunders, Philadelphia.

Siroky, M. B. 1987. Benign disorders of the prostate, p. 1995. *In* J. Noble (ed.), Textbook of general medicine and primary care. Little, Brown & Co., Boston.

Sotrel, G. 1987. Dysmenorrhea, endometriosis, chronic pelvic pain, and premenstrual syndrome, p. 2127–31. *In* J. Noble (ed.), Textbook of general medicine and primary care. Little, Brown & Co., Boston.

Walter, J. B. 1989. Pathology of human disease, p. 713, 721–23, 726, 750–58, 693. Lea & Febiger, Philadelphia.

Wilkins, E. W. (ed.). 1989. Emergency medicine: Scientific foundations and current practice, 3rd ed., p. 696–99. Williams & Wilkins, Baltimore.

CHAPTER 11

Brenner, B. M., E. L. Milford, and J. L. Seifter. 1987. Urinary tract obstruction, p. 1216. *In* E. Braunwald et al. (eds.), Principles of internal medicine, 11th ed. McGraw-Hill, New York.

Cameron, S. 1986. Kidney disease: The facts, p. 17. Oxford University Press, Oxford.

Dennis, V. W. 1985. Investigations of renal function, p. 511–13. *In* J. P. Wyngaarden and L. H. Smith (eds.), Cecil textbook of medicine. W. B. Saunders, Philadelphia.

Guyton, A. 1986. Textbook of medical physiology, p. 396. W. B. Saunders, Philadelphia.

Henry, J. B. 1974. Clinical chemistry, p. 591–94. *In* I. Davidsohn and J. B. Henry (eds.), Todd-Sanford clinical diagnosis by laboratory methods, 15th ed. W. B. Saunders, Philadelphia.

Kaloyanides, G. J. 1980. Pathogenesis and treatment of edema with special reference to the use of diuretics, p. 670–74. *In* M. H. Maxwell and C. R. Kleeman (eds.), Clinical disorders of fluid and electrolyte metabolism, 2nd ed. McGraw-Hill, New York.

Lehninger, A. L. 1982. Principles of biochemistry, p. 384. Worth Publishers, New York.

Mudge, G. H. 1985. Agents affecting volume and composition of body fluids, p. 866. *In* A. Gilman and L. Goodman (eds.), Goodman and Gilman's the pharmacological basis of therapeutics. Macmillan, New York.

Narins, R. G. et al. 1985. Metabolic acid-base disorders: Pathophysiology, classification, and treatment, p. 346. *In* A. I. Arieff and R. A. DeFronzo (eds.), Fluid, electrolyte, and acid-base disorders. Churchill Livingstone, New York.

Shaw, S. T., and E. S. Benson. 1974. Renal function and its evaluation, p. 86–89, 96, 98. *In* I. Davidsohn and J. B. Henry (eds.), Todd-Sanford clinical diagnosis by laboratory methods, 15th ed. W. B. Saunders, Philadelphia.

Straffon, R. A. 1974. Genitourinary system, p. 246, 252. *In* R. D. Judge and G. D. Zuidema (eds.), Methods of clinical examination, 3rd ed. Little, Brown & Co., Boston.

Straffon, R. A. 1973. Urine formation, p. 5–6, 5–7. *In* J. Brobeck (ed.), Best and Taylor's physiological basis of medical practice, 9th ed. Williams & Wilkins, Baltimore.

Suki, W. N. et al. 1985. Physiology of diuretic action, p. 2130, 2132, 2137–38, 2141, 2144. *In* D. W. Seldin and G. Giebisch (eds.), The kidney: Physiology and pathophysiology. Raven Press, New York.

Vick, R. 1984. Contemporary physiology, p. 644. Addison-Wesley, Menlo Park, Cal.

Wallach, J. 1986. Interpretation of diagnostic tests, p. 10, 103, 106. Little, Brown & Co., Boston.

Weiner, I. M., and G. H. Mudge. 1985. Diuretics and other agents employed in the mobilization of edema fluid, p. 887–903. *In* A. Gilman and L. Goodman (eds.), Goodman and Gilman's the pharmacological basis of therapeutics. Macmillan, New York.

CHAPTER 12

Backman, U., B. G. Danielson, and S. Ljunghall. 1985. Renal stones: Etiology, management, treatment, p. 16, 54, 84–85, 92, 95. Almqvist and Wiksell International, Stockholm.

Calabresi, P., and R. E. Parks, Jr. 1985. Antiproliferative agents and drugs used for immunosuppression, p. 1276. *In* A. Gilman and L. Goodman (eds.), Goodman and Gilman's the pharmacological basis of therapeutics. Macmillan, New York.

Coe, F. L. 1987. The patient with renal stones, p. 92. *In* R. W. Schrier (ed.), Manual of nephrology. Little, Brown & Co., Boston.

Coe, F. L., and M. J. Favus. 1987. Nephrolithiasis, p. 1213. *In* E. Braunwald et al. (eds.), Harrison's principles of internal medicine, 11th ed. McGraw-Hill, New York.

Cogan, M. G. 1985. Specific renal tubular disorders, p. 608–9, 615. *In* J. B. Wyngaarden and L. H. Smith (eds.), Cecil textbook of medicine, 17th ed. W. B. Saunders, Philadelphia.

Emmerson, B. T. 1985. Toxic nephropathy, p. 594–98, 600, 603–4. *In* J. B. Wyngaarden and L. H. Smith (eds.), Cecil textbook of medicine, 17th ed. W. B. Saunders, Philadelphia.

Garnick, M. B., and B. M. Brenner. 1987. Tumors of the urinary tract, p. 1220. *In* E. Braunwald et al. (eds.), Harrison's principles of internal medicine, 11th ed. McGraw-Hill, New York.

Jocham, D. et al. 1986. Treatment of nephrolithiasis with ESWL, p. 40–42. *In* J. S. Gravenstein and P. Klaus (eds.), Extracorporeal shock-wave lithotripsy for renal stone disease. Butterworths, Boston.

Pak, C. Y. 1985. Renal calculi, p. 631, 633. *In* J. B. Wyngaarden and L. H. Smith (eds.), Cecil textbook of medicine, 17th ed. W. B. Saunders, Philadelphia.

Rector, F. C. 1985. Obstructive nephropathy, p. 605. *In* J. B. Wyngaarden and L. H. Smith (eds.), Cecil textbook of medicine, 17th ed. W. B. Saunders, Philadelphia.

Robbins, S. L., R. S. Cotran, and V. Kumar. 1984. Pathologic basis of disease, 3rd ed., p. 984, 985, 1030, 1034, 1050, 1053. W. B. Saunders, Philadelphia.

Stamm, W. E., and M. Turck. 1987. Urinary tract infection, pyelonephritis, and related conditions, p. 1189–94. *In* E. Braunwald et al. (eds.), Harrison's principles of internal medicine, 11th ed. McGraw-Hill, New York.

Thompson, F. D., and C. R. Woodhouse. 1987. Disorders of the kidney and urinary tract, p. 86, 87, 103, 104, 109, 185. Edward Arnold, London.

Turka, L. A. 1987. Urinary tract obstruction, p. 448–49, 451. *In* B. D. Rose (ed.), Pathophysiology of renal disease, 2nd ed. McGraw-Hill, New York.

CHAPTER 13

Cameron, J. S. 1988. The nephrotic syndrome: A historical review, p. 40–43. *In* J. S. Cameron and R. J. Glascock (eds.), The nephrotic syndrome. Marcel Dekker, New York.

Couser, W. 1985. Glomerular disorders, p. 573, 577. *In* J. B. Wyngaarden and L. H. Smith (eds.), Cecil textbook of medicine, 17th ed. W. B. Saunders, Philadelphia.

Glassock, R. J. 1986. Clinical, immunologic, and pathologic aspects of human glomerular disease, p. 611, 638–39. *In* R. W. Schrier (ed.), Renal and electrolyte disorders, 3rd ed. Little, Brown & Co., Boston.

Hoffsten, P., and S. Klahr. 1983. The kidney and body fluids, p. 390, 400, 427. *In* S. Klahr (ed.), Health and disease. Plenum Medical Book Company, New York.

Hutt, M. P., and S. P. Kelleher. 1986. Proteinuria and the nephrotic syndrome, p. 576. *In* R. W. Schrier (ed.), Renal and electrolyte disorders, 3rd ed. Little, Brown & Co., Boston.

Levy, M. 1988. Infection-related proteinuric syndromes, p. 754–55. *In* J. S. Cameron and R. J. Glasscock (eds.), The nephrotic syndrome. Marcel Dekker, New York.

Rose, B. D. 1987. Pathogenesis, clinical manifestations, and diagnosis of glomerular disease, p. 146, 153–55, 161, 163. *In* B. D. Rose (ed.), Pathophysiology of renal disease, 2nd ed. McGraw-Hill, New York.

Rose, B. D., and J. B. Jacobs. 1987. Nephrotic syndrome and glomerulonephritis, p. 181, 188, 223, 225–26, 228, 254–60. *In* B. D. Rose (ed.), Pathophysiology of renal disease, 2nd ed. McGraw-Hill, New York.

Stryer, L. 1981. Biochemistry, p. 200. W. H. Freeman & Co., San Francisco.

CHAPTER 14

Coe, F. L., and S. Kathpalia. 1987. Hereditary tubular disorders, p. 1209. *In* E. Braunwald et al. (eds.), Harrison's principles of internal medicine, 11th ed. McGraw-Hill, New York.

Cogan, M. G. 1985. Specific tubular disorders, p. 608–10. *In* J. B. Wyngaarden and L. H. Smith (eds.), Cecil textbook of medicine, 17th ed. W. B. Saunders, Philadelphia.

Dunnill, M. S. 1984. Pathological basis of renal disease, 2nd ed., p. 178–89, 319. Baillière Tindall, London.

Garnick, M. B., and B. M. Brenner. 1987. Tumors of the urinary tract, p. 1218–20. *In* E. Braunwald et al. (eds.), Harrison's principles of internal medicine, 11th ed. McGraw-Hill, New York.

Hollenberg, N. K. 1987. Vascular injury to the kidney, p. 1202. *In* E. Braunwald et al. (eds.), Harrison's principles of internal medicine, 11th ed. McGraw-Hill, New York.

Humes, H. D. 1986. Disorders of sodium and water balance, p. 57–58. *In* H. D. Humes (ed.), Pathophysiology of electrolyte and renal disorders. Churchill Livingstone, New York.

Kelleher, S. P., and R. W. Schrier. 1986. The kidney in hypertension, p. 400. *In* R. W. Schrier (ed.), Renal and electrolyte disorders, 3rd ed. Little, Brown & Co., Boston.

Knauss, T. C., and H. D. Humes. 1986. Kidney disease, p. 524–26. *In* H. D. Hume (ed.), Pathophysiology of electrolyte and renal disorders. Churchill Livingstone, New York.

Linehan, W. M., W. U. Shipley, and D. L. Longo. 1989. Cancer of the kidney and ureter, p. 979–80, 982, 986. *In* V. T. DeVita, Jr., S. Hellman, and S. A. Rosenberg (eds.), Cancer: Principles and practice of oncology, 3rd ed. J. B. Lippincott Co., Philadelphia.

Narins, R. G. et al. 1985. Metabolic acid-base disorders; pathophysiology, classification, and treatment, p. 327. *In* A. I. Arieff and R. A. DeFronzo (eds.), Fluid, electrolyte, acid-base disorders. Churchill Livingstone, New York.

Ruddy, M. C., and J. A. Barone. 1986. Medical treatment of renal hypertension, p. 230. *In* J. S. Cheigh et al. (eds.), Hypertension in kidney disease. Martinus Nyhoff, Boston.

CHAPTER 15

Alfey, A. C. 1986. Chronic renal failure: Manifestations and pathogenesis, p. 473–75, 477–82. *In* R. W. Schrier (ed.), Renal and electrolyte disorders, 3rd ed. Little, Brown & Co., Boston.

Carpenter, C. B., and J. M. Lazarus. 1987. Dialysis and transplantation in the treatment of renal failure, p. 1163–64. *In* E. Braunwald et al. (eds.), Harrison's principles of internal medicine, 11th ed. McGraw-Hill, New York.

Hoffsten, P., and S. Klahr. 1983. Pathophysiology of chronic renal failure, p. 463–84, 487, 496–99. *In* S. Klahr (ed.), The kidney and body fluids in health and disease. Plenum Medical Book Co., New York.

Kliger, A. S. 1985. Complications of dialysis: Hemodialysis, peritoneal dialysis, CAPD, p. 777–81, 784, 812. *In* A. I. Arieff and R. A. DeFronzo (eds.), Fluid, electrolyte, and acid-base disorders. Churchill Livingstone, New York.

Lancaster, L. E. 1984. End stage renal disease: Pathophysiology, assessment, and intervention, p. 5. *In* L. E. Lancaster (ed.), The patient with end stage renal disease, 2nd ed. John Wiley & Sons, New York.

Lancaster, L. E. 1984. The patient receiving hemodialysis, p. 118, 121. *In* L. E. Lancaster (ed.), The patient with end stage renal disease, 2nd ed. John Wiley & Sons, New York.

Luke, R. G. 1985. Treatment of irreversible renal failure, p. 561–62. *In* J. B. Wyngaarden and L. H. Smith (eds.), Cecil textbook of medicine, 17th ed. W. B. Saunders, Philadelphia.

Martin, K. 1983. Pathophysiology of acute renal failure, p. 443–45. *In* S. Klahr (ed.), The kidney and body fluids in health and disease. Plenum Medical Book Co., New York.

Parker, S. R., and L. E. Lancaster. 1984. Access to the circulation in hemodialysis, p. 155–56, 180. *In* L. E. Lancaster (ed.), The patient with end stage renal disease, 2nd ed. John Wiley & Sons, New York.

CHAPTER 16

Coburn, J. W. 1985. Renal osteodystrophy, p. 731. *In* A. I. Arieff and R. A. DeFronzo (eds.), Fluid, electrolyte, and acid-base disorders. Churchill Livingstone, New York.

Coburn, J. W., K. Kurokowa, and F. Llach. 1980. Altered divalent ion metabolism in renal disease and renal osteodystrophy, p. 1176–77, 1180–81. *In* M. H. Maxwell and C. R. Kleeman (eds.), Clinical disorders of fluid and electrolyte metabolism, 3rd ed. McGraw-Hill, New York.

Fine, L. G., and E. P. Nord. 1985. Chronic renal failure, p. 710, 712. *In* A. I. Arieff and R. A. DeFronzo (eds.), Fluid, electrolytes, and acid-base disorders. Churchill Livingstone, New York.

Goldberger, E. 1974. A primer of water electrolyte and acid-base syndromes, 5th ed., p. 17–18. Lea & Febiger, Philadelphia.

Humes, H. D. 1986. Acute renal failure, p. 318–19. *In* H. D. Humes (ed.), Pathophysiology of electrolytes and renal disorders. Churchill Livingstone, New York.

Kelley, W. N., and T. D. Palella. 1987. Gout and other disorders of purine metabolism, p. 1623, 1625. *In* E. Braunwald et al. (eds.), Harrison's principles of internal medicine, 11th ed. McGraw-Hill, New York.

Luke, R. G. 1985. Treatment of irreversible renal failure: Dialysis, p. 562. *In* J. B. Wyngaarden and L. H. Smith, Jr. (eds.), Cecil textbook of medicine, 17th ed. W. B. Saunders, Philadelphia.

Narins, R. G. et al. 1985. Metabolic acid-base disorders: Pathophysiology, classification, and treatment, p. 277–79, 283. *In* A. I. Arieff and R. A. DeFronzo (eds.), Fluid, electrolytes, and acid-base disorders. Churchill Livingstone, New York.

Port, F. K. 1986. Chronic renal failure and uremia, p. 594. *In* H. D. Humes (ed.), Pathophysiology of electrolyte and renal disorders. Churchill Livingstone, New York.

Wallach, J. 1986. Interpretation of diagnostic tests, 4th ed., p. 105, 530, 532. Little, Brown & Co., Boston.

CHAPTER 17

Adamson, J. W. 1980. Pathophysiology of anemia, p. 10–14. *In* M. A. Lichtman (ed.), Hematology and oncology. Grune & Stratton, New York.

Adler, S., and D. S. Fraley. 1985. Acid-base regulation: Cellular and whole body, p. 255. *In* A. I. Arieff and R. A. DeFronzo (eds.), Fluid, electrolyte, and acid-base disorders. Churchill Livingstone, New York.

Bozdech, M. J. 1984. Granulocytes and myeloproliferative disorders, p. 254, 256. *In* A. MacKinney (ed.), Pathophysiology of blood. John Wiley & Sons, New York.

Chanarin, I. et al. 1984. Blood and its diseases, 3rd ed., p. 197–99. Churchill Livingstone, Edinburgh.

Gordon-Smith, E. C. 1987. Aplastic anemia and other causes of bone marrow failure, p. 19.46–19.50. *In* D. J. Weatherall et al. (eds.), Oxford textbook of medicine, 2nd ed. Oxford University Press, Oxford.

Guyton, A. 1986. Textbook of medical physiology, 7th ed., p. 44–45, 501. W. B. Saunders, Philadelphia.

Hackett, P. H. et al. 1982. Fluid retention and relative hypoventilation in acute mountain sickness. *Respiration* 43:322.

Jandl, J. H. 1987. Blood: Textbook of hematology, p. 434. Little, Brown & Co., Boston.

Lehninger, A. L. 1982. Principles of biochemistry, p. 780. Worth Publishers, New York.

Naeye, R., C. Melot, and P. Lejeune. 1986. Hypoxic pulmonary vasoconstriction and high altitude pulmonary edema. *American Review of Respiratory Disease* 134:332.

Pippard, M. J. 1987. Iron deficiency and overload, p. 19.85. *In* D. J. Weatherall et al. (eds.), Oxford textbook of medicine, 2nd ed. Oxford University Press, Oxford.

Quesenberry, P. J. 1987. Origin of the blood cells and architecture of the bone marrow, p. 5, 7, 16. *In* O. A. Thorup (ed.), Fundamentals of clinical hematology, 5th ed. W. B. Saunders, Philadelphia.

Scoggin, C. H. et al. 1977. High-altitude pulmonary edema in the children and young adults of Leadville, Colorado. *New England Journal of Medicine* 297:1269.

Sophocles, A. M. 1986. High-altitude pulmonary edema in Vail, Colorado. *Western Journal of Medicine* 144:569.

Stryer, L. 1981. Biochemistry, p. 70–71, 276. W. H. Freeman, San Francisco.

Turgeon, M. L. 1988. Clinical hematology: Theory and procedures, p. 174, 222–25. Little, Brown & Co., Boston.

Wallach, J. 1986. Interpretation of diagnostic tests, 4th ed., p. 38, 51. Little, Brown & Co., Boston.

Weatherall, D. J. 1987. Anaemia: Pathophysiology, classification, and clinical features, p. 19.67–19.69. *In* D. J. Weatherall et al. (eds.), Oxford textbook of medicine, 2nd ed. Oxford University Press, Oxford.

Weatherall, D. J. 1987. Polycythemia vera, p. 19.37, 19.38. *In* D. J. Weatherall et al. (eds.), Oxford textbook of medicine, 2nd ed. Oxford University Press, Oxford.

Weatherall, D. J. 1987. The relative and secondary polycythaemias, p. 19.152–53. *In* D. J. Weatherall et al. (eds.), Oxford textbook of medicine, 2nd ed. Oxford University Press, Oxford.

West, J. B. 1984. Human physiology at extreme altitudes on Mount Everest. *Science* 223:784–87.

Wheby, M. S. 1987. Differentiation, biochemistry, and physiology of the erythrocyte, p. 65, 81–82. *In* O. A. Thorup (ed.), Fundamentals of clinical hematology, 5th ed. W. B. Saunders, Philadelphia.

Woodson, R. D. 1980. Smoker's erythrocytosis, p. 121. *In* M. A. Lichtman (ed.), Hematology and oncology. Grune & Stratton, New York.

Chapter 18

Bennett, J. M. 1985. The classification of the acute leukemias: Cytochemical and morphologic consideration, p. 203. *In* P. H. Wiernik et al. (eds.), Neoplastic diseases of blood. Churchill Livingstone, New York.

Bozdech, M. J. 1984. Granulocyte and myeloproliferative disorders, p. 239, 257. *In* A. MacKinney (ed.), Pathophysiology of blood. John Wiley & Sons, New York.

Bunch, C. 1987. Leucocytes in health and disease, p. 19.156–61. *In* D. J. Weatherall et al. (eds.), Oxford textbook of medicine. Oxford University Press, Oxford.

Chanarin, I. et al. 1984. Blood and its diseases, 3rd ed., p. 196, 226, 231–42, 248. Churchill Livingstone, Edinburgh.

Ganick, D. J., and A. A. MacKinney. 1984. The lymphocytes, p. 266–67. In A. MacKinney (ed.), Pathophysiology of blood. John Wiley & Sons, New York.

Griffin, J. D., and I. M. Nadler. 1985. Immunobiology of chronic leukemias, p. 60. *In* P. H. Wiernik et al. (eds.), Neoplastic diseases of blood. Churchill Livingstone, New York.

Jandl, J. H. 1987. Blood: Textbook of hematology, p. 444–46, 505, 671, 676–77, 681–89, 723–41, 756–58. Little, Brown & Co., Boston.

Kanti, R. R., and A. Sawitsky. 1985. Diagnosis and treatment of chronic lymphocytic leukemia, p. 105–6. *In* P. H. Wiernik et al. (eds.), Neoplastic diseases of blood. Churchill Livingstone, New York.

Keeling, R. P. 1987. Differentiation, biochemistry, and physiology of the granulocyte, p. 90–102. *In* O. A. Thorup (ed.), Fundamentals of clinical hematology, 5th ed. W. B. Saunders, Philadelphia.

Keeling, R. P. 1987. Differentiation, biochemistry, and physiology of the monocyte and macrophage, p. 108–110. *In* O. A. Thorup (ed.), Fundamentals of clinical hematology, 5th ed. W. B. Saunders, Philadelphia.

MacKinney, A. A. 1984. Monocytes and macrophages, p. 309. *In* A. A. MacKinney (ed.), Pathophysiology of blood. John Wiley & Sons, New York.

Quesenberry, P. J. 1987. Origin of the blood cells and architecture of the bone marrow, p. 10, 12. *In* O. A. Thorup (ed.), Fundamentals of clinical hematology, 5th ed. W. B. Saunders, Philadelphia.

Chapter 19

Chanarin, I. et al. 1984. Blood and its diseases, 3rd ed., p. 183–84, 220. Churchill Livingstone, Edinburgh.

Davies, J. A., and E. C. Tuddenham. 1987. Haemostasis and thrombosis, p. 19.211. *In* D. J. Weatherall et al. (eds.), Oxford textbook of medicine, 2nd ed. Oxford University Press, Oxford.

Jandl, J. H. 1987. Blood: Textbook of hematology, p. 968–79, 983–86, 989–99, 1052–53, 1095–97, 1100–117, 1125, 1155–166. Little, Brown & Co., Boston.

Rick, M. E. 1980. Hemophilia and von Willebrand's disease, p. 216. *In* M. A. Lichtman (ed.), Hematology and oncology. Grune & Stratton, New York.

Wiernik, P. H. et al. (eds.). 1985. Neoplastic diseases of blood, p. 999–1000. Churchill Livingstone, New York.

Chapter 20

Bunn, P. A., and E. C. Ridgway. 1989. Paraneoplastic syndromes, p. 1905–6, 1916–21. *In* V. T. DeVita, Jr., S. Hellman, and S. A. Rosenberg (eds.), Cancer: Principles and practice of oncology, 3rd ed. J. B. Lippincott Co., Philadelphia.

Callender, S. T., and C. Bunch. 1987. The leukemias, p. 19.27. *In* D. J. Weatherall et al. (eds.), Oxford textbook of medicine, 2nd ed. Oxford University Press, Oxford.

Davies, J. A., and E. G. Tuddenham. 1987. Haemostasis and thrombosis, p. 19.212. *In* D. J. Weatherall et al. (eds.), Oxford textbook of medicine, 2nd ed. Oxford University Press, Oxford.

Fischback, F. 1984. A manual of laboratory diagnostic tests, 2nd ed., p. 87, 295–96, 301, 307–8, 315. J. B. Lippincott Co., Philadelphia.

Jandl, J. H. 1987. Blood: Textbook of hematology, p. 1001, 1003. Little, Brown & Co., Boston.

McKenzie, S. B. 1988. Textbook of hematology, p. 321–25. Lea & Febiger, Philadelphia.

Mosher, D. 1984. Diagnosis and treatment of hemostatic and fibrinolytic disorders, p. 228. *In* A. MacKinney (ed.), Pathophysiology of blood. John Wiley & Sons, New York.

Quesenberry, P. J. 1987. Origin of the blood cells and architecture of the bone marrow, p. 19. *In* O. A. Thorup (ed.), Fundamentals of clinical hematology, 5th ed. W. B. Saunders, Philadelphia.

Wallach, J. 1986. Interpretation of diagnostic tests, 4th ed., p. 8, 133, 386–87, 741. Little, Brown & Co., Boston.

Chapter 21

Abboud, F. 1985. Shock, p. 213. *In* J. B. Wyngaarden and L. H. Smith, Jr. (eds.), Cecil textbook of medicine, 17th ed. W. B. Saunders, Philadelphia.

Berne, R., and M. Levy (eds.), 1983. Physiology, p. 510–11, 514–15, 536, 553–57, 564–65, 570–71. C. V. Mosby Co., St. Louis.

Braunwald, E., and G. H. Williams. 1987. Alterations in arterial pressure and the shock syndrome, p. 154–55. *In* E. Braunwald et al. (eds.), Harrison's principles of internal medicine, 11th ed. McGraw-Hill, New York.

Dale, D. C., and R. G. Petersdorf. 1987. Septic shock, p. 475–76. *In* E. Braunwald et al. (eds.), Harrison's principles of internal medicine, 11th ed. McGraw-Hill, New York.

Dunagan, W. C., and M. L. Ridner (eds.), 1989. Manual of medical therapeutics, 26th ed., p. 99–100, 109. Little, Brown & Co., Boston.

Green, J. 1977. Cardiovascular and pulmonary physiology, p. 12. Lea & Febiger, Philadelphia.

Guyton, A. 1986. Textbook of medical physiology, 7th ed., p. 157, 214, 219, 225–26. W. B. Saunders, Philadelphia.

Kaplan, N. 1988. Systemic hypertension: Mechanisms and diagnosis, p. 823–27. *In* E. Braunwald (ed.), Heart disease: A textbook of cardiovascular medicine, 3rd ed. W. B. Saunders, Philadelphia.

Oparil, S. 1985. Systemic hypertension, p. 332–37, 340–54. *In* L. D. Horowitz and B. M. Groves (eds.), Signs and symptoms in cardiology. J. B. Lippincott Co., Philadelphia.

Pasternak, R. C., E. Braunwald, and J. S. Alpert. 1987. Disorders of the cardiovascular system, p. 990–91. *In* E. Braunwald et al. (eds.), Harrison's principles of internal medicine, 11th ed. McGraw-Hill, New York.

Price, T., G. Wollam, and P. Grady. 1988. Hypertensive cerebrovascular disease, p. 200–201, 209. *In* G. Wollam and W. Hall (eds.), Hypertension and management: Clinical practice and therapeutic dilemmas. Year Book Medical Publishers, Chicago.

Schick, E. 1987. Congenital heart disease, p. 1168–69. *In* J. Noble (ed.), Textbook of general medicine and primary care. Little, Brown & Co., Boston.

Sheagren, J. N. 1985. Shock syndromes related to sepsis, p. 1474. *In* J. B. Wyngaarden and L. H. Smith, Jr. (eds.), Cecil textbook of medicine, 17th ed. W. B. Saunders, Philadelphia.

Tikoff, G. 1987. Diseases of peripheral arteries and veins, p. 561. *In* J. Stein (ed.), Internal medicine, 2nd ed. Little, Brown & Co., Boston.

Vick, R. 1984. Contemporary medical physiology, p. 196–97, 200–201, 203. Addison-Wesley, Menlo Park, Cal.

Vokonas, P. 1987. Congestive heart failure, p. 1147–48, 1153. *In* J. Noble (ed.), Textbook of general medicine and primary care. Little, Brown & Co., Boston.

Weil, M., M. von Planta, and E. Rackow. 1988. Acute circulatory failure (shock), p. 561–63, 568, 575. *In* E. Braunwald (ed.), Heart disease: A textbook of cardiovascular medicine, 3rd ed. W. B. Saunders, Philadelphia.

West, J. (ed.), 1985. Best and Taylor's physiological basis of medical practice, 11th ed., p. 228–29, 263–66. Williams & Wilkins, Baltimore.

Wollam, G., W. Hall, and J. Lowdon. 1988. Approach to pharmacologic therapy for mild, moderate, severe, and resistant hypertension, p. 365. *In* G. Wollam and W. Hall (eds.), Hypertension management: Clinical practice and therapeutic dilemmas. Year Book Medical Publishers, Chicago.

Chapter 22

Berne, R. M., and M. N. Levy. 1983. Hemodynamics, p. 481. *In* R. M. Berne and M. N. Levy (eds.), Physiology. C. V. Mosby Co., St. Louis.

Dalen, J. E. 1987. Diseases of aorta, p. 1038–39. *In* E. Braunwald et al. (eds.), Harrison's principles of internal medicine, 11th ed. McGraw-Hill, New York.

Fuster, V., and B. Kottke. 1987. Atherosclerosis: Pathogenesis, pathology, and presentation of atherosclerosis, p. 961–62. *In* R. O. Brandenburg et al. (eds.), Cardiology: Fundamentals and practice. Year Book Medical Publishers, Chicago.

Gersh, B. J., and J. H. Chesebro. 1987. Acute myocardial infarction: Management and complications, p. 1157–61, 1020. *In* R. O. Brandenburg et al. (eds.), Cardiology: Fundamentals and practice. Year Book Medical Publishers, Chicago.

Henry, J. B. 1974. Clinical chemistry, p. 619–20. *In* S. Davidsohn and J. Henry (eds.), Todd-Sanford clinical diagnosis by laboratory methods, 15th ed. W. B. Saunders, Philadelphia.

Ross, R. 1986. The pathogenesis of atherosclerosis—an update. *New England Journal of Medicine* 314:488.

Ross, R. 1988. The pathogenesis of atherosclerosis, p. 1135–44. *In* E. Braunwald (ed.), Heart disease: A textbook of cardiovascular medicine. W. B. Saunders, Philadelphia.

Shabetai, R. 1985. Cardiac tamponade, p. 298. *In* L. D. Horowitz and B. M. Groves (eds.), Signs and symptoms in cardiology. J. B. Lippincott Co., Philadelphia.

Spence, A. P. 1986. Basic human anatomy, 2nd ed, p. 297–300. Benjamin-Cummings, Menlo Park, Cal.

Spittell, J. A. 1980. Raynaud's phenomenon and allied vasospastic disorders, p. 555–64. *In* J. L. Juergens, J. A. Spittell, and J. F. Fairbairn (eds.), Allen-Barker-Hines peripheral vascular disease, 5th ed. W. B. Saunders, Philadelphia.

Spittell, J. A., and R. B. Wallace. 1980. Dissecting aneurysm of the aorta, p. 404–9. *In* J. L. Juergens, J. A. Spittell, and J. F. Fairbairn (eds.), Allen-Barker-Hines peripheral vascular diseases, 5th ed. W. B. Saunders, Philadelphia.

Wheat, M. W. 1982. Acute dissecting aneurysms of the aorta, p. 211, 224. *In* E. Goldberger (ed.), Treatment of cardiac emergencies, 3rd ed. C. V. Mosby Co., St. Louis.

Chapter 23

Braunwald, E. 1987. Valvular heart disease, p. 961, 963. *In* E. Braunwald et al. (eds.), Harrison's principles of internal medicine, 11th ed. McGraw-Hill, New York.

Callahan, J., A. Tajik, and J. Seward. 1987. Echocardiography, p. 323–25. *In* R. O. Brandenburg et al. (eds.), Cardiology: Fundamentals and practice. Year Book Medical Publishers, Chicago.

Driscoll, D. 1987. Rheumatic fever, p. 1380–84. *In* R. O. Brandenburg et al. (eds.), Cardiology: Fundamentals and practice. Year Book Medical Publishers, Chicago.

Fyke, F., D. Mair, and H. Smith. 1987. Cardiac catheterization and angiography, p. 416. *In* R. O. Brandenburg, et al. (eds.), Cardiology: Fundamentals and practice. Year Book Medical Publishers, Chicago.

Gersh, B., I. Clements, and J. Chesebro. 1987. Acute myocardial infarction: Diagnosis and prognosis, p. 1116–19. *In* R. O. Brandenburg et al. (eds.), Cardiology: Fundamentals and practice. Year Book Medical Publishers, Chicago.

Gersh, B., J. Chesebro, and I. Clements. 1987. Acute myocardial infarction: Management and complications, p. 1153. *In* R. O. Brandenburg et al. (eds.), Cardiology: Fundamentals and practice. Year Book Medical Publishers, Chicago.

Gibbons, R. 1987. Nuclear cardiology, p. 340. *In* R. O. Brandenburg et al. (eds.), Cardiology: Fundamentals and practice. Year Book Medical Publishers, Chicago.

Harvey, M. 1983. Physical assessment: A tailored approach for patients in heart failure, p. 130. *In* C. R. Michaelson (ed.), Congestive heart failure. C. V. Mosby Co., St. Louis.

Horowitz, L. D., and E. W. Grogan. 1985. Congestive heart failure, p. 161–66, 269–73. *In* L. D. Horowitz and B. M. Groves (eds.), Signs and symptoms in cardiology. J. B. Lippincott Co., Philadelphia.

Lie, J. 1987. Atherosclerosis: Pathology of coronary artery disease, p. 972, 974. *In* R. O. Brandenburg et al. (eds.), Cardiology: Fundamentals and practice. Year Book Medical Publishers, Chicago.

Maron, B. J. et al. 1987. Hypertrophic cardiomyopathy. *New England Journal of Medicine* 316:780.

Polansky, B., and E. Schick, Jr. 1987. Ischemic heart disease, p. 1076. *In* J. Noble (ed.), Textbook of general medicine and primary care. Little, Brown & Co., Boston.

Pura, L. S., and C. S. Sam. 1983. Underlying causes and precipitating factors in heart failure, p. 88, 93–94, 96, 102, 107–8, 110–13. *In* C. R. Michaelson (ed.), Congestive heart failure. C. V. Mosby Co., St. Louis.

Schattenberg, T., R. Brandenburg, and P. Julsrud. 1987. The chest roentgenogram, p. 310. *In* R. O. Brandenburg et al. (eds.), Cardiology: Fundamentals and practice. Year Book Medical Publishers, Chicago.

Selwyn, A., and E. Braunwald. 1987. Ischemic heart disease, p. 980–81. *In* E. Braunwald et al. (eds.), Harrison's principles of internal medicine, 11th ed. McGraw-Hill, New York.

Shub, C. 1987. Angina pectoris and coronary heart disease, p. 1073. *In* R. O. Brandenburg et al. (eds.), Cardiology: Fundamentals and practice. Year Book Medical Publishers, Chicago.

Tsakonas, J., and A. Wasserman. 1988. Thrombolytic therapy for acute myocardial infarction. *Journal of the American Academy of Physician Assistants* 1:433.

Vick, R. L. 1984. Contemporary medical physiology, p. 462. Addison-Wesley, Menlo Park, Cal.

Vokonas, P. S. 1987. Congestive heart failure, p. 1147–54. *In* J. Noble (ed.), Textbook of general medicine and primary care. Little, Brown & Co., Boston.

Wynne, J., and E. Braunwald. 1987. The cardiomyopathies and myocarditides, p. 999–1001. *In* E. Braunwald et al. (eds.), Harrison's principles of internal medicine, 11th ed. McGraw-Hill, New York.

CHAPTER 24

Braunwald, E. 1988. Cyanosis, hypoxia, and polycythemia, p. 145–46. *In* E. Braunwald et al. (eds.), Harrison's principles of internal medicine, 11th ed. McGraw-Hill, New York.

Driscoll, D. J., V. Fuster, and D. C. McGoon. 1987. Congenital heart disease in adolescents and adults; tetralogy of Fallot, p. 1468. *In* R. O. Brandenberg et al. (eds.), Cardiology: Fundamentals and practice. Year Book Medical Publishers, Chicago.

Friedman, W. F. 1987. Congenital heart disease in infancy and childhood, p. 896–97, 901. *In* E. Braunwald et al. (eds.), Harrison's principles of internal medicine, 11th ed. McGraw-Hill, New York.

Friedman, W. F. 1987. Congenital heart disease, p. 942, 945–46, 950. *In* E. Braunwald et al. (eds.), Harrison's principles of internal medicine, 11th ed. McGraw-Hill, New York.

Guyton, A. C. 1986. Textbook of medical physiology, p. 1000. W. B. Saunders, Philadelphia.

Mair, D. D. et al. 1986. Congenital heart disease, p. 4–5, 12, 22. *In* R. L. Frye (ed.), Clinical medicine: Cardiovascular medicine, cardiovascular disease, vol. 6. Harper & Row, Philadelphia.

Schick, E. C., Jr. 1987. Congenital heart disease, p. 1163–64, 1167–70. *In* J. Noble (ed.), Textbook of general medicine and primary care. Little, Brown & Co., Boston.

Spence, A. P. 1986. Basic human anatomy, 2nd ed., p. 625. Benjamin-Cummings, Menlo Park, Cal.

Vick, R. L. 1984. Contemporary medical physiology, p. 316, 623. Addison-Wesley, Menlo Park, Cal.

CHAPTER 25

Berne, R., and M. Levy (ed.), 1983. Physiology, p. 879–80. C. V. Mosby Co., St. Louis.

Dale, D. C., and R. G. Petersdorf. 1987. Septic shock, p. 475–76. *In* E. Braunwald et al. (eds.), Harrison's principles of internal medicine, 11th ed. McGraw-Hill, New York.

Gersh, B., I. Clements, and J. Chesebro. 1987. Acute myocardial infarction: Diagnosis and prognosis, p. 1126–27. *In* R. O. Brandenburg et al. (eds.), Cardiology: Fundamentals and practice. Year Book Medical Publishers, Chicago.

Gotto, A., and J. Farmer. 1987. Risk factors for coronary artery disease, p. 1156, 1159, 1164, 1166. *In* R. O. Brandenburg et al. (eds.), Cardiology: Fundamentals and practice. Year Book Medical Publishers, Chicago.

Kaloyanides, G. 1980. Pathogenesis and treatment of edema with special reference to the use of diuretics, p. 657–59. *In* M. Maxwell and C. Kleeman (eds.), Clinical disorders of fluid and electrolyte metabolism, 3rd ed. McGraw-Hill, New York.

Polansky, B., and E. Schick, Jr. 1987. Ischemic heart disease, p. 1074–75. *In* J. Nobel (ed.), Textbook of general medicine and primary care. Little, Brown & Co., Boston.

Raine, A. et al. 1986. Atrial natriuretic peptide and atrial pressure in patients with congestive heart failure. *New England Journal of Medicine* 315: 533–37.

Shorecki, K., and B. Brenner. 1985. Edema forming states: Congestive heart failure, liver disease, and nephrotic syndrome, p. 78–79, 82, 85. *In* A. Arieff and R. DeFronzo (eds.), Fluid, electrolyte, and acid-base disorders. Churchill Livingstone, New York.

Vokonas, P. 1987. Congestive heart failure, p. 1150. *In* J. Noble (ed.), Textbook of general medicine and primary care. Little, Brown & Co., Boston.

Wallach, J. 1986. Interpretation of diagnostic tests, 4th ed., p. 202–3. Little, Brown & Co., Boston.

CHAPTER 26

Bellanti, J. 1985. Immunology III, p. 8, 18, 210, 213, 217. W. B. Saunders, Philadelphia.

Cano, R., and J. Colome. 1988. Essentials of microbiology, p. 296. West Publishing Co., St. Paul.

Haynes, B., and A. Fauci. 1987. Introduction to clinical immunology, p. 336. *In* E. Braunwald et al. (eds.), Harrison's principles of internal medicine, 11th ed. McGraw-Hill, New York.

Hess, C. 1987. The immune system: Structure, development, and function, p. 25, 52. *In* O. Thorup (ed.), Fundamentals of clinical hematology, 5th ed. W. B. Saunders, Philadelphia.

Johlik, W., and H. Willett (eds.), 1976. Zinsser microbiology, 16th ed., p. 294–95, 399–400. Appleton-Century-Crofts, New York.

Ketchum, P. 1988. Microbiology: Concepts and applications, p. 414, 422, 455, 623. John Wiley & Sons, New York.

Stryer, L. 1981. Biochemistry, 2nd ed., p. 136. W. H. Freeman & Co., San Francisco.

CHAPTER 27

Bellanti, J. 1985. Immunology III, p. 90. W. B. Saunders, Philadelphia.

Bennett, C. 1987. Immunoglobulins: Structure and genetics, p. 1150–52. *In* J. Stein (ed.), Internal Medicine, 2nd ed. Little, Brown & Co., Boston.

Ketchum, P. 1988. Microbiology: Concepts and applications, p. 423, 430–31, 442. John Wiley & Sons, New York.

Roit, I. 1984. Essential immunology, 5th ed., p. 31, 34, 37, 170–72, 163–67. Blackwell Scientific Publishing Co., Oxford.

Stobo, J. 1987. The immune response, p. 1144. *In* J. Stein (ed), Internal Medicine, 2nd ed. Little, Brown & Co., Boston.

Stryer, L. 1981. Biochemistry, 2nd ed., p. 810. W. H. Freeman & Co., San Francisco.

Van De Graaff, K. M., and S. I. Fox. 1989. Concepts of human anatomy and physiology, 2nd ed., p. 727. Wm. C. Brown Publishers, Dubuque, Ia.

Chapter 28

Bellanti, J. 1985. Immunology III, p. 230, 234, 237, 240, 352, 360–61, 398. W. B. Saunders, Philadelphia.

Lieberman, P. 1986. Anaphylaxis and anaphylactoid reactions, p. 2–12. *In* R. Slavin and G. Hunder (eds.), Medicine: Allergy and immunology, rheumatology, vol. 4. Harper & Row, Philadelphia.

Luke, R., and T. Strom. 1987. Chronic renal failure, p. 781–82. *In* J. Stein (ed.), Internal medicine, 2nd ed. Little, Brown & Co., Boston.

McConnell, I., A. Munro, and H. Waldmann. 1981. The immune system, 2nd ed., p. 254. Blackwell Scientific, Oxford.

Moreno, J., and P. Lipsky. 1987. Cellular immunity, p. 1158–63. *In* J. Stein (ed.), Internal medicine, 2nd ed. Little, Brown & Co., Boston.

Reisman, R. 1986. Serum sickness, p. 1. *In* R. Slavin and G. Hunder (eds.), Clinical medicine: Allergy and immunology, rheumatology, vol. 4. Harper & Row, Philadelphia.

Slavin, R. 1986. The allergic response: Introduction and general principles, p. 2. *In* R. Slavin and G. Hunder (eds.), Clinical medicine: Allergy and immunology, rheumatology, vol. 4. Harper & Row, Philadelphia.

Zwieman, B., and A. Levinson. 1986. Cell-mediated immunity (delayed hypersensitivity), p. 3, 9. *In* R. Slavin and G. Hunder (eds.), Clinical medicine: Allergy and immunology, rheumatology, vol. 4. Harper & Row, Philadelphia.

Chapter 29

Ammann, A. 1984. Immunodeficiency diseases, p. 392–93, 399–401, 408, 414, 418–19. *In* D. Stites et al. (eds.), Basic and clinical immunology, 5th ed. Lange Medical Publications, Los Altos, Cal.

Bennett, J. C. 1985. Rheumatoid arthritis, p. 1913. *In* J. B. Wyngaarden and L. H. Smith, Jr. (eds.), Cecil textbook of medicine, 17th ed. W. B. Saunders, Philadelphia.

Bluestein, H. 1987. Immunodeficiency diseases, p. 1229–30. *In* J. Stein (ed.), Internal medicine, 2nd ed. Little, Brown & Co., Boston.

Cassidy, J. T. 1986. Systemic lupus erythematosus, p. 1–7, chapter 34. *In* R. G. Slavin and G. G. Hunder (eds.), Clinical medicine: Allergy and immunology, rheumatology, vol. 4. Harper & Row, Philadelphia.

Clavel, F. et al. 1987. Human immunodeficiency virus type 2 infections associated with AIDS in West Africa. *New England Journal of Medicine* 316:1180.

Condemi, J. J. 1986. Miscellaneous diseases with immunologic features, p. 1–3, 12, 21, 23–24, chapter 21. *In* R. G. Slavin and G. G. Hunder (eds.), Clinical medicine: Allergy and immunology, rheumatology, vol. 4. Harper & Row, Philadelphia.

Fauci, A. 1988. Acquired immunodeficiency syndrome (AIDS), p. 1862. *In* J. Wyngaarden and L. Smith (eds.), Cecil textbook of medicine, 18th ed. W. B. Saunders, Philadelphia.

Fauci, A., and C. Lane. 1987. The acquired immunodeficiency syndrome, p. 1394. *In* E. Braunwald et al. (eds.), Harrison's principles of internal medicine, 11th ed. McGraw-Hill, New York.

Gatti, R. 1986. The immunodeficiency diseases, p. 5, chapter 18. *In* R. Slavin and G. Hunder (eds.), Clinical medicine: Allergy and immunology, rheumatology, vol. 4. Harper & Row, Philadelphia.

Harris, E. D. 1985. Systemic sclerosis, p. 1932–35. *In* J. B. Wyngaarden and L. H. Smith, Jr. (eds.), Cecil textbook of medicine, 17th ed. W. B. Saunders, Philadelphia.

Reimer, G., and E. M. Tan. 1987. Antinuclear antibodies, p. 1200. *In* J. Stein (ed.), Internal medicine, 2nd ed. Little, Brown & Co., Boston.

Sharp, J. 1986. Rheumatoid arthritis, p. 1, 7, 10–15, 18, 20, 22–26, chapter 26. *In* R. Slavin and G. Hunder (eds.), Clinical medicine: Allergy and immunology, rheumatology, vol. 4. Harper & Row, Philadelphia.

Steinberg, A. D. 1985. Systemic lupus erythematosus, p. 1924. *In* J. B. Wyngaarden and L. H. Smith, Jr. (eds.), Cecil textbook of medicine, 17th ed. W. B. Saunders, Philadelphia.

Talal, N. 1985. Sjögren's syndrome, p. 1926–67. *In* J. B. Wyngaarden and L. H. Smith, Jr. (eds.), Cecil textbook of medicine, 17th ed. W. B. Saunders, Philadelphia.

Theofilopoulos, A. 1984. Autoimmunity, p. 152–56. *In* D. Stites et al. (eds.), Basic and clinical immunology, 5th ed. Lange Medical Publications, Los Altos, Cal.

Chapter 30

Ammann, A. J. 1984. Immunodeficiency diseases, p. 418. *In* D. P. Stites et al. (eds.), Basic and clinical immunology, 5th ed. Lange Medical Publications, Los Altos, Cal.

Bluestein, H. G. 1987. Immunodeficiency diseases, p. 1229. *In* J. Stein (ed.), Internal medicine, 2nd ed. Little, Brown & Co., Boston.

Callender, S. T. 1987. Normochromic, normocytic anaemias, p. 19.91, 19.92. *In* D. J. Weatherall, J. G. Ledingham, and D. A. Warrell (eds.), Oxford textbook of medicine, 2nd ed. Oxford University Press, Oxford.

Carson, D. 1987. Rheumatoid factors, p. 1204. *In* J. Stein (ed.), Internal medicine, 2nd ed. Little, Brown & Co., Boston.

Cassidy, J. 1986. Systemic lupus erythematosus, p. 11–14, chapter 34. *In* R. G. Slavin and G. G. Hunder (eds.), Clinical medicine: Allergy and immunology, rheumatology, vol. 4. Harper & Row, Philadelphia.

Condemi, J. J. 1986. Miscellaneous diseases with immunologic features, p. 1, 7–10, chapter 21. *In* R. G. Slavin and G. G. Hunder (eds.), Clinical medicine: Allergy and immunology, rheumatology, vol. 4. Harper & Row, Philadelphia.

Fischbach, F. 1984. A manual of laboratory diagnostic tests, 2nd ed., p. 296, 301, 307, 315, 492. J. B. Lippincott Co., Philadelphia.

Geppert, T. D., and P. E. Lipsky. 1987. Evaluation of cellular immune function, p. 1188–89. *In* J. Stein (ed.), Internal medicine, 2nd ed. Little, Brown & Co., Boston.

Henry, J. 1984. Todd-Sanford Davidsohn clinical diagnosis and management, 17th ed., p. 692, 980. W. B. Saunders, Philadelphia.

Lahita, R. et al. 1984. Effects of sex hormones, nutrition, and aging on the immune response, p. 306. *In* D. Stites et al. (eds.), Basic and clinical immunology, 5th ed. Lange Medical Publications, Los Altos, Cal.

Masur, H. 1985. Pneumocystosis, p. 1796. *In* J. B. Wyngaarden and L. H. Smith, Jr. (eds.), Cecil textbook of medicine, 17th ed. W. B. Saunders, Philadelphia.

Moreno, J., and P. E. Lipsky. 1987. Cellular immunity, p. 1159. *In* J. Stein (ed.), Internal medicine, 2nd ed. Little, Brown & Co., Boston.

Nichols, S., and R. Nakamura. 1987. Autoimmune antibodies, p. 1198–99. *In* J. Stein (ed.), Internal medicine, 2nd ed. Little, Brown & Co., Boston.

Price, R. W. 1988. AIDS dementia and human immunodeficiency virus brain infection, p. 2203–4. *In* J. B. Wyngaarden and L. H. Smith, Jr. (eds.), Cecil textbook of medicine, 28th ed. W. B. Saunders, Philadelphia.

Reimer, G., and E. M. Tan. 1987. Antinuclear antibodies, p. 1200–1201, 1203. *In* J. Stein (ed.), Internal medicine, 2nd ed. Little, Brown & Co., Boston.

Ruddy, S. 1987. Complement measurement, p. 1184–85. *In* J. Stein (ed.), Internal medicine, 2nd ed. Little, Brown & Co., Boston.

Stites, D. 1984. Clinical methods for detection of antigen and antibodies, p. 331–32. *In* D. Stites, et al. (eds.), Basic and clinical immunology, 5th ed. Lange Medical Publications, Los Altos, Cal.

Stobo, J. D. 1987. The immune response, p. 1144. *In* J. Stein (ed.), Internal medicine, 2nd ed. Little, Brown & Co., Boston.

Theofilopoulos, A. 1987. Measurement of immune complexes, p. 1193. *In* J. Stein (ed.), Internal medicine, 2nd ed. Little, Brown & Co., Boston.

Wallach, J. 1986. Interpretation of diagnostic tests, 4th ed. p. 6, 7, 146, 355. Little, Brown & Co., Boston.

Wells, V., and C. Henney. 1984. Immune mechanisms, p. 142. *In* D. Stites et al. (eds.), Clinical immunology, 5th ed. Lange Medical Publications, Los Altos, Cal.

Wells, V., J. Isbister, and C. Ries. 1984. Hematologic diseases, p. 460. *In* D. Stites et al. (eds.), Clinical immunology, 5th ed. Lange Medical Publications, Los Altos, Cal.

Chapter 31

Aurbach, G., S. Marx, and A. Spiegel. 1985. Parathyroid hormone, calcitonin, and the calciferols, p. 1199, 1205. *In* J. D. Wilson and D. W. Foster (eds.), Williams textbook of endocrinology, 7th ed. W. B. Saunders, Philadelphia.

Culpepper, R. M., S. C. Hebert, and T. E. Andreoli. 1985. The posterior pituitary and water metabolism, p. 597, 635, 638–39. *In* J. D. Wilson and D. W. Foster (eds.), Williams textbook of endocrinology, 7th ed. W. B. Saunders, Philadelphia.

Daniels, G., and J. Martin. 1987. Neuroendocrine regulation and diseases of the anterior pituitary and hypothalamus, p. 1703. *In* E. Braunwald et al. (eds.), Harrison's principles of internal medicine, 11th ed. McGraw-Hill, New York.

Daughaday, W. 1985. The anterior pituitary, p. 568, 591. *In* J. D. Wilson and D. W. Foster (eds.), Williams textbook of endocrinology, 7th ed. W. B. Saunders, Philadelphia.

Fleischer, N. 1987. Disorders of the hypothalamus and anterior pituitary, p. 1895–97. *In* J. H. Steiner (ed.), Internal medicine, 2nd ed. Little, Brown & Co., Boston.

Frohman, L. A. 1985. The anterior pituitary, p. 1255–56. *In* J. B. Wyngaarden and L. H. Steiner (eds.), Cecil textbook of medicine, 17th ed. W. B. Saunders, Philadelphia.

Genuth, S. M. 1983. The hypothalamus and the pituitary gland, p. 1008. *In* R. M. Berne and M. N. Levy (eds.), Physiology. C. V. Mosby Co., St. Louis.

Hadley, M. 1984. Endocrinology, p. 7, 22, 264, 293. Prentice-Hall, Englewood Cliffs, N.J.

Ingbar, S. 1985. The thyroid gland, p. 684, 698, 743, 749, 751–56, 775–76. 707–800, 806. *In* J. D. Williams and D. W. Foster (eds.), Williams textbook of endocrinology, 7th ed. W. B. Saunders, Philadelphia.

Metz, R., and E. Larson. 1985. Blue book of endocrinology, p. 42, 83, 197–98. W. B. Saunders, Philadelphia.

Robertson, G. L. 1987. Disorders of the posterior pituitary, p. 1908–17. *In* J. H. Steiner (ed.), Internal medicine, 2nd ed. Little, Brown & Co., Boston.

Volpe, R. 1986. The thyroid, p. 4, 6, 29–34, 37, 41, 63–66. *In* J. Spittell (ed.), Clinical medicine, vol. 8. Harper & Row, Philadelphia.

Chapter 32

Benson, E. 1985. Endocrine emergencies, p. 346. *In* R. Metz and E. Larson (eds.), Blue book of endocrinology. W. B. Saunders, Philadelphia.

Bondy, P. 1985. Disorders of the adrenal cortex, p. 857. *In* J. D. Wilson and D. W. Foster (eds.), Williams textbook of endocrinology, 7th ed. W. B. Saunders, Philadelphia.

Bird, C., and A. Clark. 1986. The adrenals, p. 16–18, 22, 23. *In* J. Spittell (ed.), Clinical medicine, vol. 8. Harper & Row, Philadelphia.

Daughaday, W. 1985. The anterior pituitary, p. 596. *In* J. D. Wilson and D. W. Foster (eds.), Williams textbook of endocrinology, 7th ed. W. B. Saunders, Philadelphia.

Fujimoto, W. 1985. Disorders of glucocorticoid homeostasis, p. 45, 52, 54, 56. *In* R. Metz and E. Larson (eds.), Blue book of endocrinology. W. B. Saunders, Philadelphia.

Hadley, M. 1984. Endocrinology, p. 243. Prentice-Hall, Englewood Cliffs, N.J.

Hanna, A. 1986. Diabetes mellitus, p. 1–8, 11, 13–17, 21, 26–27. *In* J. Spittell (ed.), Clinical medicine, vol. 8. Harper & Row, Philadelphia.

Landsberg, L., and J. Young. 1987. Pheochromocytoma, p. 1775. *In* E. Braunwald et al. (eds.), Harrison's principles of internal medicine, 11th ed. McGraw-Hill, New York.

Unger, R., and D. Foster. 1985. Diabetes mellitus, p. 1041, 1053, 1057–58, 1062, 1063, 1066. *In* J. Wilson and D. Foster (eds.), Williams textbook of endocrinology, 7th ed. W. B. Saunders, Philadelphia.

Chapter 33

Alvis, R., M. Geheb, and M. Cox. 1985. Hypo- and hyperosmolar states: Diagnostic approaches, p. 199. *In* A. I. Arieff and R. A. DeFronzo (eds.), Fluid, electrolyte, and acid-base disorders. Churchill Livingstone, New York.

Benson, E. 1985. Endocrine emergencies, p. 343–44, 348. *In* R. Metz and E. Larson (eds.), Blue book of endocrinology. W. B. Saunders, Philadelphia.

Hadley, M. 1984. Endocrinology, p. 245–48. Prentice-Hall, Englewood Cliffs, N.J.

Halperin, M. L. et al. 1985. Diabetic comas, p. 939–43, 947–49, 956–58, 960–61, 964–65. *In* A. I. Arieff and R. A. DeFronzo (eds.), Fluid, electrolyte, and acid-base disorders. Churchill Livingstone, New York.

Hanna, A. 1986. Diabetes mellitus, p. 3, 7. *In* J. Spittell (ed.), Clinical medicine, vol. 8. Harper & Row, Philadelphia.

Higgins, J. T., and P. J. Mulrow. 1980. Fluid and electrolyte disorders of endocrine diseases, p. 1324–25. *In* M. H. Maxwell and C. R. Kleeman (eds.), Clinical disorders of fluid and electrolyte metabolism, 3rd ed. McGraw-Hill, New York.

Kleeman, C. R., and R. G. Narins. 1980. Diabetic acidosis and coma, p. 1362. *In* M. H. Maxwell and C. R. Kleeman (eds.), Clinical disorders of fluid and electrolyte metabolism, 3rd ed. McGraw-Hill, New York.

Narins, R. G. et al. 1985. Metabolic acid-base disorders: Pathophysiology, classification, and treatment, p. 289. *In* A. I. Arieff and R. A. DeFronzo (eds.), Fluid, electrolyte, and acid-base disorders. Churchill Livingstone, New York.

Olefsky, J. M. 1985. Diabetes mellitus, p. 1336–37. *In* J. B. Wyngaarden and L. H. Smith, Jr. (eds.), Cecil textbook of medicine, 17th ed. W. B. Saunders, Philadelphia.

Rose, D. B. 1977. Clinical physiology of acid-base and electrolyte disorders, p. 444. McGraw-Hill, New York.

Smith, J. D. et al. 1985. Clinical disorders of potassium metabolism, p. 418, 422–23, 447. *In* A. I. Arieff and R. A. DeFronzo (eds.), Fluid, electrolyte, and acid-base disorders. Churchill Livingstone, New York.

Unger, R., and D. Foster. 1985. Diabetes mellitus, p. 1018, 1036, 1045, 2039–40. *In* J. D. Wilson and D. W. Foster (eds.), Williams textbook of endocrinology, 7th ed. W. B. Saunders, Philadelphia.

Wallach, J. 1986. Interpretation of diagnostic tests, p. 80, 95–96, 422, 471–72. Little, Brown & Co., Boston.

CHAPTER 34

Crouch, J. 1982. Essential human anatomy, p. 376, 382–94. Lea & Febiger, Philadelphia.

Davis, G. 1989. Neoplasms of the stomach, p. 745, 747, 750–51, 754, 758–60. *In* M. Sleisinger and J. Fordtran (eds.), Gastrointestinal disease: Pathophysiology, diagnosis, management, 4th ed. W. B. Saunders, Philadelphia.

Friedman, S. L., and R. L. Owen. 1989. Gastrointestinal manifestations of AIDS and other transmissible diseases, p. 1250–51. *In* M. Sleisinger and J. Fordtran (eds.), Gastrointestinal disease: Pathophysiology, diagnosis, management, 4th ed. W. B. Saunders, Philadelphia.

Gosling, J. et al. 1985. Atlas of human anatomy, p. 4, 19. J. B. Lippincott Co., Philadelphia.

Hersh, T. 1988. Achalasia, p. 1333–35. *In* J. Hurst (ed.), Medicine for the practicing physician, 2nd ed. Butterworths, Boston.

Hersh, T. 1988. Gastric ulcer, p. 1361–62. *In* J. Hurst (ed.), Medicine for the practicing physician, 2nd ed. Butterworths, Boston.

Hogan, W. J., and W. J. Dodds. 1989. Gastroesophageal reflux disease (reflux esophagitis), p. 594, 596–600, 608. *In* M. Sleisinger and J. Fordtran (eds.), Gastrointestinal disease: Pathophysiology, diagnosis, management, 4th ed. W. B. Saunders, Philadelphia.

Knauer, C., and S. Silverman, Jr. 1987. Alimentary tract and liver, p. 369–70. *In* M. Krupp, S. Schroeder, and L. Tierney (eds.), Current medical diagnosis and treatment. Appleton & Lange, Norwalk, Conn.

McGuigan, J. 1987. Peptic ulcer, p. 124. *In* E. Braunwald et al. (eds.), Harrison's principles of internal medicine, 11th ed. McGraw-Hill, New York.

Richardson, C. T. 1989. Gastric ulcer, p. 879–80, 890, 893. *In* M. Sleisinger and J. Fordtran (eds.), Gastrointestinal disease: Pathophysiology, diagnosis, management, 4th ed. W. B. Saunders, Philadelphia.

Robert, A., and G. L. Kauffman, Jr. 1989. Stress ulcers, erosions, and gastric mucosal injury, p. 772, 774–75, 780–83. *In* M. Sleisinger and J. Fordtran (eds.), Gastrointestinal disease: Pathophysiology, diagnosis, management, 4th ed. W. B. Saunders, Philadelphia.

Spence, A. 1986. Basic human anatomy, 2nd ed., p. 529. Benjamin-Cummings, Menlo Park, Cal.

Wenger, J. 1988. Acute and chronic gastritis, p. 1356–60. *In* J. Hurst (ed.), Practicing physician, 2nd ed. Butterworths, Boston.

Wenger, J. 1988. Stress ulcer, p. 1359–60. *In* J. Hurst (ed.), Medicine for the practicing physician, 2nd ed. Butterworths, Boston.

Wolf, D. 1988. Dumping syndrome, p. 1372. *In* J. Hurst (ed.), Medicine for the practicing physician, 2nd ed. Butterworths, Boston.

CHAPTER 35

Bresalier, R., and Y. Kim. 1989. Malignant neoplasms of the large and small intestine, p. 1519, 1521, 1523–25, 1527–29, 1533, 1537, 1542. *In* M. Sleisinger and J. Fordtran (eds.), Gastrointestinal disease: Pathophysiology, diagnosis, management, 4th ed. W. B. Saunders, Philadelphia.

Fischbach, F. 1984. A manual of laboratory diagnostic tests, 2nd ed., p. 572, 659. J. B. Lippincott Co., Philadelphia.

Floch, M., and L. Ozick. 1988. Tropical sprue, p. 1402–3. *In* J. Hurst (ed.), Medicine for the practicing physician, 2nd ed. Butterworths, Boston.

Friedman, S. L., and R. L. Owen. 1989. Gastrointestinal manifestations of AIDS and other sexually transmissible diseases. p. 1245–48. *In* M. Sleisinger and J. Fordtran (eds.), Gastrointestinal disease: Pathophysiology, diagnosis, management, 4th ed. W. B. Saunders, Philadelphia.

Glickman, M. 1987. Inflammatory bowel disease, p. 1277. *In* E. Braunwald et al. (eds.), Harrison's principles of internal medicine, 11th ed. McGraw-Hill, New York.

Hersh, T. 1988. Celiac disease, p. 1399–1401. *In* J. Hurst (ed.), Medicine for the practicing physician, 2nd ed. Butterworths, Boston.

Klipstein, F. A. 1989. Tropical sprue, p. 1281–88. *In* M. Sleisinger and J. Fordtran (eds.), Gastrointestinal disease: Pathophysiology, diagnosis, management, 4th ed. W. B. Saunders, Philadelphia.

Knauer, C., and S. Silverman, Jr. 1987. Alimentary tract and liver, p. 371–72, 378–89. *In* M. Krupp, S. Schroeder, and L. Tierney (eds.), Current medical diagnosis and treatment. Appleton & Lange, Norwalk, Conn.

Levin, B. 1988. Ulcerative colitis, p. 753, 757. *In* J. Wyngaarden and L. Smith (eds.), Cecil textbook of medicine, 18th ed. W. B. Saunders, Philadelphia.

Levite, M. 1988. Lactase deficiency, p. 1389, 1391. *In* J. Hurst (ed.), Medicine for the practicing physician, 2nd ed. Butterworths, Boston.

Naitove, A., and T. P. Almy. 1989. Diverticular disease of the colon, p. 1419–31. *In* M. Sleisinger and J. Fordtran (eds.), Gastrointestinal disease: Pathophysiology, diagnosis, management, 4th ed. W. B. Saunders, Philadelphia.

Pinckney, C., and E. R. Pinckney. 1982. The encyclopedia of medical tests, p. 291. Facts on File Inc., New York.

Spiro, H. 1988. Chronic ulcerative colitis and proctitis, p. 1422–23. *In* J. Hurst (ed.), Medicine for the practicing physician, 2nd ed. Butterworths, Boston.

Spiro, H. 1988. Duodenal ulcer, p. 1368–69. *In* J. Hurst (ed.), Medicine for the practicing physician, 2nd ed. Butterworths, Boston.

Winawer, S. 1988. Neoplasms of the large and small intestine, p. 766–67, 770, 772. *In* J. Wyngaarden and L. Smith (eds.), Cecil textbook of medicine, 18th ed. W. B. Saunders, Philadelphia.

CHAPTER 36

Alpert, E., and K. Isselbacher. 1987. Tumors of the liver, p. 1351. *In* E. Braunwald et al. (eds.), Harrison's principles of internal medicine, 11th ed. McGraw-Hill, New York.

Bender, M., and R. Ockner. 1989. Ascites, p. 433–40. *In* M. Sleisinger and J. Fordtran (eds.), Gastrointestinal disease: Pathophysiology, diagnosis, management, 4th ed. W. B. Saunders, Philadelphia.

Boyer, T. 1988. Major sequelae of cirrhosis, p. 847–52. *In* J. Wyngaarden and L. Smith (eds.), Cecil textbook of medicine, 18th ed. W. B. Saunders, Philadelphia.

Cello, J. 1988. Carcinoma of the pancreas, p. 781, 783. *In* J. Wyngaarden and L. Smith (eds.), Cecil textbook of medicine, 18th ed. W. B. Saunders, Philadelphia.

Czaja, A. 1988. Chronic active hepatitis, p. 1288–89. *In* J. Hurst (ed.), Medicine for the practicing physician, 2nd ed. Butterworths, Boston.

Dienstag, J., J. Wands, and R. Koff. 1987. Acute hepatitis, p. 1329–30. *In* E. Braunwald et al. (eds.), Harrison's principles of internal medicine, 11th ed. McGraw-Hill, New York.

Friedman, S. L., and R. L. Owen. 1989. Gastrointestinal manifestations of AIDS and other sexually transmissible diseases, p. 1253–55. *In* M. H. Sleisinger and J. S. Fordtran (eds.), Gastrointestinal disease: Pathophysiology, diagnosis, management, 4th ed. W. B. Saunders, Philadelphia.

Galambos, J. 1988. Acute viral hepatitis, p. 1282–83. *In* J. Hurst (ed.), Medicine for the practicing physician, 2nd ed. Butterworths, Boston.

Galambos, J. 1988. Cirrhosis in medicine, p. 1295. *In* J. Hurst (ed.), Medicine for the practicing physician, 2nd ed. Butterworths, Boston.

Galambos, J. 1988. Jaundice, p. 1266. *In* J. Hurst (ed.), Medicine for the practicing physician, 2nd ed. Butterworths, Boston.

Gosling, J. et al. 1985. Atlas of human anatomy, p. 4.30, 4.31. J. B. Lippincott Co., Philadelphia.

Isselbacher, K. 1987. Disturbances of bilirubin metabolism, p. 1321. *In* E. Braunwald et al. (eds.), Harrison's principles of internal medicine, 11th ed. McGraw-Hill, New York.

Knauer, C., and S. Silverman, Jr. 1987. Alimentary tract and liver, p. 406, 408–9, 413, 420–22. *In* M. Krupp, S. Schroeder, and L. Tierney (eds.), Current medical diagnosis and treatment. Appleton & Lange, Norwalk, Conn.

Plagemann, P. 1991. Hepatitis C virus: Brief review. *Archives of Virology* 120:165–80.

Podolsky, D., and K. Isselbacher. 1987. Cirrhosis, p. 1341, 1343–45. *In* E. Braunwald et al. (eds.), Harrison's principles of internal medicine, 11th ed. McGraw-Hill, New York.

Podolsky, D., and K. Isselbacher. 1987. Derangements of hepatic metabolism, p. 1310, 1314. *In* E. Braunwald et al. (eds.), Harrison's principles of internal medicine, 11th ed. McGraw-Hill, New York.

Scharschmidt, B. 1989. Jaundice, p. 454–58. *In* M. Sleisinger and J. Fordtran (eds.), Gastrointestinal disease: Pathophysiology, diagnosis, management, 4th ed. W. B. Saunders, Philadelphia.

Scully, R., E. Mark, and B. McNeeley (eds.). 1984. Case records of Massachusetts General Hospital. *New England Journal of Medicine* 311:1170.

Smith, D. 1991. Hepatitis C update: New answers, new questions. *Postgraduate Medicine* 90:199–206.

Stryer, L. 1981. Biochemistry, p. 507. W. H. Freeman & Co., San Francisco.

Zieve, L. 1988. Hepatic encephalopathy, p. 1300–1302. *In* J. Hurst (ed.), Medicine for the practicing physician, 2nd ed. Butterworths, Boston.

CHAPTER 37

Agus, Z. S. and S. Goldfarb. 1985. Calcium metabolism: Normal and abnormal, p. 528, 546–47. *In* A. I. Arieff and R. A. DeFronzo (eds.), Fluid, electrolyte, and acid-base disorders. Churchill Livingstone, New York.

Czaja, A. 1988. Chronic active hepatitis, p. 1288, 1294. *In* J. Hurst (ed.), Medicine for the practicing physician, 2nd ed. Butterworths, Boston.

Fischbach, F. 1984. A manual of laboratory diagnostic tests, 2nd ed., p. 307. J. B. Lippincott Co., Philadelphia.

Kaplan, M. 1987. Evaluation of hepatobiliary diseases, p. 58–61. *In* J. Stein (ed.), Internal medicine, 2nd ed. Little, Brown & Co., Boston.

Knauer, C., and S. Silverman, Jr. 1987. Alimentary tract and liver, p. 407, 417, 420, 423. *In* M. Krupp, S. Schroeder, and L. Tierney (eds.), Current medical diagnosis and treatment. Appleton & Lange, Norwalk, Conn.

Levitt, M. D. 1985. Pancreatitis, p. 772–73, 781. *In* J. B. Wyngaarden and L. H. Smith (eds.), Cecil textbook of medicine, 17th ed. W. B. Saunders, Philadelphia.

Ockner, R. K. 1985. Chronic hepatitis, p. 825. *In* J. B. Wyngaarden and L. H. Smith (eds.), Cecil textbook of medicine, 17th ed. W. B. Saunders, Philadelphia.

Ockner, R. 1988. Laboratory tests in liver disease, p. 814–16. *In* J. Wyngaarden and L. Smith (eds.), Cecil textbook of medicine, 18th ed. W. B. Saunders, Philadelphia.

Parfitt, A. M., and M. Kleerekoper. 1980. Clinical disorders of calcium, phosphorous, and magnesium metabolism, p. 1032. *In* M. H. Maxwell and C. R. Kleeman (eds.), Clinical disorders of fluid and electrolyte metabolism, 3rd ed. McGraw-Hill, New York.

Reynolds, T. B. 1980. Water, electrolyte, and acid-base disorders in liver disease, p. 1252–57. *In* M. H. Maxwell and C. R. Kleeman (eds.), Clinical disorders of fluid and electrolyte metabolism, 3rd ed. McGraw-Hill, New York.

Rose, B. D. 1977. Clinical physiology of acid-base and electrolyte disorders, p. 123–34. McGraw-Hill, New York.

Rudert, C. S. 1988. Acute pancreatitis, p. 1452–53. *In* J. Hurst (ed.), Medicine for the practicing physician, 2nd ed. Butterworths, Boston.

Shorecki, K., and B. Brenner. 1985. Edema states: Congestive heart failure, liver disease, and nephrotic syndrome, p. 96. *In* A. Arieff and R. DeFronzo (eds.), Fluid, electrolyte, and acid-base disorders. Churchill Livingstone, New York.

Wallach, J. 1986. Interpretation of diagnostic tests, 4th ed., p. 47, 64, 206, 251, 253. Little, Brown & Co., Boston.

CHAPTER 38

Brust, J. C. 1989. Cerebral infarction, p. 211. *In* L. P. Rowland (ed.), Merritt's textbook of neurology, 8th ed. Lea & Febiger, Philadelphia.

Brust, J. C. 1989. Subarachnoid hemorrhage, p. 235–40. *In* L. P. Rowland (ed.), Merritt's textbook of neurology, 8th ed. Lea & Febiger, Philadelphia.

Burns, R. A. 1989. Stroke in young adults, p. 226–27, 229–30. *In* L. P. Rowland (ed.), Merritt's textbook of neurology, 8th ed. Lea & Febiger, Philadelphia.

Crouch, J. E. 1982. Essential human anatomy: A text atlas, p. 222, 347. Lea & Febiger, Philadelphia.

Fishman, R. A. 1989. Brain edema and disorders of intracranial pressure, p. 262–66. *In* L. P. Rowland (ed.), Merritt's textbook of neurology, 8th ed. Lea & Febiger, Philadelphia.

Luke, R. G. 1985. Treatment of irreversible renal failure: Dialysis, p. 561. *In* J. B. Wyngaarden and L. H. Smith (eds.), Cecil textbook of medicine, 17th ed. W. B. Saunders, Philadelphia.

Mohr, J. P. 1989. Transient ischemic attacks, p. 206. *In* L. P. Rowland (ed.), Merritt's textbook of neurology, 8th ed. Lea & Febiger, Philadelphia.

Prochop, L. D., and C. P. Shah. 1989. Disorders of cerebrospinal and brain fluids, p. 253–60. *In* L. P. Rowland (ed.), Merritt's textbook of neurology, 8th ed. Lea & Febiger, Philadelphia.

Rottenberg, D. A. 1985. Hydrocephalus, p. 2169. *In* J. B. Wyngaarden and L. H. Smith (eds.), Cecil textbook of medicine, 17th ed. W. B. Saunders, Philadelphia.

Spence, A. P. 1986. Basic human anatomy, p. 307. Benjamin-Cummings, Menlo Park, Cal.

Stein, B. M. 1989. Vascular tumors and malformations, p. 384. *In* L. P. Rowland (ed.), Merritt's textbook of neurology, 8th ed. Lea & Febiger, Philadelphia.

Toole, J. F., and M. Cole. 1988. Ischemic cerebrovascular disease, p. 1, 8–14, 17, 32, 38. *In* R. Joynt (ed.), Clinical neurology, vol. 2. J. B. Lippincott Co., Philadelphia.

Vick, R. 1984. Contemporary medical physiology, p. 261. Addison-Wesley, Menlo Park, Cal.

Walker, H. K. 1988. Cerebrovascular disease and stroke, p. 1560–61. *In* J. Hurst (ed.), Medicine for the practicing physician. Butterworths, Boston.

CHAPTER 39

Antunes, L. 1984. Gliomas, p. 237–40. *In* L. Rowland (ed.), Merritt's textbook of neurology, 7th ed. Lea & Febiger, Philadelphia.

Carter, S., and N. L. Low. 1989. Cerebral palsy and mental retardation, p. 458–62. *In* L. P. Rowland (ed.), Merritt's textbook of neurology, 8th ed. Lea & Febiger, Philadelphia.

Davenport, J. 1987. Epilepsy, p. 2143–46, 2148–49. *In* J. Stein (ed.), Internal medicine, 2nd ed. Little, Brown & Co., Boston.

Demian, P. L., and G. J. Carlucci. 1989. Two patients with histories of intermittent loss of consciousness. *Physician Assistant* 13:48–56.

Goldensohn, E., G. Glaser, and M. Goldberg. 1984. Epilepsy, p. 629, 631, 634–36. *In* L. Rowland (ed.), Merritt's textbook of neurology, 7th ed. Lea & Febiger, Philadelphia.

Greenlee, J. E. 1987. Slow virus infections, p. 2260–61. *In* J. H. Stein (ed.), Internal medicine, 2nd ed. Little, Brown & Co., Boston.

Jubelt, B., and J. R. Miller. 1989, Infections of the nervous system, p. 126–67. *In* L. P. Rowland (ed.), Merritt's textbook of neurology, 8th ed. Lea & Febiger, Philadelphia.

Katzman, R. 1989. Delirium and dementia, p. 6–7. *In* L. P. Rowland (ed.), Merritt's textbook of neurology, 8th ed. Lea & Febiger, Philadelphia.

Katzman, R. 1989. The dementias, p. 638–41. *In* L. P. Rowland (ed.), Merritt's textbook of neurology, 8th ed. Lea & Febiger, Philadelphia.

Kortyna, Ricky. 1989. Brain abscess in a 35-year-old man and a 62-year-old man. *Physician Assistant* 13:83–96.

Levin, V., G. Sheline, and P. Gutin. 1989. Neoplasms of the central nervous system, p. 1557–58, 1562–63. *In* V. DeVita, Jr., S. Hellman, and S. Rosenberg (eds.), Principles and practice of oncology, 3rd ed. J. B. Lippincott Co., Philadelphia.

Plum, F. 1988. Dementias, p. 2090. *In* J. B. Wyngaarden and L. H. Smith (eds.), Cecil textbook of medicine, 17th ed. W. B. Saunders, Philadelphia.

Price, R. W. 1988. AIDS dementia and human immunodeficiency virus brain infection, p. 2203–5. *In* J. B. Wyngaarden and L. H. Smith, Jr. (ed.), Cecil textbook of medicine, 18th ed. W. B. Saunders, Philadelphia.

Pruitt, A. 1987. Intracranial neoplasms, p. 2220–21. *In* J. Stein (ed.), Internal medicine, 2nd ed. Little, Brown & Co., Boston.

Spence, A. 1986. Basic human anatomy, 2nd ed., p. 357. Benjamin-Cummings, Menlo Park, Cal.

Stajich, J. M. 1989. Common neurologic disorders part II: Headaches. *Physician Assistant* 13:23–32.

Stein, B. 1984. Tumors of the meninges, p. 232–33. *In* L. Rowland (ed.), Merritt's textbook of neurology, 7th ed. Lea & Febiger, Philadelphia.

Walker, K. H., and R. Morris. 1988. Dementia, p. 1549. *In* J. W. Hurst (ed.), Medicine for the practicing physician. Butterworths, Boston.

CHAPTER 40

Easton, J. D. 1987. Degenerative diseases, p. 2243–44. *In* J. H. Stein (ed.), Internal medicine, 2nd ed. Little, Brown & Co., Boston.

Edmeads, J. G. 1987. Headache and facial pain, p. 2179–84. *In* J. H. Stein (ed.), Internal medicine, 2nd ed. Little, Brown & Co., Boston.

Fahn, S. 1984. Huntington disease and other forms of chorea, p. 517–19. *In* L. P. Rowland (ed.), Merritt's textbook of neurology, 7th ed. Lea & Febiger, Philadelphia.

Hart, R., and D. Easton. 1987. Demyelinating disease, p. 2245. *In* J. Stein (ed.), Internal medicine, 2nd ed. Little, Brown & Co., Boston.

McDowell, F., and J. Cedarbaum. 1988. The extrapyramidal system and disorders of movement, p. 2, 5, 8–26, 29, 32, 34–5, 57. *In* R. Joynt (ed.), Clinical neurology, vol. 3. J. B. Lippincott Co., Philadelphia.

Munsat, T. L. 1989. Adult motor neuron disease, p. 683–84. *In* L. P. Rowland (ed.), Merritt's textbook of neurology, 8th ed. Lea & Febiger, Philadelphia.

Penn, A., and L. Rowland. 1984. Neuromuscular junction, p. 561–65. *In* L. P. Rowland (ed.), Merritt's textbook of neurology, 7th ed. Lea & Febiger, Philadelphia.

Pleasure, D. E., and D. L. Schotland. 1989. Acquired neuropathies, p. 609–11. *In* L. P. Rowland (ed.), Merritt's textbook of neurology, 8th ed. Lea & Febiger, Philadelphia.

Poser, C. et al. 1984. Demyelinating diseases, p. 593–95, 601–5, 608, 610. *In* L. P. Rowland (ed.), Merritt's textbook of neurology, 7th ed. Lea & Febiger, Philadelphia.

Poser, J. B. 1988. Disorders of sensation, p. 2129–34. *In* J. B. Wyngaarden and L. H. Smith, Jr. (eds.), Cecil textbook of medicine, 18th ed. W. B. Saunders, Philadelphia.

Roses, A. 1984. Progressive muscular dystrophies, p. 577–80. *In* L. P. Rowland (ed.), Merritt's textbook of neurology, 7th ed. Lea & Febiger, Philadelphia.

Ross, G., J. Wolf, and M. Chipman. 1988. The neuralgias, p. 2–5. *In* R. Joynt (ed.), Clinical neurology, vol. 4. J. B. Lippincott Co., Philadelphia.

Rowland, L. P., and R. B. Layzer. 1986. Muscular dystrophies, atrophies, and related diseases, p. 11–12. *In* J. A. Spittell, Jr. (ed.), Clinical medicine, vol. 4. Harper & Row, Philadelphia.

Rowland, L. P. 1989. Injury to peripheral and cranial nerves, p. 419–24. *In* L. P. Rowland (ed.), Merritt's textbook of neurology, 8th ed. Lea & Febiger, Philadelphia.

Sherman, D. 1987. Muscle diseases, p. 2240–41. *In* J. Stein (ed.), Internal medicine, 2nd ed. Little, Brown & Co., Boston.

Spence, A. 1986. Basic human anatomy, 2nd ed., p. 351, 370–71, 410, 426. Benjamin-Cummings, Menlo Park, Cal.

CHAPTER 41

Adams, H., and J. Biller. 1988. Hemorrhagic intracranial vascular disease, p. 2, 11–12, 14, 18, 22, 23. *In* R. Joynt (ed.), Clinical neurology. J. B. Lippincott Co., Philadelphia.

Arieff, A. I. 1985. Effects of water, acid-base, and electrolyte disorders on the central nervous system, p. 992. *In* A. I. Arieff and R. A. DeFronzo (eds.), Fluid, electrolyte, and acid-base disorders. Churchill Livingstone, New York.

Cox, M., M. Geheb, and I. Singer. 1985. Disorders of thirst and renal water excretion, p. 133. *In* A. I. Arieff and R. A. DeFronzo (eds.), Fluid, electrolyte, and acid-base disorders. Churchill Livingstone, New York.

Davidson, I., and D. A. Nelson. 1974. Blood, p. 247. *In* I. Davidsohn and J. B. Henry (eds.), Todd-Sanford clinical diagnosis, 15th ed. W. B. Saunders, Philadelphia.

Fischbach, F. 1984. A manual of laboratory diagnostic tests, 2nd ed., p. 353. J. B. Lippincott Co., Philadelphia.

Kistler, J., A. Ropper, and J. Martin. 1987. Cerebrovascular diseases, p. 1954, 1956. *In* E. Braunwald et al. (eds.), Harrison's principles of internal medicine, 11th ed. McGraw-Hill, New York.

Mohr, J. P., and J. A. Bello. 1989. Laboratory studies in stroke, p. 193–94. *In* L. P. Rowland (ed.), Merritt's textbook of neurology, 7th ed. Lea & Febiger, Philadelphia.

Moses, A. M., S. A. Blumenthal, and D. H. Streeten. 1985. Acid-base and electrolyte disorders associated with endocrine disease: Pituitary and thyroid, p. 880. *In* A. I. Arieff and R. A. DeFronzo (eds.), Fluid, electrolyte and acid-base disorders. Churchill Livingstone, New York.

Narins, R. G. et al. 1985. Metabolic acid-base disorders: Pathophysiology, classification, and treatment, p. 307. *In* A. I. Arieff and R. A. DeFronzo (eds.), Fluid, electrolyte, and acid-base disorders. Churchill Livingstone, New York.

Ritchie, J. M. 1975. Central nervous system stimulants, p. 373. *In* L. S. Goodman and A. Gilman (eds.), Pharmacological basis of therapeutics, 5th ed. Macmillan, New York.

General

Dorland's illustrated medical dictionary, 26th ed. 1982. W. B. Saunders, Philadelphia.

Webster's new universal unabridged dictionary, 2nd ed. 1983. New World Dictionaries Simon & Schuster, New York.

Credits

From Kent M. Van De Graaff, *Human Anatomy,* 3d ed. Copyright © 1992 Wm. C. Brown Communications, Inc., Dubuque, Iowa. All Rights Reserved. Reprinted by permission. **Figure 10.6** From Kent M. Van De Graaff and Stuart Ira Fox, *Concepts of Human Anatomy and Physiology,* 3d ed. Copyright © 1992 Wm. C. Brown Communications, Inc., Dubuque, Iowa. All Rights Reserved. Reprinted by permission.

Chapter 11
Figures 11.2, 11.4, 11.6, 11.12 From John W. Hole, Jr., *Human Anatomy and Physiology,* 3d ed. Copyright © 1984 Wm. C. Brown Communications, Inc., Dubuque, Iowa. All Rights Reserved. Reprinted by permission. **Figure 11.3** From John W. Hole, Jr., *Human Anatomy and Physiology,* 4th ed. Copyright © 1987 Wm. C. Brown Communications, Inc., Dubuque, Iowa. All Rights Reserved. Reprinted by permission. **Figure 11.10** From Stuart Ira Fox, *Human Physiology,* 3d ed. Copyright © 1990 Wm. C. Brown Communications, Inc., Dubuque, Iowa. All Rights Reserved. Reprinted by permission.

Chapter 12
Figure 12.4 From John W. Hole, Jr., *Human Anatomy and Physiology,* 3d ed. Copyright © 1984 Wm. C. Brown Communications, Inc., Dubuque, Iowa. All Rights Reserved. Reprinted by permission.

Chapter 15
Figure 15.4 From Kent M. Van De Graaff, *Human Anatomy,* 3d ed. Copyright © 1992 Wm. C. Brown Communications, Inc., Dubuque, Iowa. All Rights Reserved. Reprinted by permission.

Chapter 17
Figures 17.5, 17.6, 17.7 From Stuart Ira Fox, *Human Physiology,* 3d ed. Copyright © 1990 Wm. C. Brown Communications, Inc., Dubuque, Iowa. All Rights Reserved. Reprinted by permission.

Chapter 20
Figure 20.1 From John W. Hole, Jr., *Human Anatomy and Physiology,* 5th ed. Copyright © 1990 Wm. C. Brown Communications, Inc., Dubuque, Iowa. All Rights Reserved. Reprinted by permission.

Chapter 21
Figures 21.2*b*, 21.6 From Stuart Ira Fox, *Human Physiology,* 3d ed. Copyright © 1990 Wm. C. Brown Communications, Inc., Dubuque, Iowa. All Rights Reserved. Reprinted by permission. **Figures 21.3, 21.4**

From John W. Hole, Jr., *Human Anatomy and Physiology,* 3d ed. Copyright © 1984 Wm. C. Brown Communications, Inc., Dubuque, Iowa. All Rights Reserved. Reprinted by permission. **Figure 21.8** From Kent M. Van De Graaff and Stuart Ira Fox, *Concepts of Human Anatomy and Physiology.* Copyright © 1986 Wm. C. Brown Communications, Inc., Dubuque, Iowa. All Rights Reserved. Reprinted by permission.

Chapter 23
Figures 23.2, 23.12 From John W. Hole, Jr., *Human Anatomy and Physiology,* 5th ed. Copyright © 1990 Wm. C. Brown Communications, Inc., Dubuque, Iowa. All Rights Reserved. Reprinted by permission. **Figure 23.3** From Kent M. Van De Graaff, *Human Anatomy,* 3d ed. Copyright © 1992 Wm. C. Brown Communications, Inc., Dubuque, Iowa. All Rights Reserved. Reprinted by permission.

Chapter 24
Figure 24.2 From John W. Hole, Jr., *Human Anatomy and Physiology,* 5th ed. Copyright © 1990 Wm. C. Brown Communications, Inc., Dubuque, Iowa. All Rights Reserved. Reprinted by permission.

Chapter 26
Figure 26.5 From Stuart Ira Fox, *Human Physiology,* 2d ed. Copyright © 1987 Wm. C. Brown Communications, Inc., Dubuque, Iowa. All Rights Reserved. Reprinted by permission. **Figure 26.6** From John W. Hole, Jr., *Human Anatomy and Physiology,* 5th ed. Copyright © 1990 Wm. C. Brown Communications, Inc., Dubuque, Iowa. All Rights Reserved. Reprinted by permission.

Chapter 27
Figure 27.2*a* From John W. Hole, Jr., *Human Anatomy and Physiology,* 4th ed. Copyright © 1987 Wm. C. Brown Communications, Inc., Dubuque, Iowa. All Rights Reserved. Reprinted by permission. **Figure 27.2b** From John W. Hole, Jr., *Human Anatomy and Physiology,* 3d ed. Copyright © 1984 Wm. C. Brown Communications, Inc., Dubuque, Iowa. All Rights Reserved. Reprinted by permission. **Figures 27.3, 27.5, 27.7** From L. M. Prescott, et al., *Microbiology.* Copyright © 1990 Wm. C. Brown Communications, Inc., Dubuque, Iowa. All Rights Reserved. Reprinted by permission. **Figure 27.4** From Stuart Ira Fox, *Human Physiology,* 3d ed. Copyright © 1990 Wm. C. Brown Communications, Inc., Dubuque, Iowa. All Rights Reserved. Reprinted by permission. **Figure 27.6** From *Immunology,* by I. Roitt, J. Brostoff and D. Male, Gower Medical

Publishing, London, UK, 1985. **Figure 27.8** From Stuart Ira Fox, *Human Physiology,* 2d ed. Copyright © 1987 Wm. C. Brown Communications, Inc., Dubuque, Iowa. All Rights Reserved. Reprinted by permission. **Figure 27.9** From Kent M. Van De Graaff and Stuart Ira Fox, *Concepts of Human Anatomy and Physiology,* 2d ed. Copyright © 1989 Wm. C. Brown Communications, Inc., Dubuque, Iowa. All Rights Reserved. Reprinted by permission.

Chapter 28
Figures 28.1, 28.4 From L. M. Prescott, et al., *Microbiology.* Copyright © 1990 Wm. C. Brown Communications, Inc., Dubuque, Iowa. All Rights Reserved. Reprinted by permission. **Figures 28.2, 28.3** From *Immunology,* by I. Roitt, J. Brostoff and D. Male, Gower Medical Publishing, London, UK, 1985.

Chapter 29
Figure 29.5 From Stuart Ira Fox, *Human Physiology.* Copyright © 1984 Wm. C. Brown Communications, Inc., Dubuque, Iowa. All Rights Reserved. Reprinted by permission. **Figure 29.6** From Stuart Ira Fox, *Human Physiology,* 3d ed. Copyright © 1990 Wm. C. Brown Communications, Inc., Dubuque, Iowa. All Rights Reserved. Reprinted by permission. **Figure 29.7** From L. M. Prescott, et al., *Microbiology.* Copyright © 1990 Wm. C. Brown Communications, Inc., Dubuque, Iowa. All Rights Reserved. Reprinted by permission.

Chapter 31
Figure 31.1 From John W. Hole, Jr., *Human Anatomy and Physiology,* 3d ed. Copyright © 1984 Wm. C. Brown Communications, Inc., Dubuque, Iowa. All Rights Reserved. Reprinted by permission. **Figure 31.3** From Kent M. Van De Graaff, *Human Anatomy,* 3d ed. Copyright © 1992 Wm. C. Brown Communications, Inc., Dubuque, Iowa. All Rights Reserved. Reprinted by permission. **Figures 31.4, 31.6, 31.9** From Stuart Ira Fox, *Human Physiology,* 3d ed. Copyright © 1990 Wm. C. Brown Communications, Inc., Dubuque, Iowa. All Rights Reserved. Reprinted by permission. **Figures 31.7, 31.8** From Kent M. Van De Graaff and Stuart Ira Fox, *Concepts of Human Anatomy and Physiology,* 2d ed. Copyright © 1989 Wm. C. Brown Communications, Inc., Dubuque, Iowa. All Rights Reserved. Reprinted by permission. **Figures 31.10, 31.12*b*** From John W. Hole, Jr., *Human Anatomy and Physiology,* 4th ed. Copyright © 1987 Wm. C. Brown Communications, Inc., Dubuque, Iowa. All Rights Reserved. Reprinted by permission. **Figure 31.11** From Kent M. Van De Graaff and Stuart Ira Fox, *Concepts of Human Anatomy and Physiology,* 3d ed.

Chapter 32

Figure 32.1 From John W. Hole, Jr., *Human Anatomy and Physiology,* 5th ed. Copyright © 1990 Wm. C. Brown Communications, Inc., Dubuque, Iowa. All Rights Reserved. Reprinted by permission. **Figures 32.2, 32.4, 32.7, 32.8** From Stuart Ira Fox, *Human Physiology,* 3d ed. Copyright © 1990 Wm. C. Brown Communications, Inc., Dubuque, Iowa. All Rights Reserved. Reprinted by permission. **Figure 32.3** From John W. Hole, Jr., *Human Anatomy and Physiology,* 5th ed. Copyright © 1990 Wm. C. Brown Communications, Inc., Dubuque, Iowa. All Rights Reserved. Reprinted by permission. **Figure 32.6** From John W. Hole, Jr., *Human Anatomy and Physiology,* 4th ed. Copyright © 1987 Wm. C. Brown Communications, Inc., Dubuque, Iowa. All Rights Reserved. Reprinted by permission.

Chapter 33

Figures 33.1, 33.4 From Stuart Ira Fox, *Human Physiology,* 3d ed. Copyright © 1990 Wm. C. Brown Communications, Inc., Dubuque, Iowa. All Rights Reserved. Reprinted by permission. **Figure 33.2** From Stuart Ira Fox, *Human Physiology.* Copyright © 1984 Wm. C. Brown Communications, Inc., Dubuque, Iowa. All Rights Reserved. Reprinted by permission. **Figure 33.3** Reprinted with permission. E. J. Barrett and R. A. DeFronao, "Diabetic Ketoacidosis: Diagnosis and Management." *Hospital Practice,* Volume 19, issue 4, pages 89–104. Illustration by Albert Miller.

Chapter 34

Figure 34.3 From John W. Hole, Jr., *Human Anatomy and Physiology,* 5th ed. Copyright © 1990 Wm. C. Brown Communications, Inc., Dubuque, Iowa. All Rights Reserved. Reprinted by permission.

Chapter 35

Figure 35.5 From Kent M. Van De Graaff and Stuart Ira Fox, *Concepts of Human Anatomy and Physiology,* 2d ed. Copyright © 1989 Wm. C. Brown Communications, Inc., Dubuque, Iowa. All Rights Reserved. Reprinted by permission.

Chapter 36

Figure 36.2 From John W. Hole, Jr., *Human Anatomy and Physiology,* 4th ed. Copyright © 1987 Wm. C. Brown Communications, Inc., Dubuque, Iowa. All Rights Reserved. Reprinted by permission. **Figure 36.3** From John W. Hole, Jr., *Human Anatomy and Physiology,* 3d ed. Copyright © 1984 Wm. C. Brown Communications, Inc., Dubuque, Iowa. All Rights Reserved. Reprinted by permission. **Figure 36.5** From Kent M. Van De Graaff, *Human Anatomy,* 2d ed. Copyright © 1988 Wm. C. Brown Communications, Inc., Dubuque, Iowa. All Rights Reserved. Reprinted by permission. **Figure 36.6** From Kent M. Van De Graaff and Stuart Ira Fox, *Concepts of Human Anatomy and Physiology,* 3d ed. Copyright © 1992 Wm. C. Brown Communications, Inc., Dubuque, Iowa. All Rights Reserved. Reprinted by permission. **Figure 36.7** From John W. Hole, Jr., *Human Anatomy and Physiology,* 5th ed. Copyright © 1990 Wm. C. Brown Communications, Inc., Dubuque, Iowa. All Rights Reserved. Reprinted by permission. **Figure 36.16** From Kent M. Van De Graaff, *Human Anatomy,* 3d ed. Copyright © 1992 Wm. C. Brown Communications, Inc., Dubuque, Iowa. All Rights Reserved. Reprinted by permission.

Chapter 38

Figures 38.1, 38.3, 38.6 From John W. Hole, Jr., *Human Anatomy and Physiology,* 4th ed. Copyright © 1987 Wm. C. Brown Communications, Inc., Dubuque, Iowa. All Rights Reserved. Reprinted by permission. **Figure 38.2** From John W. Hole, Jr., *Human Anatomy and Physiology,* 5th ed. Copyright © 1990 Wm. C. Brown Communications, Inc., Dubuque, Iowa. All Rights Reserved. Reprinted by permission. **Figure 38.11** From Kent M. Van De Graaff and Stuart Ira Fox, *Concepts of Human Anatomy and Physiology,* 3d ed. Copyright © 1992 Wm. C. Brown Communications, Inc., Dubuque, Iowa. All Rights Reserved. Reprinted by permission.

Chapter 39

Figures 39.1, 39.5 From John W. Hole, Jr., *Human Anatomy and Physiology,* 5th ed. Copyright © 1990 Wm. C. Brown Communications, Inc., Dubuque, Iowa. All Rights Reserved. Reprinted by permission. **Figure 39.2** From Kent M. Van De Graaff, *Human Anatomy,* 3d ed. Copyright © 1992 Wm. C. Brown Communications, Inc., Dubuque, Iowa. All Rights Reserved. Reprinted by permission. **Figure 39.4** From Stuart Ira Fox, *Human Physiology,* 3d ed. Copyright © 1990 Wm. C. Brown Communications, Inc., Dubuque, Iowa. All Rights Reserved. Reprinted by permission.

Chapter 40

Figures 40.1, 40.2 From Kent M. Van De Graaff, *Human Anatomy,* 3d ed. Copyright © 1992 Wm. C. Brown Communications, Inc., Dubuque, Iowa. All Rights Reserved. Reprinted by permission. **Figures 40.3, 40.5, 40.6** From John W. Hole, Jr., *Human Anatomy and Physiology,* 3d ed. Copyright © 1984 Wm. C. Brown Communications, Inc., Dubuque, Iowa. All Rights Reserved. Reprinted by permission. **Figure 40.4** From John W. Hole, Jr., *Human Anatomy and Physiology,* 5th ed. Copyright © 1990 Wm. C. Brown Communications, Inc., Dubuque, Iowa. All Rights Reserved. Reprinted by permission. **Figure 40.7** From John W. Hole, Jr., *Human Anatomy and Physiology,* 4th ed. Copyright © 1987 Wm. C. Brown Communications, Inc., Dubuque, Iowa. All Rights Reserved. Reprinted by permission.

PHOTOGRAPHS

Chapter 5
Figure 5.1: © Edwin Reschke

Chapter 6
Figure 6.8: Ross, M. H., Reith, E. J., and Romrell, L. J. HISTOLOGY: A TEXT AND ATLAS, 2nd edition. © Williams & Wilkins Co., 1989.

Chapter 7
Figures 7.7, 7.8: Cotran, R., Kumar, V., and Robbins, S. ROBBINS PATHOLOGIC BASIS OF DISEASE, 4th edition. © W. B. Saunders, 1989.

Chapter 8
Figures 8.2, 8.3: Courtesy Dr. M. G. Klein; **Figure 8.5:** Cotran, R., Kumar, S. and Robbins, S. ROBBINS PATHOLOGIC BASIS OF DISEASE, 4th ed. © W. B. Saunders, 1989.

Chapter 11
Figure 11.22a: © J. Hayes/Medical Images, Inc.

Chapter 12
Figure 12.1: Moran, D.T. and Rowley, J. C. VISUAL HISTOLOGY. © Lea & Febiger, 1988.

Chapter 14
Figures 14.1, 14.4: Ritchie, A. C. BOYD'S TEXTBOOK OF PATHOLOGY, 9th ed. © Lea & Febiger, 1990.

Chapter 15
Figure 15.5a and b: Janet D. Coleman.

Chapter 17
Figure 17.2: Courtesy Dorothea Zucker-Franklin, MD, Professor of Medicine at

New York University Medical Center; from Ross et al. HISTOLOGY: A TEXT-ATLAS, 2nd ed. © Williams and Wilkins Co., 1989; **Figure 17.9:** © Edwin Reschke.

Chapter 18
Figures 18.1, 18.3, 18.6: Courtesy Dr. M. G. Klein; **Figures 18.7, 18.8:** McKensie, S. B. TEXTBOOK OF HEMATOLOGY. © Lea & Febiger, 1988.

Chapter 20
Figure 20.2: Courtesy Dorothea Zucker-Franklin, MD, Professor of Medicine, New York University Medical Center; from Ross et al. HISTOLOGY: A TEXT-ATLAS, 2nd ed. © Williams and Wilkins, 1989; **Figure 20.8a and b:** Courtesy Dr. M. G. Klein.

Chapter 21
Figure 21.1: © J. Hayes, Medical Images, Inc.

Chapter 22
Figures 22.2, 22.4: Ritchie, A. C. BOYD'S TEXTBOOK OF PATHOLOGY, 9th ed. © Lea & Febiger, 1990.

Chapter 23
Figure 23.4: Cotran, R., Kumar, V., and Robbins, S. PATHOLOGIC BASIS OF DISEASE, 4th ed. © W. B. Saunders, 1989.

Chapter 27
Figure 27.1: Courtesy Dorothea Zucker-Franklin, MD, Professor of Medicine at New York University Medical Center; from Ross et al. HISTOLOGY: A TEXT AND ATLAS, 2nd ed. © Williams and Wilkins, 1989.

Chapter 29
Figures 29.1a and b, 29.2a and b, 29.3, 29.4: Centers for Disease Control.

Chapter 30
Figures 30.2a and b, 30.3, 30.4, 30.5: Centers for Disease Control; **Figure 30.6:** Dr. J. H. Crookston; from Ritchie, A. C. BOYD'S TEXTBOOK OF PATHOLOGY, 9th ed. © Lea & Febiger, 1990.

Chapter 34
Figure 34.7: Ritchie, A. C. BOYD'S TEXTBOOK OF PATHOLOGY, 9th ed. © Lea & Febiger, 1990.

Chapter 35
Figures 35.3, 35.6a and b, 35.8, 35.9: Cotran, R., Kumar, V. and Robbins, S. ROBBINS PATHOLOGIC BASIS OF DISEASE, 4th ed. © W. B. Saunders, 1989.

Chapter 36
Figure 36.4a–c: Moran, D. T. and Rowley, J. C. VISUAL HISTOLOGY. © Lea & Febiger, 1988; **Figure 36.9:** Ross, M. H., Reith, E. J., and Romrell, L. J. HISTOLOGY: A TEXT AND ATLAS, 2nd ed. © Williams & Wilkins Co., 1989.

Chapter 38
Figures 38.4, 38.5a and b: Janet D. Coleman; **Figure 38.7:** Dr. J. H. N. Deck, from Ritchie, A. C. BOYD'S TEXTBOOK OF PATHOLOGY, 9th ed. © Lea & Febiger, 1990.

Chapter 39
Figure 39.3: Courtesy of Dr. M. G. Klein.

Chapter 40
Figure 40.8: Moran, D. T. and Rowley, T. C. VISUAL HISTOLOGY. © Lea & Febiger, 1988.

Chapter 41
Figure 41.2: Dr. J. H. N. Deck, from Ritchie, A. C. BOYD'S TEXTBOOK OF PATHOLOGY, 4th ed. © Lea & Febiger, 1990.

Index

renal control, 39, 40
  urinary calcium and thiazides, 161
  and vitamin D, 39–40, 193
Calcium carbonate, 42
Calcium homeostasis, 39
Calcium ion, 40–41, 227
Calcium oxalate, 161
Calcium phosphate deposits
  uremic syndrome, 182
Calculus, 160
Calyces, 158, 161
Calyx, 138
Camphylobacter, 409
Cancer, 83
  abnormalities in clotting characteristics, 244
Candida, 393
Candida albicans, 120, 134
Capacitance system, 256
CAPD. *See* continuous ambulatory peritoneal dialysis
Capillary, 14–15, 17–18, 32, 80–81, 83–84, 92, 255, 354
  dynamics, 18
  lymphatic, 17
  permeability, 18–19
  pressure, 84
  walls, 25, 84
Capillary/capsular wall, 140
Capsular oncotic pressure, 141
Carbaminohemoglobin, 202
Carbohydrate, 54, 355
Carbohydrate side chains, 8
Carbon dioxide, 53–54, 56, 58–59, 76, 79, 81
  alveolar/capillary exchange, 49–50
  carbaminohemoglobin, 202
  dissolved in plasma, 202
  forms in plasma, 202
  and hemoglobin-oxygen affinity, 203 , 212
  hydration, 49–52, 62
  hydration in pancreatic cells, 63
  hydration in parietal cells, 53, 63
  hydration in red blood cells, 53, 201–2
  hydration in tubular cells, 146
  and hyperventilation, 49, 50
  and hypoventilation, 50
  pCO_2, 203
  source of bicarbonate, 202
Carbon dioxide narcosis, 111
Carbonic, 63
Carbonic acid, 48–49, 52–53, 56, 59
Carbonic anhydrase, 52–53, 62
  hydration of carbon dioxide, 146, 201–2
  in red blood cells, 201–2
  tubular cells, 146
Carbonic anhydrase inhibitors
  diuretics, 62, 152–53
  renal tubular acidosis, 173
Carbon monoxide blood levels
  affinity for hemoglobin, 204
  cigarette smoking, 204
  oxygen-hemoglobin dissociation, 204
Carbon tetrachloride, 163
Carbonyl group, 54
Carcinoembryonic antigen (CEA), 125, 411
Carcinoma, 360
  adenocarcinoma, 101
  anaplastic, 361
  bladder, 162
  of breast, 42
  of cervix, 122, 135
  colon, 411
  of endometrium, 135
  epithelial cell origin, 101, 122
  follicular, 361

in situ, 122
  of kidney, 42, 174–75
  large cell, 101
  oat cell, 101
  papillary, 361
  of penis, 135
  of prostate, 135
  small cell, 101
  squamous cell, 101
Cardiac arrest, 36
Cardiac arrhythmias, 36, 38, 41, 45, 284–85
Cardiac asthma, 284
Cardiac cirrhosis, 422
Cardiac conducting system, 276
Cardiac contractility, 42, 58, 280, 257
Cardiac contraction, 276, 281
Cardiac control center, 255
  of circulating blood, 255
Cardiac cycle, 252, 276
Cardiac defect, 292
Cardiac enlargement. *See* cardiomegaly
Cardiac failure, 291, 293
Cardiac function, 276
  abnormalities, 276
  terms related to, 255
Cardiac notch, 77, 81
Cardiac output, 84, 281, 254–55, 257–58
  and decreased blood flow, 205
  increased in severe anemia, 207
  and polycythemia vera, 205
  and tissue hypoxia, 200
Cardiac tamponade, 272, 281
Cardiogenic emboli, 449
Cardiogenic shock, 264
  management, 263
  manifestations, 263
Cardiomegaly, 284
  anemia, 207
Cardiomyopathies, 285
  dilated, 286
  endomyocardial fibrosis, 280
  hypertrophic, 286
  restrictive, 286
  types, 280
Cardiovascular disease, 276
Cardiovascular malformation, 290
Carotid
  arteries, 24, 294
  bodies, 56–57, 256
  sinus, 255
Carpal spasm, 43
Carrier molecule, 10
Cartilage, 78, 84
Case studies
  acute myelogenous leukemia, 242–43
  acute pancreatitis, 432–33
  adenosquamous carcinoma, 111
  adult respiratory distress syndrome, 109–10
  AIDS, 342
  alcoholic cirrhosis, 430
  autoimmune hemolytic anemia, 341
  carcinoma, 433
  chronic active hepatitis, 431
  chronic obstructive pulmonary disease, 111
  chronic renal failure, 191–92
  cirrhosis, 429, 434
  collapse and loss of consciousness, 478
  congestive heart failure, 299
  diabetes, 443
  diabetes mellitus, 381–83
  diabetic nephropathy, 194
  diarrhea, 69

discoid/systemic lupus erythematosus, 344
  end-stage renal disease, 191
  glomerulonephritis, 192
  grand mal seizure, 476
  hepatic failure, 431–32
  hepatorenal syndrome, 190
  hypertension, 433
  infectious hepatitis, 431
  insulin dependent diabetes mellitus (IDDM), 383–84
  ketoacidosis, 380–81
  loss of gastric contents, 68
  lupus/pneumonitis/nephritis, 345
  myocardial infarction, 300–301
  nephrosclerosis/glomerulonephritis, 191–92
  prostate carcinoma/DIC, 243
  pulmonary edema, 431
  pulmonary embolism, 109
  reduced renal perfusion, 190–91
  renal transplant rejection, 194
  septic shock, 302
  transient ischemic attack, 476–77
  vomiting, 68
  vomiting and diarrhea, 69
Casts
  white blood cell, 158–59
Catalyzed, 53
Catecholamines, 366
Catheterization, 158
  diagnose heart disease, 285
Cation-exchange resin, 37
Cationic proteins
  platelet secretion, 166
Cations, 32
CAT scan. *See* computerized axial tomography
CAT scanner
  photograph, 442
CCPD. *See* continuous cyclic peritoneal dialysis
Celiac sprue, 412
  comparison to tropical sprue, 408
  diagnostic tests, 407
  etiology, 406
  manifestations, 406–7
  treatment, 407
Cell, 49, 56, 64
Cell-mediated immunity (CMI), 316, 319, 320
  contact dermatitis, 326
  lymphokines, 326
  macrophage activating factor, 326
  migration inhibiting factor, 326
  role of, 320
  suppressor T cells, 326
  T-lymphocytes, 319–20
  tuberculin test, 326
  type IV hypersensitivity, 326
Cell-mediated response, 320–21
Cell membrane, 6, 8
  bilayer, 8
  fluid mosaic model, 8
  proteins, 8
  structure, 6, 8
Cellular, 20
  dehydration, 20–21, 23–25, 27, 29, 32, 45
  compartments, 64
  hydration, 20–21, 25, 29, 35, 64
  volume, 24
  water, 32
Cellulitis
  definition, 302
  septic shock, 302
Central nervous system, 32, 33

Central nervous system diseases
  SIADH, 26
Centriacinar, 92
Cephalic vein
  arteriovenous fistula, 183
Cephalosporins, 163
Cerebral, 34, 58
  hemorrhage, 32–33, 172–73, 294
  tumor, 33
Cerebral arteries
  polycystic kidney disease, 172
Cerebral arteriogram, 445, 477
Cerebral blood flow, 464
Cerebral cortex
  motor areas, 456, 469
  sensory areas, 456, 469
Cerebral edema. See edema
  acute mountain sickness, 204
Cerebral hemisphere
  frontal section, 469
Cerebral injury
  cerebral palsy, 458
  minimal brain dysfunction, 457
Cerebral palsy, 461
  ataxia, 458
  athetosis, 458
  definition of terms, 458
  diplegia, 458
  dysarthria, 458
  dyskinesia, 458
  dysphagia, 458
  etiology, 458
  hemiparesis, 458
  manifestations, 458
  paresis, 458
  spastic, 458
  tetraparesis, 458
  treatment, 458
Cerebral thrombosis, 291, 294
Cerebral ventricles, 440
Cerebrospinal fluid, 14, 32, 440
  diagram, 441
  flow of, 441
Cerebrovascular accident, 271
  and polycythemia vera, 205
  stroke, 445
Cerebrovascular accident/hypertension, 479,
    480
Cerebrovascular occlusion
  attacks, 446
  strokes caused by occlusion, 447
  transient ischemic
Cerebrum
  microscopic view, 459
Cervical carcinoma
  metastasis, 123
  renal failure, 123
  symptoms, 123
  ureteral obstruction, 123
Cervix 118–20
  squamous cell carcinoma, 122, 135
  staging of carcinoma, 123
Chelating agent
  treatment for thalassemia, 210
Chemical reaction, 49–50
Chemicals, 354
Chemoreceptor, 56–57, 255–56
Chemosensitive, 56
Chemotaxis, 311
Chemotherapy, 128, 132–33, 220
  in acute myelogenous leukemia, 242
  in polycythemia vera, 205
  stimulation of vasopressin, 358
Chest, 83, 84
  movement, 89
  pain, 293
  wall, 82, 85, 89–90

Chest X-ray
  cryptococcosis, 331
  pulmonary interstitial infiltrates, 309
  stricture in pulmonary artery, 285
Cheyne-Stokes breathing, 204, 285
  acute mountain sickness, 204
  definition, 204
CHF. See congestive heart failure
Chief cell, 62
Chills, 158
Chlamydia trachomatis, 120, 132, 134, 158
Chloride, 32, 52, 62, 65–66, 68–70, 153
  exchanged for bicarbonate, 201–2
  excretion with bicarbonate reabsorption,
    153–54
  high osmolality in renal medulla, 142–43
Chloride ions
  and vomiting, 65–66
Chloride shift, 52, 201–2
Chlorothiazide
  brand name Diuril, 152
Chlorpropamide, 358
Cholelithiasis, 425
Cholesterol, 6, 8, 39, 272–73, 366, 419
  in bile, 419
  risk for atherosclerosis, 271
Choline, 7
Choline acetyltransferase, 459
Cholinesterase, 468
Chordae tendineae, 279
Chorea, 278
Choriocarcinoma, 133, 136
Chorionic gonadotropin, 133
Christmas disease, 231
Chromosome, 220–21
Chronic active hepatitis, 431
Chronic bronchitis, 88
  and emphysema, 92
  etiology, 90
  management, 91
  obstruction, 91
  pathology, 91
  structural changes, 91
  symptoms, 90–91
Chronic intoxication
  cause of dementia, 461
Chronic liver disease
  acid-base imbalance, 430
Chronic lymphocytic leukemia, 221
  laboratory findings, 222
  management, 222
  manifestations, 222
  prognosis, 222
  symptoms, 222
Chronic myelogenous leukemia, 220
  laboratory findings, 220
  management, 220
  manifestations of, 220
  prognosis, 221
Chronic obstructive pulmonary disease
    (COPD), 88, 107, 110
  complications, 95
  cor pulmonale, 95
  hypoxemia, 95
  prognosis, 221
  respiratory acidosis, 95
Chronic renal failure, 181, 186, 190, 192,
    194
  acidosis, 181–82, 192
  alkaline phosphatase, 193–94
  ammonia, 182
  anemia, 193
  and anion gap, 192–94
  bicarbonate, 194
  BUN/creatinine, 192, 194
  calcium, 182, 193–94

case, 191–92
  causes, 190, 194
  creatinine, 181
  decreased ammonia synthesis, 192
  diabetic nephropathy, 191
  early measures to control, 183
  expected laboratory data, 194
  filtration, 150
  fluid/electrolyte imbalances, 181
  glomerular filtration rate, 181–82
  glomerulonephritis, 167–68, 191
  hepatorenal syndrome, 190
  hydrogen ion, 182
  hyperkalemia, 182
  hypernatremia, 181
  hypocalcemia, 181–82
  hyponatremia, 181
  increased ammonia, 192
  low sodium diet, 182
  magnesium, 181
  manifestations, 181, 194, 195
  mechanisms of injury, 181
  nephrosclerosis, 176, 191
  normal values, 150
  phosphorous, 181, 193–94
  polycystic kidney disease, 173
  polyuria, 181
  potassium, 181, 182
  pyelonephritis, 191
  rate of production, 150
  reabsorption, 150
  sodium, 182, 193–94
  structure, 149
  synthesis, 149
  tubular secretion, 150
  urea, 181
Chronic respiratory acidosis, 58
Chvostek's sign, 42
Chyme, 62–63
Cigarette smoking, 45, 84, 90, 92, 449
  blood carbon monoxide levels, 204
  erythrocytosis, 204
  hypoxemia, 204
  plasma volume decrease, 204
Cilia, 79, 81
Ciliated epithelium, 78
Cimetidine, 394
Circle of Willis, 448
Circulatory failure
  compensatory responses, 260
  management, 261
  manifestations, 261
  secondary effects, 260
Circulatory system, 256
  distribution of blood in, 256
Cirrhosis
  abnormal sex hormone metabolism, 424
  causes, 423
  hepatic encephalopathy, 424
  hepatocellular failure, 425
  laboratory findings, 434
  portal hypertension, 423
  prognosis, 424
Cirrhosis of liver, 84
  acid-base balance, 430
  blood chemistry, 431
  effects of hepatocellular failure, 425
  effects of portal hypertension, 424
  etiology, 423
  hematological findings, 431
  polycystic kidney disease, 173
Cirrhotic ascites
  complications, 424
  formation, 424
  treatment, 424
Cis-aconitic acid, 52

Erosive gastritis
  uremic syndrome, 182
Erythema marginatum, 278
Erythema nodosum, 409
Erythematous, 222
Erythematous rash
  butterfly pattern, 334
Erythroblast, 200, 206
Erythrocytes. *See* red blood cells
Erythrocytosis, 204, 212
  bone marrow dysfunction, 204
  malignancy, 243
  polycythemia vera, 204
  secondary causes, 204
Erythropoiesis
  nucleated red blood cells, 242
  polychromia, 242
  stages, 200
Erythropoietin, 200
  acute renal failure, 181
  and anemia, 193, 205–6, 209
  chronic renal failure, 193
  source, 200
  stimulus, 200, 204
  uremic syndrome, 182
Erythropoietin-responsive stem cell, 200
Escherichia coli, 158
Esophageal endoscopy, 392
Esophageal motility
  scleroderma, 176
Esophagus, 76, 81–82, 362
  associated with AIDS, 393
Estrogen-progestogen contraceptives, 122
Estrogen receptors, 127–28
Estrogens, 44, 45, 119, 120, 121–22, 135,
    355–56, 368
  sources, 123, 127–28
Estrogen therapy, 134
Estrone, 123
Ethacrynic acid, 42
  brand name Edecrin, 152
Ethylenediaminetetracetate (EDTA), 42,
    240
Eustachian, 76
Excretion, 36
Excretory system, 158
  anatomy, 138
Excretory urogram. *See* intravenous
    pyelogram
Exercise, 55
Exertional dyspnea
  anemia, 207
Exophthalmos
  caused by Graves' disease, 360, 363
Expectorants, 91
Expiration, 89
External iliac artery, 290
External intercostal muscles, 88
External nares, 76
External respiration, 76
Extracellular, 9, 14, 32
  compartment, 14, 24, 27, 35, 38
  dilution, 25
  expansion, 25, 29
  fluid, 9, 20, 23–24, 26, 32, 34, 36, 41, 45,
    51, 54–56, 64–66, 68, 363
  hyperosmolality, 33, 34
  osmolality, 27
  shift, 57
  solute, 24
  spaces, 354
  volume, 24–25, 29
Extracorporeal shock-wave lithotripsy, 161
Extrapyramidal disorders
  Huntington's disease, 470
  Parkinson's disease, 470

Extrapyramidal system, 469
Extrinsic pathway, 226–27, 234
  and prothrombin time, 239–40
  summary of reaction, 228
Exudation, 84

F

Fab sites (fragment antigen binding), 316,
    319–20
Facial nerve, 465–66
Facial weakness, 447
Factor I. *See* fibrinogen
Factor II. *See* prothrombin
Factor III. *See* tissue factor
Factor III:C. *See* antihemophilic factor
Factor IV. *See* calcium ion
Factor V, 233
Factor VII, 233–34
Factor VIII, 231, 233
Factor IX, 231, 233–34
Factor X, 233–34
Factor Xa, 234
Fallopian tube adenocarcinoma, 124, 135
Fallopian tubes, 118, 120–21, 134
Fanconi syndrome, 163
  renal tubular acidosis, 174
  substances in urine, 163
Fat, 54, 355
Fatigue, 293
  anemia, 207
  Graves' disease, 360
Fatty acid chains, 8
Fatty acids, 24
Fatty streak, 269
  child's aorta, 269
FDP. *See* fibrinogen/fibrin degradation
    products
Fecal occult blood, 411
Ferritin, 209
Fertilized egg, 118
Fetal circulation, 290
  aorta, 290
  aortic arch, 290
  changes at birth, 291
  ductus arteriosus, 290
  ductus venosus, 290
  foramen ovale, 290, 293
  hypogastric arteries, 290
  iliac artery, 290
  liver, 290
  pulmonary artery, 290
  sphincter, 290
  umbilical arteries, 290
  umbilical vein, 290
  urinary bladder, 290
  vena cava, 290
Fetal vessel, 295
Fever, 28, 33, 54, 84, 158
  acute mountain sickness, 204
  beta-thalassemia, 210
  low grade in anemia, 207
Fibrin, 166, 169, 226–28, 232, 241
  molecules, 228
  polymerization of, 229
Fibrinogen, 168, 227–28, 233, 241
Fibrinogen/fibrin degradation products,
    232–33, 241
  deep vein thrombosis, 242
  disseminating intravascular coagulation,
    242
  elevated in thromboembolic conditions,
    242
  myocardial infarction, 242
  pulmonary thrombosis, 242
  significance of, 232

Fibrinoid necrosis
  malignant hypertension, 176
  scleroderma, 176
Fibrinolysis, 229, 232
  overview, 226
Fibrinolytic responses, 100
Fibrinolytic therapy, 130, 233
Fibrin split products, 241
Fibrocystic disease, 126, 135
Fibroids, 121, 134
Fibromas, 125, 135
Fibrosis, 83, 121, 134–35, 168, 291, 293
  glomerulosclerosis, 176
  pericardium, 281
  scleroderma, 176
Fibrous plaque, 269
Fibrous tissue, 121
Filtration, 16, 140
Filtration/diffusion
  glomerular filtrate, 140
Fimbriae, 118
Final common clotting pathway, 241
Finger clubbing, 102
Fistula, 404
Flank pain
  diabetic nephropathy, 176
Flat plate dialyzer, 183
Flatus, 402
Fluid
  balance, 138
  deficit, 26, 34
  excess, 24
  ingestion, 23
  restriction, 183
  vascular fluid loss, 168
Fluid compartments, 14, 21
Fluid homeostasis
  and congestive heart failure, 298, 303
  failure of homeostasis, 298, 303
Fluid imbalance
  acute mountain sickness, 204
  antidiuresis, 204
  edema, 204
Fluid loss, 26–29
  cause of hypernatremia, 33
  and diabetes insipidus, 33
Fluid mosaic model, 8
Fluorescein, 167
Focal deficit, 454
Folic acid, 405
  and anemia, 206, 208
  factor red blood cell formation, 201, 208
Folic acid therapy, 405
Follicles, 134–35, 360–61
  decrease, 122
  development, 119
  layer of, 121
  maturation failure, 122
  primary, 119
  secondary, 119
Follicle stimulating hormone (FSH),
    119–22, 355–57
Follicular
  carcinoma, 361
  cysts, 121
Fontanelle
  depression, 27
Foramen of Magendie, 440
Foramen ovale, 290–91, 295
Foramina of Luschka, 440
Foreskin, 129
Fragment antigen binding sites. *See* Fab
    sites
Frank-Starling law, 283, 254–55, 257
Frontal balding, 122
FSH. *See* foliicle stimulating hormone

FSP. *See* fibrin split products
Fumaric acid, 52
Furosemide, 42
   brand name Lasix, 152

G

Gait ataxia, 472
Galactorrhea, 126, 135, 356
Gallbladder anatomy, 416
Gallstones, 68
  X-ray, 421
Gamma-aminobutyric acid, 470
Gamma globulin, 319
Gamma-glutamyl transferase, 345
  and AIDS, 344
Gas exchange, alveolar/capillary, 106
Gasserian ganglion, 465
Gastrectomy, 396
Gastric
  fluids, 62, 65, 68
  glands, 62
  juice, 62, 64–65
  mucosa, 53, 62
  sources of secretion, 390
Gastric acid secretion, 395
Gastric bleeding, 396
Gastric cancer
  possible factors, 397
Gastric carcinoma
  diagnostic tests, 397
  etiology, 397
  manifestations, 397
  treatment, 397
Gastric cells, 63
Gastric contents
  electrolytes compared to serum, 64
  electrolyte composition, 62
Gastric erosions, 395
Gastric gland
  of the mucosa, 391
Gastric mucosa, 393, 395, 398
Gastric mucosal barrier, 394, 398
  factors, 394
Gastric mucosal stress erosions, 395, 398
  central nervous system disease, 395
  cerebral or spinal cord injuries, 395
  chronic failure, 395
  conditions, 395
  diagnostic tests, 396
  etiology, 395
  extensive burns, 395
  hypotension, 395
  jaundice, 395
  manifestations, 395
  renal failure, 395
  respiratory failure, 395
  sepsis, 395
  severe trauma, 395
  treatment, 396
Gastric mucosal ulceration, 395
Gastric pit, 62
  of the mucosa, 391
Gastric ulcer
  diagnostic tests, 394
  etiology, 394
  manifestations, 394
  treatment, 394
Gastrin, 390
Gastritis, 398
  acute gastritis, 393
  chronic gastritis, 393
Gastroduodenostomy, 68
Gastroenterostomy, 396
Gastroesophageal reflux, 391
Gastroesophageal reflux disease, 391

Gastrointestinal, 39, 45, 58
  effects of hyperparathyroidism, 362
Gastrointestinal bleeding, 395
Gastrointestinal tract, 354
Gastrointestinal wall
  basic layers, 402
Generalized seizures
  absence, 455
  characteristics, 455
  tonoclonic, 455
Genetic factors, 44
Genitalia, 118, 128
Genital tract, 120
Genitourinary, 118
  structures, 162
Gentamycin, 163
Germinal cell tumors, 132, 135–36
Germinal epithelial cells, 129
GGT. *See* gamma-glutamyl transferase
Glans penis, 128, 135
Glial tumors
  astrocytoma, 456
  glioblastoma multiforme, 457
Glioblastoma, 456
Glioblastoma multiforme, 457
Gliomas
  astrocytoma, 456
  glioblastoma multiforme, 457
Globulin, 422
Glomerular basement membrane, 166, 168
Glomerular capillary walls, 166
  causes of damage, 166
  polyanionic molecules, 166
  proteoglycans, 166
  sialglycoprotein, 166
Glomerular disease
  antigens, 166
Glomerular filtrate, 28, 154, 368
  volume, 148
Glomerular filtration, 147
  chronic renal failure, 181–82
  decreased rate, 160, 192
  inhibition, 160
  rate, 150, 158
Glomerular filtration rate
  causes of decreased rate, 190
  decreased in hepatorenal syndrome, 190
  decreased rate coupled with retention of
    acid anions, 192
  and phosphorous, 193
  renal clearance tests, 150
Glomerular injury, 166
  mechanisms, 166–67, 169
Glomerulonephritis, 19, 191
  acute, 166–67, 169
  blood urea nitrogen, 148
  complement, 167
  creatinine, 167
  immunoglobulins, 167
  laboratory data, 167
  pathology, 167
  poststreptococcal, 166, 191
  prognosis, 167
  rapidly progressive, 167
  symptoms, 166
  treatment, 167
  urinalysis, 167
Glomerulosclerosis
  diabetic nephropathy, 176
Glomerulus, 27, 139–41, 159, 166
Glossitis, 405
  and anemia, 208
Glucagon, 366, 378, 403
  secretion, 373

Glucocorticoids, 42, 44–45, 168, 355, 357,
  362, 366–68
  and acute mountain sickness, 204
  catabolism, 148
  cortisol, 366
  immunosuppressive agent, 366
  and protein anti-inflammatory, 366
  therapeutic uses, 367
  treatment, 366
Glucose, 10, 17, 20, 33–34, 51, 52, 66, 356,
  358, 367
  in dialysate, 184
Glutamine
  source of ammonia, 147, 191
Gluten, 406
Gluten sensitive enteropathy, 406
Glycerol, 6
Glycoprotein, 8, 92, 231
Goblet cell, 79, 81, 91
Goiter, 361
  caused by Graves' disease, 360, 363
  iodine deficiency, 361
Gonad control, 355
Gonadoblastoma, 125, 135
Gonadotropin releasing hormone (GnRH),
  119–20, 357
Goodpasture's syndrome, 168–69, 325, 333
  anti-glomerular basement membrane
    antibodies, 168
  pulmonary hemorrhage, 168
Gout, 191
  in polycythemia vera, 205
Gouty arthritis, 191
Graafian follicle, 119–20
Grade III astrocytoma, 456
Graft vs. host, 332
Grand mal. *See* tonoclonic seizures
Granular white blood cells, 218
Granulocytes, 216
  basophils, 217
  characteristics of, 216
  development of, 216
  eosinophils, 217
  functions, 216
  hypersegmented, 208
  neutrophils, 216
Granulomas
  noncaseating epithelioid, 99
Granulosa cells, 119, 121, 135
Granulosa-theca ovarian tumors, 119, 123,
  135
Graves' disease, 359, 363
  effects, 360
  treatment, 360
  cause of hypothyroidism, 361
Growth hormone (GH), 44, 355, 356, 363,
  366
  acromegaly, 356
  deficiency, 357
  diabetes, 356
  effects of excess, 356
  extreme height, 356
  facial tissue, 356
  glucose, 356
  linear growth, 356
  mandible, 356
  from pituitary glands, 460
  protein synthesis, 356
  somatomedin, 356
  stimulus, 357
  thyrotropin releasing hormone, 356
  viscera, 356
Growth hormone releasing hormone (GRH),
  355
Guillain-Barré syndrome, 472–73

H₂ receptor antagonists, 396, 398

manifestations, 38
nephrogenic diabetes insipidus, 174
principles of treatment, 39
renal tubular acidosis, 173–74
Hypokalemic periodic paralysis, 38–39
Hyponatremia, 45, 69, 476, 479
acute renal failure, 180–81
chronic renal failure, 181
etiology, 34
manifestations, 34
and potassium ion loss, 34–35
principles of treatment, 35
Hypoparathyroidism, 43, 362–63
Hypophosphatemia, 363
Hypopituitarism
and anemia, 206
Hypotension
dehydration, 27, 29
Hypothalamic defect, 122
Hypothalamic hormone somatostatin, 363
Hypothalamic releasing factors, 122
Hypothalamo-hypophyseal portal system,
359
Hypothalamohypophyseal tract, 357
Hypothalamus, 23, 26, 33, 134, 144,
354–60, 362–63
osmoreceptors, 144
Hypothyroidism, 356–57, 360
and anemia, 206
cause of dementia, 461
Hypotonic 25, 28–29, 70
fluid, 22, 24
losses, 27
SIADH, 26
sweat, 27
Hypoventilation, 49–50, 54, 57–59, 106, 112
and blood gas patterns, 108–9
Hypovolemia
compensatory responses, 67
and vomiting, 64, 66–67
Hypovolemic shock, 264
compensatory responses, 261
management, 261
manifestations, 261
Hypoxemia, 107–9, 290
alveolar hypoventilation, 204
and asthma, 95, 110
causes, 204
and cystic fibrosis, 100
definition, 204
and emphysema, 110
impaired diffusion, 204
and pulmonary thromboembolism, 100
ventilation/perfusion abnormalities, 204
Hypoxia, 81, 84, 255, 455
causes, 200, 204
definition, 204
stimulate vasopressin, 358
stimulus for erythropoiesis, 200
Hysterectomy, 121
radical, 123

I

Ibuprofen, 121
IDDM. *See* insulin dependent diabetes
mellitus
Idiogenic osmoles, 33–34, 45
Idiopathic, 358
Idiopathic thrombocytopenic purpura, 234
Imaging techniques
excretory system, 151–52, 155
Imipramine, 132

Immediate hypersensitivity
anaphylactoid reactions, 324
anaphylaxis, 324
chemotactic factors, 324
etiology, 324
events, 324
histamine, 324
leukotriene, 324
slow-reacting substance, 324
symptoms, 324
type I response, 324
Immobilization, 42, 45
Immune complex, 166, 169
cryoglobulins, 340
Immune complex mediated antigen/
antibody complexes, 325
Arthus reaction, 325
glomerulonephritis, 326
reactions to drugs, 326
serum sickness, 325
type III hypersensitivity, 325
Immune response
uremic syndrome, 182
Immune system, 308
Immune thrombocytopenic purpura, 234
Immunity
active, 319
passive, 319
Immunoassay, 125
Immunodeficiencies, 330, 332, 337
acquired immunodeficiency syndrome, 330
selective IgA deficiency, 332
severe combined immunodeficiency
disease, 332
Immunogen, 316
Immunoglobulin, 166–67
basic structure, 316
categories, 317–18
Immunoglobulin A (IgA), 317–18
selective IgA deficiency, 332
Immunoglobulin D (IgD), 318
Immunoglobulin E (IgE), 318
Immunoglobulin Fc sites, 320
Immunoglobulin G (IgG), 317–18
Immunoglobulin M (IgM), 317–18
Immunoglobulins, 316
Immunologic characteristics of AIDS, 332
Immunologic reaction, 360
Immunologic tests
antiglobulin tests, 340
autoantibodies, 340
complement, 340
immune complexes, 340
rheumatoid factors, 340
Immunosuppressive agent, 366
Impaired diffusion
hypoxemia, 204
Implantation
abnormal, 121
ectopic, 121
tubal, 121
Impotence, 357
Inappropriate secretion of ADH. *See*
syndrome of inappropriate secretion of
ADH (SIADH)
Indigestion, 42
Indomethacin, 121
Infancy, 91
Infants, 29, 33
insensible fluid loss, 28
surface area, 28
Infarction, 270
pulmonary, 100
testicular, 130

Infection, 90–91, 360
dialysis, 186
susceptibility, 181
urinary tract, 158
Infective endocarditis, 293
Inferior lobe, 77
Infertility, 121–22, 356
Infiltrating duct carcinomas, 126
bladder, 158
cell products, 310
cervix, 120
chemicals, 310
chronic, 310–11
complement, 311
epididymis, 130–32
genitalia, 135
glans penis, 128, 132, 135
kidney, 158
peritoneum, 120
phagocytosis, 310
prostate, 131, 135
testes, 130
types of, 308
urethra, 158
vaginal, 120
vulva, 120
Inflammatory bowel disease, 120
Inflammatory reactions, 311
Inflammatory response, 309
series of events, 309
Influenza, 168
Infundibulum, 119, 354, 357
Infusion, 25
Inosine, 203
Insensible water loss, 26–28
In situ
definition, 122
formation, 121
Insomnia, 41
acute mountain sickness, 204
Graves' disease, 360
uremic syndrome, 182
Inspiration, 88–89
Inspissated, 404
Insulin, 384
effects, 378
injection, 37, 39
negative feedback, 378
secretion, 373
shock, 380
Insulin deficiency
events lead to death, 379
Insulin-dependent diabetes mellitus, 373
Integrative deficits
transient ischemic attack, 447
Intercostal muscles, 82, 88, 126
Interlobular artery, 140
Intermediate lobe (pars intermedia), 354
Intermittent peritoneal dialysis, 185
Internal nares, 76
Internal respiration, 76
Interstitial, 14, 17–18, 22, 28, 32
cells, 129
electrolyte concentration, 21
expansion, 22
fluid, 14, 17–18, 20, 23, 25, 36, 62, 83
pressure, 18
spaces, 18
Interstitial fibrosis, 83
and adult respiratory distress syndrome,
98
and sarcoidosis, 99

Meperidine, 403
Mercury, 163
Mesenchymal cells
  origin of sarcoma, 122
Mesothelium, 124
Metabolic, 54–56
Metabolic acidosis, 54–55, 379–80
  acute renal failure, 181
  and anion gap, 192–93
  chronic renal failure, 181–82, 191–92
  diuretics, 154
  hypochloremic, 65, 154
  and vomiting, 63, 66
  renal tubular acidosis, 173–74
Metabolic alkalosis, 42, 62, 64, 68, 70
  compensatory response, 64
  definition, 54
  diuretics, 154
  hypochloremic, 65, 153
  and vomiting, 63, 66
Metabolic pathways
  interconversion, 379
Metabolic waste, 49
Metabolism, 355
Metamyelocyte, 216
Metaplasia, 122
Metarterioles, 255
Metastasis, 41, 101, 122–23, 132–35, 397
  renal cell carcinoma, 174
Methoxyflurane, 163
Methyldopa, 341, 356
Methylene blue, 216
Microcytic, 206–7, 239
Microthrombi, 232
Microvilli, 79
Micturition, 139, 158
Midamor, 152
Migraine headache, 464, 473
Migrating myoelectric complexes, 397
Migration inhibiting factor, 326
Milk-alkali syndrome, 42
Mineralocorticoid, 366–68
  treatment of renal tubular acidosis, 174
Minimal brain dysfunction (MBD), 461
  etiology, 457
  manifestations, 457
  treatment, 457
Minor ischemic stroke, 447
Mithramycin, 42
Mitral regurgitation, 282
Mitral stenosis, 279
Mitral valve, 279
Mixed cell type tumors, 133
Molal solution, 24
Molecules, 49, 88
Monoblast, 218
Monoclonal antibodies, 343
  OC-125, 125
Monoclonal antibody studies
  and lymphocyte subpopulations, 343
Monocular visual problems, 447
Monocytes, 216–17, 219, 222, 229, 239, 310
Mons pubis, 118
Mons pubis, 118
Morphine, 403
  stimulate vasopressin, 358
Motor disorders
  amyotrophic lateral sclerosis, 466
  demyelinating disorders, 471
  extrapyramidal disorders, 469
  neuromuscular transmission, 468
Mucinous tumors, 124, 135
Mucociliary escalator, 79, 81, 84
Mucosa cells, 62
  glands, 81, 91
  membrane, 79, 82, 84

Mucus, 81–82, 84, 90
  membrane, 91
  secretions, 90
Multi-infarct dementia, 461
Multiple myeloma
  renal tubular acidosis, 174
Multiple sclerosis, 158, 473
  definition of terms, 472
  manifestations, 472
Mumps, 130
Muscle cells, 21
Muscles, 82–83, 88–89, 356
  contraction, 35, 39, 88
  cramps, 34
  fibers, 91
  Graves' disease, 360
  skeletal, 38
  smooth, 38, 78, 80–81, 84
  spasms, 42
  twitching, 33
  weakness, 38, 41
Muscular dystrophies
  duchenne muscular dystrophy, 467
  myotonic muscular dystrophy, 467
Muscular effect of hyperparathyroidism, 362
Muscularis mucosa, 394
Myasthenia gravis, 332–33, 473
  and autoantibodies, 340
Mycobacterium avium-intracellulare, 341–42, 425
Mycoplasma, 132
Myelin
  degeneration, 158
Myelinated axon
  electron micrograph, 472
Myelinated nerve fiber, 471
Myeloblast
  granulocytic series, 216, 242
  predominant cell in AML, 242
Myeloblastic leukemias, 220, 242
Myelocyte, 216
Myeloma, 42
Mylanta, 396
Myocardial contractility, 254–55
Myocardial disease
  cardiomyopathies, 280
  myocarditis, 280
  terms related to heart disease, 280
Myocardial infarction, 149, 276–78, 280, 285, 300
  cardiac enzymes, 300–301
  consequences, 277
  and elevated fibrin split products, 241
  leukocytosis, 302
  management, 278
  manifestations, 277
  plasma lipids, 300
Myocardial necrosis, 277
Myocarditis, 280, 286, 335
Myometrial contractility, 120
Myometrium
  invasion, 124
Myxedema, 361

N

Nails thickened, 43
Naproxen, 121
Nasal vestibule, 76
Nasogastric tube, 33
Natriuretic hormone, 298
Natural killer (NK) cells, 325
Nausea, 25, 34, 38, 41, 58, 120, 158, 167
  acute mountain sickness, 204
  acute renal failure, 181

anemia, 207
prostate carcinoma, 243
stimulates vasopressin, 358
uremic syndrome, 182
Neck vein distention, 25
Necrosis
  soft tissue, 43
Needle biopsy
  prostate, 133
Negative feedback control
  of insulin secretion, 379
Neisseria gonorrhea, 120, 132, 134, 158
Neoplasm, 122, 124, 410
  malignant, 360–61
Neoplasms of lung, 100–101
  bronchogenic carcinoma, 101–3
  malignant vs. benign, 101
Nephrectomy
  renal cell carcinoma, 174
Nephrocalcinosis, 173, 362
Nephrogenic diabetes, 163, 174, 177
  causes, 174
  insipidus, 174
Nephrolithiasis, 160, 163
  hematuria, 160
  management, 161
  obstruction, 160
  pain, 160
  polycystic kidney disease, 173
  renal tubular acidosis, 173
  urate crystals, 191
Nephron, 26–27, 38, 138–41, 159
  classification, 142
Nephron anatomy, 140, 142, 154
Nephron destruction, 192
  ammonia production, 192
  decreased buffering, 192
Nephrosclerosis, 191
  chronic renal failure, 191
  definition of, 260
  malignant hypertension, 177
  manifestations, 176
Nephrotic syndrome, 19, 84, 168–69
  causes, 168
  diabetic nephropathy, 176
  edema, 168
  findings, 168
  loss of albumin, 168
  vascular fluid loss, 168
Nephrotoxins, 158, 163, 180
  cadmium, 163
  cephalosporins, 163
  gentamycin, 163
  lead, 163
  mercury, 163
Nerve, 82–83
  cell, 354
Nerve impulse, 81
Nerve impulse transmission, 39, 41, 45
Nervousness
  Graves' disease, 360
Neuralgia, 465
Neurofibrillar tangles, 459
Neurogenic bladder
  diabetic nephropathy, 176
  dysfunction, 158, 162–63
Neuroglial cells
  astrocytes, 456
  characteristics, 456
  ependymal cells, 456
  microglia, 456
  oligodendrocytes, 456
  types, 456
Neurohypophysis, 144

Total body water. *See* water
Toxic diffuse goiter, 359
Toxic goiter, 360
Toxic multinodular goiter, 360
Toxins
  cause of acute renal failure, 180
Toxoplasma cyst, 343
TPA. *See* tissue-type plasminogen
Trabecular bone loss, 45
Trachea, 50, 76–79, 88, 358–59, 362
Tracheotomy, 78–79
Tranexamic acid, 449
Transferrin, 209
Transient ischemic attack (TIA), 446, 479
  arteritis, 477
  cerebral arteriogram, 477
  computed tomography head scan, 477
  echocardiogram, 477
  magnetic resonance imaging, 477
  manifestations, 447
Transmembrane movements, 8
Transplant
  histocompatible bone marrow, 332
Transudate, 84
Transudation, 84
Trauma, 130, 360
Treppe phenomenon, 255
Triamterene
  brand name Dyrenium, 152
Trichomonas vaginalis, 120, 134
Tricyclic antidepressants, 356
Trigeminal nerve, 465
  branches, 465
Triglyceride blood levels
  diabetes mellitus, 301
  uremic syndrome, 182
Triiodothyronine, 358–59
Trisodium citrate, 240
Trisomy 21, 459
Tropical sprue, 412
  comparison to celiac sprue, 408
  diagnostic tests, 405
  etiology, 404
  manifestations, 405
  treatment, 405–6
Trousseau's sign, 42–43
Truncal vagotomy, 396
Trypsin, 92
Tubal ectopic pregnancy, 121
Tuberculin test, 326
Tubular cells, 140
Tubular fluid, 140
Tubule
  convoluted, 140–41
Tumor markers
  alkaline phosphatase, 133
  alpha-fetoprotein, 133
  human chorionic gonadotropin, 133
Tumors, 356–57, 360
  benign, 135
  cause of acute renal failure, 180
  malignant, 42, 135
Tunica externa, 268
Tunica intima, 268
Tunica media, 268
Tunica mucosa, 402
Tunica muscularis, 402
Tunica serosa, 402
Tunica submucosa, 402
Tunica vaginalis, 131
Type I epithelial cells
  alveolar walls, 80
Type II epithelial cells
  alveolar walls, 80
Type II hypersensitivity. *See* antibody-
  dependent cellular cytotoxicity

Type III hypersensitivity. *See* immune
  complex mediated reactions
Type IV hypersensitivity. *See* cell-mediated
  immunity
Type I response. *See* immediate
  hypersensitivity

U

Ulcerative colitis, 409–10, 412
Ulcers
  and polycythemia vera, 205
Ulcer surgery
  complications, 396
Ultrafiltration, 183–84
Ultrasonography
  polycystic kidney disease, 172
Ultrasound, 130, 161
Umbilical arteries, 290
Umbilical cord artery
  arteriovenous fistula, 183
Unconjugated bilirubin
  characteristics, 420
Urea, 17, 141, 358
  body fluids, 148
  converted to ammonia, 161
  and decreased glomerular filtrate, 149
  filtration, 150
  high osmolality in renal medulla, 142, 143
  reabsorption, 149–50
  source, 190
  and water deficit, 149
Urea cycle, 148–49
Urease, 161
Uremic colitis, 182
Uremic frost, 182
Uremic stomatitis, 182
Uremic syndrome, 182–90
  anemia, 182
  anorexia, 182
  atherosclerosis, 182
  bleeding tendencies, 182
  bone resorption, 182
  calcium phosphate deposits, 182
  capillary fragility, 182
  colitis, 182
  ecchymoses, 182
  emotional lability, 182
  erythropoietin, 182
  gastritis, 182
  glomerular filtration rate, 182, 284
  growth, 182
  hydrogen ions, 183
  hypalgesia, 182
  hypertension, 183
  immune response, 182
  insomnia, 182
  magnesium, 183
  manifestations, 182
  nausea, 182
  nitrogenous wastes, 183
  parathormone, 182
  paresthesia, 182
  phosphate, 183
  platelet function, 182
  potassium, 183
  pruritis, 182
  red blood cell fragility, 182
  serous membrane effusions, 182
  skeletal defects, 182
  skin color, 182
  sodium retention, 183
  treatment, 183
  triglyceride blood levels, 182
  uremic frost, 182
  uremic stomatitis, 182
  vitamin D deficiency, 182

  vomiting, 182
  water retention, 183
Ureter, 128, 158, 161
  dilation, 131
Ureteral obstruction, 134, 160
  BUN/creatinine ratio, 150
  sodium ion balance, 160
Ureterovesical junction, 158
Urethra, 128, 138, 158, 160, 162
Urethral discharge, 130
Urethral orifice, 118
Urethral sphincter, 139
Urethritis, 158, 163
Uric acid
  and chemotherapy, 243
  and hematopoiesis, 205
  metabolized from purine, 161
  and polycythemia vera, 205
Uric acid crystals, 161
Urinalysis, 131, 158
  glomerulonephritis, 167
Urinary
  bladder carcinoma, 162
  calcium and thiazides, 161
  causes of obstruction, 162–63
  excretion, 66
  frequency, 130, 132, 162, 207
  obstruction, 163
  obstruction and pain, 160
  obstruction and atrophy of tubules, 160
  retention, 131
  stone formation inhibitors, 160
  urgency, 132
Urinary bladder, 118, 128, 131, 158, 160,
  162, 290
Urinary buffers, 146–47
Urinary tract
  infection, 131, 158, 163
  obstruction, 158, 160, 173
Urination
  frequency, 158
  painful, 158
Urine, 26, 28, 33–34, 45, 64, 140
  arginine, 162
  backflow, 139, 163
  calcium, 173, 362
  casts, 167
  composition, 138
  concentrating ability, 160
  cystine, 162
  diminished flow, 191
  flow, 138
  formation, 138–39
  hormonal influence, 138
  lysine, 162
  obstruction, 160
  ornithine, 162
  osmolality, 358
  pH, 161
  protein, 167
  specific gravity, 167
  white blood cells, 167
Urine concentration test, 151, 155
Urinometer, 151
Urobilinogen, 419–20
Uterine tubes. *See* fallopian tubes
Uterus, 118–19
  tumor, 121

V

Vaccine, 319
Vagina, 119
Vaginal inflammation, 120
Vaginal orifice, 118
Vagotomy, 68
Valve cusps, 282